Fungi Pathogenic for Humans and Animals

MYCOLOGY SERIES

Edited by

Paul A. Lemke

Department of Botany,
Plant Pathology, and Microbiology
Auburn University
Auburn, Alabama

Volume 1 Viruses and Plasmids in Fungi, *edited by Paul A. Lemke*

Volume 2 The Fungal Community: Its Organization and Role in the Ecosystem, *edited by Donald T. Wicklow and George C. Carroll*

Volume 3 Fungi Pathogenic for Humans and Animals (in three parts), *edited by Dexter H. Howard*

Volume 4 Fungal Differentiation: A Contemporary Synthesis (in preparation), *edited by John E. Smith*

Volume 5 Secondary Metabolism and Differentiation in Fungi (in preparation), *edited by Joan W. Bennett and Alex Ciegler*

Other Volumes in Preparation

Fungi Pathogenic for Humans and Animals

(IN THREE PARTS)

PART A
Biology

EDITED BY
Dexter H. Howard
Department of Microbiology and Immunology
School of Medicine, University of California
Los Angeles, California

with the assistance of
Lois F. Howard
Department of Medicine
School of Medicine, University of California
Los Angeles, California

MARCEL DEKKER, INC. New York and Basel

Library of Congress Cataloging in Publication Data

Main entry under title:

Fungi pathogenic for humans and animals.

(Mycology series ; v. 3)
Includes indexes.
Contents: pt. A. Biology.
1. Fungi, Pathogenic. I. Howard, Dexter H., [date]
II. Howard, Lois F. III. Series.
QR245.F86 589.2'0469 82-18240
ISBN 0-8247-1875-5 (pt. A)

MARCEL DEKKER, INC.

270 Madison Avenue, New York, New York 10016

Current printing (last digit):
10 9 8 7 6 5 4 3 2 1

PRINTED IN THE UNITED STATES OF AMERICA

Introduction to the Series

Mycology is the study of fungi, that vast assemblage of microorganisms which includes such things as molds, yeasts, and mushrooms. All of us in one way or another are influenced by fungi. Think of it for a moment—the good life without penicillin or a fine wine. Consider further the importance of fungi in the decomposition of wastes and the potential hazards of fungi as pathogens to plants and to humans. Yes, fungi are ubiquitous and important.

Mycologists study fungi either in nature or in the laboratory and at different experimental levels ranging from descriptive to molecular and from basic to applied. Since there are so many fungi and so many ways to study them, mycologists often find it difficult to communicate their results even to other mycologists, much less to other scientists or to society in general.

This Series establishes a niche for publication of works dealing with all aspects of mycology. It is not intended to set the fungi apart, but rather to emphasize the study of fungi and of fungal processes as they relate to mankind and to science in general. Such a series of books is long overdue. It is broadly conceived as to scope, and should include textbooks and manuals as well as original and scholarly research works and monographs.

The scope of the Series will be defined by, and hopefully will help define, progress in mycology.

Paul A. Lemke

Foreword

The current interest in the field of medical mycology is shown by the steadily increasing number of workshops, symposia, review articles, and research papers on medically important fungi. The growth in interest is not restricted to research scientists, for lectures concerned with medically important fungi are becoming an important part of the teaching curricula in biomedical sciences.

Although several fine textbooks and extensive literature on various aspects of fungal disease have accumulated, there is a lack of texts to which students can go for detailed accounts of biology and up-to-date research accomplishments on medically important fungi. This book, in a series of three parts, is devoted exclusively to meet this need.

The authors are all highly qualified, and many of them are internationally renowned scientists in the field. Professor Howard is a distinguished mycologist whose research efforts made important contributions to our knowledge of host-fungi interactions. He is also a noted teacher who has trained numerous young scientists in the field.

Part A contains the latest and most authoritative information on classification, nomenclature, morphology, and physiology of fungi pathogenic for man and animal. Publication of this volume is timely because classification and the continuous changes in the nomenclature of medically important fungi created enormous confusion among workers in this field.

Microbiologists and clinicians alike will find the taxonomic discussions of extreme interest and appreciate the kindly guidance for the identification of pathogens in each order or class. Microbiologists and biochemists will appreciate the comprehensive coverage of the morphological and physiological aspects in this volume.

K. J. Kwon-Chung
Clinical Mycology Section
Laboratory of Clinical Investigation
National Institute of Allergy and Infectious Diseases
National Institutes of Health
Bethesda, Maryland

Preface

In 1947, Walter J. Nickerson edited a volume on the *Biology of Pathogenic Fungi* which covered some aspects of the biology, physiology, and biochemistry of the fungi pathogenic for man. Since then several fine textbooks have appeared, in which are described the essential features of the clinical manifestation, epidemiology, and pathology of fungous infections together with the identifying characteristics of their etiologic agents. In addition, individual mycoses or single aspects of all of the major mycoses have received abundant monographic treatment. However, a comprehensive consideration of the biology of the fungi pathogenic to man and animals has not recently been offered. These volumes are an effort to do so. It was not my object to present a series of "selected topics" or "recent advances" but rather a thorough review of basic biology, host-parasite interactions, and current methods of detection and characterization of the zoopathogenic fungi.

Part A of this series deals with the basic biology of medically important fungi. The first section comprises a discussion of the taxonomy and classification of the zoopathogens. This section is followed by individual chapters on ultrastructural cytology, dimorphism, sporulation, blastoconidium germination, and the transport of metabolites. Some additional topics which suggested themselves for coverage could not be included in Part A because of size limitations. Other basic biology topics will be included in subsequent parts of the series. (Members of the bacterial order Actinomycetales, often included in various types of works on medical mycology, will not be considered in this series.)

So much has been accomplished in each of these fields that it would be impossible for any single author to give an in-depth presentation. Accordingly, a compilation by many individuals was attempted despite the recognized difficulties of variation and potential redundancy. The scope of these volumes is such as to attract the attention of all engaged in research on fundamental aspects of the fungi pathogenic to man and animals. However, the coverage will be of interest to students who wish an up-to-date consideration of the frontiers of investigation on these pathogens.

A special acknowledgment is required. All responsibility for necessary editing with regard to uniformity of format, proper references, and retyping of heavily altered copy was assumed by my wife, Mrs. Lois F. Howard. She was, indeed, the primary copyeditor of this work. She merits recognition for her unswerving pursuit of excellence of presentation and the indefatigable insistence on perfection of textural detail which I had hoped from the outset would characterize these volumes. Her efforts, in no small measure, and those of several of the authors, have been supported by research grant AI 16252 from the National Institute of Allergy and Infectious Diseases, National Institutes of Health, which is used to fund the Collaborative California Universities-Mycology Research Unit (CCU-MRU).

I am, of course, grateful to the splendid group of authors collaborating with me to produce these volumes.

Dexter H. Howard

Contents

Contributors

D. G. Ahearn, PhD, Georgia State University, Atlanta, Georgia

J. W. Carmichael, PhD, University of Alberta Mold Herbarium and Culture Collection, Edmonton, Alberta, Canada

Garry T. Cole, PhD, Department of Botany, The University of Texas at Austin, Austin, Texas

Nina Dabrowa, PhD, Department of Microbiology and Immunology, UCLA School of Medicine, Los Angeles, California

G. S. de Hoog, PhD, Centraalbureau voor Schimmelcultures, Baarn, The Netherlands

G. A. de Vries, PhD, Centraalbureau voor Schimmelcultures, Baarn, The Netherlands

Robert G. Garrison, PhD, The Research Service, Veterans Administration Medical Center, Kansas City, Missouri, and Department of Microbiology, University of Kansas School of Medicine, Kansas City, Kansas

Philip A. Geis, PhD,* Department of Microbiology, The University of Texas at Austin, Austin, Texas

Rishab K. Gupta, PhD, Department of Surgery, Division of Surgical Oncology, UCLA School of Medicine, Los Angeles, California

Dexter H. Howard, PhD, Department of Microbiology and Immunology, UCLA School of Medicine, Los Angeles, California

Charles W. Jacobs, Department of Microbiology, The University of Texas at Austin, Austin, Texas

Douglas S. King, PhD, Baptist Medical Center Plaza, Little Rock, Arkansas

Emil Müller, DSc, Institute of Microbiology, Swiss Federal Institute of Technology, Zurich, Switzerland

Robert A. Samson, PhD, Centraalbureau voor Schimmelcultures, Baarn, The Netherlands

M. A. A. Schipper, PhD, Centraalbureau voor Schimmelcultures, Baarn, The Netherlands

H. J. Scholer, MD, University of Basel and Pharmaceutical Research Division, F. Hoffmann-La Roche & Co., Ltd., Basel, Switzerland

Paul J. Szaniszlo, PhD, Center for Applied Microbiology and Department of Microbiology, The University of Texas at Austin, Austin, Texas

**Present affiliation:* The Proctor & Gamble Company, Cincinnati, Ohio

Contents of Part B

Fungi Pathogenic for Humans and Animals

Part One

CLASSIFICATION AND NOMENCLATURE

1

An Introduction to the Taxonomy and Nomenclature of Zoopathogenic Fungi

Dexter H. Howard / UCLA School of Medicine, Los Angeles, California

I. GENERAL COMMENTS

Medical mycology has long been encumbered with the great number of names that have been given to medically important fungi, a fact commented on by Nickerson in an early review on medical mycology [8]. Moreover, taxonomic developments have necessitated many nomenclatural changes over the years which have resulted in a bewildering synonymy for some pathogens. Since taxonomic considerations involve judgments and opinions, nomenclatural changes are often frustrating to the nonexpert and not infrequently controversial among the authorities. Thus, in planning this treatise on the biology of zoopathogenic fungi it seemed appropriate to begin with a consideration of their nomenclature and taxonomy.

The contributors to this portion of the book have chosen to approach their subjects in rather different ways. Undoubtedly, this reflects various opinions as to the type of treatment required by different groups of fungi as well as the individualism often noted among taxonomists.

II. NOMENCLATURE

The nomenclature of medically important fungi has long been a problem to scientists interested in various aspects of these microorganisms. One well-known example, frequently quoted [4], is the fact that there are over 100 synonyms for the single species *Candida albicans* [6]. This situation has been aggravated by the inexperience of investigators who were interested in aspects of the behavior of zoopathogens other than their classification.

But the problem is also related to the fact that, for the most part, these pathogens were, at the outset, known only in their imperfect states and the classification of imperfect fungi has always been subject to more uncertainty than has that of other groups. The discovery of the sexual phases of certain pathogens together with other sorts of taxonomic developments have led to many changes in nomenclature over the years.

As a starting place for a proper nomenclature of fungi pathogenic to humans and animals, I chose a recent monograph [7] on the subject compiled by a select subcommittee of the British Society for Mycopathology convened by D. W. R. Mackenzie and comprised of G. C. Ainsworth, P. K. C. Austwick, A. J. E. Barlow, C. N. D. Cruickshank, J. C. Gentles, and P. M. Stockdale. The Introduction to this monograph is an excellent account of the bases for nomenclature of the zoopathogens and other mycological matters of central importance.

III. TAXONOMIC ARRANGEMENTS

The recent overview by Ainsworth [1] of the classification of fungi presented in the introductory chapter to the two-volume treatment of fungal taxonomy which he coedited was adopted as a guide when I first began to plan these chapters on the classification of zoopathogens. Ajello [2] also followed Ainsworth's general scheme in his recent succinct consideration of the location of animal pathogens among the divisions of the fungi. More recent monographic treatments on taxonomy of the Fungi Imperfecti [3] have raised many issues for consideration, e.g., the anamorph expression of growth of teleomorphs, and these excellent volumes should be consulted for further information. Recognition of the fact that anamorphs are form-genera and form-species has been emphasized repeatedly by Carmichael [3,5].

The outline given here was constructed in May 1978 as I was outlining the chapter headings for this part of the book. I modified it in July of that year when I came across Ajello's paper on the subject [2]. His contribution is more comprehensive than mine and should be consulted for details. I further modified the key as the various chapters arrived in manuscript form. Carmichael encouraged the use of "Fungi Imperfecti" rather than "Deuteromycotina" so that the fact that the latter were form-genera and form-species would be emphasized (personal communication). I hope the presentation will be of some interest and will provide a glimpse of the position of zoopathogens in the scheme of things.

*Outline of the Zoopathogenic Fungi**

Kingdom: Fungi
- *Division*
  - I. Myxomycota (no pathogens known)
  - II. Eumycota
    - A. Mastigomycotina
      - 1. Chytridiomycetes — *Rhinosporidium seeberi* (?)

*Restricted for the most part to those genera included in Chaps. 2 to 8 (see individual chapters for lists of species). The divisions of the outline were chosen to suggest general location. Details are not presented. The divisions shown are those thought to contain pathogens of humans and animals. Prominent pathogens or potential pathogens are indicated at a generic level.

2. Oomycetes	
a. Saprolegniales	*Saprolegnia*
b. Peronosporales	*Pythium (Hyphomyces) destruens*
c. Blastocladiales	*Coelomyces*
B. Zygomycotina	
1. Trichomycetes	
2. Zygomycetes	
a. Mucorales	
i. Mucoraceae	*Absidia*
	Mucor
	Rhizomucor
	Rhizopus
ii. Cunninghamellaceae	*Cunninghamella*
iii. Mortierellaceae	*Mortierella*
iv. Saksenaeaceae	*Saksenaea*
v. Syncephalastraceae	*Syncephalastrum*
b. Entomophthorales	*Basidiobolus*
	Conidiobolus
C. Ascomycotina	
1. Hemiascomycetes	
a. Endomycetales	
i. Spermophthoraceae	*Metschnikowia*
ii. Endomycetaceae	*Endomyces*
iii. Saccharomycetaceae	*Kluyeromyces*
	Lodderomyces
	Pichia
	Saccharomyces
2. Ascoloculomycetes	
a. Dothideales	
i. Testudinaceae	*Neotestudina*
ii. Pleosporaceae	*Leptosphaeria*
iii. Piedraiaceae	*Piedraia*
3. Ascohymenomycetes	
a. Eurotiales	
i. Eurotiaceae	*Emericella*
	Eurotium
	Neosartorya
ii. Gymnoascaceae	*Ajellomyces*
	Arthroderma
	Emmonsiella
	Nannizzia
b. Sphaeriales	
i. Microascaceae	*Microascus*
	Petriellidium
	Pithoascus
ii. Ophiostomataceae	*Ceratocystis*

D. Basidiomycotina	
1. Teliomycetes	
a. Ustilaginales	
i. Filobasidiaceae	*Filobasidiella*
ii. Ustilaginaceae	*Ustilago*
2. Hymenomycetes	
a. Agaricales	
i. Coprinaceae	*Coprinus cinereus*
b. Aphyllophorales	
i. Schizophyllaceae	*Schizophyllum commune*
3. Gasteromycetes	
E. Fungi Imperfecti	
1. Blastomycetes	*Candida*
	Cryptococcus
	Loboa
	Pityrosporum
	Rhodotorula
	Torulopsis
2. Hyphomycetes	*Trichosporon*
a. "Moniliaceous"	
	Acremonium
	Arthrographis
	Aspergillus
	Beauveria
	Blastomyces
	Chrysosproium
	Coccidioides
	Epidermophyton
	Fusarium
	Fusidium
	Geotrichum
	Histoplasma
	Indiella
	Madurella
	Microsporum
	Paecilomyces
	Paracoccidioides
	Penicillium
	Prototheca
	Rhinosporidium
	Scopulariopsis
	Sporothrix
	Trichophyton
b. "Dematiaceous"	*Acrotheca*
	Alternaria
	Aureobasidium
	Cladosporium

Curvularia
Dactylaria
Drechslera
Exophiala
Ochroconis
Phialophora
Ramichloridium
Rhinocladiella
Sarcinomyces
Torula
Trichocladium
Tritirachium
Wallemia

3. Coelomycetes
   a. Sphaeropsidales — *Hendersonula*, *Macrophoma*, *Madurella*, *Phoma*, *Pyrenochaeta*

REFERENCES

1. Ainsworth, G. C. 1973. Introduction and keys to higher taxa. In G. C. Ainsworth, F. K. Sparrow, and A. S. Sussman (Eds.), *The Fungi,* Vol. IVA, Academic, New York, pp. 1-7.
2. Ajello, L. 1977. Medically important infectious fungi. Contrib. Microbiol. Immunol. *3*: 7-19.
3. Carmichael, J. W. 1979. Cross-reference names for pleomorphic fungi. In B. Kendrick (Ed.), *The Whole Fungus,* Proceedings of the Second International Mycological Conference, Vol. 1, National Museum of Natural Sciences and National Museums of Canada, Ottawa, pp. 31-42.
4. Conant, N. F., D. T. Smith, R. D. Baker, and J. L. Calloway. 1971. *Manual of Clinical Mycology,* 3rd ed., Saunders, Philadelphia, p. 340.
5. Kendrick, W. B., and J. W. Carmichael. 1973. Hyphomycetes. In G. C. Ainsworth, F. K. Sparrow, and A. S. Sussman (Eds.), *The Fungi,* Vol. IVA, Academic, New York, pp. 323-509.
6. Lodder, J. (Ed.). 1970. *The Yeasts. A Taxonomic Study,* 2nd ed., North-Holland, Amsterdam, pp. 914-919.
7. Medical Research Council. 1977. *Nomenclature of Fungi Pathogenic to Man and Animals,* Memorandum No. 23 (4th ed.), Her Majesty's Stationery Office, London.
8. Nickerson, W. J. 1953. Medical mycology. Annu. Rev. Microbiol. *7*: 245-272.

2

Mucorales

H. J. Scholer / University of Basel and Pharmaceutical Research Division, F. Hoffmann-La Roche & Co. Ltd., Basel, Switzerland

Emil Müller / Institute of Microbiology, Swiss Federal Institute of Technology, Zurich, Switzerland

M. A. A. Schipper / Centraalbureau voor Schimmelcultures, Baarn, The Netherlands

I. INTRODUCTION

This chapter is devoted to those members of the fungal order Mucorales which cause mycotic diseases in warm-blooded vertebrates, with special reference to practical, cultural identification of the species involved.

By "mycotic diseases" we exclusively mean those which are deep-seated, with the causative fungus developing in living tissues, particularly those of internal organs. Such diseases caused by the Mucorales have for nearly a century been referred to as mucormycosis (*Mycosis mucorina,* in Ref. 102), but in more recent years they were also called zygomycosis [105] or phycomycosis [78]. Since the diseases in which Mucorales are involved are fairly uniform with respect to epidemiology, clinical aspects, and pathology, and are, with very few exceptions [37], clearly distinct from the mycoses caused by other members of the Zygomycetes (particularly those of the order of Entomophthorales) or of the "Phycomycetes," we prefer mucormycosis. The use of this more specific, original designation was also recommended by Baker, to whom we owe a number of masterly descriptions of the various forms of the disease (e.g., Refs. 7-10).

Most of the Mucorales known to cause mucormycosis belong to the Mucoraceae, and within this family to a few genera: *Rhizopus (R.), Absidia, Rhizomucor (Rm.),* and *Mucor.* According to present knowledge, *R. oryzae* is the most frequent agent of the disease in humans, followed by *R. rhizopodiformis* and *A. corymbifera* [122]. The latter species seems to prevail as a pathogen in mammals other than humans. In more than 80% of the culturally proven human and animal cases one of the following four species was present: *R. oryzae, A. corymbifera, R. rhizopodiformis,* or *Rm. pusillus.* On the other hand, *R. microsporus, M. ramosissimus, M. circinelloides, M. indicus, Rm. miehei,* and the species from families other than Mucoraceae [*Mortierella (Mort.) wolfii, Cunninghamella bertholletiae, Syncephalastrum racemosum, Saksenaea vasiformis*] were isolated only from a few or even single human and/or animal cases, with the exception of *Mort. wolfii,* which was proved to be a frequent agent of mucormycotic bovine abortion in a limited area of the world (New Zealand).

Reliable cultural identification of the causative fungus was made in less than 50% of the cases of mucormycosis reported in the literature, and additional, hitherto unrecognized species may have been involved in some of the remaining cases. Additional species may also be found in the future, thanks to improved cultural identification. There are, then, reasons for discussing all of the species of Mucorales capable of growing at 37°C, including those not yet proven to cause disease, which may be regarded as "potentially pathogenic." Within the limits of this chapter, we have adopted this broader concept for the three main genera, *Rhizopus, Absidia,* and *Rhizomucor.*

The reader may still miss a number of names of Mucorales mentioned in literature and referred to as pathogenic. Some of these, e.g., *M. mucedo,* obviously were errors in diagnosis (these instances are discussed in Sec. XI), whereas most, e.g., *R. arrhizus,* are regarded as being identical with one or another of the accepted species and will be found in the respective synonymy list. For a complete presentation and discussion of synonymy, the reader must refer to the pertinent special articles.

No general keys to all the families, genera, and species of the Mucorales are given. Synoptic tables of the differential characteristics which are focused on the families and genera containing the pathogenic species are presented. It is hoped that these tables together with a selection of drawings and photographs will facilitate the diagnosis. Full species descriptions will be given only for the two most frequent agents, *R. oryzae* and *A. corymbifera.*

II. METHODS

A. Techniques

1. Morphological (Microscopic) Demonstration of Mucorales in Clinical Materials and Tissues

a. Diagnostic Significance

The positive microscopic finding of typical hyphae in tissue or materials such as sputum and pus is a proof of mucormycosis. Since the hyphae are, usually, numerous, the risk of false-negative findings is small. The high diagnostic value of microscopic morphology in tissue sections or smears is stressed by the fact that with culture, false-positive and even false-negative results may occur (Sec. II.A.2.a).

The hyphae of the Mucorales are almost always clearly distinguishable from those of other filamentous fungi, such as the Entomophthorales or *Aspergillus.* On the other hand, their microscopic morphology does not allow identification of the species involved, or even of the genus or family, except for the very rare instances when sporangia and sporangiospores are formed in vivo.

b. Direct Smear

An excellent procedure is the simple KOH (potassium hydroxide preparation) between slide and coverslip (cf. Fig. 1a).

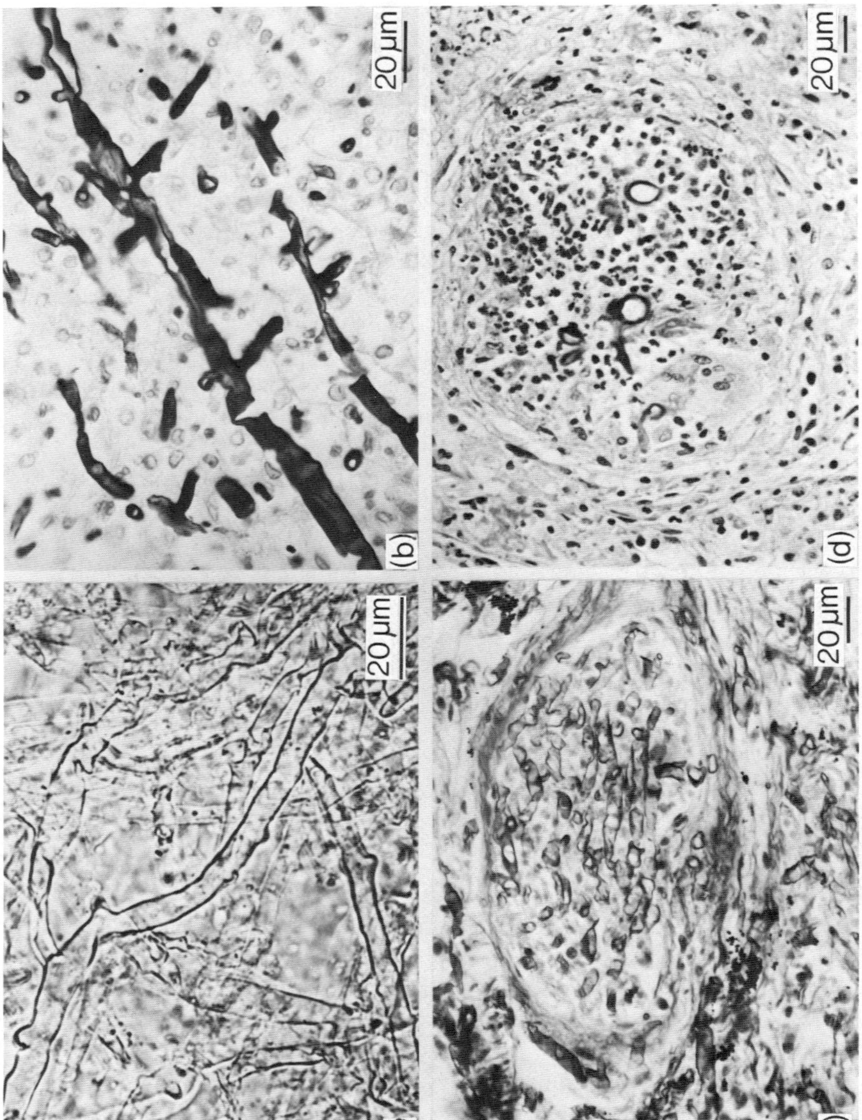

Figure 1 Morphology of Mucorales in direct smears and tissues: (a) KOH preparation from human parietal endocarditis; (b) methenamine silver-stained section from human valvular endocarditis; (c) methenamine silver-stained section from a human pulmonary arteriole (culture: *Rhizopus oryzae*); (d) PAS-stained section from spleen of a spontaneously diseased rabbit (culture: *Absidia corymbifera*). [Parts (b) and (c) from Ref. 51; part (d) from Ref. 125.]

c. Staining in Tissue Sections (see Ref. 9)

The most reliable method for demonstrating the Mucorales in tissues is the methenamine silver reaction of Grocott and Gomori by which the cell walls are stained more or less black. The intensity of the staining is, as a rule, less than in *Aspergillus* or *Candida,* and it is sometimes necessary to prolong various steps in the process by 50% or so, particularly the exposure to the chromic acid and/or methenamine silver nitrate solution. Hematoxylin or PAS (periodic acid-Schiff) are very efficient in some cases but, for unknown reasons, complete failures in others. The methenamine silver method is also applicable to direct smears, a procedure just discussed.

d. Morphological Characteristics

The hyphae of the Mucorales are thin-walled, irregularly broad with diameters from 3 to 12 μm or more, nonseptate or very rarely septate, and branching nearly at right angles or irregularly (Fig. 1a and b). They often invade the walls and lumina of blood vessels, and may give rise to thrombosis and infarction (Fig. 1c). In more chronic, granulomatous forms of mucormycosis, which are very rare in humans but obviously prevail in mammals such as cattle and horses [3], the fungal elements may become swollen, globose, or bizarre, with diameters up to 50 μm, and sometimes show yeast-like budding (Fig. 1d).

Formation of sporangiophores, sporangia, and sporangiospores was observed by the first author to describe human mucormycosis (Fürbringer, 1876 [47]) but only very exceptionally since then (some of the more recent reports are mentioned in Ref. 9). The proximity of a cavity containing air is probably required for such structures to form. With very well developed sporulation it may be possible to identify the genus and even species of the fungus [73,74]. Zygospore formation has not been found so far in vivo.

2. Primary Culture from Clinical Materials and Tissues

a. Significance

Culture is in most of the cases of mucormycosis the only method leading to species identification of the causative fungus.

Since the sporangiospores of Mucorales are quite widespread in soil, plant material, and in the air, false-positive cultures occur. This risk exists particularly in veterinary medicine, e.g., in the diagnosis of mycotic abortion. As a general rule, the fungus grown in culture should not be considered a causative agent unless there is a corresponding morphological finding from direct microscopy or histology (Sec. II.A.1).

Growth of the agents of mucormycosis in primary culture is, usually, abundant and does not require the use of particularly rich media. For unknown reasons false-negative cultures sometimes occur in the presence of numerous typical hyphae in the direct smear or histological section (e.g., Refs. 33 and 128). Perhaps the survival of the hyphae outside the body is extremely short in some cases. It is recommended that only fresh material be cultured. It was recently suggested that mechanical homogenization of the material may impair viability of the hyphae and should, therefore, be avoided [65a].

The primary culture often permits final identification of the species without the need of subculture on the special media discussed in Sec. II.A.3.b.

b. Media and Incubation Temperature

Most of the agar media used for routine primary culture of fungi are appropriate, provided that they do not contain cycloheximide, which is an inhibitor of molds, including the Mucorales. Media containing sucrose (according to older formulations such as Sabouraud's) or nitrate (e.g., Czapek's medium, which in particular is used for *Aspergillus*) are less suitable than those containing dextrose as a source of carbon, and ammonium or peptone as a source of nitrogen. The only growth factor required by some of the Mucorales (e.g., *A. corymbifera*) is thiamine, but the ingredients of ordinary media, particularly peptones, usually contain sufficient quantities, so that addition of thiamine is necessary only for semisynthetic media with purified ingredients. A satisfactory medium is the ordinary Sabouraud's dextrose agar with 1-2% dextrose, about 1% peptone, and 2-3% agar. Addition of antibacterial antibiotics (e.g., chloramphenicol and gentamycin) is recommended to prevent bacterial growth.

All agents of mucormycosis grow very well at 37°C except for some members of the "*M. circinelloides* group," for which this temperature is just at the upper limit of tolerance. Because of the potential occurrence of these rare agents of mucormycosis, an incubation temperature of 30°C is most appropriate for general use in primary culture. When one can expect the presence of one of the more frequent, thermotolerant agents such as *R. oryzae, A. corymbifera,* or *Rm. pusillus,* a higher incubation temperature, e.g., 40 or 44°C, has the advantage of being more selective since many other fungi and bacteria will not grow.

3. Cultural Identification of the Species, Including Mating Studies

a. Significance

Identification of a member of the Mucorales—with all the characteristics of various families, genera, species, and individual strains of these fungi—is essentially based on morphology, mating behavior, and dependence on environmental factors, e.g., temperature tolerance and nutrition, in culture. Thus the techniques used for culture, e.g., the composition of the media, are important. Although the morphology of the Mucorales is not markedly dependent on such variables as the ingredients of the medium, the investigator should follow the principles indicated in Sec. II.A.3.b. For the more functional tests (e.g., those of mating behavior and biochemical properties), the use of strictly standardized methods is compulsory (Secs. II.A.3.c-g).

b. Morphology in Culture (Asexual State)

A great variety of simple culture media are suitable, e.g., Sabouraud's dextrose agar, which is commonly used for primary isolation (Sec. II.A.2.b); malt agar; "synthetic mucor agar" (SMA); potato dextrose agar (PDA); and yeast extract agar (YEA). Unless stated otherwise, the descriptions given in this chapter refer to fungi grown on one of these media, which were also used in the extensive taxonomical work of C. W. Hesseltine, J. J. Ellis, and their associates, and of M. A. A. Schipper. For the composition of SMA, PDA, and YEA, see Refs. 56 and 57.

The temperature of incubation may also vary within broad limits. The temperature of 37°C, commonly used in medical laboratories, is very suitable for all Mucorales in question except those of the "*M. circinelloides* group" (cf. Sec. II.A.2.b). Somewhat lower temperatures (e.g., 28, 30, or 34°C) are more generally applicable. Ordinary room temperature is also suitable provided that it is higher than 20°C; when it is lower than 20°C, the thermophilic species, particularly those of *Rhizomucor,* will grow poorly or not at all [26,123]. A temperature within the range 25 to 37°C was used in all studies we refer to here, unless otherwise stated.

With the use of an incubator the cultures are usually kept in darkness, whereas at room temperature they are periodically exposed to daylight. Since the possible impact of a light factor is unknown for most of the species to be treated, it is generally neglected in this chapter. In the few existing studies, light was without appreciable influence on *Rm. pusillus* [117], whereas it stimulated pigment production in *M. circinelloides* [115] and *M. amphibiorum,* and production of both pigment and sporangia in *M. indicus* [117].

The study of morphology, of course, includes both macroscopical cultural appearance and microscopic morphology in mounts from the same cultures, or in slide cultures. It should be remembered that the mounting fluid may produce some structural changes (e.g., swelling of sporangiospores, disintegration of the sporangium membrane, and invagination of the columella). The microscope used for these studies must be equipped with a micrometer.

Every culture should, in addition, be examined with a good magnifying glass or even better with a stereomicroscope, which should again be equipped with a micrometer. This is the best method for the study of the various fungal structures in their undisturbed natural disposition, e.g., the sporangiophores arising from the substratum or from the aerial mycelium, their ramification and length. Not infrequently the genus, or sometimes even the species, can be diagnosed with just a stereomicroscope.

c. *Mating Studies*

The conditions for successful mating must be determined empirically for each species, but some general rules may be given. Use petri dishes with beerwort agar and/or YEA; if unsuccessful, try with another complex medium, such as whey agar, prune agar, or cherry agar [113]; inoculate the two strains as a spot each on the agar surface at a distance of a few millimeters; incubate in darkness [114,118] at a temperature that is rather below the optimum temperature of asexual growth of the species in question, and definitely (by 15-20°C) below the maximum temperature of growth of the species in question (e.g., with *Rm. pusillus,* whose maximum temperature of growth is 54-56°C, zygospores were obtained at 30-40°C, and *M. circinelloides,* which grows up to about 38°C, gave positive matings at 25°C [26,115,117]); examine with a stereomicroscope and/or a standard microscope at 2-day intervals from the time when the two colonies touch each other, with a final reading after 3 weeks. The zygospores are easily seen under the stereomicroscope when they differ in color from the sporangia (e.g., in *R. microsporus,* where they are reddish and the sporangia brown), whereas microscopic examination of smears is usually necessary when they have a similar color (e.g., in *Rm. pusillus*).

d. *Maximum Temperature of Growth*

All media mentioned above (Sec. II.A.3.b) are suitable for recording temperature tolerance, but most figures given in this chapter were obtained from SMA. Incubation temperature was increased with 2°C steps and the final reading was made after 10-14 days each. Any amount of growth was read as positive.

e. Thiamine Requirement

Comparative use of SMA without and with thiamine is suitable, provided that the asparagin contained in this basal medium is reasonably vitamin-free. Thiamine dependence is assumed when there is a marked difference of growth on the two media. In every case some restricted growth will occur in the absence of the vitamin.

f. Assimilation (Utilization) Tests

The basal agar medium must be without a carbon source for the carbon assimilation tests, and without a nitrogen source for the nitrogen (nitrate) assimilation tests. For most of the tests referred to in this chapter we used Yeast Nitrogen Base and Yeast Carbon Base from Difco, to which we added 2-3% of agar. The substrates to be tested were added at a concentration of 1% (carbon sources) and 0.5% (potassium nitrate), respectively. The inoculum should consist of a sporangiospore suspension whenever feasible, and both a negative (basal medium) and a positive control (basal medium with dextrose for carbon assimilation and with ammonium or asparagin for nitrate assimilation) should always be used. Incubation is at an optimum temperature (25-37°C according to the species) for 7-12 days. Assimilation is considered positive when growth is comparable to that of the positive control, and weakly positive when growth is less than in the positive control but still definitely more than in the negative control (growth in the latter should be very restricted or absent).

g. Fermentation Tests (Production of Gas from Carbon Sources)

The numerous inherent technical problems in fermentation tests were reviewed by Inui et al. [68]. The three most important points are:

1. Since most of the Mucorales normally grow only on the surface of liquid media, forming a membrane covered by aerial mycelia, most of the gas they produce is released directly into the air and is not easily demonstrated unless one uses gas chromatography or similar equipment [36]. One should not use Durham tubes but rather V-shaped fermentation tubes or even better the classical Einhorn tubes. Any growth appearing on the surface of the medium in the open part of the tube must periodically be transferred to the closed end of the tube by means of a loop or hook; the transferred part of mycelial mat will continue to produce gas, which is now collected in the closed end of the tube. The result of the test essentially depends on such a procedure.
2. Gas production may depend on the concentration of the substrate; e.g., it may be undetectable at 1% of dextrose but definitely positive at 5%.
3. Furthermore, gas production may require, or be stimulated, by some "vitamins"; e.g., it may be negative without, but positive with, the presence of 1% of yeast extract.

In the fermentation tests referred to here, we used essentially the techniques recommended by Inui et al. [68], but our basal medium again was Yeast Nitrogen Base (Difco) to which we added 1% of yeast extract. The concentration of the carbon sources was 2% and 5%. Basal medium plus yeast extract was used for negative controls. Einhorn tubes were inoculated with sporangiospores and incubated at 24, 30, or 37°C for up to 10-14 days. Definite gas production was read as positive fermentation (small gas bubbles, which also occurred in the negative controls, were neglected).

4. Serological Diagnosis of Mucormycosis

In spite of encouraging results obtained in some cases [69], serological diagnosis of mucormycosis has been considered difficult and unreliable [124]. Perhaps this negative statement must be revised in view of more recent findings [65a].

B. Principles of Taxonomy

1. General Considerations

The taxonomy of Mucorales is based mainly on the morphological characteristics of the spore-producing life states. Benjamin [14] distinguished sexual fructification (teleomorph), manifested exclusively by zygospores (Fig. 2), and asexual fructification (anamorph). Asexual fructification was divided into three different types: (1) endogenous formation of sporangiospores within sporangia; (2) endogenous formation of chlamydospores within vegetative hyphae or, exceptionally, sporangiophores; and (3) exogenous separation of chlamydospores, commonly called gemmae, which may show yeast-like budding (Figs. 3-5). It is not always easy to distinguish between chlamydospores and gemmae, and these two types are regarded as the same by some authors. The simplifying concept of only two classes of asexual spores—(1) the sporangiospores produced by the sporangiosporogenous apparatus (for the homologs of sporangia, see Sec. III.A), and (2) chlamydospores and gemmae essentially produced in, or from vegetative hyphae—has been adopted for the present treatise. The morphology of the purely vegetative state provides only a few characteristics which are useful in taxonomy, e.g., colony color and height of the aerial mycelium. More diversity is offered by some of the functions of the fungi, i.e., response to temperature and nutritional factors and the biochemistry of the decomposition of various substrates.

With respect to taxonomy, a hierachy of all these characteristics exists, some of them being highly significant and others less so. The actual "weight" of the characteristics is determined by their stability (reproducibility), their statistical distribution within a given group of fungi, and their association with other characteristics [100]. There is considerable variation of characteristics among strains of a given species and, on the other hand, some overlapping of the extremes from the strains of different species, although the means (e.g., those of sporangiophore length or diameter of the sporangia) may be clearly different. A certain characteristic, such as sporangiospore form or carbohydrate assimilation, may be useful for the distinction of species within one fungal group but of no use within another [123]. Overemphasis of single characteristics observed in individual strains has produced enormous confusion in the past. For example, *A. corymbifera* has no less than 48 synonyms [39,100]. Some of the problems are still unsolved in the genus *Rhizopus,* which currently awaits monographic revision. Unfortunately, the original descriptions referring to many of the existing taxa are in Japanese and cannot be read by us.

In the following, the general significance of both the most used and most misused characteristics is indicated (focused mainly on the family Mucoraceae); a detailed proof of our views is beyond the limits of this chapter.

2. Characteristics of Generally High Taxonomical Value

a. *Zygospores* (Fig. 2).

In all zygosporogenic—homothallic and heterothallic—species known to us, the morphology of the zygospores is species-specific, although the differences may be small, as are those between *Rm. miehei* and *Rm. pusillus* [120].

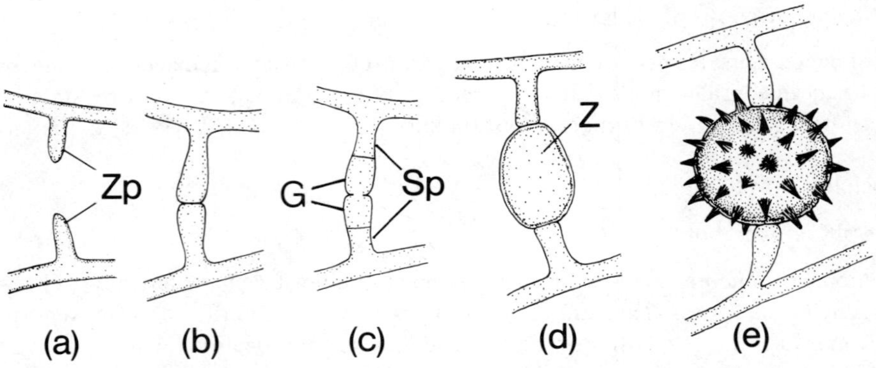

Figure 2 Mucorales. Development of the zygospore. (a) Two sexually differentiated side branches (zygophores, Zp) grow toward each other. (b) The zygophores touch; their tips swell (progametangia, still in open connection with the parent hypha). (c) The gametangia (G) are cut off by septa from the suspensors (Sp), and the cell walls dissolve at the area of contact. (d) The fused gametangia form a young zygospore (Z) with progressively thickening wall. (e) Mature, ornamented zygospore. (From Ref. 88.)

The formation of true zygospores in matings of self-incompatible, obviously similar strains is considered a proof that these strains belong to the same species. Incomplete conjugation or nonappearance of any visible reaction, however, does not necessarily provide evidence to the contrary [117].

Although zygospores have the "last word" in taxonomy, their use in practical diagnosis is limited by several facts, including the need for tester strains, the occurrence of incompatibility between (+) and (-) strains of the same species, and the consistent absence of any zygospore formation in some of the species.

b. Sporangiosporogenous Apparatus (Figs. 3–5)

Almost any family, genus, and species of Mucorales can be fully determined by the morphology of this type of asexual fructification. Grouped according to the form elements, which will be defined later, particularly in Secs. III.A, III.B, and IV.A, the most valuable characteristics are:

Sporangiophores: Length; ramification (presence, mode); rhizoids and stolons (presence); color for some genera *(Rhizopus, Rhizomucor, Thermomucor)*

Sporangia: Form; diameter; color; membrane [smooth or shiny; transparent, glittering, or opaque (sometimes depending on the age of the culture); diffluent, rupturing or persistent]

Columella: Form; diameter; color; presence of projections

Apophysis: Presence; length; form

Sporangiospores: Size; form (may indeed be variable within a given species: *A. corymbifera*); surface (smooth or, as in most members of *Rhizopus,* wrinkled = "striated"); occasionally color

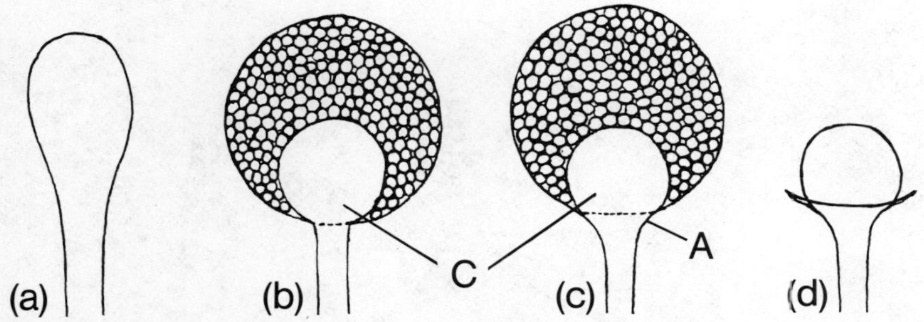

Figure 3 Mucoraceae. Development of sporangia: (a) swelling of sporangiophore tip; (b,c) sporangial differentiation with formation of columella (C), sporangiospores, and sporangium membrane without (b) and with (c) apophysis (A); (d) columella with circular remnants of the sporangium membrane (collar) after release of the sporangiospores.

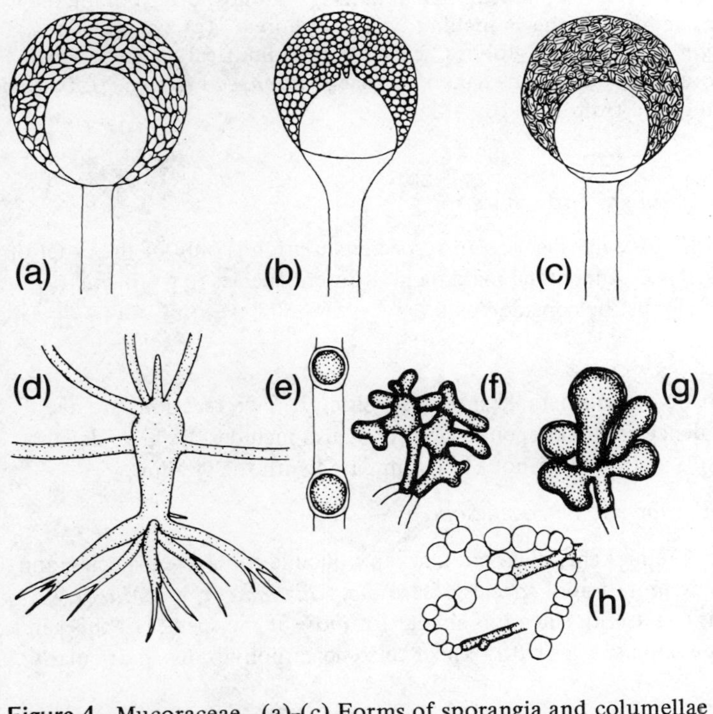

Figure 4 Mucoraceae. (a)-(c) Forms of sporangia and columellae [(b) and (c) for apophysis]: (a) is characteristic of *Mucor* and *Rhizomucor,* (b) of *Absidia,* and (c) of *Rhizopus.* (d) Rhizoid *(R. oryzae):* presence of rhizoids is characteristic of *Rhizopus* and *Rhizomucor,* to some extent also of *Absidia.* (e)-(h) Chlamydospores and/or gemmae (Sec. II.B.1): although often impressive, of comparatively little taxonomic value (from Ref. 14): (e) is *M. racemosus,* (f) and (g) are *A. corymbifera,* and (h) is *M. bainieri,* which according to Ref. 115 is a member of the *M. circinelloides* group.

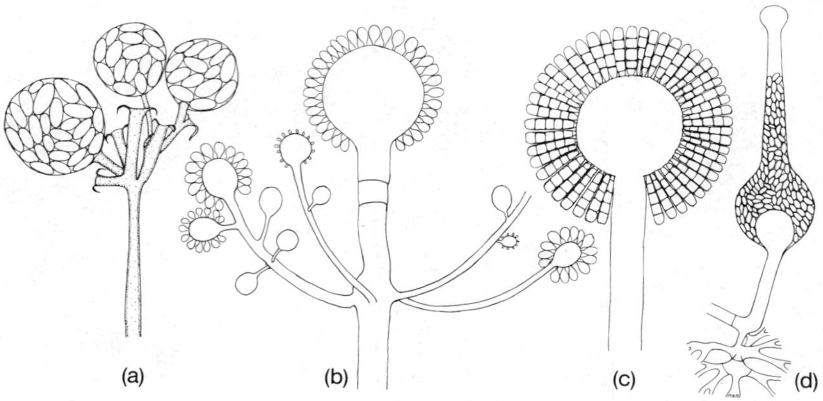

Figure 5 Sporangia and their homologs in families of Mucorales other than Mucoraceae. (a) Mortierellaceae *(Mortierella wolfii):* columella absent or very tiny. (Modified from Ref. 82.) (b) Cunninghamellacea *(Cunninghamella bertholletiae):* monosporic sporangiola ("conidia") on globose swellings of the branching "conidiophore." (c) Syncephalastraceae *(Syncephalastrum racemosum):* globose "head" with cylindrical merosporangia containing sporangiospores in one row. (d) Saksenaeaceae *(Saksenaea vasiformis);* flasked-shaped sporangium. (Modified from Ref. 109.)

c. Maximum Temperature of Growth

This is highly characteristic for each species. Intraspecies variation is only of the order of $\pm 2°C$. Indeed, the occurrence of identical maxima in different species (e.g., *R. microsporus* and *R. oligosporus*) must be considered.

d. Thiamine Requirement

This is again a characteristic of considerable species specificity. Two reactions are distinguished: (1) independence and (2) dependence, which also includes "relative dependence," in which growth is restricted without but stimulated with the vitamin.

e. Assimilation of Carbon and Nitrogen Sources

The assimilation pattern is remarkably clear-cut and reproducible for a given species, and has proven useful in the delimitation of species of *Absidia, Rhizomucor,* and *Mucor,* whereas in *Rhizopus* it is less useful, since it is similar for most of the species of this genus and may differ within the same species with respect to some carbohydrates, particularly sucrose [121].

3. Characteristics of Limited Taxonomical Value

a. Chlamydospores (Fig. 4e-h)

In most of the species the number of chlamydospores (or gemmae; see Sec. II.B.1) varies considerably from one strain to the other, in part depending on cultural conditions. For example, formation appears to be stimulated by low oxygen and high carbon dioxide tension (see Ref. 12). On the other hand, some species (e.g., *M. racemosus*) are characterized by a general abundance of chlamydospores and a few others (the psychrotolerant *R. stolonifer* and *R. sexualis*) by their virtual absence.

b. Colony Color

Although the "basic color" is typical for most of the species, the individual strains are often marked by differences in shade and, particularly, in intensity. Since in most of the species only the sporangiosporogenous apparatus is pigmented, the intensity of colony color depends essentially on the proportion fertile to sterile (aerial) mycelia (Sec. II.B.3. c). Thus, color is particularly uncharacteristic in the poorly sporulating strains, which all appear more or less "whitish."

c. Height of the Colony and Proportion Fertile (Sporangiosporogenous) to Sterile Aerial Mycelia

In a few species, particularly of *Rhizomucor,* the colony height is only a few millimeters, whereas in most others it is generally more than this, but varies according to the individual strains from about 5 to 30 mm. In many of these instances, the height is determined by sterile aerial mycelia, which are abundantly developed in the high strains but less so in the low strains. Generally, there exists a reciprocity between the development of fertile and sterile aerial mycelia, so that the strongly sporulating strains are comparatively low and the poorly sporulating high. Since the sterile aerial mycelium is generally hyaline, whereas the sporangiosporogenous apparatus is pigmented, the colony color of the low strains is often more intense than that of high strains (see Sec. II.B.3.b). This sort of "interconnected variation" is quite common in *A. corymbifera* and in the species of *Rhizopus,* where both colony height and proportion between fertile and sterile mycelia are, consequently, of little taxonomical value.

d. Fermentation Spectrum

Fermentation capacity is lacking in some species *(Rm. pusillus, Rm. miehei, A. corymbifera)* whereas it is very pronounced in others *(Rhizopus* spp. and of the "*M. circinelloides* group"). Because of the technical problems mentioned in Sec. II.A.3.g, which particularly exist with weakly fermented substrates, it is not easy to determine the exact fermentation spectrum of the fermenting species. The authors doubt whether the pattern described for some species is reliable, and they reject species that were defined only on the basis of different fermentation spectra. In the present treatise, we distinguish only between "fermenting" and "nonfermenting" species, judged mainly by their reaction with dextrose.

III. THE ORDER MUCORALES AND ITS FAMILIES

A. Systematic Position of the Order; General Characteristics; "Opportunistic" Pathogenicity

The Mucorales belong to the class Zygomycetes, which are mainly characterized by their sexual reproduction: The fusion of coenocytic gametangia, called zygogamy, results in a thick-walled zygote, called a zygospore. The vegetative mycelium is coenocytic, nonseptate or rarely septate, so that in older textbooks these fungi have been included in the broad, ill-defined group of "Phycomycetes."

The Zygomycetes include two orders that are of interest to medical mycology:

Mucorales and Entomophthorales, which differ mainly in asexual, but to some extent also sexual reproduction. In the Mucorales, nonmotile sporangiospores are formed in sporangia, or in the subdivisions of sporangia usually called merosporangia when they are elongate and contain sporangiospores in one row, and sporangioles when they are globose and contain clusters of a few (or a single) sporangiospores; single-spored sporangioles occurring in some families are recognized as conidia when they are discharged in toto and when their membrane is coalescent with the wall of the spore (see Figs. 3, 4, and 5a-c). The zygospores develop directly from the fused gametangia (Fig. 2). Ornamentation and color of the zygospore wall, equality or inequality of the size of the suspensors, and presence or absence of projections originating from the suspensors may contribute to the definition of families and genera. The pathogenic Entomophthorales are treated in Chap. 3.

The order Mucorales consists of 14 families [64]. For the families including pathogenic species, see Sec. III.B. Many of these fungi are not known to form zygospores, but their systematic position is easily established by their asexual characteristics.

Most of the Mucorales grow rapidly on simple culture media such as Sabouraud's dextrose or maltose agar, malt agar, or Synthetic Mucor Agar (SMA) and produce high and dense aerial mycelia, which often fill the tubes or plates within a few days. The natural habitat is exclusively terrestrial. The members of most families are free-living saprophytes in soil, on decaying vegetable matter such as stale bread and other food, manure from various mammals, and so on, whereas those of only a few families are, or can be, parasites on other fungi or higher plants. Many species, including those capable of producing disease in humans or animals, are widely distributed and their asexual spores are prevalent in the air. Pathogenicity to vertebrates is a very exceptional feature in the life of these fungi and is determined essentially by certain rare predispositions of the host (diabetes, malignant hemopathia, immunosuppression, etc.). In the host tissues, only a vegetative mycelium is usually formed (Fig. 1), the morphology of which is characteristic of the Mucorales and indistinguishable among all the species of the order (Sec. II.A.1.d).

B. Characteristics of the Families Known to Include Pathogenic Species (see Hesseltine and Ellis [64])

1. Mucoraceae

All sporangia are globose or pyriform and contain several, most often many sporangiospores and are provided with an infertile basal part called the columella, which has the appearance of a terminal swelling of the sporangiophore invaginated into the sporangium *sensu strictiori.* The columella may or may not show an apophysis, i.e., a basal section not surrounded by the sporangium membrane and remaining outside the sporangium *sensu strictiori.* The sporangium membrane is usually not cutinized and is persistent or diffluent; in the latter case it may leave a "collar" (Fig. 3). Merosporangia, sporangioles, or conidia are always absent. The particular morphology of the sporangium is important for the definition of genera (Fig. 4a-c), whereas the "chlamydospores" are of limited taxonomic significance (Fig. 4e-h; Sec. II.B.3.a). Twenty genera are recognized [64]. For those containing pathogenic species, see Sec. IV. The type genus is *Mucor.*

2. Mortierellaceae

Sporangia from one- to many-spored, globose, columella absent or very small (Fig. 5a). (The small, few-spored sporangia are also called sporangiola.)

Four recognized genera [64]. For the only pathogenic species, *Mort. wolfii,* see Sec. X.

3. Cunninghamellaceae

Asexual reproduction by monosporic sporangiola ("conidia"; see Sec. III.A), which are often echinulate, never in chains, and are borne on conspicuous terminal swellings of the "conidiophores" (Fig. 5b).

Four recognized genera [64]. For the only pathogenic species, *Cunninghamella bertholletiae,* see Sec. X.B.

The family was erected in 1939 by Naumov [93] as a subsection of the Choanephoraceae, to separate the genera forming only conidia. For some authors the members of *Cunninghamellaceae* (particularly those of the genus *Cunninghamella*) still belong to this family [152].

4. Syncephalastraceae

Relatively few-spored, cylindrical merosporangia produced on a terminal swelling of the sporangiophore; sporangiospores arranged in a single row, sporangium wall deliquescing (Fig. 5c). The sporangiophores are irregularly branched and are usually borne from rhizoids.

This family is monogeneric, and its single genus, *Syncephalastrum,* comprises three species [64,86,101]. For the only pathogenic species, *Syncephalastrum racemosum,* see Sec. X.C.

Formerly, *Syncephalastrum* was placed together with three genera, such as *Syncephalis* (which are mostly obligate parasites on other Mucorales), in the family Piptocephalidaceae [58].

5. Saksenaeaceae

All sporangia flask-shaped or sausage-shaped, with or without columella; sporangium wall smooth or with several spines (Fig. 5d).

Two recognized genera. For the only pathogenic species, *Saksenaea vasiformis,* see Sec. X.D.

The family was erected by Ellis and Hesseltine in 1974 [40]. Earlier, *Saksenaea* was considered a genus of Mucoraceae.

6. Summary of the Differential Characteristics of the Families Known to Include Pathogenic Species

Table 1; Figs. 3 and 5.

IV. THE GENERA OF MUCORACEAE KNOWN TO INCLUDE PATHOGENIC SPECIES

A. *Rhizopus*

The sporangiosporogenous mycelium consists at least partly of stolons extending between "knots," which bear rhizoids (Fig. 4d) fixing themselves to the substratum, and sporangiophores arising to the opposite side, single or in tufts of two to five, mostly unbranched.

Table 1 Differential Characteristics of the Families of Mucorales that Include Pathogenic Species

Family	Sporangiosporogenous apparatus	See Fig.:	Zygospores
Mucoraceae	Sporangia with columella; globose, pyriform, or ellipsoidal	3, 4a–c	Between equal or unequal suspensors
Mortierellaceae	Sporangia and sporangiola without (or with tiny) columella; globose	5a	Between tongue-shaped suspensors (not known for most of the species, including *Mort. wolfii*)
Cunninghamellaceae	Single-spored sporangiola (= "conidia") on globose or greatly elongated vesicles	5b	Between equal or unequal suspensors
Syncephalastraceae	Cylindrical merosporangia with spores in one row, on globose vesicles	5c	Between equal suspensors
Saksenaeaceae	Flask- or sausage-shaped sporangia with or without columella	5d	Not known

This "Rhizopus pattern" only exceptionally comprises the whole growth of the fungus, but in any species there also occur: (1) sporangiophores not originating from the substratum but from knots scattered along aerial hyphae, with or without the presence of rhizoids (Fig. 6), and (2) aerial hyphae which cannot be identified as stolons and which may form an extensive sterile mycelium. The term "Knoten" (English: knot) was introduced by Fischer in 1892 [43] to describe the region of the conjunction of stolons, rhizoids, and sporangiophores and does not necessarily imply a swelling.

Sporangia globose; columella globose to slightly ellipsoidal, with definite but usually inconspicuous apophysis (Fig. 4c); sporangiospores often (but not in all species) angulated and striated. Zygospores, if produced, usually with unequal, more or less globose suspensors.

Positive assimilation of ethanol as the sole source of carbon; no assimilation of nitrate; significant capacity to ferment sugars.

The number of species is difficult to estimate because of the lack of an up-to-date study of the genus, but is probably fewer than 10. The type species is *R. stolonifer.* For the pathogenic and "potentially pathogenic" species, see Sec. V.

B. *Absidia*

All sporangia are pyriform; the columella, which is always provided with a conspicuous, conical apophysis, is usually hemispherical on its top and often bears one or more projections (Fig. 4b). The sporangiophores may originate from stolons in the intervals between rhizoids or from a finely branched aerial mycelium (the latter is the case for the pathogenic species; Fig. 7). In the zygosporogenic species the suspensors may carry projec-

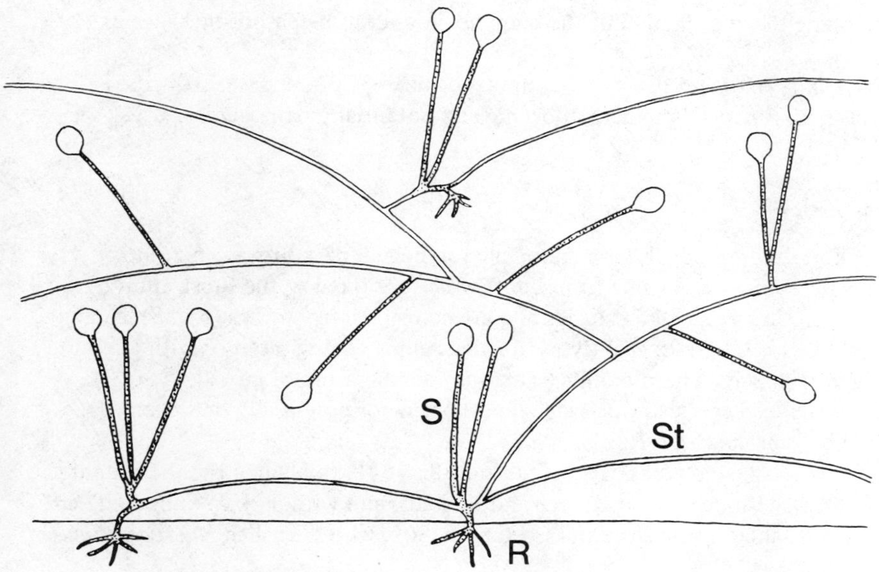

Figure 6 *Rhizopus.* "*Rhizopus* pattern" of sporangiosporogenous mycelium. R, rhizoid; St, stolon; S, sporangiophore. The knots are the points of origin of sporangiophores and typically, but not always, also of rhizoids and stolons.

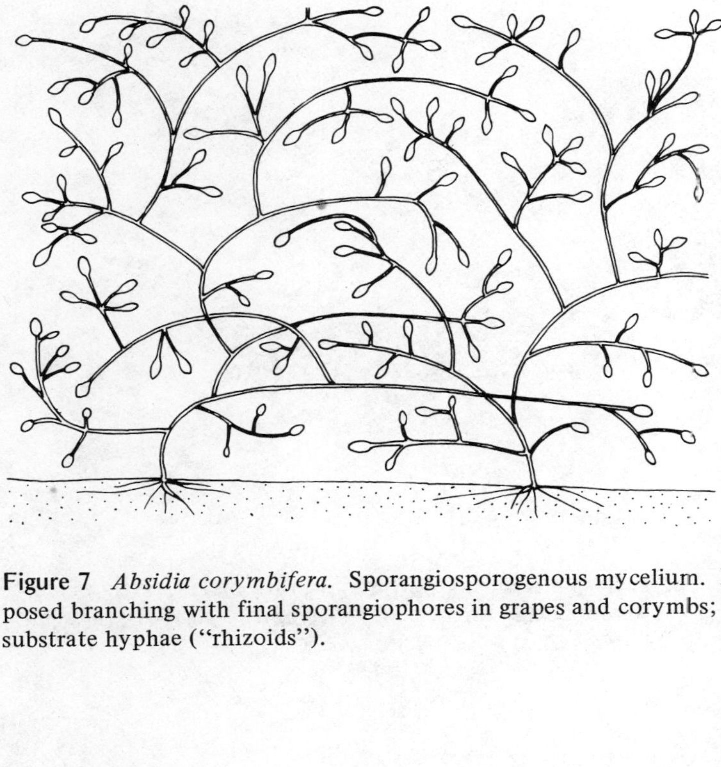

Figure 7 *Absidia corymbifera.* Sporangiosporogenous mycelium. Intensive, superimposed branching with final sporangiophores in grapes and corymbs; radially diffusing substrate hyphae ("rhizoids").

tions enveloping the zygospore, but these are not found in the pathogenic species, *A. corymbifera.*

Twenty-six species and five varieties are recognized [1,38,39,61,90,100,135]. The type species is *A. repens.* For the pathogenic and potentially pathogenic species, see Sec. VI.

C. *Rhizomucor*

Sporangiophores monopodially or sympodially branched, dark brown, in a similar way as in *Rhizopus,* sometimes arising from knots which are fixed to the substratum by weakly branched, often only spine-like rhizoids and are connected by stolons (Fig. 8); sporangia globose; columella globose to slightly pyriform or ellipsoidal without or with a minute apophysis (see Fig. 4a). Thermophilic (maximum temperature of growth > 50°C).

Three recognized species, which are all pathogenic or potentially pathogenic (see Sec. VII). The type species is *Rm. pusillus.*

This genus was reerected by Hesseltine and Ellis [64], to include the thermophilic species of *Mucor* with occasional stolons and rhizoids, and with dark sporangia. The genus name was taken from Lucet and Costantin [80], who described *Rm. parasiticus*

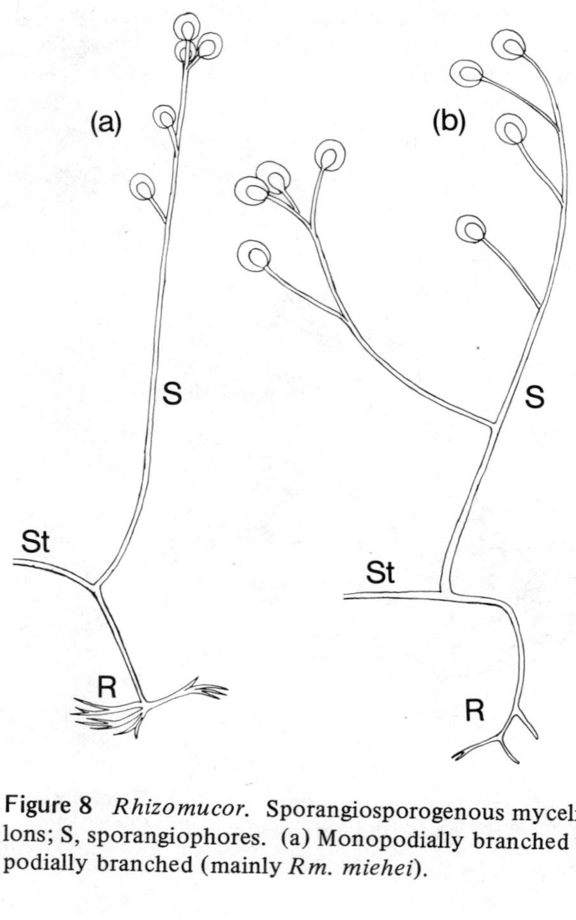

Figure 8 *Rhizomucor.* Sporangiosporogenous mycelium. R, rhizoids; St, origin of stolons; S, sporangiophores. (a) Monopodially branched (typical of *Rm. pusillus*); (b) sympodially branched (mainly *Rm. miehei*).

without being aware that the same fungus was already described as *M. pusillus* [79]. The concept of a separate genus was supported by scanning electron microscopy, proving that ornamentation of the zygospores is different from that of all other known zygospores in *Mucor* [120].

A new, monospecific genus, *Thermomucor,* which is closely related to *Rhizomucor,* was recently erected [118,132]. The single species, *T. indicae-seudaticae,* is considered potentially pathogenic and is included in Sec. VII.

D. *Mucor*

Sporangiophores not originating from stolons (distinction mainly from *Actinomucor* and *Rhizomucor*), almost always shorter than in *Phycomyces* and without metallic shine, unbranched or irregularly branched (distinction from *Dicranaphora* and *Syzygites,* the sporangiophores of which are dichotomically branched); no rhizoids (distinction mainly from *Rhizopus* and *Rhizomucor*); sporangia globose, without apophysis (see Fig. 4a; distinction from *Rhizopus* and *Spinellus*); in the zygosporogenic species suspensors without projections (distinction from *Parasitella*).

This mostly negative definition reflects the fact that the genus *Mucor* is a "torso" remaining after the other genera of Mucoraceae were separated on the basis of their particular characteristics.

Thirty-six species and 14 forms or varieties are recognized [116]. The type species is *M. caninus.* For the pathogenic and potentially pathogenic species, see Sec. VII.

E. Summary of the Differential Characteristics of the Pathogenic Genera of Mucoraceae

Table 2, Figs. 4a-c, 6-8.

V. THE PATHOGENIC SPECIES OF *RHIZOPUS*

A. *Rhizopus oryzae* Went & Prinsen Geerligs, *Verh. K. Ned. Akad. Wet. Afd. Natuurkd. Reeks 2, 4:* 16, 1895 [146]

1. Occurrence as a Pathogen

Rhizopus oryzae was recorded in roughly 60% of the cases of culturally proven human mucormycosis of any form and nearly 90% of the instances of the cranial (rhinocerebral) form of the disease [122]. On the other hand, the species was isolated only exceptionally from mucormycosis in animals [3,97].

2. Species Description

Mycelium high (1 cm or more after a few days), in shades of dirty grayish yellow to grayish brown, sometimes slightly olive, from very light (almost white) to dark, depending on the proportion of the pigmented sporangiosporogenous to the hyaline sterile mycelium; sporangia usually visible by the naked eye.

"Rhizopus pattern" of the sporangiosporogenous mycelium always present (Fig. 9) but—in a proportion which is characteristic of the individual strain—mixed up with a more irregular pattern (Sec. IV.A; see Fig. 6). *Stolons* 15-25 μm thick, hyaline but up to about 35 μm thick and slightly brown in the region of the knots. *Rhizoids* mostly

Table 2 Differential Characteristics of the Genera of Mucoraceae that Include Pathogenic Species

Genus[a]	Rhizoids and stolons	Sporangiophores	Sporangia	Columellae	Apophysis	Sporangiospores	Zygospores when present	See Fig.:
Rhizopus	Present	Single or in tufts; usually unbranched; mostly brown	Globose; gray or brown; mostly shiny	Subglobose to slightly elongated	Present (mostly inconspicuous)	Mostly angulated, and striated	Rough; mostly between unequal, globose suspensors	4c, 6
Absidia	Present but often indistinct	Finely branched, in grapes and corymbs; almost hyaline[b]	Pyriform	Hemispherical; often with projection(s)	Present; conspicuous, conical	Globose to cylindrical; smooth	Smooth to slightly roughened; with equatorial ridges; suspensors almost equal[b]	4b, 7
Rhizomucor	Present	Monopodially or sympodially branched; dark brown	Globose; gray; opaque and glittering	Subglobose; brown	Absent (or very tiny)	Globose to subglobose; small; smooth	Rough; between equal suspensors	4a, 8
Mucor	Absent	Branched or unbranched; mostly hyaline	Globose	Various forms (e.g., globose, depressed, pyriform, elongated)	Absent	Globose to cylindrical	Rough; between equal suspensors	4a

[a]Addendum: *Thermomucor* (potentially pathogenic): similar to *Rhizomucor* but with conspicuous apophysis and smooth zygospores.
[b]The characteristics indicated with an asterisk are typical of *A. corymbifera* but not of all species of the genus.

with four to eight radial branches (Fig. 4d), from almost hyaline to dark brown, often with a reddish shade; other rhizoids (particularly those on knots scattered through the aerial mycelium) with shorter, irregular branching, sometimes a single spine.

Sporangiophores single (Fig. 9) or in tufts (Fig. 10), essentially unbranched, 0.5-3.0 mm but, as a rule, 1.0-2.0 mm long, 15-25 μm thick, sometimes with knotty swellings measuring about 50 μm (bifurcation in two branches may occur near such swellings), yellowish brown to dark brown. *Sporangia* 35-280 μm in diameter, in most strains 50-200 μm, gray-beige to almost black, mostly shiny but occasionally glittering. *Columella* with 50-70% of the diameter of the sporangium, roughly globose in that the mean proportion between length and breadth is 1:1, with an individual variation from 0.9:1.0 (slightly depressed) to 1.1:1.0 (slightly elongate) (Fig. 11), very often looking like an inverted "umbrella" when the sporangium membrane has been dissolved (Fig. 12), yellowish brown to dark brown. *Apophysis* always present but short (3-12 μm in height), not angulated (see Fig. 4c), collar very short or absent. *Sporangiospores* varying in form and size both between and within the individual strains; the prevailing form is roughly rhomboidal, in fluids becoming lemon-shaped with two tapering poles, and with fine wrinkles of the surface (so-called striae) extending from one pole to the other; divergent forms are polyhedric, ovoid with only one pole, or globose (Fig. 13); the most frequent sizes are 6-8 × 4.5-6 μm with extremes from 4-5 × 3-4 μm to 9-12 × 7-8 μm or even more; slightly gray to greenish.

Zygospores have recently been obtained in certain crosses [119]; they are globose or laterally flattened, with a diameter from 60 to 140 μm, usually 80-100 μm, brown to slightly red-brown, and ornamented with flat, often lobed projections; the suspensors are very unequal, the larger is globose and almost reaches the size of the zygospore, the smaller conical to phial-shaped with a diameter of 10-20 μm (Fig. 14).

Maximum temperature of growth (40-)42-44(-46)°C.

3. Main Synonyms

a. *Proof of Synonymy Based on the Original Descriptions*

Rhizopus cambodia (Chrzaszcz) Vuillemin, *Rev. Mycol. 24*: 47, 1902 [140]
Rhizopus japonicus Vuillemin, *Rev. Mycol. 24*: 49, 1902 [140]
Rhizopus tritici Saito, *Zentralbl. Bakteriol. Parasitenkd. Infektionskr. Hyg., Abt. 2, 13*: 157, 1904 [107]
Rhizopus nodosus Namyslowski, *Bull. Int. Acad. Sci. Cracovie 208*: 682, 1906 [92]
Rhizopus batatas Nakazawa, *Zentralbl. Bakteriol. Parasitenkd. Infektionskr. Hyg., Abt. 2, 24*: 486, 1909 [91]
Rhizopus delemar Hanzawa, *Mycol. Zentralbl. 1*: 77, 1912 [54]
Rhizopus kasanensis Hanzawa, *Mycol. Zentralbl. 5*: 236, 1915 [55]
Rhizopus trubinii Hanzawa, *Mycol. Zentralbl. 5*: 236, 1915 [55]
Rhizopus usamii Hanzawa, *Mycol. Zentralbl. 5*: 237, 1915 [55]
Rhizopus maydia Bruderlein, *Bull. Soc. Bot. Geneve 9*: 108, 1917 [18]

b. *Original Description in Japanese but Synonymy Assumed on the Basis of Redescriptions Supplied by Japanese Authors in English or in German*

Rhizopus formosaenis Nakazawa, Rep. No. 2, Gov. Res. Inst. Formosa, p. 37, 1913 (fide Inui et al. [68])

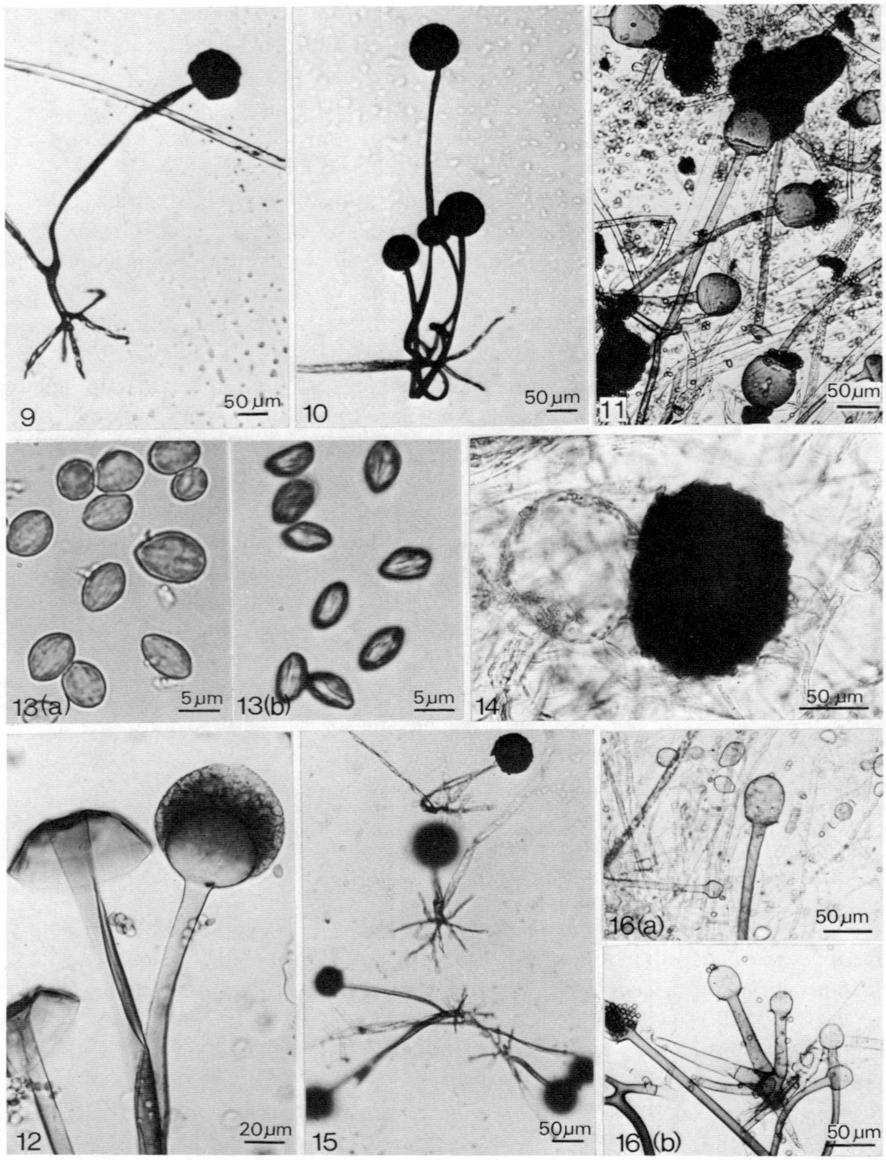

Figures 9 and 10 *Rhizopus oryzae.* Sporangia on sporangiophores originating from knots with rhizoids and stolons.
Figure 11 *R. oryzae.* Columellae.
Figure 12 *R. oryzae.* Intact sporangium (right) and inverted columella forming an "umbrella" after dissolution of the sporangium membrane (left).
Figure 13 *R. oryzae.* Sporangiospores: (a) in dry state; (b) in saline.
Figure 14 *R. oryzae.* Zygospore in a cross of two compatible strains. Note very unequal suspensors.
Figure 15 *Rhizopus rhizopodiformis.* Comparatively small sporangia on short sporangiophores originating from knots with intensively branched rhizoids and stolons.
Figure 16 *R. rhizopodiformis.* Elongate columellae with angulated ("truncate") apophysis.

Rhizopus thermosus Yamamoto, *J. Soc. Agric. For. Sapporo 17*(77): 85, 1925 (fide Yamamoto [148])

Rhizopus javanicus Takeda, *J. Agric. Chem. Soc. Jpn. 11*: 909, 1935 (fide Inui et al. [68])

c. *Doubtful Position*

Rhizopus arrhizus Fischer, Phycomycetes, in Rabenhorst's *Kryptogamenflora von Deutschland, Oesterreich und der Schweiz.* Vol. 1, Abt. IV, Kummer, Leipzig, p. 233, 1892 [43]

This name has been widely used in medical literature, and since it was published in 1892 would have the priority over *R. oryzae,* which was published in 1895. *Rhizopus arrhizus* is, however, rejected for the following reasons: In his original description, Fischer insisted on a transverse-ellipsoidal form of the columella which would be 60-100 μm in breadth and only 40-75 μm in length. During the study of many strains we did not see a *Rhizopus* with such a depressed columella. Fischer further described a peculiar arrangement of the sporangiophores, which is, however, hardly understandable for lack of a figure. No original material is available, nor is a type strain. It seems possible that Fischer was dealing with a rare *Rhizopus* species different from *R. oryzae.* On the other hand, we are convinced that the fungi referred to as *R. arrhizus* in the literature were mostly consistent with *R. oryzae.* This view was confirmed by the fact that any strain we obtained from culture collections with the label *R. arrhizus* was found to be *R. oryzae.*

Mucor norvegicus Hagem, *Untersuchungen über norwegische Mucorineen I,* Dybwad, Christiania [Oslo], p. 39, 1908 [52]

The original description would rather suit *R. microsporus,* but in 1910 Hagem [53] stated that his *M. norvegicus* was consistent with Namyslowski's *R. nodosus* [92], which is undoubtedly a synonym of *R. oryzae.*

Rhizopus suinus Nielsen, *Virchows Arch. 273*: 859, 1929 [99]

According to the original description resembling *R. rhizopodiformis* in colony color but *R. microsporus* in the form of the columella, and intermediate between these two species with respect to the form of the sporangiospores. Nielsen's original strain maintained in the CBS now resembles *R. oryzae.*

d. *Distinction from R. stolonifer (Ehrenberg ex Fries) Vuillemin, Rev. Mycol. 24: 54, 1902 [140]*

The very common, clearly nonpathogenic species *R. stolonifer* (synonym: *R. nigricans*) resembles *R. oryzae* morphologically in exhibiting similar forms but in a consistently larger size (see Table 3 and Fig. 25); in contrast to other species of the genus, *R. stolonifer* does not assimilate galactose; almost black zygospores measuring from 150 to 200 μm (or even 250 μm) are formed in crosses, with suspensors that are almost equal and rather cylindrical than globose; maximum temperature of growth is 30-32°C.

B. *Rhizopus rhizopodiformis* (Cohn) Zopf, Die Pilze, in Schenk (Ed.), *Handbuch der Botanik,* Vol. 4, Breslau, p. 587, 1890 [150]

1. Occurrence as a Pathogen

Rhizopus rhizopodiformis ranks second in frequency after *R. oryzae* as an agent of human mucormycosis, being responsible for 15% of all culturally proven cases [122]. It even ranks

Table 3 Differential Characteristics of the Pathogenic Species of *Rhizopus* (*R. stolonifer* Included for Comparison; see Fig. 23)

Species	Maximum temperature of growth (°C)	Rhizoid length[a] (μm)	Sporangiophore length (mm)	Sporangia diameter (μm)	Columellae form	Sporangiospores	Zygospores
R. oryzae	(40-)42-44(-46)	150-200 (-300)	(0.5-)1-2(-3.5)	50-200 (-250)	Subglobose	Rhomboidal to lemon-shaped variable in size; av. length 6-8 μm; striated	Only in crosses; brown (slightly reddish); av. diameter 60-140 μm; suspensors very unequal
			"Micro-*Rhizopus* group"				
R. rhizopodiformis	50-52	100-120 (intensively ramified)	0.2-1.0	40-130	Slightly elongated to pyriform; apophysis very distinct, angulated	Subglobose to slightly rhomboidal; quite equal in size; av. length 4-6 μm; smooth to slightly striated	Not known
R. microsporus	46-48	80-100	0.3-0.8	40-150	Subglobose to slightly depressed	Rhomboidal; quite equal in size; av. length 5-6 μm; striated	Only in crosses; dark red-brown; av. diameter 100 μm; suspensors unequal
R. oligosporus	46-48	50-80 (blunt, in rosettes)	0.2-0.8	30-140	Subglobose to inverse pyriform	Polymorphic; very variable in size; av. length 7-10 μm; almost smooth	Not known
R. homothallicus	46		Sporangiosporogenous apparatus poorly developed, resembling *R. microsporus*				Abundant in monosporic cultures; yellowish to bright brown; av. diameter 60-90 μm; suspensors slightly unequal, both globose
R. stolonifer (nonpathogenic!)	30-32	300-350 (intensively ramified)	1.5-4.0	150-350	Subglobose	Rhomboidal to polyhedric; very variable in size; av. length 9-11 μm; markedly striated	Only in crosses; almost black; av. diameter 150-200 μm; suspensors almost equal; conical to slightly globose

[a]Considerable polymorphism is met in some strains, but the typical figures given are found in any strain of the species.

first in the cutaneous [11,17,49] and gastrointestinal forms [95] of the disease but was only exceptionally found in the cranial (rhinocerebral) form [11], in which *R. oryzae* is by far the most prevalent species. *R. rhizopodiformis* is also among the main agents of mucormycosis of mammals, particularly of swine and of bovine abortion (see Ref. 9).

2. Main Characteristics of *Rhizopus rhizopodiformis* (see Hesseltine, in Baker et al. [11])

Mycelium high in light strains (with abundant sterile mycelium) but lower (a few millimeters) in dark strains (with the fertile mycelium prevailing), from almost white or creamish white to gray and dark gray-plumbeous (almost black); if present, the gray-plumbeous shade is of definite diagnostic value; sporangia usually not visible by the naked eye.

By its short sporangiophores and small sporangia, *R. rhizopodiformis* is a representative of the "Micro-*Rhizopus* group" (Fig. 15), which also includes the rarer species, *R. microsporus* and *R. oligosporus.* The essential differential characteristics of *R. rhizopodiformis* (see Table 3) are the comparatively long columella with angulated, truncated apophysis (Fig. 16) and the maximum temperature of growth of 50-52°C. Branching of the rhizoids is usually more intense than in *R. microsporus* and *R. oligosporus.* There are strains with small (Fig. 17) and others with somewhat larger sporangiospores (Fig. 18). In both types the form is subglobose when the spores are in a liquid, with almost no "poles" and only minimal striation being visible (Figs. 17b and 18b). In the dry state the sporangiospores appear blurred, obviously covered by an amorphous substance (Fig. 17a), or they may exhibit a more rhomboidal form and some striation (Fig. 18a), similar to but still less than *R. microsporus.*

There are no known zygospores in the species.

3. Main Synonyms

Mucor rhizopodiformis Cohn, in Lichtheim, *Zentralbl. Klin. Med. 7*: 148, 1884 [77]
Rhizopus cohnii (Cohn) Berlese & de Toni, *Syll. Fung. 7*: 213, 1888 [15]
Rhizopus equinus Costantin & Lucet, *Bull. Soc. Mycol. Fr. 19*: 202, 1903 [28]
Rhizopus chinensis Saito, *Zentralbl. Bakteriol. Parasitenkd. Infektionskr. Hyg., Abt. 2, 13*: 154, 1904 [107]

C. *Rhizopus microsporus* Van Tieghem, *Ann. Sci. Nat. 6e Ser. Bot. 1*: 83, 1875 [137]

1. Occurrence as a Pathogen

Recorded as a rare agent of gastrointestinal mucormycosis in humans [95] and in swine [50], but the possibility exists that the fungi in question were *R. rhizopodiformis* [strains from culture collections labeled *R. microsporus* and said to be isolated from some of the pertinent cases were identified as *R. rhizopodiformis* by one of us (H.J.S.)]. On the other hand, *R. microsporus* is definitely pathogenic to mice when injected intravenously [121].

2. Main Characteristics of *Rhizopus microsporus*

Rhizopus microsporus resembles *R. oryzae* by similar forms but is consistently smaller in size (Fig. 19). By the small size of the sporangiophores and sporangia it is similar to *R. rhizopodiformis* and *R. oligosporus* ("Micro-*Rhizopus* group"), from which it differs, however,

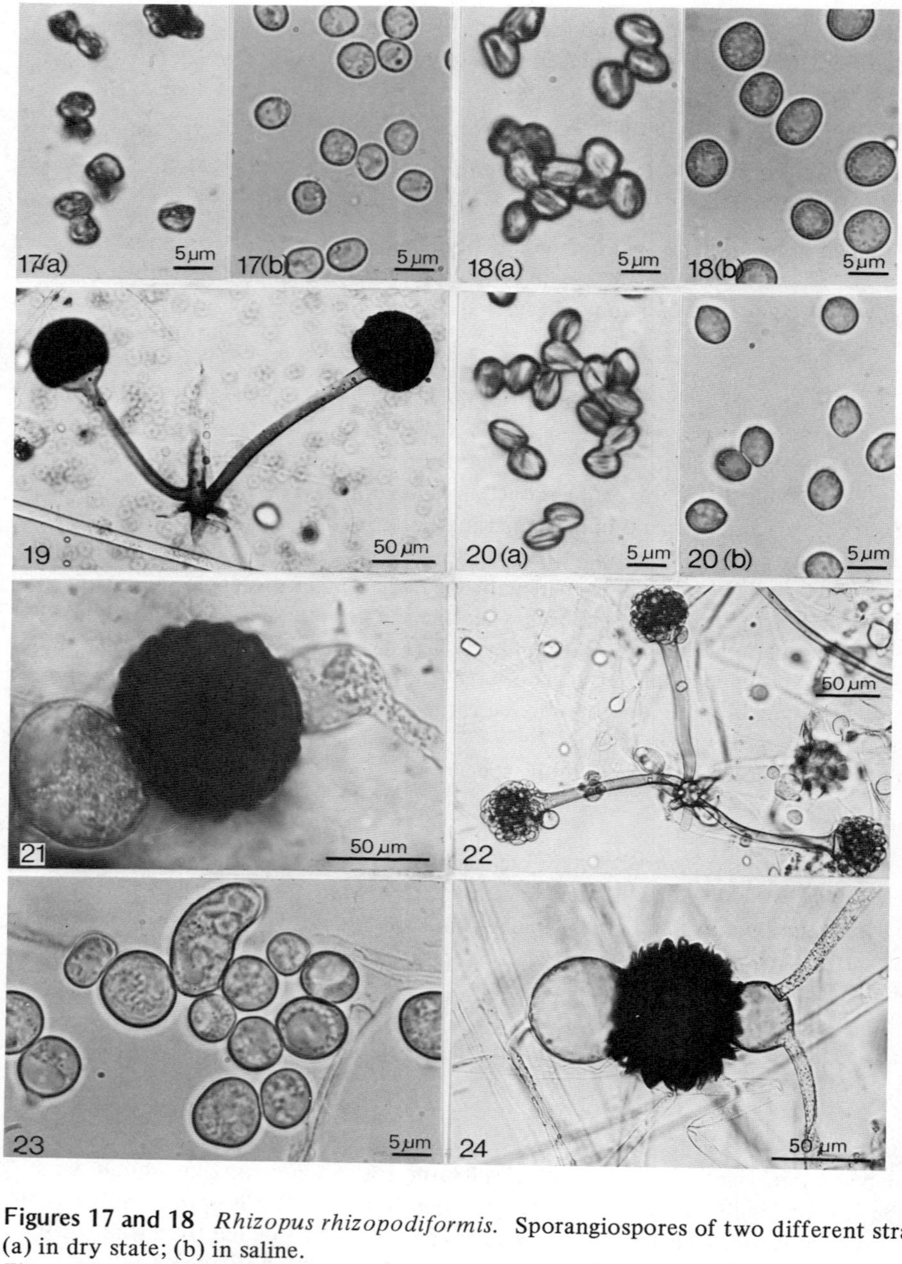

Figures 17 and 18 *Rhizopus rhizopodiformis.* Sporangiospores of two different strains: (a) in dry state; (b) in saline.

Figure 19 *Rhizopus microsporus.* Sporangia and columellae resembling *R. oryzae,* but smaller, on short sporangiophores originating from a knot with rhizoid and stolon.

Figure 20 *R. microsporus.* Sporangiospores, resembling *R. oryzae* but smaller and more regular in form and size: (a) in dry state; (b) in saline.

Figure 21 *R. microsporus.* Zygospore in a cross of two compatible strains. Suspensors unequal but less so than in *R. oryzae.*

by its essentially rhomboidal, wrinkled ("striated") sporangiospores, which are small and quite equal in size (Fig. 20). The maximum temperature of growth of 46-48°C is shared with *R. oligosporus* but is slightly different from that of *R. rhizopodiformis* (see also Table 3). Dark red-brown zygospores of *R. microsporus* with unequal suspensors (the larger globose and the smaller globose to phial-shaped) measuring from 40 to 120 μm, usually about 60 μm, were obtained in crosses (Fig. 21) [121]. They were described before by Milko et al. [85] but erroneously attributed to *R. oryzae.*

3. Synonym

Rhizopus bovinus Van Beijma thoe Kingma, *Mitteil. CBS K. Ned. Akad. Wet. Amst. 29*(2): 38, 1931 [136]

D. *Rhizopus oligosporus* Saito, *Zentralbl. Bakteriol. Parasitenkd. Infektionskr. Hyg., Abt. 2, 14*: 624, 1905 [108]

1. Occurrence as a Pathogen

So far not proved as an agent of mucormycosis but considered potentially pathogenic by its capacity of growing at 37°C and its experimental pathogenicity to mice [121].

2. Main Characteristics of *Rhizopus oligosporus* (see Hesseltine [59])

Together with *R. rhizopodiformis* and *R. microsporus* belonging to the "Micro-*Rhizopus* group." Characteristics specific for *R. oligosporus* are the extremely polymorphic sporangiospores, which are essentially large and almost smooth (Fig. 23), and the poorly developed rhizoids, with a swollen center surrounded by a rosette of short, blunt branches (Fig. 22). No known zygospores.

E. *Rhizopus homothallicus* Hesseltine & Ellis, *Mycologia 53*: 419, 1961 [60]

1. Occurrence as a Pathogen

Not proved pathogenic but considered potentially pathogenic in view of the capacity of growing at 37°C. Attempts to check experimental pathogenicity for mice have failed since an insufficient number of sporangiospores was obtained from the cultures, whereas the zygospores were too large to be injected intravenously [121].

2. Main Characteristics of *Rhizopus homothallicus* (see Hesseltine and Ellis [60])

This species is essentially characterized by its abundant zygospores, which are produced by any single strain. They are yellowish brown to bright brown and from 40 to 140 μm,

---

Figure 22 *Rhizopus oligosporus.* Comparatively small sporangia on short sporangiophores, but with large sporangiospores; rhizoid forming a rosette with short, blunt branches. (X170.)
Figure 23 *R. oligosporus.* Sporangiospores (in saline). (X1000.)
Figure 24 *Rhizopus homothallicus.* Zygospore (produced in monoculture); suspensors only slightly unequal, both globose. (X260.)

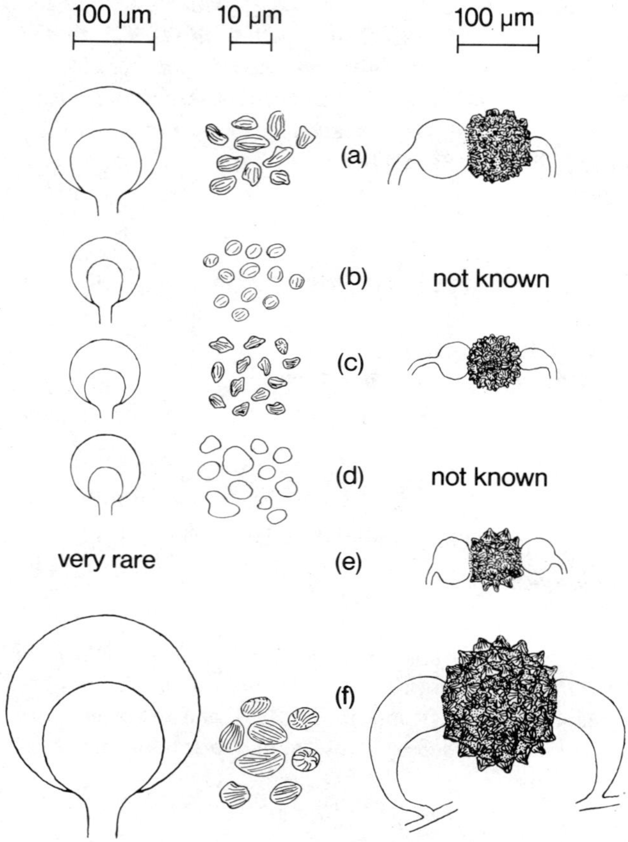

Figure 25 Differential characteristics of pathogenic species of *Rhizopus*. Left column: sporangia and columellae; middle: sporangiospores (mainly in dry state); right: zygospores. (a) *R. oryzae;* (b) *R. rhizopodiformis* (a strain with the smaller sporangiospores is represented; see Figs. 17 and 18); (c) *R. microsporus;* (d) *R. oligosporus;* (e) *R. homothallicus;* (f) for comparison, *R. stolonifer.*

usually 60-90 µm in diameter. The suspensors are globose and only slightly though clearly unequal (Fig. 24). The maximum temperature of growth of 46–48°C allows easy distinction from the only other *Rhizopus* species known as homothallic, *R. sexualis* (Smith) Callen [21], and its variety *americanus,* which do not grow at a temperature of 30°C or more (see Ref. 60). The sporangiosporogenous apparatus resembles *R. microsporus* but is poorly developed.

F. Summary of the Differential Characteristics of the Pathogenic Species of *Rhizopus*

Table 3; Fig. 25.

VI. THE PATHOGENIC SPECIES OF *ABSIDIA*

A. *Absidia corymbifera* (Cohn) Saccardo & Trotter, *Syll. Fung. 21*: 825, 1912 [106]

1. Occurrence as a Pathogen

Absidia corymbifera was probably responsible for the first reported case of human mucormycosis [47] and has remained the third most frequent agent after *R. oryzae* and *R. rhizopodiformis* found in human disease [122], where it prevails in the pulmonary form [31, 73,89,103]. It appears to be the most frequent agent of mammalian mucormycosis (including bovine mycotic abortion caused by Mucorales) and was also repeatedly isolated from mucormycosis in birds (see Ref. 3).

3. Species Description

Mycelium rapid-growing, very high, and light dirty gray to almost white in some strains but comparatively low (3-4 mm) and darker dirty grayish brown in others, reflecting different proportions between the sterile mycelium, which is hyaline, and the sporangiosporogenous apparatus, which is pigmented; the "low and dark" strains sporulate much more than the "high and light" strains; the complicated pattern with many superimposed ramifications bearing sporangia (Fig. 7) is basically the same in all strains; the fungus is fixed in the substratum by radially diffusing hyphae; it is a matter of opinion whether one should call these substrate mycelia "rhizoids," and the aerial hyphae originating from them and soon getting lost in the tangle of ramifications, "stolons"; the sporangia are not visible by the naked eye.

Sporangiophores (i.e., the ultimate branches of the aerial mycelium) finely and elegantly branched, often in form of grapes or corymbs (Fig. 26), often with one septum beneath the sporangium, subterminally slightly brownish. *Sporangia* definitely pyriform (Fig. 28), from 10 to 120 or 150 μm in diameter, transparent and almost colorless when young, later opaque (sometimes glittering) and grayish brown to greenish beige; sporangium membrane diffluent. *Columella* with a transverse diameter measuring 50-60% of that of the sporangium, hemispherical to short-ovoidal above the apophysis, often with one to two mamilliform projections (Fig. 27; see Fig. 5b), and with a long, conical apophysis (Fig. 24), grayish, collar usually visible. *Sporangiospores* smooth, slightly yellow to greenish, varying from globose to long-ellipsoidal (cylindrical) depending on the strains, with strains showing any intermediate short-ellipsoidal form: mean diameters in typical "globose strains" 3.2-3.9 × 3.1-3.5 μm; in "intermediate strains" 3.6-5.0 × 3.0-3.8 μm; and in "elongate strains" 4.1-4.9 × 2.3-2.9 μm (Fig. 29).

Zygospores formed in crosses (for the peculiar mating pattern, see Ref. 100), short-ellipsoidal in a plane perpendicular to the suspensors, 60-100 × 45-80 μm, thick-walled with very flat projections approximately 10 μm in diameter and one to three equatorial ridges, bright reddish brown; suspensors devoid of projections, equal or slightly unequal, the thicker suspensors or both often with a slightly roughened wall (Figs. 30 and 32).

No capacity to ferment sugars; assimilation of melibiose and raffinose positive, of sucrose negative [123]; thiamine-dependent.

Maximum temperature of growth 48-52°C.

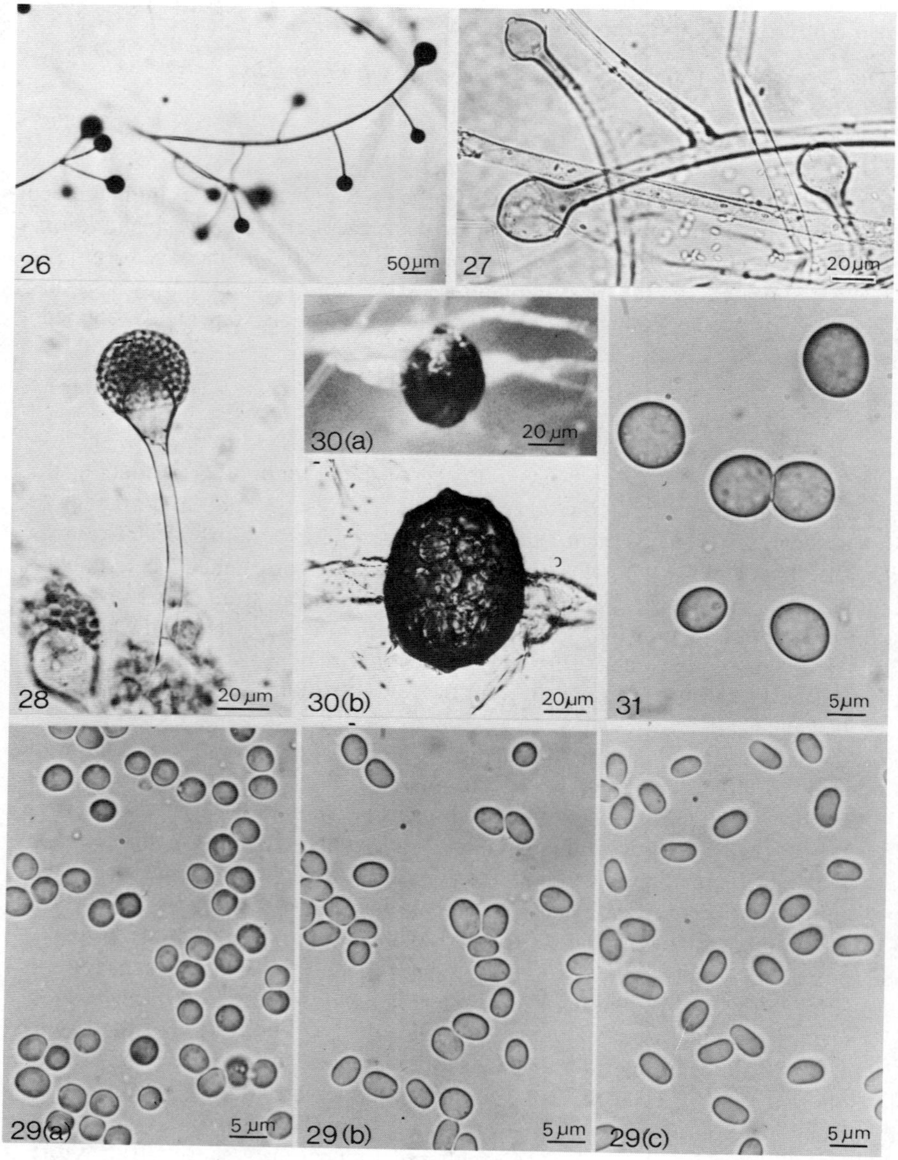

Figure 26 *Absidia corymbifera.* Sporangiophores ramified in forms of grapes and corymbs, with pyriform sporangia. (From Ref. 100.)

Figure 27 *A. corymbifera.* Columellae (upper left with mamilliform projection).

Figure 28 *A. corymbifera.* Intact, young sporangium. (From Ref. 100.)

Figure 29 *A. corymbifera.* Sporangiospores (in saline) from three different strains: (a) subglobose; (b) "intermediate"; (c) long-ellipsoidal.

Figure 30 *A. corymbifera.* Zygospores in crosses of compatible strains: (a) unmounted, incident light ; (b) in mounting fluid, transmitted light. (See also Fig. 32.) (From Ref. 100.)

Figure 31 *Absidia hyalospora.* Sporangiospores (in saline): subglobose, large (and dark in color).

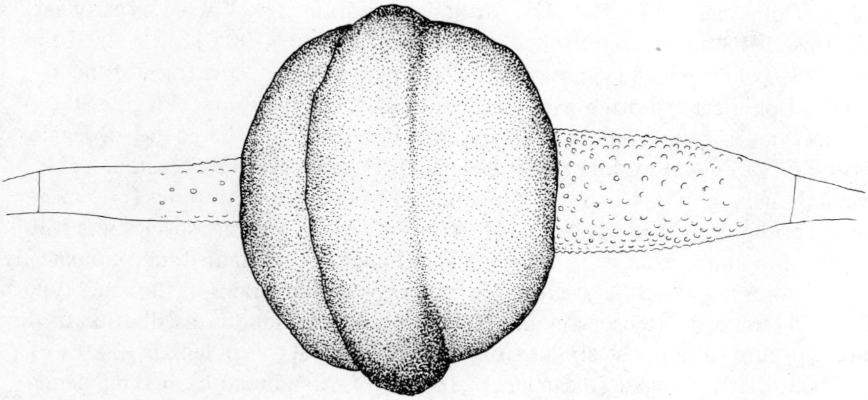

Figure 32 *Absidia corymbifera.* Zygospore in a cross of two compatible strains. (~X1000.) (See also Fig. 30).

3. Main Synonyms

a. Considered Proven

Mucor corymbifer Cohn, in Lichtheim, *Zentralbl. Klin. Med. 7*: 149, 1884 [77]
Lichtheimia corymbifera (Cohn) Vuillemin, *C. R. Acad. Sci. (Paris) 136*: 516, 1903 [141]
Mucor ramosus Lindt, *Arch. Exp. Pathol. Pharmakol. 21*: 275, 1886 [79]
Lichtheimia ramosa (Lindt) Vuillemin, *Arch. Parasitol. 8*: 563, 1903-1904 [142]
Absidia ramosa (Lindt) Lendner, *Mat. Fl. Crypt. Suisse 3*(1): 143, 1908 [75]
Absidia lichtheimii (Lucet & Costantin) Lendner, *Mat. Fl. Crypt. Suisse 3*(1): 143, 1908 [75]
Absidia truchisii (Lucet & Costantin) Lendner, *Mat. Fl. Crypt. Suisse 3*(1): 146, 1908 [75]
Absidia regnieri (Lucet & Costantin) Lendner, *Mat. Fl. Crypt. Suisse 3*(1): 146, 1908 [75]
Pseudo-Absidia vulgaris Bainier, *Bull. Soc. Mycol. Fr. 19*: 155, 1903 [6]
Absidia cornealis (Cavara & Saccardo) Dodge, *Medical Mycology,* Mosby, St. Louis, p. 114, 1935 [34]
Absidia italiana (Costantin & Perin) Dodge, *Medical Mycology,* Mosby, St. Louis, p. 112, 1935 [34]

b. Doubtful Position

Absidia dubia Bainier, Thèse, Ecole Supérieure de Pharmacie, Paris, p. 73, 1882 [5]

c. Comments on Absidia ramosa and Absidia dubia

Lindt in 1886 [79] described *A. ramosa* ("*Mucor ramosis*") as a new species: The single strain he examined had more elongate sporangiospores than the single strain named *Mucor corymbifer* by Cohn in 1884 (in Ref. 77). Although questioned by Fischer as early as 1892 [43], this distinction was confirmed by Vuillemin in 1903-1904 [142], and Lendner in 1908 [75], who even added new differential characteristics: the form of the columella (hemispherical and often with projections in *A. ramosa,* more elongate and without projections in *A. corymbifera*), the ramification and septation of the sporangiophores (both more frequent in *A. corymbifera* than in *A. ramosa*), the occurrence of rhizoids (mainly in *A. ramosa*), and the pigmentation of the sporangiospores (present in *A. ramosa* but absent in *A. corymbifera*). The concept of two separate species was still maintained by Ellis and Hesseltine in 1966 [39], although they obtained morphologically identical zygospores in crosses between the two types as within strains of the same type. In a study of 54 strains, Nottebrock et al. [100] showed a continuous distribution of the mean sporangiospore form of the strains from almost spherical (mean length 104.9% of mean breadth) to fairly elongate (mean length 167.1% of mean breadth), making demarcation impossible. The same authors also found no correlation between mating capacity and sporangiospore form of the partners, the crosses of strains with different forms yielding not less zygospores than those between strains with the same form. The lack of correlation between the sporangiospore form of the strains and some of the above-mentioned minor characteristics, such as form of the columella and septation of the sporangiophores, was also stressed (see Refs. 121 and 123).

According to Ellis and Hesseltine [39], *A. dubia,* described by Bainier in 1882 [5], would be the valid name when *A. ramosa* and *A. corymbifera* are combined as a single species, since both spherical and elongate sporangiospores were mentioned by the author. In our opinion, however, Bainier's original description is too deficient to prove the identity of the fungus he was dealing with, and is not completed by the more detailed description he gave in 1903 for "*Pseudo-Absidia dubia*" [6], since it seems too uncertain whether this newer description was based on the same material.

B. *Absidia hyalospora* (Saito) Lendner, *Mat. Fl. Crypt. Suisse 3*(1): 142, 1908 [75]

1. Occurrence as a Pathogen

Not proved as an agent of mucormycosis but considered potentially pathogenic by its capacity of growing at 37°C and its experimental pathogenicity to mice [121].

2. Main Characteristics of *A. hyalospora* (see Hesseltine and Ellis [63])

This species differs from *A. corymbifera* by its dark gray pigment, poor ramification of the sporangiophores, large sporangiospores (5-10 μm; Fig. 31), lower maximum temperature of growth (40°C), and positive assimilation of sucrose.

See Table 4 for a summary of the differential characteristics of *A. corymbifera* and *A. hyalaspora.*

Table 4 Differential Characteristics of *Absidia corymbifera* and *Absidia hyalospora*[a]

	Species	
	A. corymbifera	*A. hyalospora*
Maximum temperature of growth (°C)	48-52	40
Sporangiophores	Intensively branched, often in grapes or corymbs; slightly grayish brown (just apically, including apophysis)	Poorly branched or unbranched; gray, apically dark gray
Sporangia	Pyriform; transverse diameter up to 120(-150) μm; hyaline to grayish brown	Pyriform; transverse diameter up to 40(-80) μm; almost black
Columellae	Hemispherical to short-ovoidal; often with projections; slightly grayish brown	Hemispherical to spatulate; often with projections; dark gray
Apophysis	Long, conical	Long, conical
Sporangiospores	According to the strains from subglobose to cylindrical; av. length 3-5 μm; slightly yellowish to greenish	Subglobose; av. length 6-10 μm; gray to almost black
Zygospores	Only in crosses; with equatorial ridges; diameter 50-100 μm; bright red-brown	Not known
Assimilation of sucrose	Negative	Positive

[a]*A. hyalospora* is considered potentially pathogenic.

VII. THE PATHOGENIC SPECIES OF *RHIZOMUCOR*

A. *Rhizomucor pusillus* (Lindt) Schipper, *Studies in Mycology,* Vol. 17, p. 54, 1978 [117]

1. Occurrence as a Pathogen

There are only a few reports on human mucormycosis caused by *Rm. pusillus* (lung [96]; endocarditis [41]; disseminated forms [71,84]), but this species was more frequently isolated from diseased mammals such as cattle, horse, swine, and seal (see Ref. 3); it ranks among the more frequent agents of bovine mucormycotic abortion [97].

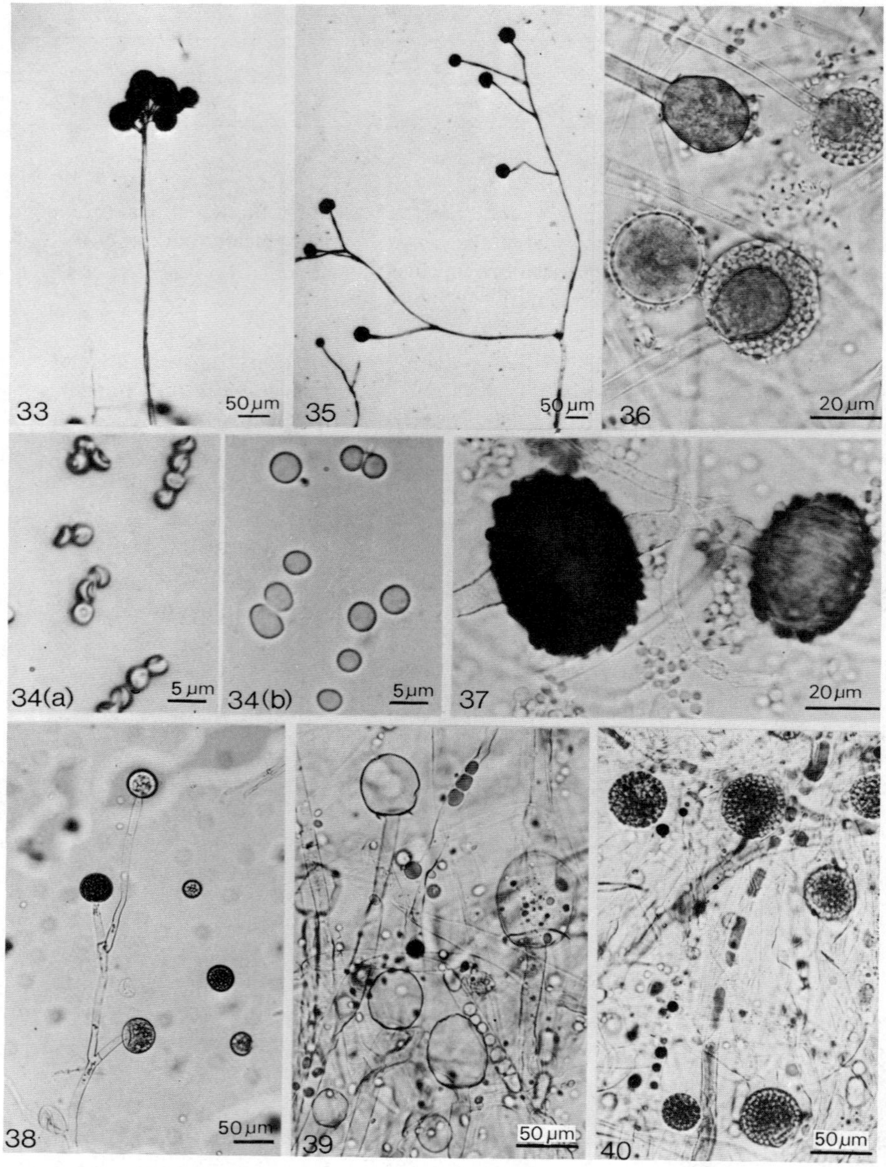

Figure 33 *Rhizomucor pusillus.* Sporangiophore with monopodial, subterminal branching in the form of a false umbel. (From Ref. 123.)

Figure 34 *Rm. pusillus.* Sporangiospores: (a) in dry state; (b) in saline. (The sporangiospores of *Rm. miehei* are indistinguishable morphologically.) (From Ref. 123.)

Figure 35 *Rhizomucor miehei.* Sporangiophores with, mainly, sympodial, loose branching.

Figure 36 *Rm. miehei.* Globose sporangia with slightly elongate to pyriform columella; absence of apophysis. (The sporangia and columellae of *Rm. pusillus* are indistinguishable morphologically.)

2. Main Characteristics of *Rhizomucor pusillus* (for full descriptions, see Cooney and Emerson [26]; Schipper [117])

Macroscopically, the cultures of this fungus are characterized by comparatively low aerial *mycelium* (up to 2 mm) and a definite hair-brown color; low-power stereomicroscopy or a loup reveals dark brown *sporangiophores* which are intensively branched (mainly monopodially, and often subterminally in the form of false umbels, Fig. 33; see also Fig. 8a), fix themselves in the medium by short *rhizoids,* and bear comparatively small *sporangia* (diameter up to 100 μm) with opaque, glittering walls; the membrane is diffluent; under the microscope one observes subglobose to slightly pyriform *columellae* (see Fig. 36) and very small *sporangiospores* (diameter 3-4 μm) which are subglobose in liquids (Fig. 34b) but reniform in the dry state (Fig. 34a); *zygospores* resembling those of *Rm. miehei* (see Fig. 37) but by most of the strains only produced in cross; maximum temperature of growth 54-58°C, minimum temperature 20°C; autoauxanotrophic; positive assimilation of sucrose; no fermentation capacity; marked pathogenicity to experimentally infected mice [121].

3. Main Synonyms (see Schipper [117])

Mucor pusillus Lindt, *Arch. Exp. Pathol. Pharmakol. 21*: 272, 1886 [79]
Rhizomucor parasiticus Lucet & Costantin, *C. R. Acad. Sci. (Paris) 129*: 1032, 1899 [80]

B. *Rhizomucor miehei* (Cooney & Emerson) Schipper, *Studies in Mycology,* Vol. 17, p. 58, 1978 [117]

1. Occurrence as a Pathogen

In their survey on animal mycoses in Great Britain, Ainsworth and Austwick [2] mentioned two strains of a "zygosporogenic *M. pusillus*" which had been isolated from bovine mastitis and from the internal organs of an aborted calf, respectively. What was probably a subculture of the latter strain (labeled "isolate from calf lung") was made available to one of us (H.J.S.) by the courtesy of P. K. C. Austwick and was identified as *Rm. miehei.* Although not proven (there are rare strains of *Rm. pusillus* behaving homothallically [17,126]), it is probable that the first-mentioned strain of Ainsworth and Austwick [2] from bovine mastitis was also *Rm. miehei.*

2. Main Characteristics of *Rhizomucor miehei* (see Cooney and Emerson [26]; Scholer and Müller [123]; Schipper [117])

Rhizomucor miehei shares most of the morphological and physiological characteristics of *Rm. pusillus* (Fig. 36) but distinguishes itself by the following. Every strain produces numer-

---

Figure 37 *Rm. miehei.* Young (right) and mature (left) zygospore produced in monoculture (homothallic). (X480.) (The zygospores of *Rm. pusillus* are similar morphologically but by most of the strains only produced in crosses.)
Figure 38 *Mucor ramosissimus* (member of the *M. circinelloides* group). Sympodial branching of sporangiophore. (X110.)
Figure 39 *Mucor circinelloides.* Subglobose to ellipsoidal columellae from larger sporangia with diffluent membrane; hyphae containing oil-like droplets. (~X200.)
Figure 40 *M. circinelloides.* Smaller sporangia with persistent membrane. (~X200.)

ous zygospores (homothallic), which are globose to slightly compressed, up to about 50 μm in diameter, reddish brown to blackish brown, with stellate warts and virtually equal, elongate, slightly conical suspensors (Fig. 37); the colony color is dirty gray rather than brown; the branching of the sporangiophores is looser than in *Rm. pusillus,* and the subterminal false umbels are absent or rare (Fig. 35; see Fig. 8a); growth is to some extent stimulated by thiamine; no assimilation of sucrose; maximum temperature of growth 54-58°C (as in *Rm. pusillus*), but minimum temperature 24°C; experimental pathogenicity to mice as marked as in *Rm. pusillus.*

C. *Rhizomucor tauricus* (Milko & Schkurenko) Schipper, *Studies in Mycology,* Vol. 17, p. 62, 1978 [117]

The description of this new species was based on the study of one strain differing from *Rm. pusillus* and *Rm. miehei* mainly by osmophily [requirement of high (10-50%) concentration of sucrose in the medium] and extremely restricted growth on ordinary media. In view of its growth at 37°C and above (up to 55°C), the species may be considered potentially pathogenic. Experimental pathogenicity was not tested.

D. Summary of the Differential Characteristic of the *Rhizomucor* Species

Table 5.

E. Appendix: *Thermomucor indicae-seudaticae* Subrahamanyam, Mehrotra & Thirumalachar, *Ga. J. Sci. 35*: 1, 1977 [132]

The new genus *Thermomucor* (with *T. indicae-seudaticae* as the only homothallic species; see Ref. 118) is closely related to *Rhizomucor* in presenting stolons, rhizoids, repeated branching of the sporangiophores, a dark pigment, and high maximum temperature of growth (55°C), but it differs by the presence of a conspicuous apophysis and by smooth zygospores; pathogenic to mice but less than *Rm. pusillus* and *Rm. miehei* (H. J. Scholer, unpublished data). The species is considered potentially pathogenic.

VIII. THE PATHOGENIC SPECIES OF *MUCOR*

Members of the genus *Mucor* were identified as agents of mucormycoses less frequently than those of *Rhizopus, Absidia,* and *Rhizomucor.* Not counting the obvious errors in diagnosis (e.g., *M. mucedo;* see Sec. XI), the names reported are concentrated on *M. circinelloides* and similar species. For the practical purpose of this chapter, and without presuming natural relationship, the respective fungi are put together in a "*M. circinelloides* group" (this is not entirely consistent with the *M. circinelloides* group of Schipper [115]). *Mucor racemosus* is dealt with in second place, although pathogenicity is subject to doubt (see Sec. VIII.A.1).

A. "*Mucor circinelloides* Group"

Mucor circinelloides Van Tieghem, *Ann. Sci. Nat. 6e Ser. Bot. 1*: 94, 1875 [137], and ibid. *4*: 390, 1876 [138]

Mucor ramosissimus Samutsevitsch, *Mater. Mycol. Phytopathol. 6*: 204, 1927 [111]

Table 5 Differential Characteristics of the Species of *Rhizomucor*

	Species		
	Rm. pusillus	*Rm. miehei*	*Rm. tauricus*[a]
Maximum temperature of growth (°C)	54-58	54-58	55
Colony color	Hair-brown	Dirty gray to deep olive	Gray-olive
Sporangiophores	Branched monopodially (often with subterminal false umbels) or sympodially	Branched mostly sympodially	Unbranched or weakly branched (subterminally)
Sporangia	Globose, diameter 40-100 (-150) μm		
Columellae	Subglobose to slightly pyriform		Globose to ovoid
Apophysis	Lacking (or very tiny)		
Sporangiospores	Globose to subglobose, length 3-4 μm		
Zygospores	Almost only in crosses; diameter up to 70 μm	In monosporic culture; diameter up to 50 μm	Not known
Assimilation of sucrose	Positive	Negative	Probably positive[b]
Thiamine requirement	Independent	Dependent (stimulated)	Not tested
Osmophily	No	No	Marked

[a]Considered potentially pathogenic.
[b]Optimum growth on beerwort agar with 30% sucrose [117].

Mucor indicus Lendner, *Bull. Soc. Bot. Geneve 21*: 258, 1929 [76]
Mucor amphibiorum Schipper, *Studies in Mycology,* Vol. *17,* p. 14, 1978 [116]

1. Occurrence as Pathogens

Mucor circinelloides was isolated from peritoneal granulomata of cattle and swine [87]. Under the synonym of *M. javanicus* (see Ref. 115) the species was found in fowl [104] and under the synonym of *M. janssenii,* in ganders [81]. Hesseltine and Ellis [62] described *M. ramosissimus* as the agent of a chronic form of human infection, essentially involving the skin, whereas the second human case due to this species presented itself as a typical cranial (rhinocerebral) mucormycosis [19]. *Mucor rouxii* (=*M. indicus*) was found in a human case of gastric mucormycosis [35]. A peculiar mycosis characterized by the production of spherules in the internal organs of various amphibia was first attributed to a variant of *M. circinelloides* [45], but the causative fungus was later described as a separate new species, *M. amphibiorum* [116].

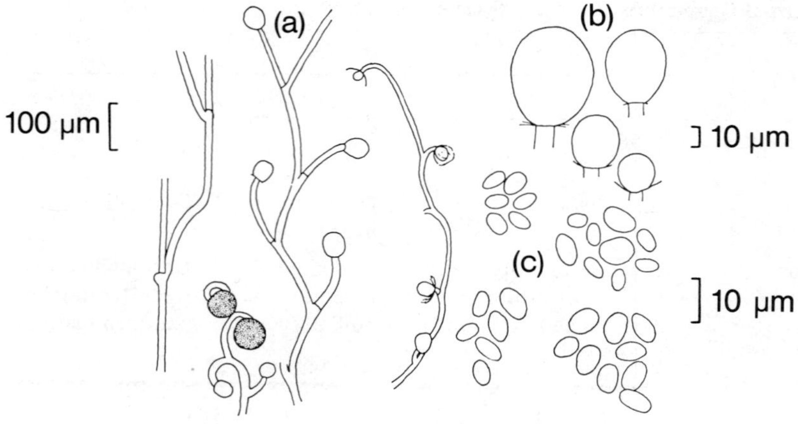

Figure 41 *Mucor circinelloides:* (a) sporangiophores; (b) columellae; (c) sporangiospores from different strains. (From Ref. 115.)

2. Main Characteristics of the *Mucor circinelloides* group and Its Species

Mucor circinelloides (with the four forms *circinelloides, lusitanicus, griseo-cyanus,* and *janssenii*), *M. ramosissimus,* and *M. indicus* were recently redescribed by Schipper [115,116], who also clarified their synonymy. In medical literature *M. indicus* was most often referred to as *M. rouxii* (Calmette) [144]. Schipper [116] also described the new species, *M. amphibiorum.* These fungi have the following characteristics in common: sporangiophores mostly sympodially branched with shorter branches sometimes circinate (Figs. 38 and 41); sporangia globose, varying in size but usually less than 100 μm in diameter, color varying from brownish gray to blackish or yellow, sometimes recurved; sporangial wall diffluent in larger sporangia (Fig. 39) but very persistent and, finally, rupturing in smaller sporangia (Fig. 40); columellae subglobose to elongate; sporangiospores less than 10 μm in length; zygospores (not observed in *M. ramosissimus*) only produced in cross, reddish brown to dark brown or black, stellate, maximum diameter 70-100 μm; chlamydospores may be numerous (but are not present in the sporangiophores); hyphae often containing oil-like yellowish brown droplets or granula; assimilation of sucrose negative; positive fermentation capacity; maximum temperature of growth 36-40°C but in *M. indicus* around 42°C; definite though weak pathogenicity to mice [121].

Mucor indicus differs from the other species by its above-mentioned higher maximum temperature of growth and further by its deep-yellow colony color, dependence on thiamine, and lack of assimilation of nitrate. *Mucor ramosissimus* shows restricted growth on any medium, extremely persistent sporangial walls, columellae that are aplanate and in smaller sporangia virtually lacking, no assimilation of ethanol. *Mucor amphibiorum* distinguishes itself by poor ramification of the sporantiophores and by definitely globose sporangiospores (see also Tables 6 and 7).

Table 6 Differential Characteristics of the Pathogenic Species of *Mucor*

Species	Maximum temperature of growth (°C)	Sporangiophores	Sporangia[a]	Columellae	Sporangiospores	Special characteristics
M. circinelloides	36-40	Sympodially branched; long branches erect, and short branches sometimes circinate	Diameter varying from 20 to 80 μm; membrane diffluent in larger but rupturing or persistent in small sporangia	Globose, in larger sporangia ellipsoidal	Ellipsoidal; length to 6-7 μm	Assimilation of ethanol and nitrate positive
M. ramosissimus	36	Sympodially branched, erect	Up to 75 μm in diameter but mostly smaller; membrane extremely persistent	Aplanate; absent in small sporangia	Ellipsoidal; length to 7-8 μm	Restricted growth; assimilation of ethanol negative, of nitrate positive
M. amphibiorum	36	Almost unbranched; erect	Up to 70-100 μm in diameter; membrane diffluent	Subglobose to ellipsoidal or pyriform	Globose; diameter 3.5-5.5 μm	Assimilation of ethanol and nitrate negative
M. indicus	42	Sympodially branched; erect or, rarely, circinate	Up to 75 μm in diameter; membrane diffluent	Subglobose to aplanate, rarely elongate	Subglobose to ellipsoidal; length to 6-7 μm	Colony color deep yellow; assimilation of ethanol positive, of nitrate negative; thiamine dependent
M. racemosus[b]	32	Sympodially or monopodially branched; erect	Up to 80-90 μm in diameter; membrane mostly diffluent but persistent in smaller sporangia	Subglobose to pyriform, often with truncate basis	Subglobose; length to 8-10 μm	Chlamydospores occurring in sporangiophores; assimilation of sucrose positive[c]

[a] Form globose or subglobose, and no apophysis, in all species.

[b] Questionable pathogenicity.

[c] Assimilation of sucrose negative in all other species listed.

Table 7 Maximum Temperature of Growth and Taxonomically Significant Biochemical Properties of the Pathogenic and Potentially Pathogenic Species of Mucoraceae

Species	Maximum temperature of growth (°C)		Assimilation of: Sucrose	Lactose	Ethanol	Nitrate	Fermentation capacity[a]	Thiamine requirement
Rhizopus oryzae	42-44	Genus *Rhizopus*						
R. rhizopodiformis	50-52		Usually					
R. microsporus	46-48		–[b]	–	+	–	+	Independent
R. oligosporus	46-48							
R. homothallicus	46							
Absidia corymbifera	48-52		–	+	–	+	–	Dependent[c]
A. hyalospora	40		+	+	–	+	–	Dependent
Rhizomucor pusillus	54-58		+	+	–	+	–	Independent
Rm. miehei	54-58		–	+	–	+	–	Dependent (weakly)
(*Thermomucor indicae-seudaticae*)	53		+	+	–	+	–	Independent
Mucor circinelloides	38-40		–	–	+	+	+	Independent
M. ramosissimus	36		–	–	–	+	+ (weak)	Independent
M. indicus	42		–	–	+	–	+	Dependent
M. amphibiorum	36		–	–	–	–	+ (weak)	Dependent
M. racemosus	32		+	–	+	+	+	Independent

[a]Fermentation of dextrose was used as the main criterion.
[b]A few sucrose-positive strains were found in *R. oryzae*, *R. microsporus*, and *R. oligosporus*.
[c]Growth restricted without, and stimulated with, the vitamin.

B. *Mucor racemosus* Fresenius, *Beiträge zur Mykologie I,* p. 12, 1850 [46]

1. Occurrence as a Pathogen

Considered doubtful by definite lack of capacity of growing at 37°C (maximum temperature of growth only 32°C), and of experimental pathogenicity in mice [121]. Some reports on natural infections are, nevertheless, quite convincing: mucormycosis in horse [44], chicken [24], and even human (pulmonary) form in a renal transplant recipient [32], occasionally also in amphibia [45].

2. Main Characteristics of *Mucor racemosus*

For a full description of the species and its three forms (*racemosus, sphaerosporus,* and *chibinensis*) and for the synonyms, see Schipper [115].

Morphologically and physiologically similar to the *M. circinelloides* group (see Sec. VIII.A.2) but with maximum temperature of growth of only 32°C, positive assimilation of sucrose; branching of the sporangiospores in a mixed sympodial and monopodial fashion; columella from subglobose to ellipsoidal and pyriform, and with a truncate basis; diameter of the sporangiospores somewhat wider than in *M. circinelloides;* chlamydospores most often abundant and occurring also in sporangiophores (Fig. 42).

C. Summary of the Differential Characteristics of the Pathogenic *Mucor* Species

Table 6.

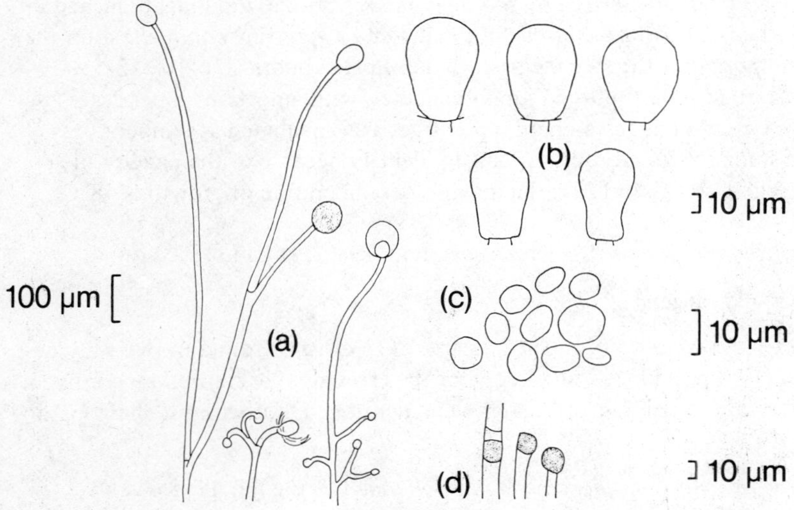

Figure 42 *Mucor racemosus:* (a) sporangiophores; (b) columellae; (c) sporangiospores; (d) chlamydospores. (From Ref. 115.)

IX. APPENDIX: TAXONOMICALLY USEFUL FUNCTIONAL CHARACTERISTICS OF THE PATHOGENIC SPECIES OF MUCORACEAE

The taxonomically significant functional characteristics (maximum temperature of growth and a selection of biochemical properties) of the pathogenic species described in Secs. V through VII are compiled in Table 7. Testing of these characteristics may facilitate diagnosis in instances of any doubts that arise with poorly sporulating isolates.

X. THE PATHOGENIC SPECIES OF MUCORALES FROM FAMILIES OTHER THAN MUCORACEAE

A. *Mortierella wolfii* Mehrotra & Baijal, *Mycopathologia 20*: 51, 1963 [82]

1. Occurrence as a Pathogen

Mortierella wolfii is not known as a human pathogen but was proved to be a common agent of bovine mycotic abortion in New Zealand [25]. A significant number of the cows were also affected by pneumonia caused by the same fungus [83]. Previously reported cases from New Zealand, which were first attributed to *Mort. alpina* and *Mort. zychae* [129], were probably due to *Mort. wolfii.* More recently, the latter species was also found as an agent of bovine mycotic abortion in Great Britain [27] and the United States [147].

2. Main Characteristics of *Mort. wolfii* (according to Mehrotra and Baijal [82]; Linnemann, in Zycha and Siepmann [152]; Gams [48]; see also Fig. 5a)

For the definition of the family Mortierellaceae, see Sec. III.B.2. The genus *Mortierella* is characterized by simple or ramified sporangiophores usually arising from the substrate mycelium with or without basal rhizoids [58]. As many as 83 species were recognized by Linnemann in Zycha and Siepmann [152], and 73 by Gams [48], who studied living cultures.

Mortierella wolfii is characterized by *sporangiophores* 80-250 μm long, branched immediately below the tip, leaving a conspicuous collar and a very tiny *columella* after the dehiscence of the *sporangia;* the *sporangiospores* are short-cylindrical, 6-10 × 3-5 μm; characteristic *chlamydospores* up to 35 μm in diameter, with amoeboid appendages are mostly present; sporulation is fostered on soil extract agar incubated at temperatures above 25°C, but some strains are still difficult to identify because of the paucity of spore formation. According to Crisan [29], the maximum temperature of growth is 48°C.

B. *Cunninghamella bertholletiae* Stadel, Dissertation, Kiel, 1911 [131]

1. Occurrence as a Pathogen

Cunninghamella bertholletiae was (under the name of *C. elegans*) found in two fully described human cases of pulmonary, and in one of the cases also generalized mucormycosis [67,72], and three additional human isolates were mentioned, but without clinicopathological details [144].

2. Main Characteristics of *Cunninghamella bertholletiae* (cf. the family characteristics of Cunninghamellaceae given in Sec. III.B.3 and Fig. 5b)

Cunninghamella bertholletiae shows a rapid-growing, high, light gray to brownish *mycelium* resembling that of a *Rhizopus* sp.; the wall of its *conidia,* which are hyaline and measure

9 × 7 μm on the average, is slightly roughened; *Mucor*-like *zygospores* are produced in crosses [130] ; maximum temperature of growth was 46°C in the two mucormycosis isolates studied by us (from the cases of Hutter [67] and Kwon-Chung et al. [72]). This is in agreement sith Weitzman and Crist [145], whose human isolates all "grew well at 42°C with some growth at 45°C," whereas according to Cutter [30], the species "sometimes grows at 40°C."

Cunninghamella bertholletiae was considered conspecific with *C. elegans* by several authors [75,110,151], whereas by others it was recognized as a separate species [16,93]. The latter was recently confirmed by Weitzman and Crist [145], who also found that all previous clinical isolates were *C. bertholletiae* rather than *C. elegans.*

C. *Syncephalastrum racemosum* Cohn, in Schröter, *Kryptogamenflora von Schlesien,* Vol. 1, p. 217, 1886 [127]

1. Occurrence as a Pathogen

Found only in one case of bovine mycotic abortion [133]. "*Syncephalastrum* sp." was additionally reported as being responsible for one case of human cutaneous mucormycosis [70], and also for gastric mucormycosis in a calf [94].

2. Main Characteristics of *Syncephalastrum racemosum* (see Benjamin [13])

Since *S. racemosum* is the type species of *Syncephalastrum,* which is the only genus in the Syncephalastraceae, the main characteristics of this species are obviously identical with those of this family (see Sec. III.B.4 and Fig. 5c). This is another rapid-growing fungus with abundant aerial mycelium superficially resembling a *Rhizopus* sp. Production of *Mucor*-like zygospores in crosses [152]; maximum temperature of growth 40°C. For synonyms, see Ref. 13.

D. *Saksenaea vasiformis* Saksena, *Mycologia 45*: 434, 1953 [109]

1. Occurrence as a Pathogen

This species was recorded only once as an agent of human cranial (orbital) mucormycosis [4]. "*Saksenaea* sp." was found responsible for one additional human case of generalized mucormycosis [22].

2. Main Characteristics of *Saksenaea vasiformis* (according to Saksena [109]; Ellis and Hesseltine [40])

For the definition of Saksenaeaceae, see Sec. III.B.5 and Fig. 5d. The genus *Saksenaea* with its only species, *Saks. vasiformis,* is characterized by flask-shaped *sporangia* up to 50 μm long and up to 20 μm thick with a definite, globose *columella,* and by small, smooth, elongate *sporangiospores* (3-4 × 1.5-2 μm); the *mycelium* is rapid growing but not very high; sporulation is often obtained only with difficulty (e.g., when pieces of potato dextrose agar permeated with mycelium are floated on sterile distilled water [109]); maximum temperature of growth is 44°C.

XI. SPECIES OF MUCORALES WITH DOUBTFUL PATHOGENICITY

A. General Considerations

There are two possible sources of errors with the reports on Mucorales cultured from pathological material: (1) the fungus may in fact have caused the disease in question but the diagnosis of the species was incorrect, and (2) the fungus, although correctly identified, was not the cause of the disease but was a contaminant. The latter risk exists because of the rather ubiquitous occurrence of the spores of many Mucorales and must particularly be considered when there is no morphological proof of typical hyphae in the material or when the clinicopathological picture does not fit one of the usual forms of mucormycosis.

An error of any of the two types was, as a rule, assumed with species whose maximum temperature of growth is clearly below 37°C. With some species we checked experimental pathogenicity to mice and found them nonpathogenic [121].

B. An Annotated List of Species Not Accepted as Pathogenic (with Reference to the Case Reports Claiming Pathogenicity, and with Maximum Temperature of Growth)

Rhizopus nigricans = *R. stolonifer* [e.g., 20,23,42,65,112,143]. Maximum temperature of growth 30-32°C.

Mucor mucedo [47,149]. In the first-mentioned case [47], the fungus was *A. corymbifera* according to the author's drawings. Maximum temperature of growth 22°C.

Mucor hiemalis [143]. Maximum temperature of growth 32°C.

"Mucor spinosus" [98], synonym of *M. plumbeus* see Ref. 115). Maximum temperature of growth 30°C.

"Mucor dispersus" [134]. Maximum temperature of growth 32°C. *M. dispersus* is considered a synonym of *Backusella lamprospora* (see Ref. 116).

Mortierella alpina and *Mort. zychae* Smith, 1966 [129]; these fungi were probably *Mort. wolfii.* Maximum temperature of growth 30-32°C.

Mortierella polycephala [126,139]. Maximum temperature of growth 22°C.

Mortierella hygrophila [66]. Maximum temperature of growth 30°C.

REFERENCES

1. Agareval, S. C. 1974. *Absidia ushtrina* sp. nov. from Indian alkaline soils. Kavaka *2*: 47-48.
2. Ainsworth, G. C., and P. K. C. Austwick. 1955. A survey of animal mycoses in Britain: mycological aspects. Trans. Br. Mycol. Soc. *38*: 369-389.
3. Ainsworth, G. C., and P. K. C. Austwick. 1973. *Fungal Diseases of Animals.* Commonwealth Agricultural Bureaux, Farnham Royal, Slough, England.
4. Ajello, L., D. F. Dean, and R. S. Irwin. 1976. The zygomycete *Saksenaea vasiformis* as a pathogen of humans with a critical review of the etiology of zygomycosis. Mycologia *68*: 52-62.
5. Bainier, G. 1882. Etude sur les Mucorinées. Thèse, Ecole Supérieure de Pharmacie, Paris.
6. Bainier, G. 1903. Sur quelques espèces de Mucorinées nouvelles ou peu connues. Bull. Soc. Mycol. Fr. *19*: 153-172.
7. Baker, R. D. 1956. Pulmonary mucormycosis. Am. J. Pathol. *32*: 287-313.

8. Baker, R. D. 1957. Mucormycosis, a new disease? JAMA *163*: 805-808.
9. Baker, R. D. 1971. Mucormycosis (opportunistic phycomycosis). In R. D. Baker (Ed.), *The Pathologic Anatomy of Mycoses,* Vol. 3, Part 5, *Handbuch der speziellen pathologischen Anatomie und Histologie,* Springer-Verlag, Berlin, pp. 830-918.
10. Baker, R. D., D. E. Bassert, and E. Ferrington. 1957. Mucormycosis of the digestive tract. Arch. Pathol. *63*: 176-182.
11. Baker, R. D., J. H. Seabury, and J. D. Schneidau. 1962. Subcutaneous and cutaneous mucormycosis and subcutaneous phycomycosis. Lab. Invest. *11*: 1091-1102.
12. Bartnicki-García, S. 1963. Mold-yeast dimorphism of *Mucor.* Bacteriol. Rev. *27*: 293-304.
13. Benjamin, R. K. 1959. The merosporangiferous Mucorales. Aliso *4*: 321-433.
14. Benjamin, R. K. 1979. Zygomycetes and their spores. In B. Kendrick (Ed.), *The Whole Fungus,* Proceedings of the Second International Mycological Conference, Vol. 2, National Museum of Natural Sciences and National Museums of Canada, Ottawa, pp. 573-616.
15. Berlese, A. N., and J. B. de Toni. 1888. Phycomyceteae. Syll. Fung. *7*: 181-220.
16. Blakeslee, A. F., J. L. Cartledge, and D. S. Welch. 1921. Sexual dimorphism in *Cunninghamella.* Bot. Gaz. (Crawfordsville) *72*: 185-219.
17. Bottone, E. J., I. Weitzman, and B. A. Hanna. 1979. *Rhizopus rhizopodiformis*: emerging etiological agent of mucormycosis. J. Clin. Microbiol. *9*: 530-537.
18. Bruderlein, J. 1917. Le *Rhizopus maydis,* n.sp. Bull. Soc. Bot. Genève. *9*: 108-112.
19. Bullock, J. D., L. M. Jampol, and A. J. Fezza. 1974. Two cases of orbital phycomycosis with recovery. Am. J. Ophthalmol. *78*: 811-815.
20. Burrow, G. N., R. B. Salmon, and J. P. Nolan. 1963. Successful treatment of cerebral mucormycosis with amphotericin B. JAMA *183*: 370-372.
21. Callen, E. O. 1940. The morphology, cytology, and sexuality of the homothallic *Rhizopus sexualis* (Smith). Ann. Bot. *4*: 791-818.
22. Campbell, C. K., W. C. Marshall, and R. J. Hay. 1980. A fatal infection with *Saksenaea.* Paper presented at the 16th Annual Meeting British Society for Mycopathology, Liverpool, March 31-April 2.
23. Campion, C. K., and T. M. Johnson. 1969. Rhino-orbito-cerebral phycomycosis. Mich. Med. News *68*: 807-810.
24. Caretta, G., and E. Piontelli. 1971. Mucormicosi polmonare avicola da *Mucor racemosus* Fres. Atti Ist. Bot. Univ. Lab. Crittog., Pavia *7*: 33-39.
25. Carter, M. E., D. O. Cordes, M. E. di Menna, and R. Hunter. 1973. Fungi isolated from bovine mycotic abortion and pneumonia with special reference to *Mortierella wolfii.* Res. Vet. Sci. *14*: 201-206.
26. Cooney, D. G., and R. Emerson. 1964. *The Thermophilic Fungi,* Freeman, San Francisco.
27. Corbel, M. J., and S. M. Eades. 1977. Cerebral mucormycosis following experimental inoculation with *Mortierella wolfii.* Mycopathologia *60*: 129-133.
28. Costantin, J., and A. Lucet. 1903. Sur un *Rhizopus* pathogène. Bull. Soc. Mycol. Fr. *19*: 200-216.
29. Crisan, E. V. 1973. Current concepts of thermophilism and the thermophilic fungi. Mycologia *65*: 1171-1198.
30. Cutter, V. M. 1946. The genus *Cunninghamella* (Mucorales). Farlowia *2*: 321-343.
31. Darja, M., and M. I. Davy. 1963. Pulmonary mucormycosis with cultural identification. Can. Med. Assoc. J. *89*: 1235-1238.
32. Davies, R. R. 1972. Protothecosis and opportunistic fungal infections. Trans. St. John's Hosp. Dermatol. Soc. *58*: 38-42

32. Dillon, M. L., W. C. Sealy, and B. L. Fetter. 1958. Mucormycosis of the bronchus successfully treated by lobectomy. J. Thorac. Surg. *35*: 464-468.
34. Dodge, C. W. 1935. *Medical Mycology,* Mosby, St. Louis.
35. Douvin, D., Y. Lefichoux, and C. Huguet. 1975. Phycomycose gastrique. Arch. Anat. Pathol. *23*: 133-138.
36. Drozd, J., and J. Novak. 1979. Headspace gas analysis by gas chromatography. J. Chromatogr. *165*: 141-165.
37. Dworzack, D. L., A. S. Pollock, G. R. Hodges, W. G. Barnes, L. Ajello, and A. Padhye. 1978. Zygomycosis of the maxillary sinus and palate caused by *Basidiobolus haptosporus.* Arch. Intern. Med. *138*: 1274-1276.
38. Ellis, J. J., and C. W. Hesseltine. 1965. The genus *Absidia:* globose-spored species. Mycologia *57*: 222-235.
39. Ellis, J. J., and C. W. Hesseltine. 1966. Species of *Absidia* with ovoid sporangiospores II. Sabouraudia *5*: 59-77.
40. Ellis, J. J., and C. W. Hesseltine. 1974. Two new families of Mucorales. Mycologia *66*: 87-95.
41. Erdos, M. S., K. Butt, and L. Weinstein. 1972. Mucormycotic endocarditis of the pulmonary valve. JAMA *222*: 951-953.
42. Ferstenfeld, J. E., S. H. Cohen, H. D. Rose, and M. W. Rytel. 1977. Postgrad. Med. J. *53*: 337-342.
43. Fischer, A. 1892. Phycomycetes. In L. Rabenhorst (Ed.), *Kryptogamenflora von Deutschland, Oesterreich und der Schweiz,* Vol. 1, Abt. IV, Kummer, Leipzig, reprinted by Cramer, Weinheim, 1963.
44. Frank, A. 1890. Eine mykotische Neubildung am Widerrist des Pferdes. Wochenschr. Tierheilk. Viehz. *34*: 13-15.
45. Frank, W., U. Roester, and H. J. Scholer. 1974. Sphaerulen-Bildung bei einer Mucorspezies in inneren Organen von Amphibien. Zentralbl. Bakteriol. Parasitenkd. Infektionskr. Hyg., Abt. 1: Orig. Reihe A, *226*: 405-417.
46. Fresenius, G. 1850. *Beiträge zur Mykologie I,* Frankfurt.
47. Fürbringer, P. 1876. Beobachtungen über Lungenmycose beim Menschen. Virchows Arch. Pathol. Anat. Physiol. *66*: 330-365.
48. Gams, W. 1977. A key to the species of *Mortierella.* Persoonia *9*: 381-391.
49. Gartenberg, G., E. J. Bottone, G. T. Kensch, and I. Weitzman. 1978. Hospital-acquired mucormycosis *(Rhizopus rhizopodiformis)* of skin and subcutaneous tissue. N. Engl. J. Med. *299*: 1115-1118.
50. Gitter, M., and P. K. C. Austwick. 1959. Mucormycosis and moniliasis in a litter of suckling pigs. Vet. Rec. *71*: 6-11.
51. Gloor, F., A. Löffler, and H. J. Scholer. 1961. Mucormykosen. Pathol. Microbiol. *24*: 1043-1064.
52. Hagem, O. 1908. *Untersuchungen über norwegische Mucorineen I.* Dybwad, Christiania [Oslo].
53. Hagem, O. 1910. Neue Untersuchungen über norwegische Mucorineen. Ann. Mycol. *8*: 266-286.
54. Hanzawa, J. 1912. Zur Morphologie und Physiologie von *Rhizopus Delemar,* dem Pilz des neueren Amylo-Verfahrens. Mycol. Zentralbl. *1*: 77-91.
55. Hanzawa, J. 1915. Studien über einige *Rhizopus*-Arten. I. Systematisches. Mycol. Zentralbl. *5*: 230-247.
56. Haynes, W. C., L. J. Wickerham, and C. W. Hesseltine. 1955. Maintenance of cultures of industrially important microorganisms. Appl. Microbiol. *3*: 361-368.
57. Hesseltine, C. W. 1954. The section *Genevensis* of the genus *Mucor.* Mycologia *46*: 358-366.

58. Hesseltine, C. W. 1955. Genera of the Mucorales with notes on their synonymy. Mycologia *47*: 344-363.
59. Hesseltine, C. W. 1965. Industrial mycology. A millennium of fungi, food and fermentation. Mycologia *57*: 149-194.
60. Hesseltine, C. W., and J. J. Ellis. 1961. Notes on Mucorales, especially *Absidia*. Mycologia *53*: 406-426.
61. Hesseltine, C. W., and J. J. Ellis. 1964. The genus *Absidia: Gongronella* and cylindrical-spored species of *Absidia*. Mycologia *56*: 568-601.
62. Hesseltine, C. W., and J. J. Ellis. 1964. An interesting species of *Mucor, M. ramosissimus*. Sabouraudia *3*: 151-154.
63. Hesseltine, C. W., and J. J. Ellis. 1966. Species of *Absidia* with ovoid sporangiospores I. Mycologia *58*: 761-785.
64. Hesseltine, C. W., and J. J. Ellis. 1973. Mucorales. In G. C. Ainsworth, F. K. Sparrow, and A. S. Sussman (Eds.), *The Fungi,* Vol. 4B; *Basidiomycetes and Lower Fungi,* Academic, New York, pp. 187-217.
65. Hoagland, R. J., J. Sube, R. H. Bishop, and B. F. Holding. 1961. Mucormycosis. Am. J. Med. Sci. *242*: 415-422.
65a. Hohl, P. E., E. Prevost, F. J. John, and L. Kaufman. 1980. Zygomycosis: clinical spectrum, mycology, and serological findings. CDC reference No. 80/8034, Center for Diseases Control, Atlanta.
66. Hörter, R., and F. Hunsteger. 1962. *Mortierella*-Mykose bei Vögeln. Dtsch. Tierärztl. Wochenschr. *69*: 49-51.
67. Hutter, R. V. P. 1959. Phycomycetous infection (mucormycosis) in cancer patients, a complication of therapy. Cancer (N.Y.) *12*: 330-350.
68. Inui, T., Y. Takeda, and H. Iizuka. 1965. Taxonomical studies on genus *Rhizopus*. J. Gen. Appl. Microbiol. *11* (Suppl.): 1-121.
69. Jones, K. W., and L. Kaufman. 1978. Development and evaluation of an immunodiffusion test for diagnosis of systemic zygomycosis (mucormycosis), a preliminary report. J. Clin. Microbiol. 7: 97-103.
70. Kamalam, A., and A. S. Thambiah. 1980. Cutaneous infection by *Syncephalastrum*. Sabouraudia *18*: 19-20.
71. Kramer, B. S., A. D. Hernandez, R. L. Reddick, and A. S. Levine. 1977. Cutaneous infection, manifestation of disseminated mucormycosis. Arch. Dermatol. *113*: 1075-1076.
72. Kwon-Chung, K. J., R. C. Young, and M. Orlando. 1975. Pulmonary mucormycosis caused by *Cunninghamella elegans* in a patient with chronic myelogenous leukemia. Am. J. Clin. Pathol. *64*: 544-548.
73. Lang, F. J., and F. Grubauer. 1923. Über *Mucor*- und *Aspergillus*-Mykose der Lunge. Virchows Arch. Pathol. Anat. Physiol. *245*: 480-512.
74. La Touche, C. J., T. W. Sutherland, and M. Telling. 1963. Histopathological and mycological features of a case of rhinocerebral mucormycosis (phycomycosis) in Britain. Sabouraudia *3*: 148-150.
75. Lendner, A. 1908. Les Mucorinées de la Suisse. Mat. Fl. Crypt. Suisse *3*(1): 1-77.
76. Lendner, A. 1929. Détermination de Mucorinées (deux Mucors nouveaux). Bull. Soc. Bot. Genève *21*: 256-263.
77. Lichtheim, L. 1884. Über pathogene Mucorineen und die durch sie erzeugten Mykosen des Kaninchens. Zentralbl. Klin. Med. 7: 140-177.
78. Lie-Kian-Joe, N. I. T. Eng, S. Tjokronegoro, S. Schaafma, and C. W. Emmons. 1959. Phycomycosis of the central nervous system, associated with diabetes mellitus. Am. J. Clin. Pathol. *32*: 62-70.
79. Lindt, W. 1886. Mitteilungen über einige neue pathogene Schimmelpilze. Arch. Exp. Pathol. Pharmakol. *21*: 269-298.

80. Lucet, A., and J. Costantin. 1899. Sur une nouvelle Mucorinée pathogène. C. R. Acad. Sci. (Paris) *129*: 1031-1034.
81. Marjanková, K., K. Křivanec, and J. Zajiček. 1978. Mass occurrence of necrotic inflammation of the penis in ganders caused by phycomycetes. Mycopathologia *66*: 21-26.
82. Mehrotra, B. S., and U. Baijal. 1963. Species of *Mortierella* from India III. Mycopathologia *20*: 49-54.
83. Menna di, M. E., M. E. Carter, and D. O. Cordes. 1972. The identification of *Mortierella wolfii* isolated from cases of abortion and pneumonia in cattle and a search for its infection source. Res. Vet. Sci. *13*: 439-442.
84. Meyer, R. D., M. H. Kaplan, M. Ong, and D. Armstrong. 1973. Cutaneous lesions in disseminated mucormycosis. JAMA *225*: 737-738.
85. Milko, A. A., T. C. Kirilenko, and V. A. Shkurenko. 1966. On some species of the genus *Rhizopus* Ehrenberg (in Russian). Mikrobiol. Zh. *28*: 30-35.
86. Misra, P. C. 1975. A new species of *Syncephalastrum*. Mycotaxon *3*: 51-54.
87. Morquer, R., C. Lombard, M. Berthelon and L. Lacoste. 1965. Pouvoir pathogène des Mucorales dans le règne animal. Une nouvelle mycose chez les bovidés et les porcins. C. R. Acad. Sci. (Paris) *260*: 6173-6176.
88. Müller, E., and W. Loeffler. 1976. *Mycology, an Outline for Science and Medical Students.* Thieme, Stuttgart.
89. Murphy, J. D., and S. Bornstein. 1950. Mucormycosis of the lung. Ann. Intern. Med. *33*: 442-453.
90. Naganishi, H., and S. Hirahara. 1970. Two new species of the *Absidia* genus isolated from "Kurone" developed during the manufacture of soy sauce Koji. Bull. Hiroshima Jogakuin Coll. *20*: 13-17.
91. Nakazawa, R. 1909. *Rhizopus batatas,* ein neuer Pilz aus dem Koji des Bataten-branntweins von der Insel Hachijo (Japan). Zentralbl. Bakteriol. Parasitenkd. Infektionskr. Hyg., Abt. 2, *24*: 482-487.
92. Namyslowski, B. 1906. *Rhizopus nigricans* et les conditions de la formation de ses zygospores. Bull. Int. Acad. Sci. Cracovie *208*: 676-692.
93. Naumov, N. A. 1939. *Clés des Mucorinées (Mucorales),* Lechevalier, Paris.
94. Nayak, B. C., A. G. Rao, S. K. Ray, and S. K. Chanda. 1975. Mycotic ruminitis in a calf. Indian Vet. J. *52*: 56-57.
95. Neame, P., and D. Rayner. 1960. Mucormycosis, report on twenty-one cases. Arch. Pathol. *70*: 261-268.
96. Nicod, J. L., C. Fleury, and J. Schlegel. 1952. Mycose pulmonaire double à *Aspergillus fumigatus* Fres. et à *Mucor pusillus* Lindt. Schweiz. Z. Allg. Pathol. Bakteriol. *15*: 307-321.
97. Nicolet, J., S. Lindt, and H. J. Scholer. 1966. L'avortement mycosique de la vache. Considérations sur le diagnostic de routine. Pathol. Microbiol. *29*: 644-655.
98. Nielsen, N. 1902-1903. Skimmelvegetation i Kindhulen hos en Hest. Maanedsskr. Dyrlaeg. *14*: 622-624.
99. Nielsen, N. 1929. Mucormykose beim Schwein II. Mitteilung: Beschreibung der isolierten Pilze. Virchows Arch. *273*: 859-863.
100. Nottebrock, H., H. J. Scholer, and M. Wall. 1974. Taxonomy and identification of mucor-mycosis-causing fungi. 1. Synonymity of *Absidia ramosa* with *A. corymbifera.* Sabouraudia *12*: 64-74.
101. Novobranova, T. J. 1972. Species novae fungorum imperfectorum e regione Alma-Ataënsi. Novitates Systematicae Plantarum non Vasularium (in Russian) *9*: 181-183.
102. Paltauf, A. 1885. Mycosis mucorina. Ein Beitrag zur Kenntnis der menschlichen Fadenpilzerkrankungen. Virchows Arch. Pathol. Anat. Physiol. *102*: 543-564.

103. Podack, M. 1899. Zur Kenntnis des sogenannten Endothelkrebses der Pleura und der Mucormykosen im menschlichen Respirationsapparate. Dtsch. Arch. Klin. Med. *63*: 1-73.
104. Porges, N., J. F. Muller, and L. B. Lockwood. 1935. A mucor found in fowl. Mycologia *27*: 330-331.
105. Reinhardt, D. J., W. Kaplan, and L. Ajello. 1970. Experimental cerebral zygomycosis in alloxan-diabetic tablets. 1. Relationship of temperature tolerance of selected zygomycetes to pathogenicity. Infect. Immun. *2*: 404-413.
106. Saccardo, P. A., and A. Trotter. 1912. Phycomycetae de Bary. Syll. Fung. *21*: 813-863.
107. Saito, K. 1904. Eine neue Art der "Chinesischen Hefe." Zentralbl. Bakteriol. Parasitenkd. Infektionskr. Hyg., Abt. 2, *13*: 154-160.
108. Saito, K. 1905. *Rhizopus oligosporus,* ein neuer technischer Pilz Chinas. Zentralbl. Bakteriol. Parasitenkd. Infektionskr. Hyg., Abt. 2, *14*: 623-627.
109. Saksena, S. B. 1953. A new genus of the Mucorales. Mycologia *45*: 426-436.
110. Samson, R. A. 1969. Revision of the genus *Cunninghamella.* Proc. K. Ned. Akad. Wet. Amst., C 1, *72*: 322-325.
111. Samutsevitsch, M. M. 1927. On the question of the fungal flora in the soil. Mater. Mycol. Phytopathol. *6*: 204-213.
112. Sandler, R., C. B. Tallman, D. G. Keany, and W. R. Irving. 1971. Successfully treated rhinocerebral phycomycosis in well controlled diabetes. N. Engl. J. Med. *285*: 1180-1182.
113. Schipper, M. A. A. 1969. Zygosporic stages in heterothallic *Mucor.* Antonie Leeuwenhoek J. Microbiol. Serol. *35*: 189-208.
114. Schipper, M. A. A. 1973. A study on variability in *Mucor hiemalis* and related species. *Studies in Mycology,* Vol. 4, Centraalbureau voor Schimmelcultures, Baarn, pp. 1-41.
115. Schipper, M. A. A. 1976. On *Mucor circinelloides, M. racemosus* and related species. *Studies in Mycology,* Vol. 12, Centraalbureau voor Schimmelcultures, Baarn, pp. 1-40.
116. Schipper, M. A. A. 1978. On certain species of *Mucor* with a key to all accepted species. *Studies in Mycology,* Vol. 17, Centraalbureau voor Schimmelcultures, Baarn, pp. 1-52.
117. Schipper, M. A. A. 1978. On the genera *Rhizomucor* and *Parasitella. Studies in Mycology,* Vol. 17, Centraalbureau voor Schimmelcultures, Baarn, pp. 53-71.
118. Schipper, M. A. A. 1979. *Thermomucor* (Mucorales). Antonie Leeuwenhoek J. Microbiol. Serol. *45*: 275-280.
119. Schipper, M. A. A. 1980. Progress report 1979, Centraalbureau voor Schimmelcultures. Verh. K. Ned. Akad. Wet. Afd. Natuurkd. Reeks 2, *75*: 4-5.
120. Schipper, M. A. A., R. A. Samson, and J. A. Stalpers. 1975. Zygospore ornamentation in the genera *Mucor* and *Zygorhynchus.* Persoonia *8*: 321-328.
121. Scholer, H. J. 1970. Mucormykosen bei Mensch und Tier: Taxonomie der Erreger, Chemotherapie im Tierexperiment und in der Klinik. Habilitationsschrift, Medical Faculty, University of Basel. (Copy can be borrowed from University Library, Schönbeinstrasse 18-20, Basel, Switzerland.)
122. Scholer, H. J. 1980. Mucormykosen und Mycormykose-Erreger. Paper presented at the 15th Scientific Conference of the German Speaking Mycological Society, Aachen, May 2-4.
123. Scholer, H. J., and E. Müller. 1966. Beziehungen zwischen biochemischer Leistung und Morphologie bei Pilzen aus der Familie der Mucoraceen. Pathol. Microbiol. *29*: 729-741.

124. Scholer, H. J., and L. Peter. 1980. Serology of mucormycosis-causing fungi and of mucormycosis. In H. J. Preusser (Ed.), *Medical Mycology*. Zentralbl. Bakteriol. Parasitenkd. Infektionskr. Hyg., Abt. 1, Suppl. 8, pp. 183-191.
125. Scholer, H. J., and R. Richle. 1965. Spontane Aspergillose und Mucormykose des Kaninchens. In H. Grimmer and H. Rieth (Eds.), *Krankheiten durch Schimmelpilze bei Mensch und Tier,* Springer-Verlag, Berlin, pp. 111-119.
126. Scholz, H. D., and L. Meyer. 1965. *Mortierella polycephala* als Erreger einer Lungenmykose beim Rind. Berl. Münch. Tieräerztl. Wochenschr. *78*: 27-30.
127. Schröter, J. 1886. *Kryptogamenflora von Schlesien,* Vol. 1, Kernis Verlag, Breslau.
128. Siebenmann, R., and T. Wegmann. 1968. Generalisierte Mucormykose. Schweiz. Med. Wochenschr. *98*: 537-543.
129. Smith, J. M. B. 1966. An interesting bovine mycotic complex in New Zealand. N.Z. Vet. J. *14*: 226.
130. Spalla, G. 1963. Research on sexual reproduction in Mucorales. Sexual reproduction in *Cunninghamella blakesleeana* and *C. elegans.* Riv. Patol. Veg. *3*: 107-128.
131. Stadel, O. 1911. Über einen neuen Pilz, *Cunninghamella bertholletiae.* Dissertation, Kiel.
132. Subrahamanyam, A., B. S. Mehrotra, and M. J. Thirumalachar. 1977. *Thermomucor,* a new genus of Mucorales. Ga. J. Sci. *35*: 1-4.
133. Turner, P. D. 1964. *Syncephalastrum* associated with bovine mycotic abortion. Nature (Lond.) *204*: 399.
134. Turner, P. D. 1965. Simultaneous infection of a bovine foetus by two fungi. Nature (Lond.) *205*: 300-301.
135. Váňová, M. 1968. Contribution to the taxonomy of the genus *Absidia* (Mucorales). 1. *Absidia macrospora* sp. nov. Česk. Mykol. *22*: 296-300.
136. Van Beyma thoe Kingma, F. H. 1931. Über einen neuen *Rhizopus, Rhizopus bovinus* nov. spec. Mitteil. CBS K. Ned. Akad. Wet. Amst. *29*(2): 38-40.
137. Van Tieghem, P. 1875. Nouvelles recherches sur les Mucorinées. Ann. Sci. Nat. 6. Ser. Bot. *1*: 5-175.
138. Van Tieghem, P. 1876. Troisième mémoire sur les Mucorinées. Ann. Sci. Nat. 6. Ser. Bot. *4*: 312-398.
139. Van Ulsen, F. W. 1955. Schimmelabortus bij runderen. Tijdschr. Diergeneeskd. *80*: 1081-1088
140. Vuillemin, P. 1902. Recherches sur les Mucorinées saccharifiantes (Amylomyces). II. Série des *Rhizopus.* Rev. Mycol. *24*: 46-60.
141. Vuillemin, P. 1903. La série des Absidées. C. R. Acad. Sci. (Paris) *136*: 514-516.
142. Vuillemin, P. 1903-1904. Le *Lichtheimia ramosa* (*Mucor ramosus* Lindt) champignon pathogène, distinct du *L. corymbifera.* Arch. Parasitol. *8*: 562-572.
143. Watson, K. C., and P. B. Neame. 1960. In vitro activity of amphotericin B on strains of Mucoraceae pathogenic to man. J. Lab. Clin. Med. *56*: 251-257.
144. Wehmer, C. 1900. Die "Chinesische Hefe" und der sogenannte *Amylomyces* (= *Mucor rouxii*). Zentralbl. Bakteriol. Parasitenkd. Infektionskr. Hyg., Abt. 2, *6*: 353-365.
145. Weitzman, I., and M. Y. Crist. 1979. Studies with clinical isolates of *Cunninghamella.* I. Mating behaviour. Mycologia *61*: 1024-1033.
146. Went, F. A. F. C., and H. C. Prinsen Geerligs. 1895. Beobachtungen über die Hefearten und zuckerbildenden Pilze der Arackfabrikation. Verh. K. Ned. Akad. Wet. Afd. Natuurkd. Reeks 2, *4*: 1-31.
147. Wohlgemuth, K., and W. V. Knudtson. 1977. Abortion associated with *Mortierella wolfii* in cattle. J. Am. Vet. Med. Assoc. *171*: 437-439.

148. Yamamoto, Y. 1930. Ein Beitrag zur Kenntnis der Gattung *Rhizopus* I und II. J. Fac. Agric. Hokkaido Imp. Univ. *28*: 1-327.
149. Zaitsev, A. P., and A. F. Kuteinikova. 1977. Mycological investigation of the bronchial aspirate of patients with chronic pulmonary diseases (in Russian). Zh. Mikrobiol. Epidemiol. Immunobiol. *12*: 94-96.
150. Zopf, W. 1890. Die Pilze. In A. Schenk (Ed.), *Handbuch der Botanik,* Vol. 4, Breslau.
151. Zycha, H. 1935. Pilze II: Mucorineae. In *Kryptogamenflora der Mark Brandenburg,* Vol. 4a, Bornträger, Leipzig.
152. Zycha, H., and R. Siepmann. 1969. Mucorales; eine Beschreibung aller Gattungen und Arten dieser Pilzgruppe. Cramer, Lehre.

3
Entomophthorales

Douglas S. King / Baptist Medical Center Plaza, Little Rock, Arkansas

I. INTRODUCTION

Basidiobolus haptosporus Drechsler, *Conidiobolus coronatus* (Cost.) Batko, and *C. incongruus* Drechsler are the only species of the Entomophthorales that are known to cause human disease. Of the various names that have been used to refer to the diseases caused by the first two, those applied by Emmons et al. [14] were the most consistent and descriptive. For *B. haptosporus,* this was entomophthoramycosis basidiobolae, and for *C. coronatus,* entomophthoramycosis conidiobolae. Only one instance of human infection involving *C. incongruus* has been documented [31].

Basidiobolus haptosporus and *C. coronatus* are widespread and usually occur as saprophytes; the latter is especially abundant in plant detritus [6,26]. However, each causes disease in humans only in the warmer regions of the world. In addition to infecting humans, *C. coronatus* also infects other mammals and insects over a broader geographical range than that in which it infects humans. *Conidiobolus incongruus* has been isolated only from the United States: twice from plant detritus and once from humans in West Virginia [28].

The Entomophthorales is a well-defined order of the Zygomycetes in which only one family (the *Entomophthoraceae*) and 10 genera are currently recognized [30]. The largest genus is *Entomophthora* Fresenius, with over 100 species. Members of this genus have been isolated only as pathogens of invertebrates, even though several can be easily cultured in vitro. The separation of *Entomophthora* and *Conidiobolus* Brefeld is a long-standing taxonomic problem which is discussed further in Sec. III. *Massospora* Peck has 11 species that are all obligate pathogens of cicadas. Except for *Basidiobolus,* only one to three species are recognized for each of the remaining genera. The two species of *Stronwellsea* Batko & Weiser infect only flies. Species of *Meristacrum* Drechsler and *Zygnemomyces* Miura infect nematodes, whereas those of *Completoria* Lohde, *Ancylistes* Pfitzer, and *Ballocephala* Drechsler infect fern gametophytes, desmid algae, and tardigrades, respectively.

The variety of spores produced and the ecological relationships at the genus and species levels give rise to taxonomic problems within the order. A brief review of the morphology of the order will be helpful in considering the systematic relationships of the species pathogenic for mammals.

II. MORPHOLOGY OF THE ENTOMOPHTHORALES

A. Vegetative Growth

Thaxter [38] coined the term "hyphal body" to describe the short hyphal segments that are prominent during the vegetative growth phase of many *Entomophthora* species. This mode of growth is found in several genera, including *Basidiobolus* and *Conidiobolus,* and has been considered characteristic of vegetative growth in the order by several workers (e.g., Refs. 1 and 40). However, the production of hyphal bodies should be considered only one phase of development, since it intergrades with truly mycelial development (e.g., Refs. 2 and 20). In two species of *Conidiobolus,* the vegetative phase is represented by only a coenocytic mycelium [28].

B. Conidia

The asexual spores of all members of the order are produced singly at the apex (apices) of simple or branched conidiophores. These spores may be evolutionarily derived from sporangia, but except for *Basidiobolus,* evidence for their sporangial nature is sparse, and they usually function as conidia [30]. Except for *Massospora,* all species produce two types of conidia, and a single species may produce four types (e.g., *C. coronatus*). Each type of conidium may germinate by one or more germ tubes to produce a vegetative mycelium, and in most instances, may produce another conidium. Species of *Massospora* are unusual in that they do not produce forcibly discharged conidia, and are not considered in the remainder of this discussion.

Primary conidia are produced atop simple or branched conidiophores that arise from vegetative elements. They may be one of several shapes and uni- to quadri- or multinucleate. They are forcibly discharged toward the strongest source of light. Discharge of all types of forcibly discharged conidia is usually effected through pressure developed within the conidium (e.g., *Conidiobolus*), and due to eversion of the conidium wall, where it was attached to the conidiophore, each conidium bears a more or less prominent basal papillum following discharge. In several species, including *Entomophthora muscae* (the type species of the order) and all species of *Basidiobolus,* discharge is effected through pressure developed within the conidiophore, which ruptures in the discharge process.

Except in the genus *Entomophthora,* the primary conidia in each genus are uniform with regard to conidial shape, number of nuclei, and mechanism of discharge. The taxonomic implications of the heterogeneity in *Entomophthora* have been discussed elsewhere [30].

A primary conidium may produce a forcibly discharged (i.e., secondary) conidium that is also phototropic. Except for their smaller size, replicative conidia usually resemble the primary conidia from which they arise, although they have a different shape in a few species. Also, in a few species of *Entomophthora* and *Conidiobolus,* replicative conidia of two shapes are produced: one that resembles the primary conidia and one that has a different shape.

A replicative conidium may produce subsequent generations of like conidia. In several species of *Entomophthora* and *Conidiobolus,* and in all but one species of *Basidiobolus,* primary and replicative conidia also may produce capillospores. These are elongate, passively detached conidia borne atop very thin, long conidiophores. Multiplicative (i.e., micro) conidia of different types are produced by several species of *Conidiobolus* and *B. microsporus,* and villose conidia are unique to *C. coronatus.* These spore types are discussed under the respective genera.

C. Resting Spores

Resting spores that are important for surviving unfavorable conditions are common in the order. Chlamydospores are one form of commonly encountered resting spore. They develop from preexisting vegetative elements by the production of a single thick wall, with or without alterations in shape and/or size. Zygospores are produced in many species following conjugation of equal or unequal gametangia derived from the same thallus (heterothallism has not been reported in the order). They may be formed within one of the gametangia as in *Conidiobolus* and *Basidiobolus,* or exogenous to the gametangia as in several species of *Entomophthora.* Azygospores are formed in many species of *Entomophthora* in which gametangial copulation has not been observed.

In both zygospores and azygospores the spore wall is formed de novo rather than by modification of preexisting walls as in the Mucorales. Both usually have two thick walls and are morphologically indistinguishable. Whether they are equivalent from the aspect of genetic recombination has not been determined.

III. PATHOGENIC GENERA AND SPECIES

A. *Basidiobolus*

1. Biology

The genus *Basidiobolus* (Figs. 1-5) is distinguished primarily by the developmental processes that lead to the production of forcibly discharged conidia, the manner in which these conidia are discharged, and the developmental sequence in the production of zygospores. In all species of *Basidiobolus,* the replicative conidia (Fig. 2) are like the primary conidia (Fig. 1) except that they are produced atop conidiophores that arise from other globose conidia rather than from vegetative elements. In producing a forcibly discharged conidium, the apical portion of the conidiophore first swells to form what will be the subconidial vesicle, and the conidium is blown out from the top of this swelling. In the discharge process, the subconidial vesicle ruptures and contracts in a well-defined manner,

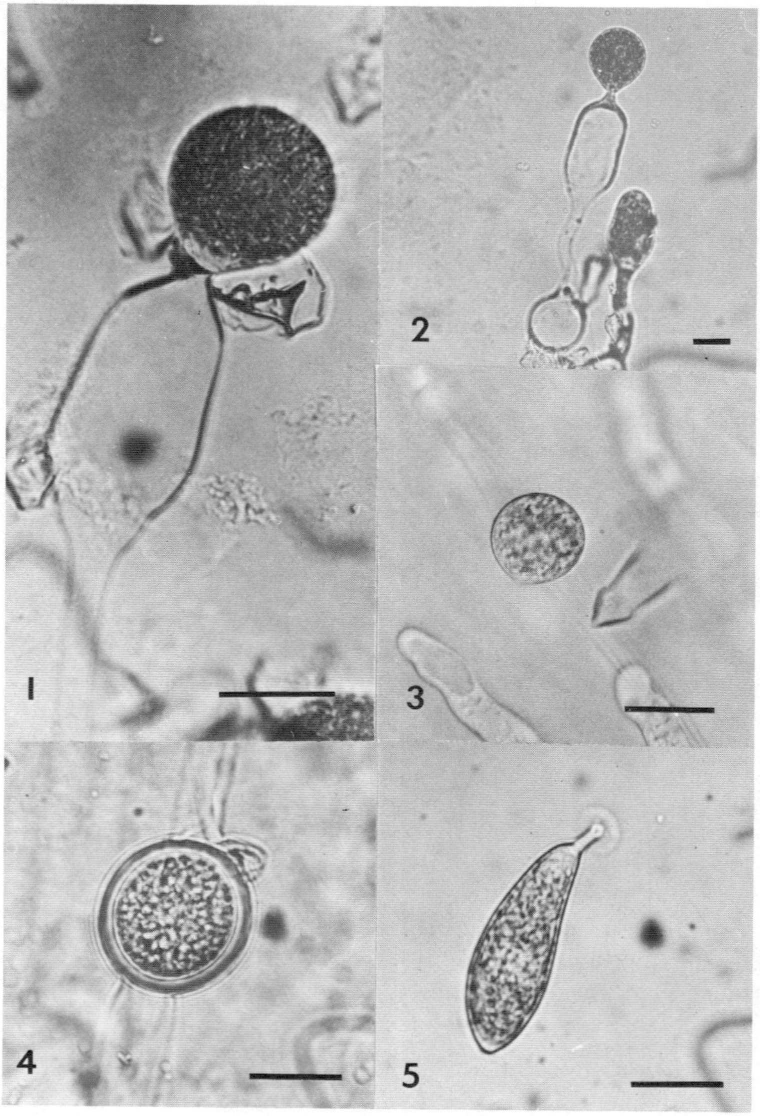

Figures 1-5 *Basidiobolus haptosporus.* Bars equal 20 μm. (1) Primary conidium on conidiophore. (2) Replicative conidium on conidiophore arising from another globose (either primary or replicative) conidium. (3) Discharged globose conidium and conidiophore apex. (From Ref. 28a.) (4) Mature zygospore with "beaks." (5) Capillospore with adhesive material at apex.

and pressure developed within the conidiophore projects away the conidiophore apex and conidium (Fig. 3), which usually separate in flight. Pressure within the conidiophore also provides the power for conidial discharge in several species of *Entomophthora,* but only in *Basidiobolus* is the apical portion of a subconidial vesicle projected away with the conidium.

The gametangia that copulate to produce the distinctive "beaked" *Basidiobolus* zygospore (Fig. 4) are usually derived from contiguous segments of the same hypha, but conjugation may occur between cells of separate hyphae, or even between the two cells produced from a single division of a conidium. During conjugation, each gametangium produces a protuberance at a right angle to the axis of the gametangium, but in apposition with the protuberance of the other gametangium. The single nucleus in each gametangium migrates into its protuberance, divides, one of the daughter nuclei migrates back into the gametangium, and a septum delimits the protuberance with its nucleus and associated cytoplasm. The contents of one gametangium flows into the other, and a spherical, thick-walled zygospore that rests within the persistent gametangial wall is formed. The cytoplasm in the persistent protuberances, or "beaks," degenerates.

Except for *B. microsporus,* each species of *Basidiobolus* produces capillospores (Fig. 5) that are distinguished from those produced by other members of the order by the presence of an apical beak covered with adhesive material. The microspores of *B. microsporus* are somewhat similar morphologically, but differ in size, mode of origin, and lack of an adhesive apex.

Species of *Basidiobolus* often are considered to be characteristically associated in nature with the gut contents of amphibians and reptiles. The extent of this relationship has been adequately discussed and demonstrated [3,23,33]. However, this is not an obligate association. All species grow well as saprophytes on a variety of common mycological media, and all except *B. microsporus* have been isolated from plant detritus. *Basidiobolus microsporus* has been found only in association with the dung and nests of rodents, and *B. magnus* has been isolated only from plant detritus [24].

All species of *Basidiobolus* are prone to lose the ability to produce conidia and zygospores after being maintained in culture for a relatively short time. Isolates of *Basidiobolus* possibly can be identified to genus level by staining the nuclei in hyphal segments [34]. However, for reliable identification to genus, conidia or zygospores must be observed, and except for *B. microsporus,* zygospores are necessary for identification to species. Therefore, a strain should be examined morphologically as soon as possible after it has been isolated. On any of several common mycological media, conidia should be produced within a few days and mature zygospores should be produced within 2 weeks.

2. Taxonomy

Currently, only one family, the Entomophthoraceae, is recognized in the order. Several workers have considered *Basidiobolus* sufficiently distinct to warrant recognition of a separate family to accommodate it. Criteria that distinguish the genus include the morphology and development of forcibly discharged conidia and zygospores, and associated structures, discussed above; the unique nature of the nuclei; the report of cellulose in the cell walls [4]; and the production of spores through internal cleavage in conidia. The distinctive nature of *Basidiobolus* nuclei have been indicated by several studies [17,34,37], but sufficiently detailed studies to make comparisons have not been reported for other members of the Entomophthorales. The presence of cellulose in the cell walls has been refuted [19].

In several species of *Basidiobolus,* both globose conidia and capillospores may cleave internally to form sporangiospores [7-9, 11, 35], termed "meristospores" by Drechsler [7]. In *B. microsporus,* globose conidia cleave internally to produce uninucleate cells, each of which gives rise to a single elongate microspore. The production of these spore types indicates the possible origin of conidia in the Entomophthorales from sporangia through evolutionary reduction, although internal cleavage of conidia has not been observed in any other genus of the order.

The criteria for separating the species of *Basidiobolus* are not entirely satisfactory. Only two species, *B. ranarum* Eidam and *B. magnus* Drechsler, are usually characterized as producing a *Streptomyces*-like odor; both also are usually characterized as producing undulate-walled zygospores, and as differing in the size of morphological structures. *Basidiobolus microsporus* is also characterized as producing undulate-walled zygospores. Three species, *B. haptosporus* Drechsler, *B. meristosporus* Drechsler, and *B. philipinensis* Josue & Quimio, were described as producing only smooth-walled zygospores (see Fig. 4). *Basidiobolus heterosporus* Srinivasan & Thirumalachar was described as producing both smooth- and undulate-walled zygospores.

In examining 169 strains of *Basidiobolus* isolated from lower vertebrates and eight previously identified strains from the American Type Culture Collection, Hutchison et al. [24] found much variability and a lack of correlation among characters previously used to separate the species. Strain ATCC 16109, a human isolate, and 25% of the lower vertebrate isolates produced both smooth- and undulate-walled zygospores. Six of these lower vertebrate isolates also produced the *Streptomyces*-like odor. One strain that produced only smooth-walled zygospores also produced the distinctive odor. Six strains, including *B. magnus* strain ATCC 15379, lost the ability to produce the odor on subculturing. Also, the character of the zygospore walls of strains ATCC 15379 and 14448 (the type strain of *B. haptosporus*) varied with the culture medium. Strain ATCC 14450, the type strain of *B. meristosporus,* produced both smooth- and undulate-walled zygospores on each of three media. The first strain isolated from humans, ATCC 14052, produced only smooth-walled zygospores, but also produced the odor.

The only species of *Basidiobolus* that can be unequivocally identified based on its original description is *B. microsporus* because it produces microspores. Because a lack of correlation among the characters previously used to separate the species has been demonstrated, only the character of the zygospore wall and the production of microspores should be used until further studies elucidate the validity of additional criteria.

The current consensus is that only one species of *Basidiobolus* causes entomophthoramycosis basidiobolae:

Basidiobolus haptosporus Drechsler, *Bull. Torrey Bot. Club 74*: 403-413, 1947
= *Basidiobolus meristosporus* Drechsler, *J. Wash. Acad. Sci. 45*: 49-56, 1955

From its description [25], *B. philipinensis* appears to be indistinguishable from *B. haptosporus.*

Lie-Kian-Joe et al. [32] identified the isolate from the first case as *B. ranarum* and published photographs of undulate-walled zygospores in support of their identification. However, this isolate is now considered referable to *B. haptosporus* [14]. Drechsler [7] derived the specific epithet of *B. meristosporus* from the fact that conidia of this fungus cleaved internally to produce meristospores when deposited on Raciborski's medium, but this has been demonstrated in other species. Attempts were made to separate *B. haptosporus* and *B. meristosporus* by differences in rate of growth at 37°C versus 25°C [8,16], but Srinivasan and Thirumalachar [35] judged this to be significant only at the strain level, and other work [24] supported their contention.

Although *B. haptosporus* has been isolated in many parts of the world since it was first reported from plant detritus from the United States [5], it is quite restricted as a disease agent for mammals. Except for one case in a horse [39], it has been proven by culture to infect only humans between the tropics of cancer and capricorn in the eastern hemisphere.

B. *Conidiobolus*

1. Biology

The systematics of *Conidiobolus* have been recently considered in detail [26-28]. Each of the 27 species produces its multinucleate primary and replicative conidia atop unbranched conidiophores (Figs, 6, 7, and 11). Each subspherical conidium is discharged due to pressure developed within the conidium, and each bears a more or less prominent papillum after discharge (Figs. 8 and 12) due to eversion of the wall against the top of the conidiophore. Capillospores lacking an adhesive apex are produced in four species. Multiplicative conidia (Fig. 13) have been observed in 10 species. These are small, subspherical conidia produced from a primary conidium on radially projecting sterigmata. As many as 40 may be produced, they are forcibly discharged but not phototropic, and cleavage of the mother conidium does not occur. Villose conidia (Fig. 9; i.e., villose resting spores) are produced only by *C. coronatus* from primary and replicative conidia by the production of a thick wall and growth of villose appendages. The appendages are quite variable in number and length. Zygospores (Fig. 14) are known in 19 species. These arise from the conjugation of equal or unequal gametangia, and the zygospore, which usually has two thick walls and an eccentric oil globule, forms within one of the gametangia. Chlamydospores (Fig. 10) have been reported in several species, including several isolates of *C. coronatus.*

Species of *Conidiobolus* are most commonly encountered in plant detritus, and all except six species were originally isolated from plant detritus. *Conidiobolus utriculosis, C. coronatus, C. brefeldianus,* and *C. mycophagus* were originally isolated from fruiting bodies of higher fungi, and *C. stromoideus* and *C. pseudococcus* were isolated from insects. Including *C. coronatus,* at least three species have been shown to infect insects in nature, and at least three others have been isolated from insects infected with *Entomophthora.* A few species are abundant and cosmopolitan (e.g., *C. coronatus*), but most have been isolated only one to a few times (e.g., *C. incongruus*). Most isolates have been reported from the United States and India, but this undoubtedly reflects efforts to isolate members of the genus rather than their distribution in nature.

Although isolates of *Conidiobolus* do not tend to lose the ability to sporulate as readily as isolates of *Basidiobolus,* this can still present a problem. The ability to produce multiplicative conidia appears to be the most easily lost. They can usually be observed by allowing the conidia from a young culture to be deposited on freshly prepared 2% water agar, and observing them for up to 20 h. Most strains of *C. coronatus* readily produce villose conidia, but their production is difficult to induce in a small percentage of strains. This appears to be particularly true for human isolates [29]. Examination of conidia

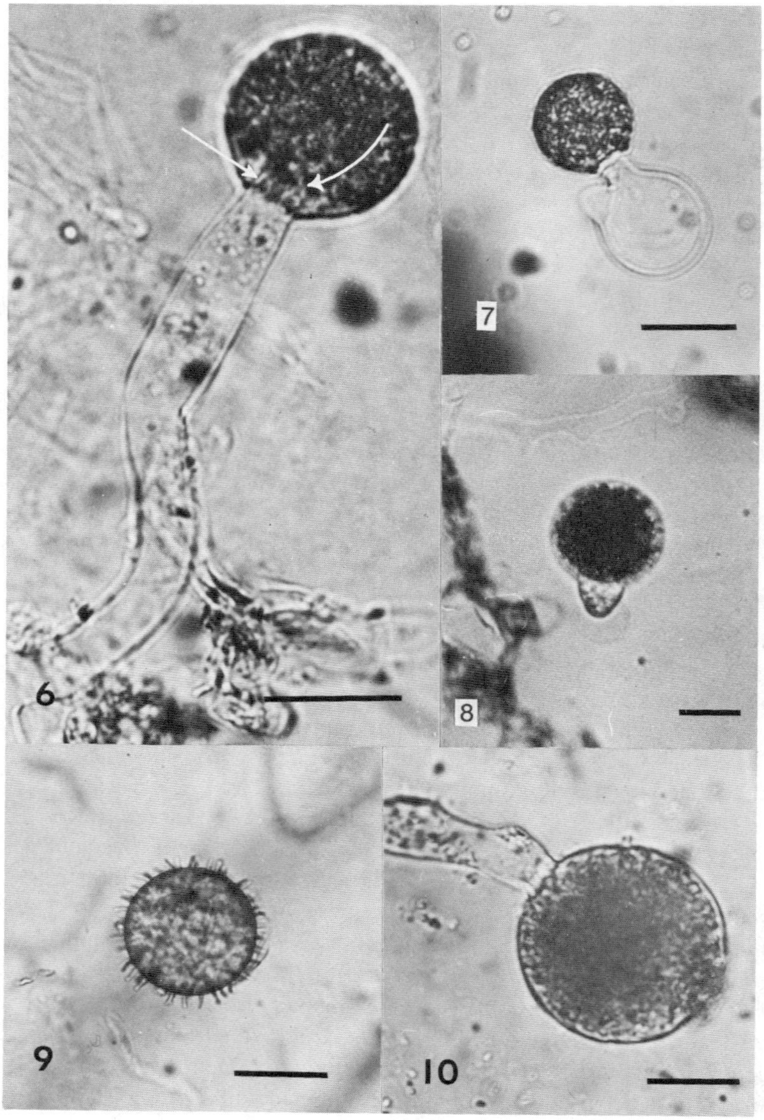

Figures 6-10 *Conidiobolus coronatus*. Bars equal 20 μm. (6) Primary conidium on conidiophore arising from vegetative mycelium. Arrows indicate dome-shaped septum separating conidium and conidiophore which splits during discharge of the conidium. (7) Replicative conidium on conidiophore arising from a globose (either primary or replicative) conidium. (8) Discharged globose conidium. Note rounded shape of papillum (see Fig. 12). (9) Relatively thin-walled villose conidium. (10) Chlamydospore arising from hyphae in agar.

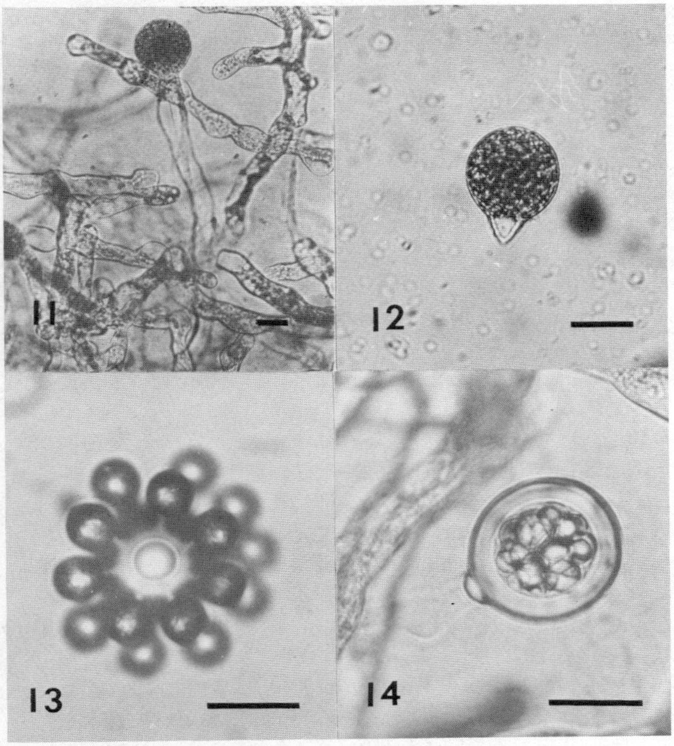

Figures 11-14 *Conidiobolus incongruus.* Bars equal 20 μm. (11) Almost mature primary conidium on conidiophore arising from vegetative mycelium. (12) Globose discharged conidium. Note pointed shape of papillum (see Fig. 8). (13) Multiplicative conidia on radially projecting sterigmata (= microconidiophores) arising from a discharged primary conidium. (14) Mature zygospore. The zygospore lies within the persistent wall of one of the gametangia. The other gametangium is not visible.

deposited on a petri dish lid above a culture, or on top of a coverglass placed on the surface of the culture, will usually reveal at least a few of these unique spores within 1 to 2 weeks.

2. Taxonomy

Although the unique villose conidia readily distinguish *C. coronatus* as a distinct species, there has been much confusion as to the correct name and systematic position of this species. This has resulted from the larger problem of separating the genera *Entomophthora* and *Conidiobolus.* For many years the criterion used to separate these genera was the parasitism of insects by *Entomophthora* and the saprophytic habit of *Conidiobolus.* Because of this distinction, *C. coronatus* was placed in *Entomophthora* for many years. However, this distinction is no longer tenable. Not only are more *Conidiobolus* species being found in association with insects, but more *Entomophthora* species are being axenically cultured.

A separation of *Entomophthora* and *Conidiobolus* based on the character of the primary conidiophore has been proposed [36]: macronemous in *Entomophthora* and micronemous in *Conidiobolus.* However, several species of *Entomophthora* have conidiophores that can be characterized as micronemous, and two *Conidiobolus* species have macronemous conidiophores. *Conidiobolus* is much more homogeneous than *Entomophthora* with respect to primary conidia and conidiophores, culturability, and mode of formation of zygospores. However, several *Entomophthora* species conform to the current generic concept of *Conidiobolus* [27], and thus these genera cannot yet be satisfactorily separated. This situation was alluded to over a decade ago, when Gustafsson [18] noted that his group c and Hutchison's [22] "apiculata type" *Entomophthora* species were similar to *Conidiobolus.*

The correct nomenclature for the etiologic agent of entomophthoramycosis conidiobolae is:

Conidiobolus coronatus (Cost.) Batko, *Entomophaga, Mem. Hors. Ser. 2*: 129, 1964
≡ *Boudierella coronata* Costantin, *Bull. Soc. Mycol. Fr. 13*: 40, 1897
≡ *Delacroixia coronata* (Cost.) Sacc. & Syd., *Syll. Fung. 14*: 457, 1899
≡ *Entomophthora coronata* (Cost.) Kevork., *J. Agric. Univ. P. R. 21*: 198, 1937
≡ *Conidiobolus coronatus* (Cost.) Srin. & Thirum., *Mycopathol. Mycol. Appl. 24*: 296, 1964
≡ *Conidiobolus* (subgen. *Delacroixia*) *coronatus* (Cost.) Tyrrell & MacLeod, *J. Invert. Pathol. 20*: 12, 1972
= *Conidiobolus villosus* Martin, *Bot. Gaz. (Crawfordsville) 80*: 317, 1925

This species is an occasional pathogen of insects worldwide, and has been isolated from horses as far north as southern Texas [13] and as far south as Australia [21]. However, infections of humans have been confined to the tropics around the world.

All strains of *C. coronatus* produce villose conidia, and about 75% of the strains isolated from plant detritus produce multiplicative conidia [26]. A few strains of *C. coronatus* produce abundant spherical chlamydospores submerged in an agar medium. Emmons and Bridges [13] speculated that these might represent zygospores, but no evidence of conjugation has been reported. They have a single thick wall and relatively homogeneous contents (Fig. 10), in marked contrast to the structure of *Conidiobolus* zygospores (Fig. 14).

The nomenclature and systematics of *C. incongruus* is much simpler than the other two disease agents:

Conidiobolus incongruus Drechsler, *Am. J. Bot. 47*: 370, 1960
= *Conidiobolus gonimodes* Drechsler, *Mycologia 53*: 292, 1961

The single human isolate was first identified as *Entomophthora* species [12,15]; it was properly identified as *C. incongruus* by King and Jong [31].

Zygospores and multiplicative conidia must be observed to identify a strain as *C. incongruus.* The similarities of *C. incongruus* to three other species has been commented upon [10,29]. These are *C. polytocus* Drechsler, *C. pseudococcus* (Speare) Tyrrell & MacLeod, and *C. brefeldianus* Couch. *Conidiobolus polytocus* does not produce zygospores. The separation from *C. brefeldianus* rests on the fact that the hyphal segments in the older portions of a colony of *C. incongruus* become distended, whereas in *C. brefeldianus* they do not [27]. This distinction is somewhat less than satisfactory, but a conservative approach dictates that a larger number of strains than are currently available should be examined in comparative cultures before any taxonomic revision is proposed.

IV. RECENT DEVELOPMENTS

Several publications on the generic taxonomy of the Entomophthorales have appeared since this chapter was prepared. The taxonomic positions of the three species considered here have not been affected. However, substantial revisions in the criteria for delimiting genera have been proposed [41-45].

REFERENCES

1. Alexopoulos, C. J. 1962. *Introductory Mycology,* 2nd ed. Wiley, New York.
2. Brobyn, P. J., and N. Wilding. 1977. Invasive and developmental processes of *Entomophthora* species infecting aphids. Trans. Br. Mycol. Soc. *69*: 349-366.
3. Coremans-Pelseneer, J. 1974. Biologie des champignons du genre *Basidiobolus* Eidam 1886. Saprophytisme et pouvoir pathogène. Acta Zool. Pathol. *60*: 1-143.
4. Couch, J. N. 1939. A new *Conidiobolus* with sexual reproduction. Am. J. Bot. *26*: 119-130.
5. Drechsler, C. 1947. A *Basidiobolus* producing elongated secondary conidia with adhesive beaks. Bull. Torrey Bot. Club *74*: 403-413.
6. Drechsler, C. 1952. Widespread distribution of *Delacroixia coronata* and other saprophytic Entomophthoraceae in plant detritus. Science *115*: 575-576.
7. Drechsler, C. 1955. A southern *Basidiobolus* forming many sporangia from globose and from elongated adhesive conidia. J. Wash. Acad. Sci. *45*: 49-56.
8. Drechsler, C. 1956. Supplementary developmental stages of *Basidiobolus ranarum* and *Basidiobolus haptosporus.* Mycologia *48*: 655-676.
9. Drechsler, C. 1958. Formation of sporangia from conidia and hyphal segments in an Indonesian *Basidiobolus.* Am. J. Bot. *45*: 632-638.
10. Drechsler, C. 1960. Two new species of *Conidiobolus* found in plant detritus. Am. J. Bot. *47*: 368-377.
11. Drechsler, C. 1964. An odorous *Basidiobolus* often producing conidia plurally and forming some diclinous sexual apparatus. Am. J. Bot. *51*: 770-777.
12. Eckert, H. L., G. H. Khoury, R. S. Pore, E. F. Gilbert, and J. R. Gaskell. 1972. Deep *Entomophthora* phycomycotic infection reported for the first time in the United States. Chest *61*: 392-394.
13. Emmons, C. W., and C. H. Bridges. 1961. *Entomophthora coronata,* etiologic agent of a phycomycosis of horses. Mycologia *53*: 307-312.
14. Emmons, C. W., C. H. Binford, J. P. Utz, and K. J. Kwon-Chung. 1977. *Medical Mycology,* 3rd ed. Lea & Febiger, Philadelphia.
15. Gilbert, E. F., G. H. Khoury, and R. S. Pore. 1970. Histopathological identification of *Entomophthora* phycomycosis. Arch. Pathol. *90*: 583-587.
16. Greer, D. L., and L. Friedman. 1966. Studies on the genus *Basidiobolus* with reclassification of the species pathogenic for man. Sabouraudia *4*: 231-241.
17. Gull, K., and A. P. J. Trinci. 1974. Nuclear division in *Basidiobolus ranarum.* Trans. Br. Mycol. Soc. *63*: 457-460.
18. Gustafsson, M. 1965. On species of the genus *Entomophthora* Fres. in Sweden. I. Classification and distribution. Lantbrukshoegsk. Ann. *31*: 103-212.
19. Hoddinott, J., and O. A. Olson. 1972. A study of the carbohydrates in the cell walls of some species of the Entomophthorales. Can. J. Bot. *50*: 1675-1679.
20. Humber, R. A. 1976. The systematics of the genus *Strongwellsea* (Zygomycetes: Entomophthorales). Mycologia *68*: 1042-1060.
21. Hutchins, D. R., and K. G. Johnston. 1972. Phycomycosis in the horse. Aust. Vet. J. *28*: 269-278.

22. Hutchison, J. A. 1963. The genus *Entomophthora* in the western hemisphere. Bull. Kans. Acad. Sci. *66*: 237-264.
23. Hutchison, J. A., and M. A. Nickerson. 1970. Comments on the distribution of *Basidiobolus ranarum*. Mycologia *62*: 585-587.
24. Hutchison, J. A., D. S. King, and M. A. Nickerson. 1972. Studies on temperature requirements, odor production and zygospore wall undulation of the genus *Basidiobolus*. Mycologia *64*: 467-474.
25. Josue, A. R., and T. H. Quimio. 1976. A Philippine *Basidiobolus* isolated from lizard dung. Nova Hedwigia *27*: 483-492.
26. King, D. S. 1976a. Systematics of *Conidiobolus* (Entomophthorales) using numerical taxonomy. I. Biology and cluster analysis. Can. J. Bot. *54*: 45-65.
27. King. D. S. 1976b. Systematics of *Conidiobolus* (Entomophthorales) using numerical taxonomy. II. Taxonomic considerations. Can. J. Bot. *54*: 1285-1296.
28. King, D. S. 1977. Systematics of *Conidiobolus* (Entomophthorales) using numerical taxonomy. III. Descriptions of recognized species. Can. J. Bot. *55*: 718-729.
28a. King, D. S. 1978. Basidiobolus haptosporus. In Melvin S. Fuller (Ed.), *Lower Fungi in the Laboratory,* Dept. of Botany, University of Georgia, Athens.
29. King, D. S. 1979. Systematics of fungi causing entomophthoramycosis. Mycologia *71*: 731-745.
30. King, D. S., and R. A. Humber. 1982. Identification: Entomophthorales. In H. D. Burges (Ed.), *Microbial Control of Insects, Mites, and Plant Diseases,* Vol. 2, Academic, New York.
31. King, D. S., and S. C. Jong. 1976. Identity of the etiological agent of the first deep entomophthoraceous infection of man in the United States. Mycologia *68*: 181-183.
32. Lie-Kian-Joe, Njo-Injo Tjoei Eng, A. Pohan, H. van der Meulen, and C. W. Emmons. 1956. *Basidiobolus ranarum* as a cause of subcutaneous mycosis in Indonesia. Arch. Dermato-syphiligr. Clin. Hop. Saint Louis *74*: 378-383.
33. Nickerson, M. A., and J. A. Hutchison. 1971. The distribution of the fungus *Basidiobolus ranarum* Eidam in fish, amphibians, and reptiles. Am. Midl. Nat. *86*: 500-502.
34. Robinow, C. F. 1963. Observations on cell growth, mitosis, and division in the fungus *Basidiobolus ranarum*. J. Cell Biol. *17*: 123-152.
35. Srinivasan, M. C., and M. J. Thirumalachar. 1967. Studies on *Basidiobolus* species from India with discussion on some of the characters used in the speciation of the genus. Mycopathol. Mycol. Appl. *33*: 56-64.
36. Srinivasan, M. C., M. J. Narasimhan, and M. J. Thirumalachar. 1964. Artificial culture of *Entomophthora muscae* and morphological aspects for differentiation of the genera *Entomophthora* and *Conidiobolus*. Mycologia *56*: 683-691.
37. Sun, N. C., and C. C. Bowen. 1972. Ultrastructural studies of nuclear division in *Basidiobolus ranarum* Eidam. Caryologia *25*: 471-494.
38. Thaxter, R. 1888. The Entomophthorae of the United States. Mem. Boston Soc. Nat. Hist. *4*: 133-201.
39. Van Overeem, C. 1925. Beitrage zur pilzflora von niederländisch indien. 10. Uber ein merkwürdges vorkommen von *Basidiobolus ranarum* Eidam. Bull. Jard. Bot. Buitenzorg, Ser. 3, *7*: 423-431.
40. Waterhouse, G. M. 1973. Entomophthorales. In G. C. Ainsworth, F. K. Sparrow, and A. S. Sussman (Eds.), *The Fungi,* Vol. IVB, Academic, New York, pp. 219-229.
41. Remaudière, G., and G. L. Hennebert. 1980. Mycotaxon *11*: 269-321.
42. Remaudière, G., and S. Keller. 1980. Mycotaxon *11*: 323-338.
43. Humber, R. A. 1981. Mycotaxon. *13*: 191-240.
44. Humber, R. A., and I. Ben-Ze'ev. 1981. Mycotaxon *13*: 506-516.
45. Tucker, B. E. 1981. Mycotaxon *13*: 481-505.

4

The Yeasts

D. G. Ahearn / Georgia State University, Atlanta, Georgia

I. INTRODUCTION

To many scientists, the term "yeast" implies a unicellular budding organism that anaerobically metabolizes sugars to alcohol and carbon dioxide, i.e., a fungus typified by *Saccharomyces cerevisiae.* This concept was derived from the late-seventeenth-century use of the term "yeast" to denote the single-cell microscopic entities associated with the effervescence of beer. These unicells were later observed to form endogenous spores. A concept evolved for yeasts as fungi that fermented sugars, permanently maintained a unicellular growth form by budding, and produced ascospores. False yeasts or *Torula* grew as single cells that reproduced by budding but formed neither mycelium nor ascospores. These false yeasts were distinguished as often causing undesirable fermentations in brewing. The false yeasts, which were classified in the imperfect fungi, generally were thought to be ascogenous forms that had lost their sporulating capacity, or ascogenous forms for which an opposite mating type had not been discovered. Transitional fungi, with or without ascospores, e.g., *Endomyces, Monilia (Candida),* which formed both mycelium and single-budding cells, frequently were termed yeast-like, and considered separate from the yeasts [8,27]. The names *Torula* and *Monilia* are illegitimate if applied to yeasts, yet in current industrial and patent literature, *Candida utilis* is frequently referred to as *Torula* yeast and in some medical literature candidiasis is still termed moniliasis.

A few yeasts, notably the Sporobolomycetaceae, had long been considered by many mycologists to be related to basidiomycetes due to their production of conidia (ballistospores) forcibly ejected from the tip of a pedicel by a drop excretion mechanism. This is an asexual method of reproduction, but the sexual stage later observed in *Sporobolomyces* supported a basidiomycetous relationship [1].

The sparcity of discrete morphological structures for recognition of yeasts led to an early reliance on physiological properties, mainly the assimilation and fermentation of carbohydrates for species recognition. The morphological flexibility of the vegetative growth form (ability to produce filaments) was at first important in classification of yeasts but later downgraded as it became evident that environmental conditions regulated the interconversion of the single cell and hyphal stages of numerous fungi. The physiological properties of yeasts are amazingly stable and in conjunction with morphological properties have provided characteristics for the practical distinction of many yeast species. However, certain of these phenotypic characters are unstable and their unquestioned use has resulted in confusion and long lists of synonymies for certain species.

II. ASCOMYCETOUS YEASTS

The majority of yeasts are classified with the hemiascomycetes or, if asexual, recognized as having ascomycetous affinities. There are approximately 35 genera of ascosporogenous yeasts. These are distinguished mainly on the mode of formation and shape of asci and ascospores, but for certain genera physiological properties such as vigorous fermentation *(Saccharomyces)* or nitrate assimilation *(Hansenula)* are of prime importance. An individual species or strain may be homothallic or heterothallic. Yeast taxonomy currently is in a state of flux, with various authors suggesting somewhat different classification schemes [4,23,27,38].

III. BASIDIOMYCETOUS YEASTS

That some of the traditional yeasts were nonascogenous was established first by Banno [2,3]. He found that compatible yeast cells of *Rhodotorula glutinis* conjugated and produced dicaryotic mycelium with clamp connections and dark resting spores. The thick-walled, smut-like resting spores germinated to produce germ hyphae bearing typically haploid sporidia that reproduced as the initial yeast isolates. Fell and co-workers [5,6] reported similar life cycles among candidas and other rhodotorulas. These genera, with teleomorphs (sexual stages) in *Rhodosporidium* Banno and *Leucosporidium* Fell, Hunter and Phaff, include fungi with both homothallic and heterothallic life cycles. In *Rhodosporidium toruloides* the mating between two haploid cells is controlled by pheromone-like materials secreted from each mating type. A lipopeptidyl substance secreted from one mating type induces the opposite mating type to produce a germ tube and a second lipopeptide. This second lipopeptide induces germ tube formation in the first cell type and conjugation occurs [33,34].

The morphology of the teleomorph, i.e., lateral sporidia borne on a four-celled promycelium, found within *Rhodosporidium* and *Leucosporidium* is comparable to that of the Ustomycota. Moore [20,21,23] considered these fungi to be neither Ascomycetes nor Basidiomycetes and proposed two new classes, Ustomycetes and Sporidiomycetes, within the division Ustomycota of the superdivision Basidiomycotina. Moore [20] characterized the Ustomycota as follows: "Resistant cell a ustospore; germination producing one or more germ hyphae that may develop directly but more usually produce several to many sporidia. Assimilative phase: yeasts, yeast-like, pseudomycelial, or mycelial with or without clamps. Asexual reproduction: budding, fragmentation, or ballistospores. Septa: abscissional or perforate, ultrastructurally lamellate, lacking Woronin bodies or parenthesomes. Lenticular bodies may develop on the inner surface of the wall." Moore [23] grouped the basidiomyceteous yeasts of the genera *Sporidiobolous, Aessosporon* (possible synonym of *Sporidiobolous*), *Rhodosporidium,* and *Leucosporidium* in the Sporidiomycetes.

Kreger-van Rij [11] suggested that the similarities between *Cryptococcus* and *Bullera* (a ballistospore producer) linked the former to the basidiomycetes. This relationship of the cryptococci to the basidiomycetes was supported further by analyses of their relatively high guanine plus cytosine (G + C) content of the DNA (49-61%), their extracellular polysaccharide composition, and their enzymatic organization within the tryptophan pathway [10,18,28,32]. Olive [25] proposed the name *Filobasidium floriforme* for a fungus that produced filiform basidia bearing 6-8 sessile basidiospores. Rodrigues de Miranda [29] demonstrated similar basidial structures from a cross of strains of *Torulopsis capsuligena* and described *Filobasidium capsuligenum.* Moore and Kreger-van Rij [24] pointed out that *F. floriforme* Olive had anamorphs in *Cryptococcus albidus.* Shadomy [31] first observed clamp connections in hyphal forms of *C. neoformans* and suggested that the species be classified in the genus *Leucosporidium* [6]. Kurtzman [13] obtained conjugation between mixtures of strains of *Cryptococcus laurentii,* but definitive sexual structures were not observed. Kwon-Chung [14,15] obtained subglobose to flask-shaped basidia with sessile, catenulate basidiospores from crosses of strains of *C. neoformans* and erected the genus *Filobasidiella* for two teleomorph stages. Presumably, the induction of conjugation tubes by lipopeptidyl substances among the Filobasidiaceae is analogous to that observed in *Rhodosporidium* since opposite mating types of nonfertile crosses can induce hyphal production. The genera *Filobasidium* and *Filobasidiella* were included in the order Tremellales of the class Phragmobasidiomycetes by Moore [21]. The Filobasidiaceae, in common with *Tremella,* differ from the other basidiomycetes by producing dolipore septa with striated material in each orifice and parenthesomes that are vesiculate or absent [22].

IV. IMPERFECT YEASTS

The classification of yeasts with no known sexual stage always has been a problem. These yeasts are grouped in the form-class Fungi Imperfecti. The philosophy of the form-genera is based on the need for a practical identification system. In many instances the form-genera include species of divergent phylogeny. For example, the genus *Candida* contains anamorphs of such ascomycetous genera as *Saccharomyces, Pichia, Hansenula, Debaryomyces,* and *Clavispora* and of the basidiomycetous genus *Leucosporidium.* Species of unknown teleomorph, but of basidiomycetous affinity as determined by cell wall character, are also classified in the form-genera *Candida* and *Torulopsis.* Further heterogeneity in these genera is indicated from analysis of their coenzyme Q systems, which varies in each genus from Q_6 to Q_{10} [40]. To date, the Q_{10} system has been found only in yeasts with basidiomycete affinities.

Recently, Yarrow and Meyer [41] reiterated that the genus name *Torulopsis* was an illegitimate homonym of *Torula.* They proposed transferring the species of *Torulopsis* to the genus *Candida.* Acceptance of this proposal in the medical literature probably will be slow because species of the respective genera of clinical significance (e.g., *Candida albicans* and *Torulopsis glabrata*) are phenotypically distinct. However, there is little justifi-

Table 1 General Morphological and Physiological Differences Between Anamorphs of Ascomycetous and Basidiomycetous Yeasts

	Ascomycetous	Basidiomycetous
Diazonium B blue reaction	No color change	Red color reaction
Cell walls[a]	Two or three layers, lacking xylose, chitin content relatively low, mannan high	Laminar, xylose, galactose, or fucose present,[b] chitin content relatively high, mannan low
G + C	30-60%	49-70+%
Urease reaction	Generally lacking	Usually strong
Coenzyme Q	Q_6-Q_9	Q_8-Q_{10}
Mitosis[a]	Intranuclear usually in neck of bud	Nuclear membrane partially breaks down, bud mitosis
Fermentative metabolism	Often present	Infrequent

[a]Limited number of observations.
[b]Xylose found in Filobasidiaceae; galactose and fucose found in Sporobolomycetaceae.

cation for maintenance of the two genera, particularly since no practical means of distinguishing many other species of the two genera is currently available [4].

As additional information on a form-species is obtained, including ideally a determination of its life cycle, it is reclassified. For example, the ability to assimilate inositol, a strictly oxidative assimilation, and failure to produce hyphae were the basis for provisionally assigning two yeasts to *Cryptococcus: C. cereanus* and *C. lactativorus.* These yeasts were noted to differ from other species of the genus in their mode of budding and by their failure to produce extracellular starch and to demonstrate urease. Their G + C content (about 49 and 46%, respectively) was within the range of both ascomycetous (30-60%) and basidiomycetous species (49-70%), but their mode of budding suggested an ascomycetous relationship [26]. Later, Rodrigues de Miranda [30] found that *C. lactativorus* represented a heterothallic ascomycete and *C. cereanus* a homothallic ascomycete. Both species produced thick-walled ascospores and were reclassified in the genus *Sporopachydermia.*

A variety of ultrastructural and biochemical properties have been employed for the distinction of ascomycetous and basidiomycetous yeasts (Table 1). Of these the apparent correlation of the occurrence of the laminar cell wall with the formation of a red color reaction upon exposure of the cells to diazonium B blue provides the most practical procedure for distinguishing the asexual yeasts of ascomycetous and basidiomycetous affinities [7,9,12,35]. Weijman [39] recently has used cell wall data as a basis for reclassification of the genera *Trichosporon* and *Geotrichum.* He proposed that the genus *Geotrichum* be reserved for ascomycetous anamorphs with cell walls characterized by a predominance of galactomannans and *Trichosporon* be reserved for xylose-containing species. Von Arx and Weijman [37] have recommended a similar type of separation for *Candida,* limiting the genus to ascomycetous anamorphs. Heterobasidiomycetous yeasts with cell walls containing xylose were transferred to *Apiotrichum,* whereas those containing fucose or galactose were placed in *Rhodotorula.*

The gradual recognition that the traditional yeasts included phylogenetically diverse fungi has resulted in a broadening of the concept of what a yeast is. Lodder and Kreger-van Rij [17] recognized yeasts as eucaryotic heterotrophs that, under routine cultural conditions, generally existed as hyaline to lightly pigmented, nonmotile, single cells, globose to spherical in shape. Vegetative reproduction was usually by budding, but certain groups reproduced by fission. Unicellular fungi that produced obvious melanin pigmentation were excluded arbitrarily from the yeasts. Current taxonomic treatments recognize that a precise definition of yeasts does not include all fungi to which the term had been historically applied. Lodder [16] stated: "Yeast may be defined as microorganisms in which the unicellular form is conspicuous and which belong to the fungi." Phaff et al. [27] state: "Yeasts are usually defined as fungi that do not produce asexual spores (conidia) borne on distinct aerial structures (conidospores) and that spend at least part of the vegetative cycle as single cells." Von Arx [36] suggested that fungi without buds but that produce asci which form on the hyphae also be considered yeasts.

The unicell concept for at least part of the life cycle appears most suitable for a definition of yeasts. It is this unicell stage that provides a maximum ratio of cell volume to surface area which optimizes diffusion and facilitates the utilization of substrates. This shape, in conjunction with the mode of growth and size (from about 2 to 25 μm in diameter dependent on species), favors the practical use of yeasts in many industrial processes. There have been considerable advances in yeast taxonomy during the past 20 years that are responsible for the present unstable state of yeast classification. I have reviewed some of these developments mainly on a generic level. The impact of molecular studies, particularly DNA sequence comparisons, is resulting in numerous reclassifications at the species level [19,28]. The increasing industrial and medical importance of yeasts can be expected to promote further developments and changes in yeast taxonomy and classification.

REFERENCES

1. Bandoni, R. J., K. J. Lobo, and S. A. Brezden. 1971. Conjugation and chlamydospores in *Sporobolomyces odorus.* Can. J. Bot. *49*: 683-686.
2. Banno, I. 1963. Preliminary report on cell conjugation and mycelial stage in *Rhodotorula* yeasts. J. Gen. Appl. Microbiol. *9*: 249-251.
3. Banno, I. 1967. Studies on the sexuality of *Rhodotorula.* J. Gen. Appl. Microbiol. *13*: 167-196.
4. Barnett, J. A., R. W. Payne, and D. Yarrow. 1979. *A Guide to Identifying and Classifying Yeasts,* Cambridge University Press, Cambridge.
5. Fell, J. W. 1974. Heterobasidiomycetous yeasts *Leucosporidium* and *Rhodosporidium.* Their systematics and sexual incompatibility systems. Trans. Mycol. Soc. Jpn. *15*: 316-323.
6. Fell, J. W., A. Statzell, I. Hunter, and H. J. Phaff. 1969. *Leucosporidium* gen. nov. The heterobasidiomycetous stage of several yeasts of the genus *Candida.* Antonie Leeuwenhoek J. Microbiol. Serol. *35*: 433-462.
7. Hagler, A. N., and D. G. Ahearn, 1981. Rapid diazonium blue B test to detect basidiomycetous yeasts. Int. J. Syst. Bacteriol. *31*: 204-208.

8. Henrici, A. T. 1930. *Molds, Yeasts and Actinomycetes,* Wiley, New York.
9. Hopsu-Havu, V. K., E. Tunnela, and D. Yarrow. 1978. *Candida sonckii* sp. nov. Antonie Leeuwenhoek J. Microbiol. Serol. *44*: 435-438.
10. Hutter, R., and J. A. DeMoss. 1967. Organization of the tryptophan pathway: a phylogenetic study of the fungi. J. Bacteriol. *94*: 1896-1907.
11. Kreger-van Rij, N. J. W. 1964. The genus *Cryptococcus.* Ann. Soc. Belge Med. Trop. *44*: 601-610.
12. Kreger-van Rij, N. J. W., and M. Beenhuis. 1971. A comparative study of the cell wall structure of basidiomycetous and related yeasts. J. Gen. Microbiol. *68*: 87-95.
13. Kurtzman, C. P. 1973. Formation of hyphae and chlamydospores by *Cryptococcus laurentii.* Mycologia *65*: 388-395.
14. Kwon-Chung, K. J. 1975. A new genus *Filobasidiella,* the perfect state of *Cryptococcus neoformans.* Mycologia *67*: 1197-1200.
15. Kwon-Chung, K. J. 1976. A new species of *Filobasidiella,* the sexual state of *Cryptococcus neoformans.* Mycologia *68*: 821-833.
16. Lodder, J. (Ed.). 1970. *The Yeasts. A Taxonomic Study,* North-Holland, Amsterdam.
17. Lodder, J., and N. J. W. Kreger-van Rij. 1952. *The Yeasts. A Taxonomic Study,* North-Holland, Amsterdam.
18. Meyer, S. A., and H. J. Phaff. 1970. Taxonomic significance of the DNA base composition in yeast. In D. G. Ahearn (Ed.), *Recent Trends in Yeast Research,* Vol. 1, Spectrum (Georgia State University), Atlanta, pp. 1-29.
19. Meyer, S. A., M. T. Smith, and F. P. Simione. 1978. Systematics of *Hanseniaspora* Zike and *Kloeckera* Janke. Antonie Leeuwenhoek J. Microbiol. Serol. *44*: 79-96.
20. Moore, R. T. 1972. Ustomycota, a new division of higher fungi. Antonie Leeuwenhoek J. Microbiol. Serol. *38*: 567-584.
21. Moore, R. T. 1978. Taxonomic significance of septal ultrastructure with particular reference to the jelly fungi. Mycologia *70*: 1007-1024.
22. Moore, R. T. 1979. Septal ultrastructure in *Sirobasidium magnum* and its taxonomic implications. Antonie Leeuwenhoek J. Microbiol. Serol. *45*: 113-118.
23. Moore, R. T. 1980. Taxonomic proposals for the classification of marine yeasts and other yeast-like fungi including the smuts. Bot. Mar. *23*: in press.
24. Moore, R. T., and N. J. W. Kreger-van Rij. 1972. Ultrastructure of *Filobasidium* Olive. Can. J. Microbiol. *18*: 1949-1951.
25. Olive, L. S. 1968. An unusual new heterobasidiomycete with tilletic-like basidia. J. Elisha Mitchell Sci. Soc. *84*: 261-266.
26. Phaff, H. J., M. W. Miller, M. Miranda, W. B. Heed, and W. T. Starmer. 1974. *Cryptococcus cereanus,* a new species of the genus *Cryptococcus.* Int. J. Syst. Bacteriol. *24*: 486-490.
27. Phaff, H. J., M. W. Miller, and E. M. Mrak. 1978. *The Life of Yeasts,* 2nd ed., Harvard University Press, Cambridge, Mass.
28. Price, C. W., G. B. Fuson, and H. J. Phaff. 1978. Genome comparison in yeast systematics: delimitation of species within the genera *Schwanniomyces, Saccharomyces, Debaryomyces,* and *Pichia.* Microbiol. Rev. *42*: 161-193.
29. Rodrigues de Miranda, L. 1972. *Filobasidium capsuligenum* nov. comb. Antonie Leeuwenhoek J. Microbiol. Serol. *38*: 91-99.
30. Rodrigues de Miranda, L. 1978. A new genus: *Sporopachydermia.* Antonie Leeuwenhoek J. Microbiol. Serol. *44*: 439-450.
31. Shadomy, H. J. 1970. Clamp connections in two strains of *Cryptococcus neoformans.* In D. G. Ahearn (Ed.), *Recent Trends in Yeast Research,* Vol. I, Spectrum (Georgia State University), Atlanta, pp. 67-72.

32. Storch, R., C. J. Alexopoulos, and H. J. Phaff. 1969. Nucleotide compositon of deoxyribonucleic acid of some species of *Cryptococcus, Rhodotorula* and *Sporobolomyces.* J. Bacteriol. *98*: 1069-1072.
33. Tsuchiya, E., K. Abe, and S. Fukui. 1978. Sequence of events in early stage of sexual differentiation caused by rhodotorucine A, a fungal sex hormone: biological determination using matingless mutants derived from *Rhodosporidium toruloides* M 1057. J. Gen. Appl. Microbiol. *24*: 291-294.
34. Tsuchiya, E., and S. Fukui. 1978. Binding of rhodolorucine A, a lipopeptidyl mating hormone, to *a* cells of *Rhodosporidium toruloides* for induction of sexual differentiation. Biochem. Biophys. Res. Commun. *85*: 473-479.
35. van der Walt, J. P., and V. K. Hopsu-Havu. 1976. A colour reaction for the differentiation of ascomycetous and hemibasidiomycetous yeasts. Antonie Leeuwenhoek J. Microbiol. Serol. *42*: 157-163.
36. von Arx, J. A. 1979. Propagation in the yeasts and yeast-like fungi. In B. Kendrick (Ed.), *The Whole Fungus,* Proceedings of the Second International Mycological Conference, Vol. 2, National Museum of Natural Sciences and National Museums of Canada, Ottawa, pp. 555-571.
37. von Arx, J. A., and A. C. M. Weijman. 1979. Conidiation and carbohydrate composition in some *Candida* and *Torulopsis* species. Antonie Leeuwenhoek J. Microbiol. Serol. *45*: 547-555.
38. von Arx, J. A., L. Rodrigues de Miranda, M. T. Smith, and D. Yarrow. 1977. The genera of yeasts and yeast-like fungi. *Studies in Mycology,* Vol. 14, Centraalbureau voor Schimmelcultures, Baarn.
39. Weijman, A. C. M. 1979. Carbohydrate composition and taxonomy of *Geotrichum, Trichosporon* and allied genera. Antonie Leeuwenhoek J. Microbiol. Serol. *45*: 119-127.
40. Yamada, Y., and K. Kondo. 1972. Taxonomic significance of the coenzyme Q system in yeast and yeast-like fungi. In G. Terui (Ed.), *Fermentation Technology Today,* Proceedings of the IVth International Fermentation Symposium, Society of Fermentation Technology, Japan, pp. 781-784.
41. Yarrow, D., and S. A. Meyer. 1978. A proposal for amendment of the diagnosis of the genus *Candida* Berkhout nom. cons. Int. J. Syst. Bacteriol. *28*: 611-615.

5

Ascomycetes: Eurotiales, Sphaeriales, and Dothideales

G. A. de Vries / Centraalbureau voor Schimmelcultures, Baarn, The Netherlands

I. CLASSIFICATION OF THE ASCOMYCETES PATHOGENIC TO HUMANS AND ANIMALS AFTER VON ARX [4]

Class Ascomycetes

Subclass I. Ascohymenomycetidae: ascus wall one-layered (unitunicate), thin or rather thick, sometimes deliquescing at an early stage, with or without apical structures

Order Eurotiales: ascocarps globose, nonostiolate; asci nonoperculate, small, with evanescent wall, irregularly distributed in the ascocarp, without apical structures

Family Eurotiaceae: peridium pseudoparenchymatous, closed; ascocarp a cleistothecium *sensu stricto*

Family Gymnoascaceae (Onygenaceae): peridium composed of an open network of peridial hyphae; ascocarp a gymnothecium (cleistothecium pro parte)

Order Sphaeriales: ascocarps globose or flask-shaped, with apical ostiolum or canal; asci irregularly distributed or differently arranged in the ascocarps, often with apical structures

Family Microascaceae

Subclass II. Ascoloculomycetidae: ascus wall two-layered (bitunicate); inner membrane expanding after rupture of the outer membrane; ascocarps developing from cells that divide in all directions

Order Dothideales.

Family Pleosporaceae

Family Testudinaceae

Family Piedraiaceae

II. EUROTIALES

A. Gymnoascaceae (Onygenaceae)

1. Introduction

Within the Gymnoascaceae (Onygenaceae) the two very closely related genera *Ajellomyces* McDonough & Lewis and *Emmonsiella* Kwon-Chung can be separated from other pathogenic, spiral-forming genera such as *Arthroderma* Berk. and *Nannizzia* Stockdale because the ascocarps (gymnothecia, cleistothecia) of the former have spirals which do not arise from the peridial hyphae but from the ascogonium or the hyphae encircling the ascogonium [28,37]. Quite recently, McGinnis and Katz [39] considered *Emmonsiella* Kwon-Chung to be a synonym of *Ajellomyces* McDonough & Lewis. They emended the diagnosis of the latter genus as follows: "Cleistothecia globose to stellate, tan; peridial hyphae composed of thick-walled, septate spirals arising from a common center, giving forth numerous secondary, thin-walled, branching hyphae from their sides; asci subglobose, clavate to pyriform, 8-spored, deliquescent; ascospores globose, hyaline." This emended genus comprises two species. For practical reasons the original classification with two separate genera is followed here.

2. *Ajellomyces* McDonough & Lewis, *Mycologia 60*: 76-83, 1968.

Gymnothecia globose, small with thick-walled, closely coiled hyphae radiating from a common center. *Peridial hyphae* repeatedly branched, with irregularly shaped cells, constricted at the cross walls. At least some of the clusters of asci originate from lateral branches of the coils. *Asci* 8-spored, evanescent. *Ascospores* globose, hyaline, or brownish.

a. *Ajellomyces dermatitidis McDonough & Lewis, loc. cit., 1968 (Fig. 1) Conidial state synonymy*

Blastomyces dermatitidis Gilchrist & Stokes, *J. Exp. Med. 3*: 53-78, 1898

Oidum dermatitidis (Gilchrist & Stokes) Ricketts, *J. Med. Res. 6*: 373-547 1901

Cryptococcus dermatitidis (Gilchrist & Stokes) Castellani & Chalmers, *Manual of Tropical Medicine,* Baillière, Tindall, and Cox, London, p. 769, 1913

Mycoderma dermatitidis (Gilchrist & Stokes) Brumpt (as *M. dermatitis*), *Précis de Parasitologie,* Masson, Paris, p. 934, 1913

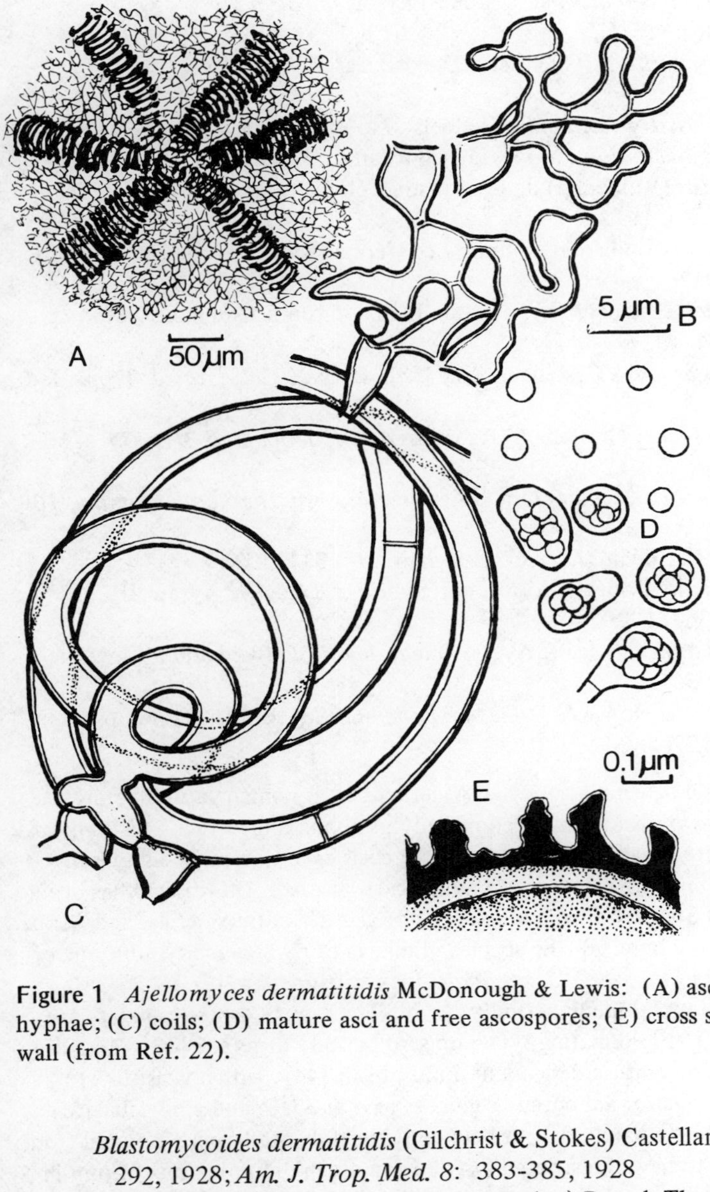

Figure 1 *Ajellomyces dermatitidis* McDonough & Lewis: (A) ascocarp; (B) peridial hyphae; (C) coils; (D) mature asci and free ascospores; (E) cross section of ascospore wall (from Ref. 22).

Blastomycoides dermatitidis (Gilchrist & Stokes) Castellani, *Am. Med. 34*: 291-292, 1928; *Am. J. Trop. Med. 8*: 383-385, 1928

Geotrichum dermatitidis (Gilchrist & Stokes) Basgal, Thesis, Inst. Oswaldo Cruz, Rio de Janeiro, p. 29, 1931

Endomyces dermatitidis (Gilchrist & Stokes) Moore, *Ann. Mo. Bot. Gard. 20*: 49-118, 1933

Torulopsis dermatitidis (Gilchrist & Stokes) Almeida, *An. Fac. Med. Univ. São Paulo 9*: 10, 1933

Gilchristia dermatitidis (Gilchrist & Stokes) Redaelli & Ciferri, *J. Trop. Med. 37*: 281, 1934

Zymonema dermatitidis (Gilchrist & Stokes) Dodge, *Medical Mycology*, Mosby, St. Louis, p. 168, 1935.

Chrysosporium dermatitidis (Gilchrist & Stokes) Carmichael, *Can. J. Bot. 40*: 1154, 1962

Cryptococcus gilchristi Vuillemin, in Gedoelst, *Les champignons parasites de l'homme et des animaux domestiques,* Joseph Van In, Lierre, p. 45, 1902

Zymonema gilchristi (Vuillemin) de Beurmann & Gougerot, *Les Nouvelles Mycoses,* Masson, Paris, p. 34, 1912

Mycoderma gilchristi (Vuillemin) Jannin, *Les Mycoderma,* These de Nancy, 1001, pp. 248-253, 1913

Glenospora gammeli Pollacci & Nannizzi, in Blankenhorn & Gammel, *J. Clin. Invest. 4*: 471-484, 1927

Blastomycoides tulasnensis Castellani, *Am. Med. 4*: 292, 1928; *Am. J. Trop. Med. 8*: 386-387, 1928

Monosporium tulasnense (Castellani) Agostini, *Am. J. Trop. Med. Hyg. 35*: 266-269, 1932

Aleurisma tulasnense (Castellani) Ota & Kawatsure, *Arch. Dermatol. Syphilol. 169*: 156-160, 1933

Glenospora brevis Castellani, *J. Trop. Med. Hyg. 36*: 311, 1933

Endomyces capsulatus Dodge & Ayres, in Rewbridge, Dodge & Ayres, *Am. J. Pathol. 5*: 349-364, 1929

Zymonema capsulatum (Dodge & Ayers) Dodge, *Medical Mycology,* Mosby, St. Louis, p. 167, 1935

Endomyces capsulatus Dodge & Ayers var. *isabellinus* MacBryde & Thompson, *Arch. Dermatol. 27*(1): 49-69, 1933

There is still no general agreement on what generic name is nomenclaturally correct for the conidial state. *Blastomyces* is most commonly used but is incorrect. *Ajellomyces dermatitidis* is the causative organism of a systemic mycosis called Gilchrist's disease, blastomycosis, North American blastomycosis, or Chicago disease. This disease has been reported from the United States, Canada, Mexico, Central and South America, and Africa. *Ajellomyces dermatitidis* is a heterothallic species. Linkage of pathogenicity with one of the two mating types, as in *Emmonsiella capsulata,* does not appear to occur [27,38]. Mating is controlled by a bipolar incompatibility [27]. The monascospore isolates vary widely in pathogenicity. The two mating types are serologically different [43]. In culture the fungus shows a temperature-dependent dimorphism [44], with a yeast-like (Y) form developing on blood agar or Sabouraud's glucose agar at 37°C, and a mycelial (M) form which grows on these media at 30°C or below. It has only rarely been isolated from soil [14] and is supposed to have a restricted distribution on an unknown substratum in soil or organic debris [14].

(1) Ascigerous State

Gymnothecia globose, dark to reddish brown, 200-350 μm. The ascocarp formation initiates with fusion of nearly globose gametangia of unequal size. Tightly coiled, 12-40 μm broad, septate spirals with walls of unequal thickness grow out radially from the ascogonium. Anastomosing, perpendicularly branched, *peridial hyphae* arise from these spirals. The triangular to quadrangular peridial cells are connected by narrow isthmuses. Clusters of asci are formed on lateral branches at the proximal end of the spirals. The 8-spored *asci* are globose to subglobose, 3.5-6 × 3.7-7.5 μm. The hyaline to very pale brown *ascospores* are globose, 1.5-2 μm in diameter, smooth-walled when observed under the regular microscope but with peg-like projections up to 1600 Å under the electron microscope [22].

(2) Conidial State

Tissue form: In mammalian tissues the fungus grows as a yeast with globose to subglobose cells, 8-15 μm, rarely up to 30 μm in diameter, with a rather thick wall (0.5-0.75 μm). The young blastoconidia are broadly attached to the parent cells, the point of attachment of bud to parent measuring 4-5 μm in diameter, and remain so rather tenaciously. The budding process is uni-, sometimes bipolar, and very rarely multilateral. The multilateral budding type is never like the "pilot wheel" type seen in *Paracoccidioides brasiliensis.* The majority of the cells are multinucleate. Some strains show hyphae between their blastoconidia. A microform, which can be distinguished from the tissue form of *Histoplasma capsulatum* only by the fact that it is multinucleate instead of uninucleate, has been observed [1].

Yeast-like form in culture: The fungus grows rather slowly, developing wrinkled, waxy, whitish colonies on Sabouraud's glucose agar, brain heart infusion agar, beef infusion agar, and so on. Microscopical examination shows a mixture of yeast-like cells, similar to those seen in the tissues, and short mycelial fragments. The cells remain attached to each other even more tenaciously than in the tissues.

Mycelial form in culture: Conversion to the mycelial form takes place at 30°C or lower. Young *colonies* are usually moist, sometimes have synnemata (coremia: "prickly stage"), and develop a variable amount of white or brownish aerial mycelium. Ascocarp development is directly related to the formation of buff to yellowish-brown pigment in each strain and is independent of the mating factors [27]. A diffusible brown pigment may be produced. The *hyphae* are delicate, 2-4 μm wide, septate, thin-walled. *Conidia* of aleuriospore type are numerous in mature colonies. They are cream-colored, lateral or terminal, sessile or pedicellate, globose to oval or pyriform, smooth-walled, and 2-10 μm in diameter.

3. *Emmonsiella* Kwon-Chung, *Science 177*: 368-369, 1972

Gymnothecia globose to subglobose with closely coiled hyphae which grow out radially from hyphae that encircle the ascogonium. *Peridial hyphae* anastomosing, septate, arising from the spirals. *Asci* pear- to club-shaped, 8-spored. *Ascospores* globose.

a. *Emmonsiella capsulata Kwon-Chung, loc. cit., 1972 (Fig. 2) Conidial state synonymy*

Histoplasma capsulatum Darling, *JAMA 46*: 1283-1285, 1906

Cryptococcus capsulatus (Darling) Castellani & Chalmers, *Manual of Tropical Medicine,* Baillière, Tindall, and Cox, London, p. 1076, 1919

Torulopsis capsulatus (Darling) Almeida, *An. Fac. Med. Univ. São Paulo 9*: 10, 1933

Posadasia capsulata (Darling) Moore, *Ann. Mo. Bot. Gard. 21*: 347-348, 1934

Posadasia pyriformis Moore, loc. cit., 1934

Histoplasma pyriforme (Moore) Dodge, *Medical Mycology,* Mosby, St. Louis, p. 155, 1935

Emmonsiella capsulata is the causative organism of histoplasmosis (Darling's disease, reticuloendothelial cytomycosis, or reticuloendotheliosis). It is a heterothallic fungus with a heterothallism of the two-allele (one-locus) type. Its pathogenicity appears to be strongly but not completely linked with the (-) mating type, the ratio of (+):(-) being 1:7 for iso-

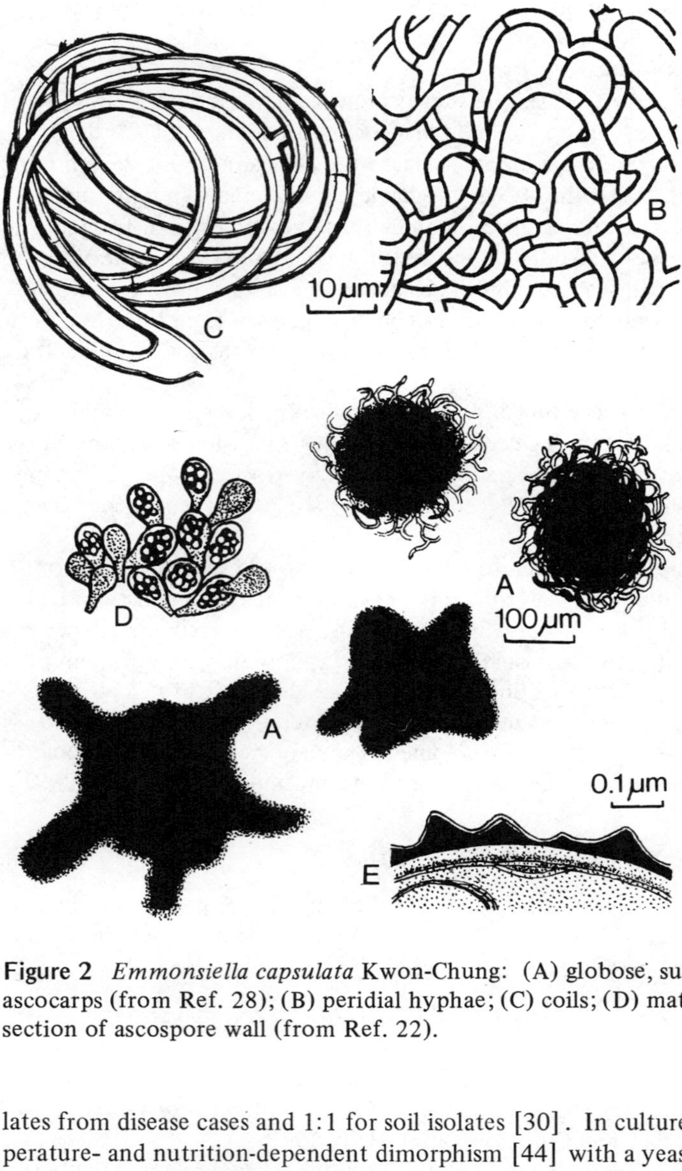

Figure 2 *Emmonsiella capsulata* Kwon-Chung: (A) globose, subglobose, and stellate ascocarps (from Ref. 28); (B) peridial hyphae; (C) coils; (D) mature asci; (E) cross section of ascospore wall (from Ref. 22).

lates from disease cases and 1:1 for soil isolates [30]. In culture the fungus shows a temperature- and nutrition-dependent dimorphism [44] with a yeast-like (Y) form developing on Francis' glucose cysteine blood agar, in tissue culture media, and other special media at 37°C and a mycelial (M) form on routine media at 30°C or below.

(1) Ascigerous State

The yellowish or buff-colored *gymnothecia* are globose to subglobose, becoming irregularly stellate with age, 80-250 μm in diameter. The ascocarp formation is initiated with fusion of two morphologically undifferentiated hyphae of opposite type. One of the hyphae winds several times around the other. The tip of the winding hypha and the site of its contact with the other hypha swell into two knob-like structures, which become encircled by hyphae that arise from the larger knob. The 2 to 10, but usually 3 to 5 conspicuous spirals that give the aging ascocarp its stellate shape grow out from the outer encircling hyphae to a length of 30-100 μm. The dichotomously branched, anastomosing, sinuous *peridial hyphae* with their rather uniform diameter originate from the outer wall of coils at the proximal end of the spiral. The outer wall of the coiled hypha is 0.8-1 μm thick, the inner wall only 0.3-0.5 μm. The 8-spored, shortly pedicellate *asci* are club- to pear-shaped, 10-16 × 5 μm. The hyaline *ascospores* are globose, 1.5-2 μm, smooth-walled when observed under the regular microscope but with minute, triangular warts under the electron microscope [22].

(2) Conidial State

Tissue form: In mammalian tissues the fungus grows as a yeast with oval cells, 2-3 × 3-4 μm in size, which multiply by budding at the narrow end of the cells. In stained preparations the young bud is often attached to the parent cell by a very delicate and rather long filament. The cells are uninucleate. The (-) mating type strains are more easily converted into the Y-form than are the (+) type strains [30].

Yeast-like form in culture: On media appropriate for the development of the Y-form the fungus produces moist, creamy, whitish colonies of oval budding cells, similar to those of the tissue form, mixed with a smaller or greater amount of irregular hyphae. The maximum size (7.6 μm) of the yeast-like cells is larger in vitro than it is in the tissues.

Mycelial form in culture: Conversion to the M-form takes place at a temperature of 30°C or lower. The *colonies* are downy or hairy, white at first, buff to brown later, grossly identical to colonies of *Ajellomyces dermatitidis* but with a somewhat slower growth rate. Vegetative *hyphae* are branched, septate, 2-5 μm wide. Racquet hyphae are scarce. Reproduction is by two types of *conidia* (aleuriospores): (1) globose or pear-shaped microconidia, 2-6 μm in diameter, sessile or formed at the tips of short conidiophores branching off at right angles from the supporting vegetative hyphae, and (2) globose, rarely pear-shaped macroconidia, 8-25 μm (usually 10-14 μm) in diameter, with walls covered with many blunt spines or finger-like appendages.

b. *Histoplasma duboisii Vanbreuseghem, Ann. Soc. Belge Med. Trop. 32: 578, 1952*

H. duboisii, the causative fungus of African histoplasmosis, has often been regarded as a variety of *H. capsulatum* (*H. capsulatum* var. *duboisii* Drouhet, 1957; Ciferri, 1960). Kwon-Chung's [31] discovery that *H. duboisii* strains are compatible with *H. capsulatum* strains and can produce ascocarps, indistinguishable from *Emmonsiella capsulata,* supports the variety concept. However, as no viable ascospores have been formed, the separate species concept cannot be ruled out. Histoplasmosis is a cosmopolitan deep mycosis. African histoplasmosis has a greater tendency to be restricted to the skin and subcutaneous tissues and is confined to the African continent.

c. *Histoplasma farciminosum (Rivolta) Redaelli & Ciferri, Boll. Sez. Ital. Soc. Int. Microbiol. 6: 376-379, 1934*

Histoplasma farciminosum is the etiological agent of epizootic lymphangitis in Equicae. The taxonomic position of this fungus is not yet settled. An ascigerous state is unknown.

B. Eurotiaceae

Eurotiaceae Clements & Shear, *The Genera of Fungi,* Hafner, New York, p. 50, 1931. Ascocarps usually formed as true cleistothecia with a closed, more or less firm peridium, soft, sometimes poorly delimited, and confluent; *peridium* pseudoparenchymatous or prosenchymatous; *asci* arising from croziers on the ascogenous hyphae which originate from coiled initials or from anastomoses of apparently undifferentiated, vegetative hyphae; *ascospores* usually oblate, one-celled without germ pores hyaline, yellow, red, or violet, often with equatorial crests.

Conidial states: *Aspergillus, Paecilomyces,* and *Penicillium.*

According to Malloch and Cain [36], the name of Eurotiaceae Clements & Shear is a later synonym for the Trichocomataceae Fischer, and the name Aspergillaceae is illegitimate when it is applied to ascigerous states. As the name Trichocomataceae is not yet applied in medical mycological literature, the better known name Eurotiaceae is used here.

1. *Eurotium* Link ex Fr., *Syst. Mycol. 3*: 331, 1829

Ascocarps arising from coiled initials, globose, nonostiolate, usually yellow, rarely white, glabrous; *peridium* pseudoparenchymatous, composed of a single layer of cells; *asci* irregularly positioned, 8-spored, subglobose to globose, evanescent; *ascospores* oblate, usually with equatorial crests or furrows, smooth- or rough-walled, hyaline to pale yellow.

Conidial state: *Aspergillus glaucus* group with phialides borne directly on the vesicle.

Raper and Fennell [41] describe 15 species and one variety as members of their *Aspergillus glaucus* group. Austwick (in Ref. 41) mentioned several reports of pathogenicity of *Aspergillus amstelodami, A. montevidensis,* and *A. repens* in cases varying from cerebral abscess and mycetoma to keratitis and otomycosis. Since *A. amstelodami* was reported most frequently in these cases, a short description is given below.

a. *Eurotium amstelodami Mangin, Ann. Sci. Nat. Bot. Biol. Veg., Ser. 9, 10*: *361, 1909 (Figs. 3 and 4)*

Eurotium repens (Corda) Sacc. var. *amstelodami* Vuillemin, *Bull. Soc. Mycol. Fr. 16*: 131, 1920

Conidial state: *Aspergillus amstelodami* (Mangin) Thom & Church, *The Aspergilli,* Williams & Wilkins, Baltimore, p. 113, 1926; Raper and Fennell, *The Genus Aspergillus,* Williams & Wilkins, Baltimore, pp. 166-169, 1965.

(1) Ascigerous State

Ascocarps globose or subglobose, yellow, 100-160 μm, glabrous; *asci* 10-13 μm, 8-spored; *ascospores* oblate, hyaline, rough-walled, with a distinct, equatorial furrow which is flanked by irregular ridges on both sides, 4.5-5 × 3.5-3.8 μm.

(2) Conidial state

Conidial heads radiate to columnar, usually 120-150 μm in diameter, occasionally wider; *conidiophores* hyaline to pale yellow-green or pale olivaceous, smooth-walled, 250-350 μm in length, 10-12 μm wide near the vesicle; *vesicles* subglobose, 10-25 μm in diameter; *phialides* uniseriate, 5-11 × 2-4 μm; *conidia* subglobose to elliptical, usually with flattened ends, finely spinulose, 3.5-5.2 μm in length, but generally 4.5-5 × 3.5-4 μm.

This species grows well at 37°C. Strains isolated from sputum, lungs, bronchi, and external ear canal were studied in the division of medical mycology of the Centraalbureau voor Schimmelcultures at Baarn.

2. *Emericella* Berk. & Br., in Berk., *Introduction to Cryptogamic Botany,* H. Baillière, London, p. 340, 1857

Ascocarps arising from coiled initials, globose, nonostiolate, yellowish red or dark brown, solitary or clustered, within masses of Hülle cells; *asci* irregularly positioned, 8-spored,

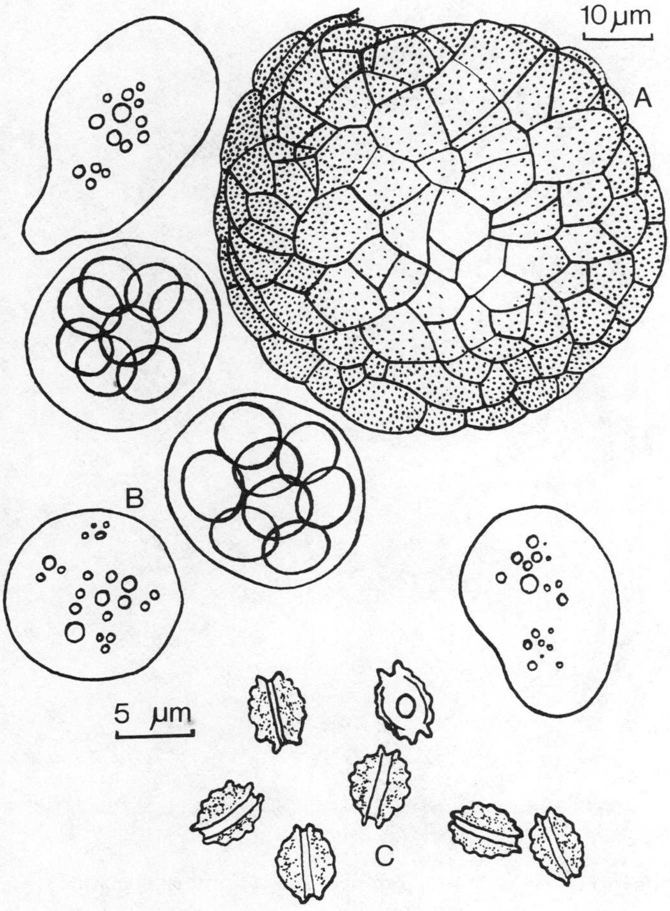

Figure 3 *Eurotium amstelodami* Mangin (CBS 518.65, neotype): (A) cleistothecium; (B) asci; (C) ascospores.

subglobose to globose, evanescent; *ascospores* oblate, usually with equatorial crests, smooth or rough, orange to purple or bluish violet.

Conidial state: *Aspergillus nidulans* group with phialides borne on metulae.

At present there are about 12 species of *Emericella*, most of which are treated as species of *Aspergillus* in the *Aspergillus nidulans* group by Raper and Fennell [41] and Christensen and Raper [13].

a. Emericella nidulans (Eidam) Wint. var. nidulans as Emericella nidulans (Eidam) Wint., in Rabenhorst, Krypt.-Fl. 1(2): 62, 1884 (Figs. 5 and 6)

Sterigmatocystis nidulans Eidam, in Cohn, *Beitr. Biol. Pflanz. 3*: 392-411, 1883

Conidial state: *Aspergillus nidulans* (Eidam) Wint., loc. cit., 1884.

(1) Ascigerous State

Ascocarps globose, blackish brown, developing separately, 100-200 µm, covered by a closely appressed network of anastomosing hyphae which bear a dense aggregation of pale yellow, thick-walled, globose to subglobose Hülle cells up to 25 µm wide; wall of

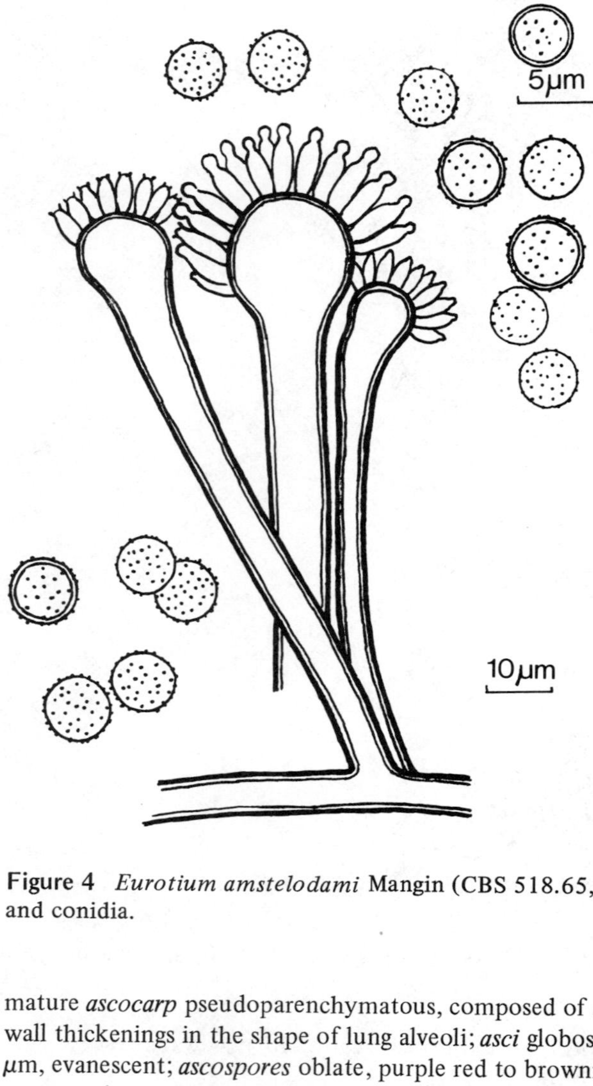

Figure 4 *Eurotium amstelodami* Mangin (CBS 518.65, neotype). Conidial structures and conidia.

mature *ascocarp* pseudoparenchymatous, composed of a single layer of cells with external wall thickenings in the shape of lung alveoli; *asci* globose to subglobose, 8-spored, 7-12 µm, evanescent; *ascospores* oblate, purple red to brownish violet, smooth-walled, with two equatorial crests, 3.8-4.5 X 3.5-4 µm excluding the ±0.5-µm-high crests.

(2) Conidial State

Conidial heads short, columnar; *conidiophores* mostly sinuous, smooth-walled, cinnamon brown, up to 100 µm long and about 5 µm wide, usually narrowing toward the foot cell; vesicle hemispherical, cinnamon brown, up to 13 µm in diameter; *metulae* 4-7 X 2-3.5 µm; *phialides* 5-9 X 2-3 µm; *conidia* globose, rugulose, 3-4 µm in diameter, green in mass.

Aspergillus nidulans has five varieties, which differ primarily in ascospore morphology [13]. It grows well at 37°C and is an occasional pathogen for humans and animals, either alone or associated with other potential pathogens, such as *Aspergillus fumigatus, A. flavus,* and *A. niger* [15,42,53]. Its pathogenicity for rabbits was experimentally proved by Drake [17].

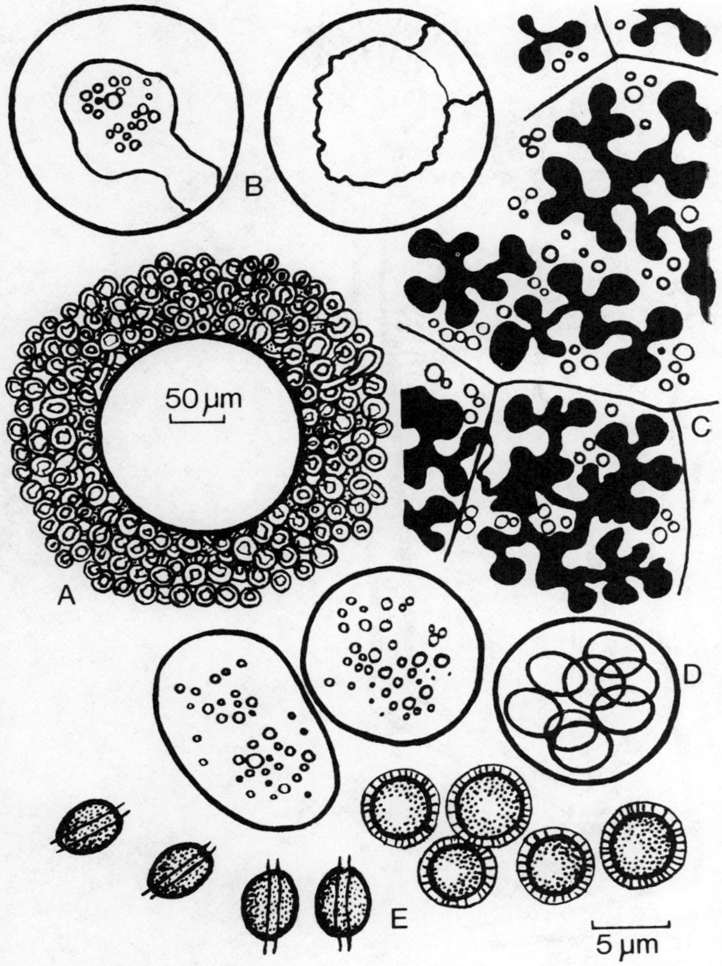

Figure 5 *Emericella nidulans* (Eidam) Wint. var. *nidulans* (CBS 121.35, type): (A) cleistothecium with mantle of Hülle cells; (B) Hülle cells; (C) cells of peridium; (D) asci; (E) ascospores.

3. *Neosartorya* Malloch & Cain, *Can. J. Bot. 50*: 2620, 1972

Ascocarps arising from coiled initials, globose, glabrous to tomentose, nonstromatic, with a multilayered, pseudoparenchymatous peridium; *asci* irregularly positioned, subglobose to globose, evanescent; *ascospores* oblate, smooth to variously sculptured, with two or four equatorial crests, hyaline.

Conidial state: *Aspergillus fumigatus* group with phialides borne directly on the vesicles.

At present there are eight species of *Neosartorya* with four varieties. Raper and Fennell [41] classified their species in the *Aspergillus fischeri* series, the ascigerous coun-

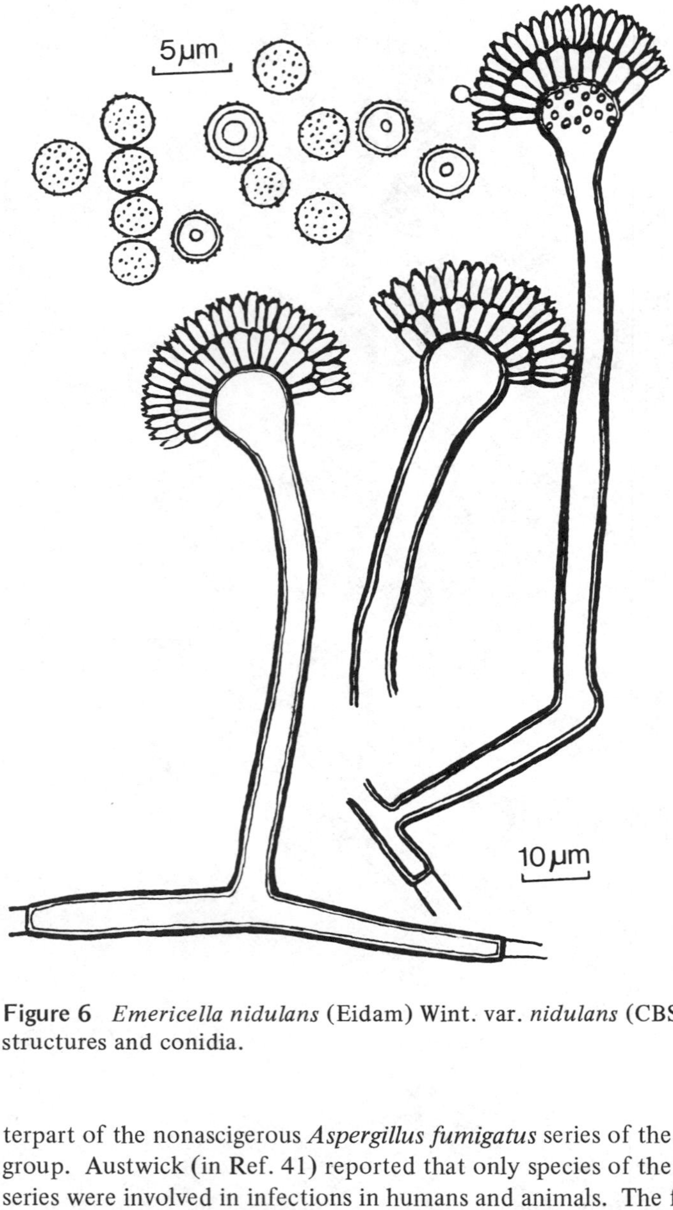

Figure 6 *Emericella nidulans* (Eidam) Wint. var. *nidulans* (CBS 121.35, type). Conidial structures and conidia.

terpart of the nonascigerous *Aspergillus fumigatus* series of the *Aspergillus fumigatus* group. Austwick (in Ref. 41) reported that only species of the *Aspergillus fumigatus* series were involved in infections in humans and animals. The first report of pathogenicity to humans of a member of the other series, namely *A. fischeri* var. *spinosus,* was from Gerber et al. [21] in 1973.

a. Neosartorya fennelliae Kwon-Chung & Kim

A heterothallic species isolated from the eyeballs of laboratory rabbits, which was proved to be nonpathogenic, although it was regarded as a potential pathogen in patients with compromised immune defenses [29].

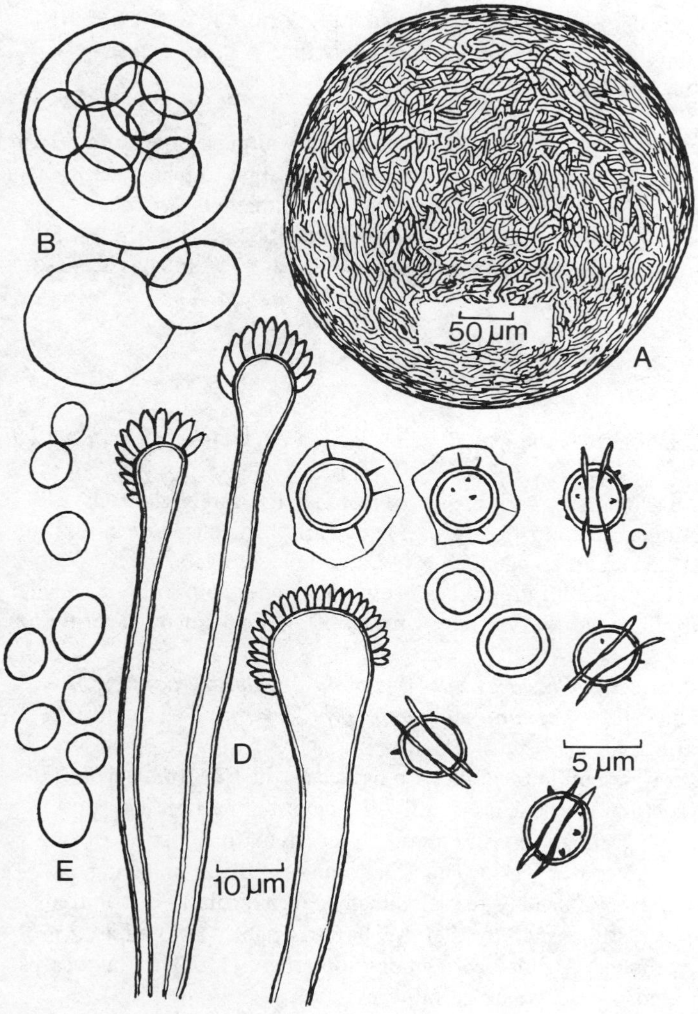

Figure 7 *Neosartorya fischeri* (Wehmer) Malloch & Cain var. *spinosa* (Raper & Fennell) Malloch & Cain: (A) cleistothecium; (B) asci; (C) ascospores; (D) conidial structures; (E) conidia.

b. Neosartorya fischeri (Wehmer) Malloch & Cain var. spinosa (Raper & Fennell) Malloch & Cain, Can. J. Bot. 50: 2621, 1972 (Fig. 7)

Conidial state: *Aspergillus fischeri* Wehmer var. *spinosus* Raper & Fennell, *The Genus Aspergillus,* Williams & Wilkins, Baltimore, p. 256, 1965.

(1) Ascigerous State

Ascocarps globose, pale yellow, developing separately, 200-300 µm, surrounded by a loose network of hyphae; wall of mature ascocarps fragile, consisting of two to three layers of irregularly flattened cells; *asci* globose to subglobose, 8-spored, 10-12 × 8-10 μm, evanescent; *ascospores* globose to subglobose, hyaline, with 0.5-2 μm long spines, 4.5 μm in diameter excluding the 2-μm-high equatorial crests.

(2) Conidial State

Conidial heads loosely columnar, 100-150 × 40-50 μm or nearly globose and up to 50 μm in diameter, pale greenish gray; *conidiophores* smooth-walled, hyaline to pale green, 100-150 μm long; *vesicles* flask-shaped, grayish green, up to 18 μm in diameter; *phialides* hyaline to pale grayish green, 5-8 × 2-5 μm; *conidia* globose to subglobose, smooth, 2-4.5 × 2.4-3.5 μm, pale grayish green. *N. fischeri* var. *spinosa* grows well at 37°C and has been isolated from soil.

III. SPHAERIALES

A. Microascaceae

Microascaceae Luttrell ex Malloch, *Mycologia 62*: 734-735, 1970; von Arx, *Persoonia 10*: 23-31, 1978

Ascocarps globose to irregular in shape, dark brown to black, hairy, rarely glabrous, arising from coiled initials, ostiolate or nonostiolate; *asci* irregularly distributed or in vertical rows, arising singly or in chains on an ascogenous system, without croziers, ovoid to globose, evanescent; *ascospores* straw-colored, reddish brown or copper-colored, one-celled, dextrinoid when young, usually smooth, normally with an often inconspicuous germ pore at one or both ends.

Conidial states: *Scopulariopsis (Phaeoscopulariopsis, Masoniella, Wardomyces), Doratomyces, Scedosporium (Monosporium),* and *Graphium- (Sporocybe-, Dendrostilbella-)* or *Sporothrix*-like forms.

The Microascaceae closely resemble the Ophiostomataceae, the Chaetomiaceae, and the Eurotiaceae. The Ophiostomataceae can be easily distinguished by their colorless ascospores without germ pores [35] and by the pseudoparenchymatous instead of filamentous central tissue of the ascocarps [34]. The Chaetomiaceae differ in having croziers, nondextrinoid ascospores, and usually fascicled instead of irregularly distributed asci. The Eurotiaceae can be distinguished by their hyaline to brightly colored ascocarps and ascospores, and their complex phialide-bearing conidiophores [41]. Their ascocarps are nonostiolate and their ascospores have no germ pores.

Species belonging to the genera *Microascus, Petriella, Petriellidium,* and *Pithoascus* have been reported, usually under the names of their conidial states, as pathogens on humans and animals.

1. *Microascus Zukal, Verh. Zool. Bot. Ges. Wien 35*: 339, 1885; Morton & Smith, *Commonw. Mycol. Inst. Mycol. Pap. 86*: 1-96, 1963; von Arx, *Persoonia 8*: 191-192, 1975

Colonies spreading, becoming brown by pigmented aerial mycelium; *ascocarps* superficial or immersed, subglobose to globose, papillate or long-necked, ostiolate, dark brown to black, carbonaceous to membranaceous, glabrous or more often hairy to setose, especially around the ostiolum, with a pseudoparenchymatous wall composed of three to eight layers of isodiametric or flattened, dark cells; *asci* numerous, clustered or catenulate, often in vertical rows, ovoid or broadly clavate, evanescent, 8-spored; *ascospores* irregularly disposed in the ascus, asymmetrical, often reniform, heart-shaped or triangular, yellowish or straw-colored when mature, with a single, small but often prominent germ pore, dextrinoid when young, smooth.

Conidial states: *Scopulariopsis (Phaeoscopulariopsis, Masoniella, Wardomyces), Doratomyces* and *Graphium- (Sporocybe-, Dendrostilbella-)* or *Sporothrix*-like forms. Barron et al. [7] distinguished 14 species of *Microascus.* Morton and Smith (loc. cit.) and von Arx (loc. cit.) reduced the number to 10, but have a mutually diverging opinion about which species should be accepted in the genus.

Microascus species are rather common soil fungi [7], but they are also quite often isolated from plant seeds and various kinds of animal dung. The following species have occasionally been reported as pathogens on humans and animals.

a. Microascus trigonosporus Emmons & Dodge, Mycologia 23: 317, 323, 1931; Barron et al., Can. J. Bot. 39: 1616-1617, 1961; Morton & Smith, Commonw. Mycol. Inst. Mycol. Pap. 86: 32-33, 1963 (Fig. 8)

(1) Ascigerous State

Ascocarps globose, 120-375 μm, with or without one or more papillae or up to 250-μm-long necks, carbonaceous, black, glabrous or with scattered, flexuous hairs; *asci* globose, ovoid to pyriform, 8-12 × 6-9 μm; *ascospores* triangular in planar view, broadly rounded at the corners and normally with all three sides concave, oblong in side view, 3.5 × 2.5-3.5 μm, very pale brown when seen under the microscope, rusty red in mass.

(2) Conidial State

Scopulariopsis trigonospora Emmons & Dodge, loc. cit., 1931. *Conidiogenous cells* solitary or in whorls, occasionally in branched, penicillioid structures, mostly 10-15 × 2-3 μm, with basal swelling, elongating percurrently; *conidia* globose, subglobose or lemon-shaped, 3.5-5.5 × 2.5-3.5 μm, with a truncate base.

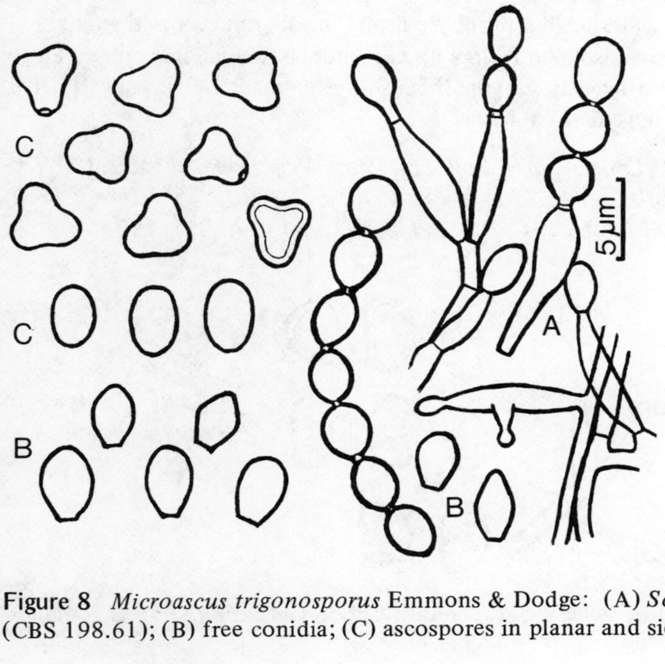

Figure 8 *Microascus trigonosporus* Emmons & Dodge: (A) *Scopulariopsis* conidial state (CBS 198.61); (B) free conidia; (C) ascospores in planar and side view (CBS 272.49).

There is no substantial evidence of pathogenicity of this species. The type culture, which was isolated from a skin lesion caused by *Epidermophyton floccosum* in a Puerto Rican, was regarded as a contaminant.

b. *Microascus cinereus (Emile-Weil & Gaudin) Curzi, Boll. Stn. Patol. Veg. Roma 11*: *60, 1931; von Arx, Persoonia 8*: *194, 1975; Morton & Smith, Commonw. Mycol. Inst. Mycol. Pap. 86*: *27-29, 1963; Udagawa, J. Gen. Appl. Microbiol. 8*: *44-46, 1962 (Fig. 9)*
M. griseus Mathur et al., *Sydowia 16*: 47, 1962 (publ. 1963)
M. lunasporus Jones, *Mycologia 28*: 503, 1936.
M. pedrosoi Fuentes & Wolf, *Mycologia 48*: 63, 446, 1956

Conidial state:
Scopulariopsis cinerea Emile-Weil & Gaudin, *Arch. Med. Exp. Anat. Pathol. 28*: 460, 1919
S. oidiospora Zach, *Österr, Bot. Z. 83*: 182, 1934
Sporotrichum arabicum Massee, *J. Bot. Br. Foreign 34*: 153, 1896

(1) Ascigerous State
Ascocarps 150-350 μm, black, with one or more, rarely up to 100-μm-long necks; *asci* ovate, 10-14 × 7-12 μm; *ascospores* roughly triangular, with one concave side, or reniform to broadly lunate, 4.5-6.5 × 3-5 μm.

(2) Conidial State
Colony dark violaceous gray, grayish brown to black; *conidiogenous cells* borne singly along the hyphae or on short conidiophores in small verticils or penicillioid structures, 3-15 μm long, cylindrical or flask-shaped, elongating percurrently into a 1.5-2.5 μm wide, annellated zone; *conidia* ovate, with rounded or pointed apex and a truncate base, 3.5-7 × 2.5-4.5 μm, dark gray to black in mass.

Microascus cinereus is a widespread saprophyte found in soil, dung, stored grains, and so on. The type culture was isolated from a diseased human toenail in France. This single observation does not sufficiently support the assumption made by Udagawa [50] that the fungus is mildly pathogenic to humans.

c. *Microascus manginii (Loub.) Curzi, Boll. Stn. Patol. Veg. Roma 11*: *60, 1931 (Fig. 10)*
Nephrospora manginii Loubière, *C. R. Acad. Sci. (Paris) 177*: 211, 1923

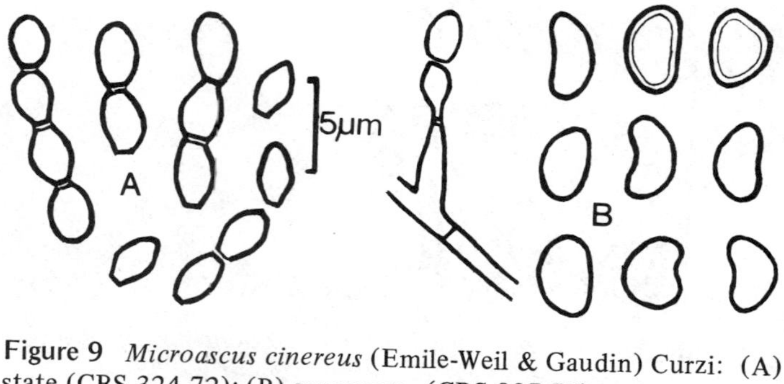

Figure 9 *Microascus cinereus* (Emile-Weil & Gaudin) Curzi: (A) *Scopulariopsis* conidial state (CBS 324.72); (B) ascospores (CBS 807.73).

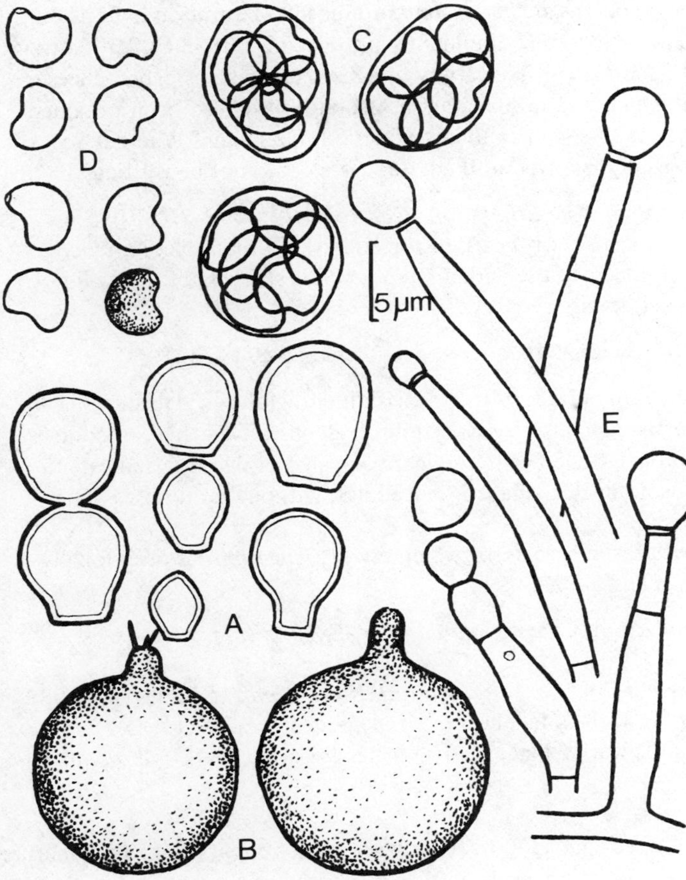

Figure 10 *Microascus manginii* (Loub.) Curzi [(A)-(D): CBS 816.73; (E): CBS 506.66]: (A) conidia; (B) ascocarps; (C) mature asci; (D) ascospores; (E) conidiogenous cells.

Conidial state:

Scopulariopsis candida (Guéguen) Vuillemin, *Bull. Soc. Mycol. Fr. 27*: 143, 1911

S. alboflavescens Zach, *Österr, Bot. Z. 83*: 177, 1934

(1) Ascigerous State

Ascocarps pyriform to globose, black, 85-150 × 90-175 μm; *asci* globose, ± 10 μm, or ovate and up to 29 μm long; *ascospores* reniform to lunate, 4-5 × 2.5-4 μm.

(2) Conidial State

Colonies white to pale buff; *conidiogenous cells* borne singly along the hyphae or in groups on short, rarely branched conidiophores, 10-25 μm long, with a cylindrical or slightly swollen base, 2.5-4.5 μm in width and a cylindrical, equally wide or slightly narrower annellated zone; *conidia* subglobose to broadly ovate, with a more or less rounded apex and a truncate base with a prominent collar, 5-9 × 4-7 μm, white to very pale buff in mass; solitary aleuriospores (chlamydospores) occur occasionally. This species is usually isolated in its conidial state. Mature ascocarps are very rarely produced. Maturation may take up to 2 years [54], but fertile ascocarps were abundantly produced by strain CBS 506.66 on malt agar after 2 weeks. It occurs on a wide variety of substrates, such as paper, cheese, wood, soil, milled rice, bat dung, and in hospital air. Isolates have also been obtained from dentine, a toenail, tinea capitis, and an unknown lesion in humans. Whether *Scopulariopsis candida* was etiologically involved in the latter cases could not be verified.

d. Microascus cirrosus Curzi, Boll. Stn. Patol. Veg. Roma 11: 302-310, 1930
This species is reported in connection with cattle abortion [6]. It resembles *M. cinereus* in its cultural characters, but differs in the darker color and less stable habit. It will not be described here. For further details, see Ref. 7.

2. *Pithoascus* von Arx, *Proc. K. Ned. Akad. Wet. Ser. C, 76*: 295, 1973

Ascocarps superficial or partly immersed, solitary or confluent, globose, glabrous, black, nonostiolate or with an abortive, inconspicuous ostiolum; ascocarp wall thick, pseudoparenchymatous, composed of flattened cells; *asci* barrel-shaped or ovoid, often catenulate, 8-spored; *ascospores* fusoid or boat-shaped, unicellular, without germ pores.
Conidial state: *Arthrographis.*

Pithoascus is a genus with seven species, of which one, *P. langeronii,* is occasionally isolated from diseased human and animal tissues.

a. Pithoascus langeronii von Arx, Persoonia 10: 24-26, 1978 (Fig. 11)
Conidial state:
Arthrographis langeronii Cochet, *Ann. Parasitol. Hum. Comp. 17*: 98-101, 1939
Oidiodendron kalrai Tewari & Macpherson, *Mycologia 63*: 602-611, 1971
Arthrographis kalrai (Tewari & Macpherson) Sigler & Carmichael, *Mycotaxon 4*: 360-363, 1976

(1) Ascigerous State

Ascocarps immersed, rarely superficial, globose, yellowish when young, black when mature, nonostiolate, 75-160 μm in diameter; ascocarp initials composed of a central cell surrounded by inflated, 5-7 μm wide cells, soon covered by radial hyphae; ascocarp wall composed of 2-4 μm broad cells, with an inner layer of flattened, 7-13 μm wide cells and an outer wall which later becomes dark, encrusted, and nearly amorphous; *asci* numerous, in radiating, clustered rows, ellipsoid, clavate or cylindrical, sessile, sometimes stalked, thin-walled, rather persistent, 8-spored, 8-13 × 4-7 μm; *ascospores* fusiform to boat-shaped, dextrinoid when young, straw-colored to yellow-brown when mature, smooth, nonostiolate, 3.5-5 × 2-2.5 μm.

(2) Conidial State

Good growth at 25-37°C. *Colony* velutinous or floccose, whitish to pale buffish brown, with yellowish to brown reverse; vegetative hyphae 1.3-4.8 μm wide; *arthroconidia* in branched chains, smooth, cylindrical, hyaline, 3.5-8.5 × 2-3 μm, *chlamydospores* lateral, terminal, or intercalary, globose to subglobose, smooth, 3.3-4.5 μm. Colony on brain heart infusion agar at 37°C cream-colored, yeast-like, with oval to elongate, budding cells, 2.5 × 3.6 μm in diameter, which gradually become mixed with arthroconidia.

Pithoascus langeronii is usually seen in its conidial state. Ascocarps occur very rarely. Tewari and Macpherson [48] and Swenberg et al. [47] have demonstrated its potential pathogenicity for mice. It is reported to be keratinolytic [46], but the present author did not observe any change in inoculated human hair. *Pithoascus langeronii* has been isolated

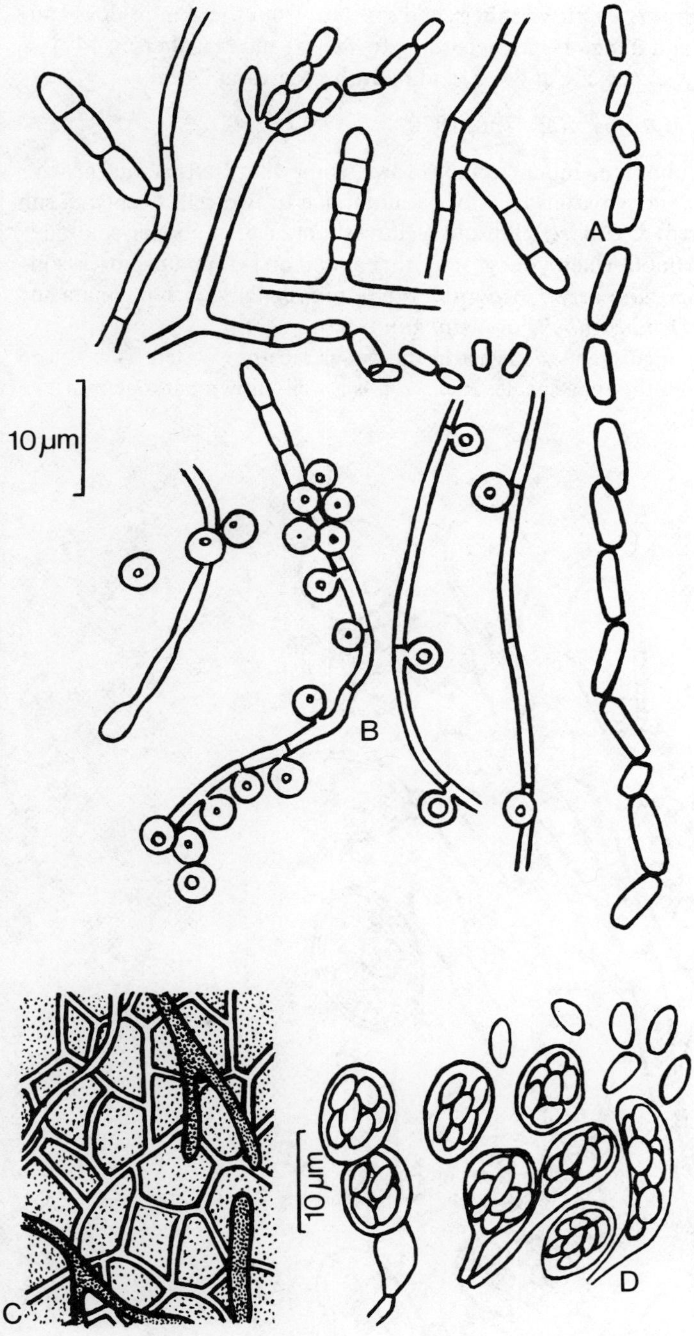

Figure 11 *Pithoascus langeronii* von Arx: (A) arthroconidia; (B) chlamydospores; (C) ascocarp wall; (D) mature asci and ascospores. [Parts (C) and (D) from Ref. 49.]

from human skin, nails, cornea, gastric washings, and sputum; from the skin of dogs and horses; and from air, soil, and dung. It has, according to Tewari and Macpherson [49], a strong caseinolytic activity, expecially at 37°C, and hydrolyses urea in 24 h.

3. *Petriellidium* Malloch, *Mycologia 62*: 738, 1970

Ascocarps nonostiolate, globose, membranaceous, brown, somewhat transparent, nearly smooth; hyaline pseudoparenchymatous layer of peridium one to two cells deep; *asci* subglobose to globose, 8-spored; *ascospores* ellipsoid, yellowish, orange, or copper-colored, with two germ pores, dextrinoid when young, smooth; *conidia* on sympodulae or as solitary aleuriospores (*Scedosporium* or *Monosporium* type), occasionally on synnemata and aggregated in a wet mass (*Dendrostilbella* or *Graphium* type).

Von Arx (loc. cit.) distinguished six species in this genus mostly isolated from soil or plant material. Only one, the type-species *P. boydii,* is a well-known pathogen on humans and animals.

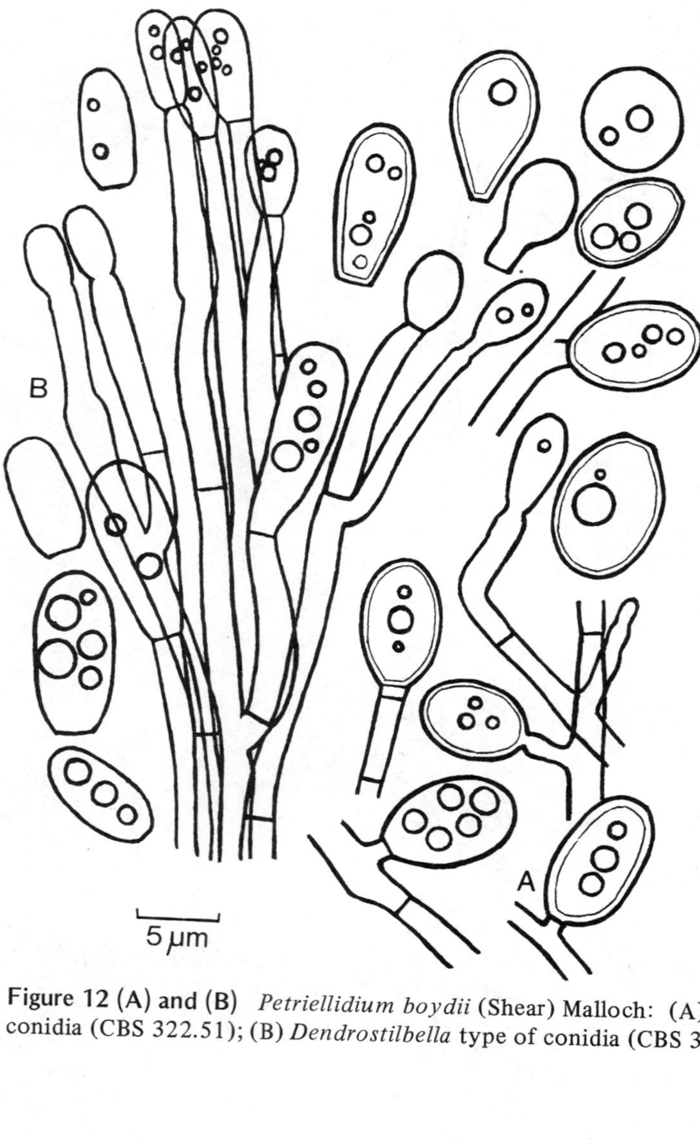

Figure 12 (A) and (B) *Petriellidium boydii* (Shear) Malloch: (A) *Scedosporium* type of conidia (CBS 322.51); (B) *Dendrostilbella* type of conidia (CBS 322.51).

a. *Petriellidium boydii (Shear) Malloch, loc. cit., 1970; von Arx, loc. cit., 1973 (Fig. 12)*
Allescheria boydii Shear, *Mycologia 14*: 242, 1922

Conidial states:
Cephalosporium boydii Shear, loc. cit., 1922
Dendrostilbella boydii Shear, loc. cit., 1922
Monosporium apiospermum Sacc., *Ann. Mycol. 9*: 254, 1911 (nom. ill.)
Scedosporium apiospermum (Sacc.) Sacc., 1919, not validly published by Dodge, *Medical Mycology*, Mosby, St. Louis, p. 838, 1935.

(1) Ascigerous State

Ascocarps usually submerged in the agar, globose, nonostiolate, membranaceous, dark brown, 140-200 μm in diameter, with a 4-6-μm-thick wall, composed of two to three layers of meanderingly interwoven (textura epidermoidea), flattened, 2-6 μm wide, dark brown, hyphoid cells, often covered with brown, thick-walled, septate, 2-3-μm-wide hyphae; ascocarp initials as in *Pithoascus* [3; J. A. von Arx, personal communication];

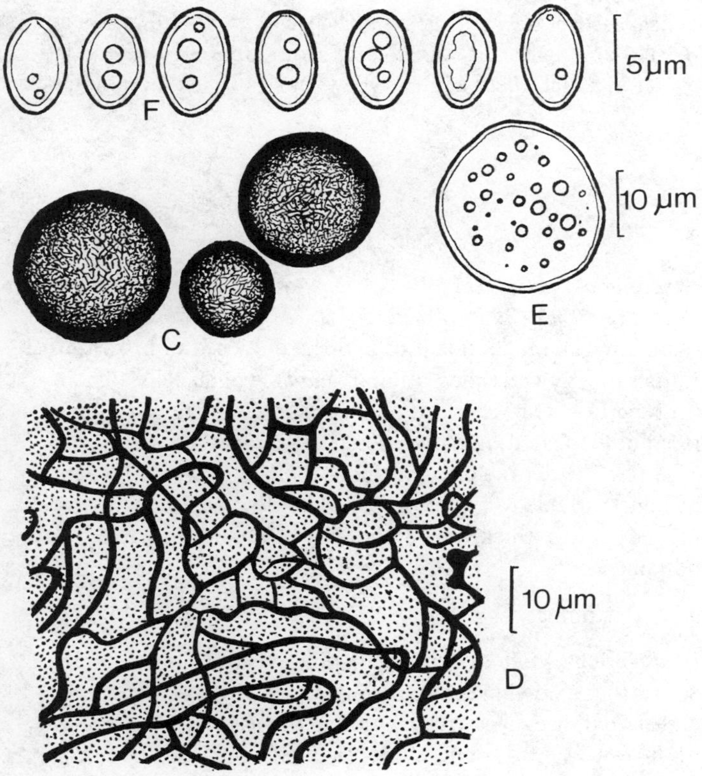

Figure 12 (C), (D), (E), and (F) *Petriellidium boydii* (Shear) Malloch: (C) asoccarps; (D) ascocarp wall; (E) immature ascus; (F) asocspores.

asci globose to ellipsoid, 12-20 × 9-13 μm, evanescent, 8-spored; *ascospores* smooth, ellipsoid, straw-colored, 6-7 × 3.5-4 μm, dextrinoid when young, with two germ pores.

(2) Conidial State

Colony lanose, white or usually pale to dark grayish, with straw-colored, dark olivaceous brown to almost black reverse. *Conidia* of the *Scedosporium (Monosporium)* type, chlamydospore-like, broadly clavate or obovate, 6-12 × 3.5-6 μm, with a rather thick, brown wall and a truncate base, sessile on the vegetative hyphae or more or less aggregated on the distal part of undifferentiated conidiophores; *Dendrostilbella (Graphium)*-type conidia clavate or nearly cylindrical, with a truncate base, hyaline to very pale brown, 5-7 × 2-3 μm, borne as a wet mass in short sympodulae on elongating conidiogenous cells, which usually arise in erect synnemata.

Petriellidium boydii is an etiological agent of white grain maduromycosis (mycetoma pedis) in humans and (rarely) animals [19]. It was also involved in cases of chronic otomycosis, keratomycosis, septicaemia, meningitis, cerebral mycosis, and pulmonary and maxillary sinus infections [12,25]. It is a cosmopolitan, homothallic, rather common soil fungus, especially in temperate and subtropical climates. The conidial states of several *Petriella* species are hardly if at all distinguishable from the conidial states of *Petriellidium boydii*. It is therefore not unlikely that among the many nonascigerous strains of *P. boydii* described as pathogens, strains occur which belong to *Petriella* Curzi. Representatives of this rather small genus differ from *P. boydii* by their ostiolate ascocarps and darker, usually reddish brown, asymmetrical ascospores [3,8]. *Petriella* species are found in soil and dung. One species, *P. setifera* (Schmidt) Curzi, was isolated from lobomycotic skin lesions in dolphins and from the water in which these animals were kept [40]. It did not appear to be etiologically involved.

IV. DOTHIDEALES

A. Testudinaceae

Testudinaceae von Arx, *Persoonia 6*: 366, 1971
Zopfiaceae Arnaud, *Bull. Soc. Mycol. Fr. 29*: 259, 1913 (as "Zopfiacées")
Colonies spreading, dark, aerial mycelium abundant or scarce, composed of brown, thick-walled hyphae; ascocarp initials mostly consisting of intercalary, swollen cells, dividing into all directions; ascocarp tuberiform or spherical, nonostiolate; peridial wall usually divided into plates, composed of elongated, radially arranged, thick-walled, dark cells; *asci* irregularly arranged or in fascicles, borne on ascogenous hyphae, clavate, obovate, or nearly spherical, with a thickened double or a thin single membrane; *ascospores* two-celled, brown, glabrous or ornamented, mostly thick-walled; paraphyses absent or scarce, filamentous, often septate and ramose.

1. *Neotestudina* Segretain & Destombes, *C. R. Acad. Sci. (Paris) 253*: 2579, 1961

Ascocarps scattered, rarely gregarious, globose, without ostioles, usually immersed; *peridium* carbonaceous almost glabrous, showing some similarity to a turtle's shell, by its translucency; *asci* globose, 8-spored, irregularly distributed in the ascocarp; *ascospores* ellipsoidal, two-celled, dark brown. Von Arx and Müller (loc. cit.) distinguished three species in this genus. *Neotestudina* species have been isolated from soil. One of them, the type species *N. rosatii*, is an occasional pathogen in humans.

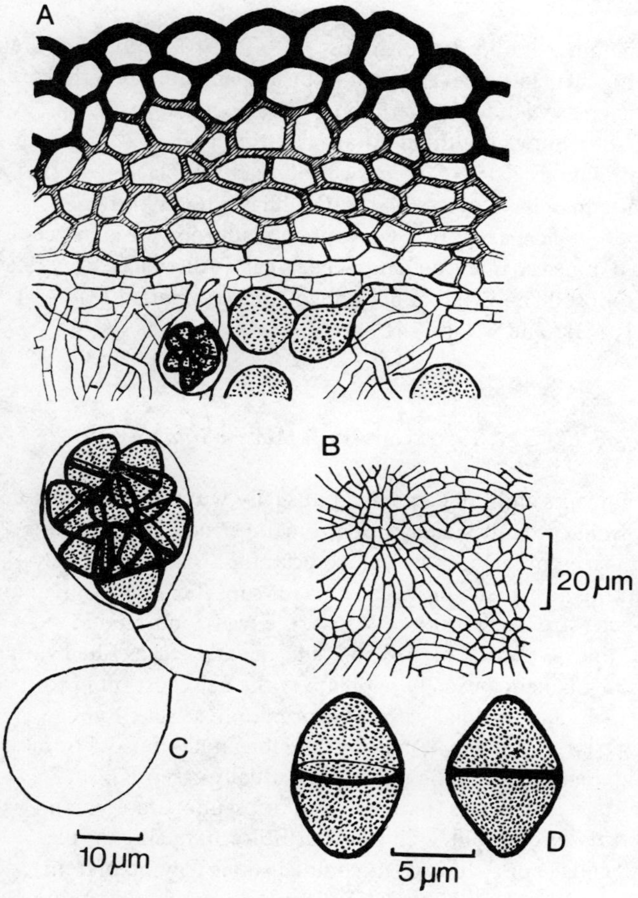

Figure 13 *Neotestudina rosatii* Segretain & Destombes: (A) cortical, subcortical, and ascogenous tissues; (B) surface view of peridium; (C) asci; (D) ascospores.

a. Neotestudina rosatii Segretain & Destombes, loc. cit., 2577-2579, 1961 (Fig. 13)
Zopfia rosatii (Segretain & Destombes) Hawksworth & Booth, *Commonw. Mycol. Inst. Mycol. Pap.* 27-31, 1974
Pseudophaeotrichum sudanense Aue, Müller & Stoll, *Nova Hedwigia 17*: 84, 1969
Pseudodelitschia coriandri Kapoor, Bahl & Lal, *Kavaka 3*: 127, 1976

Ascocarps nonostiolate, superficial, scattered, black, ellipsoid to globose, carbonaceous, smooth, 180-600 μm outer layer of peridium pseudoparenchymatous, divided into plates, which are composed of radially arranged, thick-walled, dark-brown to black, mainly 4-10-μm-wide cells; cells of inner layers less heavily sclerotized, paler, imperceptibly merging into the thin-walled, hyaline, polygonal, pseudoparenchymatous, mainly 5-7-μm-wide cells of the rest of the ascocarp; cells in the center deliquescing as the ascogenous hyphae and pseudoparaphyses develop; pseudoparaphyses rather sparse, branched, entwined, hyaline, mainly 1-2 μm wide; *asci* lateral on wide ascogenous hyphae, bitunicate, with a thick inner membrane, obovate to globose, 12-40 × 10-35 μm, 8-spored; *ascospores* brown, two-celled, irregular in shape, ellipsoid to bicampanulate, sometimes asymmetrical, constricted at the cross wall, with a germ pore at each end at maturity, smooth-walled, 9-12.5 × 4.5-8 μm. Slow growth on potato dextrose agar at 20°C; better growth at 30-37°C, with an optimum at 33°C. *Colonies* velvety, grayish black to brownish black, sometimes with a white margin. Reverse black. Conidial state unknown.

Neotestudina rosatii is the etiological agent of a white-grain maduromycosis (mycetoma pedis) in Africa (Somalia, Guinea, Cameroon, and Senegal) and Australia. In Africa and India it was also isolated from soil and living or decaying plants (*Arachis, Vigna sinensis,* and *Coriandrum*) [23], bark, and wood [24].

B. Pleosporaceae

Pleosporaceae Luttrell, *Mycologia 47*: 521, 1955; von Arx & Müller, *Stud. Mycol. 9*: 64-65, 1975

Saprophytes or parasites on plants and fungi; *mycelium* usually within the host tissue, hyaline or pigmented, often forming pseudostromatic or stromatic structures or superficial subicula; *ascocarps* medium-sized or large, spherical, conical, flask-shaped or rarely almost cylindrical, immersed in the host tissue or in a stroma, or superficial in a subiculum of dark hyphae or on an erumpent stroma, smooth or setose, the wall composed of isodiametric or flattened, thin- or thick-walled, dark or light cells, in early stages filled with often vertically elongated cells or filaments, usually papillate at the apex; ostiolum formed as an apical pore or channel, which may be lined with setae or periphyses-like hyphae; *asci* numerous, arising from the base of the loculi, cylindrical, elongated ellipsoid or saccate, bitunicate, with thickened membranes in the upper part, usually 8-spored, surrounded by hyaline, usually septate paraphyses (paraphysoids or pseudoparaphyses), which are often numerous; *ascospores* ellipsoid, cylindrical or filiform, hyaline or pigmented, with or without an appendage or a slime coat; conidial states may be present; *conidia* formed in pycnidia (e.g., form-genera *Phoma, Hendersonula,* etc.) or on free conidiophores (e.g., form-genera *Drechslera, Curvularia, Alternaria, Stemphylium*).

1. *Leptosphaeria* Ces. & De Not., *Comment. Soc. Crittog. Ital. 1*: 234, 1863; Müller, *Sydowia 4*: 206, 1950

Ascocarps perithecium-like, globose, ellipsoid, very rarely lenticular, solitary or in small, more or less dense groups, often nearly coalescing, rarely in elongated rows, exceptionally more than 500 μm in diameter, with a cellular, often crustose cortex and a generally apical ostiolum, which is developed either as a group of thinner-walled cells or as an open canal with ingrowing periphyses-like hyphae; *asci* cylindrical, clavate, ovoid, oblong, or elongated ellipsoid, 8-spored, arising from thin-walled, often elongated cells full of protoplasm in the basal or sometimes also lateral parts of the ascocarp, surrounded by cellular or filiform paraphyses-like cells; *ascospores* fusoid, elongated or clavate, two- or multiseptate, hyaline or pigmented, often with an inflated cell, or with hyaline appendages at both ends. Von Arx and Müller [5] distinguished about 50 species, which are usually saprophytes, but occasionally parasites of plants. The following two species are reported as causative agents of black-grain maduromycosis in humans in West Africa (Senegal and Mauritania).

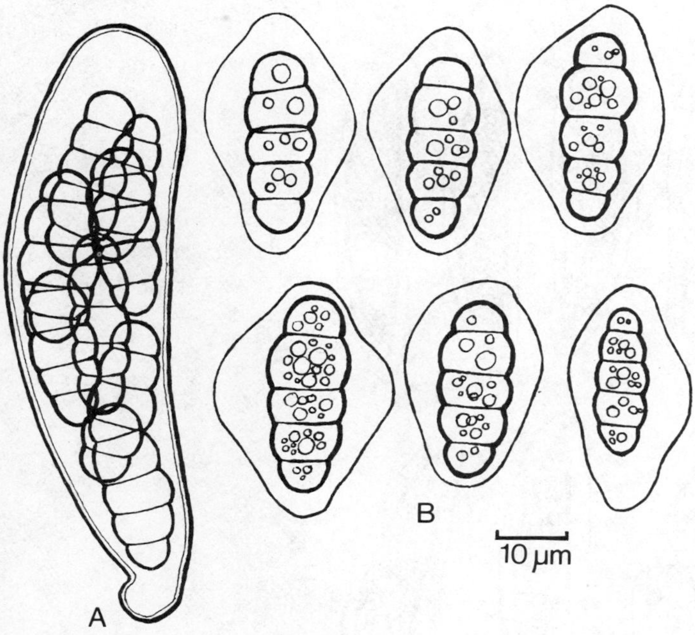

Figure 14 *Leptosphaeria senegalensis* Segretain et al. (IP 614): (A) mature ascus; (B) ascospores.

a. Leptosphaeria senegalensis Segretain et al., C. R. Acad. Sci. (Paris) 248: 3730-3772, 1959 (Fig. 14)

Ascocarps solitary, black, in the middle and basal parts covered with short, gray hyphae, subglobose, nonostiolate, 100-300 µm; *peridium* made up of several layers of isodiametric cells, those of the cortex with thick, dark brown walls; *asci* abundant, bitunicate, clavate, 8-spored, 80-160 × 17-28 µm, arising from a stroma, mixed with hyaline, filiform, sparingly branched, septate paraphyses; *ascospores* uni- or biseriate, hyaline or pale brown, oval or ellipsoid, 23-36 × 8-13.4 µm, usually five-celled, rarely up to eight-celled at maturity, with a thick, hyaline, occasionally pale brown slime coat, which usually has its greatest dimension toward the upper end of the spore, sometimes with a detachable, hyaline, irregular structure at the basal end [5].

b. Leptosphaeria tompkinsii El-Ani, Mycologia 58: 406-411, 1966 (Fig. 15)

Ascocarps solitary, black, covered with dark brown, flexuous hyphae, globose to subglobose, often compressed, nonostiolate, 214-535 µm; peridium 15-34 µm thick, thinner toward the top, made up of several layers of isodiametric cells, those of the cortex with thick, dark brown walls, the attenuated upper part underlain with a mass of polygonal, thin-walled, uninucleate cells; *asci* abundant, bitunicate, clavate to cylindrical, 8-spored, 80-115 × 20-25 µm, mixed with many filiform, non- or only sparingly branched, septate paraphyses; *ascospores* tri- or biseriate, hyaline, fusoid, straight or slightly curved, second or third cell from apex inflated, five to nine (usually seven to nine) celled, 32-45 × 8.8-11 µm, with a rather narrow, hyaline, slime coat.

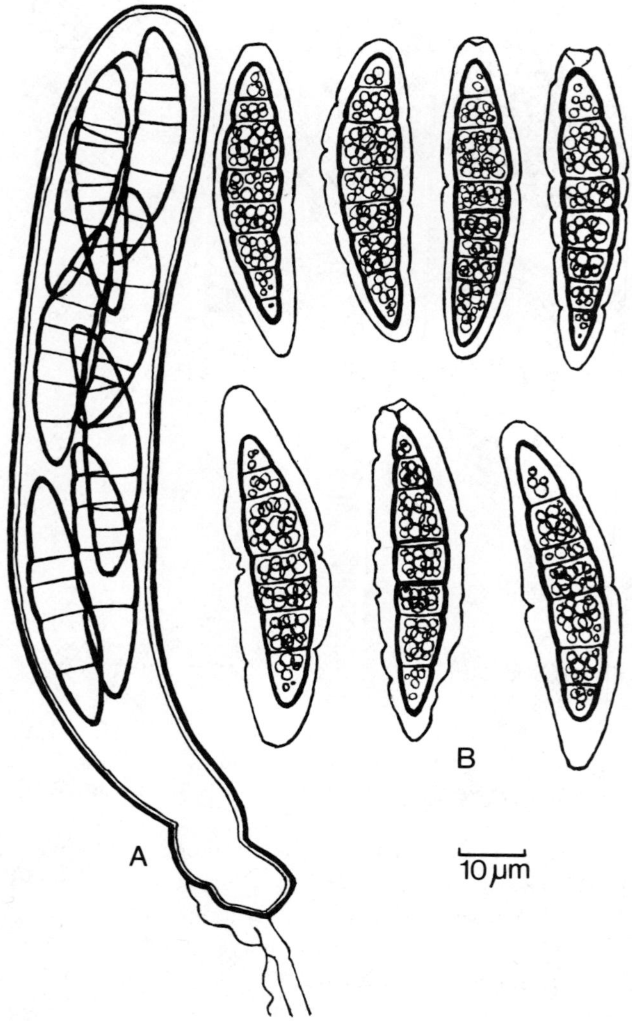

Figure 15 *Leptosphaeria tompkinsii* El-Ani (IP 1151.76): (A) mature asci; (B) ascospores.

The rough grains in the host tissue of the two species are indistinguishable from each other. They are black, solid or hollow, ±1 mm, sometimes forming larger clumps. Microscopically, a hyaline or pale brown center of scattered, inflated cells mixed with some hyphae is surrounded by a dark brown periphery, where the inflated cells, embedded in a dark brown cement, are close together. In culture both species grow more quickly at 37°C than at 30°C. They show a rather great variability in growth rate. Both produce velutinous to lanose, pale and later dark gray to dark olivaceous brown colonies with an almost black reverse. Brownish soluble pigments are formed in some media. Conidia are not formed.

According to Emmons et al. [20], overlapping of ascospore size, shape, and septation occurs. They therefore regarded the two species as similar, but still treated them as separate. A decision concerning their eventual synonymy should be postponed until more is known about the morphology of the species in their natural habitat. *Leptosphaeria tompkinsii* has been reported on thorns and wood [45].

According to El-Ani and Gordon [18], the classification of *Leptosphaeria senegalensis* needs to be reconsidered. However, a reclassification would be premature considering the instability of the taxonomy of the taxa in and around the Pleosporaceae, as well as of the genus *Leptosphaeria* itself.

C. Piedraiaceae

Piedraiaceae Viegas ex Ciferri et al., *Inst. Micol. Bras. Publ. 45*: 7, 1956; Boedijn, *Persoonia 2*: 65, 1961; von Arx & Müller, *Stud. Mycol. 9*: 130-131, 1975 (Fig. 16)

Keratinophilic, growing in or on hairs of humans and animals; *ascocarps* pulvinate, often surrounding the hair shaft, subcuticular when young, dark brown to black, pseudoparenchymatous; *asci* borne singly or in a small number in loculi which have an inconspicuous, ostiolar pore, ellipsoid or obovate, bitunicate, with an apically thickened inner membrane; *ascospores* botuliform or fusiform, one-celled, with or without an appendage at one or both ends, hyaline or slightly pigmented.

1. *Piedraia* da Fonseca & Leão, *Mem. Inst. Oswaldo Cruz, Suppl. 4*: 124, 1928

For a description of *Piedraia,* the only genus of the family, see the description of *Piedraiaceae.*

The following eight species have been described in this genus: *Piedraia hortai* (Brumpt) da Fonseca & Leão, 1928 (type species), *P. paraguayensis* Delamare & Gatti, 1928, *P. sarmentoi* Pereira fil., 1929, *P. venezuelensis* Brumpt & Langeron, 1934, *P. surinamensis* Dodge, 1935, *P. columbiana* Dodge, 1935, *P. javanica* Boedijn & Verbunt, 1938, and *P. quintanilhai* van Uden et al., 1963. Ciferri et al. [11] suggested that there is only one species, *P. hortai,* and that all other names were synonyms (except the later described *P. quintanilhai*). In 1959, however, Boedijn [9] expressed his doubts about this synonymy in a paper in which he described *P. javanica* as a new species differing from *P. hortai* in having thick-walled ascospores with rounded ends in contrast to the thin-walled ascospores with strongly acuminate ends of the latter species. The renewed study of sporulating material, which Boedijn regarded as necessary for the solution of the problem, has never been carried out. *Piedraia* usually grows on and not inside the hair. Occasionally, however, the cortex and medulla of the hair are invaded, resulting in hair destruction [26]. This was also observed by van Uden et al., who described *P. quintanilhai* as a species with ascospores without appendages, found on but also in the hairs of monkeys, squirrels, genets, and civets from Central Africa [51]. Since it has become known that the asci of the Piedraiaceae are bitunicate [2], there is no longer an objection to their classification in the subclass Ascoloculomycetidae.

The Piedraiaceae have been classified in the following orders: Moniliales [10], Plectascales [52], Endomycetales [16], Microthyriales [32,33], Myriangiales [11], and Dothideales [5]. The Dothideales are here considered as the most appropriate order for accommodating this family.

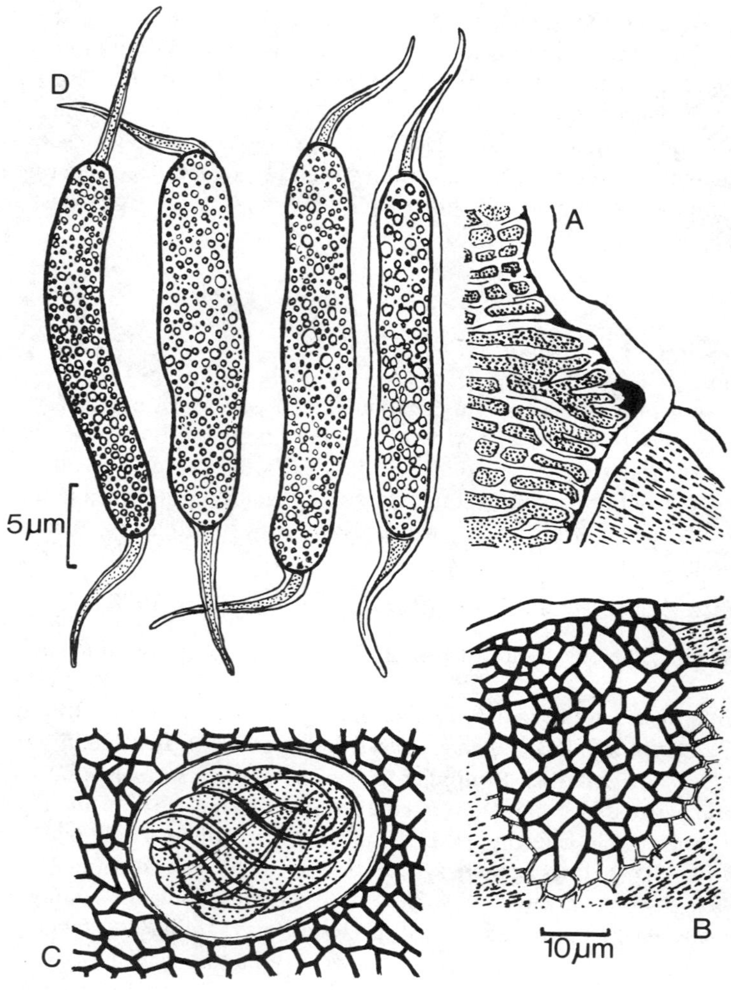

Figure 16 *Piedraia hortai* (Brumpt) da Fonseca & Leão: (A) border of subcuticular zone of a mature ascostroma; (B) young ascostroma on human hair; (C) mature ascus; (D) ascospores.

a. Piedraia hortai (Brumpt) da Fonseca & Leão, loc. cit., 1928

Trichosporum hortai Brumpt, Précis de parasitologie, Masson, Paris, p. 951, 1913

Ascocarps (ascostromata) in vivo very variable in size and shape, black, rounded-elongate, with dotted or irregular surface, up to 1000 μm long and up to 330 μm high, with a thin subcuticular margin, usually surrounding the hair shaft, composed of vertical rows of thick-walled, dark brown, polygonal cells, 2-10 × 2.5-6.5 μm; loculi irregularly distributed throughout the ascostroma, but mostly found at about the same level in the upper part, globose or ovate, 30-60 μm, without a proper wall, opening to the surface by an inconspicuous, ostiolar pore, containing one, rarely more asci; *asci* ellipsoid or obovate, 30-60

μm, with an evanescent wall, 2 to 8-spored; *ascospores* fusiform or botuliform, curved, hyaline, with one or very rarely more than one appendage of variable (4-30 μm) length at one or both ends, 30-47 X 5.5-10 μm (without appendages).

REFERENCES

1. Almeida, F. de. 1954. Formas pequeñas de *Paracoccidioides brasiliensis, Blastomyces dermatitidis* e *Histoplasma capsulatum* nos tecidos. An. Fac. Med. Univ. São Paulo *28*: 141-148.
2. Arx, J. A. von. 1963. Die Gattungen der Myriangiales. Persoonia *2*: 421-475.
3. Arx, J. A. von. 1973. The genera *Petriellidium* and *Pithoascus* (Microascaceae). Persoonia *7*: 367-375.
4. Arx, J. A. von. 1976. *Pilzkunde,* 3rd ed., Lubrecht and Cramer, Vaduz, pp. 91-92.

4a. Arx, J. A. von. 1978. Notes on Microascaceae with the description of two new species. Persoonia *10*: 23-31.

5. Arx, J. A. von, and E. Müller. 1975. A re-evaluation of the bi-tunicate ascomycetes with keys to families and genera. *Studies in Mycology,* Vol. 9, Centraalbureau voor Schimmelcultures, Baarn, p. 76.
6. Austwick, P. K. C., and J. A. J. Venn. 1962. Mycotic abortion in England and Wales 1954-1960. Proceedings of the Fourth International Congress on Animal Reproduction, The Hague.
7. Barron, G. L., R. F. Cain, and J. C. Gilman. 1961. The genus *Microascus.* Can. J. Bot. *39*: 1609-1631.
8. Barron, G. L., R. F. Cain, and J. C. Gilman. 1961. A revision of the genus *Petriella.* Can. J. Bot. *39*: 837-845.
9. Boedijn, K. B. 1959. On black and white piedra. Mycopathologia *11*: 354-358.
10. Brumpt, E. 1913. *Précis de parasitologie,* 2nd ed., Masson et Cie, Paris, pp. 949, 951.
11. Ciferri, R., A. Chaves Batista, and S. Campos. 1956. Taxonomy of *Piedraia hortai* and systematic position of the Piedraiaceae family. Inst. Micol. Bras. Publ. *45*: 1-9.
12. Conant, N. F., D. T. Smith, R. D. Baker, and J. L. Callaway. 1971. *Manual of Clinical Mycology,* 3rd ed., Saunders, Philadelphia, p. 469.
13. Christensen, M., and K. B. Raper. 1978. Synoptic key to *Aspergillus nidulans* group species and related *Emericella* species. Trans. Br. Mycol. Soc. *71*: 177-191.
14. Denton, J. F., E. S. McDonough, L. Ajello, and R. J. Auscherman. 1961. Isolation of *Blastomyces dermatitidis* from soil. Science *133*: 1126-1127.
15. Doby, J. M., and M. Kombila-Favry. 1978. Présence de formes sexuées (cleistothèces et Hülle cells) dans un cas humain d'Aspergillose du sinus maxillaire chez *Aspergillus nidulans* associé à *Aspergillus fumigatus.* Mycopathologia *64*: 157-163.
16. Dodge, C. W. 1935. *Medical Mycology,* Mosby, St. Louis, pp. 131-134.
17. Drake, C. H. 1947. The pathogenicity of *Aspergillus nidulans.* Mycopathologia *4*: 103-119.
18. El-Ani, A. S., and M. A. Gordon. 1965. The ascospore sheath and taxonomy of *Leptosphaeria senegalensis.* Mycologia *57*: 275-278.
19. Emmons, C. W. 1944. *Allescheria boydii* and *Monosporium apiospermum.* Mycologia *36*: 188-193.
20. Emmons, C. W., C. H. Binford, J. P. Utz, and K. J. Kwon-Chung. 1977. *Medical Mycology,* 3rd ed., Lea & Febiger, Philadelphia, pp. 356, 460.
21. Gerber, J., J. Chomicki, J. W. Braunsberg, R. Jones, and K. J. Hammerman. 1973. Pulmonary aspergillosis caused by *Aspergillus fischeri* var. *spinosus.* Am. J. Clin. Pathol. *60*: 861-866.

22. Glick, A. D., and K. J. Kwon-Chung. 1973. Ultrastructural comparison of coils and ascospores of *Emmonsiella capsulata* and *Ajellomyces dermatitidis*. Mycologia *65*: 216-220.
23. Hawksworth, D. L. 1979. Ascospore sculpturing and generic concepts in the Testudinaceae (syn. Zopfiaceae). Can. J. Bot. *57*: 91-99.
24. Hawksworth, D. L., and C. Booth. 1974. A revision of the genus *Zopfia* Rabenh. Commonw. Mycol. Inst. Mycol. Pap. *135*: 1-38.
25. Hecht, R., and J. Z. Montgomerie. 1978. Maxillary sinus infection with *Allescheria boydii (Petriellidium boydii)*. Johns Hopkins Med. J. *142*: 107-109.
26. Kaplan, W. 1959. The occurrence of black piedra in primate pelts. Trop. Geogr. Med. *11*: 115-126.
27. Kwon-Chung, K. J. 1971. Genetic analysis on the incompatibility system of *Ajellomyces dermatitidis*. Sabouraudia *9*: 231-238.
28. Kwon-Chung, K. J. 1973. Studies in *Emmonsiella capsulata*. I. Heterothallism and development of the ascocarp. Mycologia *65*: 109-121.
29. Kwon-Chung, K. J., and S. J. Kim. 1974. A second heterothallic *Aspergillus*. Mycologia *66*: 628-638.
30. Kwon-Chung, K. J., R. J. Weeks, and H. H. Larsh. 1974. Studies in *Emmonsiella capsulata (Histoplasma capsulatum)*. II. Distribution of the two mating types in 13 endemic states of the United States. Am. J. Epidemiol. *99*: 44-49.
31. Kwon-Chung, K. J. 1975. Perfect state *(Emmonsiella capsulata)* of the fungus causing large-form African histoplasmosis. Mycologia *67*: 980-990.
32. Langeron, M. 1929. Les Asterinées parasites de l'homme. La Piedra. Ann. Parasitol. Hum. Comp. *7*: 309-324.
33. Langeron, M. 1945. *Précis de mycologie*. Masson et Cie, Paris.
34. Luttrell, E. S. 1951. Taxonomy of the Pyrenomycetes. Univ. Mo. Stud. *24*(3): 1-121.
35. Malloch, D. 1970. New concepts in the Microascaceae illustrated by two new species. Mycologia *62*: 727-740.
36. Malloch, D., and R. F. Cain. 1972. The Trichocomataceae: Ascomycetes with *Aspergillus, Paecilomyces* and *Penicillium* imperfect states. Can. J. Bot. *50*: 2613-2628.
37. McDonough, E. S., and A. L. Lewis. 1968. The ascigerous stage of *Blastomyces dermatitidis*. Mycologia *60*: 76-83.
38. McDonough, E. S., W. J. McNamara, D. M. Chan, and M. Wallenfang. 1973. Geographic distribution of "+" and "–" isolates of *Blastomyces (Ajellomyces) dermatitidis* in North America. Am. J. Epidemiol. *98*: 63-67.
39. McGinnis, M. R., and B. Katz. 1979. *Ajellomyces* and its synonym *Emmonsiella*. Mycotaxon *8*: 157-164.
40. Poelma, N. F., G. A. de Vries, E. A. Blythe-Russell, and M. H. F. Luykx. 1974. Lobomycosis in an Atlantic bottle-nosed dolphin in the Dolphinarium Harderwijk. Aquat. Mamm. *2*: 11-15.
41. Raper, K. B., and D. I. Fennell. 1965. *The Genus Aspergillus*, Williams & Wilkins, Baltimore, pp. 491-542.
42. Redmond, A., I. J. Carré, J. D. Biggart, and D. W. R. Mackenzie. 1965. Aspergillosis *(Aspergillus nidulans)* involving bone. J. Pathol. Bacteriol. *89*: 391-395.
43. Rippon, J. W., D. N. Anderson, S. Jacobsohn, M. Soo Hoo, and E. D. Garber. 1977. Blastomycosis: specificity of antigens reflecting the mating types of *Ajellomyces dermatitidis*. Mycopathologia *60*: 65-72.
44. Romano, A. H. 1966. Dimorphism. In G. C. Ainsworth and A. S. Sussman (Eds.), *The Fungi. An Advanced Treatise*, Vol. 2, Academic, New York, pp. 181-209.

45. Segretain, G., M. Andre, H. Sarrat, and P. Destombes. 1974. *Leptosphaeria tompkinsii,* agent de mycétomes au Sénégal. Bull. Soc. Fr. Mycol. Med. *3*: 71-74.
46. Sigler, L., and J. W. Carmichael. 1976. Taxonomy of *Malbranchea* and some other Hyphomycetes with arthroconidia. Mycotaxon *4*: 349-488.
47. Swenberg, J. A., A. Koestner, and R. P. Tewari. 1969. Experimental mycotic encephalitis. Acta Neuropathol. Berl. *13*: 75-90.
48. Tewari, R. P., and C. R. Macpherson. 1968. Pathogenicity and neurologic effects of *Oidiodendron kalrai* for mice. J. Bacteriol. *95*: 1130-1139.
49. Tewari, R. P., and C. R. Macpherson. 1971. A new dimorphic fungus, *Oidiodendron kalrai:* morphological and biochemical characteristics. Mycologia *63*: 602-611.
50. Udagawa, S. 1962. *Microascus* species new to the mycoflora of Japan. J. Gen. Appl. Microbiol. *8*: 39-51.
51. Uden, N. van, A. de Barros-Machado, and R. Costelo-Branco. 1963. On black piedra in central African mammals caused by the ascomycete *Piedraia quintanilhai* nov. sp. Rev. Biol. *3*: 271-276.
52. Viegas, A. P. 1943. *Piedraia hortai* (Brumpt) Fonseca & Area Leão. Bragantia *3*: 37-40.
53. Welsh, R. A., and J. M. Buchness. 1955. *Aspergillus* endocarditis, myocarditis and lung abscesses. Report of a case. Am. J. Clin. Pathol. *25*: 782-786.
54. Zach, F. 1934. Untersuchungen über einige neuen Arten der Gattung *Scopulariopsis* Bainier. Österr. Bot. Z. *83*: 173-186.

6

Ascomycetes: The Dermatophytes

Dexter H. Howard / UCLA School of Medicine, Los Angeles, California

I. INTRODUCTION

The term dermatophyte is used to designate a large, closely related group of keratinophilic fungi that cause infections of the skin, the hair, and the nails. Most of these fungi were originally described as Hyphomycetes but several are now known to have perfect states (teleomorphs) in the family Gymnoascaceae of the order Eurotiales. Nevertheless, in keeping with custom, the dermatophytes are being treated separately in this book.

The infections produced by the dermatophytes, called dermatophytoses (syn. tinea, ringworm), are generally restricted to the nonliving, cornified layers of the skin and its appendages [29,76], but occasionally the dermis and the subcutaneous tissue may be involved in a form of the disease known as Majocchi's granuloma, and mycetomas have been described [105]. Although the penetration of tissue by the dermatophytes is usually quite superficial, adsorption of products from the fungi leads to sensitization of the host, an event that is manifested by specific delayed hypersensitivity and other sorts of allergic responses.

II. HISTORICAL ANTECEDENTS

The development of knowledge about the ringworm fungi parallels that of medical mycology in general, for these microorganisms were among the first to be identified as human and animal pathogens [6]. The story, with variations, has been told on several occasions [1,6,47]. Briefly, recognition of the dermatophytes began in 1837 with the work of Robert Remak, who found arthroconidia and hyphae in the hair shafts of a child with favus. Remak did not publish his findings at this time but allowed them to be incorporated into the thesis of a close friend of his, Xavier Hube [1,6]. It has been said that Remak did not recognize the importance of his discovery [1] and that it remained for Lucas Schoenlein to claim favus as a mycotic disease [85]. In doing this Schoenlein was strongly influenced by Bassi's classic work on the fungal nature of the muscardine disease of silkworms [1,14]. In 1840, Remak published his observations on the "plantlike nature" of the structures he had observed in the hair shafts of patients with favus [72], and he subsequently showed that the infection was contagious by successfully inoculating himself [73]. In 1845, he described the fungus of favus more completely and named it in honor of Schoenlein [1,74]. Some have interpreted this as an effort to curry favor with Schoenlein in order to obtain a university chair, an appointment which at that time was generally denied to Jews [12]. If this were the motivation for the somewhat unusual behavior of Robert Remak, the effort was not totally successful, for he was appointed "extraordinary professor" (but not full professor, which he wanted) at the University of Berlin. To his credit he eschewed the obvious alternative of a certificate of baptism to achieve his goal [1].

The Hungarian-born eccentric David Gruby also refused to buy a university chair with a certificate of baptism [1]. Gruby published his thesis in Vienna and then settled down in Paris. From 1841 to 1844 he published a series of papers which are generally acknowledged to establish him as the founder of medical mycology [6,17,47]. In the first two papers of this series Gruby described, independently of Remak and Schoenlein, the fungal nature of favus [38,39]. The next year he gave an account of an ectothrix dermatophytosis of the beard [40], and although he did not name the fungus, it was subsequently described as *Microsporon (Trichophyton) mentagrophytes* by Charles Robin [80]. In 1843, Gruby [42] published his account of classical epidemic ringworm of the scalp and named the etiologic agent *Microsporum audouinii* in honor of Victor Audouin, director of the Paris Natural History Museum. His final published work on dermatophytes was one in which he established the etiologic relation of *T. tonsurans* Malmsten, 1848 to endothrix dermatophytosis [43]. Although extraneous to the subject of this chapter, it should be noted that Gruby also described the fungal nature of thrush [41]. Interestingly, Gruby made no further mycological observations after 1844, but concentrated on his medical practice in Paris and attended many notables, including George Sand, Alexander Dumas (both father and son), Frederic Chopin, Franz Liszt, and Heinrich Heine [54].

In the middle of the nineteenth century there arose a concept which caused considerable confusion for a number of years. The work of Anton De Bary [25] and that of the Tulasne brothers [96] had established the fact of polymorphism among certain fungi, notably the parasites of cereals. The term polymorphism denoted that one fungus could appear in different forms in the various stages of its life cycle. This concept came to be widely and often inappropriately applied with the disastrous consequences that all fungi and bacteria were thought to be much the same and simply developmental stages of one another [55].

The result of this extension of the concept of polymorphism to those fungi in which it played no role for a time obscured the specific etiology of infectious diseases [55]. The one person who was important in rescuing the idea of the specificity of cause in the mycoses was Raimond Sabouraud, who insisted that there were different species of dermatophytes, which he placed in four genera: *Achorion, Epidermophyton, Microsporum,* and *Trichophyton* [83]. However, in spite of the firm foundation laid by Sabouraud, the literature on dermatophytes subsequently became confused by investigators who went too far in the opposite direction and based species identification on small variations in clinical appearance of lesions or on slight differences in colonial morphology. The situation became so muddled that in 1935 some 118 species of dermatophytes were described by C. W. Dodge in his monographic consideration of medical mycology [27].

Emmons proposed the basis for the current taxonomic treatment of the dermatophytes [28]. He adopted a rigorous approach, based on accepted rules of nomenclature, to the classification of the ringworm fungi and established three genera *(Microsporum, Trichophyton,* and *Epidermophyton)* to contain all the species considered legitimate. Emmons's proposal is widely accepted as valid [56], although alternative schemes are suggested from time to time [97].

The existence of a sexual phase of growth among certain dermatophytes was indicated by the observations of a number of early workers (see Ref. 89 for a review of the literature). Nannizzi [59] described cleistothecia (gymnothecia) in a strain of *Sabouraudites (Microsporum) gypseum.* He named the isolate *Gymnoascus gypseus.* Nannizzi's work was generally discredited because it was thought that he dealt with contaminated substrates. However, his observations were proven to be correct by Griffin [37] and by Stockdale [89] and the existence of a sexual form of growth was also demonstrated for species in the genus *Trichophyton* by the results of studies by Dawson and Gentles [23, 24].

The rediscovery of a sexual phase of growth among the dermatophytes led to a flurry of taxonomic and genetic studies. Few other groups of medically important fungi have been subjected to review as often as have the dermatophytes [53]. Accordingly, in this chapter the major emphasis is placed on directing the reader to appropriate reviews in which is presented more comprehensive coverage of a given topic.

III. DERMATOPHYTE MEMBERS OF THE FAMILY GYMNOASCACEAE

A. General Considerations

1. Definitions

The dermatophytes comprise a large, closely related group of fungi that cause infections of the hair, skin, and nails [29,76]. The word tinea is used to designate these infections, known colloquially as ringworm, and an adjective is appended to that word to indicate the primary area of involvement. For example, tinea capitis is ringworm of the scalp, tinea barbae is ringworm of the beard, tinea corporis is ringworm of the general body surfaces, tinea cruris is ringworm of the groin, tinea pedis is ringworm of the feet, and tinea unguium is ringworm of the nails [15]. The diseases may also be termed collectively dermatophytoses [56].

The species of dermatophytes covered in this chapter are listed in Table 1.

2. Identification

The dermatophytes are identified primarily on the basis of their appearance in materials from the host and in culture.

Table 1 The Dermatophytes

Anamorphs	Teleomorphs
Epidermophyton	Unknown
E. floccosum	Unknown
E. stockdaleae	Unknown
Microsporum	*Nannizzia*
M. amazonicum	*N. borellii*
M. audouinii	Unknown
M. boullardii	Unknown
M. canis	*N. otae*
M. cookei	*N. cajetani*
M. distortum	Unknown
M. equinum	Unknown
M. ferrugineum	Unknown
M. fulvum	*N. fulva*
M. gallinae	Unknown
M. gypseum	*N. gypsea, N. incurvata*
M. magellanicum	Unknown
M. nanum	*N. obtusa*
M. persicolor	*N. persicolor*
M. praecox	Unknown
M. racemosum	*N. racemosa*
M. ripariae	Unknown
M. vanbreuseghemii	*N. grubyia*
Trichophyton	*Arthroderma*
T. ajelloi	*A. uncinatum*
T. concentricum	Unknown
T. equinum	Unknown
T. flavescens	*A. flavescens*
T. georgiae	*A. ciferrii*
T. gloriae	*A. gloriae*
T. gourvilii	Unknown
T. longifusum	Unknown
T. megninii	Unknown
T. mentagrophytes	*A. benhamiae, A. vanbreuseghemii*
T. phaseoliforme	Unknown
T. rubrum	Unknown
T. schoenleinii	Unknown
T. simii	*A. simii*
T. soudanense	Unknown
T. terrestre	*A. insingulare, A. quadrifidum* *A. lenticularum*
T. tonsurans	Unknown
T. vanbreuseghemii	*A. gertleri*
T. verrucosum	Unknown
T. violaceum	Unknown
T. yaoundei	Unknown

a. The Appearance of Dermatophytes in the Tissues of the Host

Direct examination of materials from a host infected with a dermatophyte not infrequently reveals the fungus. Specimens of skin, nails, and hair from suspicious lesions are placed on a slide in a drop of 10-20% KOH. A coverslip is placed over the drop and the preparation examined microscopically. The various species of dermatophytes look alike in skin scales and nail scrapings [29]. The appearance is that of septate, branching hyphae which may break up into barrel-shaped or rounded arthroconidia [29]. Invasion of the hair may be either of the ectothrix or endothrix sort [29]. Details of the appearance of different species of dermatophytes in and on the surface of hair will be given with the description of individual species.

b. Wood's Light

A Wood's lamp emits ultraviolet light at approximately 3650 Å. Hairs invaded by some species of *Microsporum* emit a greenish yellow fluorescence. Hairs involved by most species of *Trichophyton* do not fluoresce but *T. schoenleinii*-infected hairs emit a dull greenish fluorescence. *Epidermophyton* spp. do not invade hair. Infected skin scales and nails do not fluoresce.

c. Cultures

Specimens from lesions suspected of being tinea are cultured. Several recipes for suitable media are available [48]. Some variation of glucose peptone agar with inhibitors is generally employed. The colonial appearance will form a part of the description to be given for each species considered.

Generally, the dermatophytes are identified on the basis of their colonial and microscopic morphology, and this will be the approach taken in the description of taxonomy to follow. However, physiological tests may also be of great value in their identification. For example, it was recognized early on that the inability of *M. audouinii* to grow on rice grains served to distinguish it from *M. canis* and *M. gypseum* [33]. The classic work of Lucille Georg and her colleagues led eventually to the development of a series of nutritional tests which could be used to augment species recognition among the *Trichophyton* spp. [15,33,48]. Efforts to extend this sort of approach to include patterns of assimilation of carbon and nitrogen sources and to expand the approach to identification of all dermatophytes have been made (e.g., Refs. 67 and 87).

Other endeavors to augment the morphological approach to dermatophyte taxonomy have included: (1) comparative observations of proteins by disc electrophoresis [4,8,84,86]; (2) description of antigenic differences among strains [69]; (3) study of DNA homologies among isolates [22]; (4) analysis by gas-liquid chromatography [18]; and (5) the ability to perforate hair in vitro [64a].

3. Epidemiology

The natural occurrence of dermatophytes varies among the species. Typical associations are generally recognized, although it is not always easy to describe a species as belonging to only one category [29,76]. *Anthropophilic* species are obligate parasites of humans and commonly show reduced sporulation. *Zoophilic* species occasionally infect humans but are predominantly found on other animals (cattle, cats, dogs, etc). *Geophilic* species commonly found in soil are of limited pathogenicity to humans and animal, and mostly have abundant sporulation [15]. Among the last are species of *Microsporum, Epidermophyton,* and *Trichophyton* which have not been found to cause disease. Their morpho-

Table 2 Geographic Distribution of Dermatophytes

Anthropophilic	Zoophilic	Geophilic
	Cosmopolitan	
E. floccosum	*M. canis*	*M. gypseum*
M. audouinii	*M. gallinae*	*M. cookei*
T. mentagrophytes[a]	*T. mentagrophytes*	*T. terrestre*
T. rubrum	*T. verrucosum*	*T. ajelloi*
T. tonsurans		*M. fulvum*
T. violaceum		*M. nanum*
M. ferrugineum		*M. persicolor*
		T. vanbreuseghemii
	Limited	
T. gourvilii	*M. distortum*	*E. stockdaleae*
T. concentricum	*T. equinum*	*M. vanbreuseghemii*
T. megninii	*M. amazonicum*	*T. flavescens*
T. soudanense		*T. georgiae*
T. yaoundei		*T. gloriae*
T. schoenleinii		*T. simii*
M. praecox		*M. boullardii*
		T. longifusum
		M. racemosum
		T. phaseoliforme
		M. magellanicum
		M. ripariae

[a]See Note Added in Proof, p. 142.

logical similarity to pathogens, however, has led to inclusion in this chapter (Tables 2 and 3).

Some of the dermatophytes have a very restricted geographic distribution, whereas others are cosmopolitan. Examples of each are recorded in Table 2. An excellent review of this topic has appeared recently [66,68].

4. Taxonomy

Prior to 1961 the taxonomy of the dermatophytes was based on the anamophic forms. The practical needs of the clinical laboratory also involve observations on this stage for identification. Thus, the customary arrangement of a discussion of the dermatophytes is based on the conidia of the asexual phase of growth even when comprehensive coverage is also given the teleomorphic forms [7,9,15,71].

In 1934, Emmons proposed that the dermatophytes be classified in three genera: *Epidermophyton, Microsporum,* and *Trichophyton* [28]. These genera were distinguished by the morphology of the large, multiseptate macroconidia which each produced. A second type of conidia, the small unicellular, spherical, oval, or pyriform microconidia was used to some degree for differentiation of species. The scheme was widely adopted and remains the dominant one in the field today [3,15,28,71].

The morphological features that distinguish the three genera are shown in Fig. 1. The macroconidia of *Epidermophyton* are smooth-walled, clavate, and borne in clusters of

Table 3 Arrangements of Dermatophytes

I. Pathogenic Species of Dermatophytes
- A. Teleomorph (anamorph)
  - 1. *Arthroderma (Trichophyton)*
    - *A. benhamiae (Trichophyton mentagrophytes)*
    - *A. gertleri (T. vanbreuseghemii)*
    - *A. simii (T. simii)*
    - *A. uncinatum (T. ajelloi)*
    - *A. vanbreuseghemii (T. mentagrophytes)*
  - 2. *Nannizzia (Microsporum)*
    - *N. borellii (Microsporum amazonicum)*[a]
    - *N. cajetani (M. cookei)*
    - *N. fulva (M. fulvum)*
    - *N. grubyia (M. vanbreuseghemii)*
    - *N. gypsea (M. gypseum)*
    - *N. incurvata (M. gypseum)*
    - *N. obtusa (M. nanum)*
    - *N. otae (M. canis)*
    - *N. persicolor (M. persicolor)*
    - *N. racemosa (M. racemosum)*
- B. Anamorphs—no teleomorphs yet described
  - 1. *Epidermophyton*
    - *E. floccosum*
  - 2. *Microsporum*
    - *M. audouinii*
    - *M. distortum*
    - *M. equinum*
    - *M. ferrugineum*
    - *M. gallinae*
    - *M. praecox*
  - 3. *Trichophyton*
    - *T. concentricum*
    - *T. equinum*
    - *T. gourvilii*
    - *T. megninii*
    - *T. rubrum*
    - *T. schoenleinii*
    - *T. soudanense*
    - *T. tonsurans*
    - *T. verrucosum*
    - *T. violaceum*
    - *T. yaoundei*

II. Saprophytic Species of Dermatophytes
- A. Teleomorph (anamorph)
  - 1. *Arthroderma (Trichophyton)*
    - *A. ciferrii (T. georgiae)*
    - *A. gloriae (T. gloriae)*

(continued)

Table 3 (continued)

1. *Arthroderma (Trichophyton)* (continued)
A. flavescens (T. flavescens)
A. insingulare (T. terrestre)
A. lenticularum (T. terrestre)
A. quadrifidum (T. terrestre)
B. Anamorphs—no teleomorphs yet described
1. *Microsporum*
M. boullardii
M. magellanicum
M. ripariae
2. *Epidermophyton*
E. stockdaleae
3. *Trichophyton*
T. longifusum
T. phaseoliforme

[a]The pathogenicity of this species is not established (see the text).

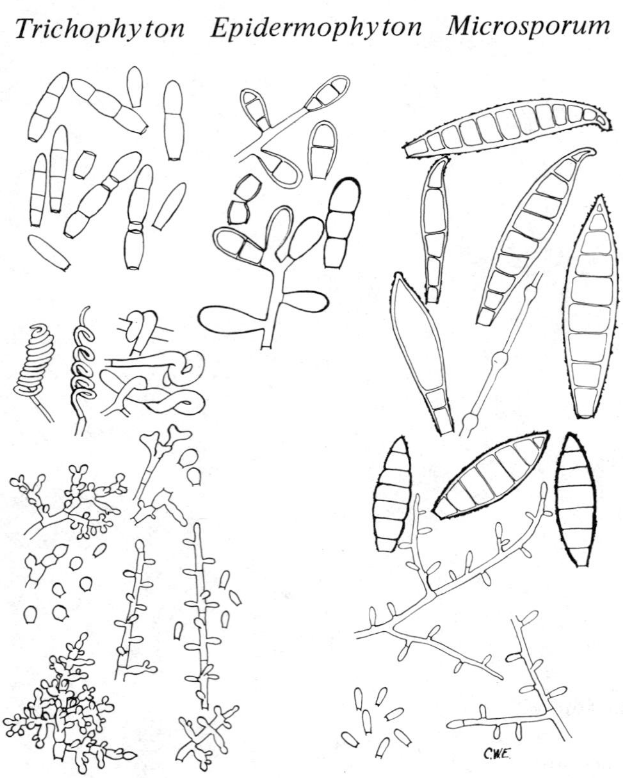

Figure 1 Spore types in the three genera of dermatophytes. (From Ref. 28, C. W. Emmons, Dermatophytes, *Arch. Dermatol. Syphilol. 30*: 337, © 1934, American Medical Association.)

two or three; those of *Microsporum* are rough-walled, mostly spindle-shaped, and usually borne singly; and those of *Trichophyton* are smooth-walled, mostly cylindrical, and borne singly.

The current list of authenticated species dates from that originally prepared by Ajello in 1968 [4] and includes those listed in Table 1. There is reasonably good general agreement about the list [8,15,71], but disagreements may be raised about the authenticity of a few of those discussed in this chapter and perhaps some divergence of opinion about some not herein included. I have tried to evaluate such points under the discussion section of each taxon described.

A taxonomic scheme based on the teleomorph stage of the dermatophytes became possible with the published descriptions by Stockdale [89] and by Dawson and Gentles [23,24]. The mating behavior and genetic studies of the dermatophytes have served to clarify the taxonomic status of these fungi [99]. A study of sexual phase of growth of zoopathogenic fungi is not usually essential to their identification in a clinical laboratory. Thus, most practical discussions of the classification of the dermatophytes are based on features of the anamorphs [15,48]. In keeping with the treatment in this volume of other gymnoascaceous fungi (Chap. 5), I will develop the following discussion on the basis of the teleomorph stage of the dermatophytes. Only certain interesting or controversial synonyms have been cited. More comprehensive lists of synonomy are available in standard texts [21,29,76]. A summary of the arrangement of the dermatophytes as developed in the remainder of the chapter is shown in Table 3.

B. Pathogenic Species

1. Introduction

As pointed out above, the morphology of the asexual phase of growth of the dermatophytes suggested to early investigators that these fungi were closely related to the Gymnoascaceae. The rediscovery of the sexual phase of growth of some of them confirmed this early assumption, and it is now accepted that the dermatophytes are Ascomycetes in the family Gymnoascaceae of the order Eurotiales. Within the Gymnoascaceae the two genera *Ajellomyces* McDonough & Lewis and *Emmonsiella* Kwon-Chung are distinguished from the other pathogenic, spiral-forming genera, *Arthroderma* Berk. and *Nannizzia* Stockdale, by the fact that the spirals of the ascocarps (gymnothecia, cleistothecia) of the former do not arise from the peridial hyphae but from the ascogonium or the hyphae encircling the ascogonium (see Chap. 5).

2. *Arthroderma* Berkeley, *Outlines of British Fungology,* London, p. 357, 1860

Gymnothecia small, whitish to pale yellow to buff. *Peridial hyphae* repeatedly branched to produce a densely packed network, straight or curved, hyaline, septate. *Cells* symmetrically dumbbell-shaped, densely asperulate except in constricted portion. *Appendages* long, slender, thin-walled, smooth, septate, spiral hyphae terminal on the peridial branches. Some species produce slender, straight, septate appendages in addition to the spirals. *Asci* globose to subglobose, evanescent, 8-spored. *Ascospores* small, smooth, oblate, 2.5-3.5 × 1.5-2.5 μm, yellowish in mass. Type species: *Arthroderma curreyi* Berk. 1860.

a. Arthroderma benhamiae Ajello & Cheng, Sabouraudia, 5: 232, 1967

(1) Conidial State

Trichophyton mentagrophytes (Robin) Blanchard, *sensu lato, Traité de pathologie générale,* Vol. 2, Masson et Cie, p. 811, Paris, 1896

Distinctive varieties:

Trichophyton interdigitale Priestley, *Med. J. Aust. 4*: 417-475, 1917

Trichophyton quinckeanum (Zopf) MacLeod & Muende, *Practical Handbook of the Pathology of the Skin,* 2nd ed., H. K. Lewis & Co. Ltd., London, p. 383, 1940

Trichophyton erinacei (Smith & Marples) Padhye & Carmichael, *Sabouraudia 7*: 178-181, 1969

Trichophyton proliferans English & Stockdale, *Sabouraudia 6*: 267-270, 1961 (see Ref. 81).

(2) Description

Heterothallic. Gymnothecia globose, pale white, 250-450 μm in diameter, excluding appendages. *Peridial hyphae* hyaline, septate, dichotomously branched, thin-walled. *Cells* dumbbell-shaped, echinulate, asymmetrically constricted, 8-12 X 4.5-5.2 μm. *Appendages* of two sorts: (1) elongated, smooth-walled hyphae tapering at the apex, 60-200 μm long; (2) smooth-walled, spiral hyphae. *Asci* globose to ovate, 4.2-7.2 X 3.6-6 μm, thin-walled, evanescent, 8-spored. *Ascospores* hyaline, smooth, oblate, 1.5-1.8 X 2.5-2.8 μm, yellow in mass.

Colonies develop rapidly on glucose peptone agar; white, floccose to granular, yellowish to peach-colored. Reverse pale yellow to brown to reddish brown. *Microconidia* clavate, borne laterally on undifferentiated hyphae in floccose strains or nearly spherical on conidiophores, forming clusters in granular strains. *Macroconidia* rare or abundant in granular forms, clavate, smooth-walled, 20-50 X 6-8 μm, with 3-5 septa. Spiral hyphae and nodular organs are frequently present.

Discussion

This species is a common cause of tinea pedis, tinea corporis, and tinea unguium. It may also cause tinea barbae and tinea capitis. The hairs show large-spored ectothrix involvement. *Arthroderma benhamiae* also infects cattle, horses, dogs, cats, sheep, pigs, rabbits, squirrels, monkeys, chinchillas, silver foxes, laboratory rats, mice, and other animals [15]. The fungus which is worldwide in distribution has also been isolated from soil [15,62]. The microconidia of highly granular cultures are infectious and laboratory-acquired infections may occur (D. H. Howard, unpublished observation).

The discovery of the sexual phase of growth of *T. mentagrophytes* has helped to sort out some of the confusion about the status of species and varieties of the genus which had been based on morphological variations. For example, the organism causing mouse favus, *T. quinckeanum,* was shown to be conspecific with *T. mentagrophytes* by Ajello et al. [11], and many isolates of *T. interdigitale* are compatible with tester strains of *A. benhamiae* [60]. Weitzman [99] has summarized the work that eventually led to recognition of *T. erinacei* as a variety of *T. mentagrophytes* and she emphasized that a large number of tester strains of diverse origin may be necessary in order to reveal true relationships. Weitzman et al. [104] made a detailed cytological study of *A. benhamiae* and have determined the chromosome number to be 4 [102]. Some investigators maintain varietal status for especially distinctive anamorphs of *A. benhamiae* [30,32,81] but reducing some of them to synonymy is also recommended [15,62].

Among the dermatophytes *A. benhamiae* is well suited to genetic studies because: (1) under properly controlled conditions it abundantly produces uninucleate microconidia which are hardy enough for easy manipulation in the laboratory, (2) it is an important human pathogen, (3) it is easily mated, and (4) the ascospores are reasonably easy to manipulate and give a good percentage of viability. Yet except for the aforementioned mating studies which have helped to resolve taxonomic questions, *A. benhamiae* has not been widely used for the development of mutants useful in studies on the basic biology and pathogenesis of disease [45]. Kwon-Chung has critically reviewed early reports of reversion of pleomorphic mutants of dermatophytes and of a putative suppressor mutation in *T. mentagrophytes* [53]. Krammer and Knight [49] studied the virulence for guinea pigs of various auxotrophic mutants of *T. mentagrophytes,* each requiring a single amino acid, vitamin, purine, and pyrimidine. All mutants tested infected guinea pigs and the relative severity of the infections was within normal variation. The nutritional requirements of the isolates were manifest in isolates from the lesions. Rippon [75] and Rippon and Garber [77] reported a difference in extracellular elastase production and virulence between one strain each of the *A* and *a* mating types of *A. benhamiae.* Chu-Cheung and Maniotis showed that the locus controlling elastase segregated independently of the mating type locus but was closely linked to that governing colonial morphology [19]. Regrettably, the authors did not assess virulence in the segregants.

b. Arthroderma gertleri Böhme, Mykosen 10: 251, 1967

(1) Conidial State

Trichophyton vanbreuseghemii Rioux, Jarry & Juminer, *Nat. Monspel. Ser. Bot. 16*: 153-162, 1964.

(2) Description

Heterothallic. Gymnothecia globose, pale yellow, 200-600 μm in diameter, excluding appendages. *Peridial hyphae* hyaline, septate, uncinately branched, curving over the gymnothecium. *Cells* dumbbell-shaped, echinulate, usually asymmetrically but occasionally symmetrically constricted. *Appendages* septate spirals varying in length and number of turns, borne terminally. Multiseptate, thin-walled, smooth, cylindrical macroconidia are sometimes attached to the pyridial hyphae. *Asci* subglobose, thin-walled, evanescent, 4 × 5 μm, 8-spored. *Ascospores* hyaline smooth, oblate, 1.5-2.8 × 2-2.5 μm, yellow in mass.

Colonies flat or with radial grooves, floccose at first, becoming velvety (described as the texture of "fine glove leather" [71] with a finely granular center, white to buff. Reverse creamy white to pale buff. *Microconidia* pyriform, subglobose, ovoid, thin-walled, unicellular, sessile or on short pedicels, borne laterally or terminally, 2-7 × 1.5-2.5 μm. *Macroconidia* numerous, thin-walled, smooth, cylindrical, multiseptate, borne singly, 30-55 × 6-8 μM.

Discussion

This dermatophyte is probably worldwide in distribution and has been isolated from a tinea of the hand [71]. Large-spored ectothrix involvement of guinea pig hair is reported [15]. The fungus has been isolated from soil [62,71].

c. Arthroderma simii Stockdale, Mackenzie & Austwick, Sabouraudia 4: 113, 1965

(1) Conidial State

Trichophyton simii (Pinoy) Stockdale, Mackenzie & Austwick, *Sabouraudia 4*: 114, 1965.

Synonyms:

Epidermophyton simii Pinoy, *C. R. Soc. Biol. (Paris) 72*: 59, 1912

Pinoyella simii (Pinoy) Castellani & Chalmers, *Manual of Tropical Medicine,* 3rd ed., Baillière, Tindall & Cox, London, p. 1023, 1919

(2) Description

Heterothallic. Gymnothecia globose, pale buff to buff, 200-700 μm, thin excluding appendages. *Peridial hyphae* pale buff, septate, with somewhat thin, verruculose walls. Up to three secondary branches may arise from the apex of the same cell of a peridial hypha. Distal branches curve over the gymnothecium, and the ultimate branches are rarely composed of more than two or three cells. *Cells* of the inner parts of the peridial hyphae are 12.5-21 × 3-4.5 μm, not constricted but swelling toward the apex from which the branches arise, 5-7 μm in diameter. *Cells* of the outer branches are asymmetrically constricted, 6.5-12.5 × 4.2-6.7 μm at their widest part. *Appendages* slender, smooth-walled, septate, spiral hyphae. Smooth-walled, cylindrical to fusiform macroconidia are produced by the peridial hyphae. *Asci* subglobose, thin-walled, evanescent, 5-6.7 μm in diameter, 8-spored. *Ascospores* hyaline, smooth, oblate, 2.5-3.5 × 1.5-2.5 μm, yellow in mass.

Colonies velvety with a finely granular uniform surface and a fluffy asteroid margin. The surface is pale buff and the reverse straw to salmon. *Macroconidia* numerous, borne terminally on complex branched hyphae, hyaline, smooth-walled, fusiform, occasionally cylindrofusiform, 30-80 × 6-11 μm, with 4-7 (up to 10) septa. In older cultures the macroconidia become constricted at the septa as a result of an increase in the diameter of individual cells forming intercalary, thick-walled chlamydospores. *Microconidia* rare in young cultures, more abundant in old cultures, clavate to pyriform, 2-6.5 × 1.5-4 μm, sessile or on short pedicels, and borne singly along the sides of the hyphae. Spiral hyphae are moderately abundant on the vegetative mycelium of old cultures.

Discussion

Arthroderma simii is of restricted geographic distribution as most of the cases have been reported in or from India [15,71]; also reported from Brazil and Guinea [60]. The fungus produces tinea in monkeys, poultry, and dogs. Several human cases have been reported and the type of hair involvement is endothrix [15]. The organism also occurs in the soil.

A fascinating attribute of *A. simii* is that it induces a form of stimulated growth of the mycelium of several different species of dermatophytes [92]. This stimulation occurs in strains of mating type opposite that of the tester *A. simii* type. The interaction, of course, never goes on to formation of fertile ascospores within asci except in opposite mating types of *A. simii* itself. But the observation of stimulated growth has been used to presume the juxtaposition of opposite mating types. In some instances isolates of species not known to engage in a sexual form of growth have been identified as being of a single mating type. For example, all isolates of *T. rubrum* were shown in this manner to be the *a* mating type [106]. Such observations strengthen the argument that the reason certain species cannot be induced to reveal a sexual phase of growth is that they are all of one mating type. Thus, if virulence were associated with mating type, one could imagine the evolutionary disappearance of a mating type. This would be especially likely to happen if the environment were highly restricted. It is interesting to reflect on the fact that none of the anthropophilic species has been shown to have a perfect state [99].

The chromosome number of *A. simii* is 4 [102] and the incompatibility system has been subjected to a comprehensive study [52]. At one time it was suggested that *A. simii* was one of the "three" teleomorphs of the *T. mentagrophytes* complex [94]. Although this concept has been reiterated on one recent occasion [44], it would appear that the results of careful studies affirm the distinctness of *T. simii* as the single anamorph of *A. simii* [95, 99, 100]. The relationships are, however, close [99,100].

d. Arthroderma uncinatum Dawson & Gentles, Sabouraudia 1: 55, 1961

(1) Conidial State

Trichophyton ajelloi (Vanbreuseghem) Ajello, *Sabouraudia 6: 148, 1968.*

Synonym:

Keratinomyces ajelloi Vanbreuseghem, *Bull. Acad. R. Med. Belg. 38*: 1076, 1952.

(2) Description

Heterothallic. Gymnothecia globose, pale buff, 300-900 μm in diameter, excluding appendages. *Peridial hyphae* pale yellow, hyaline, septate, uncinately branched usually to the outside of the main hyphae. *Cells* thick-walled, strongly echinulate, symmetrically dumbbell-shaped, 7-11 × 4-7 μm. *Appendages* septate, smooth-walled, spirals which vary considerably in length and number of turns, borne terminally. Sometimes smooth-walled, multiseptate, fusiform macroconidia are produced by peridial hyphae. *Asci* subglobose, thin-walled, evanescent, 5.4-7.2 × 5-6.5 μm, 8-spored. *Ascospores* hyaline, smooth, oblate, 2.5-3.2 × 1.8-3 μm, yellow in mass.

Colonies flat to folded, with a downy to powdery surface, cream to orange tan. Reverse with or without diffusable, vinaceous red pigment which becomes bluish black with age. *Macroconidia* produced in abundance, cylindrical to fusiform, multiseptate (8-12 septa), thick-walled, smooth, 18-60 × 8-12 μm. *Microconidia* present in some strains, pyriform to obovate, 3-9 × 2-5 μm [60].

Discussion

It is to be noted that there are two colonial forms of *T. ajelloi*: one producing a diffusible red pigment, the other not. In the original description of *A. uncinatum* it was recorded that nonpigmented varieties did not mate with pigmented ones [24]. Moreover, the size of the macroconidia of the two forms has been reported to differ [90]. Thus, the suggestion was made that the nonpigmented isolates might represent a distinct taxon [90]. However, Weitzman et al. [103], and Padhye and Carmichael [61,62] examined a larger number of strains, observed fertile crosses between pigmented and nonpigmented isolates, and concluded that the lack of pigment did not denote a separate taxon.

Weitzman and Silva-Hutner [101] studied the genetics of pigmentation in *A. uncinatum* and concluded from the segregation pattern that there was a single gene difference between pigmented and nonpigmented parents. *Arthroderma uncinatum* has a haploid complement of four chromosomes [102]. A variety of *T. ajelloi* with very small macroconidia has been described as *T. ajelloi* var. *nanum* [4,10].

e. Arthroderma vanbreuseghemii Takashio, Ann. Soc. Belge. Med. Trop. 53: 427, 1973

(1) Conidial State

Trichophyton mentagrophytes (Robin) Blanchard, *Traité de pathologie générale,* Vol. 2, Masson et Cie, Paris, p. 811, 1896.

(2) Description

Heterothallic. Gymnothecia are globose, 300-650 μm in diameter, straw- or buff-colored. *Peridial hyphae* are hyaline or slightly buff, septate, interwoven, dichotomously branched, and thin-walled. *Cells* dumbbell-shaped, echinulate, asymmetrically constricted, 8-12 × 4.5-5.2 μm. *Appendages* of two sorts: (1) elongate, smooth-walled hyphae tapering at the apex, 60-200 μm long; (2) smooth-walled spiral hyphae with up to 15 turns. *Asci* globose to ovate, 4.2-7.2 × 3.6-6 μm, thin-walled, evanescent, with 8 ascospores. *Ascospores* are hyaline, smooth, oblate, 1.5-1.8 × 2.5-2.8 μm, yellow in mass.

Colonies described under *A. benhamiae.*

Discussion

This species is the second perfect state of the *T. mentagrophytes* complex (see above regarding thoughts on *A. simii*). The differences are the larger ascospores of *A. vanbreuseghemii* and no interspecific crosses. Mating studies in the complex have been comprehensively recorded by Takashio [93,95]. Variations in the colonial morphology have been presented [95]. The disease pattern and distribution have been described under *A. benhamiae.*

3. *Nannizzia* Stockdale, *Sabouraudia 1*: 45-46, 1961

Gymnothecia globose. *Peridium* consisting of a network of hyaline, septate, verticillately and/or dichotomously branched hyphae. *Cells* moderately thick-walled, densely asperulate, more or less symmetrically constricted. *Peridial hyphae* with numerous free ends. *Appendages* of three kinds: (1) elongate, slender, smooth-walled, septate, occasionally branching, straight or loosely coiled hyphae; (2) elongate, slender, smooth-walled, septate, occasionally branching, tightly coiled hyphae; (3) macroconidia. *Asci* globose to ovate, hyaline, 8-spored. *Ascospores* lenticular, yellow in mass. Type species: *Nannizzia incurvata* Stockdale, 1961.

It should be noted that Stockdale describes the macroconidium as a kind of appendage to the peridial hyphae, whereas Ajello and co-workers prefer not to so consider them [58]. I have chosen to use the descriptions given by the authors cited, usually without major modification.

a. Nannizzia borellii Moraes, Padhye & Ajello, Mycologia 67: 1112, 1975

(1) Conidial State

Microsporum amazonicum Moraes, Borelli & Feo, *Med. Cutanea, 2*: 201-205, 1967.

(2) Description

Heterothallic. Gymnothecia globose, dull white or pale yellow, 300-500 μm in diameter, excluding appendages. *Peridial hyphae* are hyaline, septate, and dichotomously branched. *Cells* are thin-walled, finely echinulate, slightly constricted in the middle, and swollen near the septa. *Appendages* are of two kinds: (1) elongate, septate, smooth-walled, tapering hyphae up to 150 μm in length, 2-3 μm in diameter at the base and 1.2-1.5 μm at the tip; (2) elongate, smooth-walled, slender hyphae coiled into spirals; macroconidia may be at the end of some hyphal appendages. *Asci* are hyaline, subglobose, 4-5 μm in diameter with 8 ascospores. *Ascospores* hyaline, oblate, 2-3 × 2.5 μm in diameter, yellow in mass.

Colonies on glucose peptone agar are fluffy to powdery with a distinctive gray-olive buff color [71]. *Macroconidia* are symmetrical, spindle-shaped, echinulate, thick-walled, with 1-4 septa. *Microconidia* not observed.

Discussion

The fungus was isolated from hair of spiny rats *(Proechimys guyannensis)* and from hair of a rat belonging to the genus *Oryzymys.* Although the organism is associated with animals, it perhaps is a saprophyte since naturally occurring lesions have not been described.

b. Nannizzia cajetani Ajello, Sabouraudia 1: 173-177, 1961

(1) Conidial State

Microsporum cookei Ajello, *Mycologia 51*: 69-70, 1959.

(2) Description

Heterothallic. Gymnothecia globose, pale yellow, 368-686 μm in diameter. *Peridial hyphae* hyaline, septate, verticillately branched. *Cells* echinulate, thin-walled, and slightly constricted at the site of septations. *Appendages* of two kinds: (1) elongate, slender, tapered, smooth hyphae up to 480 μm in length, 3.6 μm in diameter at the base, 2.4 μm at midlength, and 1.2 μm at the tip; (2) elongate, smooth-walled, slender hyphae coiled into spirals. *Asci* globose to ovate, 6-9 μm in diameter. *Ascospores* ovate, smooth-walled, and golden, 3-3.6 × 1.8 μm.

Colonies on glucose peptone agar grow rapidly and are flat and spreading, with a rather powdery surface which is yellowish, buff, or dark tan. The reverse is deep purplish red. *Macroconidia* are oval to ellipsoidal, thick-walled, and echinulate, 31-50 × 10-15 μm, with 6 septa. *Microconidia* obovate, are produced abundantly.

Discussion

This species is geophilic and has a worldwide distribution. The fungus may be isolated from animals but clinical disease is not observed in them [15]. Tinea corporis has been reported in humans [15]. Genetic analysis of the incompatibility system in *N. cajetani* has been accomplished by Kwon-Chung [50].

c. *Nannizzia fulva Stockdale, Sabouraudia 3: 120-212, 1963*

(1) Conidial State

Microsporum fulvum Uriburu, *Argent. Med. 1*: 563-582, 1909.

(2) Description

Heterothallic. Gymnothecia globose, pale buff, 500-1250 μm in diameter, excluding appendages. *Peridial hyphae* are hyaline, septate, branched hyphae with thin, densely verrucose walls. Up to four branches arising in succession at the apex of the same cell; the branches are straight or curved over toward the gymnothecium. *Appendages* of three kinds: (1) straight, slender, smooth-walled, septate hyphae up to 250 μm long, 3-4.5 μm in diameter at the base, tapering to 1.5-2 μm at the tip; (2) slender, smooth-walled, septate, spiral hyphae, not branched or much branched, each branch loosely to tightly coiled with up to 15 turns; (3) moderately thick-walled, verrucose macroconidia, usually cylindrical, tapering toward the apex and the base, or clavate, sometimes ellipsoid or fusiform, 30-50 × 9-12 μm, with up to 5 septa. *Asci* subglobose, evanescent, thin-walled, 5-7 μm in diameter, with 8 ascospores. *Ascospores* smooth-walled, lenticular, 1.5-2 × 2-5.4 μm, yellow in mass.

Colonies on glucose peptone agar are dense, downy to granular, pale buff to rosy buff. The reverse is rosy buff to amber. *Macroconidia* are predominantly cylindrical, slightly tapering toward both ends, and with a rounded apex or clavate, occasionally ellipsoid or fusiform, 25-28 × 5-12 μm, with up to 5 septa and moderately thick, verrucose walls. Whip-like appendages (see *N. gypsea*) rare. Macroconidia borne singly or on branched conidiophores. *Microconidia* 1.7-3.3 × 3.3-8.3 μm, clavate, smooth-walled or slightly rough-walled, sessile or on short pedicels, borne laterally along the hyphae. Spiral hyphae similar to those seen on gymnothecia are formed abundantly on the vegetative mycelium.

Discussion

This species produces tinea corporis and tinea capitis in humans. Hair involvement is of the ectothrix sort but there is no fluorescence under a Wood's lamp. The fungus is also reported from dogs and occasionally from other animals. *Nannizzia fulva* is a geophilic dermatophyte found throughout the world.

Genetic studies on the relationship between the enzyme elastase and the mating type of *N. fulva* have been reviewed by Kwon-Chung [53].

d. Nannizzia grubyia Georg, Ajello, Friedman & Brinkman, Sabouraudia 1: 189-196, 1962

(1) Conidial State

Microsporum vanbreuseghemii Georg, Ajello, Friedman & Brinkman, *Sabouraudia 1*: 189-196, 1962.

(2) Description

Heterothallic. Gymnothecia globose, white to pale buff, 150-600 μm in diameter, exclusive of appendages. *Peridial hyphae* are hyaline, septate, branching dichotomously and uncinately, mostly curved to one side, the outside of the main hypha. *Cells* are moderately thick-walled, densely echinulate, and moderately constricted in the center. *Appendages* terminate bluntly or taper into smooth-walled, loosely coiled hyphae with two to three turns or into elongate, thin, smooth-walled, loosely coiled, septate hyphae tapering from 3.0 μm in diameter at the base to 1.5 μm at the tip, with 30-50 tight coils. *Macroconidia,* thick-walled, cylindrofusiform, multiseptate, and densely echinulate, borne laterally and terminally on the peridial hyphae. *Asci* are globose, thin-walled, evanescent, 4.8-6.0 μm in diameter, with 8 ascospores. *Ascospores* hyaline, pale yellow, smooth-walled, ovate, 2.4 × 3.0 μm, yellow in mass.

Colonies on glucose peptone agar are rapid-growing, flat with powdery to fluffy surface. Surface color white to yellowish; pink to deep rose colonies have also been described [36]. The reverse is light yellow to bright lemon yellow (some strains are colorless). *Macroconidia* are numerous, cylindrofusiform with thick, echinulated walls, and with 5-12 septa. The size range is 58.8-61.7 × 10.4-10.6 μm. *Macroconidia* are numerous, pyriform to obovate, 9.2 × 4.0 μm, borne laterally on the hyphae, sessile, or on pedicels.

Discussion

This species was originally described from tinea lesions in a Malabar squirrel, a dog, and two humans. The fungus has been isolated from soil [15]. Experimental infections in guinea pigs resulted in hyphal elements within the hair shafts and large numbers of spores on the surface of the hair. The hairs did not fluoresce under a Wood's lamp. However, in a natural human infection the hairs did fluoresce [36].

e. Nannizzia gypsea (Nannizzi) Stockdale, Sabouraudia 3: 119, 1963

(1) Conidial State

Microsporum gypseum (Bodin) Guiart & Grigorakis, *sensu lato Lyon Med. 141*: 369-378, 1928.

(2) Description

Heterothallic. Gymnothecia are globose, pale buff, 500-1250 μm in diameter, excluding appendages. *Peridial hyphae* are hyaline, pale buff, septate, branched with thin, densely verruculose walls. Up to four branches arising in succession at the apex of the same cells. *Appendages* are of three kinds: (1) straight, slender, smooth-walled, septate hyphae, up to 250 μm long, 2.5-4.0 μm at the base tapering to 1.5-2.0 μm; (2) slender, smooth-walled, septate, spiral hyphae, rarely branched, 2.5-3.5 μm at the base tapering to 1.5-2 μm, loosely to tightly coiled with a variable number of turns; (3) moderately thick-walled, verrucose, ellipsoid or fusiform macroconidia, 35-55 × 10-13.5 μm, with up to 5 septa. *Asci* are subglobose, thin-walled, evanescent, 5-7 μm in diameter, with 8 ascospores. *Ascospores* smooth, lenticular, 1.5-2 × 2.5-4 μm, yellow in mass.

Colonies on glucose peptone agar are coarsely granular, radiating, with arachnoid edge, surface rosy buff color, while the reverse is buff to cinnamon. *Macroconidia* ellipsoid to fusiform, 25-58 × 8.5-15 μm, with moderately thick, verrucose walls. Slender whip-like appendages, up to 30 μm long borne on the apical cell of the conidia, are seen in some isolates but not in others. *Microconidia* are clavate, 1.7-3.5 × 3.3-8.3 μm, unicellular, smooth-walled, or slightly roughened, sessile or on pedicels, borne laterally on the hyphae. Spiral hyphae similar to those borne on the gymnothecium are also formed on the vegetative mycelium.

Discussion

This species is worldwide in distribution. It is found in the soil and causes an inflammatory tinea corporis and tinea capitis in humans. Small-spore ectothrix involvement of hairs is seen but generally no fluorescence. The fungus also infects a variety of animals [15]. A genetic study of a phenomenon of pleomorphism has been made by means of induced mutants of this fungus and of *N. incurvata* (see below). The results of such studies have been comprehensively reviewed by Kwon-Chung [53].

f. Nannizzia incurvata Stockdale, Sabouraudia 1: 46-47, 1961

(1) Conidial State

Microsporum gypseum (Bodin) Guiart & Grigoraksi, *sensu lato Lyon Med. 141*: 369-378, 1928.

(2) Description

Heterothallic. Gymnothecia globose, pale buff to yellow, 350-650 μm in diameter, excluding appendages. *Peridial hyphae* pale buff, hyaline, septate, verticillately branched with up to five branches, which usually curve in toward their main axis. *Cells* moderately thick-walled, densely asperulate, with one to three more or less symmetrical constrictions, usually up to 7-8 μm in diameter but occasionally reaching 11 μm. Free ends numerous and appendages of three kinds: (1) elongate, slender, smooth-walled, septate, occasionally branched hyphae, up to 300 μm long, 3.0-4.5 μm in diameter at the base tapering to 1.5-2.0 μm in diameter, straight or loosely coiled; (2) elongate, slender, smooth-walled, septate, occasionally branched hyphae, 2.5-3.5 μm in diameter at the base tapering to 1.5-2.0 μm in diameter, tightly coiled with up to 30 spirals; (3) macroconidia thick-walled, asperulate spindle-shaped, 40-57 × 10-12.5 μm, with 1-5 septa. *Asci* globose to ovate, 5-7 μm in diameter, 8-spored. *Ascospores* yellow, smooth-walled, lenticular, 2.8-3.5 × 1.5-2.0 μm.

Colonies grow rapidly on most laboratory media, becoming powdery and buff to cinnamon brown in color. Reverse pale yellow to tan, occasionally red in some isolates. *Microconidia* clavate, thin-walled, and sessile on the hyphae. *Macroconidia* numerous, large, echinulate, thick-walled, ellipsoid, with 3-9, commonly 4-6 septa, 8-12 × 30-50 μm.

Discussion

This species produces tinea corporis or tinea capitis. Hair involvement is of the small-spored ectothrix type and the hairs usually do not fluoresce. The fungus, which also infects animals, is found in the soil throughout the world.

A description of the morphogenesis of the sexual phase of growth was included with the original description [89] and has been reexamined in detail by Kwon-Chung [51].

g. Nannizzia obtusa Dawson & Gentles, Sabouraudia 1: 56, 1961

(1) Conidial State

Microsporum nanum (Fuentes, Aboulafia & Vidal) Fuentes, *Mycologia 48*: 613-614, 1956.

(2) Description

Heterothallic. Gymnothecia globose, pale buff, 250-450 μm in diameter, excluding appendages. *Peridium* about 50 μm thick. *Peridial hyphae* pale yellow, hyaline, septate, branching mostly dichotomous, occasionally verticillate, angle between branch and main hypha usually obtuse. Cells thick-walled, echinulate, cylindrical, 8-20 μm. Average 13 × 4-7 μm, may have one or two slight constrictions. *Appendages* of two sorts: (1) septate, smooth-walled, tightly coiled, lateral or terminal; (2) elongate, slender, septate, hyphae up to 450 μm long, terminal. *Asci* subglobose, thin-walled, evanescent, 5.5-6.5 × 5-6 μm, 8-spored. *Ascospores* hyaline, smooth or finely roughened, lenticular, 2.7-3.2 × 1.5-2 μm, yellow in mass.

Colonies white, cottony, and spreading, becoming granular and buff-colored. Reverse red to brown. *Macroconidia* small, clavate, thick-walled, echinulate, generally with 1 septum but a few spores may have 2 or even 3 septa, 12-18 × 4-7.5 μm.

Discussion

This species is worldwide in distribution. The fungus causes tinea capitis, in which the hairs display the small-spored ectothrix type of involvement but do not fluoresce. *Nannizzia obtusa* is primarily a pathogen of pigs.

h. Nannizzia otae Hasegawa & Usui, Jpn. J. Med. Mycol. 16: 148-153, 1974

(1) Conidial State

Microsporum canis Bodin, *Arch. Parasitol. 5*: 5-30, 1902.

(2) Description

Heterothallic. Gymnothecia globose, white becoming buff with age, 280-700 μm in diameter, exclusive of appendages. *Peridial hyphae* hyaline, septate, echinulate, constricted at the cell junction, usually dichotomously but occasionally verticillately branched. The tips of peridial hyphae are blunt and curved toward the gymnothecium. *Appendages* of three kinds: (1) long (up to 150 μm), slender, straight, smooth-walled, and septate hyphae; (2) spiral or long, coiled hyphae with 10-15 turns; and (3) hyphal appendages with thick-walled, echinulate macroconidia. *Asci* are subglobose, thin-walled, evanescent, 5-7 μm in diameter, with 8 ascospores. *Ascospores* are smooth, lenticular, 2-2.5 × 2.5-4.8 μm.

Colonies on glucose peptone agar grow fairly rapidly, forming a cottony or woolly mycelium, white to buff in color. The reverse is yellow to orange-brown. *Macroconidia* are numerous, large, thick-walled, spindle-shaped, echinulate, 8-20 × 40-150 μm, with 6-15 septa. *Microconidia* usually scarce, clavate to elongate, sessile, or on short pedicels borne laterally on the hyphae. Racquet hyphae, pectinate hyphae, nodular bodies, and chlamydospores may be seen.

Discussion

The fact that *N. otae (M. canis)* grows on polished rice grains whereas *M. audouinii* does not was one of the earliest uses of nutritional tests in identification of dermatophytes [33]. *Nannizzia otae* causes tinea capitis and tinea corporis. The source of infection is almost invariably an animal. The infected hairs fluoresce under a Wood's lamp and the small-spored ectothrix type of hair shaft involvement is seen. The fungus is zoophilic being frequently isolated from dogs and cats and occasionally from monkeys, horses, rabbits, and pigs.

i. Nannizzia persicolor Stockdale, Sabouraudia 5: 357-358, 1967

(1) Conidial State

Microsporum persicolor (Sabouraud) Guiart & Grigorakis, *Lyon Med. 141*: 369-378, 1928.

(2) Description

Heterothallic. Gymnothecia are globose, 350-900 μm in diameter, pale buff to buff, and composed of a central mass of asci surrounded by a peridium of loosely interwoven hyphae. *Peridial hyphae* are hyaline, pale buff, septate, branching, with up to four branches arising from the apex of the same cell. The distal branches curve back over the gymnothecium, and the ultimate branches are composed of up to five cells. *Cells* of the peridial hyphae have rather thin, verruculose walls and are 5.5-27.5 μm long × 2.5-7.5 μm in diameter at their widest part. The cells of the distal branches are dumbbell-shaped, with one slight constriction in the middle; they are symmetrical in both face and side views. *Appendages* are numerous, branched, spirally coiled hyphae. *Asci* subglobose, thin-walled, evanescent, 4.5-6.0 × 5.0-7.0 μm in diameter, and contain 8 ascospores. *Ascospores* are hyaline, smooth-walled, lenticular, 2.5-3.3 × 1.6-2.1 μm, pale yellow.

Colonies on glucose peptone agar are fast-growing, flat at first becoming fluffy and folded, yellowish buff to pink. The reverse is variable from peach, rose, to deep ochre. *Macroconidia* are not abundantly produced, rather thin-walled, clavate to fusiform, echinulate at the tip, usually with 6 septa. *Microconidia* produced in abundance, clusters, clavate, fusiform, or globose. Stalked, elongate, clavate conidia are distinctive.

Discussion

This species is worldwide in distribution, producing tinea infections in wolves, bats, shrews, and mice, and more rarely, infection of humans [65]. At one time the fungus was considered to be a *Trichophyton, T. persicolor,* closely related or, in some people's view, synonymous with *T. mentagrophytes* (see Ref. 71). Stockdale established that *Nannizzia persicolor* was a distinct taxon and expressed surprise that the conidial state should be *Trichophyton* [91]. She recorded the slightly verruculose walls of the macroconidia, a feature used subsequently to support inclusion of the species in *Microsporum* as *M. persicolor* (see Ref. 7).

j. Nannizzia racemosa Rush-Munro, Smith & Borelli, Mycologia 62: 856-859, 1970

(1) Conidial State

Microsporum racemosum, Borelli, *Acta Med. Venez. 12*: 148-151, 1965.

(2) Description

Heterothallic. Gymnothecia globose, pale buff, 300-700 μm in diameter, exclusive of appendages. *Peridial hyphae* hyaline, septate, verticillately and dichotomously branched. Up to four branches arising in succession at the apex of the same cell, many branches curve back over the gymnothecium. *Cells* are thin-walled, echinulate, often more or less symmetrically constricted with one to three constrictions, up to 8 μm in diameter. *Appendages* of three kinds: (1) slender, smooth-walled, septate hyphae, 2.3-3.0 μm in diameter at the base tapering to 1.5-2.0 μm at the tip, coiled with up to 15 turns; (2) straight, slender, smooth-walled hyphae, 3.0-4.0 μm in diameter at the base tapering to 1.7-2.0 μm at the tip, up to 380 μm long; (3) delicate, smooth-walled hyphae, 0.7-0.9 μm, which are up to 200 μm long. *Asci* globose or oval, 4.4-5.5 μm in diameter, containing 8 spores. *Ascospores* hyaline, smooth-walled, oval, 2.5-3.0 × 1.2-1.5 μm, yellow in mass. The distinctive feature of the gymnothecia of this fungus is the flagellum-like appendages (type 3) [71].

Colonies are very fast-growing on glucose peptone agar. Said to be the fastest-growing dermatophyte [71]. The surface of the colony is powdery cream-white. The reverse is grape-red. *Macroconidia* spindle-shaped, echinulate, rather thick-walled, up to 60 μm long, 5-10 septa. The macroconidia frequently have a terminal filament. *Microconidia* are quite distinctive. They are club-shaped, mostly stalked, and are borne in large, wand-like clusters (racemes) [16].

Discussion

This fungus has been reported from soils of South America and Romania [13] but is probably worldwide. The organism produces lesions experimentally [82].

4. Related Anamorphs for Which a Teleomorph Is Presently Unknown

a. Introduction

As explained at the outset, the dermatophytes were most often first described in their asexual form of growth. With the rediscovery of sexual reproduction among them in the 1960s the teleomorphs of a number of the dermatophytes have been described (Table 1). Yet for a large number of them, no such stage is presently known. These anamorphs will now be described.

b. Pathogenic Species

(1) *Epidermophyton* Sabouraud, *Les Teignes,* Masson et Cie., Paris, 1910

The genus is characterized on the basis of the macroconidia, which are 6-10 × 8-15 μm in size, clavate, smooth-walled, with 0-4 (usually 2-3) septa. No microconidia are found. The type species is *Epidermophyton floccosum* (Harz) Langeron & Milochevitch, 1930.

Epidermophyton floccosum (Harz) Langeron & Milochevitch, *Ann. Parasitol. Hum. Comp. 8:* 495-497, 1930

Colonies. The fungus grows slowly on glucose peptone agar. The colonies have a velvety to powdery surface which is gently folded in a number of radiating furrows and is khaki-yellow in color. The reverse is yellow to tan. Usually, isolates mutate early and almost invariably, with the production tufts of white, sterile hyphae which soon overgrow the entire colony. *Macroconidia* clavate, smooth, fairly thick-walled, with 0-4 (usually 2-3) septa; the macroconidia are borne in characteristic clusters of twos and threes. *Microconidia* are not produced. Chlamydospores, racquet hyphae, and nodular bodies are often abundantly produced.

Discussion

This species is worldwide in distribution. The fungus produces tinea pedis and tinea cruris. There are rare reports of tinea unguium [15]. A dog has been reported to have yielded *E. floccosum* but animals are generally not infected.

A second species of *Epidermophyton, E. stockdaleae,* has been described. It will be discussed together with other saprophytes which are related to the dermatophytes (see Sec. III.C).

(2) *Microsporum* Gruby, *C. R. Acad. Sci. (Paris) 17*: 301-303, 1843

The genus is characterized on the basis of the macroconidia, which are large, thick-walled, echinulate, fusiform to obovate conidia with 3-15 septa, generally borne singly. Smaller one-celled microconidia are produced on short pedicels or are sessile on the hyphae. Pectinate hyphae, racquet hyphae, nodular bodies, coils, and chlamydospores may be present. Type species is *Microsporum audouinii,* Gruby, 1843.

Microsporum audouinii Gruby, *C. R. Acad. Sci. (Paris) 17*: 301-303, 1843

Synonyms

Sabouraudites (Microsporum) langeronii Vanbreuseghem, *Ann. Parasitol. Hum. Comp. 25*: 409-417, 1950.

Microsporum rivalieri Vanbreuseghem, *Sabouraudia 2*: 215-224, 1965.

Colonies on glucose peptone agar grow rather slowly, form a matted, velvety surface with straggly edges, and are light tan to brown in color. The reverse is buff-salmon to orange-brown. *Macroconidia* are rarely found in cultures. When observed they are distorted and very irregular with thick echinulate or smooth walls. Some may be nearly spindle-shaped. The number of septa is from 2 to 8. *Microconidia,* when present, are sessile or borne on short pedicels along the hyphae, clavate, and single-celled. Racquet hyphae, pectinate hyphae, nodular bodies, and chlamydospores may be present and are not infrequently the only microscopic structures visible. As pointed out earlier, the poor growth of *M. audouinii* on rice grain serves to distinguish the species from *M. canis (N. otae)* and *M. gypseum (N. gypsea* and *N. incurvata).*

Discussion

This species produces classical, epidemic tinea capitis in children. The hairs show ectothrix involvement and fluoresce under a Wood's lamp. The fungus was originally endemic in Europe but now occurs in a much wider area. *Microsporum audouinii* is one of the forms of ringworm most susceptible to griseofulvin therapy, and the incidence of disease has been remarkably reduced wherever the drug is used. A few animal infections have been reported but *M. audouinii* is clearly anthropophilic.

Microsporum distortum diMenna & Marples, *Trans. Br. Mycol. Soc. 37*: 372-374, 1954

Colonies on glucose peptone agar grow rather rapidly and are flat, velvety to fluffy, white to tan. The reverse is colorless or yellow to tan. *Macroconidia* thick-walled, echinulate, with 6-10 septa. The spores are basically spindle-shaped but all bend and are grossly distorted. Their size is roughly similar to that of *M. canis (N. otae)* but difficult to record usefully because of the distortions. *Microconidia* are clavate and sessile on the hyphae.

Discussion

This species produces tinea capitis. The hairs show ectothrix involvement and fluoresce under a Wood's lamp. Cases have been reported in monkeys, dogs, horses, and pigs. The fungus occurs in the United States, Australia, and New Zealand.

Microsporum equinum (Delacroix & Bodin) Gueguen, *Les champignons parasites de l'homme et des animaux domestiques,* p. 144, Joseph Van In & Cie, Lierre, 1904

Colonies on glucose peptone agar are downy to powdery, pale buff to pinkish buff. The reverse is yellow to buff amber. *Macroconidia* elliptical to fusiform, thick-walled, echinulate, 18-62 × 5-13.5 μm, with 4-8 septa and borne terminally in clusters [20,64]. *Microconidia* sessile or on short pedicels, pyriform to clavate, 3-8 × 2-3.5 μm. Spiral hyphae and chlamydospores have been seen [20]. The description of this species in the early literature [20] varies in detail from that more recently published [64].

Discussion

This species is now included in most standard lists of *Microsporum* spp. [7-9] but it was not always so [32,71] and modern descriptions of it are somewhat different from each other [20,64]. It was originally reported in Europe, Java, and Uruguay [4,5] and said to cause tinea of horses. Small-spored ectothrix involvement is described [20]. The fungus is said to infect humans [64]. The results of a recent study support the view that *M. equinum* is a distinct species and not a synonym of *M. canis* [64].

Microsporum ferrugineum Ota, *Bull. Soc. Pathol. Exot. 15*: 588-596, 1922

Synonym

Trichophyton ferrugineum (Ota) Langeron & Milochevitch, *Ann. Parasitol. Hum. Comp. 8*: 422-436, 1930.

Colonies on glucose peptone agar are slow-growing, heaped with many deep furrows, glabrous and waxy, deep yellow to orange in color. Reverse is buff to brownish. A white velvety cover may form over the colony. Sectoring occurs with great variations in color intensities. *Macroconidia* are rarely observed. However, such structures have been reported when isolates are cultured on a potato dextrose charcoal agar or diluted Sabouraud's agar [15,98]. The spores so generated were typical for the genus *Microsporum,* i.e., spindle-shaped, echinulate, thick-walled, 40.5-47 × 8.5-10.5 μm, with 2-8 septa. The mycelium is comprised of hyphae some of which have prominent cross walls ("bamboo" hyphae). Chlamydospores may be seen.

Discussion

The fungus causes tinea capitis in children. Hair involvement is of the small-spored ectothrix sort. The hairs fluoresce under a Wood's lamp. The organism is strictly anthropophilic and is found in middle Europe, Asia, the USSR, and Africa.

Microsporum gallinae (Megnin) Grigorakis, *Ann. Dermatol. Syphilol. 10*: 18-53, 1929

Synonym

Trichophyton gallinae Silva & Benham, *J. Invest. Dermatol. 18*: 432-472, 1952.

Colonies on glucose peptone agar grow moderately rapidly, are flat at first with radial folds. Edges of the colonies may be irregular. The reverse of the colony is deep strawberry-red. The pigment diffuses into the medium. Macroconidia relatively thin-walled, echinulate at the tips, 6-8 × 15-50 μm, frequently curved, with 2-10 septa. *Microconidia* pyriform to clavate, smooth-walled, borne laterally on the hyphae.

Discussion

This species is a very rare incitant of tinea capitis, tinea corporis, and tinea cruris in humans. The ectothrix type of hair involvement is seen. More typically seen as a cause of tinea in chickens and other fowl. The distribution is worldwide.

Microsporum praecox Rivalier, *Ann. Inst. Pasteur 86*: 276-284, 1953

Colonies on glucose peptone agar are fast-growing and similar in appearance to those of *M. gypseum* (see *N. gypsea* and *N. incurvata*). Powdery buff with a yellowish orange reverse. *Macroconidia* long narrow, lanceolate, with fairly thick, roughened walls, 60-65 × 8-10 μm, with 6-9 septa. *Microconidia* pyriform in shape [79].

Discussion

This species was isolated from a pustular, vesicular tinea lesion on the wrist from an adult [71].

(3) *Trichophyton* Malmsten, *Trichophyton tonsurans* harskarärde Mögel, 1845 (Transl.), *Arch. Anat. Physiol. Wiss. Med.* pp. 1-19, 1848

The genus is characterized on the basis of the macroconidia, which are elongate, clavate, to fusiform, generally thin-walled, smooth, and have 0-10 septa. The size ranges from 8-50 × 4-8 μm. Smaller one-celled microconidia are produced. These cells are globose, 2.5-4 μm, or clavate to pyriform, and 2-3 × 2-4 μm. Spiral hyphae and chlamydospores are produced. The type species is *Trichophyton tonsurans* Malmsten, 1845.

Trichophyton concentricum Blanchard, *Traité de pathologie générale,* Vol. 2, Masson et Cie, Paris, pp. 811-926, 1895

Colonies on glucose peptone agar are very slow-growing and are raised deeply folded, smooth and white, becoming cream to amber or brown, and covered with short, gray hyphae. The reverse is cream to brown. *Macroconidia* not observed. Microscopically similar to *T. schoenleinii* (see below). The hyphae are swollen, bear chlamydospores and aborted branches. *Microconidia* have been reported [71].

Discussion

The fungus causes tinea imbricata (i.e., a form of tinea corporis in which concentric rings of scales occur on the skin [75]). The disease has been reported in the South Pacific islands, in Guatemala, southern Mexico, and central Brazil.

Trichophyton equinum (Matruchot & Dassonville) Gedoelst, *Les champignons parasites de l'homme et des animaux domestiques,* p. 88, Joseph Van In & Cie, Lierre, 1902

Colonies grow rapidly on glucose peptone agar and are flat, developing folds with age, surface white, cottony with yellow color in the edge around the new growth. The reverse is bright yellow becoming dark pink to brown with age. *Macroconidia* are rarely found in cultures grown on glucose peptone agar. A few such spores are found when strains are grown on wort agar [35]. They are slightly clavate, thin-walled, smooth, with 3-4 septa. *Microconidia* are abundantly produced on wort agar. They are nearly spherical to slightly pyriform, borne laterally along the hyphae, sessile or on short pedicels and in clusters.

Discussion

This species considered by some as being synonymous with or a variety of *T. mentagrophytes* was reevaluated as a separate taxon [35]. The fungus is rare in humans [15] and commonly causes ringworm in horses [34]. The large-spored ectothrix type of hair involvement is seen. There is a variety recorded from New Zealand and Australia with less exacting nutritional requirements than customary. This form has been referred to as *T. equinum* var. *autotrophicum* [88].

Trichophyton gourvilii Cantanei, *Bull. Soc. Pathol. Exp. 26*: 377-381, 1933

Colonies on glucose peptone agar are waxy with a heaped and folded surface, becoming velvety with age. The color is lavender to deep garnet red. The reverse is reddish purple. The colonies are said to resemble those of *T. violaceum* and *T. soudanense* [15]. *Macroconidia* smooth-walled, cylindrical. *Microconidia* pyriform, smooth-walled, borne laterally on the hyphae. Not all isolates produce micro- and macroconidia (see Ref. 32).

Discussion

This fungus is found in Africa, where it produces tinea capitis and tinea corporis. Endothrix type of hair involvement is observed.

Trichophyton megninii Blanchard, *Traité de pathologic générale,* Vol. 2, Masson et Cie, Paris, pp. 811-926, 1896

Synonym

Trichophyton kuryangei Vanbreuseghem & Rosenthal, *Ann. Parasitol. Hum. Comp. 36*: 797-802, 1961 (see Ref. 46).

Colonies grow slowly on glucose peptone agar and are cottony to velvet, white at first becoming pink with age. A nondiffusible rose to red pigment develops on the reverse side. *Macroconidia* are scarce, clavate, thin and smooth-walled, with 2-10 septa, said to resemble those of *T. rubrum* [71]. *Microconidia* are numerous, small, pyriform to clavate, borne singly or in clusters along the hyphae.

Discussion

Produces tinea barbae predominantly and to a lesser extent tinea corporis and tinea capitis. Hair involvement of the large-spored ectothrix type. Predominantly anthropophilic, it has been reported on dogs [15]. The fungus is found in Europe and Africa.

Trichophyton rubrum (Castellani) Sabouraud, *Br. J. Dermatol. 23*: 389-390, 1911

Synonym

Trichophyton fluviomuniense Pereiro Miguens, *Sabouraudia 6*: 312-317, 1968 (see Ref. 71).

Colonies on glucose peptone agar are cottony and white, later becoming velvety. The reverse side is reddish to rose purple. *Macroconidia* elongate, cylindrical, thin and smooth-walled, with 3-8 septa. Generally, these are scarce on glucose peptone agar but are formed in cultures on heart infusion tryptose agar [15,57]. *Microconidia* numerous, clavate, 2-3 × 3-5 μm, borne in clusters or singly along the hyphae. Chlamydospores, racquet hyphae and nodular bodies may be seen in primary cultures. A wide variety of colonial types has been described (see Ref. 32).

Discussion

This species is worldwide in distribution and clinically is among the most important dermatophyte pathogens. It produces tinea corporis, tinea pedis, tinea cruris, and tinea unguium. Nail lesions are very refractory to treatment. The fungus rarely causes disease in animals but instances are recorded [15]. *Trichophyton rubrum* does not perforate autoclaved hair, a test useful in distinguishing isolates of it from *T. mentagrophytes (Arthroderma benhamiae, A. vanbreuseghemii)* [64a].

Mating studies with *A. simii* [106] have revealed that all tested isolates of *T. rubrum* were the same mating type *a*. It may be that virulence or some other selective factor has eliminated the *A* mating type of *T. rubrum.*

Trichophyton schoenleinii (Lebert) Langeron & Milochevitch, *Ann. Parasitol. Hum. Comp. 8*: 465-508, 1930

Colonies are very slow-growing on glucose peptone agar, heaped with many irregular folds, waxy smooth, later becoming velvety white. *Macroconidia* have never been reported. *Microconidia* are virtually never seen. Antler-like terminal branching ("favic chandeliers") is the single most obvious morphological feature. Chlamydospores may be seen.

Discussion

The fungus is widespread in Eurasia and North Africa, where it produces favus or tinea favosa, which is a chronic, cicatricial form of tinea capitis distinguished by the formation of scutula, crusts, scarring, atrophy, and permanent hair loss. Hair involvement is quite distinctive. Within the hair, hyphae and air spaces accompanied by degenerate hyphae are seen. A dull greenish fluorescence of the hairs under a Wood's lamp has been described. Tinea corporis has been observed, as have infections in animals.

Trichophyton soudanense Joyeux, *C. R. Soc. Biol. (Paris) 73*: 15-16, 1912

Synonym

Langeronia soudanensis Vanbreuseghem, *Ann. Parasitol. Hum. Comp. 25*: 493-508, 1950.

Colonies are very slow-growing, flat to folded, suede-like, yellow, dried-apricot colored, often with a fringed border [78]. Sterile rapidly growing hyphae develop readily in culture. *Macroconidia* not seen. *Microconidia* are seen in cultures grown on potato dextrose agar and are pyriform, smooth-walled, borne laterally on hyphae, sessile or on short pedicels [78]. The mycelium is characterized by hyphae which break up into segments or arthroconidia and by reflexive branching.

Discussion

This species produces tinea capitis and tinea corporis. It is found principally in Africa, but cases have been reported in England, Brazil, and the United States [15].

Trichophyton tonsurans Malmsten, 1845. *Trichophyton tonsurans* harskärande Mögel, Stockholm (transl.) *Arch. Anat. Physiol. Wiss. Med.*, pp. 1-99, 1848.

Colonies grow fairly slowly on glucose peptone agar, flat, somewhat powdery, yellowish surface. The reverse is mahogany red. A sulfureum variety produces pale to deep yellow colonies. *Macroconidia* are not frequently seen, but when found are somewhat thick-walled, club-shaped with 3-5 septa. Some more abundantly sporulating strains have been described [26]. *Microconidia* are numerous, clavate, 2-5 × 3-7 μm, and are borne laterally along the hyphae on pedicels of various lengths. Chlamydospores and racquet hyphae may be seen.

Discussion

This species is worldwide in distribution. It produces black-dot tinea capitis and occasionally tinea corporis and tinea pedis. Hair shaft involvement is of the endothrix sort. Animal infections have been rarely reported.

Trichophyton verrucosum Bodin, *Les champignons parasites de l'homme,* Masson et Cie, Paris, p. 121, 1902

Colonies are very slow-growing on glucose peptone agar, heaped, deeply folded, glabrous and waxy, sometimes with a fine white velvety surface. Colonial variants have been described and named (var. *ochraceum,* var. *album,* and var. *discoides* [71]). These variants were originally described as species [2] but are treated as varieties by most [32]. On glucose peptone agar generally only chlamydospores and hyphae are observed microscopically. On enriched media *microconidia* are produced. They are clavate to pyriform, sessile, borne laterally along the hyphae. *Macroconidia* are rare and are "rat-tailed" or "string bean"-shaped with 3-5 septa, smooth, thin-walled.

Discussion

This species produces highly inflammatory tinea capitis, tinea barbae, and tinea corporis. The hairs show ectothrix type of involvement. The fungus is worldwide in distribution. Characteristically, it is a zoophilic species found primarily in cattle, and has also been recorded in donkeys, dogs, goats, sheep, and horses [15].

Trichophyton violaceum Bodin, *Les champignons parasites de l'homme,* Masson et Cie, Paris, p. 113, 1902

Colonies are slow-growing on glucose peptone agar, waxy, wrinkled, heaped up or verrucous, with a deep purplish red pigmentation. Colonies tend to produce sector variants. Generally, only torturous, tangled hyphae are observed. Pyriform *microconidia* borne laterally may be seen but only rarely. Clavate *macroconidia* [71] with 3-5 septa, and smooth, thin walls have been reported.

Discussion

This species produces tinea corporis and tinea capitis of the black-dot variety. Hair invasion is of the endothrix sort. Worldwide in distribution although unusual in the United States. Reported to infect animals other than humans [15].

Trichophyton yaoundei, Cochet & Doby-Dubois, *Sem. Hop. Paris 33*: 26-30, 1957

Colonies are very slow-growing on glucose peptone agar, glabrous, raised, folded, white to cream in color at first becoming tan to brown with age. The pigment diffuses into the medium. Sectoring variants devoid of diffusible pigment on the reverse occur regularly. *Macroconidia* are not formed. *Microconidia* rare but when seen are pyriform and borne laterally on the hyphae. Chlamydospores are present.

Discussion

This species produces endothrix-type tinea capitis and is found in equatorial Africa.

C. Saprophytic Species

1. Introduction

A number of the species of *Arthroderma* have conidial states in the genus *Trichophyton*. In addition, a few species of *Microsporum* and one of *Epidermophyton* have been isolated and described as saprophytes. These forms are generally included among the dermatophytes, even though they are not known to cause ringworm in humans or other animals. Such forms are discussed here because it is customary to include them in any consideration of the dermatophytes (Table 1).

2. *Arthroderma* Berkeley, 1860 (see description in Sec. III.B)

a. Arthroderma ciferrii Varsavsky & Ajello, Riv. Patol. Veg. 4: 358-359, 1964

(1) Conidial State

Trichophyton georgiae Varsavsky & Ajello, *Riv. Pathol. Veg. 4*: 357-358, 1964.

(2) Description

Homothallic. Gymnothecia globose, pale buff to pale brownish vinaceous, 450-700 μm in diameter, excluding appendages. *Peridial hyphae* hyaline, septate, uncinately branched with curled end. *Cells* of the peridial hyphae dumbbell-shaped, symmetrically or asymmetrically constricted, densely asperulate, 8-12 μm long, 7-8.5 μm at enlarged ends in width. *Appendages* of two kinds: (1) slender, smooth-walled spiral hyphae, borne terminally; and (2) slender, smooth-walled, septate, tapering hyphae extending outward, 60-96 μm long. *Asci* subglobose, thin-walled, evanescent, 4.8 × 6 μm, 8-spored. *Ascospores* hyaline, smooth, oblate, 2.5-3 × 1.4-2.8 μm, yellow in mass.

Colonies on glucose peptone agar are flat and granular with an umbonate center. Center pale vinaceous and downy. Flat area surrounding umbonate center powdery to granular, vinaceous brown. Reverse irregularly spotted with dark red or dark vinaceous brown. *Macroconidia* thought not to occur [32], but some of the spores identified as microconidia have septa (see below; Ref. 62) and might be regarded as macroconidia [71]. *Microconidia* variable in size and shape, elongate, clavate, occasionally pyriform to subglobose, borne laterally or terminally, singly on short pedicels, or sessile on the hyphae. *Microconidia* smooth-walled, usually nonseptate, occasionally with 2 or 3 septa, 4-6.5 × 2-2.5 μm.

Discussion

This fungus is found in soil, on hair and feathers, or in birds' nests or excreta [62]. The organism has been reported from the United States, Venezuela, Australia, Czechoslovakia, and Yugoslavia [62].

b. Arthroderma gloriae Ajello, Mycologia 59: 257, 1967

(1) Conidial State

Trichophyton gloriae Ajello, *Mycologia 59*: 257, 1967.

(2) Description

Heterothallic. Gymnothecia globose, pale yellow, 250-450 μm in diameter, excluding appendages. *Peridial hyphae* thin-walled, septate, hyaline, uncinately branched. *Cells* dumbbell-shaped, echinulate, symmetrically constricted, 7.5 × 5 μm. *Appendages* moderately slender, thin-walled hyphae pointed at the end, 150 μm long. *Asci* subglobose, thin-walled, evanescent, 3.5 × 4.6 μm, 8-spored. *Ascospores* hyaline, thin-walled, smooth, oblate, 2-2.5 × 1.5-2 μm, yellow in mass.

Colonies downy, slightly folded, pale cream with a brownish center. Reverse yellow with yellow brown center. *Microconidia* pear-shaped, borne singly on short pedicels or sessile, 1.5-6 X 1.5-2.5 μm. *Macroconidia* numerous, borne singly, more often in clusters of 4-30, cylindrical, multiseptate, with 1-10 septa, smooth walls up to 10 μm thick, 9-60 X 3-7 μm.

Discussion

This fungus has been in the soil and on animal hair from several states of the United States [71].

c. Arthroderma flavescens Rees, Sabouraudia 5: 206-207, 1967

(1) Conidial State

Trichophyton flavescens Padhye & Carmichael, *Can. J. Bot. 9*: 1535, 1971.

(2) Description

Heterothallic. Gymnothecia globose, pale buff, 450-650 μm in diameter, excluding appendages. *Peridial hyphae* hyaline, septate, with two inwardly curving branches developed from apex of some peridial cells. *Cells* dumbbell-shaped, thick-walled, asymmetrically constricted, echinulate, 11-16.5 X 3-4.8 μm. *Appendages* smooth-walled, septate, spiral hyphae usually terminal. Smooth-walled, septate macroconidia are rarely attached to peridial hyphae. *Asci* subglobose, thin-walled, evanescent, 5.9-8.3 X 4.8-7.5 μm, 8-spored. *Ascospores* hyaline, smooth, oblate, with equatorial rim, 3.8-4 X 1.8-2 μm, yellow in mass.

Colonies on glucose peptone agar white to pale yellow with buff center. Reverse bright yellow darkening to yellow brown. *Microconidia* borne laterally or terminally, smooth, thin-walled, sessile or on short pedicels, unicellular, occasionally two-celled, pyriform, ovate, 5-16 X 4-8 μm. *Macroconidia* numerous, hyaline, usually 2-6 septa, smooth, thin-walled, cylindrical with rounded apices, 26-86 X 8-14 μm.

Discussion

This species is found in Australia, where it was isolated from feathers.

d. Arthroderma insingulare Padhye & Carmichael, Sabouraudia 10: 47, 1972

(1) Conidial State

Trichophyton terrestre, sensu lato Durie and Frey, *Mycologia 49*: 401, 1957.

(2) Description

Heterothallic. Gymnothecia globose, white to pale yellow, 250-500 μm in diameter, excluding appendages. *Peridial hyphae* pale yellow, hyaline, septate, uncinately branched, usually the outside of the main hypha. *Cells* thick-walled, echinulate, asymmetrically dumbbell-shaped, 8-12 μm long, 5-6 μm wide at enlarged ends and 3-4 μm wide at the internode. *Appendages* spiral, septate, smooth-walled, terminal hyphae which vary considerably in length and number of turns. *Asci* subglobose, thin-walled, evanescent, 4-6 X 3.5-5 μm, 8-spored. *Ascospores* hyaline, smooth, oblate, 2.5-3 X 2.2-2.5 μm, yellow in mass.

Colonies on glucose peptone agar are white and fluffy, becoming pale yellow, and downy to granular. Reverse yellowish brown. Some strains develop a red pigment on the reverse. *Microconidia* elongate, pyriform, borne singly or in groups, 3-6.5 X 1.5-3.5 μm. *Macroconidia* cylindrical or rarely clavate, smooth, thin-walled, sessile, 8-52 X 4-5 μm, with 2-6 septa. Intermediate forms present. Terminal or intercalary chlamydospores, spirals, racquet hyphae, antler-like hyphae, and nodular bodies occasionally seen. The colonial description of *T. terrestre* given at this point for convenience should more properly have been introduced under *Arthroderma quadrifidum,* where it was first used.

Discussion

This species has been isolated from soil, hair, and feathers. The fungus is known to occur in Canada, the United States, Hungary, and Czechoslovakia.

e. Arthroderma lenticularum Pore, Tsao & Plunkett, Mycologia, 57: 970-971, 1965

(1) Conidial State

Trichophyton terrestre, sensu lato Durie & Frey, *Mycologia 49*: 401, 1957.

(2) Description

Heterothallic. Gymnothecia globose, pale buff, 300-600 μm in diameter, excluding appendages. *Peridial hyphae* pale yellow, hyaline septate, uncinately branched usually to one side, the outside of the main hyphae. *Cells* thick-walled, echinulate, dumbbell-shaped, symmetrically constricted, 5.5-8.5 × 7-10 μm. *Appendages* septate, spiral hyphae borne terminally. *Asci* subglobose, thin-walled, evanescent, 4-4.8 × 5-5.6 μm, 8-spored. *Ascospores* hyaline, smooth, oblate, 2.2-3 × 1.5-7.5 μm, yellow in mass.

Colonies (see description of *A. singulare* above). This fungus is found in soil in the United States [15].

f. Arthroderma quadrifidum Dawson & Gentles, Sabouraudia 1: 56-57, 1961

(1) Conidial State

Trichophyton terrestre, sensu lato Durie & Frey, *Mycologia 49*: 401, 1957.

Synonym

Trichophyton thuringiense Koch, *Mykosen 12*: 287-290, 1969 (see Ref. 63).

(2) Description

Heterothallic. Gymnothecia globose, pale buff, 400-700 μm in diameter, excluding appendages. *Peridial hyphae* pale yellow, hyaline, septate, uncinately branched usually to one side, the outside of the main hypha. *Cells* thick-walled, strongly echinulate, dumbbell-shaped when young, but when mature resembling a short humerus bone with condyles much accentuated and formed on one face only, 8-13 × 5-9 μm. *Appendages* septate spirals varying considerably in length and number of turns, borne terminally. *Asci* subglobose thin-walled, evanescent, 4-6 × 3.5-5 μm, 8-spored. *Ascospores* hyaline, smooth, oblate, 2.5-3 × 2.2-3 μm, yellow in mass.

Colony (see description of *A. singulare* above).

Discussion

This fungus is of worldwide occurrence and is found in soil, on feathers, and on animal hair [15].

Pore and Plunkett [70] reported three more teleomorphs in the *T. terrestre* complex, but these new species were not named or described by the authors and so have not been included in compilations of the *T. terrestre* complex [15,24,32,62,71].

3. *Epidermophyton* Sabouraud, 1910 (see description in Sec. III.B)

a. Epidermophyton stockdaleae Prochacki & Englehardt-Zasada, Mycopathol. Mycol. Appl. 54: 341-345, 1974

Colonies grow fairly rapidly on glucose peptone agar, flat, with a slightly umbonate center, powdery to granular, and buff cream-colored. The reverse is yellow becoming ferruginous to blush ferruginous. *Macroconidia* numerous, smooth-walled, clavate, or cylindrical, borne laterally or terminally, 20-50 × 3.8-12.9 μm, with 2-9 septa. *Microconidia* not formed. Degenerate macroconidia like those seen in *A. simii* are observed.

Discussion

This species was isolated from soil in Szczecin, Poland.

4. *Microsporum* Gruby, 1843 (see description in Sec. III.B)

a. Microsporum boullardii Dominik & Majchrowicz, Ecol. Pol. Ser. A, 13: 415-447 1965

Colonies on glucose peptone agar are rapid-growing, flat with a dense downy to floccose chamois-like surface, tawny buff to pink buff in color. Reverse red to purplish, resembles some isolates of *M. fulvum* (see *N. fulva* above). *Macroconidia* are fairly thick-walled, echinulate, clavate to bullet-shaped, 24-58 $\times$ 7.5-12 μm, 4-6 septa, borne singly. *Microconidia* single-celled, clavate, and abundant.

Discussion

This fungus has been isolated from soil of Guinea, Africa [15,71]. Seemingly, this is the only report of its isolation [5]. Its ability to produce disease is unproven.

b. Microsporum magellanicum Coretta & Piontelli, Sabouraudia 15: 1-10, 1977

Colonies on glucose peptone agar are finely granular, powdery, cream to yellowish in color; reverse yellow to ochre. *Macroconidia* are hyaline, ovate or clavate, rarely cylindrical, measuring 14.4-21.6 $\times$ 4.8-7.2 μm, verrucous and pedicillate-walled also verruculose, with 1-2 or rarely 4-6 septa, borne singly. *Microconidia* not seen but young macroconidia are pyriform or clavate with a truncate base, lightly verruculose, measuring 2.4-3 $\times$ 4.8-5.8 μm, sessile or pedunculate often with a septum. Chlamydospores observed rarely.

Discussion

This keratinophilic fungus was isolated from soil collected in the extremity of Chile, in southern Shetland, and the Antarctic continent. It has only been described in this saprophytic situation.

c. Microsporum ripariae Hubalek & Rush-Munro, Sabouraudia 11: 287-289, 1973

Colonies on glucose peptone agar are densely powdery to velvety and floccose with some zonate growth. Reverse is light yellow or lemon yellow and brownish in the center. *Macroconidia* are produced in limited numbers, borne singly, hyaline, relatively thin-walled, verrucose to verruculose, fusiform to cigar-shaped or almost cylindrical, 27-47 $\times$ 6.3-11 μm, with 4-6 septa. *Microconidia* are numerous, unicellular, hyaline, smooth, thin-walled, and sessile or borne on short lateral branches, often in bunches. They are pyriform and on the average 3.2 $\times$ 2.1 μm. Arthroconidia and terminal chlamydospores are occasionally observed.

Discussion

This fungus has only been reported one time. It was isolated from feathers and nests of the sand martin *(Riparia riparia)* in central Europe. Although isolated from a seemingly saprophytic source, it is capable of causing experimental infections in guinea pigs and humans.

5. *Trichophyton* Malmsten, 1848 (see description in Sec. III.B)

a. Trichophyton longifusum (Florian & Galgoczy) Ajello, Sabouraudia 6: 148, 1968 (see Ref. 10)

Colonies on glucose peptone agar are fluffy to powdery, and white to pale yellow in color. The reverse is yellowish brown. *Macroconidia* are long and cylindrical, 7-9 μm in diameter and up to 300 μm long, occurring in clusters and characteristically may arise from one another and merge into terminal hyphae. *Microconidia* not reported in original description [4].

Discussion

Isolated from soil in Hungary [31]. The species was transferred to the genus *Trichophyton* by Ajello [4] and otherwise has seemingly received scant attention.

b. Trichophyton phaseoliforme Borelli & Feo, Acta Med. Venez. 13: 176-177, 1966
Colonies on glucose peptone agar are powdery, white to bright cinnamon colored. *Microconidia* are cylindrical and produced in terminal clusters. Microconidia are curved and "cashew-nut" shaped, borne laterally on the hyphae and in large numbers on thickened and enlarged hyphae in the ascocarp-like (pycnidia?) structures.

Discussion

This species deserves further work to establish various morphological features which might alter its current taxonomic treatment. It was isolated from soil in Romania, from an apparently normal rodent in Venezuela, and is perhaps worldwide in distribution [71].

ACKNOWLEDGMENT

I am grateful to Dr. A. A. Padhye, C.D.C., Atlanta, Georgia, for reviewing and criticizing the manuscript.

NOTE ADDED IN PROOF

It will be noted that *Trichophyton mentagrophytes* is listed under both the Anthropophilic and the Zoophilic dermatophytes in Table 2 (p. 118). This seeming uncertainty indicates a widely acknowledged observation that there is a zoophilic form of *T. mentagrophytes* which often causes only mild infections in animals but elicits highly inflammatory responses in humans, while there is also an anthropophilic form of *T. mentagrophytes* which produces relatively mild, chronic infections in humans and spreads characteristically from human to human [5,77,76]. This ecological observation is sometimes indicated by the use of varietal names: *T. mentagrophytes* var. *mentagrophytes* for the zoophilic forms and *T. mentagrophytes* var. *interdigital* for the anthropophilic forms.

REFERENCES

1. Ainsworth, G. C. 1976. *Introduction to the History of Mycology*, Cambridge University Press, Cambridge.
2. Ainsworth, G. C., and L. K. Georg. 1954. Nomenclature of the faviform trichophytons. Mycologia *46*: 9-11.
3. Ajello, L. 1962. Present day concepts of the dermatophytes. Mycopathol. Mycol. Appl. *17*: 315-324.
4. Ajello, L. 1968. A taxonomic review of the dermatophytes and related species. Sabouraudia *6*: 147-159.
5. Ajello, L. 1974. Natural history of the dermatophytes and related fungi. Mycopathol. Mycol. Appl. *53*: 93-110.
6. Ajello, L. 1977. Milestones in the history of medical mycology: the dermatophytes. In K. Iwata (Ed.), *Recent Advances in Medical and Veterinary Mycology*, Proceedings of the Sixth Congress of the International Society for Human and Animal Mycology, University of Tokyo Press, Tokyo, pp. 3-11.
7. Ajello, L. 1977. Taxonomy of the dermatophytes: a review of their imperfect and perfect states. In K. Iwata (Ed.), *Recent Advances in Medical and Veterinary Mycology*, Proceedings of the Sixth Congress of the International Society for Human and Animal Mycology, University of Tokyo Press, Tokyo, pp. 289-297.

8. Ajello, L. 1977. Medically important infectious fungi. Contrib. Microbiol. Immunol. *3*: 7-19.
9. Ajello, L. 1978. Gegenwärtige Kenntnisse über die imperfekten und perfekten Formen der *Epidermophyton-*, *Mikrosporum-* und *Trichophytonarten.* Hautarzt *29*: 6-9.
10. Ajello, L., and A. A. Padhye. 1979. An orthographic correction. Mycotaxon *8*: 383-384.
11. Ajello, L., L. Bostick, and S.-L. Cheng. 1968. The relationship of *Trichophyton quinckeanum* to *Trichophyton mentagrophytes.* Mycologia *60*: 1185-1189.
12. Alkiewicz, J. A. 1967. On the discovery of *Trichophyton schoenleinii (Achorion schoenleinii).* Mycopathol. Mycol. Appl. *33*: 28-32.
13. Alteras, I., and R. Evolceanu. 1969. First Isolation of *Microsporum racemosum*—Dante Borelli, 1965 from Romanian soil. New data on its pathogenic properties. Mykosen *12*: 223-230.
14. Bassi, A. 1835. *Del mal del segno calcinaccio o moscardion: malattia che affligge i bachi de seta e sul modo di liberarne le bigattaje anche le piu infestate.* Lodi, Italy.
15. Beneke, E. S., and A. L. Rogers. 1980. *Medical Mycology Manual,* Burgess, Minneapolis.
16. Borelli, D. 1965. *Microsporum racemosum* nova species. Acta Med. Venez. *12*: 148-151.
17. Bullock, W. 1938. *The History of Bacteriology,* Oxford University Press, London. Reprint: Dover, New York, 1960.
18. Carmichael, J. W., A. S. Sekhon, and L. Sigler. 1972. Classification of some dermatophytes by pyrolysis-gas-liquid chromatography. Can. J. Microbiol. *19*: 403-407.
19. Chu-Cheung, S., and J. Maniotis. 1973. A genetic study of extracellular elastin-hydrolysing protease in the ringworm fungus *Arthroderma benhamiae.* J. Gen. Microbiol. *74*: 299-304.
20. Conant, N. F. 1937. Studies in the genus *Microsporum.* III. Taxonomic studies. Arch. Dermatol. Syphilol. *36*: 781-808.
21. Conant, N. F., D. T. Smith, R. D. Baker, and J. L. Calloway. 1971. *Manual of Clinical Mycology,* Saunders, Philadelphia.
22. Davison, F. D., D. W. R. Mackenzie, and R. J. Owen. 1980. Deoxyribonucleic acid base compositions of dermatophytes. J. Gen. Microbiol. *118*: 465-470.
23. Dawson, C. O., and J. C. Gentles. 1959. Perfect stages of *Keratinomyces ajelloi.* Nature (Lond.) *183*: 1345-1346.
24. Dawson, C. O., and J. C. Gentles. 1961. The perfect states of *Keratinomyces ajelloi* Vanbreuseghem, *Trichophyton terrestre* Durie and Frey, and *Microsporum nanum* Fuentes. Sabouraudia *1*: 49-57.
25. De Bary, A. 1884. *Vergleichende Morphologie und Biologie die Pilze, Mycetozoen, und Bacterien,* Leipzig; English transl., Oxford, 1887.
26. De Vries, G. A. 1971. Observations on *Trichophyton tonsurans.* Sabouraudia *9*: 1-5.
27. Dodge, C. W. 1935. *Medical Mycology,* Mosby, St. Louis.
28. Emmons, C. W. 1934. Dermatophytes. Natural grouping based on the form of the spores and accessory organs. Arch. Dermatol. Syphilol. *30*: 337-362.
29. Emmons, C. W., C. H. Binford, J. P. Utz, and K. J. Kwon-Chung. 1977. *Medical Mycology,* Lea & Febiger, Philadelphia.
30. English, M. P., and P. M. Stockdale. 1968. *Trichophyton proliferans* sp. nov., a human pathogen. Sabouraudia *6*: 267-270.
31. Florian, E., and J. Galgoczy. 1964. *Keratinomyces longifusus* sp. nov. from Hungary. Mycopathol. Mycol. Appl. *24*: 73-80.
32. Frey, D., R. I. Oldfield, and R. C. Bridger. 1979. *Color Atlas of the Pathogenic Fungi,* Year Book Medical Publishers, Chicago.

33. Georg, L. K., and L. B. Camp. 1957. Routine nutritional tests for the identification of dermatophytes. J. Bacteriol. *74*: 113-121.
34. Georg, L. K., W. Kaplan, and L. B. Camp. 1957. Equine ringworm with special reference to *Trichophyton equinum*. Am. J. Vet. Res. *18*: 798-810.
35. Georg, L. K., W. Kaplan, and L. B. Camp. 1957. *Trichophyton equinum*—a reevaluation of its taxonomic status. J. Invest. Dermatol. *29*: 27-37.
36. Georg, L. K., L. Ajello, L. Friedman, and S. A. Brinkman. 1962. A new species of *Microsporum* pathogenic to man and animals. Sabouraudia *1*: 189-196.
37. Griffin, D. M. 1960. A perfect stage of *Microsporum gypseum*. Nature (Lond.) *186*: 94-95.
38. Gruby, D. 1841. Mémoire sur une végétation qui constitute la vraie teigne. C. R. Acad. Sci. (Paris) *13*: 72-75.
39. Gruby, D. 1841. Sur les mycodermes qui constituent la teigne faveuse. C. R. Acad. Sci. (Paris) *13*: 309-311.
40. Gruby, D. 1842. Sur une espèce de mentagre contagieuse résultante du développement d'un nouveau cryptogame dans la racine depoils de la barbe de l'homme. C. R. Acad. Sci. (Paris) *15*: 512-513.
41. Gruby, D. 1842. Recherches anatomiques sur une plante cryptogame qui constitue le vrai muguet des enfants. C. R. Acad. Sci. (Paris) *14*: 634-636.
42. Gruby, D. 1843. Recherches sur la nature, la siège et le développement du porrigo décalvans ou phytoalopécie. C. R. Acad. Sci. (Paris) *17*: 301-303.
43. Gruby, D. 1844. Recherches sur les cryptogames qui constituent la maladie contagieuse du cuir chevelu décrit sous le nom de teigne tondante (Maton), Herpes tonsurans (Cazenave). C. R. Acad. Sci. (Paris) *18*: 583-585.
44. Heronga, M., and S. Watanabe. 1980. Mating behavior of 334 Japanese isolates of *Trichophyton mentagrophytes*. Mycologia *72*: 1159-1170.
45. Howard, D. H., and N. Dabrowa. 1979. Mutants of *Arthroderma benhamiae*. Sabouraudia *17*: 35-50.
46. Jeffries, C. D., E. Reiss, A. Padhye, and L. Ajello. 1981. The relationship between *Trichophyton kuryangei* and *T. megninii*. Abstr. Annu. Meet. Am. Soc. Microbiol., 1981, F 50, p. 321.
47. Keddie, F. M. 1969. Medical mycology, 1841-1870. In *Medicine and Science in the 1860s*, Wellcome Institute of the History of Medicine, London, pp. 137-140.
48. Koneman, E. W., G. D. Roberts, and S. F. Wright. 1978. *Practical Laboratory Mycology*, Williams & Wilkins, Baltimore.
49. Krammer, H., and S. G. Knight. 1959. Studies on the infectivity of nutritional mutants of *Trichophyton mentagrophytes*. J. Invest. Dermatol. *32*: 621.
50. Kwon-Chung, K. J. 1967. Studies on the sexuality of *Nannizzia*. I. Heterothallism vs. fertile isolates. Sabouraudia *6*: 5-13.
51. Kwon-Chung, K. J. 1969. Studies on the sexuality of *Nannizzia*. II. Morphogenesis of gametangia in *N. incurvata*. Mycologia *61*: 593-605.
52. Kwon-Chung, K. J. 1972. Genetic study of the incompatibility system in *Arthroderma simii*. Sabouraudia *10*: 74-78.
53. Kwon-Chung, K. J. 1974. Genetics of fungi pathogenic for man. Crit. Rev. Microbiol. *3*: 115-133.
54. Leleu, L. 1908. *Le Dr. Gruby*. P.-V. Stock, Editeur, Paris.
55. Loeffler, F. 1887. Vorlesungen über die geschichtliche Entwickelung der Lehre von den Bakterien. F. C. W. Vogel, Leipzig (English transl.: D. H. Howard, Coronado Press, Kansas City, 1982).
56. Medical Research Council. 1977. *Nomenclature of Fungi Pathogenic to Man and Animals*. Memorandum No. 23 (4th ed.), Her Majesty's Stationery Office, London.

57. Miguens, M. P. 1968. Un nuevo dermatofito de origin tropical; *Trichophyton fluviomuniense* sp. nov. Sabouraudia *6*: 312-317.
58. Moraes, M., A. A. Padhye, and L. Ajello. 1975. The perfect state of *Microsporum amazonicum*. Mycologia *67*: 1109-1113.
59. Nannizzi, A. 1927. Richerche sull'origine saprofitica del funghi delle tigne. 2. *Gymnoascus gypseum* sp. n. forma ascofora del *Sabouraudites* (*Achorion*) *gypseum* (Bodin) Ota et Langeron. Atti Accad. Fisiocr. Siena *10*: 89-97.
60. Padhye, A. A., and J. W. Carmichael. 1969. Mating behavior of *Trichophyton mentagrophytes* varieties paired with *Arthroderma benhamiae* mating types. Sabouraudia *7*: 178-181.
61. Padhye, A. A., and J. W. Carmichael. 1970. Mating reactions of pigmented and non-pigmented isolates of *Arthroderma uncinatum*. Sabouraudia *8*: 112-115.
62. Padhye, A. A., and J. W. Carmichael. 1971. The genus *Arthroderma* Berkeley. Can. J. Bot. *49*: 1525-1540.
63. Padhye, A. A., and J. W. Carmichael. 1973. The mating reaction in the *Trichophyton terrestre* complex. Sabouraudia *11*: 64-69.
64. Padhye, A. A., I. Weitzman, and L. Ajello. 1979. Mating behavior of *Microsporum equinum* with *Nannizzia otae*. Mycopathologia *69*: 87-90.
64a. Padhye, A. A., C. N. Young, and L. Ajello. 1980. Hair perforation as a diagnostic criterion in the identification of *Epidermophyton, Microsporum,* and *Trichophyton* species. Proceedings of the Fifth International Conference on the Mycoses. Pan Am. Health Organ. Sci. Publ. No. 396, Washington, D.C., pp. 115-120.
65. Padhye, A. A., F. Blank, P. J. Koblenzer, S. Spatz, and L. Ajello. 1973. *Microsporum persicolor* infection in the United States. Arch. Dermatol. *108*: 561-562.
66. Philpot, C. M. 1977. Some aspects of the epidemiology of tinea. Mycopathologia *62*: 3-13.
67. Philpot, C. M. 1977. The use of nutritional tests for the differentiation of dermatophytes. Sabouraudia *15*: 141-150.
68. Philpot, C. M. 1978. Geographical distribution of the dermatophytes: a review. J. Hyg. *80*: 301-313.
69. Philpot, C. M. 1978. Serological differences among the dermatophytes. Sabouraudia *16*: 247-256.
70. Pore, P. S., and O. A. Plunkett. 1967. Biological species and variations in *Arthroderma*. Mycopathol. Mycol. Appl. *31*: 225-241.
71. Rebell, G., and D. Taplin. 1970. *Dermatophytes: Their Recognition and Identification,* University of Miami Press, Coral Gables (see 4th printing, 1979, for citations in this chapter).
72. Remak, R. 1840. Zur kenntniss von der pflanzlichen Natur der Porrigo lupinosa. Med. Ztg. *9*: 73-74.
73. Remak, R. 1842. Gelungene Impfung des Favus. Med. Ztg. *11*: 137.
74. Remak, R. 1845. *Diagnostische und pathogenetische Untersuchungen in der Klinik des Hern Geh. Raths D. Schoenlein auf dessen Veranlassung angestellt und mit Benutzung andersweitiger Beobachtungen veroffentlecht.* Hirschwald, Berlin.
75. Rippon, J. W. 1967. Elastase: production by ringworm fungi. Science *157*: 947.
76. Rippon, J. W. 1974. *Medical Mycology,* Saunders, Philadelphia.
77. Rippon, J. W., and E. D. Garber. 1969. Dermatophyte pathogenicity as a function of mating type and associated enzymes. J. Invest. Dermatol. *53*: 445.
78. Rippon, J. W., and M. Medenica. 1964. Isolation of *Trichophyton soudanense* in the United States. Sabouraudia *3*: 301-302.
79. Rivalier, E. 1953. Description de *Sabouraudites praecox* nova species suivie de remarques sur le genre *Sabouraudites*. Ann. Inst. Pasteur *86*: 276-284.

80. Robin, C. 1853. *Histoire naturelle des végétaux parasites qui croissent sur l'homme et sur les animaux vivants,* Paris.
81. Rush-Munro, F. M., and J. M. B. Smith. 1971. Further observations on *Trichophyton erinacei* and *T. proliferans.* Sabouraudia *9*: 61-64.
82. Rush-Munro, F. M., J. M. B. Smith, and D. Borelli. 1970. The perfect state of *Microsporum ramosum.* Mycologia *62*: 856-859.
83. Sabouraud, R. J. A. 1910. *Les Teignes,* Masson et Cie, Paris.
84. Schechter, Y., J. W. Landau, N. Dabrowa, and V. D. Newcomer. 1966. Comparative disc electrophoretic studies of proteins from dermatophytes. Sabouraudia *5*: 144-149.
85. Schoenlein, J. L. 1839. Zur Pathogenie der Impetigines. Arch. Anat. Physiol., p. 82.
86. Sekhon, A. S., A. A. Padhye, and J. Carmichael. 1974. Disc-gel electrophoresis of + and – strains and of gymnothecial cultures of *Arthroderma tuberculatum.* Sabouraudia *12*: 12-17.
87. Shadomy, H. J., and C. M. Philpot. 1980. Utilization of standard laboratory methods in the laboratory diagnosis of problem dermatophytes. Am. J. Clin. Pathol. *74*: 197-201.
88. Smith, J. M. B., R. D. Jolly, L. K. Georg, and M. D. Connole. 1968. *Trichophyton equinum* var. *autotrophicum;* its characteristics and geographical distribution. Sabouraudia *6*: 296-304.
89. Stockdale, P. M. 1961. *Nannizzia incurvata* gen. nov., sp. nov., a perfect state of *Microsporum gypseum* (Bodin) Guiart et Grigorakis. Sabouraudia *1*: 41-48.
90. Stockdale, P. M. 1964. Personal observations on the production of sexual forms of dermatophytes. Ann. Soc. Belge Med. Trop. *44*: 821-830.
91. Stockdale, P. M. 1967. *Nannizzia persicolor* sp. nov., the perfect state of *Trichophyton persicolor.* Sabouraudia *5*: 355-359.
92. Stockdale, P. M. 1968. Sexual stimulation between *Arthroderma simii* Stockdale, Mackenzie and Austwick and related species. Sabouraudia *6*: 176-181.
93. Takashio, M. 1973. Etude des phénomènes de reproduction liés au vieillissement et au rajeunissement des cultures de champignons. Ann. Soc. Belge Med. Trop. *53*: 427-580.
94. Takashio, M. 1977. The *Trichophyton mentagrophytes* complex. In K. Iwata, (Ed.), *Recent Advances in Medical and Veterinary Mycology,* Proceedings of the Sixth Congress of the International Society for Human and Animal Mycology, University of Tokyo Press, Tokyo, pp. 271-276.
95. Takashio, M. 1979. Taxonomy of dermatophytes based on their sexual states. Mycologia *71*: 968-976.
96. Tulasne, L. R., and C. 1861-1865. *Selecta fungorum carpologia,* 3 vols., Paris (English transl. by W. B. Grove, edited by A. H. R. Buller, 1931, Oxford).
97. Vanbreuseghem, R. 1958. *Mycoses of Man and Animals* (English transl. by J. Wilkerson), Thomas, Springfield, Ill.
98. Vanbreuseghem, R., C. DeVroey, and M. Takashio. 1970. Production of macroconidia by *Microsporum ferrugineum* Ota 1922. Sabouraudia *7*: 252-256.
99. Weitzman, I. 1980. Genetic studies of mating reactions of zoopathogenic fungi. In E. S. Kuttin and G. L. Baum (Eds.), *Human and Animal Mycology,* Excerpta Medica, Amsterdam, pp. 251-254.
100. Weitzman, I., and A. A. Padhye. 1976. Is *Arthroderma simii* the perfect state of *Trichophyton quinckeanum*? Sabouraudia *14*: 65-74.
101. Weitzman, I., and M. Silva-Hutner. 1970. Genetic studies on the segregation of pigmentation in *Arthroderma uncinatum (=Trichophyton ajelloi).* Abstr. Annu. Meet. Am. Soc. Microbiol., 1970, MM 12, p. 120.

102. Weitzman, I., P. W. Allderdice, and M. Silva-Hutner. 1970. Chromosome numbers in species of *Nannizzia* and *Arthroderma*. Mycologia *62*: 89-97.

103. Weitzman, I., I. Kozma, and M. Silva-Hutner. 1969. Some observations on *Arthroderma uncinatum*. Sabouraudia *7*: 216-218.

104. Weitzman, I., P. W. Allderdice, M. Silva-Hutner, and O. J. Miller. 1968. Meiosis in *Arthroderma benhamiae (=Trichophyton mentagrophytes)*. Sabouraudia *6*: 232-237.

105. West, B. C., and K. J. Kwon-Chung. 1980. Mycetoma caused by *Microsporum audouinii*. Am. J. Clin. Pathol. *73*: 447-454.

106. Young, C. N. 1968. Pseudocleistothecia in *Trichophyton rubrum*. Sabouraudia *6*: 160-162.

7

On the Potentially Pathogenic Dematiaceous Hyphomycetes

G. S. de Hoog / Centraalbureau voor Schimmelcultures, Baarn, The Netherlands

I. INTRODUCTION

Much has been published about the medical significance of various groups of dematiaceous hyphomycetes. However, the amount of definite knowledge, particularly of the opportunistic Dematiaceae, is not commensurate with the number of pages written on the subject, since most articles either present no more than suppositions or notable correlations, or are taxonomically ambiguous.

In recent years, modern revisions of *Aureobasidium* [141], *Drechslera* [98], *Exophiala* [81], *Rhinocladiella* [81], *Sarcinomyces* [141], and *Tritirachium* [79] have become available. For some larger genera, however, such treatments are overdue. The taxonomic problems are particularly pressing in *Alternaria, Cladosporium,* and *Phialophora.* Notwithstanding the latter, most medical mycologists use fungal names as though everyone had a clear vision of what these stand for. In only a few cases can the identity of a particular fungus be checked later by study of either a good description and illustration, or of preserved material. The disadvantage of this poor level of reproducibility is particularly obvious where opportunistic fungi are concerned. When one is not sure of the identity of a reported fungus, no further research on this species can be carried out, and any later report of a strain involved in a mycosis under the same name does not yield corroborative evidence. For example, the strains involved in alternariosis probably belong to several species; consequently, it cannot be known whether *Alternaria alternata* is really an opportunistic fungus, or whether it has only been found growing in a lesion once or twice, which can be said of almost any ubiquitous fungus.

The problems are hardly less significant in the case of well-recognized diseases. Often, the fungi were identified by means of the clinical picture, leading to circular reasoning in the quest for the etiologic agent of the disease. Some disorders may possibly be less common than they are generally thought to be. Too often, doubtful literature data have been uncritically copied by later authors. For example, the long-supposed existence of American and Asian varieties to tinea nigra palmaris, mentioned by numerous authors, is based on a single, probably erroneous, publication. When related fungi may cause different diseases, uncritical use of names may even lead to complete chaos. For example, now that the black yeasts have been sorted out, it has become evident that under such names as *Phialophora dermatitidis* a host of different fungi have been included.

In summary, in most articles the case reports have been presented as singular, giving insufficient foundation on which to base later, inductive research. When data are not comparable, reviews are necessarily untrustworthy. In this chapter the author does not claim to contribute anything fundamental toward solving this problem. However, to avoid presenting too many unsure statements, type or authentic material of the described species has mostly been used when possible. This material was used regardless of whether it corresponded with the said pathogens in the literature, and hence it is not claimed that all the fungi described are potential pathogens, or that all potential dematiaceous pathogens are listed here. Nevertheless, I have tried to correlate the literature data to the taxonomy in the discussions on most species. As a consequence, these are rather long where the data are very ambiguous, but short for a few relatively well-established diseases.

II. DESCRIPTION OF GENERA AND SPECIES

A. *Alternaria* Nees ex Fr.

Alternaria Nees, *Das System der Pilze und Schwämme,* Würzburg, p. 72, 1816; ex Fr., *Syst. Mycol. 1*: 46, 1821
Macrosporium Fr., *Syst. Mycol. 3*: 373, 1832
Rhopalidium Fr. & Mont., in Mont., *Sylloge Generum Specierumque Cryptogamarum,* Paris, p. 297, 1856

Colonies spreading, velvety or lanose, gray, brown, or black. *Hyphae* hyaline, olivaceous or brown. *Conidiophores* differentiated, ascending or erect, usually solitary, unbranched or loosely branched, brown, septate, producing one to several conidia by sympodial growth. *Conidiogenous cells* terminal, integrated, usually becoming intercalary, with one to several dark conidial scars. *Conidia* catenate or solitary, dry, olivaceous to brown, smooth- or rough-walled, with several transverse and often also some longitudinal septa, ovoidal, ellipsoidal, obclavate or cylindrical, mostly beaked.

Type species: *Alternaria alternata* (Fr.) Keissl.

Discussion

Alternaria is a large and pluriform genus [98,99,155,217]. Three roughly defined groups can be distinguished. One consists mostly of plant pathogenic species with single, obclavate conidia with long apical beaks. A second, heterogeneous group comprises species with noncatenate, unbeaked conidia. The third group has species with catenate conidia with short beaks. This group contains several ubiquitous saprophytes, occasionally reported as human pathogens. Although Simmons [275] mentioned basic differences in conidiogenesis of the third *Alternaria* group and *Ulocladium* Preuss, the distinction between the two genera, when seen with the light microscope, is gradational. *Ulocladium* species usually have conidia with a pointed base at a very young stage, later becoming dark brown and coarsely warted. The conidia are noncatenate or produce secondary conidiophores; conidial scars on conidiophores are relatively inconspicuous. Young conidia of comparable species of *Alternaria,* on the contrary, mostly have a more or less rounded base, and are medium brown, smooth-walled or verruculose at maturity. The conidia are usually produced in chains; conidial scars on conidiophores are relatively conspicuous. Some strains, however, are accommodated with difficulty in either genus.

1. *Alternaria alternata* (Fr.) Keissl. Species Complex (Fig. 1)

Alternaria tenuis Nees, *Das System der Pilze und Schwämme,* Würzburg, p. 72, 1816, ex Pers. *Mycologia Europaea,* Erlangen, p. 20, 1822
= *Torula alternata* Fr., *Syst. Mycol. 3*: 500, 1832 (name change)
= *Alternaria alternata* (Fr.) Keissl., *Beih. Bot. Zentralbl. 29*: 434, 1912

Colonies under near-ultraviolet (UV) light rapidly expanding, appearing glassy with minute black specks of conidia on water agar with wheat straw; dark gray, powdery on hay infusion agar; black with lanose, dull gray aerial mycelium on oatmeal agar. *Hyphae* hyaline to pale olivaceous, 2.5-3 μm wide. *Conidiophores* usually arising solitarily, laterally from undifferentiated hyphae, simple, straight or flexuous, usually with one terminal conidial scar, occasionally elongating sympodially; conidiophores smooth- and rather thick-walled with 3-7 thinner septa, pale olivaceous brown, often darker near the apex, 20-50 μm long, 3-5 μm wide. *Conidia* formed in long, often branched chains, pale to dark olivaceous brown, apical cells somewhat lighter; conidia becoming verruculose but apical cell remaining smooth-walled; conidia with 3-5 (up to 7) transverse, often also with 1 (up to 3) longitudinal or oblique septa, obovoidal to obclavate, slightly constricted at the septa, abruptly tapering toward a unicellular, cylindrical beak which measures one-sixth to one-fourth (up to one-third) of the spore body; overall length (12-)23-34(-45) μm, width 7-10 μm.

Discussion

Alternaria alternata auct. probably comprises several, minutely distinct but highly variable [153,154] species. Since these species have not yet been sorted out, the description above has been based on CBS 603.78 = QM 9553, which is regarded by E. G. Simmons (personal communication) as an isolate close to his concept of typical *A. alternata.* Any synonymy at this stage is premature.

The *Alternaria* strains reported as human or animal pathogens clearly belong to several species. Pribram [240] repeatedly isolated a mold close to the *Alternaria* state of *Pleospora infectoria* Fuckel from skin lesions. The same species was found to cause maxillary osteomyelitis in an otherwise healthy human patient [119]. An *Alternaria* strain with

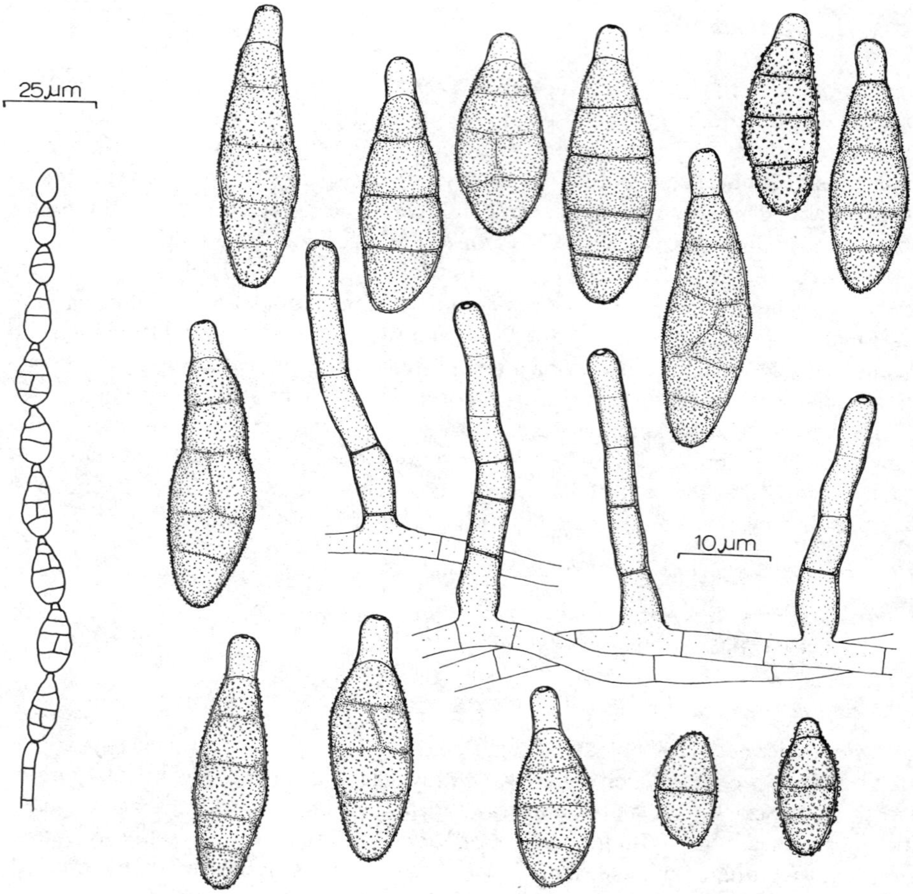

Figure 1 *Alternaria alternata,* CBS 603.78.

relatively large conidia was described as causing psoriasis-like skin lesions in cats and other mammals [107]. Specimens close to *Ulocladium* were reported from lacrymal ducts [115], from onychomycosis [84], and from pulmonary granuloma [180]. Pathogenic strains morphologically similar to the *A. alternata* above were reported several times [109,113, 200,224,238]. Numerous authors mentioned *A. alternata* or an unidentified *Alternaria* species as an occasional pathogen; from these reports the identity of the causative agents cannot be established.

English [105] demonstrated the ability to disintegrate human hairs in a variety of normally saprophytic species, among which was *Alternaria.* Botticher [41] found *Alternaria* species to be able to grow in keratinous structures. When in vivo, they are mainly found in hosts with a primary debilitating disease or an injury [109,200,232,263] or in association with another pathogenic fungus [142,261,267]. When such conditions are supposed to be absent, the fungus is mainly found producing nonspecific corneal ulcers [17,113], or human or animal skin lesions with ulcers, crusted papules, or scales [40-42, 84,93,109,200,233]. Hyphal fragments are often observed in the tissue. Experimental infection of small mammals frequently resulted in superficial eczematous lesions [92,227, 233,247]. In contradiction to this general picture, Ohashi [224] reported an epidemic internal disease caused by *A. alternata;* his inoculation experiments resulted in a high death rate among guinea pigs.

B. *Aureobasidium* Viala & Boyer

Aureobasidium Viala & Boyer, *C. R. Acad. Sci. (Paris) Ser. D, 112*: 1149, 1891
= *Aureobasis* Clem. & Shear, *The Genera of Fungi,* Hafner, New York, p. 343, 1931 (name change) (for synonymy, see Hermanides-Nijhof [141])

Colonies spreading, smooth, often covered with slimy masses of conidia, usually with sparse aerial mycelium; light brown, yellow, pink, or black. *Hyphae* with cells often wider than long, hyaline, frequently disarticulating into separate cells or becoming brown and thick-walled. *Conidiogenous cells* undifferentiated, lateral, terminal, or intercalary. *Conidia* blowing out simultaneously in dense groups, hyaline, smooth-walled, one-celled, rather variable in shape and size. Secondary conidia common; endoconidia often present.
Type species: *Aureobasidium pullulans* (de Bary) Arnaud.

Discussion

In her revision of the genus, Hermanides-Nijhof [141] restricted *Aureobasidium* to species with blastic conidia formed synchronously on hyaline, usually undifferentiated cells. In this sense *Aureobasidium* comprises mainly plant pathogens that were formerly classified in *Kabatiella.* Five more generic synonyms were added. She classified similar species that form conidia percurrently on intercalary cells in *Hormonema* Lagerberg & Melin. In the older literature the differences in conidiogenesis were often not recognized, so that the genera were frequently confused. Any data on the pathogenicity of *Aureobasidium* or *Pullularia* species should therefore be regarded with precaution. For similar reasons, the vast amount of literature on fungi under the names given above will not be dealth with here.

1. *Aureobasidium pullulans* (de Bary) Arnaud (Fig. 2)

Dematium pullulans de Bary, *Morphologie und Physiologie der Pilze Flechten und Myxomyceten,* Engelmann, Leipzig, p. 182, 1866
= *Aureobasidium pullulans* (de Bary) Arnaud, *Ann. Mycol. 8*: 475, 1910
= *Pullularia pullulans* (de Bary) Berkhout, *Schimmelgesl. Monilia, Oidium, Oospora en Torula, Dissertation,* Utrecht, p. 55, 1923
= *Hormonema pullulans* (de Bary) Robak, *Nytt Mag. Naturvidensk. 71*: 255, 1932

Colonies on malt agar rapidly expanding, attaining a diameter of about 40 mm in 8 days, appearing smooth, soon covered with a slimy mass due to abundant sporulation, yellow, cream, light pink or light brown, often becoming blackish due to chlamydospore production. *Hyphae* hyaline, smooth- and thin-walled, 2-16 μm wide, with cells often wider than long, forming a tough, compact mycelium. *Conidiogenous cells* undifferentiated, intercalary or terminal on hyaline hyphae, or arising as short lateral branches, producing conidia blastically in dense groups; at conidial detachment an indistinct scar or a denticle up to 1 μm in length remains. *Conidia* hyaline, smooth-walled, one-celled, rather variable in shape and size, usually 8-12 × 4-6 μm, often producing smaller, secondary conidia. Endoconidia, measuring about 6 × 3 μm, sometimes present. Thick-walled, dark brown chlamydospore-like cells common in aging cultures.
Neotype strain: CBS 584.75.

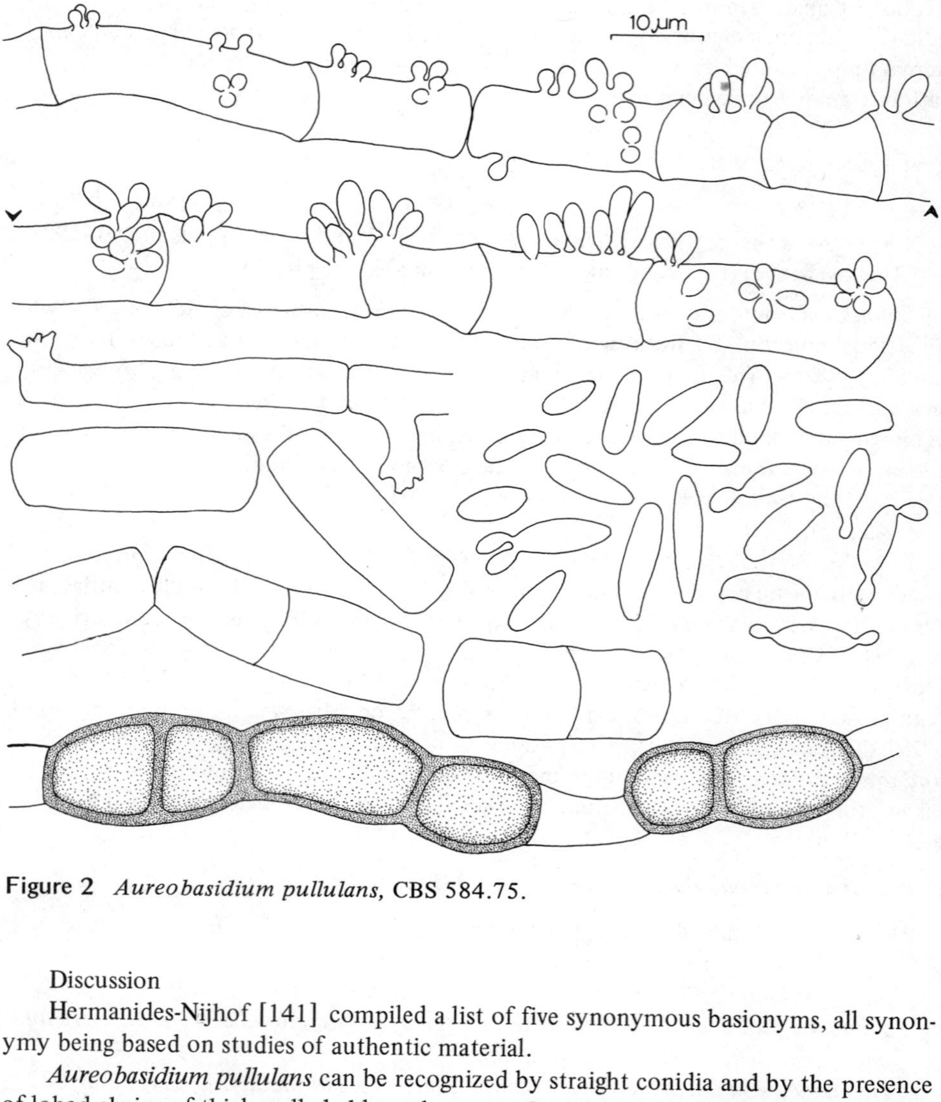

Figure 2 *Aureobasidium pullulans,* CBS 584.75.

Discussion

Hermanides-Nijhof [141] compiled a list of five synonymous basionyms, all synonymy being based on studies of authentic material.

Aureobasidium pullulans can be recognized by straight conidia and by the presence of lobed chains of thick-walled chlamydospores. In most strains chlamydospores occur rather infrequently. For practical reasons, Hermanides-Nijhof [141] described a variety, *melanigenum* Hermanides-Nijhof, which rapidly turns black due to abundant chlamydospore production.

Aureobasidium pullulans is a common saprophyte, particularly occurring on plant material, rotten wood, paper, foodstuffs, and similar substrates. It is also frequently found on the surface of phanerogams. Cooke [71,72] and Ciferri et al. [64] also presented an extensive review on its habitat, although it should be noted that part of their data refer to other species, incorrectly placed in synonymy with *A. pullulans.*

Species under the name *A. pullulans* are rather frequently isolated from humans and animals, only rarely being found to be pathogens [63,246]. Wynne and Gott [310] isolated strain CBS 298.56 from a lymph node of a patient with Hodgkin's granuloma. Cases of onychomycosis were reported by Vieira [302] and by Marill et al. [202]. Vermeil et al. [301] and Gordeff et al. [123] described an *A. pullulans* as the causative agent of dry, verrucose skin lesions on humans. Sarrat et al. [264] mentioned a strain that caused a small, ulcerating skin lesion. Salkin et al. [262] repeatedly isolated the species from a skin lesion of a porcupine. Akagi et al. [5] reported it as the cause of a pulmonary mycosis with sepsis, and Jones [156] isolated it from a corneal ulcer. Pathogenicity of isolated strains to mice was established by several authors [5,123,301]. Bulman and Stretton [48] reported experimentally induced disease in Dutch rabbits by means of CBS 298.56. In spite of these data, Ajello [2] suspected that *A. pullulans* occurs almost only as a contaminant.

The strain isolated from eczematous dermatitis and described by Leone [176] as *A. pullulans* was probably an *Exophiala* species.

C. *Cladosporium* Link ex Fr.

Cladosporium Link, *Mag. Ges. Naturforsch. Freunde, Berl. 7*: 37, 1815; ex Fr., *Syst. Mycol. 1*: 46, 1821

Colonies spreading or restricted, velvety, powdery or lanose; olivaceous green, brown, or gray. *Hyphae* pale brown, occasionally forming irregular chlamydospores or small sclerotial bodies. *Conidiophores* conspicuously differentiated or not, usually erect, unbranched to profusely branched in the apical region, brown, septate, producing conidial chains in more or less sympodial order. *Conidiogenous cells* with one to three dark, slightly prominent conidial scars, cylindrical and integrated or morphologically intermediate between conidia and hyphal cells, often being converted into ramoconidia or arthroconidia. *Conidia* catenate, dry, olivaceous to brown, smooth- or rough-walled, with 0-3 transverse septa, globose, fusiform or subcylindrical, each with one to three dark, slightly prominent scars.

Lectotype species: *Cladosporium herbarum* (Pers.) Link ex S.F. Gray.

Discussion

The generic delimitation is questionable. On the basis of different opinions on the circumscription of the genus, Ellis [98] and von Arx [15] presented different lists of generic synonyms. Many hundreds of species have been reported, but it is doubtful whether these all represent separate taxa; the genus urgently needs revision.

In medical literature the name *Hormodendrum* auct. has often been used for *Cladosporium*-like strains, although *Hormodendrum* Bonord. was based on a *Penicillium* species. The pathogenic taxa described as *Hormodendrum* were revised by de Hoog and Hermanides-Nijhof [82].

Several *Cladosporium* species are ubiquitous saprophytes, occurring most abundantly in the air. As a result, they are frequently isolated as culture contaminants and also from human and animal parts that are exposed to air. Few are pathogenic, causing quite different diseases. Supposed pathogenicity of strains similar to *C. cladosporioides* (Fres.) de Vries is mainly found in older literature, and has been doubted by later authors. However, the possibility of this species causing or being involved in human mycoses may not be excluded. Polack et al. [237] reported a dual infection of a traumatic cornea by *Phialophora verrucosa* Medlar and *C. cladosporioides.* Forster et al. [113] and Gugnani et al. [128] mentioned a case of corneal ulcer caused by *Cladosporium oxysporum* Berk. & Curt.

Recently, Borelli and Marcano [38] described a *Cladosporium castellanii* sp. nov., which was said to cause human tinea nigra in Venezuela. McGinnis and Padhye [190] found that the type strain, IMI 183818 = CBS 105.75, is identical to *Stenella araguata* Syd. Another strain from the same source, CBS 106.75 = Borelli 7441, was sent to the CBS with Borelli's statement that it was the best of the series of strains isolated from the tinea nigra case, including the type. This CBS 106.75, however, represents a species different from the type of *C. castellanii.* Since it is probable that the original specific concept was based on two different species, the name *C. castellanii* should be rejected.

KEY TO THE POTENTIALLY PATHOGENIC SPECIES OF *CLADOSPORIUM*

1a. Conidiophores with few septa, conspicuously nodose . . . . *C. herbarum*
1b. Conidiophores regularly septate in the apical region, smoothly constricted at the septa 2
2a. Conidia nearly all globose, a few fusiform . . . . *C. sphaerospermum*
2b. Conidia all fusiform or ellipsoidal, never globose 3
3a. Colonies powdery with irregularly folded center; conidia with rather inconspicuous scars about 0.5 μm wide . . . . *C. carrionii*
3b. Colonies velvety to hairy, regularly cushion-shaped; conidia with conspicuous scars . . . . *C. bantianum*

1. *Cladosporium herbarum* (Pers.) Link ex S. F. Gray (Fig. 3)

Dematium herbarum Pers., *Synopsis Methodica Fungorum,* H. Dieterich, Gottingae, p. 699, 1801
= *Acladium herbarum* (Pers.) Link, *Mag. Ges. Naturforsch. Freunde, Berl. 3*: 12, 1809
= *Cladosporium herbarum* (Pers.) Link, *Mag. Ges. Naturforsch. Freunde, Berl. 7*: 37, 1815
= *Byssus herbarum* (Pers.) DC., *Fl. Fr., 5*: 11, 1815
= *Cladosporium herbarum* (Pers.) Link ex S. F. Gray, *Nat. Arr. Br. Pl. 1*: 556, 1821
= *Sporocladium herbarum* (Pers.) Chev., *Fl. Gen. Environ. Paris 1*: 36, 1826
Perfect state: *Mycosphaerella tassiana* (de Not.) Joh.

Colonies restricted, powdery to velvety, regular; dark green, reverse greenish black with narrow white margin. Submerged *hyphae* pale olivaceous, 2.5-3.5 μm wide, with thin septa, often verruculose; sclerotial bodies composed of dark, thick-walled cells may be present. *Conidiophores* differentiated, usually simple, smooth-walled or basally verrucose, rather thick-walled, forming thin septa in a later stage, brown, 3-4.5 μm wide, up to 200 μm long; conidial chains are produced more or less sympodially in groups of two to six from one to three terminal or intercalary swellings 5-8 μm in width. *Conidia* formed in rather long, often branched, chains, finely echinulate, pale brown, lower conidia cylindrical to elongate, higher conidia broadly fusiform to ellipsoidal, continuous, 5.5-12 × 3-6 μm, or one-septate, 10-22 × 3-7 μm, with one to three black, slightly protuberant scars 0.8-1.5 μm in width.

Discussion

Hughes [147] compiled a list of 19 synonymous basionyms, all synonymy being based on studies of authentic material. The relationships with similar species have been elaborated by de Vries [85] and Ellis [98]. The conidiogenesis was described in detail by Hashmi et al. [138].

Cladosporium herbarum is an ubiquitous saprophyte, mainly isolated from plant material and frequently also from air and soil. The literature on its habitat has been re-

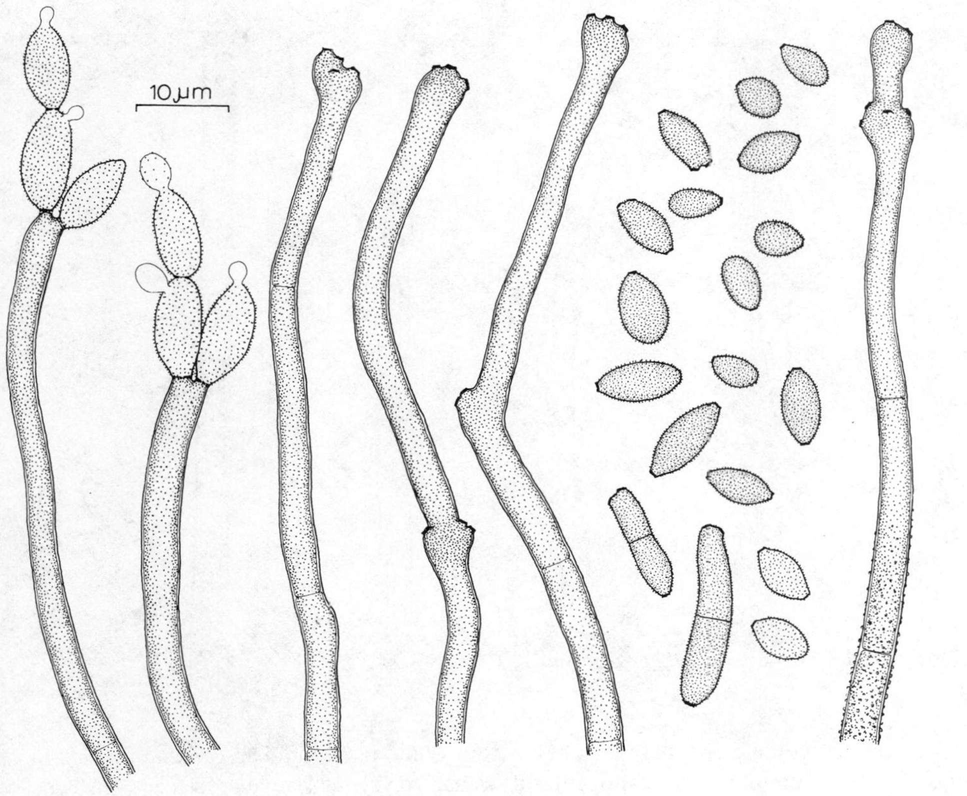

Figure 3 *Cladosporium herbarum,* CBS 177.71.

viewed by Domsch and Gams [89]. It rarely occurs as a pathogen on waterfowl and flamingos, forming black sclerotia on the outermost feathers, rupturing the cortex and causing break-off of distal portions [23,24]. Hubálek [145] experimentally proved *C. herbarum* to be a secondary pathogen on the skin of house sparrows. Bulman and Stretton [48] demonstrated that the species is not pathogenic in experimentally infected rabbits.

2. *Cladosporium sphaerospermum* Penz. (Fig. 4)

Cladosporium sphaerospermum Penz., *Michelia 2*: 473, 1882
Hormodendrum langeronii da Fonseca, Leão & Penido, *Sci. Med. 5*: 563, 1927
= *Cladosporium langeronii* (da Fonseca et al.) Vuill., *Les Champignons Parasites et les Mycoses de l'Homme,* Lechevalier, Paris, p. 78, 1931

Colonies restricted, slightly domed, often furrowed, powdery; dark grayish green, reverse greenish black with narrow white margin. Submerged *hyphae* pale olivaceous to brown, 2-4 μm wide, often verruculose, forming a compact mycelium. *Conidiophores* somewhat differentiated, usually simple, smooth-walled or verruculose, rather thick-walled, stalks about 3-6 μm wide and remotely septate at the base, slightly narrower toward the apex, apical part septate about every 10 μm, gradually narrower at the septa, the upper

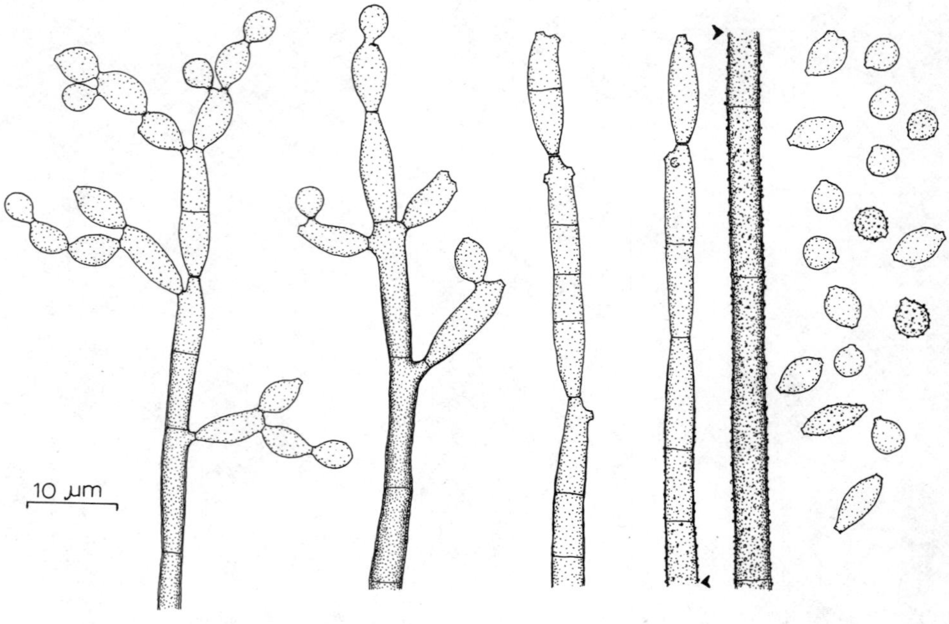

Figure 4 *Cladosporium sphaerospermum,* CBS 193.54.

cells later becoming converted into one- to four-celled ramoconidia. *Conidia* formed in rather long, often branched chains, smooth-walled to verrucose, pale olivaceous brown, often becoming distinctly warted and dark brown in aging cultures; terminal and subterminal conidia continuous, usually globose, 3-4 µm in diameter, lower conidia subglobose to broadly fusiform, up to about 15 µm in length, often septate, with one to three conspicuous, pigmented, often slightly protuberant scars about 1 µm in width.
Neotype strain: CBS 193.54.

Discussion

The species is easily distinguished from other common species of *Cladosporium* by the abundance of globose conidia. *Cladosporium cladosporioides* is similar, but distinct by more elongate conidia which are smooth-walled and rather pale.

Cladosporium sphaerospermum is an ubiquitous saprophyte, occurring on all sorts of substrates. It was originally described from *Citrus* leaves, but the neotype strain, CBS 193.54, was isolated from human nails. Male and Trappeiner [197] isolated the species from a case of nail infection following traumatic severance of the nerves. Silva and Carvalho [273] suspected it to be the etiologic agent in a case of chromomycosis. A *C. sphaerospermum*-like strain described by Otcenásek et al. [228] as *Hormodendrum cladosporioides* (Fres.) Sacc. was involved in psoriasis-like, later eczematous skin lesions on humans. The original strain of *H. langeronii,* CBS 189.54, reidentified as *C. sphaerospermum* by several authors [82,85], was isolated from human nodular skin lesions.

3. *Cladosporium carrionii* Trejos (Fig. 5)

Cladosporium carrionii Trejos, *Rev. Biol. Trop. 2*: 106, 1954

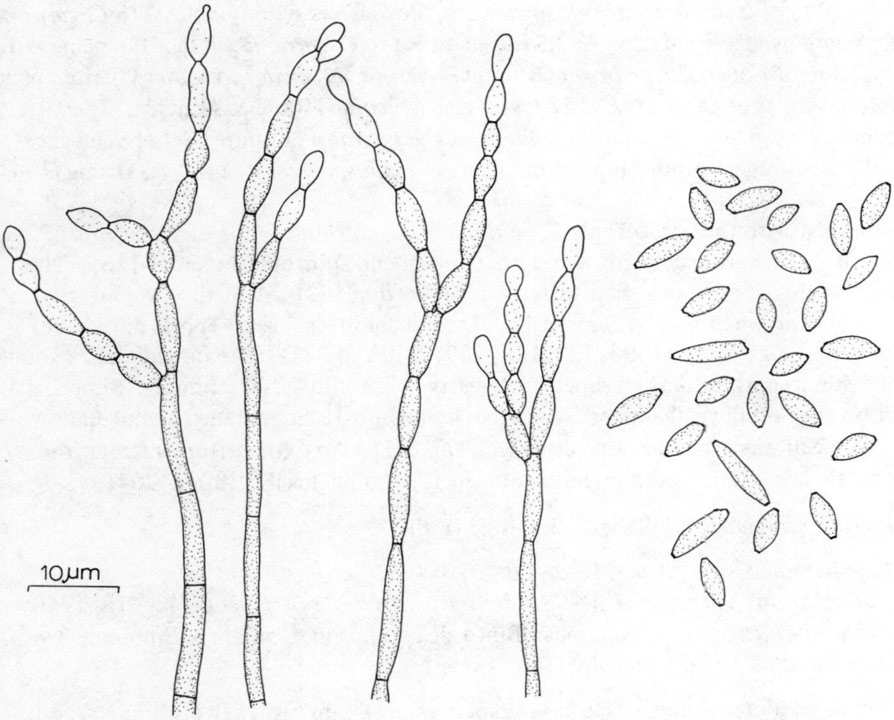

Figure 5 *Cladosporium carrionii,* CBS 165.54.

Colonies restricted, appearing powdery with elevated, slightly irregularly folded center, and smooth margin; olivaceous green, with olivaceous black reverse. *Hyphae* pale olivaceous, 1.5-2.5 μm wide, with thin septa, forming a compact mycelium. *Conidiophores* slightly or not differentiated, simple or loosely branched, with smooth, slightly thickened walls, distinctly septate, pale olivaceous, 2-3 μm wide, in the apical part the conidiophores become more profusely septate with constrictions at the septa, the upper cells are often converted into ramoconidia. *Conidia* formed in long, often branched chains, one-celled, smooth-walled or minutely warted, fusiform, 4.5-8.5 × 2.2-2.6 μm, with one to three dark, rather narrow scars.

Lectotype strain: CBS 165.54.

Discussion

The species is close to *Cladosporium bantianum* (Sacc.) Borelli, but markedly smaller on average; a statistical comparison has been carried out by Trejos [295]. In addition, the colonies are rather flat, often powdery, whereas those of *C. bantianum* are cushion-shaped and velvety to hairy. Differentiation of the two species by means of fluorescent-antibody techniques has been described by Al-Doory and Gordon [9]. Some strains have a rather strong resemblance to *Rhinocladiella pedrosoi* (Brumpt) Schol-Schwarz, being mainly distinguished by undifferentiated conidiophores and longer conidial chains. The general taxonomic position has been discussed on the basis of cell wall composition [289] and the ultrastructure of the septa [73].

Trejos [295] listed *Fonsecaea cladosporium* "Powell" as synonymous with *C. carrionii,* but although Powell's strain, CBS 162.54, is identical to *C. carrionii,* in fact the name is a recombination of *Fonsecaea pedrosoi* (Brumpt) Negroni var. *cladosporioides* Carrión, which was based on the type of *Phialoconidiophora guggenheimia* Moore & Almeida. The latter name is now known as a synonym of *Rhinocladiella pedrosoi* (Brumpt) Schol-Schwarz [81,269]. Most older authors applied the name *F. cladosporium* to a species fitting *C. carrionii.*

In contrast with the neurotropic *C. bantianum, C. carrionii* is one of the agents of chromycosis in humans, occurring mainly in tropical and subtropical regions [46]. The fungi causing this disease were first sorted out by Carrión [53]. A further, well-documented study was presented by Simson [276]; some additional case reports have been published later (e.g., Refs. 11, 94, 112, 117, 209, 239, and 277). The fungus causes granulomatous skin lesions, leading to superficial, warty to cauliflower-like tumors. Superficial inoculation may produce characteristic dermal lesions [75]. Subcutaneous inoculation gives a nonspecific reaction or none at all [35,277,295]. An extract from a strain from human cutaneous lesions had no significant effects on animal cell cultures [205].

4. *Cladosporium bantianum* (Sacc.) Borelli (Fig. 6)

Torula bantiana Sacc., *Ann. Mycol. 10*: 320, 1912
= *Cladosporium bantianum* (Sacc.) Borelli, *Riv. Anat. Patol. Oncol. 17*: 618, 1960
Cladosporium trichoides Emmons in Binford, Thompson, Gorham & Emmons, *Am. J. Clin. Pathol. 22*: 540, 1952

Colonies restricted, circular, cushion-shaped, up to 6 mm high, velvety to hairy; olivaceous green, with olivaceous black reverse. *Hyphae* olivaceous brown, straight and rather firm, 1.5-4 μm wide, regularly septate. *Conidiophores* slightly differentiated or undifferentiated, simple or loosely branched, with smooth, slightly thickened walls, pale olivaceous to brown, 1.5-3.5 μm wide; in the apical part the conidiophores becoming more profusely septate with constrictions at the septa; the upper cells are often converted into cylindrical to fusiform ramoconidia. *Conidia* formed in long, often branched chains, one-celled, smooth-walled, pale olivaceous, occasionally subhyaline, elongate to fusiform, with pointed or truncate ends, occasionally nearly cylindrical, 6-12.5 × 2.5-4 μm; with 1-2 (up to 3) pale pigmented, 0.5-1 (up to 2) μm wide scars.

Discussion

The original description of *Torula bantiana* and *Cladosporium trichoides* are quite different. Saccardo [258] described *T. bantiana* with conidia 8-11 × 5 μm. According to Binford et al. [29], *C. trichoides* has conidia measuring 4-7 × 2-2.5 μm. The type strain of the latter, CBS 173.52 = ATCC 10958, now has conidia which resemble hyphal cells, often being subhyaline, nearly cylindrical, and measuring about 8-12 μm in length. A more recent isolate deposited in the CBS collection, CBS 328.65, closely fits the authentic description of *C. trichoides,* except that the conidia are mostly slightly larger. It should be stated that the micrographs of CBS 173.52 given by Emmons et al. [101] with an indication of the magnification, show several conidia measuring over 10 μm and only a few less than 5 μm. It is therefore likely that all strains belong to a single species which tends to produce more cylindrical conidia when it is transferred onto artificial media. The suggestion of Emmons et al. [102] that *T. bantiana* is a nomen confusum, being based on the descriptions of Saccardo [258] and Borelli [34], is incorrect since any fungus name refers to the protolog.

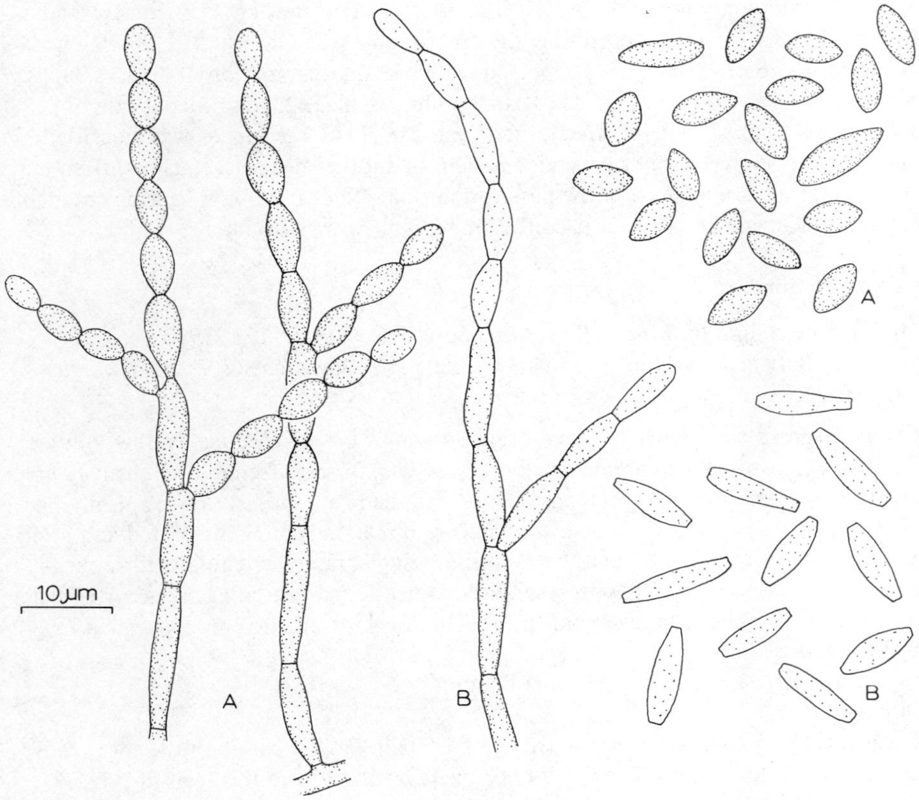

Figure 6 *Cladosporium trichoides*: (A) CBS 173.52; (B) CBS 328.65.

Cladosporium bantianum causes a human, rarely animal [149] cerebral phaeohyphomycosis which is usually fatal. Multiple, granulomatous brain abscesses are formed, and the disorder is often accompanied by chronic meningitis. The disease is well documented now; about 20 confirmed cases have been described in the literature (e.g., Refs. 19, 29, 69, 77, 95, 252, 272, 288, 294, and 306). The majority of these reports has been reviewed by Bennett et al. [25]. The pathologic histology has been studied by Araviyskiy and Aronson [13] in particular.

The brain disease caused by *C. bantianum* is cosmopolitan, though rare. The species has twice been isolated from soil [168,204] and once from bark [87]; otherwise, it is known almost only from nervous tissue. The manner of entry has not been determined convincingly. The virulence and neurotropism of the fungus has been established experimentally by several authors [12,110,272,294]. Similar brain abscesses are occasionally provoked by *Rhinocladiella pedrosoi* (Ref. 181; as *C. trichoides,* Refs. 116 and 290). In one of these cases [116] the patient had also a dermal chromomycosis caused by the same fungus. *Rhinocladiella pedrosoi* is only able to provoke brian abscesses experimentally when inoculated intracranially [110].

Maheswari Amma et al. [195] reported a phaeohyphomycosis on the foot clinically resembling mycetoma. The isolated *C. bantianum* strains showed neurotropy in animal tests. Once a pulmonary infection by *C. bantianum* was reported [181]. However, all artificial pulmonary infections performed by Mandal and Mohapatra [199] showed no significant result. Saito et al. [259] reported widespread inflammation of air sacs, lungs, and brains of South American parrots. Nsanzumuhire et al. [223] ascribed a case of chromomycosis to *C. bantianum.* In another such case [114] the fungus was identified by means of histologic preparations only; *C. carrionii* or another dematiaceous pathogen may have been involved. Judging from the description of a said *C. trichoides* causing cutaneous ulcers given by Gugnani et al. [129], confusion with *C. carrionii* is likely.

D. *Curvularia* Boedijn

Curvularia Boedijn, *Bull. Jard. Bot. Buitenzorg, Ser. III, 13*: 123, 1933
Malustela Batista & de Lima in Batista, de Lima & de Vasconcelos, *Publ. Inst. Micol. Recife 263*: 5, 1960

Colonies spreading, velvety or hairy; gray, brown, or black. *Hyphae* hyaline, olivaceous or brown. Stromata sometimes present. *Conidiophores* differentiated, usually erect, solitary, unbranched, brown, septate, producing conidia in sympodial order, apically becoming flexuose or nodose. *Conidiogenous* cells terminal, becoming integrated and intercalary, with numerous dark conidial scars. *Conidia* single, rarely catenulate, dry, olivaceous to brown, smooth-walled, with several transverse septa, cells at each end often paler than those in the middle; conidia ovoid to subcylindrical, often curved due to an asymmetrical median cell.
Type species: *Curvularia lunata* (Wakker) Boedijn.

Discussion

The genus *Curvularia* was recently treated by Ellis [97-99], based on the specimens available at the CMI. About 30 described taxa were omitted from this revision. The generic delimitation is quite clear. The few catenulate species are close to *Alternaria* but have asymmetric conidia. *Drechslera* Ito comprises some species with *Curvularia*-like conidia, but these have pseudosepta. Species with cylindrical conidia resemble *Embellisia* Simmons, but the comparable species of the latter genus differ by having conidia with markedly thickened, occasionally oblique septa, and distinctive conidial scars.

Most *Curvularia* species occur on living or dead plant material. Some are commonly isolated from soil or other substrates. Occasional reports are available of species being parasitic on humans and animals. The etiologic strains are usually identified to be the most common *Curvularia* species without protuberant hila, *C. geniculata* and *C. lunata,* but according to Forster et al. [113] *C. pallescens* Boedijn, *C. senegalensis* (Speg.) Subram., and *C. verruculosa* Tandon & Bilgrami ex M. B. Ellis may also cause nonspecific corneal ulcers. *Curvularia pallescens* was also described as the cause of a pulmonary mycetoma with cerebral metastasis in a human patient [170].

KEY TO THE POTENTIALLY PATHOGENIC SPECIES OF *CURVULARIA*

1a. Conidia predominantly 4-septate . . . . *C. geniculata*
1b. Conidia predominantly 3-septate . . . . *C. lunata*

1. *Curvularia geniculata* (Tracey & Earle) Boedijn (Fig. 7)

Helminthosporium geniculatum Tracey & Earle, *Bull. Torrey Bot. Club 23*: 207, 1896

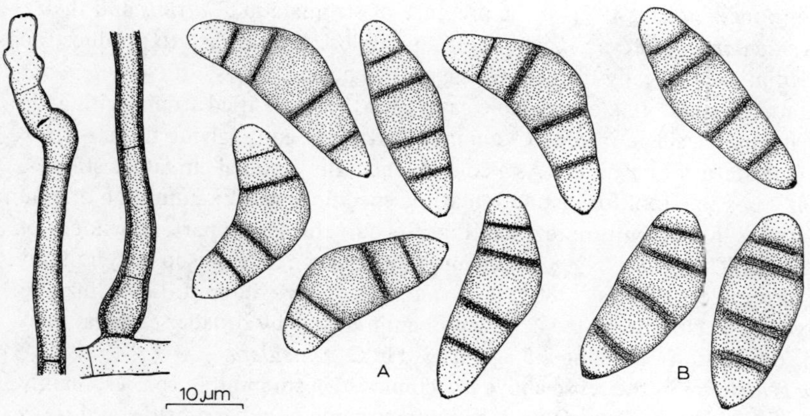

Figure 7 *Curvularia geniculata*: (A) CBS 220.52; (B) CBS 185.49.

= *Curvularia geniculata* (Tracey & Earle) Boedijn, *Bull. Jard. Bot. Buitenzorg, Ser. III, 13*: 129, 1933

Curvularia affinis Boedijn, *Bull. Jard. Bot. Buitenzorg, Ser. III, 13*: 130, 1933

Brachysporium sesami Sawada, *Spec. Publ. Coll. Agric. Nat. Taiwan Univ. 8*: 198, 1959

= *Curvularia sesami* (Sawada) Matsushima, *Icones Microfungorum a Matsushima lectorum,* Kobe, Japan, p. 42, 1975 (incidentally mentioned)

Perfect state: *Cochliobolus geniculatus* Nelson.

Colonies on PCA under near-UV light rapidly expanding, appearing dark gray, with cottony to lanose, grayish white to gray aerial mycelium. *Hyphae* hyaline to pale brown, (1-)2-4 µm wide, intermingled with hyphae up to 8 µm wide. *Conidiophores* arising singly, laterally from undifferentiated hyphae, simple, straight, apical portion flexuose with conidial scars on the edges; conidiophores smooth- and thick-walled, particularly near the base, with 3-15 thinner septa, brown, lighter toward the apex, 100-300 µm long, 4.5-8 µm wide, basal cell often swollen up to 12 µm. *Conidia* single, formed in sympodial order, smooth and rather thick-walled and brown, end cells with thinner walls and somewhat paler; conidia usually with (3-)4 septa, the median cell being the largest, conidia curved to nearly straight, fusiform with rounded ends; five-celled conidia 22-35 × 8-14 µm.

Discussion

In his treatment of the genus, Ellis [97] described *C. geniculata* as having distinctly curved conidia. *Brachysporium sesami,* which was enlisted into synonymy, had conidia with similar curvature. *Curvularia affinis* was described as being distinguished mainly by straight or nearly straight conidia. Groves and Skolko [126], however, when examining the type of *C. geniculata,* found that it had nearly straight conidia. Their species concept was adopted by most subsequent authors [47,204,281]. Strain CBS 333.64, sent by the discoverer of the perfect state, R. R. Nelson, also has nearly straight conidia. Matsushima [204] described the conidia of his *C. affinis* as being slightly larger. However, these measurements practically match those given by Groves and Skolko [126] of the type of *C. geniculata.* The type culture of *C. affinis,* CBS 154.34, now sporulates poorly; another authentic strain, CBS 185.49, has conidia in the same range. Another difference between the species, mentioned by Ellis [97], is the presence of stromata in *C. affinis* and their absence in *C. geniculata.* However, most strains rapidly lose their ability to produce such structures. Boedijn [32] already mentioned the occurrence of "races" of *C. geniculata* with and without stromata. In a comparable case, Ellis [97] classified strains with and without stromata all in a single species, *C. lunata* (Wakker) Boedijn, giving those with stromata the mere status of a variety. As a consequence, the conidial curvature still remains as the only possible tool for distinction at the specific level. Examination of a number of similar strains in pure culture revealed that the curvature is, in part, dependent on cultural conditions. Therefore, *C. geniculata* and *C. affinis* are here placed in synonymy, as was already proposed by Somal [280]. *Curvularia palmarum* Subram. [284] may be a further synonym. The closely related *C. fallax* Boedijn differs by broader conidia, the average length to breadth ratio being 2.5 against 3.0 in *C. geniculata.*

Curvularia geniculata in the sense above is a common, cosmopolitan species, mainly occurring saprophytically on phanerogams, sometimes being weakly parasitic, and is occasionally isolated from soil. Georg [120] reported it from a mycotic keratitis in humans. Although the identity of this strain was confirmed by Nelson, the photographs show a species with mainly 3-septate conidia, similar to *C. lunata.* The same holds true for the strains described by Nityananda et al. [222], also isolated from mycotic keratitis, and Brodey et al. [45], from a maduromycotic mycetoma with black granules in a dog. Similar tubefactions in a dog's foot were reported by Bridges [43]. The conidia of the etiologic agent were 4-septate. Another case of black-grain maduromycosis caused by a said *C. geniculata* was described by Baylet et al. [22]. In a similar case in horses [33], the causative fungus was not cultured. Photographs of the etiologic agent, given by Baylet et al. [22], showed 3-septate conidia only. Curiously, however, Nityananda et al. [222] referred to this strain as *C. lunata* in their literature review, although their own strain, depicted as being 3-septate, was referred to as *C. geniculata.* Kaufman [164] reported a said *C. geniculata* on a homografted aortic valve.

2. *Curvularia lunata* (Wakker) Boedijn (Fig. 8)

Acrothecium lunatum Wakker, in Wakker & Went, *Ziekten Suikerriet Java,* Brill, Leiden, p. 196, 1898

= *Curvularia lunata* (Wakker) Boedijn, *Bull. Jard. Bot. Buitenzorg, Ser. III, 13*: 127, 1933

Colonies on PCA under near-UV light rapidly expanding, appearing glassy with minute black specks of conidia, often with dark gray to black, lanose, aerial mycelium. *Hyphae* hyaline to pale brown, 1-5 μm wide. *Conidiophores* arising singly, laterally from undifferentiated hyphae, simple, straight, apical portion flexuose or somewhat nodose, with conidial scars on the edges; conidiophores smooth- and thick-walled, particularly near the base, with 3-15 thinner septa, brown, lighter toward the apex, 100-300 μm long, 4-6 μm wide, basal cell often swollen up to 12 μm. *Conidia* single, formed in sympodial order, smooth- and rather thick-walled, olivaceous brown to dark brown but end cells somewhat paler and with thinner walls; conidia usually with 3 septa, the subapical cell being the largest; conidia mostly curved, if straight slightly asymmetrical by septa of subapical cell not being parallel; conidia obovoidal to broadly clavate, 21-31 × 8.5-12 μm.

Discussion

Although the type is probably lost [230], the circumscription of the species is unambiguous (e.g., Refs. 47, 126, 171, 204, and 284). It has been amply described in connec-

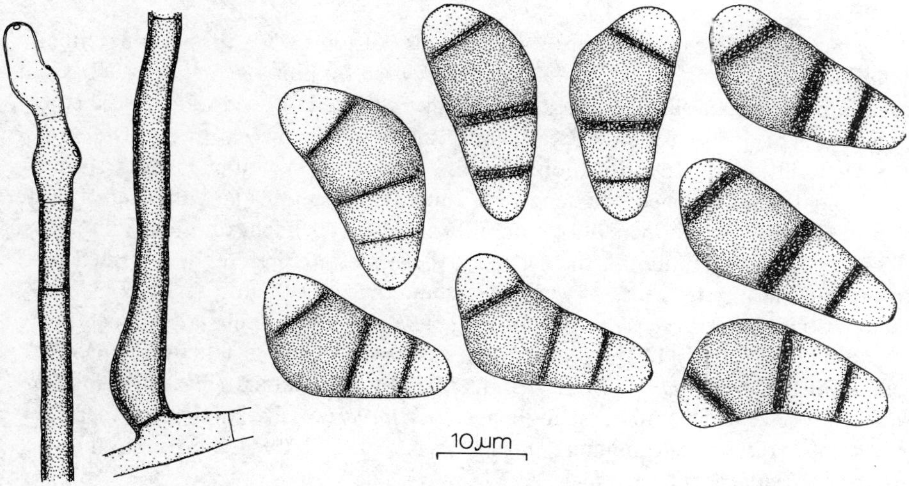

Figure 8 *Curvularia lunata,* CBS 159.33.

tion with similar species by Ellis [97,98], who accommodated strains with distinct stromata in the variety *aeria* (Batista et al.) M. B. Ellis. *Curvularia clavata* Jain is similar but most conidia of that species are straight; in *C. pradisii* R. L. & B. L. Mathur four-celled conidia are also common.

The species is ubiquitous, occurring most frequently on phanerogams and plant remains. It has sometimes been reported as causing human ulcerative keratitis [10,179, 221]. A *Curvularia* species close to *C. lunata,* also causing keratitis, was mentioned by Forster et al. [113]. Maghoub [196] described a case of black-grain mycetoma in humans, caused by *C. lunata.* Harris and Downham [135] described a *Curvularia* sp. close to *C. lunata* from a human lymph node. It should be stressed here that in the medical literature *C. lunata* may have been confused with *C. geniculata.* Wind and Polack [309] ascribed the etiologic agent in a case of keratitis to *C. lunata,* but a micrograph in their article shows a species with straight, multiseptate conidia.

E. *Drechslera* Ito

Angiopoma Lév., *Ann. Sci. Nat., Ser. 2, 16*: 235, 1841
Drechslera Ito, *Proc. Imp. Acad. Jpn. 6*: 355, 1930 (nom. cons.)
Bipolaris Shoemaker, *Can. J. Bot. 37*: 882, 1959

Colonies spreading, usually hairy; gray, brown, or black. *Hyphae* hyaline, olivaceous or brown. Stromata sometimes present. *Conidiophores* differentiated, usually erect, solitary, unbranched or loosely branched, brown, septate, producing conidia in sympodial order, apically becoming flexuose or nodose. *Conidiogenous cells* terminal, becoming integrated and intercalary, with numerous dark conidial scars. *Conidia* single, rarely catenulate or forming short secondary conidiophores, dry, pale to dark brown, smooth-walled, with several transverse pseudosepta, occasionally with some true septa, ellipsoidal, clavate or fusiform, often curved.
Lectotype species: *Drechslera tritici-vulgaris* (Nisikado) Ito ex Hughes.

Discussion

The genus *Drechslera* now comprises the majority of the species formerly described as *Helminthosporium* Link ex Fr. It was recently revised by Ellis [98,99] on the basis of the herbarium material available at the CMI. Species cultured on seed were described by Chidambaram et al. [59]. An older treatment of many species now known as *Drechslera* is that of Putterill [242]. Identification of strains cultivated on artificial media is sometimes problematic, since the conidia tend to become shorter and wider with every transfer. The sizes of conidia and numbers of pseudosepta are also dependent on the medium used [133,134,159,257]. To minimize these effects, the strains are preferably described on PCA or on water-agar with pieces of wheat-straw under near-UV light.

The relationships of *Drechslera* with other genera and the perfect state connections were discussed by Luttrell [183]. *Exserohilum* Leonard & Suggs differs by having dark, protuberant hila. *Curvularia* Boedijn is distinguished by the conidial septa, all being true. Several authors have advocated the maintenance of *Bipolaris* as a separate genus, partly on the basis of perfect state characters, but in absence of the latter the distinction of *Drechslera* and *Bipolaris* is often vague.

Most *Drechslera* species are pathogens of Gramineae, having a restricted host range. Some common saprophytes, however, are occasionally known to cause mycoses, mainly in animals. The closely related species *Exserohilum rostratum* (Drechsler) Leonard & Suggs [= *Drechslera rostrata* (Drechsler) Richardson & Fraser = *D. halodes* (Drechsler) Subram. & Jain] was occasionally reported from human corneal ulcers [113,157].

The *Helminthosporium* possible etiologic agent of bovine nasal granuloma is probably a *Drechslera* or *Exserohilum* species. McKenzie and Connole [191] and Pritchard et al. [241] isolated *E. rostrata* from several such cases. Patton [231], however, reported *D. spicifera* from a similar case. Numerous older reports of pathogenic *"Helminthosporium"* species are available, but their identity cannot be established with certainty.

KEY TO THE POTENTIALLY PATHOGENIC SPECIES OF *DRECHSLERA*

1a. Conidia almost exclusively 3-septate . . . . *D. spicifera*
1b. Conidia predominantly 5-septate . . . . *D. hawaiiensis*

1. *Drechslera spicifera* (Bain.) von Arx (Fig. 9)

Brachycladium spiciferum Bain., *Bull. Trimest. Soc. Mycol. Fr. 24*: 81, 1908
= *Curvularia spicifera* (Bain.) Boedijn, *Bull. Jard. Bot. Buitenzorg, Ser. III, 13*: 127, 1933
= *Helminthosporium spiciferum* (Bain.) Nicot, *Oesterr. Bot. Z. 100*: 482, 1953
= *Drechslera spicifera* (Bain.) von Arx, *The Genera of Fungi Sporulating in Pure Culture,* Cramer, Vaduz, p. 222, 1970
= *Bipolaris spicifera* (Bain.) Subram., *Hyphomycetes,* Indian Council of Agricultural Research, New Delhi, p. 756, 1971
Helminthosporium tetramera McKinney, *U.S. Dept. Agric. Bull. 1347*: 33, 1925
= *Curvularia tetramera* (McKinney) Gilman, *A Manual of Soil Fungi,* Iowa State College Press, Ames, Iowa, p. 303, 1945
= *Drechslera tetramera* (McKinney) Subram. & Jain, *Curr. Sci. 35*: 355, 1966
= *Bipolaris tetramera* (McKinney) Shoemaker, *Can. J. Bot. 37*: 884, 1969
Perfect state: *Cochliobolus spicifer* Nelson.

Colonies on PCA under near-UV light rapidly expanding, appearing glassy with minute black specks of conidia, with some central, gray aerial mycelium. *Hyphae* hyaline to pale

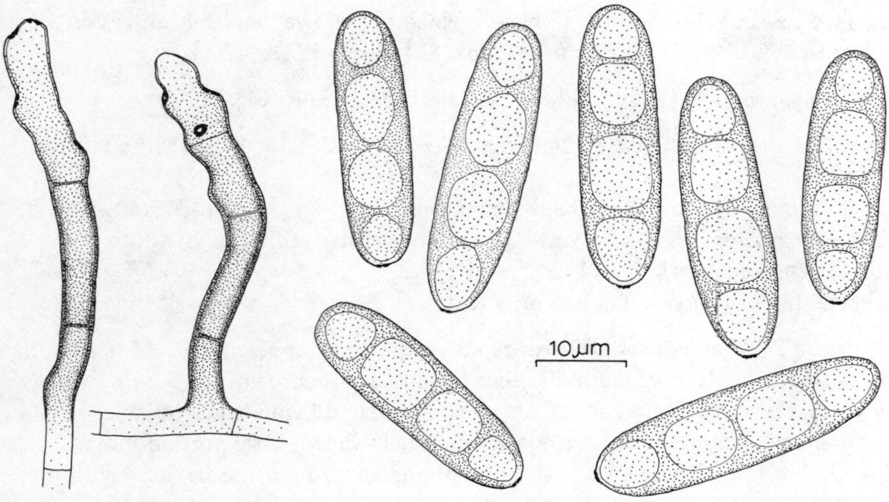

Figure 9 *Drechslera spicifera,* CBS 274.52.

brown, 3-6 μm wide. *Conidiophores* arising singly, laterally from undifferentiated hyphae or terminally from short side branches, simple, straight, smooth- and thick-walled, basal portion short, straight, with 0-4 septa, basal cell occasionally inflated; apical portion flexuose with conidial scars on the edges, of indefinite length, 4-6 μm wide, becoming multiseptate. *Conidia* single, formed in sympodial order, smooth- and rather thin-walled, brown except for a narrow subhyaline zone at the base and sometimes also at the tip; conidia invariably with 3 pseudosepta, cylindrical with rounded ends, 24-35 × 8-12 μm.

Discussion

Drechslera spicifera has almost exclusively 3-pseudoseptate conidia, a character that segregates it from nearly all other *Drechslera* species. *Drechslera australiensis* (Bugn.) Subram. & Jain ex M. B. Ellis is similar but has narrower conidia, usually below 10 μm in width; in *D. nicotiae* Mouchacca the conidia are wider, usually over 15 μm. The species was illustrated by Ellis [98], Nicot [218], Chidambaram et al. [59], and Matsushima [203].

The species is a common, weak parasite of phanerogams and is also found in soil and saprophytically on a variety of substrates. Muller et al. [215] reported multiple subcutaneous nodules in a cat, caused by *D. spicifera.* Granulomatous nodules, partly ulcerating, on a horse in debilitated condition, were found by Schauffler [265]. Human cutaneous phaeohyphomycosis was reported by Estes et al. [106]. Intraperitoneal inoculation of a pathogenic strain in mice remained without significant reaction [265]. Similar findings were given by Kaplan et al. [162], using a strain isolated from black, papulate cutaneous plaques on a horse. Granulomatous nodules in nostrils and other parts of the body of a cow were described by Patton [231]. Zapater et al. [314] and Forster et al. [113] mentioned *D. spicifera* as the etiologic agent of human corneal ulcers. Bridges and Beasley [44] described maduromycotic tumefactions with ulcerating areas occurring on a cat's foot. Hall [131] reported a *Helminthosporium* sp. causing similar nodular swellings on a dog. Judging from his photographs, probably *D. spicifera* was involved. Two cases of chronic pneumonia, of which one showed multiple abscesses, were said to be caused by a "*Helminthosporium*" sp. [88] close to *D. spicifera.*

2. *Drechslera hawaiiensis* (Bugn.) Subram. & Jain ex M.B. Ellis (Fig. 10)

Helminthosporium hawaiiense Bugn., *Rev. Gen. Bot. 62*: 238, 1955 (without Latin diagnosis)
= *Drechslera hawaiiensis* (Bugn.) Subram. & Jain, *Curr. Sci. 35*: 354, 1966; ex M. B. Ellis, *Dematiaceous Hyphomycetes,* Commonwealth Mycological Institute, Kew, Surrey, England, p. 415, 1971
Perfect state: *Cochliobolus hawaiiensis* Alcorn.

Colonies on PCA under near-UV light rapidly expanding, appearing blackish, smooth, or with gray, lanose aerial mycelium. *Hyphae* hyaline to pale brown, 1.5-3.5 μm wide. *Conidiophores* arising singly, laterally from undifferentiated hyphae or terminally from short side branches, simple, straight, smooth- and thick-walled, basal portion short, straight, with 0-4 septa, basal cell occasionally inflated; apical portion flexuose with conidial scars on the edges, up to 80 μm long, 3-5 μm wide, becoming multiseptate. *Conidia* single, formed in sympodial order, smooth- and rather thick-walled, brown, young conidia often with a subhyaline or pale brown zone at the base; conidia with (3-)5(-7) pseudosepta, cylindrical to nearly fusiform, with rounded or slightly acuminate ends, 18-35 × 6-9 μm.

Discussion

Recent descriptions of *D. hawaiiensis* are those of Ellis [98] and Chidambaram et al. [59]. The species is a weak parasite on a variety of phanerogams and is also isolated from soil, textiles, and other substrates, mainly in tropical and subtropical countries. A case of fatal human meningoencephalitis has been described by Fuste et al. [117]. Experimental infection of mice did not result in abnormalities in the brain, but the fungus did affect the liver and spleen. Some isolates from plants were also able to invade internal organs. A case of nasal obstruction and subsequent bone erosion in a human patient in apparent good health was described by Young et al. [312]. A *Helminthosporium* sp. was isolated

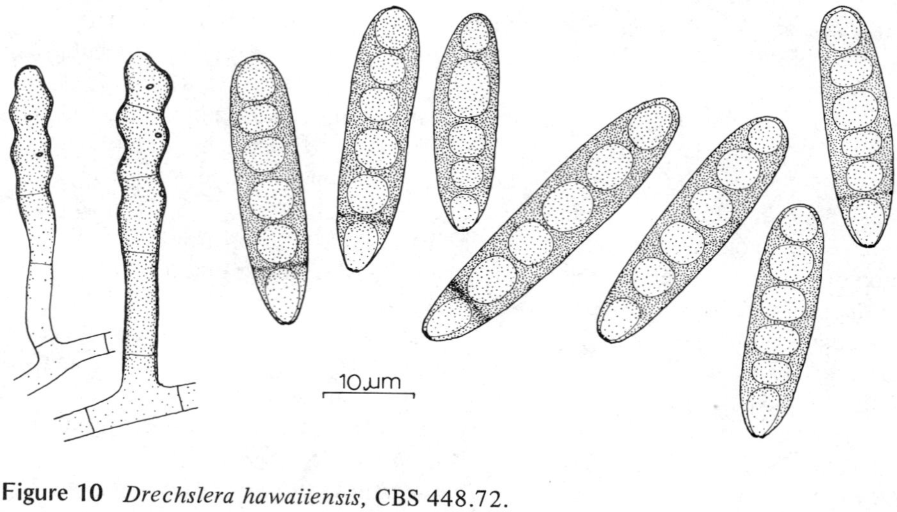

Figure 10 *Drechslera hawaiiensis,* CBS 448.72.

by Krachmer et al. [169] from several corneal ulcers. Judging from the illustration, they may have been dealing with a *D. hawaiiensis.*

Roberts et al. [254] described *Helminthosporium anomalum* Gilman & Abbott as the etiologic agent of granulomatous proliferations of the nasal mucosa in cattle. Their photograph shows a fungus superficially resembling *Drechslera hawaiiensis. Helminthosporium anomalum* was originally described as having large, multiseptate conidia. The type strain, CBS 161.27, is now sterile. Although Agnihotrudu [1] depicted under this name a species with protuberant hila, identity with the *Drechslera* state of *Cochliobolus carbonus* Nelson is more likely.

F. *Exophiala* Carmichael

Foxia Castell., *J. Trop. Med. Hyg. 11*: 261, 1908 (without description)
Nadsoniella Issatchenko, *Annu. Exped. Sci. Ind. Mourmansk 1906,* p. 273, 1914
Exophiala Carmichael, *Sabouraudia 5*: 122, 1966
Wangiella McGinnis, *Mycotaxon 5*: 354, 1977

Colonies restricted, appearing smooth or velvety, often initially mucous at the center; black, olivaceous green, or dark brown. *Hyphae* pale olivaceous, torulose at the center. *Conidiogenous cells* undifferentiated or somewhat differentiated, intercalary or free, often forming part of a more or less circumscribed, unbranched or brush-like conidiophore; subglobose to cylindrical, with one to three small butts or scars which produce conidia repeatedly. *Conidia* mucous, usually one-celled hyaline to pale olivaceous. Sometimes an accompanying *Rhinocladiella, Phaeococcomyces,* or *Phialophora* state is present. Sclerotial bodies sometimes occur.
Type species: *Exophiala salmonis* Carmichael.

Discussion

Exophiala is a form-genus of black yeast-like hyphomycetes in which conidia are formed in basipetal order from narrow, annellated zones. In addition, sympodial conidiogenous structures, classifiable in *Rhinocladiella* Nannf. may be present, and mere budding cells, for which the form-genus *Phaeococcomyces* de Hoog (= *Phaeococcus* de Hoog non Borzi) is available. Occasionally, phialides with collarettes (*Phialophora* Medlar) are present.

McGinnis [186] classified one *Exophiala* species in the newly erected genus *Wangiella* McGinnis. The conidiogenous cells of this genus were referred to as phialides without collarettes, in contrast to the annellides of *Exophiala.* However, the concerned species only differs from *Exophiala* in having a higher percentage of annellations which are too short to be seen with the light microscope. Scanning electron microscope (SEM) pictures clearly demonstrate their presence in the type strain, CBS 207.35 (Fig. 11). In strain CBS 525.76, classified by McGinnis [186] in *Wangiella,* the annellated zones are often even more distinct, being visible with the light microscope [81].

Exophiala was recently revised by de Hoog [81] and McGinnis [185-189]. The differences between these revisions mainly concern *Hormiscium dermatitidis* Kano, the type species of *Wangiella.* All strains assigned to this species by McGinnis [186] were classified by de Hoog [81] in *Exophiala.* However, I regard the type strain, CBS 207.35, as distinct from all other strains. The distinctive features seen in cultures grown on Sabouraud's glucose agar were the wide, hardly elongating annellated zones [66] and the local occurrence of phialides, as recently delimited by Cole and Samson [68]. In the remaining strains,

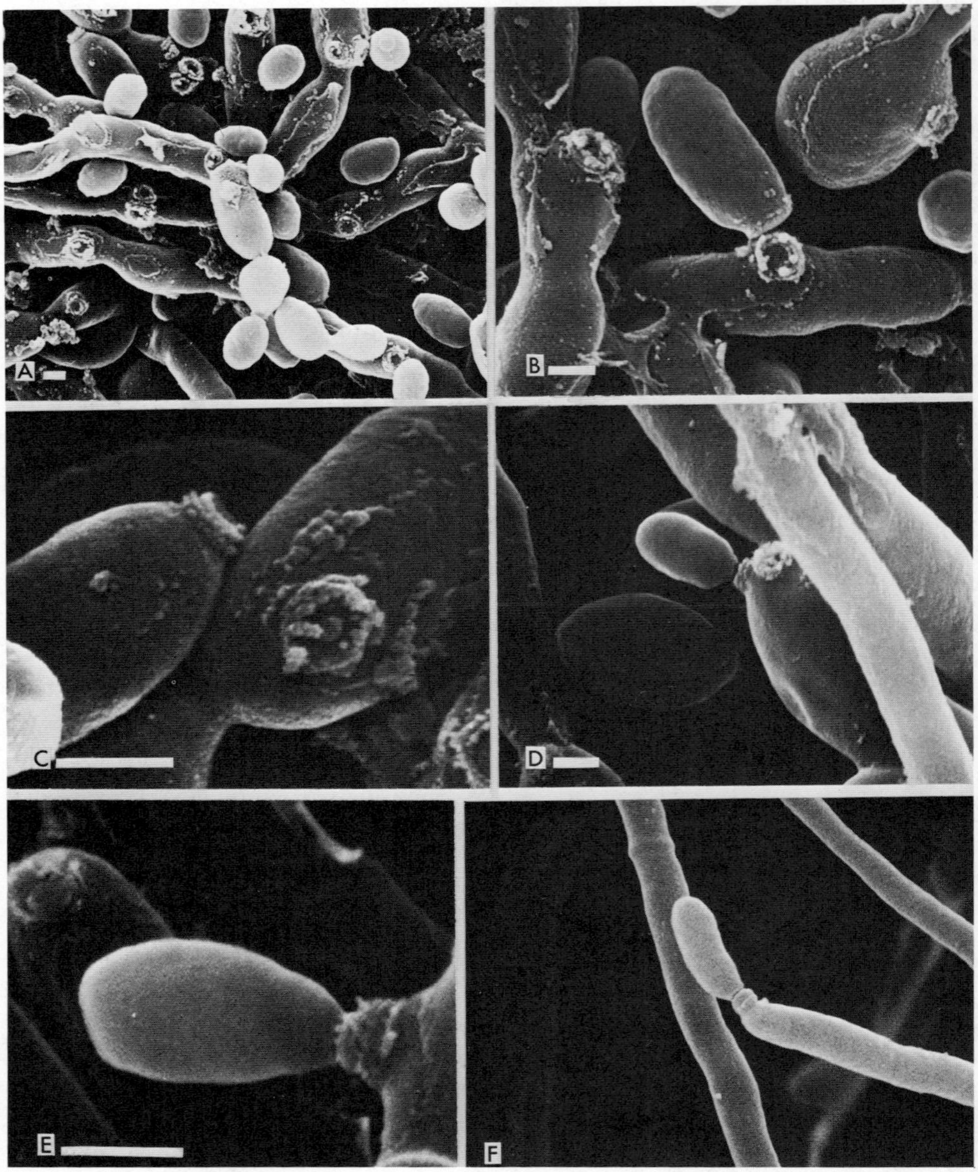

Figure 11 Scanning electron micrographs of conidiogenous cells of some *Exophiala* species: (A)-(C) *E. dermatitidis*, CBS 207.35; (D) *E. mansonii*, CBS 525.76; (E)-(F) *E. jeanselmei*, CBS 529.76. The markers indicate 1 μm.

the annellated zones, when present, show distinct annellations and do not taper toward the tip. These strains were classified in *E. mansonii* (Castell.) de Hoog. In this species the annellations may either be not or easily observable with the light microscope within the same strain, and are comparable to those of some red yeasts [315]. *Exophiala jeanselmei* (Langer.) McGinnis & Padhye has regular, tapering annellated zones and more spreading colonies than *E. mansonii*. For convenience, the large and variable species, *E. jeanselmei*, was subdivided into three varieties, according to their rates of differentiation. A summary of de Hoog's paper [81] is given below.

Several generic synonyms are known. One is *Foxia* Castell., based on *F. mansonii* (Castell.) Castell. (= *Exophiala mansonii*). To date no generic description has been found, the specific diagnosis being published separately; it should therefore be regarded as invalidly published, although Art. 42 of the Code of Botanical Nomenclature [283] is somewhat ambiguous on this point. *Nadsoniella nigra* Issatchenko, the only species described in *Nadsoniella* Issatchenko, is probably an earlier synonym of *Exophiala jeanselmei*. The black yeasts, being more or less intermediate between the yeasts and the hyphomycetes, have usually been omitted by nomenclatural indices and ignored by reviewers of the respective groups, so that the early descriptions of these fungi can only be found by chance. To prevent repetitive nomenclatural changes, and in the light of the uncertainty described above, it is felt that any name change should be postponed.

In the medical literature the species now known as *Exophiala* are quite confused, having been described as the etiologic agents of chromomycosis, mycetomas, phaeohyphomycosis, keratomycosis, tinea nigra palmaris, and cerebral chromomycosis. Most species cause human chromomycoses; others are pathogenic to fish. Though case reports of these diseases clearly indicate an *Exophiala* causative agent, identifications at the species level are often doubtful due to persistent taxonomic confusion. For this reason a literature review is omitted here.

KEY TO THE POTENTIALLY PATHOGENIC SPECIES OF *EXOPHIALA*

1a. Annellated zones 1-2 μm wide; conidia usually becoming septate, darkening with age . . . . *E. werneckii*
1b. Annellated zones narrower; conidia continuous or septate, hyaline 2
2a. Conidiogenous cells predominantly on torulose mycelium 3
2b. Conidiogenous cells also on unswollen hyphae or conidiophores 4
3a. Colonies attaining a diameter less than 6 mm in 14 days; annellated zones cylindrical, often with distinct annellations . . . . *E. mansonii*
3b. Colonies attaining a diameter of over 6 mm in 14 days; annellated zones tapering toward the tip, annellations indistinct . . . *E. jeanselmei* var. *heteromorpha*
4a. Free conidiogenous cells absent or very rare . . . . . . *E. jeanselmei* var. *lecanii-corni*
4b. Free conidiogenous cells common 5
5a. Free conidiogenous cells little inflated, often somewhat thick-walled and darker than the supporting hyphae 6
5b. Free conidiogenous cells usually inflated, ellipsoidal to subglobose, not darker than the supporting hyphae 7
6a. Conidiophores one- to two-celled; conidia ellipsoidal, rounded at the base, hyaline. . . . . *E. jeanselmei* var. *jeanselmei*
6b. Conidiophores two- to eight-celled; conidia subglobose, slightly narrowed toward the base, subhyaline. . . . . *E. spinifera*

7a. Conidia without discernible hilum; usually less than 4 μm long . . . . *E. dermatitidis*
7b. Conidia with a distinct, unpigmented hilum; over 5 μm long 8
8a. Conidia frequently 1-3-septate . . . . *E. salmonis*
8b. Conidia rarely septate . . . . *E. pisciphila*

1. *Exophiala salmonis* Carmichael (Fig. 12)

Exophiala salmonis Carmichael, *Sabouraudia, 5*: 122, 1966
= *Aureobasidium salmonis* (Carmichael) Borelli, *Med. Cutanea 3*: 588, 1969
Exophiala brunnea Papendorf, *Trans. Br. Mycol. Soc. 52*: 487, 1969

Colonies on oatmeal agar attaining a diameter of 8-14 mm in 14 days, appearing smooth, dry, soon becoming depressed lanose; gray to dull olivaceous green. *Budding cells* absent. *Hyphae* 2-3 μm wide. *Conidiogenous cells* slightly differentiated or undifferentiated, intercalary or free, the latter flask-shaped to subcylindrical, 7-21 × 2-3.8 μm; annellated zones up to 3 × 0.8 μm, with vague annellations. *Conidia* hyaline, ellipsoidal to short cylindrical, zero to one (up to four)-celled; continuous conidia 5.6-6.8 × 2-2.4 μm, two-celled conidia 6.5-8.5 × 2.2-3.5 μm.
Type strain: CBS 157.57.

Discussion

The fungus was originally described as the etiologic agent of an epidemic cerebral mycetoma of trout [49]. The disease showed a high rate of mortality, but the manner of entry of the fungus could not be determined in the experiment. Richards et al. [249] found *E. salmonis* forming granulomas in Atlantic salmon, particularly in the posterior kidneys. Further isolates are known from leaf litter and rotten wood. The process of conidiogenesis was said to be phialidic by Cole [65], but this statement was based on a misidentification [66].

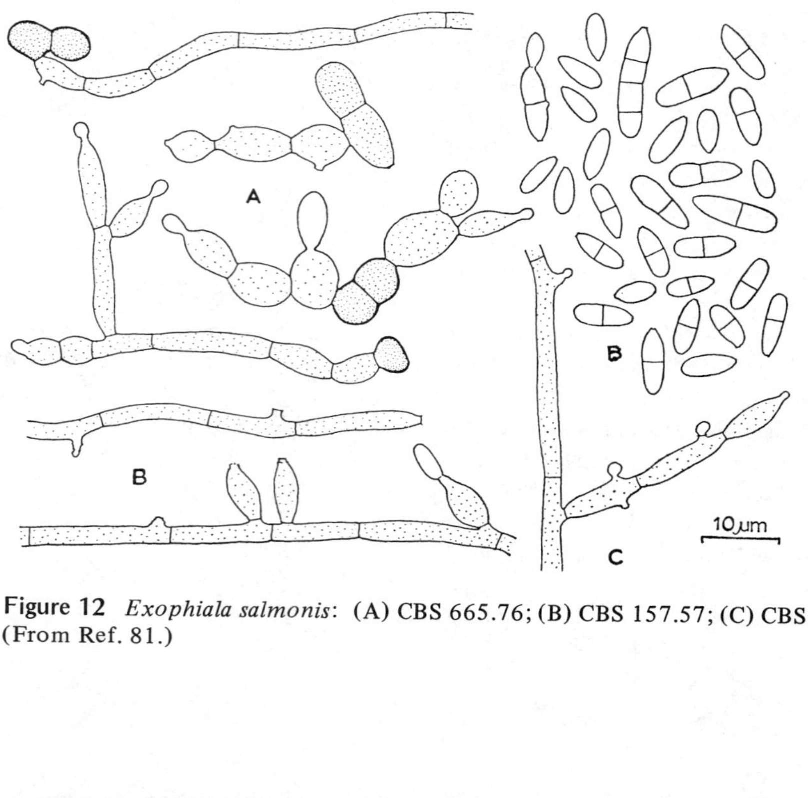

Figure 12 *Exophiala salmonis*: (A) CBS 665.76; (B) CBS 157.57; (C) CBS 665.76. (From Ref. 81.)

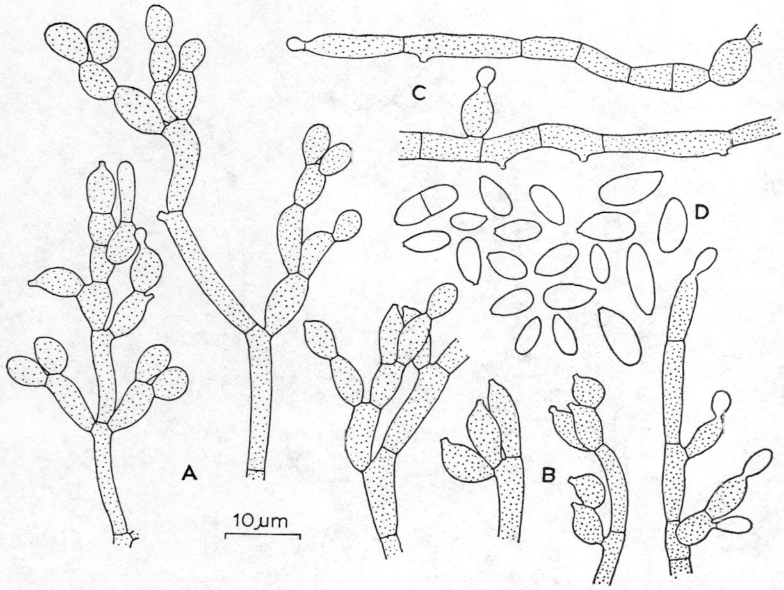

Figure 13 *Exophiala pisciphila* (A) CBS 100.68; (B,C) CBS 537.73; (D) various strains. (From Ref. 81.)

2. *Exophiala pisciphila* McGinnis & Ajello (Fig. 13)

Exophiala pisciphila McGinnis & Ajello, *Mycologia 66*: 518, 1974

Colonies on oatmeal agar attaining a diameter of 12-18 mm in 14 days, appearing smooth, dry, soon becoming thinly floccose; gray to blackish olivaceous. *Budding cells* absent. *Hyphae* 2-2.5 μm wide. *Conidiogenous cells* slightly or undifferentiated, usually free, ellipsoidal, 6-18 × 2-4.5 μm; annellated zones often very short, occasionally up to 2 μm long, 0.8-1.2 μm wide, with vague annellations. *Conidia* hyaline, broadly ellipsoidal to fusiform, rarely slightly curved, continuous, occasionally two-celled, 6-8 × 2.5-4 μm. Type strain: CBS 537.73.

Discussion

The fungus was originally described (Ref. 111) as *E. salmonis,* causing a systemic mycosis in catfish. Necrotic and granulomatous reactions were observed in several genera of captive fishes by Blazer and Wolke [31]. The disease was reproduced in three other genera by intraperitoneal injection of conidia. De Hoog [81] mentioned isolates from many different sources. The fungus is particularly common in soil, but due to its slow growth it is often overlooked (W. Gams, personal communication).

3. *Exophiala spinifera* (Nielsen & Conant) McGinnis (Fig. 14)

Phialophora spinifera Nielsen & Conant, *Sabouraudia 6*: 228, 1968
= *Exophiala spinifera* (Nielsen & Conant) McGinnis, *Mycotaxon 5*: 337, 1977
= *Rhinocladiella spinifera* (Nielsen & Conant) de Hoog, *Stud. Mycol. 15*: 93, 1977

Colonies on oatmeal agar attaining a diameter of 6-16 mm in 14 days, appearing slimy at the center, black, outer part with flocculose aerial mycelium, grayish olivaceous. *Budding*

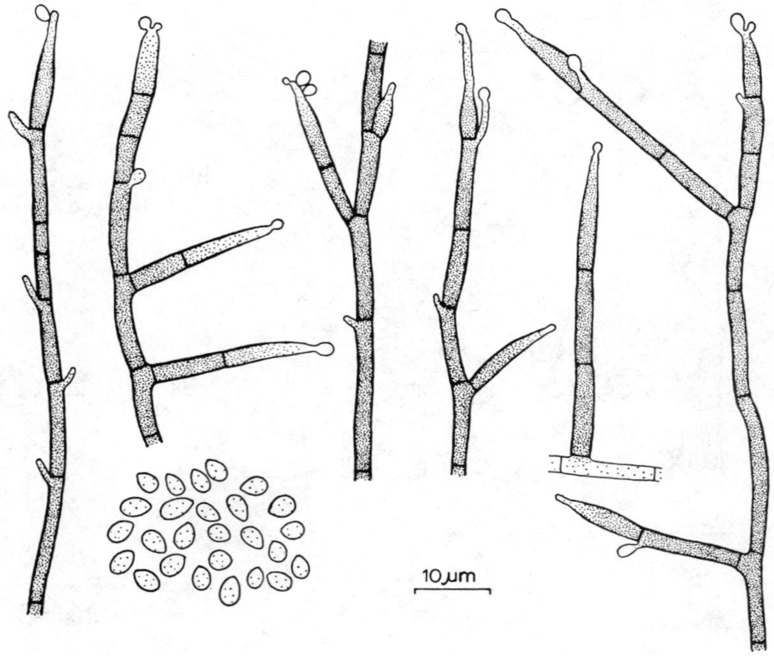

Figure 14 *Exophiala spinifera,* CBS 899.68. (From Ref. 81.)

cells abundant. *Hyphae* 1.5-2.5 µm wide. *Conidiogenous cells* differentiated, intercalary or free, composing suberect conidiophores; terminal conidiogenous cells cylindrical to acicular, 8-13 × 1.8-2.5 µm, apically with a narrow rhachis up to 10 µm long which forms conidia in basipetal order and occasionally sympodially; annellations narrow, vague. *Conidia* subhyaline, continuous, subglobose to ellipsoidal, 2-4 × 1.8-2.8 µm.
Type strain: CBS 899.68.

Discussion

The species is intermediate between *Rhinocladiella atrovirens* Nannf. and *Exophiala jeanselmei* in having suberect conidiophores with long rhachides, and mainly percurrent conidiogenesis. De Hoog [81] classified it in *Rhinocladiella,* but in accordance with other literature in medical mycology, *Exophiala* is preferred. In addition to differentiated conidiophores, budding cells and micronematous conidiogenous cells are common.

Rajam et al. (Ref. 243, as *Fonsecaea dermatitidis*) isolated CBS 194.61 from a fatal case of human chromomycosis. Initially, verrucous skin lesions were formed; later multiple cystic and nodular swellings developed and the mucous membranes were affected. CBS 899.68 was isolated from a human nasal granuloma [220]. The fungus has also been isolated from nests and burrows [81].

4. *Exophiala jeanselmei* (Langer.) McGinnis & Padhye (Figs. 15-17)

Torula jeanselmei Langer., *Ann. Parasitol. Hum. Comp. 6*: 396, 1928
= *Pullularia jeanselmei* (Langer.) Dodge, *Medical Mycology,* Mosby, St. Louis, p. 675, 1935
= *Phialophora jeanselmei* (Langer.) Emmons, *Arch. Pathol. 39*: 368, 1945

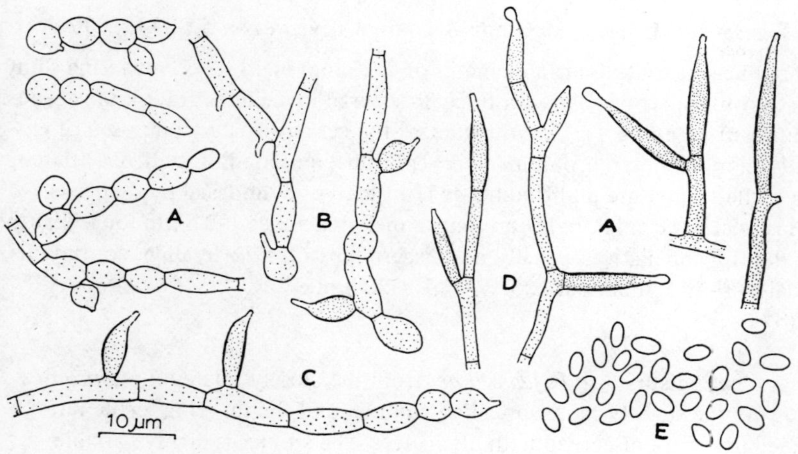

Figure 15 *Exophiala jeanselmei* var. *jeanselmei*: (A) CBS 528.76; (B) CBS 538.76; (C) CBS 537.76; (D) CBS 265.49; (E) CBS 528.76 and 537.76. (From Ref. 81.)

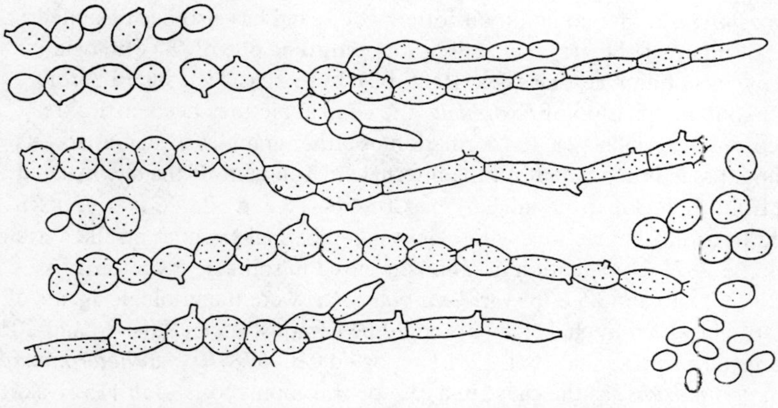

Figure 16 *Exophiala jeanselmei* var. *heteromorpha*, various strains. (From Ref. 81.)

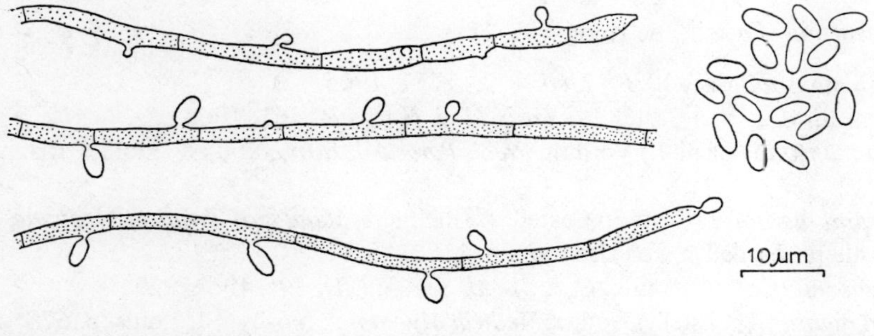

Figure 17 *Exophiala jeanselmei* var. *lecanii-corni*, CBS 123.33. (From Ref. 81.)

= *Exophiala jeanselmei* (Langer.) McGinnis & Padhye, *Mycotaxon 5*: 345, 1977

Colonies on oatmeal agar attaining a diameter of 7-20 mm in 14 days, appearing slimy at the center, black, outer part usually with flocculose aerial mycelium, grayish olivaceous. *Budding cells* abundant. *Hyphae* 1.5-2.5 μm wide, at the center usually composed of globose cells 4.7-5.9 μm in diameter. *Conidiogenous cells* differentiated or undifferentiated, intercalary or free, the former are undifferentiated, inflated or cylindrical hyphal cells, the latter are ellipsoidal to acicular, 6-15 μm long; annellated zones 1-2.5 μm long, slightly tapering toward the tip, annellations usually not discernible. *Conidia* hyaline, continuous, broadly ellipsoidal to short cylindrical, 2.6-5.9 × 1.2-2.5 μm.
Type strain: CBS 664.76.

Discussion

The main features to distinguish *E. jeanselmei* from the closely related *E. mansonii* (Castell.) de Hoog are the slightly more spreading colonies and the tapering fertile butts on which the annellations are discerned with difficulty. The species is highly variable. For this reason, de Hoog [81] subdivided it into three varieties, characterized respectively by prevalence of dark, acicular conidiogenous cells (var. *jeanselmei,* Fig. 15), budding cells and strongly torulose mycelium [var. *heteromorpha* (Nannf.) de Hoog, Fig. 16; type: CBS 232.33] and prevalence of unswollen, undifferentiated conidiogenous cells [var. *lecaniicorni* (Benedek & Specht) de Hoog, Fig. 17; type: CBS 123.33].

Exophiala jeanselmei is rather common on rotten wood, and has also been isolated from several kinds of liquids [81]. It is generally supposed to be one of the etiologic agents of chromomycosis and mycetoma (e.g., Ref. 102), but as has been stated before, with the modern taxonomic revision of *Exophiala,* the clinical pictures need critical reexamination. According to Ajello [2], the formation of either granules or dematiaceous mycelium in the host tissue is dependent on the internal environment of the site invaded, whether resulting from trauma or from inhalation. CBS 664.76 and 528.76 (var. *jeanselmei*) were reported to form tumefactions with black, subepidermal granules on the extremities [100,151,173,286,287]. Inoculation in rabbits resulted in solitary, mycetoma-like nodules [286]. CBS 529.76 and 657.76 (var. *heteromorpha*) were the etiologic agents of subcutaneous nodules in which hyphal elements could be demonstrated (Refs. 86 and 201, both as *Phialophora gougerotii*). Tsai et al. [296] reported CBS 578.76 (var. *heteromorpha,* as *Hormodendrum dermatitidis*) as the cause in a case of chromomycosis. Jen [152] isolated CBS 577.76 (var. *heteromorpha,* as *Hormodendrum dermatitidis*) from a cervical lymph node, and Jotisankasa et al. [158] described CBS 579.76 (var. *heteromorpha,* as *Phialophora dermatitidis*) as causing a brain abscess in a human patient. The latter strain appeared to be neurotropic when experimentally inoculated.

5. *Exophiala mansonii* (Castell.) de Hoog (Fig. 18)

Microsporum mansonii Castell., *Br. Med. J. 2*: 1271, 1905
= *Foxia mansonii* (Castell.) Castell., *J. Trop. Med. Hyg. 11*: 261, 1908
= *Malassezia mansonii* (Castell.) Verdun, *Précis Parasitol. Hum.*, 2nd ed., Doin, Paris, p. 698, 1912
= *Cladosporium mansonii* (Castell.) Castell. & Chalmers, *Manual of Tropical Medicine,* 3rd ed., Ballière, Tindall & Cox, London, p. 1100, 1919
= *Torula mansonii* (Castell.) Vuill., *C. R. Acad. Sci., Ser. D, 89*: 406, 1929
= *Dematium mansonii* (Castell.) Dodge, *Medical Mycology,* Mosby, St. Louis, p. 678, 1935
= *Aureobasidium mansonii* (Castell.) W. B. Cooke, *Mycopathol. Mycol. Appl. 17:* 34, 1962

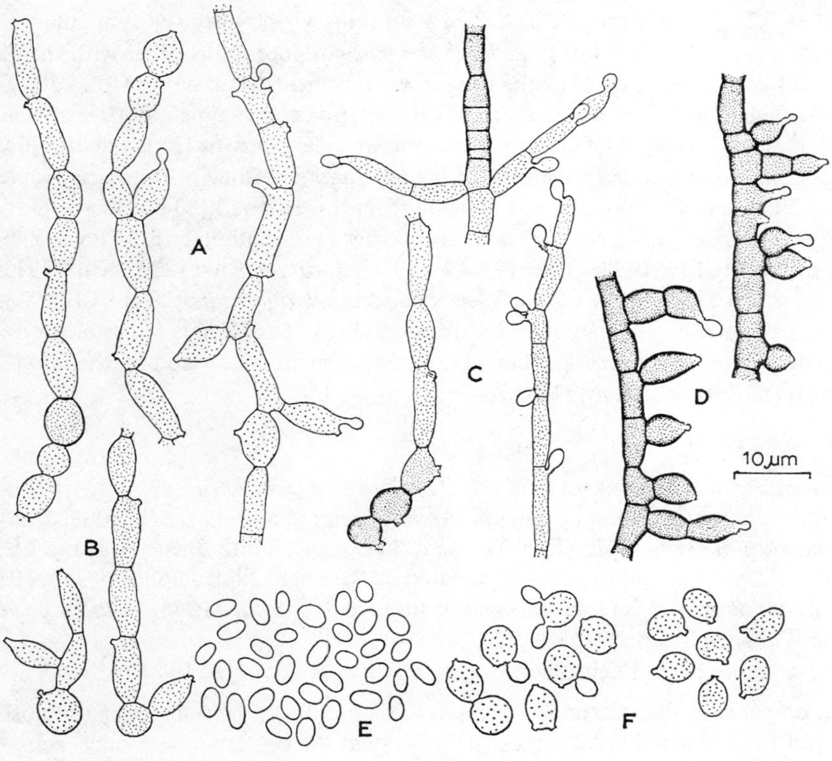

Figure 18 *Exophiala mansonii*: (A) CBS 158.58; (B) CBS 352.52; (C) CBS 525.76; (D) CBS 101.67; (E,F) various strains. (From Ref. 81.)

= *Rhinocladiella mansonii* (Castell.) Schol-Schwarz, *Antonie Leeuwenhoek J. Microbiol. Serol. 34*: 122, 1968

= *Exophiala mansonii* (Castell.) de Hoog, *Stud. Mycol. 15*: 114, 1977

Perfect state: *Dictyotrichiella mansonii* Schol-Schwarz.

Colonies on oatmeal agar attaining a diameter of 2-6 mm in 14 days, appearing smooth or depressed flocculose, slimy at the center, grayish to olivaceous black. *Budding cells* abundant. *Hyphae* 1.5-2.5 µm, at the center usually composed of globose cells. *Conidiogenous cells* undifferentiated, intercalary or free, the latter subglobose to ellipsoidal, 5-8.5 × 2-3.5 µm; annellated zones cylindrical, 0.5-1.5 µm wide, of variable length, often hardly elongating; annellations frequently discernible. *Conidia* hyaline, ellipsoidal to short cylindrical, 2.9-4.7 × 1.8-2.6 µm.

Neotype strain: CBS 158.58.

Discussion

De Hoog [81] listed four synonymous basionyms, among which was *Sporotrichum gougerotii* Matruchot (based on neotype, CBS 526.76). On the basis of the symptoms on the host provoked by the original *Microsporum mansonii* [56], McGinnis [188] considered the species to be synonymous to the hyaline yeast *Malassezia furfur* (Robin) Baillon. Since, however, the described fungus was not necessarily the causative organism of the disease, de Hoog [81] did not include described symptoms when newly typifying the species. Castellani's poor description [56,57] of *M. mansonii* does not conflict with the neotype strain, CBS 158.58. The color of the fungus was not mentioned, but as CBS 158.58, to which A. Castellani in his correspondence to CBS referred to as "type" of *M. mansonii,* is black [58], it seems unlikely that his 1905 strain was hyaline. Thus, de Hoog's neotypification [81] of *E. mansonii* is in accordance with the International Code of Botanical Nomenclature [283]. The species is occasionally isolated from rotten wood [81]. Most confirmed isolates of *E. mansonii,* however, came from cases of chromomycosis (CBS 232.39, as *Hormodendrum negronii* [234]; CBS 292.49, as *Mycotorula schawii* [235]; CBS 351.53, as *Torula bergeri* [26,27]; CBS 581.76, as *Hormodendrum dermatitidis* [297]. CBS 523.76 originates from an undescribed case of endocarditis, and CBS 526.76 was the etiologic agent of a cystic lesion on a human patient [313]. Experimental infection by the latter strain gave no significant reaction [313].

6. *Exophiala dermatitidis* (Kano) de Hoog (Fig. 19)

Hormiscium dermatitidis Kano, *Aichi Ika Daigaku Igakukwai Zasshi 41*: 11, 1934
= *Fonsecaea dermatitidis* (Kano) Carrión, *Arch. Dermatol. Syphilol. 61*: 1008, 1950
= *Hormodendrum dermatitidis* (Kano) Conant, in Conant, Smith, Baker, Gallaway & Martin, *A Manual of Clinical Mycology,* 2nd ed., Saunders, Philadelphia, p. 276, 1954
= *Phialophora dermatitidis* (Kano) Emmons, Binford & Utz, *Medical Mycology,* Lea & Febiger, Philadelphia, p. 291, 1963
= *Exophiala dermatitidis* (Kano) de Hoog, *Stud. Mycol., 15*: 118, 1977

Colonies on oatmeal agar attaining a diameter of 5 mm in 14 days, appearing smooth with some local tufts of aerial mycelium, slightly slimy at the center, grayish olivaceous to black. *Budding cells* abundant. *Hyphae* 1.5-3 μm wide, somewhat torulose at the center. *Conidiogenous cells* undifferentiated, arising from ascendent hyphae, intercalary or free, composing a slightly differentiated conidial apparatus; free conidiogenous cells subglobose to broadly ellipsoidal, 4-5 × 3.5 μm, with one to three broad scars on which the conidia are basipetally produced. *Conidia* hyaline, broadly ellipsoidal, 2.6-4 × 2-3 μm. Locally subglobose phialides with broad collarettes are formed on torulose branches. Type strain: CBS 207.35.

Discussion

Despite the frequent usage of the epithet *dermatitidis* for medically important black yeast-like fungi, the species in the sense described above is hitherto only definitely known from the type. It was reported by Kano [161] as the etiologic agent in a case of verrucose skin lesions on humans. The strain appeared to be pathogenic in experimental infection [161].

7. *Exophiala werneckii* (Horta) von Arx (Fig. 20)

Cladosporium werneckii Horta, *Rev. Med. Cir. Braz. 22*: 269, 1921
= *Dematium werneckii* (Horta) Dodge, *Medical Mycology,* Mosby, St. Louis, p. 677, 1935
= *Pullularia werneckii* (Horta) de Vries, *Contribution to the Knowledge of the Genus Cladosporium,* Hollandia, Baarn, p. 101, 1952
= *Exophiala werneckii* (Horta) von Arx, *The Genera of Fungi Sporulating in Pure Culture,* Cramer, Vaduz, p. 180, 1970

Cryptococcus metaniger Castell., *Arch. Dermatol. Syphilol. 16*: 402, 1927
= *Cladosporium metaniger* (Castell.) Ferrari, *Atti Ist. Bot. Univ. Lab. Crittog. Pavia, Ser. 4, 3*: 183, 1932

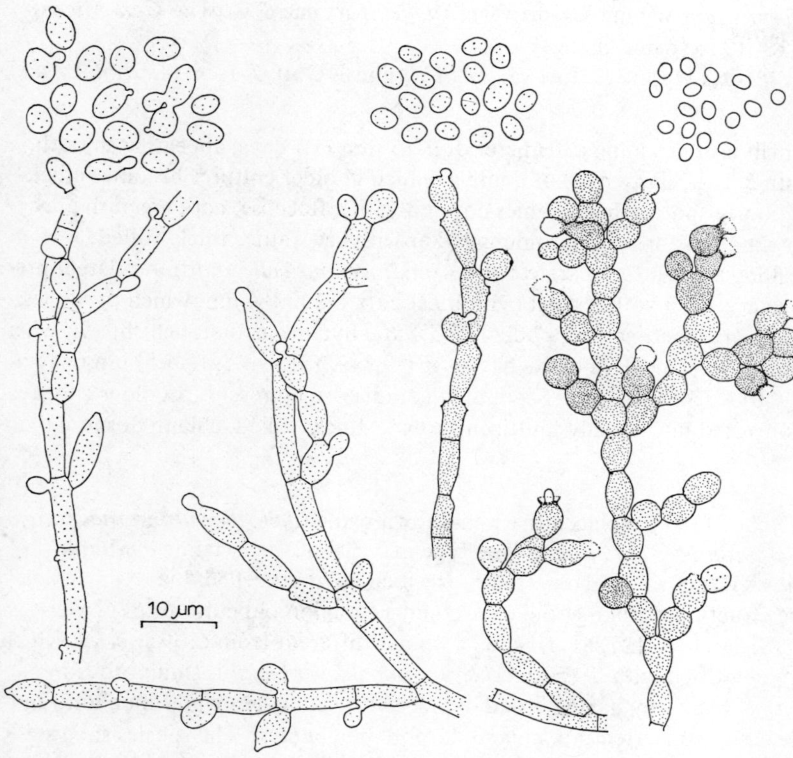

Figure 19 *Exophiala dermatitidis*, CBS 207.35. (From Ref. 86.)

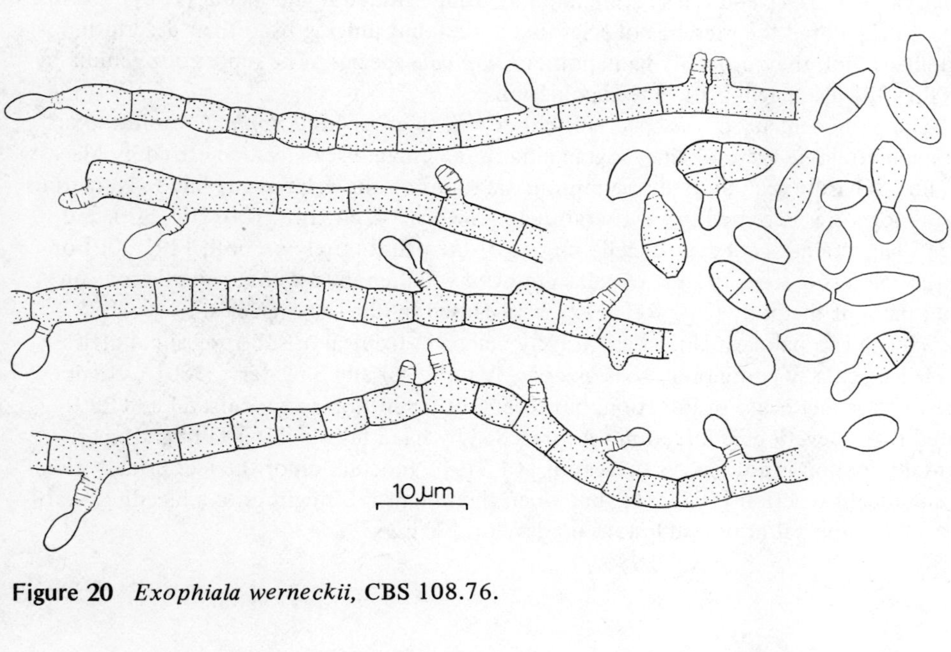

Figure 20 *Exophiala werneckii*, CBS 108.76.

= *Pullularia fermentans* Wynne & Gott var. *castellanii* Wynne & Gott, *J. Gen. Microbiol. 14*: 518, 1956 (name change)
Pullularia fermentans Wynne & Gott var. *laeoi* Wynne & Gott, *J. Gen. Microbiol. 14*: 517, 1956

Colonies on malt agar attaining a diameter of 9-15 mm in 7 days, appearing smooth, black, covered with a dark, slimy mass of conidia, which in older cultures becomes crustaceous. Sterile or poorly sporulating colonies covered with a floccose, dark greenish gray mycelial mat. *Hyphae* light brown, becoming darker with age, rather thick-walled, 3-10 μm wide, easily falling apart into separate cells. *Conidiogenous cells* undifferentiated, predominantly intercalary, each with a short cylindrical butt 1-2 × 1-4 μm, which produces conidia percurrently or sometimes sympodially. *Conidia* hyaline at first, light brown when older, smooth-walled, ellipsoidal, one-celled (5-)7-8(-9.5) × (2.5-)3.5-4(-4.5) μm or two-celled (7-)8-9.5(-10) × (3.5-)4.5 μm. Secondary conidia often present. In older cultures the conidia are converted into usually multicelled, dark, thick-walled, chlamydospore-like bodies.

Discussion

Hermanides-Nijhof [141] replaced the well-known name *Cladosporium werneckii* by the older name *Sarcinomyces crustaceus.* The type material of the latter at the Berlin herbarium (B) shows rather small, dark structures which are multicelled due to cross-wall formation. These structures resemble the dark conidia found in old cultures of *C. werneckii.* Recently, Sigler et al. [272a], isolated a fungus different from *C. werneckii,* which they convincingly identified with *S. crustaceus. Exophiala werneckii* is similar to *Hormonema* Lagerberg & Melin, by having broad hyphae with cells often wider than long and conidia which are easily converted into chlamydospore-like bodies. The species shows abundant polar budding on several media; some strains consist merely of budding cells. The conidiogenesis and ultrastructure were described by Gustafson et al. [130], McGinnis [184] and Cole [66]. The mold-yeast interconversion is particularly dependent on nutritional factors [144] and is also stimulated by using agitated liquid media [132]. Volcán et al. [303] noted the presence of sclerotial bodies, but judging from their description and illustration they probably had another *Exophiala* species. The report of pycnidia by Borelli [37] also needs critical reexamination.

The pathogenicity of the species is rather well established. *Exophiala werneckii* is the main etiologic agent of tinea nigra palmaris, a human disease characterized by blackish, not or hardly scaly, usually asymptomatic stains on the palm of the hand. Numerous case reports are presented in the literature (e.g., Refs. 14, 54, 148, 160, 207, 300, and 307). The strains occur superficially on one of the palms, rarely on both [291,300] or on the sole [166,266,311]. It remains doubtful whether all cases of similar lesions on other parts of the body (e.g., Ref. 212) were caused by the same species.

Tinea nigra palmaris almost exclusively occurs in tropical or subtropical countries, and is particularly common in coastal areas. Van Velsor and Singetary [300] were not able to trace any cause of infection, but several other authors (e.g., Refs. 54 and 229) stated that in every case infection was preceded by hand injuries after falling. Experimental infection is not always successful [14,279]. Inoculation of the feet usually gives no significant reaction [165,251], but when the fungus is brought onto a bleeding, scarified area of the palm, typical lesions do develop [251,253].

G. *Ochroconis* de Hoog & von Arx

Ochroconis de Hoog & von Arx, *Kavaka 1*: 57, 1973

Colonies restricted, appearing farinose, velvety or funiculose, brown to olivaceous. *Hyphae* smooth- or somewhat rough-walled, pale olivaceous. *Conidiophores* slightly or conspicuously differentiated, cylindrical, often flexuose, producing conidia on scattered, cylindrical to conical denticles. After detachment an inconspicuous frill often remains both on the denticle and on the conidium base. *Conidia* one- to four-celled, pale olivaceous brown, smooth- or rough-walled, ellipsoidal, cylindrical, clavate or cuneiform. Type species: *Ochroconis constricta* (Abbott) de Hoog & von Arx.

Discussion

Ochroconis was recently introduced [83] for the majority of the species formerly classified in *Scolecobasidium* Abbott, which was restricted to species with T- or Y-shaped conidia formed on short, inflated conidiophores. *Scolecobasidiella* M. B. Ellis can be distinguished by conidiophores which often disarticulate into separate cells. *Johnstonia* M. B. Ellis differs by the regular branching pattern of the conidiophores. *Veronaea* Cif. & Mont. is distinct because of its blunt conidial denticles. *Dactylaria* Sacc. has hyaline or subhyaline, rather fragile conidia and cylindrical denticles. The latter genus, however, is circumscribed unsatisfactorily and revision is overdue.

Many species now classified in *Ochroconis* were treated by Barron and Busch [21] and Ellis [98]. They are frequently isolated from soil and rotten plant material, occasionally occurring on animals.

KEY TO THE POTENTIALLY PATHOGENIC SPECIES OF *OCHROCONIS*

1a. Conidia usually four-celled and verrucose . . . . *O. tshawytschae*
1b. Conidia usually two-celled and smooth-walled 2
2a. Conidia distinctly cuneiform, upper cell being markedly wider than basal cell . . . . *O. gallopavum*
2b. Conidia cylindrical to clavate, upper cell not markedly wider than basal cell . . . . *O. humicola*

1. *Ochroconis gallopavum* (W. B. Cooke) de Hoog, Comb. Nov. (Fig. 21)

Diplorhinotrichum gallopavum W. B. Cooke, in Georg, Brierer & W. B. Cooke, *Sabouraudia 3*: 241, 1964 (basionym)
= *Dactylaria gallopava* (W. B. Cooke) Bhatt & Kendrick, *Can. J. Bot. 46*: 1257, 1968

Colonies on malt agar at room temperature attaining a diameter of about 10 mm in 10 days, restricted, appearing flat or with radial furrows, with some powdery or velvety aerial mycelium; dark brown, with blackish brown center; reverse blackish brown. A reddish brown pigment is released into the medium. *Hyphae* subhyaline to brown, often with some encrusted pigment, thin- to rather thick-walled, 2-3.5 μm wide, forming a compact mycelium. *Conidiophores* slightly differentiated, arising orthotropically from undifferentiated hyphae, initially subulate, later elongating sympodially to up to 60 μm, 2-4 μm wide, finally laying down 1-2 (up to 4) thin septa; conidia are produced on one to five remote, cylindrical denticles. After detachment a frill usually remains both on the denticle and on the conidial base. *Conidia* usually two-celled, rarely up to four-celled,

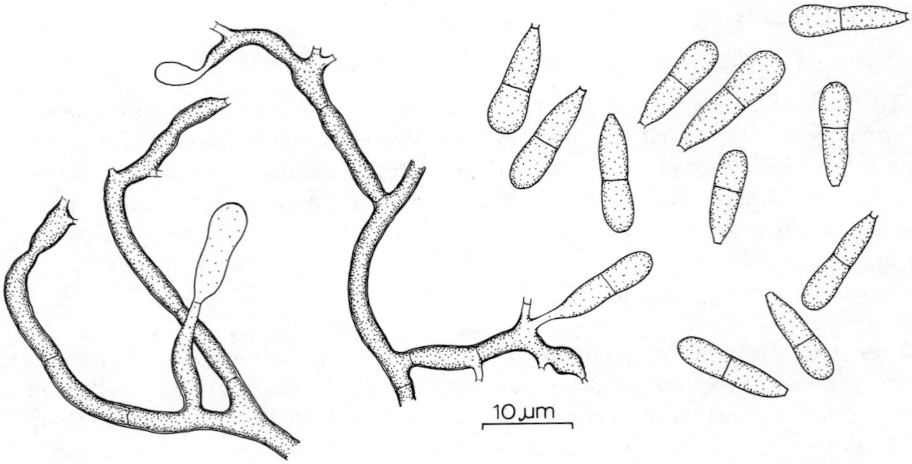

Figure 21 *Ochroconis gallopavum,* CBS 437.64.

smooth-walled to verruculose, pale brown, clavate to cuneiform, conspicuously constricted at the septa, 11-18 × 2.5-4.5 µm.
Type strain: CBS 437.64.

Discussion

The species has hitherto been referred to as belonging to the *Dactylaria* complex. However, the conidia are released rhexolytically, frills remaining on both the denticles and on the conidial bases, as is also the case in *Ochroconis* and *Scolecobasidium.* This similarity has already been pointed out by Sumner and Evans [285]. The species classified in *Dactylaria, Diplorhinotrichum* Höhnel and *Eriomycopsis* Speg. all have more or less cylindrical, flat-topped denticles without frills. Other characteristics pointing to *Ochroconis* are the color of the colony, the slight differentiation of the conidiophores, the scattered, plagiotropic denticles, and the pigmentation and warticulation of the conidia. The species is close to *Ochroconis humicola* and *O. constrictum,* differing from the former by constricted, from the latter by less verrucose conidia.

The species is known as the causative agent of acute encephalitis in turkey poults and chicks [30,121,244,304]. Granulomatous brain lesions are visible macroscopically; the mycelium is often demonstrable in the brain tissue. Experimental, intratracheal inoculation in turkey poults caused death within 2 weeks, showing mortality percentages comparable to those found in the field outbreaks [30]. Lesions on various organs could also be induced by inoculating spore suspensions in the thoracic air sacs of chicks, suggesting that natural infection occurs by inhalation of the fungus [244].

In nature, the species occurs in geothermal habitats and on self-heated substrates [108,292] and also on broiler-house litter [304]. In some such strains, the pathogenicity to turkey poults was established [292]. In vitro, the species is thermophilic, with an optimum near 40°C.

2. *Ochroconis humicola* (Barron & Busch) de Hoog & von Arx (Fig. 22)

Scolecobasidium humicola Barron & Busch, *Can. J. Bot. 40*: 83, 1962
= *Ochroconis humicola* (Barron & Busch) de Hoog & von Arx, *Kavaka 1*: 57, 1973

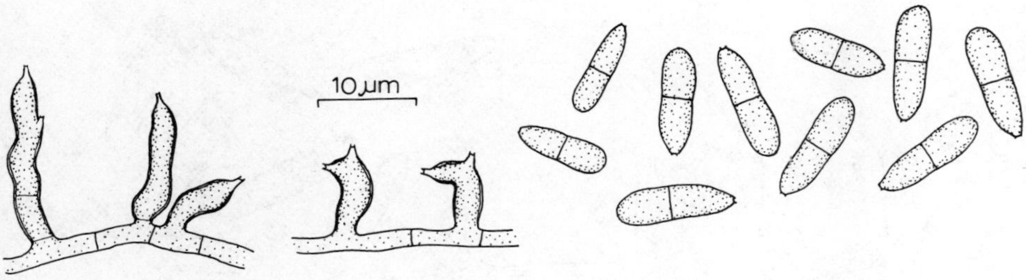

Figure 22 *Ochroconis humicola,* CBS 172.74.

Colonies on maize meal agar attaining a diameter of about 7 mm in 10 days, restricted, slightly domed at the center, otherwise flat, farinose to velvety; brown, brownish gray at the center, with regular and sharp margin and umber reverse. *Hyphae* subhyaline to pale olivaceous, usually smooth- and rather thick-walled, 1.5-2.5 μm wide; aerial hyphae often strongly flexuose; small sclerotial bodies may be present in the submerged mycelium. *Conidiophores* slightly differentiated, arising orthotropically from undifferentiated hyphae, cylindrical, often flexuose, up to about 100 μm long, 2-3 μm wide, with 0-4 (up to 8) thin septa; conidia are produced on one to five scattered, cylindrical to conical, 1-2 μm long denticles. After detachment an inconspicuous frill often remains both on the denticle and on the conidial base. *Conidia* usually two-celled, smooth-walled, pale olivaceous brown, cylindrical with rounded ends or somewhat clavate, not or hardly constricted at the septa, 7-15 × 2.5-4 μm.

Discussion

The species was described from soil, e.g., by Barron and Busch [21], Barron [20], Ellis [98], Domsch and Gams [89], and Matsushima [203,204]. Recently, the fungus was found to affect the kidney and other internal organs of coho salmon [256]. In experimental infection via the food, the fungus appeared to be only weakly pathogenic. Elkan and Philpot [96] isolated the species from diseased frogs, but were not able to establish whether *O. humicola* was the etiologic agent. Ajello et al. [3] reported kidneys and other internal organs of rainbow trout as being invaded by *O. humicola* mycelium. Intraperitoneal inoculation resulted in death of the fish within 2 weeks.

3. *Ochroconis tshawytschae* (Doty & Slater) Kirilenko & All-Achmed (Fig. 23)

Heterosporium tshawytschae Doty & Slater, *Am. Midl. Nat. 36*: 663, 1946
= *Scolecobasidium tshawytschae* (Doty & Slater) McGinnis & Ajello, *Trans. Br. Mycol. Soc. 63*: 202, 1974
= *Ochroconis tshawytschae* (Doty & Slater) Kirilenko & All-Achmed, *Mykrobiol. Zh. 39*: 305, 1977
Scolecobasidium variabile Barron & Busch, *Can. J. Bot. 40*: 83, 1962
= *Ochroconis variabilis* (Barron & Busch) de Hoog & von Arx, *Kavaka 1*: 58, 1973
Scolecobasidium macrosporum Roy, Dwivedi & Mishra, *Lloydia 25*: 164, 1962

Colonies on malt agar attaining a diameter about 15 mm in 10 days, appearing slightly domed, felt-like to velvety; olivaceous brown to olive, with regular and sharp margin and evenly dark olivaceous reverse. *Hyphae* subhyaline to pale olivaceous, usually smooth-

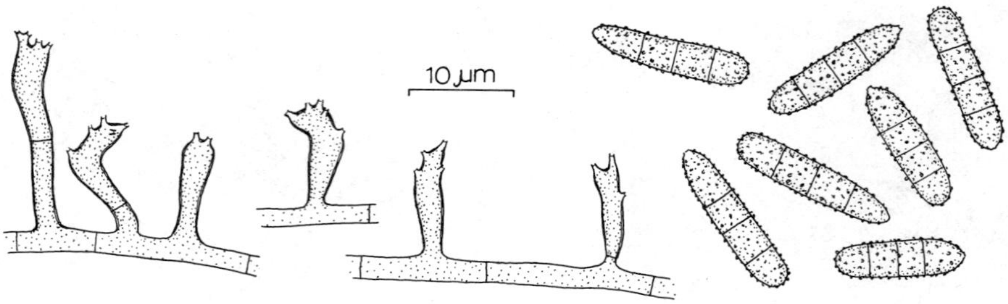

Figure 23 *Ochroconis tshawytschae,* CBS 332.65.

and rather thick-walled, 1.5-2.5 μm wide, forming a compact mycelium. *Conidiophores* slightly differentiated, arising orthotropically from undifferentiated hyphae, cylindrical, often somewhat inflated or flexuose, up to about 30 μm long, 2-3.5 μm wide, with 0-2 thin septa; conidia are produced on one to five scattered, cylindrical denticles in the apical region. After detachment an inconspicuous frill often remains both on the denticle and on the conidial base. *Conidia* usually four-celled, verrucose, pale brown, cylindrical with rounded ends or somewhat clavate, not or hardly constricted at the septa, 12-25 × 3.5-5.5 μm.

Discussion

The species is close to *O. anelii* (Graniti) de Hoog & von Arx, but differs by narrower conidia. Recent descriptions are, for example, those of Barron and Busch [21], Matsushima [203,204], and Mouchacca [214].

Ochroconis tshawytschae is mainly found in soil. Doty and Slater [91] isolated the fungus from a case of phaeohyphomycosis in chinook salmon. Experimental infection remained unsuccessful.

H. *Phialophora* Medlar

Phialophora Medlar, *Mycologia* 7: 202, 1915

Colonies restricted, appearing lanose, floccose or nearly smooth; usually olivaceous, brown, gray, or blackish; colonies when smooth mostly pink or subhyaline. *Hyphae* hyaline to brown. *Conidiogenous cells* phialidic, undifferentiated, some intercalary, or slightly differentiated and arising in more or less dense brushes from short lateral branches, or conspicuously differentiated and forming part of a loosely branched conidial apparatus; collarettes inconspicuous and cylindrical, or more distinct and cup-, funnel-, or vase-shaped or flaring. *Conidia* produced in slimy heads or chains, one-celled, smooth-walled, usually hyaline, globose to allantoid; two types of conidia may be produced within the same colony. Chlamydospores, when present, one-celled and dark brown or forming small, ill-defined, olivaceous brown sclerotial bodies.

Type species: *Phialophora verrucosa* Medlar.

Discussion

Phialophora is a heterogeneous assemblage of about 45 species, some of which are known to have perfect states in such widely different ascomycete genera as *Ascocoryne* Groves & Wilson, *Mollisia* (Fr.) Karst., *Pyrenopeziza* Fuckel (Helotiales), *Lasiosphaeria* Ces. & de Not., *Coniochaeta* (Sacc.) Massee (Sordariaceae), and *Gaeumannomyces* von

Arx & Olivier (Diaporthaceae). Gams and Holubová-Jechová [118] introduced a section *Catenulatae* W. Gams for species producing conidia in chains. Cole and Kendrick [67] made another subdivision by proposing that the genus *Margarinomyces* Laxa be maintained for *Phialophora* species with undifferentiated, often intercalary phialides with inconspicuous collarettes. Domsch et al. [90] opposed this subdivision by pointing to the presence of more distinct collarettes in aging cultures of *P. bubakii* (Laxa) Schol-Schwarz. It remains difficult to comprehend how all these markedly different species can be maintained in a single genus. Revision of *Phialophora sensu lato* is long overdue and any proposed synonymy would be premature at this stage. Contributions toward a monographic treatment have been made by van Beyma [298], Cole and Kendrick [67], Gams and Holubová-Jechová [118], Moreau [213], and Schol-Schwarz [270]. A poorly illustrated review of the species of medical importance was given by McGinnis [184].

Phialophora species with undifferentiated, hyaline phialides resemble *Acremonium* Link ex Fr., but the latter has smaller collarettes and subulate phialides in the aerial mycelium. Dark *Phialophora* species with poorly differentiated phialides are structurally similar to those of *Exophiala* Carmichael, but these produce conidia from annellated zones. Sometimes the annellations in *Exophiala* are too narrow to be discerned with the light microscope, but the collarettes are definitely absent. Some *Exophiala* species may also produce a *Phialophora* state. Similar states are occasionally found in *Rhinocladiella* Nannf. Still, Cooper and Schneidau [74] found *P. verrucosa* to be serologically closer to *Cladosporium carrionii* than to *Rhinocladiella pedrosoi. Phialophora* species with brushes of phialides are close to *Myrioconium* Syd., but this genus has hyaline phialides with funnel-shaped collarettes in sporodochia. *Phialophora* species with differentiated conidiogenous cells can be distinguished from *Chloridium* Link ex Fr. and *Exochalara* W. Gams & Hol.-Jech. by the absence of unbranched, dark, septate conidiophores. *Chalara* (Corda) Rabenh. usually has tubular collarettes, but distinction from some *Phialophora* species of the section *Catenatae* is sometimes problematic [118].

Phialophora species are common in rotten wood and soil, and may also be isolated from foodstuffs. Some are parasitic on phanerogams. The species that are similar to *Exophiala* are known to be potential etiologic agent of chromomycosis in humans. A few species, most of which are not well known, may be involved in various disorders in man.

KEY TO THE POTENTIALLY PATHOGENIC SPECIES OF *PHIALOPHORA*

1a. Phialides with funnel- or vase-shaped, never flaring collarettes which are darker than the rest of the phialide 2
1b. Phialides with collarettes of other shape, not markedly darker than the rest of the phialide 3
2a. Collarettes funnel-shaped, not longer than wide. . . . . *P. verrucosa*
2b. Collarettes vase-shaped, usually longer than wide . . . . *P. americana*
3a. Phialides arising in dense, more or less radiating brushes from short lateral branches . . . . *P. repens*
3b. Phialides arising in loose, not radiating brushes or directly from loosely branched, little differentiated, or fasciculate hyphae 4
4a. Phialides thin-walled, hyaline; colonies smooth, moist 5
4b. Phialides rather thick-walled, brown; colonies floccose, dry. 6
5a. Chlamydospores present. . . . . *P. mutabilis*
5b. Chlamydospores absent . . . . *P. hoffmannii*

6a. Conidia of two types: hyaline, allantoid, and brown, globose; collarettes wide, flaring . . . . *P. richardsiae*
6b. Conidia all hyaline, reniform to allantoid; collarettes narrow, nearly tubular . . . . *P. parasitica*

1. *Phialophora verrucosa* Medlar (Fig. 24)

Phialophora verrucosa Medlar, *Mycologia* 7: 202, 1915

Colonies attaining a diameter of about 15 mm in 10 days, appearing flat, nearly smooth with some thin, appressed aerial mycelium; olivaceous gray, with regular and sharp margin; reverse olivaceous gray, olivaceous black at the center. *Hyphae* subhyaline to pale brown, 1.5-2.5 μm wide, locally lateral branches having inflated cells of up to 6 μm. *Conidiogenous cells* pale brown, arising orthotropically from rather broad hyphae, often also arising plagiotropically in small groups from short lateral branches which are frequently inflated, flask-shaped, 7-15 μm long, widest (3-5 μm) near the middle; upper third of the cell is made up of a brown, funnel-shaped collarette 2.5-3 μm long, 1.5 μm wide at the base, and up to 4 μm wide at the tip. *Conidia* hyaline, smooth- and thin-walled, broadly ellipsoidal, 2.5-4 × 1.5-2.8 μm, often adhering in mucous balls at the tip of the phialide.

Discussion

Within *Phialophora verrucosa sensu lato,* several morphological types can be distinguished. A few strains have collarettes which are trapezoid in outline and nearly of equal length and width at the apex. The majority of the isolates, however, have vase-shaped collarettes with parallel walls which flare terminally, usually being much longer than wide. In a third type, most phialides are intercalary. More study is required to determine whether these different types each warrant a taxonomic status. According to Domsch et al. [90], the strains with vase-shaped collarettes should be classified in *Phialophora americana.* A strain considered by W. Gams to be typical for *P. americana* is described below. Since in the medical literature, however, this distinction has not been recognized, *P. americana* being placed in synonymy with *P. verrucosa,* the data on these taxonomic units have been merged here under *P. verrucosa.* According to Zweibel and Wang [316] *P. verrucosa* is usually isolated from other natural substrates. In experimental infection, however, both species proved to be pathogenic to mice.

Descriptions of *P. verrucosa* in the broad sense have been given, for example, by Schol-Schwarz [270], Moreau [213], MacKinnon [193], Matsushima [204], and Cole and Kendrick [67]. The ultrastructure of the septa was studied by Cooper et al. [73], and the conidiogenesis by Cole [66]. The influence of trace elements on sporulation and chlamydospore production was demonstrated by Reiss and Nickerson [248]. Sclerotial bodies bearing septae were found by Ajello and Runyan [4].

Phialophora verrucosa sensu lato is rather frequently isolated from rotten wood, but is also known as a rare [46] etiologic agent of chromomycosis on humans. The reports of a species occurring on a horse [6,7] were not confirmed by culturing. Usually, nodular to granulomatous lesions are formed, mainly on the extremities and occasionally on other parts of the body [70,216]. Several case reports have been presented in the literature (e.g., Refs. 136, 172, 206, 210, and 211).

In most descriptions of chromomycosis, *Rhinocladiella pedrosoi* is taken to be the causative agent, sometimes even when the fungus is not unambiguously identified in culture; the role of *Phialophora* in this disease is therefore not clearly understood. A case of

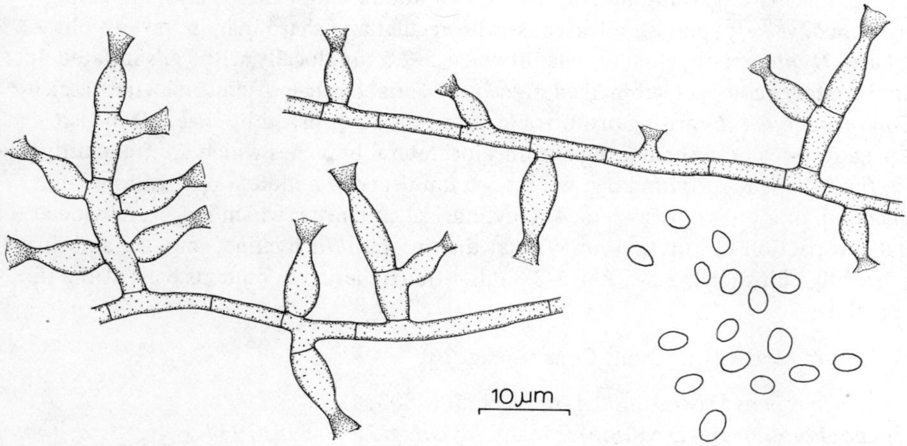

Figure 24 *Phialophora verrucosa*, CBS 138.67.

keratomycosis caused by *P. verrucosa* was reported by Wilson et al. [308]. Polack et al. [237] described a dual infection of a human cornea after trauma by *P. verrucosa* and *Cladosporium cladosporioides* (Fres.) de Vries.

The virulence of *P. verrucosa* was tested experimentally by several authors and was mostly found to be weak [95,110] or to vary from strain to strain [192,210,274]. Nevertheless, in some cases typical lesions could be reproduced [192]. Inoculation onto rabbits' eyes gave minimal inflammation [237], but in a monkey the reaction was significant [308].

2. *Phialophora americana* (Nannf.) Hughes (Fig. 25)

Cadophora americana Nannf., in Melin & Nannf., *Sven. Skogvardsfoeren. Tidskr. 32*: 412, 1934
= *Phialophora americana* (Nannf.) Hughes, *Can. J. Bot. 36*: 795, 1958
Phialophora macrospora Moore & Almeida, *Ann. Mo. Bot. Gard. 23*: 545, 1936
= *Fonsecaea pedrosoi* (Brumpt) Negroni var. *phialophorica* Carrión, *Mycologia 34*: 432, 1942 (name change)

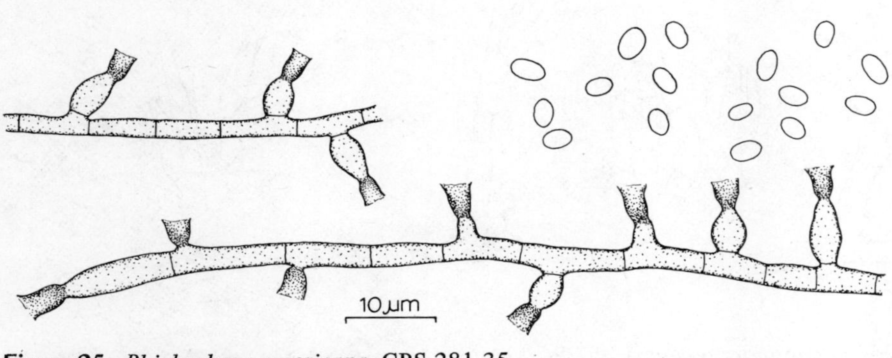

Figure 25 *Phialophora americana*, CBS 281.35.

Colonies on malt agar attaining a diameter of about 8 mm in 10 days, appearing cushion-shaped, velvety; grayish olivaceous, with regular and sharp margin; reverse olivaceous black. *Hyphae* subhyaline to pale brown, 1.5-2.5 μm, locally with cells inflated up to 6 μm, forming a compact submerged mycelium, aerial hyphae often somewhat fasciculate. *Conidiogenous cells* arising orthotropically from rather broad hyphae, often also arising plagiotropically in small groups from short lateral branches which are frequently inflated, flask-shaped, 7-20 μm long, widest 3-5 μm near the middle; upper third of the cell is made up of a brown, vase-shaped to cylindrical collarette, which is 1.5 μm wide at the basal constriction and up to 4 μm wide at the tip. *Conidia* hyaline, smooth- and thin-walled, broadly ellipsoidal, 2.5-5 × 1.5-2.8 μm, often adhering in mucous balls at the tip of the phialide.

3. *Phialophora repens* (Davidson) Conant (Fig. 26)

Cadophora repens Davidson, *J. Agric. Res. 50*: 803, 1935
= *Phialophora repens* (Davidson) Conant, *Mycologia 29*: 598, 1937

Colonies on malt agar attaining a diameter of about 10 mm in 10 days, appearing smooth, with some farinose aerial mycelium; grayish brown, pale brownish gray toward the vague margin, with similarly colored reverse. *Hyphae* subhyaline to pale brown, thin- or rather thick-walled, 2.5-5 μm wide, often fasciculate. *Conidiogenous cells* hyaline, usually arising plagiotropically in dense brushes from profusely branched, orthotropic, basally pale brown side branches, 6-15 μm long, 2-3.5 μm wide just above the base, flask-shaped to elongate, apically with a short, inconspicuous collarette of about 1 μm width. *Conidia* hyaline, thin-walled, cylindrical to reniform, 3-4 × 1.2-1.8 μm.
Type strain: CBS 294.39.

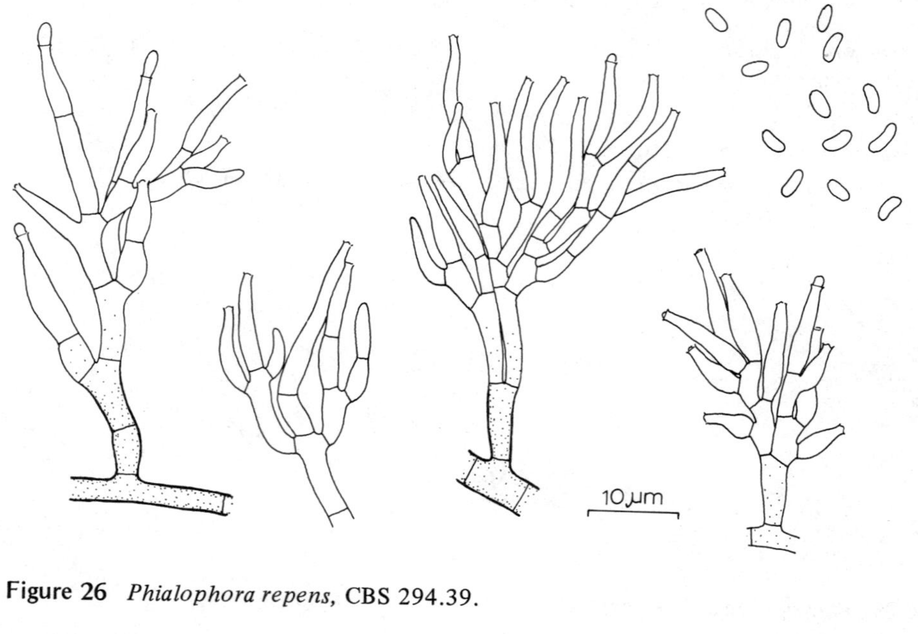

Figure 26 *Phialophora repens,* CBS 294.39.

Discussion

The species was originally isolated from rotten pine lumber. Recently, Meyers et al. [208] described CBS 423.73 as the etiologic agent of a granulomatous nodule in the scalp of a black with advanced lepromatous leprosy. This strain does not show penicillate brushes of phialides; whether the occurrence of this branching pattern has taxonomic significance still needs to be established.

4. *Phialophora hoffmannii* (van Beyma) Schol-Schwarz Species Complex (Fig. 27)

Margarinomyces hoffmann van Beyma, *Zentralbl. Bakteriol. Parasitenkd. Infektsrk. Hyg., Abt. II, 99*: 386, 1939
= *Phialophora hoffmannii* (van Beyma) Schol-Schwarz, *Persoonia 6*: 79, 1970
Phialophora aurantiaca van Beyma, *Antonie Leeuwenhoek J. Microbiol. Serol. 6*: 277, 1940
Perfect state (pro parte): *Coniochaeta ligniaria* (Grev.) Cooke.

Colonies on malt agar attaining a diameter of about 20 mm in 10 days, appearing flat and even, smooth, at the very center with some fluffy aerial mycelium, pink, with regular and sharp margin; reverse pale pink. *Hyphae* hyaline to subhyaline, smooth- and thin-walled, 1-2.5 μm wide, often fasciculate, forming a tough and compact mycelium. *Conidiogenous cells* arising orthotropically from undifferentiated hyphae or in dense *Microdochium*-like clusters from short and inflated lateral branches; often also intercalary on undifferentiated hyphae; conidiogenous cells hyaline to subhyaline; scattered, orthotropic cells elongate, 7-20 μm long, cylindrical to flask-shaped, the latter inflated up to 3 μm just below the middle; clustered phialides subglobose to ovoidal, about 3-7 × 2-3 μm; intercalary phialides cylindrical with one or several scattered collarettes; collarettes unpigmented, cylindrical, about 1.5 μm wide. *Conidia* hyaline, smooth- and thin-walled, broadly ellipsoidal to cylindrical, sometimes slightly curved, usually 3-3.5 × 1.5-2.5 μm.
Type strain: CBS 245.38.

Discussion

The name *Phialophora hoffmannii* probably stands for several, closely related taxa, which can hardly be distinguished. *Phialophora mutabilis* was reduced to synonymy by

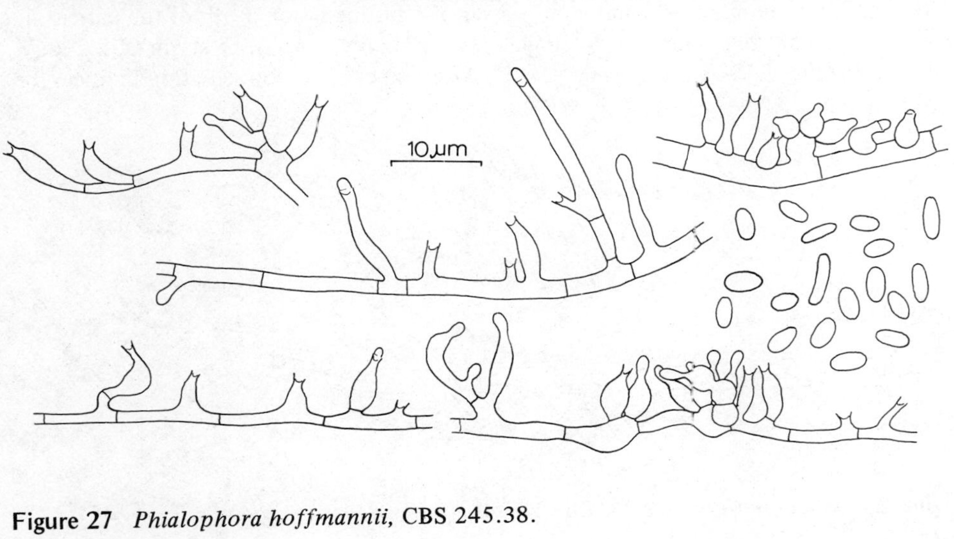

Figure 27 *Phialophora hoffmannii*, CBS 245.38.

McGinnis [184], although the type strain is different from that of *P. hoffmannii* by having dark brown, one-celled chlamydospores.

Isolates more or less fitting the concept described above are infrequently isolated from rotten wood, soil, foodstuffs, and similar substrates. McGinnis's [184] listing of the species as a human pathogen was based on the *P. hoffmannii*-like CBS strain isolated from a case of human keratitis. Another, similar strain, CBS 151.52, was isolated from a nodular swelling on a cat's tongue, and CBS 550.76 from a human maduromycosis.

5. *Phialophora mutabilis* (van Beyma) Schol-Schwarz (Fig. 28)

Margarinomyces mutabilis van Beyma, *Antonie Leeuwenhoek J. Microbiol. Serol. 10*: 48, 1944/45
= *Phialophora mutabilis* (van Beyma) Schol-Schwarz, *Persoonia 6*: 80, 1970

Colonies on malt agar attaining a diameter of about 20 mm in 10 days, appearing flat, with central, ascending to erect, radiating, up to 2 mm high hyphal fascicles; colonies pink, later becoming blackish brown due to abundant formation of chlamydospores; margin regular and sharp, reverse pale pink, centrally becoming dull brown. *Hyphae* hyaline to subhyaline, smooth- and thin-walled 1-3 μm wide, strongly fasciculate, forming a tough and compact mycelium. *Conidiogenous cells* arising ortho- or plagiotropically from hyphal strands, or being intercalary on fasciculate hyphae with collarettes pointing sideward; conidiogenous cells hyaline to subhyaline, flask-shaped to acicular, flexuose, 6-12 μm long when having basal septum, 1.5-3.5 μm wide, phialidic protrusions of intercalary cells up to 8 μm long; collarettes unpigmented, cylindrical, about 1.5 μm wide. *Conidia* hyaline, smooth- and thin-walled, subcylindrical to cylindrical, sometimes slightly curved, usually 4-6 × 1.8-2.5 μm. *Chlamydospores* free or intercalary, thick- and smooth-walled, brown, broadly ellipsoidal to broadly clavate, about 7 × 4.5 μm.
Type strain: CBS 157.44.

Discussion

The species was originally described from river water; further isolates, reported by Schol-Schwarz [270], were from soil. Strain CBS 897.72 was isolated as the causative agent of a human endocarditis, described by Pierach et al. [236]. The fungus had formed a luxuriant growth on a transplanted mitral valve, resulting in the death of the patient due to obstruction of the circulation. A similar case, probably concerning strain CBS 206.73, was described by Slifkin and Bowers [278]. After experimental inoculation in mice, the

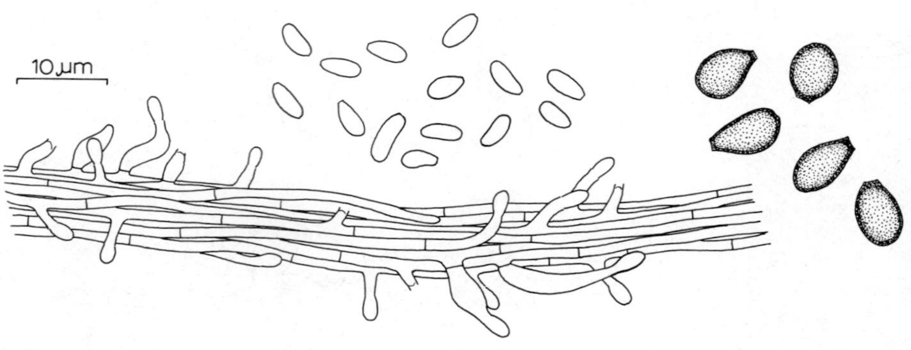

Figure 28 *Phialophora mutabilis*, CBS 157.44.

fungus could be reisolated from several internal organs, where hyphae were visible in the tissues.

6. *Phialophora richardsiae* (Nannf.) Conant (Fig. 29)

Cadophora richardsiae Nannf., in Melin & Nannf., *Sven. Skogsfoeren. Tidskr. 32*: 421, 1934
= *Phialophora richardsiae* (Nannf.) Conant, *Mycologia 29*: 598, 1937
Cadophora brunnescens Davidson, *J. Agric. Res. 50*: 803, 1935
= *Phialophora brunnescens* (Davidson) Conant, *Mycologia 29*: 598, 1937
Phialophora calyciformis G. Smith, *Trans. Br. Mycol. Soc. 45*: 391, 1962

Colonies on malt agar attaining a diameter of about 30 mm in 10 days, appearing velvety to hairy or slightly funiculose; grayish brown, reverse grayish brown to olivaceous black with whitish margin. *Hyphae* subhyaline to brown, brown hyphae usually with rather thick walls, 2.5-4 μm wide, often fasciculate. *Conidiogenous cells* brown, usually arising orthotropically from brown, often fasciculate hyphae, occasionally also arising in loose, plagiotropic brushes from ascending hyphal tips, flask-shaped to acicular, often with a septum just above the base, 10-30 μm long, 2.5-4 μm wide at the base, tapering toward the 1-2-μm-wide tip, which bears a flaring phialidic collarette up to 4 μm in width. *Conidia* of two types: hyaline, thin-walled, short-cylindrical, reniform to allantoid, 3-6 × 1.5-2.5 μm, and pale brown, with rather thick walls, 2.8-3.5 × 2-2.8 μm.
Type strain: CBS 270.33.

Discussion

Phialophora richardsiae can easily be recognized by the dark, acicular phialides with flaring collarettes and by the presence of brown, globose conidia in addition to allantoid

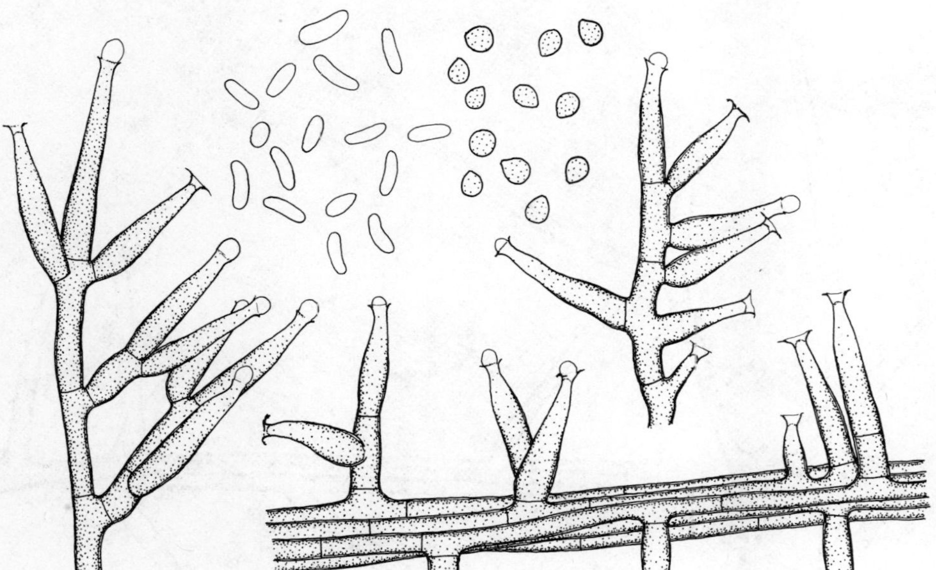

Figure 29 *Phialophora richardsiae*, CBS 270.33.

conidia. Detailed descriptions were given by Domsch and Gams [89], Cole and Kendrick [67], Schol-Schwarz [270], Moreau [213], Nicot and Caillat [219], and McGinnis [184]. The ultrastructure and conidiogenesis were studied by Oláh and Reisinger [225].

The fungus is mainly isolated from wood [219,305], where it may cause soft rots [182]. It occasionally occurs in soil [90]. Schwartz and Emmons [271] described a strain that caused a subcutaneous, cystic granuloma on humans. The pathology was not tested experimentally. Strain CBS 723.74 was reported [178] to be involved in a case of human dacryocystitis. CBS 271.66 was isolated from the tissue of a prostate gland.

7. *Phialophora parasitica* Ajello, Georg & Wang (Fig. 30)

Phialophora parasitica Ajello, Georg & Wang, in Ajello, Georg, Steigbigel & Wang, *Mycologia 66*: 493, 1974

Colonies on malt agar attaining a diameter of about 20 mm in 10 days, appearing powdery to velvety, somewhat domed and floccose at the center, remaining nearly smooth at the regular margin; dull brown; reverse dull brown at the center, with an ochraceous orange tinge near the margin, becoming darker with age. *Hyphae* subhyaline to pale brown, thin- but later rather thick-walled, often becoming verrucose, 2-3.5 μm wide; most aerial hyphae later converted into a more or less differentiated conidial apparatus, sometimes fasciculate. *Conidiogenous cells* arising orthotropically from undifferentiated hyphae or plagiotropically from suberect, regularly cymosely branched hyphae, acicular to linear, 15-30 μm long, 2-3.5 μm wide at the base, tapering toward the 1-1.5-μm-wide tip, which bears an unpigmented, funnel-shaped to cylindrical phialidic collarette, up to 2.5 μm in width. *Conidia* thin-walled, hyaline, reniform, 3.5-5 × 1-2 μm.
Type strain: CBS 860.73.

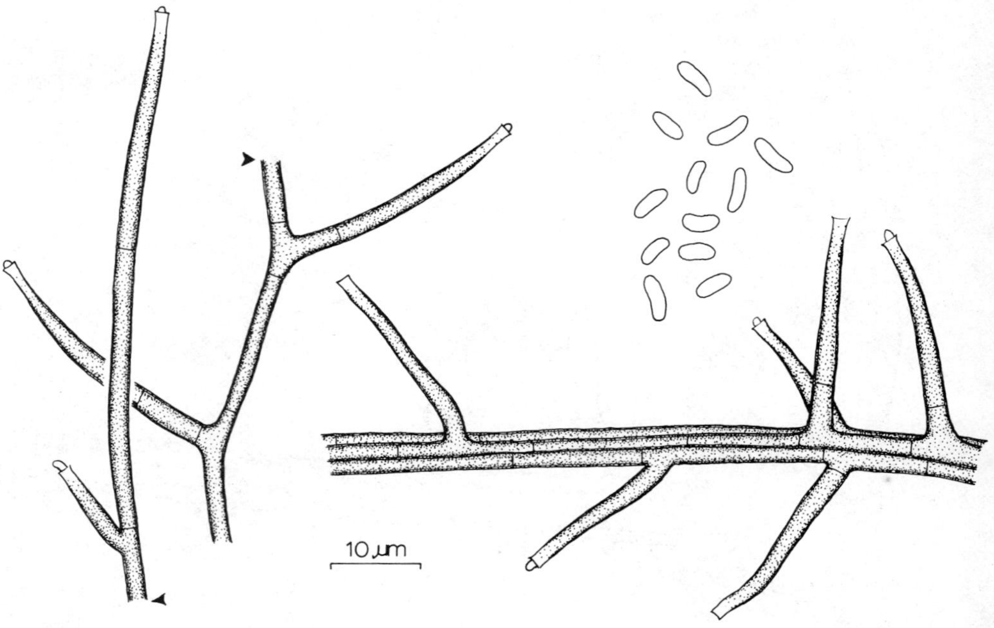

Figure 30 *Phialophora parasitica*, CBS 860.73.

Discussion

The species was originally isolated from a subcutaneous, granulomatous abscess on a human patient undergoing immunosuppressive maintenance therapy. Hyphae were demonstrated in the tissue. It has been obtained repeatedly from human skin lesions at C.D.C., Atlanta (A. A. Padhye, personal communication). Hawksworth et al. [140] and Hawksworth and Gibson [139] described some isolates from various kinds of diseased trees. *Phialophora parasitica* was possibly one of the causative agents of these wilts.

I. *Ramichloridium* Stahel ex de Hoog

Ramichloridium Stahel, *Trop. Agric. (Trinidad) 14*: 44, 1937 (without Latin diagnosis); ex de Hoog, *Stud. Mycol. 15*: 59, 1977

Colonies spreading, appearing smooth, farinose or lanose, brown to olivaceous green; often a yellow or orange pigment being exuded into the medium. *Hyphae* smooth- and thin-walled, hyaline. *Conidiophores* differentiated, usually unbranched, erect, with thin septa. *Conidiogenous cells* terminally on conidiophores, brown, cylindrical, with conidium-bearing denticles in the apical region. *Conidia* hyaline or subhyaline, usually one-celled. Type species: *Ramichloridium apiculatum* (Miller et al.) de Hoog.

Discussion

The genus was recently [81] reintroduced for hyphomycetes with erect, denticulate conidiophores producing one-celled conidia. A taxonomic discussion of the genus was given by de Hoog [81]; 13 species were described. *Pleurophragmium* Cost. is distinguished by having multicelled conidia, and *Veronaea* Cif. & Montemartini by broad and flat conidial scars, which are not prominent.

1. *Ramichloridium cerophilum* (Tubaki) de Hoog (Fig. 31)

Acrotheca cerophila Tubaki, *J. Hattori Bot. Lab. 20*: 143, 1958
= *Cladosporium cerophilum* (Tubaki) Matsushima *Icones Microfungorum a Matsushima lectorum,* Kobe, Japan, p. 34, 1975
= *Ramichloridium cerophilum* (Tubaki) de Hoog, *Stud. Mycol. 15*: 74, 1977
Acrotheca aquaspersa Borelli, *Acta Cient. Venez. 23*: 195, 1972

Colonies on malt agar attaining a diameter of 10-20 mm in 14 days, appearing velvety to hairy; dark olivaceous gray, with sharp margin and dark olivaceous reverse. *Hyphae* pale olivaceous, smooth- and slightly rough-walled, 1.5-3 μm wide. *Conidiophores* differentiated, straight, unbranched, thick-walled with 1-5 thin septa, dark brown, up to 50 μm high, cells 8-15 μm long, 2.2-2.8 μm wide. *Conidiogenous cells* terminally on conidiophores, cylindrical, 9-23 μm long, later often becoming septate, in the apical region with crowded, slightly prominent, unpigmented or faintly pigmented conidial scars about 0.5 μm in width. *Conidia* subhyaline, smooth- and thin-walled, one- or occasionally two-celled, fusiform to clavate, 4.5-7.5 × 1.5-2.4 μm, often giving rise to one to four short series of smaller, secondary conidia.
Type strain: CBS 103.59.

Discussion

As mentioned by de Hoog [81], some slight differences have been found between the type strains of *R. cerophilum* and *Acrotheca aquaspersa,* CBS 313.73. The former almost without exception has only lateral conidiophores, usually giving rise to conidia in short chains, whereas in the latter terminal conidiophores are also common, and secondary

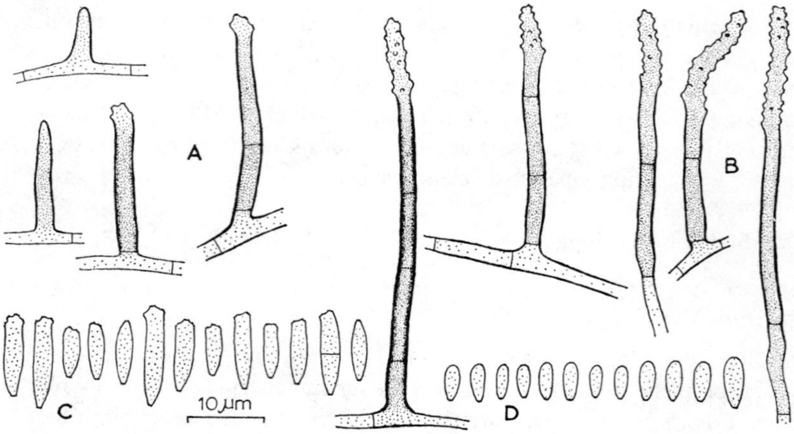

Figure 31 *Ramichloridium cerophilum*: (A) CBS 103.59; (B) CBS 313.73; (C) catenate conidia; (D) terminal conidia, both of CBS 103.59. (From Ref. 81.)

conidia are rare. Some additional differences are presented by the cultural aspects of the strains. Further study is needed to establish whether these differences warrant maintenance of *Acrotheca aquaspersa* as a separate species.

The original isolate of *R. cerophilum* came from decaying leaves in Japan. Matsushima [204] described an isolate from soil. CBS 313.73 was an etiologic agent of inflammatory, suppurative skin lesions in man [36]. Intradermal inoculation in human skin provoked similar lesions; when the fungus was inoculated intraperitoneally in mice, internal organs were affected [36].

J. *Rhinocladiella* Nannf.

Rhinocladiella Nannf., in Melin & Nannf., *Sven. Skogsvardsfoeren. Tidskr. 32*: 461, 1934

Colonies restricted, appearing velvety, lanose or nearly smooth; gray, green, or olivaceous brown. Budding cells and accompanying *Exophiala* and *Phialophora* states may be present. *Hyphae* hyaline to pale olivaceous. *Conidiophores* slightly differentiated, suberect, usually branched, pale to dark brown. *Conidiogenous cells* intercalary or free, cylindrical, grow apically with pointed, blunt or flat conidium-bearing denticles with unpigmented scars. *Conidia* hyaline to subhyaline, one-celled, smooth-walled, single or in short chains.

Type species: *Rhinocladiella atrovirens* Nannf.

Discussion

The genus was recently revised by de Hoog [81]. He reduced five genera to synonymy with *Rhinocladiella*, among which was *Fonsecaea* Negroni. In cultures of the type species, budding cells are common, and an accompanying *Exophiala* state, showing basipetal conidiogenesis on slightly differentiated conidiogenous cells, is often present. *Exophiala spinifera* is intermediate between both genera, having erect conidiophores with basipetal conidiogenesis.

KEY TO THE POTENTIALLY PATHOGENIC SPECIES OF *RHINOCLADIELLA*

1a. Conidiogenous cells grow apically with prominent, butt-shaped denticles . . . . *R. pedrosoi*
1b. Conidiogenous cells grow apically with flat scars . . . . *R. compacta*

1. *Rhinocladiella pedrosoi* (Brumpt) Schol-Schwarz (Fig. 32)

Hormodendrum pedrosoi Brumpt, *Précis Parasitologie,* 3rd., Masson, Paris, p. 1105, 1922
= *Acrotheca pedrosoi* (Brumpt) de Fonseca & de Arẽa Leão, *C. R. Soc. Biol. (Paris) 89*: 762, 1923
= *Trichosporium pedrosoi* (Brumpt) Langer., *Ann. Parasitol. Hum. Comp.* 7: 149, 1929
= *Gomphinaria pedrosoi* (Brumpt) Dodge, *Medical Mycology,* Mosby, St. Louis, p. 850, 1935
= *Fonsecaea pedrosoi* (Brumpt) Negroni, *Rev. Inst. Bacteriol. (Buenos Aires)* 7: 424, 1936
= *Hormodendroides pedrosoi* (Brumpt) Moore & Almeida, *Ann. Mo. Bot. Gard. 23*: 547, 1936
= *Carrionia pedrosoi* (Brumpt) Briceño-Iragorry, *Rev. Clin. Luis Razetti 1*: 108, 1939
= *Phialophora pedrosoi* (Brumpt) Red. & Cif., *Tratt. Micopathol. Umana 5*: 592, 1942

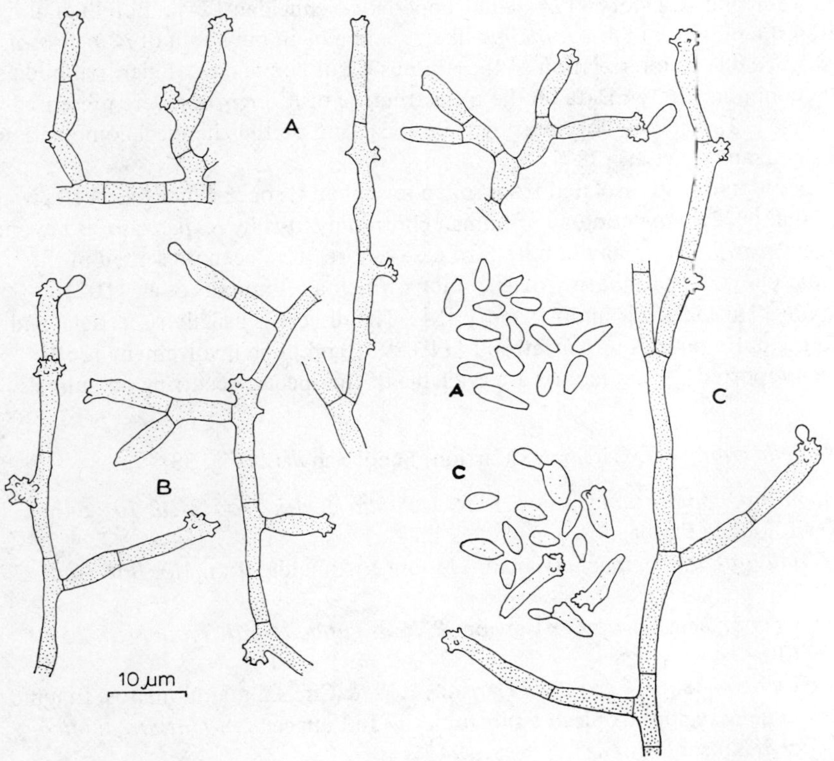

Figure 32 *Rhinocladiella pedrosoi*: (A) CBS 271.37; (B) CBS 272.37; (C) CBS 274.66. (From Ref. 81.)

Colonies on oatmeal agar attaining a diameter of 6-12 mm in 14 days, appearing lanose or velvety, olivaceous. Reverse olivaceous black. *Budding cells* absent. *Hyphae* olivaceous, smooth-walled, 1.5-3 μm wide. *Conidiophores* slightly differentiated, ascending to suberect, branched, pale brown, rather thick-walled, with septa similarly thickened. *Conidiogenous cells* integrated, cylindrical, 8-16 × 2-2.5 μm, later often becoming septate, apically bearing some butt-shaped denticles 0.5-1.5 μm in length with unpigmented scars 0.3-0.8 μm wide. Fertile parts of intercalary cells extend from the upper half of the cells; rhachides rarely intercalary. *Conidia* hyaline or pale brown, thin- and smooth-walled, short-cylindrical, with truncate bases, 3.4-4.8 × 1.6-2 μm, often giving rise to one or several short series of secondary conidia.
Neotype strain: CBS 271.37.

Discussion

Rhinocladiella pedrosoi can easily be recognized by the large, prominent conidium-bearing denticles and the conidia often arising in short chains. Some strains, however, may resemble *R. atrovirens,* which has pimple-shaped denticles on a distinct rhachis, and single conidia. In culture, *Exophiala* states or budding cells are absent from *R. pedrosoi* but often present in *R. atrovirens.* Phialides may rarely occur in either species [103,104]. *Cladosporium carrionii* is also similar, but has undifferentiated conidiogenous cells, flat scars, and long conidial chains. Dark cultures of *R. pedrosoi* may have the appearance of a *Cladosporium* species. Serologic differentiation of these species was elaborated, for example, by Gordon and Al-Doory [124] and Cooper and Schneidau [74]. Borelli et al. [39] reported the presence of *Macrophoma*-like pycnidia in an old strain of *R. pedrosoi,* but their evidence on the interrelation of these states is not convincing. Sclerotial bodies, however, are common [163]. Data on the ultrastructure of *R. pedrosoi* were presented by Cooper et al. [73], on conidiogenesis by Cole [66], and on the chemical composition of cell walls by Szaniszlo et al. [289].

The species is frequently isolated from soil; a review on its occurrence has been given by Gezuela et al. [122]. The etiology of human chromomycosis by *R. pedrosoi* is beyond all doubt (see *C. carrionii*): many hundreds of case reports have been presented in the literature. Many reviews have been written, among which are Emmons et al. [102], Carrión and Silva-Hutner [55], and Al-Doory [8]. The disease is usually superficial and restricted to the extremities, but neurotropy [116,290] and bone involvement [260] have also been reported. Some reports are available of the species occurring on animals [60,250].

2. *Rhinocladiella compacta* (Carrión ex Carrión) Schol-Schwarz (Fig. 33)

Hormodendrum compactum Carrión, *P. R. J. Public Health Trop. Med. 10*: 543, 1935 (without Latin diagnosis)

= *Phialoconidiophora compacta* (Carrión) Moore & Almeida, *Ann. Mo. Bot. Gard. 23*: 549, 1936

= *Fonsecaea compacta* Carrión ex Carrión, *P. R. J. Public Health Trop. Med. 15*: 359, 1940

= *Phialophora compacta* (Carrión ex Carrión) Red. & Cif., Le granulomatosi fungine dell'uomo nelle regioni tropicali e subtropicali. In Pollacci, G., *Trattato di Micopatologia,* 5, Sansoni, Firenze, p. 595, 1942

= *Rhinocladiella compacta* (Carrión ex Carrión) Schol-Schwarz, *Antonie Leeuwenhoek J. Microbiol. Serol. 34*: 141, 1968

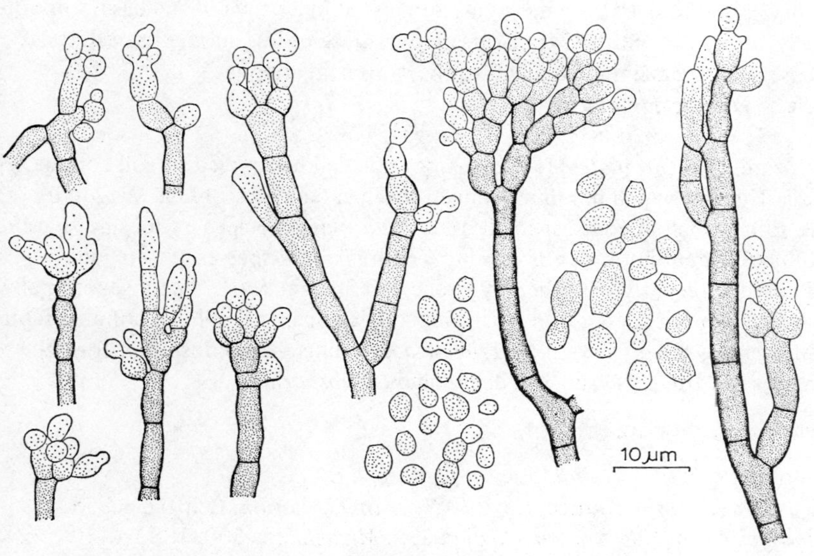

Figure 33 *Rhinocladiella compacta,* CBS 285.47. (From Ref. 81.)

Colonies on oatmeal agar attaining a diameter of 2 mm in 14 days, slightly domed, powdery to velvety, olivaceous black. Reverse olivaceous black. *Hyphae* smooth- and rather thick-walled, flexuose, brown, 3-4 μm wide, forming a compact mycelium. *Conidiophores* slightly differentiated, suberect, with numerous short branches densely aggregated in the apical region. *Conidiogenous cells* integrated, ellipsoidal to doliform, about 8 × 4 μm, with flat, pale pigmented scars, the conidia little differentiated, remaining attached to the supporting cells for a long time. Terminal *conidia* rather thin-walled, subhyaline, truncate, about 2.5-4.5 × 2-3.5 μm.

Neotype strain: CBS 285.47.

Discussion

The taxonomic relationships of *R. compacta* have been discussed by de Hoog [81]. The species has been described as a rare [46,198] etiologic agent of human chromomycosis [50-52], occasionally occurring on animals [197]. The species has mostly been reported along with other cases of chromomycosis, caused by *R. pedrosoi* in particular, and due to the absence of authentic materials it is unclear whether the few strains denominated *R. compacta* were mere deviating strains of *R. pedrosoi.*

K. *Trichocladium* Harz

Trichocladium Harz, *Bull. Soc. Imp. Mosc. 44*: 125, 1871
= *Harziella* O. Kuntze, *Revisio Generum Plantarum,* Leipzig, p. 855, 1891 (name change)
Culcitalna Meyers & Moore, *Am. J. Bot. 47*: 348, 1960

Colonies spreading, gray to black, sometimes exuding a citron yellow pigment into the agar. *Hyphae* hyaline or subhyaline. *Conidiogenous cells* undifferentiated, intercalary or arising laterally from undifferentiated hyphae, hyaline or subhyaline, cylindrical.

Conidia single, dry, olivaceous to brown, usually somewhat paler toward the base, smooth-walled to coarsely verrucose, with 1-5 transverse septa, more or less clavate in outline, with truncate base, often remaining attached to the supporting hyphae.
Lectotype species: *Trichocladium asperum* Harz.

Discussion

The genus is similar to *Cirrenalia* Meyers & Moore, which has curved to coil conidia, and to *Taeniolella* Hughes, which has much longer, often catenulate conidia. *Endophragmia brevis* M. B. Ellis, which in contrast to all other *Endophragmia* species does not have percurrently elongating conidiophores, is close to *Trichocladium* species with regularly balloon-shaped conidia, recently described by Crane and Shearer [76]. Many superficially similar genera are known but these differ by having longitudinal and oblique conidial septa. Some widely different hyphomycetes have *Trichocladium*-like chlamydospores, for example, some species of *Idriella* Nelson & Wilhelm and *Cladosporium*.

1. *Trichocladium asperum* Harz (Fig. 34)

Trichocladium asperum Harz, *Bull. Soc. Imp. Mosc. 44*: 125, 1871
= *Harziella aspera* (Harz) O. Kuntze, *Revisio Generum Plantarum,* Leipzig, p. 855, 1891
Trichocladium asperum Harz var. *charticola* Sacc., *Michelia 2*: 560, 1882
= *Trichocladium charticola* (Sacc.) Sacc., *Syll. Fung. 15*: 424, 1901
= *Dicoccum asperum* (Corda) Sacc. var. *charticola* (Sacc.) Lindau, in Rabenhorst, *Krypt.-Fl.,* 2nd ed., I, *8*: 764, 1907

Colonies on PCA under near-UV light spreading, appearing glassy to dirty white with minute black specks due to abundant sporulation. *Hyphae* hyaline, smooth- and thin-walled, 1.5-3.5 (up to 8) μm wide, loosely branched. *Conidiogenous cells* undifferentiated, intercalary or arising laterally from undifferentiated hyphae, cylindrical; intercalary cells bearing one to two conidia, lateral cells cylindrical, often slightly clavate, up to 20 μm long, nonseptate or with some septa in the apical region. *Conidia* single, usually with one transverse septum, thick-walled, coarsely verrucose, dark brown, broadly cuneiform to ellipsoidal, distinctly constricted at the septum, 15-30 × 10-15 μm, with truncate base, 2-3 μm in width.

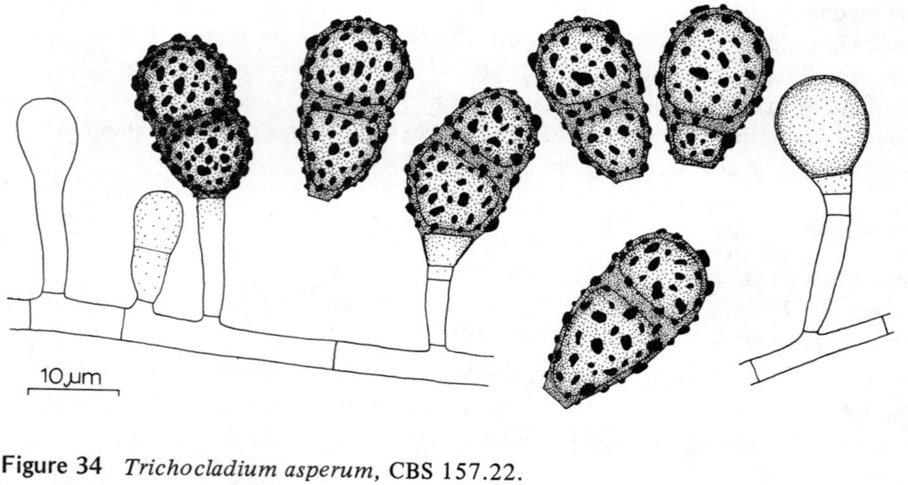

Figure 34 *Trichocladium asperum,* CBS 157.22.

Discussion

The species is easily distinguished from other *Trichocladium* species by its rough-walled conidia. It was described and depicted, for example, by Groves and Skolko [127], Hughes [146], Ellis [98], Barron [20], Domsch and Gams [89], and Matsushima [204]. Matsushima [204] described an additional *Scopulariopsis* state in culture, which Domsch and Gams [89] referred to as a "state with phialospores."

The species is rather common on plant remains and is frequently isolated from soil [89]. Bietti [28] described a case of superficial keratomycosis in humans; this case was recently listed by Hoffmann [143].

L. *Tritirachium* Limber

Spirotrichum Saito, *J. Ferment. Technol., Osaka 27*: 1, 1939 (without Latin diagnosis); ex van Beyma, *Antonie Leeuwenhoek Microbiol. Serol. 6*: 286, 1940

Tritirachium Limber, *Mycologia 32*: 26, 1940

Colonies restricted, velvety; ochraceous, cinnamon or purple. *Hyphae* pale, thin-walled. *Conidiophores* differentiated but basally with gradational transition into the vegetative mycelium, smooth- and rather thick-walled, bearing several whorls of conidiogenous cells or verticillate side branches. *Conidiogenous cells* consisting of an acicular to subulate basal part and a narrow, regularly flexuose, inconspicuously cicatrized rhachis. *Conidia* one-celled, hyaline to subhyaline, smooth- and thin-walled, globose to ellipsoidal.

Type species: *Tritirachium dependens* Limber.

Discussion

The genus was revised by de Hoog [79,80] and restricted to species with verticillately arranged conidiogenous cells, bearing regularly flexuose, inconspicuously cicatrized rhachides. *Acrodontium* de Hoog is similar, but has denticulate rhachides.

1. *Tritirachium oryzae* (Vincens) de Hoog (Fig. 35)

Beauveria oryzae Vincens, *Rev. Pathol. Veg. Entomol. Agric. Fr. 10*: 122, 1910

= *Tritirachium oryzae* (Vincens) de Hoog, *Stud. Mycol. 1*: 22, 1972

Beauveria brumptii Langer., *Bull. Acad. Nat. Med. Paris 111*: 133, 1934

= *Tritirachium brumptii* (Langer.) Langer., *Ann. Parasitol. Hum. Comp. 22*: 94, 1947

Spirotrichum purpureum Saito, *J. Ferment. Technol. Osaka 27*: 1, 1939 (without Latin diagnosis)

= *Tritirachium purpureum* Saito ex van Beyma, *Antonie Leeuwenhoek J. Microbiol. Serol. 8*: 117, 1942

Spirotrichum musae van Beyma, *Antonie Leeuwenhoek J. Microbiol. Serol. 6*: 286, 1940

= *Tritirachium musae* (van Beyma) van Beyma, *Antonie Leeuwenhoek J. Microbiol. Serol. 8*: 117, 1942

Tritirachium roseum van Beyma, *Antonie Leeuwenhoek J. Microbiol. Serol. 8*: 118, 1942

Colonies on malt agar growing slowly, attaining a diameter of 3-5 mm in 8 days, appearing velvety, rosy vinaceous to lilac. Reverse vinaceous brown. *Hyphae* hyaline to pale brown, smooth- and thin-walled, 1-2 μm wide, forming a compact submerged mycelium. *Conidiophores* reddish-brown, smooth- and rather thick-walled, suberect, basally gradationally merging into the vegetative mycelium, 200-1500 μm long, in the upper part

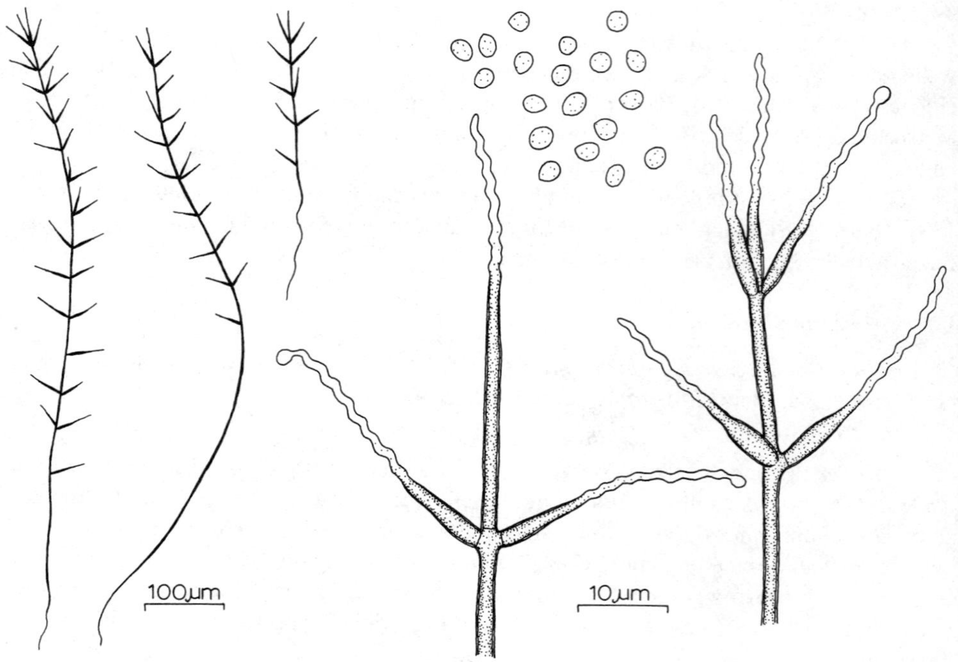

Figure 35 *Tritirachium oryzae*, CBS 182.42.

bearing 5-20 whorls of one to three conidiogenous cells. *Conidiogenous cells* consisting of an elongate basal part, slightly swollen below the middle, 9-20 × 1.5-3 µm, and a geniculate, inconspicuously cicatrized rhachis, up to 165 µm long and regularly 1 µm wide. *Conidia* hyaline, smooth- and thin-walled, globose to ellipsoidal, (1.5-)2-3(-3.5) × 1.5-3 µm.

Discussion

The species can easily be recognized by its purplish colonies and the regularly flexuose conidiiferous rhachides. It is fairly frequently isolated from air and soil. Langeron [174, 175] described a specimen from corneal ulcers in humans. Rodrigues et al. [255] mentioned a similar case, where *T. oryzae* occurred after corneal trauma. Srinivasa Rao and Ramakrishnan [282] isolated *T. oryzae* from three corneal ulcers, but their attempts to reproduce infection in rabbits' eyes remained unsuccessful. *Verticillium rubrum* Baquis [18], treated as a doubtful species by de Hoog [79] but possibly close to *T. oryzae*, was also described from an ocular mycosis.

M. *Wallemia* Johan-Olsen

Wallemia Johan-Olsen, *Forh. Videnskselsk. Krist. 12*: 6, 1887
Hemispora Vuill., *Bull. Trimest. Soc. Mycol. Fr. 22*: 125, 1906

Colonies restricted, irregular, powdery to velvety, brown to purplish brown; osmophilic. *Hyphae* hyaline or subhyaline, forming a compact mycelium. *Conidiophores* erect, in dense, parallel arangement, cylindrical, slightly constricted at the apex, producing a cylindrical meristematic cell which forms arthric conidia. *Conidia* one-celled, initially cubical, soon becoming globose.

Type species: *Wallemia sebi* (Fr.) von Arx.

Discussion

The probably monotypic genus is characterized by its process of conidiogenesis [137, 194,293]. From the tip of a smooth-walled conidiogenous cell a verruculose meristematic cell is produced which retains some extension capacity, even after delimitation. Later it disarticulates into separate cells. In other hyphomycete genera with meristem arthroconidia, such as *Basipetospora* Cole & Kendrick, *Oidium* Sacc., *Briosia* Cavara, and *Ojibwaya* Sutton, no fixed conidiogenous cell can be recognized. The colony aspect of *Wallemia* shows a marked resemblance to that of *Oidiodendron* Robak.

1. *Wallemia sebi* (Fr.) von Arx (Fig. 36)

Torula epizoa Corda in Sturm, *Deutschl. Fl., Abt. III, 2*: 97, 1829
= *Sporendonema epizoum* (Corda) Red. & Cif., *J. Trop. Med. Hyg. 37*: 169, 1934
Sporendonema sebi Fr., *Syst. Mycol. 3*: 435, 1832
= *Wallemia sebi* (Fr.) von Arx, *The Genera of Fungi Sporulating in Pure Culture,* Cramer, Vaduz, p. 166, 1970
Torula sacchari Corda, *Icon. Fung. 4*: 23, 1840
Wallemia ichtyophila Johan-Olsen, *Forh. Videnskselsk. Krist. 12*: 6, 1887
Hemispora stellata Vuill., *Bull. Trimest. Soc. Mycol. Fr. 22*: 125, 1906
Sporotrichum navale Joly, *Rev. Mycol. 26*: 98, 1961

Colonies restricted on malt agar with 40% additional sugar, attaining a diameter of 3 mm in 10 days, appearing slightly domed, powdery, with sharp, somewhat irregular margin; ochraceous brown to purplish brown. *Hyphae* hyaline, smooth- and thin-walled, 1.5-2.5 μm wide, forming a compact mycelium. *Conidiophores* erect, in dense, parallel arrangement, unbranched, only slightly differentiated, subhyaline, smooth-walled, slightly constricted at the apex, producing a cylindrical, verruculose meristematic cell which disarticulates into arthric conidia. *Conidia* dry, one-celled, pale brown, verruculose, initially cubical, soon becoming globose, 2-3.5 μm in diameter.

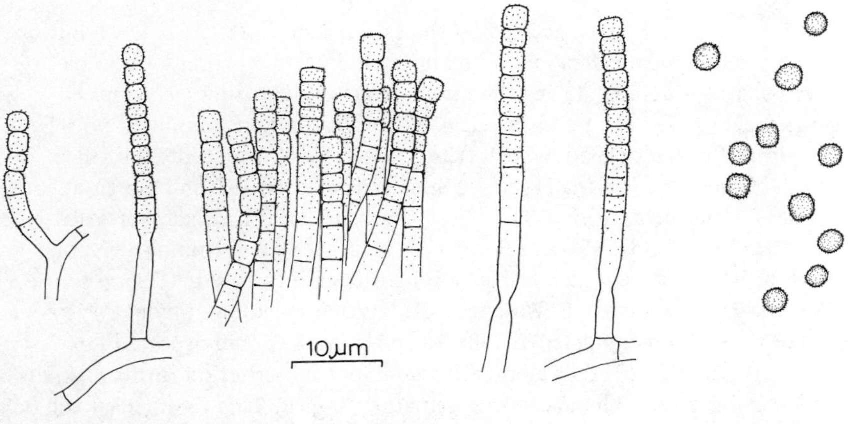

Figure 36 *Wallemia sebi,* CBS 103.10.

Discussion

The synonyms listed above are generally considered to be identical with *Wallemia sebi,* although critical type studies have not been carried out. Redaelli and Ciferri [245] and Ciferri [61] listed many additional synonyms, which all need to be confirmed. The authentic strain of *Hemispora stellata,* CBS 103.10, still sporulates well; the description above was largely based on this strain. CBS 200.33 is possibly an authentic *Oospora d'agatae* Sacc.

The species is frequently isolated from substrates with a high osmotic pressure, particularly salted fish [167], sugar-containing materials, and foodstuffs. Its osmophilia has been firmly established by Ormerod [226] and others; according to Van de Lustgraaf [299], it may be xerophilic.

Wallemia sebi may also be isolated from the skin of healthy and diseased humans and animals. The numerous early reports in the literature on disorders supposedly caused by this species led Ciferri and Redaelli [62] to describe the clinical picture as variable. Janke [150], however, presented a critical review and accepted only those reports in which the pathogenicity was unambiguously demonstrated. From this selection, the appearance of the disease was more stable, and could be summarized as verrucose lesions with subcutaneous, tuberculoid infiltrates, leading to ulceration. Accepted cases included those of Gougerot and Caraven [125], Auvrey [16], de Beurmann et al. [78] and Schrameck and Weidenfeld [268].

Epicutaneous inoculation of small animals gave no significant reaction [125], but subcutaneous inoculation gave subcutaneous nodules [150], although the virulence was low. The pathogenicity to rabbits was established by Ota et al [227].

ACKNOWLEDGMENTS

The author wishes to express his gratitude to Drs. J. A. von Arx, M. B. Ellis, W. Gams, W. Jooste, E. G. Simmons, and G. A. de Vries for giving their comments on the manuscript or parts thereof. Miss C. A. N. van Oorschot is acknowledged for correcting the English text and Miss J. Elings for technical assistance.

NOTE ADDED IN PROOF

The manuscript for this chapter was received by the Editor in May 1979. A few additions were made in July 1981. The advances recorded in the literature on this group of pathogenic fungi have been considerable during the intervening period of time. New potentially pathogenic genera have been published, *e.g., Cladophialophora* Borelli [Borelli, D. 1980. Causal agents of chromoblastomycosis (Chromomycetes). Proceedings of the Fifth International Conference on the Mycoses, Superficial, Cutaneous and Subcutaneous Infections. Caracas, Venezuela, April 27-30, 1980. Pan Am. Health Organ. Sci. Publ. No. 396, Washington, D.C., pp. 334-335] and *Botryomyces* de Hoog and Rubio (de Hoog, G.S., and C. Rubio. 1982. A new dematiaceous fungus from human skin. *Sabouraudia 20*:15-20.). Recently, M. Hironago, S. Watanabe, K. Nishimura, and M. Miyaji (Mycologia *73*:118-1183, 1981) reported elongated annellated zones on the conidiospores of *exophiala dermatitidis* (CBS 207.35), thus eliminating an important criterion for its distinction from *Exophiala mansonii.* The taxonomic structure of *Exophiala* Carmichael will be described in light of recent discoveries in Chapter 11, Part C of this series.

REFERENCES

1. Agnihotrudu, V. 1959. Fungi isolated from rhizosphere, V. J. Madras Univ. Ser. B *29*: 155-181.
2. Ajello, L. 1978. The black yeasts as disease agents: historical perspective. In *The Black and White Yeasts,* Proceedings of the Fourth International Conference on the Mycoses, Pan Am. Health Organ. Sci. Publ. No. 356, Washington, D.C., pp. 9-16.
3. Ajello, L., M. R. McGinnis, and J. Camper. 1977. An outbreak of phaeohyphomycosis in rainbow trout caused by *Scolecobasidium humicola.* Mycopathologia *62*: 15-22.
4. Ajello, L., and L. Runyan. 1973. Abortive "perithecial" production by *Phialophora verrucosa.* Mycologia *45*: 947-950.
5. Akagi, M., T. Fujino, S. Hayashi, Y. Yamamoto, and T. Akabe. 1958. A case of pulmonary infection caused by a black yeast-like fungus. Med. J. Osaka Univ. *8*: 585-599.
6. Akün, R. 1953. Chromoblastomycose beim Pferde. Dtsch. Tieraertzl. Wochenschr. *60*: 519.
7. Akün, R. 1953. Eine chromoblastomycosisähnliche Pilzkrankheit beim Pferde. Zentralbl. Allg. Pathol. Pathol. Anat. *90*: 294-297.
8. Al-Doory, Y. 1972. *Chromomycosis,* Mountain Press, Missoula, Mont.
9. Al-Doory, Y., and M. A. Gordon. 1963. Application of fluorescent antibody procedures to the study of pathogenic dematiaceous fungi. I. Differentiation of *Cladosporium carrionii* and *Cladosporium bantianum.* J. Bacteriol. *86*: 332-338.
10. Anderson, B., S. S. Roberts, C. Gonzalez, and E. W. Chick. 1959. Mycotic ulcerative keratitis. Arch. Ophthalmol. *62*: 169-179.
11. Aranello, F. 1967. Experiencias en cuatro casos de cromomicosis y su tratamiento. Rev. Soc. Peru. Dermatol. *1*: 139-144.
12. Araviyskiy, R. A. 1968. Suktsinoksidaznaya i shchelochnofosfataznaya aktivost' pri éksperimental'ny̌kh mikozakh mozga. Byull. Eksp. Biol. Med. *65*: 114-118.
13. Araviyskiy, R. A., and V. B. Aronson. 1968. Comparative histopathology of chromomycosis and cladosporiosis in the experiment. Mycopathol. Mycol. Appl. *36*: 322-340.
14. Arêa Leão, A. E., A. Cury, and J. Martins Ferreira. 1945. "Tinea nigra" (Keratomycosis "nigricans palmaris"): observacão e estudo de um caso. Rev. Bras. Biol. *5*: 165-177.
15. Arx, J. A. von 1970. *The Genera of Fungi Sporulating in Pure Culture,* Cramer, Lehre.
16. Auvrey, M. 1909. A propos d'une nouvelle mycose observée chez l'homme. Suppuration cervicale due à l'*Hemispora stellata.* Bull. Mem. Soc. Chir. Paris *20*: 686.
17. Azar, P., J. V. Aquavella, and R. S. Smith. 1957. Keratomycosis due to an *Alternaria* species. Am. J. Ophthalmol. *79*: 881-883.
18. Baquis, E. 1905. Una nuova forma di Cheratomicosi. Ann. Ottalmol. Clin. Ocul. *34*: 945-947.
19. Barnola, J., and A. A. Ortega. 1961. Cladosporiosis profunda. Mycopathol. Mycol. Appl. *15*: 422-428.
20. Barron, G. L. 1968. *The Genera of Hyphomycetes from Soil,* Williams & Wilkins, Baltimore.
21. Barron, G. L., and L. V. Busch. 1962. Studies on the soil hyphomycete *Scolecobasidium.* Can. J. Bot. *40*: 77-84.
22. Baylet, J., R. Camain, and G. Segretain. 1959. Identification des agents des madaromycoses du Sénégal et de la Mauritanie. Description d'une espèce nouvelle. Bull. Soc. Pathol. Exot. *52*: 448-477.

23. Beer, J. V. 1969. Isolation of pathogenic fungi from waterfowl. J. Appl. Bacteriol. Tech. Ser. *3*: 103-111.
24. Beer, J. V., and J. Kear. 1973. Fungal infections of the plumage in flamingos. Avic. Mag. *79*: 163-164.
25. Bennett, J. E., H. Bonner, A. E. Jennings, and R. I. Lopez. 1973. Chronic meningitis caused by *Cladosporium trichoides*. Am. J. Clin. Pathol. *59*: 398-407.
26. Berger, L., M. Beaudry, and E. Gaumond. 1945. Chromoblastomycosis due to a new species of fungus. Can. Med. Assoc. J. *53*: 138-143.
27. Berger, L., and M. Langeron. 1949. Sur un type nouveau de chromomycose observé au Canada (*Torula bergeri* n.sp.). Ann. Parasitol. Hum. Comp. *24*: 574-599.
28. Bietti, A. 1922. Tre casi di forma benigna di cheratomicosi con ifomiceti non ancora rinvenuti nella cornea. Ann. Ottalmol. Clin. Ocul. *50*: 301.
29. Binford, C. H., A. K. Thompson, and M. E. Gorham. 1952. Mycotic brain abscess due to *Cladosporium trichoides*. Report of a case. With mycologic report by C. W. Emmons. Am. J. Clin. Pathol. *22*: 535-542.
30. Blalock, H. G., L. K. Georg, and W. T. Durieux. 1973. Encephalitis in turkey poults due to *Dactylaria (Diplorhinotrichum) gallopava*—a case report and its experimental reproduction. Avian Dis. *17*: 197-204.
31. Blazer, V. S., and R. E. Wolke. 1979. An *Exophiala*-like fungus as the cause of a systemic mycosis of marine fish. J. Fish Dis. *2*: 145-152.
32. Boedijn, K. B. 1933. Uber einige phragmosporen Dematiazeen. Bull. Jard. Bot. Buitenzorg, Ser. III, *13*: 120-134.
33. Boomker, J., J. A. W. Coetzer, and D. B. Scott. 1977. Black grain mycetoma (maduromycosis) in horses. Onderstepoort J. Vet. Res. *44*: 249-252.
34. Borelli, D. 1960. *Torula bantiana,* agents di un granuloma cerebrale. Riv. Anat. Patol. Oncol. *17*: 615-622.
35. Borelli, D. 1972. A method for producing chromomycosis in mice. Trans. R. Soc. Trop. Med. Hyg. *66*: 793-794.
36. Borelli, D. 1972. *Acrotheca aquaspersa* nova species agente de cromomicosis. Acta Cient. Venez. *23*: 193-196.
37. Borelli, D. 1978. *Aureobasidium werneckii*: acervuli, sporodochia, and pycnidia. In *The Black and White Yeasts,* Proceedings of the Fourth International Conference on the Mycoses, Pan Am. Health Organ. Sci. Publ. No. 356, Washington, D.C., pp. 79-90.
38. Borelli, D., and C. Marcano. 1973. *Cladosporium castellanii* nova species agente de tinea nigra. Castellania *1*: 151-154.
39. Borelli, D., C. Marcano, and M. Feo. 1974. Informe sobre las actividades de la sección de micología durante el año de 1972. Gac. Med. Caracas *82*: 133-149.
40. Borsook, M. E. 1933. Skin infection due to *Alternaria tenuis* (with the report of a case). Can. Med. Assoc. J. *29*: 479-482.
41. Botticher, W. W. 1966. *Alternaria* as a possible human pathogen. Sabouraudia *4*: 256-258.
42. Bourlond, A., J. Decroix, F. Dobbelaere, and A. Lissoir. 1974. Alternariose dermique. Ann. Dermatol. Syphiligr. *101*: 413-415.
43. Bridges, C. H. 1957. Maduromycotic mycetomas in animals. *Curvularia geniculata* as an etiologic agent. Am. J. Pathol. *33*: 411-427.
44. Bridges, C. H., and J. N. Beasley. 1960. Maduromycotic mycetomas in animals. *Brachycladium spiciferum* as an etiologic agent. J. Am. Vet. Med. Assoc. *137*: 192-201.
45. Brodey, R. S., H. F. Schrijver, M. J. Deubler, W. Kaplan, and L. Ajello. 1967. Mycetoma in a dog. J. Am. Vet. Med. Assoc. *151*: 442-451.
46. Brygoo, E. R., and P. Destombes. 1976. Epidémiologie de la chromoblastomycose humaine. Bull. Inst. Pasteur (Paris) *74*: 219-243.

47. Bugnicourt, M. F. 1950. Les espèces du genre *Curvularia* isolées des semences de riz. Rev. Gen. Bot. *57*: 65-77.
48. Bulman, R. A., and J. Stretton. 1974. Lesions experimentally produced by fungi implicated in extrinsic allergic alveolitis. J. Hyg. *73*: 369-374.
49. Carmichael, J. W. 1966. Cerebral mycetoma of trout due to a *Phialophora*-like fungus. Sabouraudia *5*: 120-123.
50. Carrión, A. L. 1935. Chromoblastomycosis. Preliminary report on a new clinical type of the disease caused by *Hormodendrum compactum,* nov. sp. P. R. J. Public Health Trop. Med. *10*: 543-545.
51. Carrión, A. L. 1936. Chromoblastomycosis. A new clinical type caused by *Hormodendrum compactum.* P. R. J. Public Health Trop. Med. *11*: 663-682.
52. Carrión, A. L. 1940. The specific fungi of chromomycosis. P. R. J. Public Health Trop. Med. *15*: 340-361.
53. Carrión, A. L. 1942. Chromoblastomycosis. Mycologia *34*: 424-441.
54. Carrión, A. L. 1950. Yeastlike dematiaceous fungi affecting the human skin. Special reference to so-called *Hormiscium dermatitidis.* Arch. Dermatol. Syphilol. *61*: 996-1009.
55. Carrión, A. L., and M. Silva-Hutner. 1970. Taxonomic criteria for the fungi of chromoblastomycosis with reference to *Fonsecaea pedrosoi.* Int. J. Dermatol. *10*: 35-43.
56. Castellani, A. 1905. Tropical forms of Pityriasis versicolor. Br. Med. J. *2*: 1271-1272.
57. Castellani, A., and A. J. Chalmers. 1913. *Manual of Tropical Medicine,* Baillière, Tindall & Cox, London.
58. Castellani, A., and R. De Silva. 1957. Tinea nigra. A general account based on far eastern experience. Mykosen *1*: 41-49.
59. Chidambaram, P., S. B. Mathur, and P. Neergaard. 1973. Identification of seed-borne *Drechslera* species. Friesia *10*: 165-207.
60. Cicmanec, J. L., D. H. Ringler, and E. S. Beneke. 1973. Spontaneous occurrence and experimental transmission of the fungus *Fonsecaea pedrosoi,* in the marine toad, *Bufo marinus.* Lab. Anim. Sci. *23*: 43-47.
61. Ciferri, R. 1958. *Mauginiella* a synonym of *Sporendonema.* Atti Ist. Bot. Univ. Lab. Crittog. Pavia, Ser. 5, *15*: 126-133.
62. Ciferri, R., and P. Redaelli. 1934. *Sporendonema epizoum* (Corda) Cif. et Red.; an entity including *Hemispora stellata* and *Oospora d'agatae.* J. Trop. Med. Hyg. *37*: 167-170.
63. Ciferri, R., and P. Redaelli. 1957. *Trichosporium mantagazzae* Poll. e *Cladosporium chlamydeum* n. sp. Mycopathol. Mycol. Appl. *8*: 183-188.
64. Ciferri, R., M. Ribaldi, and A. Corte. 1956. Revision of 23 strains of *Aureobasidium pullulans* (de By.) Arn. (= *Pullularia pullulans*). Atti Ist. Bot. Univ. Lab. Crittog. Pavia, Ser. *5, 14*: 3-15.
65. Cole, G. T. 1976. Conidiogenesis in pathogenic hyphomycetes. I. *Sporothrix, Exophiala, Geotrichum* and *Microsporum.* Sabouraudia *14*: 81-98.
66. Cole, G. T. 1978. Conidiogenesis in the black yeasts. In *The Black and White Yeasts,* Proceedings of the Fourth International Conference on the Mycoses, Pan Am. Health Organ. Sci. Publ. No. 356, Washington, D.C., pp. 66-78.
67. Cole, G. T., and B. Kendrick. 1973. Taxonomic studies of *Phialophora.* Mycologia *65*: 661-688.
68. Cole, G. T., and R. A. Samson. 1979. *Patterns of Development in Conidial Fungi,* Pitman, London.
69. Collomb, H., M. Dumas, P.-L. Girard, and I.-P. N'Diaye. 1973. Dématiomycose cérébrale (cladosporiose cérébrale). Un cas chez un noir sénégalais. Sem. Hop. Paris *49*: 2857-2862.

70. Conti-Díaz, I. A., R. Vignale, and E. Peña de Pereira. 1969. Cromoblastomicosis tratada con thermoterapia local. Med. Cutanea (Barcelona) *3*: 383-386.
71. Cooke, W. B. 1959. An ecological life history of *Aureobasidium pullulans* (de Bary) Arnaud. Mycopathol. Mycol. Appl. *12*: 1-45.
72. Cooke, W. B. 1962. A taxonomic study of the "black yeasts." Mycopathol. Mycol. Appl. *17*: 1-43.
73. Cooper, B. H., S. Grove, C. Mims, and P. Szaniszlo. 1973. Septal ultrastructure in *Phialophora pedrosoi, Phialophora verrucosa* and *Cladosporium carrionii.* Sabouraudia *11*: 127-130.
74. Cooper, B. H., and J. D. Schneidau. 1970. A serological comparison of *Phialophora verrucosa, Fonsecaea pedrosoi* and *Cladosporium carrionii* using immunodiffusion and immunoelectrophoresis. Sabouraudia *8*: 217-226.
75. Coulanges, P., and E. R. Brygoo. 1970. Chromoblastomycose expérimentale de la souris (technique de Shepard). Bull. Soc. Pathol. Exot. *63*: 194-201.
76. Crane, J. L., and C. A. Shearer. 1978. Two new species of *Trichocladium* (Hyphomycetes) from submerged wood. Mycologia *70*: 866-874.
77. Cricklow, D. K., F. T. Enrile, and M. Y. Memon. 1973. Cerebellar abscess due to *Cladosporium trichoides (bantianum)*: case report. Am. J. Clin. Pathol. *60*: 416-421.
78. de Beurmann, L., M. Clair, and H. Gougerot. 1909. Une nouvelle mycose, l'hémisporose. Un cas d'hémisporose de la verge. Bull. Mem. Soc. Med. Hop. Paris *3*: 917-919.
79. de Hoog, G. S. 1972. The genera *Beauveria, Isaria, Tritirachium* and *Acrodontium* gen. nov. *Studies in Mycology,* Vol. 1, Centraalbureau voor Schimmelcultures, Baarn, pp. 1-41.
80. de Hoog, G. S. 1973. Additional notes on *Tritirachium.* Persoonia *7*: 437-441.
81. de Hoog, G. S. 1977. *Rhinocladiella* and allied genera. *Studies in Mycology,* Vol. 15, Centraalbureau voor Schimmelcultures, Baarn, pp. 1-140.
82. de Hoog, G. S., and E. J. Hermanides-Nijhof. 1977. Survey of the black yeasts and allied fungi. *Studies in Mycology,* Vol. 15, Centraalbureau voor Schimmelcultures, Baarn, pp. 178-221.
83. de Hoog, G. S., and J. A. von Arx. 1973. Revision of *Scolecobasidium* and *Pleurophragmium.* Kavaka *1*: 55-60.
84. Delacrétaz, J., D. Grigoriu, and A. Grigoriu. 1970. *Alternaria tenuis* en pathologie cutanée humaine. Ann. Dermatol. Syphiligr. *97*: 15-20.
85. de Vries, G. A. 1952. *Contribution to the Knowledge of the Genus Cladosporium Link ex Fr.,* Hollandia, Baarn.
86. Di Salvo, A. F., and W. H. Chew. 1968. *Phialophora gougerotii*: an opportunistic fungus in a patient treated with steroids. Sabouraudia *6*: 241-245.
87. Dixon, D. M., H. J. Shadomy, and S. Shadomy. 1977. Isolation of *Cladosporium trichoides* from nature. Mycopathol. Mycol. Appl. *62*: 125-127.
88. Dolan, C. T., L. A. Weed, and D. E. Dines. 1970. Bronchopulmonary Helminthosporiosis. Am. J. Clin. Pathol. *53*: 235-242.
89. Domsch, K. H., and W. Gams. 1970. *Pilze aus Agrarböden,* Fischer Verlag, Stuttgart.
90. Domsch, K. H., W. Gams, and H. Anderson. 1980. *Compendium of Soil Fungi,* Academic, London.
91. Doty, M. S., and D. W. Slater. 1946. A new species of *Heterosporium* pathogenic on young chinook salmon. Am. Midl. Nat. *36*: 663-665.
92. Dovgich, N. A. 1962. Ekzema spiny Sobak-mikoticheskoe zabolevanie. Veterinariya *39*: 31-32.
93. Dovgich, N. A. 1965. K voprosu ob épidermofitii Sobak. Veterinariya *41*: 37-38.
94. Dumas, M., I. P. N'Diaye, and P.-L. Girard. 1973. Mycoses nerveuses au Sénégal. S. Afr. J. Med. Sci. *4*: 231-238.

95. Duque, O. 1963. Cladosporiosis cerebral experimental. Rev. Latinoam. Anat. Patol. *7*: 101-110.
96. Elkan, E., and C. M. Philpot. 1973. Mycotic infections in frogs due to a *Phialophora*-like fungus. Sabouraudia *11*: 99-105.
97. Ellis, M. B. 1966. Dematiaceous Hyphomycetes. VII. *Curvularia, Brachysporium,* etc. Commonw. Mycol. Inst. Mycol. Pap. *106*: 1-57.
98. Ellis, M. B. 1971. *Dematiaceous Hyphomycetes,* Commonwealth Mycological Institute, Kew, Surrey, England.
99. Ellis, M. B. 1976. *More Dematiaceous Hyphomycetes.* Commonwealth Mycological Institute, Kew, Surrey, England.
100. Emmons, C. W. 1945. *Phialophora jeanselmei* comb. n. from mycetoma of the hand. Arch. Pathol. *39*: 364-368.
101. Emmons, C. W., C. H. Binford, and J. P. Utz. 1963. *Medical Mycology,* 2nd ed., Henry Kimpton, London.
102. Emmons, C. W., C. H. Binford, J. P. Utz, and K. J. Kwon-Chung. 1977. *Medical Mycology,* 3rd ed., Lea & Febiger, Philadelphia.
103. Emmons, C. W., and A. L. Carrión. 1936. *Hormodendrum pedrosoi,* an etiological agent in chromoblastomycosis. P. R. J. Public Health Trop. Med. *11*: 639-650.
104. Emmons, C. W., and A. L. Carrión. 1936. The *Phialophora* type of sporulation in *Hormodendrum pedrosoi* and *Hormodendrum compactum.* P. R. J. Public Health Trop. Med. *11*: 703-710.
105. English, M. P. 1965. The saprophytic growth of non-keratinophilic fungi on keratinized substrate, and a comparison with keratinophilic fungi. Trans. Br. Mycol. Soc. *48*: 219-235.
106. Estes, S. A., W. G. Merz, and L. G. Maxwell. 1977. Primary cutaneous phaeohyphomycosis caused by *Drechslera spicifera.* Arch. Dermatol. *113*: 813-815.
107. Euzeby, J., C. Laclaire, C. Chauve, and A. Eghbali. 1973. Investigations mycologiques. I. Recherches sur quelques pseudo-dermatophytes. Bull. Soc. Sci. Vet. Med. Comp. Lyon *75*: 355-360.
108. Evans, H. C. 1971. Thermophilous fungi from coal spoil tips. I. Taxonomy. Trans. Br. Mycol. Soc. *57*: 241-254.
109. Farmer, S. G., and R. A. Komorowski. 1976. Cutaneous microabscess formation from *Alternaria alternata.* Am. J. Clin. Pathol. *66*: 565-569.
110. Felger, C. E., and L. Friedman. 1962. Experimental cerebral chromoblastomycosis. J. Infect. Dis. *3*: 1-7.
111. Fijan, N. 1969. Systemic mycosis in channel catfish. Bull. Wildl. Dis. Assoc. *5*: 109-110.
112. Findley, G. H. 1957. Chromoblastomycosis caused by Simson species of *Hormodendrum.* S. Afr. Med. J. *31*: 538-540.
113. Forster, R. K., G. Rebell, and L. A. Wilson. 1975. Dematiaceous fungal keratitis. Clinical isolates and management. Br. J. Ophthalmol. *59*: 372-376.
114. Fountain, R. B. 1968. The chromomycoses in England. Trans. Ann. Rep. St. John's Hosp. Dermatol. Soc. (Lond.) *54*: 187-196.
115. François, J., M. Elewaut-Rysselaere, and E. De Vos. 1966. Mycoses et pseudomycoses des voies lacrymales. Ann. Ocul. *199*: 1129-1142.
116. Fukushiro, R., S. Kagawa, S. Nishiyama, and H. Takahashi. 1957. Un cas de chromoblastomycose cutanée avec métastase cérébrale mortelle. Presse Med. *65*: 2142-2143 (original not seen).
117. Fuste, F. J., L. Ajello, R. Threlkeld, and E. R. Henry. 1973. *Drechslera hawaiiensis*: causative agent of fatal fungal meningo-encephalitis. Sabouraudia *11*: 59-63.
118. Gams, W., and V. Holubová-Jechová. 1976. *Chloridium* and some other dematiaceous hyphomycetes growing on decaying wood. *Studies in Mycology,* Vol. 13, Centraalbureau voor Schimmelcultures, Baarn, pp. 1-99.

119. Garau, J., R. D. Diamond, L. B. Lagrotteria, and S. A. Kabins. 1977. *Alternaria* osteomyelitis. Ann. Intern. Med. *86*: 748-749.
120. Georg, L. K. 1964. *Curvularia geniculata*, a cause of mycotic keratitis. J. Med. Assoc. State Ala. *31*: 234-236.
121. Georg, L. K., B. W. Brierer, and W. B. Cooke. 1964. Encephalitis in turkey poults due to a new fungus species. Sabouraudia *3*: 239-244.
122. Gezuela, E., J.-E. MacKinnon, and I. A. Conti-Díaz. 1972. The frequent isolation of *Phialophora verrucosa* and *Phialophora pedrosoi* from natural sources. Sabouraudia *10*: 266-273.
123. Gordeff, A., M.-J. Leroux, C. Vermeil, O. Morin, and M. Bouc. 1969. Peut-on parler, en Bretagne, d'une Blastomycose cutanée possible à *Aureobasidium pullulans*? Bull. Soc. Fr. Mycol. Med. *16*: 19-20.
124. Gordon, M. A., and Y. Al-Doory. 1965. Application of fluorescent-antibody procedures to the study of pathogenic dematiaceous fungi. II. Serological relationships of the genus *Fonsecaea*. J. Bacteriol. *89*: 551-556.
125. Gougerot, H., and M. Caraven. 1909. Mycose nouvelle: l'hémisporose, ostette humaine primitive du tibia due à l'*Hemispora stellata* (note préliminaire). C. R. Soc. Biol. (Paris) *11*: 474.
126. Groves, J. W., and A. J. Skolko. 1945. Notes on seed-borne fungi. III. *Curvularia*. Can. J. Res., Ser. C, *23*: 94-104.
127. Groves, J. W., and A. J. Skolko. 1946. Notes on seed-borne fungi. IV. *Acremoniella, Chlamydomyces,* and *Trichocladium*. Can. J. Res., Ser. C, *24*: 74-80.
128. Gugnani, H. C., S. Gupta, and R. S. Talwar. 1978. Role of opportunistic fungi in ocular infections in Nigeria. Mycopathologia *65*: 155-166.
129. Gugnani, H. C., A. V. Suseelan, C. Nwokolo, and A. N. U. Jnoku-Obi. 1977. Cutaneous cladosporiosis due to *Cladosporium trichoides*. J. Trop. Med. Hyg. *80*: 177-178.
130. Gustafson, R. A., R. V. Hardcastle, P. J. Szaniszlo. 1975. Budding in the dimorphic fungus *Cladosporium werneckii*. Mycologia *67*: 942-951.
131. Hall, J. E. 1965. Multiple maduromycotic mycetomas in a colt caused by *Helminthosporium*. Southwest. Vet. *18*: 233-234.
132. Hardcastle, R. V., and P. J. Szaniszlo. 1974. Characterization of dimorphism in *Cladosporium werneckii*. J. Bacteriol. *119*: 294-302.
133. Harding, H. 1975. Effect of D-amino acids on conidium size and numbers of pseudosepta per conidium in isolates of *Bipolaris sorokiniana*. Can. J. Bot. *53*: 600-603.
134. Harding, H. 1975. Effect of pH and sucrose concentration on conidium size and septation in four *Bipolaris* species. Can. J. Bot. *53*: 1457-1464.
135. Harris, J. J., and T. F. Downham. 1978. Unusual fungal infections associated with immunologic hyporeactivity. Int. J. Dermatol. *17*: 323-330.
136. Harwood-Nash, D. C. F. 1962. A case of chromoblastomycosis in a coloured male. S. Afr. Med. J. *36*: 647-650.
137. Hashmi, M. H., and G. Morgan-Jones. 1973. Conidium ontogeny in hyphomycetes. The meristem arthrospores of *Wallemia sebi*. Can. J. Bot. *51*: 1669-1671.
138. Hashmi, M. H., G. Morgan-Jones, and B. Kendrick. 1973. Conidium ontogeny in hyphomycetes. The blastoconidia of *Cladosporium herbarum* and *Torula herbarum*. Can. J. Bot. *51*: 1089-1091.
139. Hawksworth, D. L., and I. A. S. Gibson. 1977. *Phialophora parasitica*. CMI Descr. Pathog. Fungi Bact. 54-540.
140. Hawksworth, D. L., I. A. S. Gibson, and W. Gams. 1976. *Phialophora parasitica* associated with disease conditions in various trees. Trans. Br. Mycol. Soc. *66*: 427-431.
141. Hermanides-Nijhof, E. J. 1977. *Aureobasidium* and allied genera. *Studies in Mycology,* Vol. 15, Centraalbureau voor Schimmelcultures, Baarn, pp. 141-177.

142. Higashi, N., and Y. Asada. 1973. Cutaneous alternariosis with mixed infection of *Candida albicans.* Report of a patient responding to natamycin. Arch. Dermatol. *108*: 558-560.
143. Hoffmann, D. H. 1965. Pilzinfektionen des Auges. Systematik, Klinik, Erkennung und Behandlung. Fortschr. Augenheilkd. *16*: 63-217.
144. Houston, M. R., K. H. Meyer, N. Thomas, and F. T. Wolf. 1969. Dimorphism in *Cladosporium werneckii.* Sabouraudia *7*: 195-198.
145. Hubálek, Z. 1974. Cutaneous staphylococcosis and secondary infection of house sparrow with the fungus *Cladosporium herbarum.* Folia Parasitol. *21*: 59-66.
146. Hughes, S. J. 1952. *Trichocladium* Harz. Trans. Br. Mycol. Soc. *35*: 152-157.
147. Hughes, S. J. 1958. Revisiones hyphomycetum aliquot cum appendice de nominibus rejiciendis. Can. J. Bot. *36*: 727-836.
148. Itani, Z. 1972. Tinea nigra palmaris. Mykosen *15*: 27-34.
149. Jang, S. S., E. L. Biberstein, M. G. Rinaldi, A. M. Hennes, G. A. Boorman, and R. F. Taylor. 1977. Feline brain abscesses due to *Cladosporium trichoides.* Sabouraudia *15*: 115-123.
150. Janke, D. 1950. Zur Kenntniss der Hemisporose. Arch. Dermatol. Syphilol. *190*: 95-113.
151. Jeanselme, E., L. Huet, and M. Lotte. 1927. Nouveau type de mycétome à grains noirs dû à une *Torula* encore non décrite. Bull. Soc. Fr. Dermatol. Syphiligr. *35*: 369-375.
152. Jen, T. M. 1966. Mycological studies of *Hormodendrum dermatitidis (Fonsecaea dermatitidis)* isolated from the cervical lymphnode of a Chinese in Taiwan (Formosa). J. Formosan Med. Assoc. *65*: 66-68.
153. Joly, P. 1959. Variations morphologiques et notion d'espèce chez le genre *Alternaria* (Nees) Wiltshire. Bull. Trimest. Soc. Mycol. Fr. *75*: 149-158.
154. Joly, P. 1962. Recherches sur les genres *Alternaria* et *Stemphylium.* III. Action de la lumière et des ultra-violets. Rev. Mycol. *27*: 1-16.
155. Joly, P. 1964. Le genre *Alternaria.* Encycl. Mycol. *33*: 1-250.
156. Jones, B. R. 1974. *Pullularia* corneal ulcer. Arch. Ophthalmol. *92*: 529-530.
157. Jones, B. R. 1975. Principles in the management of oculomycosis. Trans. Am. Acad. Ophthalmol. Otolaryngol. *79*: 15-53.
158. Jotisankasa, V., H. S. Nielsen, and N. F. Conant. 1970. *Phialophora dermatitidis,* its morphology and biology. Sabouraudia *8*: 98-107.
159. Kafi, A., and S. A. J. Tarr. 1966. Growth, sporulation and conidial characteristics of five graminicolous species of *Helminthosporium.* I. Effect of nutrients. Trans. Br. Mycol. Soc. *49*: 327-337.
160. Kane, J., B. Birkett, and J. B. Fischer. 1976. Tinea nigra infection in Canada. Sabouraudia *14*: 327-330.
161. Kano, K. 1937. Uber die Chromoblastomykose durch einen noch nicht als Pathogen beschriebenen Pilz. Arch. Dermatol. *176*: 282-294.
162. Kaplan, W., F. W. Chandler, L. Ajello, R. Gauthier, R. Higgins, and P. Cayouette. 1975. Equine phaeohyphomycosis caused by *Drechslera spicifera.* Can. Vet. J. *16*: 205-208.
163. Karrer, H., and N. F. Conant. 1953. A cleistothecium-like structure found in *Hormodendrum pedrosoi.* Mycologia *45*: 947-950.
164. Kaufman, S. M. 1971. *Curvularia* endocarditis following cardiac surgery. Am. J. Clin. Pathol. *56*: 466-470.
165. Keddie, F. 1964. *Cladosporium werneckii* infection and in vitro culture. Arch. Dermatol. *89*: 432-435.
166. Kerdel-Vegas, F., and M. C. Bastardo de Alvornoz. 1966. Tinea nigra. Report of five cases seen in Venezuela. Dermatologica *132*: 320-330.
167. Klebahn, H. 1919. Die Schädlinge des Klippfisches. Ein Beitrag zur Kenntniss der salzliebenden Organismen. Mitt. Inst. Allg. Bot. Hamberg *4*: 11-69.

168. Klite, P. D., H. B. Kelley, and F. H. Diercks. 1965. A new soil sampling technique for pathogenic fungi. Am. J. Epidemiol. *81*: 124-130.
169. Krachmer, J. H., R. L. Anderson, P. S. Binder, G. O. Waring, J. J. Rowsey, and E. S. Meck. 1978. *Helminthosporium* corneal ulcers. Am. J. Ophthalmol. *85*:666-670.
170. Lampert, R. P., J. H. Hulto, W. H. Donnelly, and S. T. Shulman. 1977. Pulmonary and cerebral mycetoma caused by *Curvularia pallescens*. J. Pediatr. *91*: 603-605.
171. Lam-Quang-Bach. 1964. *Curvularia lunata* (Wakker) Boedijn. Fiches Phytopathol. Trop. *15*: 1-7.
172. Lane, C. G. 1915. A cutaneous lesion caused by a new fungus *(Phialophora verrucosa)*. J. Cutaneous Dis. *33*: 840-846.
173. Langeron, M. 1928. Mycétome à *Torula jeanselmei* Langeron. Nouveau type de mycétome à grains noirs. Ann. Parasitol. Hum. Comp. *6*: 385-403.
174. Langeron, M. 1934. Mycose oculaire primitive due au "*Beauveria brumpti.*" Bull. Acad. Nat. Med. (Paris) *111*: 133-137.
175. Langeron, M. 1947. *Tritirachium brumpti* (Langeron et Lichaa 1934) Langeron 1947 et le genre *Tritirachium* Limber 1940. Ann. Parasitol. Hum. Comp. *22*: 94-99.
176. Leone, R. 1977. Su una forma di "*Aureobasidium pullulans*" (De Bary) Arnaud var. "*pullulans*" isolata dall'uomo. Allionia *22*: 65-79.
177. Limsila, T., T. Stitnimankarn, and P. Thasnakorn. 1970. Pulmonary cladosporoma. Report of a case. J. Med. Assoc. Thail. *53*: 586-592.
178. Listemann, H. 1975. Die kulturelle Untersuchung eines Tränensteines mit Isolierung des Pilzes *Phialophora richardsiae*. E. Rodenwaldt Arch. *2*: 45-52.
179. Llamozas, G. P., V. Suprani, and M. De Albornoz. 1966. Queratomicosis por *Curvularia lunata*. Acta Med. Venez. *13*: 178-181.
180. Lobritz, R. W., T. H. Roberts, R. V. Marraro, P. K. Carlton, and P. J. Thorp. 1979. Granulomatous pulmonary disease secondary to *Alternaria*. JAMA *241*: 596-597.
181. Lucasse, C., J. Chardome, P. Magis, and R. Vanbreuseghem. 1954. Mycose cérébrale par *Cladosporium trichoides* chez un indigène du Congo Belge. Note mycologique sur *Cladosporium trichoides* Emmons 1952. Ann. Soc. Belge Med. Trop. *34*: 475-484.
182. Lundström, H. 1972. Microscopic studies of cavity formation by soft rot fungi *Allescheria terrestris* Apinis, *Margarinomyces luteo-viridis* v. Beyma and *Phialophora richardsiae* (Nannf.) Conant. Stud. For. Suec. *98*: 1-18.
183. Luttrell, E. S. 1977. Correlations between conidial and ascigerous state characteristics in *Pyronema, Cochliobolus* and *Setosphaeria*. Rev. Mycol. *41*: 271-279.
184. McGinnis, M. R. 1978. Human pathogenic species of *Exophiala, Phialophora,* and *Wangiella*. In *The Black and White Yeasts,* Proceedings of the Fourth International Conference on the Mycoses, Pan Am. Health Organ. Sci. Publ. No. 356, Washington, D.C., pp. 37-59.
185. McGinnis, M. R. 1977. *Exophiala spinifera,* a new combination for *Phialophora spinifera*. Mycotaxon *5*: 337-340.
186. McGinnis, M. R. 1977. *Wangiella,* a new genus to accommodate *Hormiscium dermatitidis*. Mycotaxon *5*: 353-363.
187. McGinnis, M. R. 1978. Taxonomy of *Exophiala jeanselmei* (Langeron) McGinnis and Padhye. Mycopathologia *65*: 79-87.
188. McGinnis, M. R. 1979. Taxonomy of *Exophiala werneckii* and its relationship to *Microsporum mansonii*. Sabouraudia *17*: 145-154.
189. McGinnis, M. R., and A. A. Padhye. 1977. *Exophiala jeanselmei,* a new combination for *Phialophora jeanselmei*. Mycotaxon *5*: 341-352.

190. McGinnis, M. R., and A. A. Padhye. 1978. *Cladosporium castellanii* is a synonym of *Stenella araguata.* Mycotaxon 7: 415-418.
191. McKenzie, R. A., and M. D. Connole. 1977. Mycotic nasal granuloma in cattle. Aust. Vet. J. *53*: 268-270.
192. MacKinnon, J.-E. 1934. Estudio del primer caso Uruguayo de cromoblastomicosis y "revista crítica" sobre la enfermedad. Arch. Urug. Med. Cir. Espec. *5*: 201-226.
193. MacKinnon, J.-E. 1936. Description d'une souche de *Phialophora verrucosa* Thaxter (Medlar, 1915) isolée du premier cas de dermatite verrucueuse observé en Uruguay. Ann. Parasitol. Hum. Comp. *14*: 78-84.
194. Madelin, M. F., and S. Dorabjee. 1974. Conidium ontogeny in *Wallemia sebi.* Trans. Br. Mycol. Soc. *63*: 121-130.
195. Maheswari Amma, S., C. K. J. Paniker, P. T. Iype, and S. Rangaswamy. 1979. Phaeohyphomycosis caused by *Cladosporium bantianum* in Kerala (India). Sabouraudia *17*: 419-423.
196. Mahgoub, E. S. 1973. Mycetomas caused by *Curvularia lunata, Madurella grisea, Aspergillus nidulans* and *Nocardia brasiliensis* in Sudan. Sabouraudia *11*: 179-182.
197. Male, O., and J. Trappeiner. 1965. Nagelveränderungen durch Schimmelpilze. Dermatol. Wochenschr. *151*: 212-221.
198. Malkina, A. Y. 1967. Significance of soil in the epidemiology of dermatophytes. Mikol. Fitopatol. *1*: 330-333.
199. Mandal, A., and L. N. Mohapatra. 1973. Experimental cladosporiosis. I. Artificial respiratory infection. Indian J. Microbiol. *13*: 205-208.
200. Mård, P.-A., and T. Hallberg. 1978. *Alternaria alternata* as a cause of opportunistic fungal infections in man. Scand. J. Infect. Dis. Suppl., *16*: 36-40.
201. Mariat, F., G. Segretain, P. Destombes, and H. Darrasse. 1967. Kyste sous-cutané mycosique (phaeo-sporotrichose) à *Phialophora gougerotii* (Matruchot 1910) Borelli 1955, observé au Sénégal. Sabouraudia *5*: 209-219.
202. Marill, F. G., M. Liautaud, B. Liautaud, J. Vodov, and F. Mariat. 1973. Dermatitie verniqueuse à *Aureobasidium pullulans* (de Bary). Bull. Soc. Fr. Mycol. Med. *20*: 73-74.
203. Matsushima, T. 1971. *Microfungi of the Solomon-Islands and Papua-New Guinea,* Kobe, Japan.
204. Matsushima, T. 1975. *Icones microfungorum a Matsushima lectorum,* Kobe, Japan.
205. Mayoux, A. 1974. Action d'un extract soluble de *Cladosporium carrionii* sur des cultures cellulaires in vitro. Arch. Inst. Pasteur Madagascar *42*: 89-92.
206. Medlar, E. M. 1915. A cutaneous infection caused by a new fungus, *Phialophora verrucosa* with a study of the fungus. J. Med. Res. *32*: 507-522.
207. Merwin, C. F. 1965. Tinea nigra palmaris. Review of the literature and case report. Pediatrics *36*: 537-541.
208. Meyers, W. M., J. R. Dooley, and K. J. Kwon-Chung. 1975. Mycotic granuloma caused by *Phialophora repens.* Am. J. Clin. Pathol. *64*: 549-555.
209. Mohapatra, L. N., V. K. Sood, and H. L. E. Grueber. 1967. Chromoblastomycosis in India and Nepal. Mykosen *10*: 309-314.
210. Moore, M. 1942. The virulence of strains of *Phialophora verrucosa* determined by inoculating chlorio-allantoic membranes of chick embryos. J. Invest. Dermatol. *5*: 411-422.
211. Moore, M., Z. K. Cooper, and R. S. Weiss. 1943. Chromomycosis (Chromoblastomycosis): report of two cases. JAMA *122*: 1237-1243.
212. Morales, J. 1956. Nota sobre un caso de tinea nigra causado por *Cladosporium werneckii.* Rev. Med. Exp. Lima *10*: 53-57.

213. Moreau, C. 1963. Morphologie comparée de quelques *Phialophora* et variations du *P. cinerescens* (Wr.) van Beyma. Rev. Mycol. *28*: 260-276.
214. Mouchacca, J. 1971. Remarques sur la taxinomie des espèces de *Scolecobasidium* à conidies pleuriseptées. Rev. Mycol. *36*: 298-305.
215. Muller, G. H., W. Kaplan, L. Ajello, and A. A. Padhye. 1975. Phaeohyphomycosis caused by *Drechslera spicifera* in a cat. J. Am. Vet. Med. Assoc. *166*: 150-154.
216. Nakamura, T., J. A. Grant, R. Threlkeld, and L. Wible. 1971. Primary chromoblastomycosis of the nasal septum. Am. J. Clin. Pathol. *58*: 365-370.
217. Neergaard, P. 1945. *Danish Species of Alternaria and Stemphylium,* Elnar Munksgaard, Copenhagen.
218. Nicot, J. 1953. Un Helminthosporium saprophyte du sol: *Helminthosporium spiciferum* (Bain.) nov. comb. Oesterr. Bot. Z. *100*: 478-485.
219. Nicot, J., and M. Caillat. 1967. Etude morphologique d'une souche africaine de *Phialophora richardsiae* (Nannf.) Conant. Rev. Mycol. *32*: 28-40.
220. Nielsen, H. S., and N. F. Conant. 1966. A new human pathogenic *Phialophora.* Sabouraudia *6*: 228-231.
221. Nityananda, K., P. Sivasubramaniam, and L. Ajello. 1962. Mycotic keratitis caused by *Curvularia lunata.* Case report. Sabouraudia *2*: 35-39.
222. Nityananda, K., P. Sivasubramaniam, and L. Ajello. 1964. A case of mycotic keratitis caused by *Curvularia geniculata.* Arch. Ophthalmol. *71*: 456-458.
223. Nsanzumuhire, H., D. Vollum, and A. Poltera. 1974. Chromomycosis due to *Cladosporium trichoides* treated with 5-fluorocytosine. Am. J. Clin. Pathol. *61*: 257-263.
224. Ohashi, Y. 1960. On a rare disease due to *Alternaria tenuis* (alternariasis). Tohoku J. Exp. Med. *72*: 78-82.
225. Oláh, G. M., and O. Reisinger. 1974. Etude ultrastructurale et cytochimique de l'appareil sporifère chez *Phialophora richardsiae.* Can. J. Bot. *52*: 2473-2480.
226. Ormerod, J. G. 1967. The nutrition of the halophilic mould *Sporendonema epizoum.* Arch. Mikrobiol. *56*: 31-39.
227. Ota, M., P.-T. Huang, and T. Okamura. 1935. Jpn. J. Dermatol. *37*: 95 (original not seen).
228. Otcenásek, M., Z. Hubálek, J. Dvorák, and M. Sabatová. 1968. Ein weiterer Chromomykose-Fall in der Tschechoslowakei? Mykosen *11*: 719-724.
229. Pardo-Castello, V., and F. Trespalacios. 1959. Superficial and deep mycoses in Cuba: a report based on 1,174 cases. South. Med. J. *52*: 7-14.
230. Parmelee, J. A. 1956. The identification of the *Curvularia* parasite of *Gladiolus.* Mycologia *48*: 558-567.
231. Patton, C. S. 1977. *Helminthosporium spiciferum* as the cause of dermal and nasal maduromycosis in a cow. Cornell Vet. *67*: 236-244.
232. Pedersen, N. B., P.-A. Mårdh, T. Hallberg, and N. Johnsson. 1976. Cutaneous alternariosis. Br. J. Dermatol. *94*: 201-209.
233. Percebois, G., M. F. Biava, and L. Kures. 1978. Discussion du pouvoir pathogène de certaines espèces d'*Alternaria.* A propos de trois observations. Bull. Soc. Fr. Mycol. Med. *7*: 15-18.
234. Pereira, M. J. 1938. *Mycologia de las Cromomicoses,* São Paulo (original not seen).
235. Pereira, M. J. 1948. *Mycotorula schawi* n.sp. Rev. Med. Rio Grande do Sul. *5*: 61-69.
236. Pierach, C. A., G. Gülmen, G. J. Dhar, and J. C. Kiser. 1973. *Phialophora mutabilis* endocarditis. Ann. Intern. Med. *79*: 900-901.
237. Polack, F. M., C. Siverio, and R. H. Bresky. 1976. Corneal chromomycosis:

double infection with *Phialophora verrucosa* (Medlar) and *Cladosporium cladosporioides* (Fresenius). Ann. Ophthalmol. *8*: 139-144.

238. Pollacci, G. 1922. Miceti del corpo umana e degli animali (Nota seconda). Contributo botanico all studio delle dermatomicosi. Riv. Biol. *4*: 313-328.
239. Powell, R. E. 1952. A survey of chromoblastomycosis in Queensland. Aust. J. Dermatol. *1*: 214-222.
240. Pribram, E. 1930. Ringworm of the scalp caused by *Alternaria tenuis* Nees. J. Infect. Dis. *47*: 11-15.
241. Pritchard, D., B. F. Chick, and M. D. Connole, 1977. Eumycotic mycetoma due to *Drechslera rostrata* infection in a cow. Aust. Vet. J. *53*: 241-244.
242. Putterill, K. M. 1954. Some graminicolous species of *Helminthosporium* and *Curvularia* occurring in South Africa. Bothalia *6*: 347-378.
243. Rajam, R. V., K. C. Kandhari, and M. J. Thirumalachar. 1958. Chromoblastomycosis caused by a rare yeast-like dematiaceous fungus. Mycopathol. Mycol. Appl. *9*: 5-19.
244. Ranck, F. M., L. K. Georg, and D. H. Wallace. 1974. Dactylariosis—a newly recognized fungus disease of chickens. Avian Dis. *18*: 4-20.
245. Redaelli, P., and R. Ciferri. 1934. Studio comparativo di ventun ceppi di *Sporendonema epizoum* (Corda) Nobis (= *Hemispora stellata* Vuillemin; = *Oospora d'Agatae* Saccardo; ecc.). Atti Ist. Bot. Univ. Lab. Crittog. Pavia Ser. IVa *5*: 145-198.
246. Redaelli, P., and R. Ciferri, 1949. Un quadriennio di attivita' del centro di micologia umana e comparata di Pavia. Mycopathol. Mycol. Appl. *4*: 25-47.
247. Reddy, B. D., D. C. Kelley, H. C. Minocha, and H. D. Anthony. 1974. Pathogenicity of *Alternaria alternata* and its antibody production in experimental animals. Mycopathol. Mycol. Appl. *54*: 385-390.
248. Reiss, E., and W. J. Nickerson. 1971. Control of dimorphism in *Phialophora verrucosa.* Sabouraudia *12*: 202-213.
249. Richards, R. H., A. Holliman, and S. Helgason. 1978. *Exophiala salmonis* infection in Atlantic salmon *Salmo salar* L. J. Fish Dis. *1*: 357-368.
250. Rieth, H., H. Koch, and A. Y. El-Fiki. 1959. Zum Nachweis hautpathogener Pilze im Laboratoriums-Tierstallen. Arch. Klin. Exp. Dermatol. *209*: 258-269.
251. Rietmann, B. 1930. Note préliminaire sur une épidermomycose palmaire noire observée dans l'état Bahia au Brésil. Bull. Soc. Fr. Dermatol. Syphiligr. *37*: 202-207.
252. Riley, O., and S. H. Mann. 1960. Brain abscess caused by *Cladosporium trichoides.* Review of 3 cases and report of a fourth case. Am. J. Clin. Pathol. *33*: 525-553.
253. Ritchie, E. B., and T. E. Taylor. 1964. A study of tinea nigra palmaris. Report of a case and inoculation experiments. Arch. Dermatol. *89*: 601-605.
254. Roberts, E. D., H. A. McDaniel, and E. A. Carbey. 1963. Maduromycosis of the bovine nasal mucosa. J. Am. Vet. Med. Assoc. *142*: 42-48.
255. Rodrigues, M. M., P. Laibson, and W. Kaplan. 1975. Exogenous corneal ulcer caused by *Tritirachium roseum.* Am. J. Ophthalmol. *80*: 804-806.
256. Ross, A. J., and W. T. Yasutake. 1973. *Scolecobasidium humicola,* a fungal pathogen of fish. J. Fish. Res. Board Can. *30*: 994-995.
257. Ruppel, E. G. 1974. Factors affecting conidial dimensions of a *Drechslera* species. Mycologia *66*: 803-807.
258. Saccardo, P. A. 1912. Notae mycologicae. 1. Fungi ex Gallia, Abyssinia, Japonia, Mexico, Canada, Amer. bor. et centr. Ann. Mycol. *10*: 310-322.
259. Saito, Y., T. Tsuchiya, Y. Nomura, M. Takenaka, and K. Tabuchi. 1967. Three cases of mycosis in *Amazona aestiva.* Bull. Azabu Vet. Coll. *1*: 47-53.

260. Salgado, U., R. Macedo, and A. Louriero. 1967. Cromomicosis: observación de lesiones en tres casos. Dermatologia (Mex.) *11*: 293-297.
261. Salkin, I. F., M. A. Gordon, and W. B. Stone. 1975. Dual infection of a white-tailed deer by *Dermatophilus congolensis* and *Alternaria alternata*. J. Am. Vet. Med. Assoc. *167*: 571-573.
262. Salkin, I. F., M. A. Gordon, and W. B. Stone. 1976. Cutaneous infection of a porcupine (*Erethizon dorsatum*) by *Aureobasidium pullulans*. Sabouraudia *14*: 47-49.
263. Salkin, I. F., and W. B. Stone. 1974. Subcutaneous mycotic infection of a white-tailed deer. J. Wildl. Dis. *10*: 34-38.
264. Sarrat, H., H. Mathe, J. P. Marchand, and I. Faye. 1973. Agents étiologiques rares de mycoses sous-cutanés. Bull. Soc. Pathol. Exot. *66*: 615-620.
265. Schauffler, A. F. 1972. Maduromycotic mycetoma in an aged mare. J. Am. Vet. Med. Assoc. *160*: 998-1000.
266. Scherwitz, C., M. Knickenberg, W. N. Meigel, and G. Rassner. 1976. Zum Krankheitsbild der Tinea nigra. Mykosen *19*: 429-437.
267. Schnapka, O. 1956. Onychomycosis nigricans. Beitrag zum Krankheitsbild der "schwarzen Nägel." Arch. Klin. Exp. Dermatol. *202*: 45-50.
268. Schrameck, M., and S. Weidenfeld. 1912. Iconogr. Dermatol. *6*.
269. Schol-Schwarz, M. B. 1968. *Rhinocladiella,* its synonym *Fonsecaea* and its relation to *Phialophora.* Antonie Leeuwenhoek J. Microbiol. Serol. *34*: 119-152.
270. Schol-Schwarz, M. B. 1970. Revision of the genus *Phialophora* (Moniliales). Persoonia *6*: 59-94.
271. Schwartz, I. S., and C. W. Emmons. 1968. Subcutaneous cystic granuloma caused by a fungus of wood pulp (*Phialophora richardsiae*). Am. J. Clin. Pathol. *49*: 500-505.
272. Segretain, G., F. Mariat, and E. Drouhet. 1955. Sur *Cladosporium trichoides* isolé d'une mycose cérébrale. Ann. Inst. Pasteur (Paris) *89*: 465-470.
272a. Sigler, L., A. Tsuneda, and J. W. Carmichael. 1981. *Phaeotheca* and *Phaeosclera,* two new genera of dematiaceous hyphomycetes, and a redescription of *Sarcinomyces* Lindner. Mycotaxon *12*: 449-467.
273. Silva, D., and T. Carvalho. 1974. Cromoblastomicose por *Cladosporium sphaerospermium.* Med. Cutanea Ibero-Latinoam. *2*: 197-200.
274. Silva, M. 1957. The parasitic phase of the fungi of chromoblastomycosis: development of sclerotic cells in vivo and in vitro. Mycologia *49*: 318-331.
275. Simmons, E. G. 1978. *Alternaria*—an exercise in diversity. Proc. Int. Symp. Taxon. Fungi Madras *1*: 130-135.
276. Simson, F. W. 1946. Chromoblastomycosis, some observations on the types of the disease in South Africa. Mycologia *38*: 432-449.
277. Simson, F. W., C. Harington, and J. Barnetson. 1943. Chromoblastomycosis: a report of six cases. J. Pathol. Bacteriol. *55*: 191-198.
278. Slifkin, M., and H. M. Bowers. 1975. *Phialophora mutabilis* endocarditis. Am. J. Clin. Pathol. *63*: 120-130.
279. Smith, J. G., W. M. Sams, and F. J. Roth. 1958. Tinea nigra palmaris. A disorder easily confused with junction nevus of the palm. JAMA *167*: 312-314.
280. Somal, B. S. 1976. A key to the species of *Curvularia.* Indian J. Mycol. Plant Pathol. *6*: 59-64.
281. Sprague, R. 1950. *Diseases of Cereals and Grasses in North America,* Ronald Press, New York.
282. Srinivasa Rao, P. N., and T. S. Ramakrishnan. 1968. Studies on the fungus diseases affecting man in and around Manipal. III. Fungi and diseases of the eye. Indian J. Pathol. Bacteriol. *11*: 53-60.

283. Stafleu, F. A. (Ed.) 1978. International code of botanical nomenclature. Regnum Veg. *97*: 1-457.
284. Subramanian, C. V. 1953. Fungi imperfecti from Madras. V. *Curvularia*. Proc. Indian Acad. Sci. Sect. B *38*: 27-37.
285. Sumner, J. L., and H. C. Evans. 1971. The fatty acid composition of *Dactylaria* and *Scolecobasidium*. Can. J. Microbiol. *17*: 7-11.
286. Symmers, D. 1945. Experimental reproduction of maduromycotic lesions in rabbits. Arch. Pathol. *39*: 358-363.
287. Symmers, D., and A. Sporer. 1944. Maduromycosis of the hand, with special reference to heretofore undescribed foreign body granulomas formed around disintegrated chlamydospores. Arch. Pathol. *37*: 309-318.
288. Symmers, W. S. C. 1960. A case of cerebral chromoblastomycosis (cladosporiosis) occurring in Britain as a complication of polyarteritis treated with cortisone. Brain *83*: 37-51.
289. Szaniszlo, P. J., B. H. Cooper, and H. S. Voges. 1972. Chemical compositions of the hyphal walls of three chromomycosis agents. Sabouraudia *10*: 94-102.
290. Takahashi, H., A. Sasaki, T. Arai, Y. Tsukamoto, O. Sato, and K. Sano. 1973. Chromoblastomycosis in the cisterna magna and the spinal aubarachnoid space. Case report. J. Neurosurg. *38*: 506-509.
291. Tamayo, L., R. Ruiz-Maldonado, and V. Vasquez. 1974. Tiña negra palmar bilateral. Dermatologia (Mex.) *18*: 164-169.
292. Tansey, M. R., and T. D. Brock. 1973. *Dactylaria gallopava*, a cause of avian encephalitis, in hot spring effluents, thermal soils and self-heated coal waste piles. Nature (Lond.) *242*: 202-203.
293. Terracina, F. C. 1974. Fine structure of the septum in *Wallemia sebi*. Can. J. Bot. *52*: 2587-2590.
294. Tommasi, M., J. Coudert, R. Fau, R. Chateau, and A. Goutelle. 1964. Mycose cérébrale à "*Cladosporium trichoides*" (un cas lyonnais avec étude anatomique et mycologique). Pathol. Biol. Paris Sect. Arch. Anat. Pathol. *12*: 48-50.
295. Trejos, A. 1954. *Cladosporium carrionii* n.sp. and the problem of cladosporia isolated from chromoblastomycosis. Rev. Biol. Trop. *2*: 75-112.
296. Tsai, C. Y., Y. C. Lü, L.-T. Wang, T. L. Hsu, and J.-L. Sung. 1966. Systemic chromoblastomycosis due to *Hormodendrum dermatitidis* (Kano) Conant. Report of the first case in Taiwan. Am. J. Clin. Pathol. *46*: 103-114.
297. Urabe, H., K. Yasumoto, and K. Nakashima. 1967. A case of chromoblastomycosis probably involvement of the central nervous system. Jpn. J. Dermatol. Urol. *29*: 1012-1021.
298. van Beyma thoe Kingma, J. F. H. 1943. Beschreibung der im Centraalbureau voor Schimmelcultures vorhandenen Arten der Gattungen *Phialophora* Thaxter und *Margarinomyces* Laxa, nebst Schlüssel zu ihrer Bestimmung. Antonie Leeuwenhoek J. Microbiol. Serol. *9*: 51-67.
299. Van de Lustgraaf, B. 1976. Xerophilic fungi in mattress dust. Mykosen *20*: 101-106.
300. Van Velsor, H., and H. Singetary. 1964. Tinea nigra palmaris. Arch. Dermatol. *90*: 59-61.
301. Vermeil, C., A. Gordeff, M.-J. Leroux, O. Morin, and M. Bouc. 1971. Blastomycose chéloidienne à *Aureobasidium pullulans* (De Bary) Arnaud en Bretagne. Mycopathol. Mycol. Appl. *43*: 35-39.
302. Vieira, J. R. 1961. Onicomycose por *Aureobasidium pullulans* (De By.) Arnaud. Proc. Int. Congr. Trop. Med. Malaria *4*: 768-777.

303. Volcán, G., G. A. Godoy, F. Battistini, and A. Alvarez. 1976. Fruiting organs of *Cladosporium werneckii.* Sabouraudia *14*: 115-122.
304. Waldrip, D. W., A. A. Padhye, L. Ajello, and M. Ajello. 1974. Isolation of *Dactylaria gallopava* from broiler-house litter. Avian Dis. *18*: 445-451.
305. Wang, C. J. K. 1965. Fungi of pulp and paper in New York. Tech. Publ. N.Y. State Univ. Coll. For. *87*: 1-115.
306. Watson, K. C. 1962. Cerebral chromomycosis. J. Pathol. Bacteriol. *84*: 233-237.
307. Whiting, D. A., and E. A. Bisset. 1973. Tinea nigra. S. Afr. Med. J. *47*: 1659-1661.
308. Wilson, L. A., R. R. Sexton, and D. Ahearn. 1966. Keratomycosis. Arch. Ophthalmol. *76*: 811-816.
309. Wind, C. A., and F. M. Polack. 1970. Keratomycosis due to *Curvularia lunata.* Arch. Ophthalmol. *84*: 694-696.
310. Wynne, E. S., and C. L. Gott. 1956. A proposed revision of the genus *Pullularia.* J. Gen. Microbiol. *14*: 512-519.
311. Yaffee, H. S., and I. A. Grots. 1965. Tinea nigra "plantaris," Arch. Dermatol. *91*: 153-154.
312. Young, C. N., J. G. Swart, D. Ackermann, and K. Davidge-Pitts. 1978. Nasal obstruction and bone erosion caused by *Drechslera hawaiiensis.* J. Laryngol. Otol. *92*: 137-143.
313. Young, J. M., and E. Ulrich. 1953. Sporotrichosis produced by *Sporotrichum gougeroti.* Arch. Dermatol. *67*: 44-52.
314. Zapater, R. C., E. J. Albesi, and G. H. Garcia. 1975. Mycotic keratitis by *Drechslera spicifera.* Sabouraudia *13*: 295-298.
315. Zsolt, J., and E. K. Novák. 1977. Some remarks on the vegetative reproduction of yeasts. Proc. Int. Spec. Symp. Yeasts *5*: 39-40.
316. Zweibel, S. M., and C. J. K. Wang. 1978. Reexamination of *Phialophora verrucosa.* In *The Black and White Yeasts,* Proceedings of the Fourth International Conference on the Mycoses, Pan Am. Health Organ. Sci. Publ. No. 356, Washington, D.C., pp. 91-100.

8

Miscellaneous Hyphomycetes and Coelomycetes

J. W. Carmichael / University of Alberta Mold Herbarium and Culture Collection,
Edmonton, Alberta, Canada

I. HYPHOMYCETES

A. Introduction

The Hyphomycetes are characterized by conidia produced diffusely on the mycelium, on mononematous or synnematous conidiophores, or from conidiogenous cells aggregated on the surface of a sporodochium. In addition, sterile mycelia and molds producing only bulbils or sclerotia are also usually disposed here. The form-genera of the Hyphomycetes are defined primarily on the shape, septation and color of the conidia, and on their ontogeny and arrangement in chains or heads.

In this chapter the stress is placed on directing the reader to the most appropriate sources for checking the identity and nomenclature of human pathogens. The form-genera of Hyphomycetes have been reviewed and illustrated by Carmichael et al. [16], and references for the original descriptions of form-genera and their type species can be found therein. Frequent reference is made here to the medical mycology texts published by Rippon [43] and Emmons et al. [22], but the same topics are usually included in other current medical mycology texts.

Some of the genera listed below are not Hyphomycetes, but are included here for convenience.

B. Form-Genera with Pathogenic Species

1. *Acremonium* Link, 1809

Lectotype species: *A. alternatum* Link (fide Hughes, 1958)
Synonym: *Cephalosporium* auct., non Corda 1839
Connected teleomorphs: *Emericellopsis, Nectria, Chaetomium, Niesslia, Cordyceps,* etc. [16,23]

For many years fungi producing slimy balls of conidia at the tips of long, tapering phialides were referred to the form-genus *Cephalosporium* Corda, 1839. However, the lectotype species, *C. acremonium,* is of doubtful identity from the published descriptions, and no preserved material has been discovered [23, p. 12].

The form-genus *Acremonium* Link 1809, was originally described as having solitary conidia, but mounts from the type specimen show occasional clumps and short chains of spores. Thus, when Gams [23] monographed the *Cephalosporium*-like fungi, he rejected *Cephalosporium* in favor of the earlier name *Acremonium.* Since *A. alternatum* produces conidia in chains which tend to collapse into slimy masses, Gams included in *Acremonium* those species with persistently catenate conidia as well as species with initially slimy conidia. Because of the broad spectrum of Ascomycota that produce relatively undifferentiated, elongate phialides, *Acremonium* includes a rather heterogeneous assemblage of anamorphs and the limits of the form-genus are difficult to define. In his treatment of *Acremonium,* Gams included the monophialidic species of *Paecilomyces* and all the species of *Gliomastix.* Most workers have accepted the *Paecilomyces* species in *Acremonium,* but many still recognize *Gliomastix* as a separate form-genus with darkly pigmented conidia and phialides. The paucity of morphological characters makes most species of *Acremonium* difficult to identify. In his monograph, Gams [23] gives keys to the species of *Acremonium* and some similar form-genera (*Septofusidium, Monocillium, Verticillium,* and *Aphanocladium*).

Most species of *Acremonium* are found in soil, on decaying plant material, or on other fungi. A few species are pathogenic for insects or humans. The most common human infection is mycetoma. Koshi et al. [31] give a case report on mycetoma due to *A. recifei* with illustrations of the fungus and a review of the literature on mycetomas caused by species of *Acremonium.* A few species of *Acremonium* have been reported from skin or other infections; some species only once and others but a few times. Some of the supposed *Acremonium* (or *Cephalosporium*) species reported from skin and nail infections have now been reidentified as species of *Fusarium* (see Ref. 23, p. 233).

Madurella mycetomi occasionally produces phialides and conidia which resemble *Acremonium.*

2. *Arthrographis* Cochet ex Sigler & Carmichael, 1976

Type species: *A. kalrae* (Tewari & Macpherson) Sigler & Carmichael
= *Arthrographis* state of *Pithoascus langeronii* von Arx
Connected teleomorphs: *Pithoascus* [4]

Arthrographis was first reported by Cochet [19] from a case of onychomycosis. Since then it has been reported occasionally from skin and other clinical specimens [46].

3. *Aspergillus* Link, 1809

Lectotype species: *A. glaucus* Link (fide Clements & Shear, 1931)
Connected teleomorphs: *Emericella, Eurotium, Fennellia, Neosartorya,* etc. [16]

Aspergillus was grown on melon slices by Micheli in 1718 and was one of the first molds to be described and illustrated [12]. The aspergilli are among the commonest molds found in soil and on decaying organic material of all kinds. An exemplary monograph of the genus was prepared by Raper and Fennell in 1965 [42]. A compilation of the species published since 1965 has been prepared recently by Samson [45].

Aspergillus infections of insects, birds, and mammals (including humans) were recognized during the nineteenth century (see Austwick in Ref. 42). Current medical mycology texts (e.g., Ref. 43) give details of the various clinical types of infection produced by aspergilli and also allergic conditions and toxicoses (see also Part B, Chap. 10).

Although there are more than 100 species of *Aspergillus,* there are only a few of medical importance. *A. fumigatus* is by far the most pathogenic and frequently encountered. Members of the *A. niger* and *A. flavus* groups cause infections less commonly. A variety of other species are encountered from time to time. *A. flavus* and related species may produce a carcinogenic toxin (aflatoxin) when growing under certain conditions on peanuts, corn, or other substances (see Part B, Chap. 10).

Species of *Aspergillus* usually are identified readily by using Raper and Fennell's monograph [42]. The most pathogenic species, *A. fumigatus,* is particularly easy to recognize by its crowded phialides borne directly on a greenish vesicle that tapers gradually into the greenish conidiophore. The conidia are globose or ovoidal and form compact green or grayish green columsn.

4. *Beauveria* Vuillemin, 1912

Type species: *B. densa* (Link) Picard
= *B. bassiana* (Balsamo) Vuillemin
Connected teleomorphs: none known

Beauveria densa is a cosmopolitan and omnivorous parasite of insects [33]. It is probably also a frequent saprophyte on dead insects in soil, on dung, and on other substrates. Since it is somewhat variable, both on the host and in culture, *B. densa* has received many different names [20]. Probably most, or all, of the beauverias described from insects belong to this species. Some strains produce synnemata on the host and in culture.

Beauveria densa has been reported as a pathogen of the tortoise and humans [43]. It is frequently isolated from rodent lungs [14] and occasionally from human sputum or skin scrapings.

5. *Cephalosporium* Corda, 1839

Lectotype species: *C. acremonium* Corda (fide Clements & Shear, 1931)

No type specimen of *C. acremonium* is available and the description by Corda suggests a *Mortierella* [23]. For pathogenic species of *"Cephalosporium,"* see under *Acremonium.*

6. *Chrysosporium* Corda, 1833

Type species: *C. merdarium* (Link) Carmichael
Synonyms: *Blastomyces* Costantin & Rolland 1888, non Gilchrist & Stokes 1898; *Emmonsia* Ciferri & Montemartini 1959
Connected teleomorphs: *Ajellomyces, Arthroderma, Ctenomyces, Thielavia, Keratinophyton, Anixiopsis, Renispora, Phanerochaete,* and others [16]

The name *Chrysosporium* was brought into modern usage for fungi with single-celled aleurioconidia by Hughes [27] and Carmichael [15]. Before that, medical workers had referred fungi with aleurioconidia to *Sporotrichum, Aleurisma,* or other form-genera.

Many species of *Chrysosporium* are anamorphs of Gymnoascaceae and thus this form-genus is of great importance in medical mycology. The agents of histoplasmosis, blastomycosis, and ringworm have *Chrysosporium* anamorphs (see Chaps. 5 and 6). However, single-celled aleurioconidia are also produced by a variety of unrelated fungi. *C. pruinosum* is the conidial anamorph of a Basidiomycete *(Phanerochaete). Chrysosporium merdarium* and some other species have no known teleomorph.

The simple morphology of *Chrysosporium* makes it difficult to separate from similar anamorph genera. *Sporotrichum* has similar aleurioconidia, but the hyphae and conidia are colored yellow, orange, or brown and are usually encrusted with granular material. The type species of *Geomyces* was transferred to *Chrysosporium* by Hughes [27] but later returned to *Geomyces* by Sigler and Carmichael [46]. Van Oorschot [38] has proposed that *Myceliophthora,* based on *M. lutea,* be maintained separate from *Chrysosporium.* The form-genus *Emmonsia,* proposed by Ciferri and Montemartini [17], does not seem to have any characters that would separate it from *Chrysosporium* or *Myceliophthora.*

7. *Coccidioides* Stiles, in Rixford & Gilchrist, 1896

Only species: *C. immitis* Rixford & Gilchrist
Connected teleomorphs: none known

The endosporulating spherules of the tissue phase of *Coccidioides* have long puzzled mycologists. At first, the spherules were taken to be part of the life cycle of a protozoan [44]. When the mold stage was discovered in culture, the spherules were taken to be asci or sporangia (see, for example, Ref. 21). Recent work has filled in many details of the endosporulation process and the conversion from arthroconidia to spherules. However, the taxonomic significance of endosporulation is still uncertain [47,48].

The saprophytic mold phase of *C. immitis* has arthroconidia whose morphology fits the anamorph genus *Malbranchea* [46]. This suggests a connection with the Gymnoascaceae, but a firm taxonomic disposition of *C. immitis* cannot be made without further information.

8. *Fusarium* Link, 1809

Type species: *F. roseum* Link
Connected teleomorphs: *Nectria, Giberella, Calonectria,* and others [16]

Species of *Fusarium* have long been known as plant pathogens and litter decomposers. In recent years, *Fusarium solani* and a few other species have been reported as agents of mycotic keratitis, usually following corneal trauma and treatment with antibacterial antibiotics and/or steroids [22]. Species of *Fusarium* are also isolated on occasion from the crusts and debris of severely burned patients, but pathogenicity in this circumstance is doubtful [43]. Occasional reports of nail infections with *Fusarium* have also appeared. Several species of *Fusarium* are capable of producing toxins when growing on grain or forage (see Part B, Chap. 10).

Some species of *Fusarium* produce an *Acremonium* microconidial stage in young cultures or on the high-sugar media favored by medical mycologists. Later production of the characteristic canoe-shaped macroconidia permits identification as a *Fusarium.* Solitary microconidia are produced in *Fusarium sporotrichoides* and a few other species.

Identification of *Fusarium* species can be attempted using the recent books published by Booth [8,9]. The genus *Cylindrocarpon* is similar to *Fusarium* and may be confused with it [7].

9. *Fusidium* Link, 1809

Lectotype species: *F. griseum* Link (fide Hughes, 1958)
Connected teleomorphs: *Biostictus* [30]

Fusidium griseum grows on senescent or shed leaves and has not been connected with human or animal disease nor grown in culture. *Fusidium griseum* produces elongate blastoconidia in branched, dry chains [6]. The limits of the form-genus and the number of species are not known. Phialidic species originally described in *Fusidium* have now been transferred to *Acremonium* [23]. Yeast-like fungi with elongate, fusiform budding cells have been placed in *Fusidium,* as have some fungi with very narrow hyphae [32].

10. *Indiella* Brumpt, 1906

Type species: *I. mansoni* Brumpt [18]

This genus was described by Brumpt, apparently as a hyaline counterpart of *Madurella* [21]. However, there are no cultures available of any of the original species. Therefore, *Indiella* is a doubtful genus. *Indiella americana,* a later species, has been identified as the *Scedosporium* state of *Petriellidium boydii* [21, p. 838].

11. *Loboa* Ciferri et al., 1956

Only species: *L. loboi* (Fonseca & Area Leão) Ciferri et al.

Loboa loboi is a yeast-like organism seen in the lesions of keloidal blastomycosis or Lobo's disease of humans and dolphins *(Tursiops).* It is well described in recent medical mycology texts [22]. The organism has never been grown in culture. Early reports of its cultural characters were based on other organisms [11]. Since the organism is known only in the form of budding cells without hyphae, it belongs in the Blastomycetes, not the Hyphomycetes (assuming that it is a fungus).

12. *Madurella* Brumpt, 1905

Type species: *M. mycetomi* (Laveran) Brumpt
Connected teleomorphs: none known

The form-genus *Madurella* was based on hyphae in the grains from a case of mycetoma. It must therefore be taken as a form-genus based on sterile mycelium. Two species are common agents of mycetoma: *M. mycetomi* and *M. grisea.* They are described in medical mycology texts. Some strains of *M. mycetomi* produce black sclerotia or minute phialoconidia when growing on appropriate media. *M. grisea* may be the hyphal state of *Pyrenochaeta romeroi* [3].

13. *Paracoccidioides* Almeida, 1930

Only species: *Zymonema brasiliense* Splendore 1912
= *P. brasiliensis* (Splendore) Almeida
Connected telemorphs: none known

Paracoccidioides brasiliensis causes a chronic granulomatous disease in humans that is geographically restricted to Central and South America [43]. At room temperature the fungus forms slow-growing white colonies composed of fine mycelium. Occasional strains produce a few conidia which could be accommodated in the form-genus *Chrysosporium* [10]. At 37°C, the fungus converts to a multiple-budding yeast, similar to the tissue phase.

The original descriptions of *Paracoccidioides* and *Z. brasiliensis* mention only budding cells and sterile mycelium. Thus *Paracoccidioides* could be taken either as a form-genus of the Blastomycetes or as a member of the Mycelia Sterilia.

14. *Penicillium* Link, 1809

Lectotype species: *P. expansum* Link [25]
Connected teleomorphs: *Eupenicillium* and *Talaromyces* [40]

Species of *Penicillium* are among the most common and widespread fungi on decaying organic material of all kinds. Fortunately, human or animal infections with penicillia are extremely rare [43]. Only *P. marneffei* is pathogenic for experimental animals [22].

The genus was monographed by Raper and Thom [41], but species identification is difficult. Perhaps the new monograph by Pitt [40] will remedy this situation.

15. *Prototheca* Kruger, 1894

Type species: *P. zopfii* Kruger

Prototheca is a genus of colorless, yeast-like organisms which morphologically resemble the green alga *Chlorella.* Reproduction is by internal division into a few to several endospores (called autospores by algologists).

Prototheca infections in humans and other animals have been reviewed recently by Kaplan [28]. Identification procedures for *Prototheca* species are given by Padhye et al. [39].

16. *Rhinosporidium* Seeber, 1912

Type species: *R. seeberi* (Wernicke) Seeber

Rhinosporidium seeberi causes the production of granulomatous polyps from the mucous membranes of humans and other animals. The organism grows in the tissue as endosporulating spherules, somewhat similar to those of *Coccidioides immitis,* but larger and thicker-walled [22,43].

Rhinosporidium seeberi has not been grown in culture. Since no hyphae have been seen, it is not a Hyphomycete. Indeed, its cytology in tissue, and the fact that experimental transmission is difficult, both suggest protozoal affinity, rather than fungal.

17. *Sporothrix* Hektoen & Perkins, 1900

Type species: *S. schenckii* Hektoen & Perkins
Connected teleomorphs: *Ceratocystis, Ceratocystiopsis,* and others [16]

Sporothrix schenckii causes a chronic, ulcerative wound infection in humans and other animals [1,43]. Despite considerable variation in colonial morphology, most clinical isolates are considered to belong to the same species. A somewhat different organism was described by Ajello and Kaplan [2] as *S. schenckii* var. *luriei*.

Numerous other isolates of morphologically similar fungi have been made from wood, dung, soil, and other substrates [26]. Whether some of these represent relatively avirulent strains of *S. schenckii* is not yet settled. The ascomycete genera *Ceratocystis* and *Ceratocystiopsis* contain a number of species with *Sporothrix* conidial states [37,51]. It has been suggested that *S. schenckii* may be the conidial state of *Ceratocystis stenoceras* (see Ref. 34 and references therein). However, the connection of *C. stenocenas* with sporotrichosis and with *S. schenckii* is still under investigation. No comprehensive monograph of *Ceratocystis* has yet been published. *Sporothrix* and some similar form-genera have been treated by de Hoog [20a], but a comprehensive account of *Sporothrix* anamorphs is still needed.

II. COELOMYCETES

A. Introduction

The Coelomycetes are characterized by conidia produced inside a conidioma [29]. Pycnidial conidiomata are formed entirely by the fungus; acervular conidiomata are formed partly from fungus hyphae and partly from host tissue. Most Coelomycetes are parasitic or saprophytic on higher plants. Their conidia are usually exuded from the conidioma in a slimy mass.

Identification of Coelomycetes, even to form-genus, is difficult because the taxonomy of this group is still in an unsatisfactory state. Many of the form-genera have never been illustrated. Sutton [49] prepared keys to 52 form-genera with acervuli and 145 form-genera with pycnidia. He also published [50] an annotated list of the 1350 generic names proposed for Coelomycetes. Illustrations of 200 form-genera of Coelomycetes have been published in the *Icones generum Coelomycetum* [35].

B. Form-Genera with Pathogenic Species

Only a few Coelomycetes have been encountered in human disease processes, mostly in wound infections.

Pyrenochaeta romeroi has been reported several times from black-grain mycetoma. *Pyrenochaeta romeroi* may be the pycnidial state of *Madurella grisea* [3].

Hendersonula toruloidea has recently been recognized as a cause of dermatophyte-like infections of the skin and nails in tropical regions [24]. The *Scytalidium* state of *H. toruloidea*, which is the state usually seen in culture, has been described and illustrated by Campbell [13], Sigler and Carmichael [46], and Nicot [36].

Species of *Phoma* and *Macrophoma* have been reported from a skin lesion [5] and ocular keratitis [43].

Punithalingam [40a] has recently published an extensive review of 12 species of fungi belonging to the Sphaeropidales isolated from humans. This comprehensive coverage includes many illustrations and keys to the species.

REFERENCES

1. Ainsworth, G. C., and P. K. C. Austwick. 1973. *Fungal Diseases of Animals,* 2nd ed., Commonwealth Agricultural Bureaux, Farnham Royal, Slough, England.
2. Ajello, L., and W. Kaplan. 1969. A new variant of *Sporothrix schenckii.* Mykosen *12*: 633-644.
3. Alilou, M. 1977. *Madurella grisea* est-il synonyme de *Pyrenchaeta romeroi*? Bull. Soc. Mycol. Med. *6*: 227-230.
4. Arx, J. A. von. 1978. Notes of Microascaceae with the description of two new species. Persoonia *10*: 23-26.
5. Bakerspigel, A. 1970. The isolation of *Phoma hibernica* from a lesion on a leg. Sabouraudia *7*: 261-264.
6. Barron, G. L. 1968. *The Genera of Hyphomycetes from Soil,* Williams & Wilkins, Baltimore.
7. Booth, C. 1966. The genus *Cylindrocarpon.* Commonw. Mycol. Inst. Mycol. Pap. *104*: 1-56.
8. Booth, C. 1971. *The Genus Fusarium,* Commonwealth Mycological Institute, Kew, Surrey, England.
9. Booth, C. 1977. *Fusarium, Laboratory Guide to the Identification of the Major Species,* Commonwealth Mycological Institute, Kew, Surrey, England.
10. Borelli, D. 1955. Las aleurias de *Paracoccidioides brasiliensis.* VI Cong. Venez. Cien. Med. *4*: 2241-2253.
11. Borelli, D. 1968. Lobomicosis: nomenclatura de su agente. Med. Cutanea *3*: 151-156.
12. Buller, A. H. R. 1915. Micheli and the discovery of reproduction in fungi. Trans. R. Soc. Can. Sect. IV, Ser. III, *9*: 1-25 and Pls. I-IV.
13. Campbell, G. K. 1974. Studies on *Hendersonula toruloides* isolated from human skin and nail. Sabouraudia *12*: 150-156.
14. Carmichael, J. W. 1961. Fungi from Alberta rodents. Mycopathologia *14*: 129-135.
15. Carmichael, J. W. 1962. *Chrysosporium* and some other aleuriosporic Hyphomycetes. Can. J. Bot. *40*: 1137-1173 and pls. I-VIII.
16. Carmichael, J. W., W. B. Kendrick, I. L. Conners, and L. Sigler. 1980. *Genera of Hyphomycetes,* University of Alberta Press, Edmonton.
17. Ciferri, R., and A. Montemartini. 1959. Taxonomy of *Haplosporangium parvum.* Mycopathologia *10*: 303-316.
18. Clements, F. E., and C. L. Shear. 1931. *The Genera of Fungi,* Hafner, New York.
19. Cochet, G. 1939. Sur un nouveau champignon arthrospore *(Arthrographis langeroni)* agent pathogène d'une onychomycose humain. Ann. Parasitol. Hum. Comp. *17*: 98-101 and pl. 3.
20. de Hoog, G. S. 1972. The genera *Beauveria, Isaria, Tritirachium* and *Acrodontium* gen. nov. *Studies in Mycology,* Vol. 1, Centraalbureau voor Schimmelcultures, Baarn, pp. 1-41.

20a. de Hoog, G. S. 1974. The genera *Blastobotrys, Sporothrix, Calcarisporium* and *Calcarisporiella* gen. nov. *Studies in Mycology,* Vol. 7, Centraalbureau voor Schimmelcultures, Baarn, pp. 1-84.

21. Dodge, C. W. 1935. *Medical Mycology,* Mosby, St. Louis.
22. Emmons, C. W., C. H. Binford, J. P. Utz, and K. J. Kwon-Chung. 1977. *Medical Mycology,* 3rd ed., Lea & Febiger, Philadelphia.
23. Gams, W. 1971. *Cephalosporium-artige schimmelpilze (Hyphomycetes),* Fischer, Stuttgart.

24. Gentles, J. C., and E. G. V. Evans. 1970. Infection of the feet and nails with *Hendersonula toruloidea.* Sabouraudia *8*: 72-75.
25. Hawksworth, D. L., J. I. Pitt, and B. C. Sutton. 1976. Typification of the genus *Penicillium.* Taxon *25*: 665-670.
26. Howard, D. H., and G. F. Orr. 1963. Comparison of strains of *Sporotrichum schenckii* isolated from nature. J. Bacteriol. *85*: 816-821.
27. Hughes, S. J. 1958. Revisiones Hyphomycetum aliquot. Can. J. Bot. *36*: 727-836.
28. Kaplan, W. 1977. Protothecosis and infections caused by morphologically similar green algae. In *The Black and White Yeasts,* Proceedings of the Fourth International Conference on the Mycoses, Pan Am. Health Organ. Sci. Publ. No. 356, Washington, D.C., pp. 218-232.
29. Kendrick, B., and T. R. Nag Raj. 1979. Morphological terms in Fungi Imperfecti. In B. Kendrick (Ed.), *The Whole Fungus,* Proceedings of the Second International Mycological Conference, Vol. 1, National Museum of Natural Sciences and National Museums of Canada, Ottawa, pp. 43-61.
30. Korf, R. P. 1973. Discomycetes and Tuberales. In G. C. Ainsworth, F. K. Sparrow, and A. S. Sussman (Eds.), *The Fungi,* Vol. IVA, Academic, New York, p. 282.
31. Koshi, G., A. A. Padhye, L. Ajello, and F. W. Chandler. 1979. *Acremonium recifei* as an agent of mycetoma in India. Am. J. Trop. Med. Hyg. *28*: 692-696.
32. Lechevallier, H., M. P. Lechevallier, D. A. Handley, B. K. Ghosh, and J. W. Carmichael. 1977. Strains of *Fusidia* which can be mistaken for Actinomycetes. Mycologia *69*: 81-95.
33. MacLeod, D. M. 1954. Investigations on the genera *Beauveria* Vuill. and *Tritirachium* Limber. Can. J. Bot. *32*: 818-890.
34. Mariat, F. 1975. Observations sur l'écologie de *Sporothrix schenckii* et de *Ceratocystis stenoceras.* Sabouraudia *13*: 217-225.
35. Nag Raj, T. R., and F. DiCosmo. 1978. *Icones generum Coelomycetum X,* University of Waterloo Biology Series No. 19, Waterloo, Canada.
36. Nicot, J. 1977. Charactères systématiques et biologiques d'*Hendersonula toruloidea.* Bull. Soc. Mycol. Med. *6*: 245-250.
37. Olchowecki, A., and J. Reid. 1974. Taxonomy of the genus *Ceratocystis* in Manitoba. Can. J. Bot. *52*: 1675-1711.
38. Oorschot, C. A. N. van. 1977. The genus *Myceliophthora.* Persoonia *9*: 401-408.
39. Padhye, A. A., J. G. Baker, and R. F. D'Amato. 1979. Rapid identification of *Prototheca* species by the API20C system. J. Clin. Microbiol. *10*: 579-582.
40. Pitt, J. I. 1979. *The genus Penicillium and Its Teleomorphic States Eupenicillium and Talaromyces,* Academic, New York.
40a. Punithalingam, E. 1979. Sphaeropsidales in cultures from humans. Nova Hedwigia *31*: 119-158.
41. Raper, K. B., and C. Thom. 1949. *A Manual of the Penicillia,* Williams & Wilkins, Baltimore.
42. Raper, K. B., and D. I. Fennell. 1965. *The Genus Aspergillus,* Williams & Wilkins, Baltimore.
43. Rippon, J. W. 1982. *Medical Mycology,* Saunders, Philadelphia.
44. Rixford, E., and T. C. Gilchrist. 1896. Two cases of protozoan (Coccidioidal) infection of the skin and other organs. Johns Hopkins Hosp. Rep. *1*: 209-268.
45. Samson, R. A. 1979. A compilation of the aspergilli described since 1965. *Studies in Mycology,* Vol. 18, Centraalbureau voor Schimmelcultures, Baarn, pp. 1-38.
46. Sigler, L., and J. W. Carmichael. 1976. Taxonomy of *Malbranchea* and some other Hyphomycetes with arthroconidia. Mycotaxon *4*: 349-488.

47. Sun, S. H., and M. Huppert. 1976. A cytological study of morphogenesis in *Coccidioides immitis.* Sabouraudia *14*: 185-198.
48. Sun, S. H., S. S. Sekhon, and M. Huppert. 1979. Electron microscopic studies of saprobic and parasitic forms of *Coccidioides immitis.* Sabouraudia *17*: 265-273.
49. Sutton, B. C. 1973. Coelomycetes. In G. C. Ainsworth, F. K. Sparrow, and A. S. Sussman (Eds.), *The Fungi,* Vol. IVA, Academic, New York.
50. Sutton, B. C. 1977. Coelomycetes VI. Nomenclature of generic names proposed for Coelomycetes. Commonw. Mycol. Inst. Mycol. Pap. *141*: 1-253.
51. Upadhyay, H. P., and W. B. Kendrick. 1975. Prodromus for a revision of *Ceratocystis* (Microascales, Ascomycetes) and its conidial states. Mycologia *67*: 798-805.

Part Two

MORPHOLOGY AND PHYSIOLOGY

9

Ultrastructural Cytology of Pathogenic Fungi

Robert G. Garrison / The Research Service, Veterans Administration Medical Center, Kansas City, Missouri, and University of Kansas School of Medicine, Kansas City, Kansas

I. INTRODUCTION

In recent years a substantial literature has accumulated on the fine structure of a variety of fungal organisms. Included in this literature are numerous studies of fungi pathogenic for the mammalian host. As might be expected, such studies at the ultrastructural level reveal the fungi to be a diverse group of organisms. Of the many thousands of known fungi, only about 55 are commonly pathogenic for humans and animals. There are no unique cytological entities which characterize or otherwise distinguish the pathogenic fungi from nonpathogenic species [12].

Fungi pathogenic for humans and animals are found in each of the five subdivisions of the Eumycota, the true fungi. Most zoopathogens are in the subdivision Ascomycotina and Fungi Imperfecti, but several important pathogens occur in the Mastigomycotina, Zygomycotina, and the Basidiomycotina. All of the true fungal pathogens are typically eucaryotic cells. This term implies the utilization of a highly organized system of cell parts, thus allowing each cytological component to have distinct functions. Certain of the organisms traditionally included in discussions of medical mycology are procaryotic cells; the pathogenic species of the genera *Actinomyces* and *Nocardia* may be mentioned. The procaryotic cell utilizes the plasma membrane and structures derived from it to perform a host of cellular functions without isolation into specialized individual units.

Since the early days of mycology as a science, the light microscope has been the single most useful tool for the identification of the fungal pathogen and study of mycotic disease. Inasmuch as the identification of fungi is quite often one of morphological attributes, microscopic examination of either fixed or slide-culture preparations is routinely employed in the medical mycology or pathology laboratory. The light microscope has many advantages: the initial cost, simplicity and ease of maintainance, ease of specimen preparation, and the ability to examine living cells. Before the late 1950s when the electron microscope became generally available, morphological and cytological investigations of fungi were done with methods of light microscopy. However, the electron microscope has one basic advantage which the light microscope cannot possibly achieve. It permits a clear, sharp resolution of objects and structures up to 1000 times smaller than that obtainable with even the best of light instruments. The light microscope can define small detail, but the resolving power of the light microscope is limited to structures larger than one-half the wavelength of visible light or about 0.25 μm.

Second-generation transmission electron microscopes are now widely available to the mycologist. Rapid advances have been made in techniques of fixation, staining, embedding, and sectioning. The scanning electron microscope (SEM) is now being used extensively for study of the surface morphology of the fungi. Scanning electron microscopy permits examination of the three-dimensional details of exposed surfaces, e.g., fungal conidia, sexual spores, and complex structures such as fruiting bodies. With the availability of higher-resolution instruments, scanning electron microscopy can make unique contributions in areas of fungal morphology and cytology where its use is appropriate. As yet, the scanning transmission electron microscope (STEM) has not been used extensively in study of the fungi. However, the greatly enhanced image contrast and information-processing facilities offered by this type of instrument have potential in the study of fungal ultrastructure and cytology.

Electron microscopes operating within the megavolt range are not yet widely available to biologists in general. Only isolated studies have been reported on aspects of fungal cytology as observed by high-voltage electron microscopy [36,49,50,90]. Nevertheless, the application of this advanced technique to the detailed study of the fungi is indeed an exciting prospect. Megavolt electron microscopy (MEM) has a number of entirely unique advantages: its penetrating power, the control of chromatic aberration, and decreased radiation damage. Indeed, it may well be possible to study the living fungal cell at high magnification with this class of electron microscope [49]. The implications of this concept to the study of fungal cytology need not be emphasized.

The present review is not intended to be a comprehensive summary of generalized fungal cytology and ultrastructure, but is related primarily to aspects of the ultrastructural cytology of those fungal organisms implicated in mycotic disease of humans and animals. Previous reviews concerned with more general concepts of fungal ultrastructure include those by Beckett et al. [9], Bracker [12], Hawker [96], and Moore [162]. The review by Matile et al. [157] is recommended to those interested in the cytology of yeasts. A study of the ultrastructure of the Actinomycetales can be found in the review by Williams et al. [232], while the fine structure of the genus *Actinomyces* and other filamentous bacteria has been reviewed by Slack and Gerencser [205]. Greenhalgh and Evans [85] have reviewed many of the basic techniques applicable to the preparation of fungal materials for examination by electron microscopy.

II. THE CELL WALL

A. Extracellular Components

A few pathogenic fungi can produce capsular or slime layer materials which are located exterior to the outermost cell wall surface. Under the conditions of conventional fixation, these mucoid substances may be lost during preparation of the fungus for electron microscopy or are otherwise electron transparent on subsequent examination in thin section. Many strains of *Cryptococcus neoformans* are heavily encapsulated. The capsule contains a network of delicate microfibrils that measure 30-40 Å in diameter and which radiate from the cell wall to coil and intertwine in various directions [54]. These thin, elongate, uniformly structured, electron-opaque filaments are believed to represent complex polysaccharides. Their exact function is unknown, but it was suggested that they may not really constitute separate components of the capsule proper [28]. Acidified dialyzed iron stains the mucoid capsule, converting it from a lucent to an extremely dense layer, thus demonstrating its content of acid mucosubstance. Acid phosphatase activity was reported to be associated with this capsular layer in *C. neoformans* [146].

Osmiophilic microfibrillar material is apparent at the outermost cell wall surface of yeast-like cells of *Sporothrix schenckii,* but is much less apparent or even lacking at the outer cell walls of hyphal cells of the corresponding mycelial phase [71,130] (Fig. 1a). The electron opacity of this microfibrillar material is markedly enhanced on staining with acidified dialyzed iron and by the PATAg method of Thiéry [220] (Fig. 1b).

By use of fluorescein-concanavalin A conjugates and an electron cytochemical method using horseradish peroxidase, it was observed that yeast-like forms of *S. schenckii* reacted more strongly with concanavalin A than did the mycelial forms [221], and such reactivity is probably due to the presence of peptidorhamnomannans. In a somewhat similar manner, the outermost wall layer of many strains of *Candida albicans* blastoconidia is often microfibrillar in appearance (Fig. 1c). Such microfibril components may be discontinuous and sometimes diminished or even absent in buds or very young cells. This filamentous layer ranges in thickness from 550 to 800 Å and is composed of anastomosing filaments of increased electron opacity. Electron-opaque granules measuring about 25-30 Å in thickness are regularly embedded in the filament network [27,52]. Scanning electron microscopy of *C. albicans* blastoconidia indicates smooth as well as roughened outer surfaces [3]. It is possible that these rough surfaces might represent conspicuous microfibrillar components at the outermost cell wall layers. Microfibrillar material located at and along the outermost wall layers has been reported for the cells of *Torulopsis glabrata* [16] and the arthro-

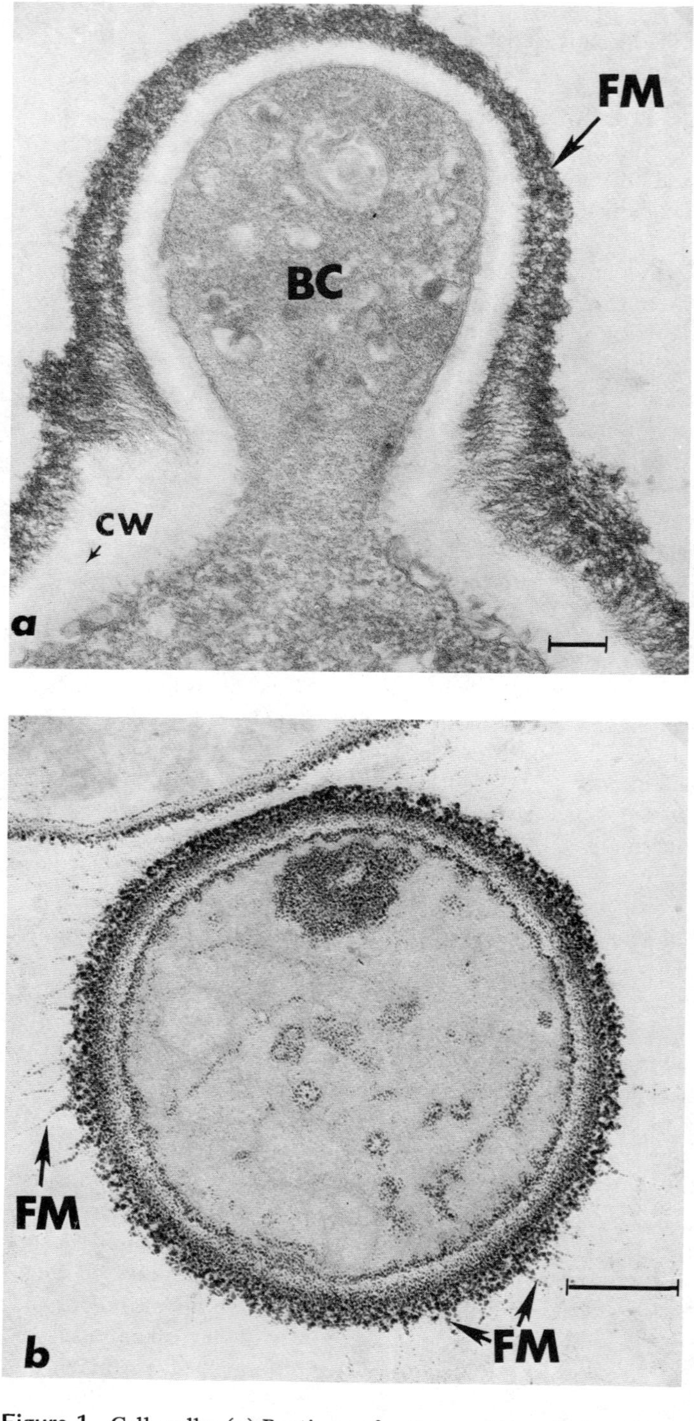

Figure 1 Cell walls. (a) Portions of a thin section of a budding yeast-like cell of *Sporothrix schenckii* stained with acidified dialyzed iron. Note the increased electron opacity of the filamentous material (FM) at the outermost cell wall (CW) surfaces of the bud cell (BC) due to the binding of the dialyzed iron stain. Glutaraldehyde-dialyzed iron-osmium tetroxide. Bar equals 0.1 μm. (b) Thin section of a yeast-like cell of *S. schenckii* stained by the PATAg method of Thiéry [220]. Note the electron-opaque deposits of silver granules at the cell wall and the delicate strands of filamentous material (FM) extending out from the outer-

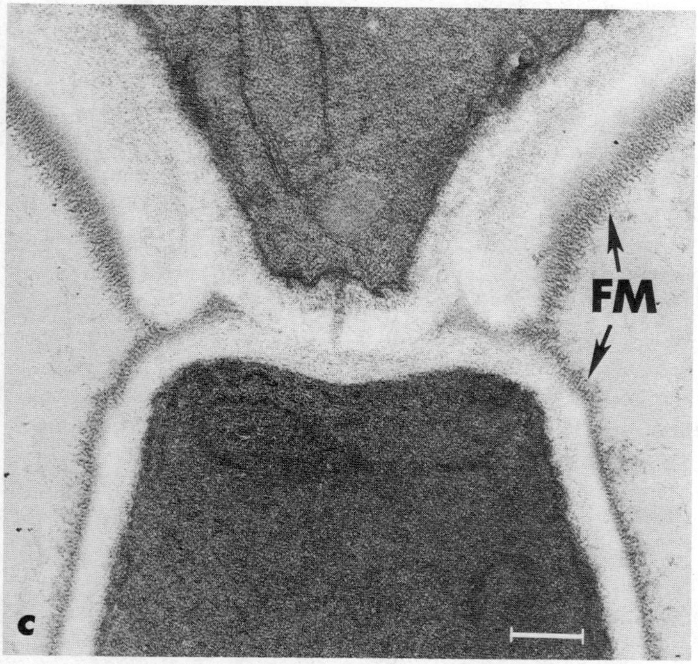

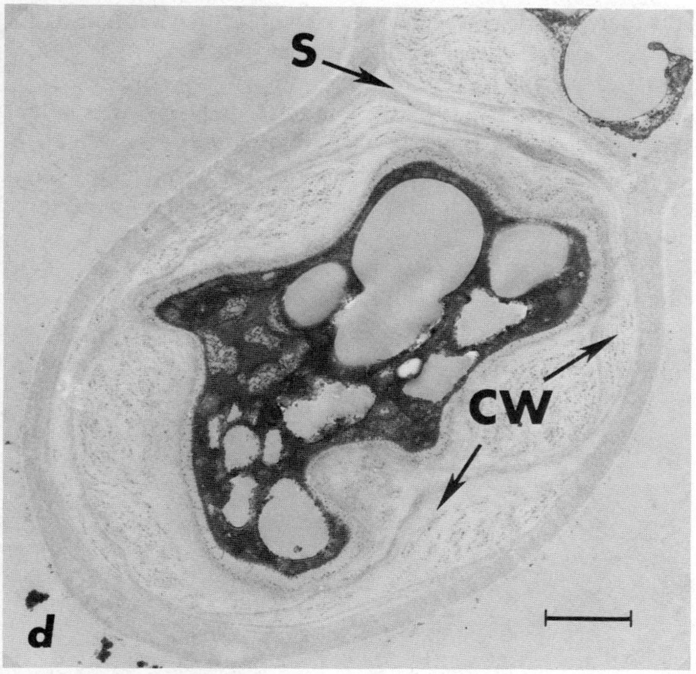

most cell wall surface. Glutaraldehyde-osmium tetroxide PATAg. Bar equals 0.25 μm. (c) Thin section of a septal area of *Candida albicans* undergoind germ tube formation. The outermost cell wall surface shows the presence of filamentous material (FM). Glutaraldehyde-osmium tetroxide. Bar equals 0.25 μm. (d) Portions of a thin section of an arthroconidium-like cell of *Aspergillus parasiticus* grown in a manganese-deficient medium. Note the septal area (S) and the hypertrophied cell wall (CW). Glutaraldehyde-osmium tetroxide. Bar equals 1.0 μm. (From Ref. 68.)

conidia of *Geotrichum candidum* [208]. Thus, it is evident that microfibrillar material can occur in a variety of pathogenic fungi having taxonomic heterogeneity. Whether this microfibrillar material can be regarded as a constitutive cell wall layer, or whether it represents a true extracellular component having a specific function, remains to be determined.

A number of the pathogenic dermatophytic fungi produce pigments which appear to be associated with the cell wall. It has been shown that the walls of pigmented strains and nonpigmented mutants of *Trichophyton violaceum* differ in their ultrastructural appearance in thin section. Pigmented colonies of this dermatophyte are characterized by the formation of a deep violet pigmentation which may ultimately be lost on serial transfer of the fungus. The outermost wall surface of pigmented strains is covered by a rough, aggregated substance 10 to 40 nm in thickness which consists of a finely granular to filamentous material. This material may be continuous with a similar granular substance seen within the hyphal cytoplasm [109,110]. In a somewhat analogous manner, certain of the pathogenic dematiaceous fungi comprising the genera *Cladosporium, Fonsecaea,* and *Phialophora* may possess demonstrable electron-opaque material at the outer hyphal wall surfaces. Electron microscopy of thin sections of glutaraldehyde-osmium tetroxide-fixed hyphal cells of certain strains of *C. carrionii* and *F. pedrosoi* indicates the presence of numerous, irregular aggregates of increased electron opacity which occur along the entirety of the outer wall surface of hyphal cells and perhaps conidia as well. At least in some species, these electron-opaque aggregates may represent an osmium-binding substance responsible for the pigmentation of vegetative hyphal filaments which occur exterior to the wall proper. *Penicillium marneffei* is characterized by the production of a brilliant water-soluble red pigment, but electron microscopy of thin sections of the hyphae of this potentially pathogenic fungus does not reveal the presence of electron-opaque, aggregated material at the outer wall [67]. Thus, it is likely that the chemical nature of some fungal pigments is such that they are not rendered visible following conventional fixation procedures used for electron microscopy.

Fine hairlike appendages on the cell walls of certain asocmycetous and heterobasidiomycetous yeastlike fungi have been described [40,41,182]. These hairlike structures were termed fungal fimbriae because of their close similarity to the fimbriae (pili) found on certain gram-negative bacteria. Fungal fimbriae have been observed under the electron microscope by negative staining with phosphotungstate, shadow casting with tungsten oxide, and freeze etching. These structures appear to be involved in certain cellular functions, such as conjugation and flocculation. Fungal fibriae or their equivalents have not been reported from any of the pathogenic fungi, but their possible occurrence in some species cannot be discounted.

B. The Cell Wall Proper

Reference to fungal walls and septa as being electron translucent and not having fibrillar or granular structure is common in the literature. It would seem that such wall material has not been contrasted adequately to illustrate fine detail. The walls of many pathogenic fungi fixed with glutaraldehyde-osmium tetroxide appear electron translucent on thin section, or at best only slightly stained. Some cell walls appear electron opaque only at the outermost surface regions. Prolonged staining with aqueous heavy metal salts, such as uranyl acetate or lead citrate, generally does little to increase the electron opacity, and considerable detail may thus be overlooked [107].

Most studies on the fine structure of the yeast cell wall have been done on *Saccharomyces cerevisiae* and *Candida albicans* [157]. The walls of these yeasts, fixed with potassium permanganate, appear almost identical with a dense outer and a broad inner electron-transparent layer. Hoch [107] found that barium permanganate rather than the potassium salt yielded greater detail of cell walls of some fungi that normally do not stain well with uranyl acetate-lead citrate procedures. Djaczenko and Cassone [52] employed a primary fixative containing acrolein and tris(1-aziridinyl)phosphine oxide (TAPO) in conjunction with osmium tetroxide as a postfixative. They distinguished the presence of five layers in the walls of mature blastoconidia of *C. albicans* and of four layers in the walls of buds and young cells. Cells of *Cryptococcus neoformans* fixed by the TAPO method showed a highly electron-opaque capsule with fibrillar and granular structures and a cell wall organized in two main layers. The outer wall layer was electron transparent and contained material of low to medium electron opacity. The inner wall layer was lamellar, with two sublayers of varying compactness [28]. Techniques employing mild oxidation of the wall with periodic acid followed by the use of osmiophilic reagents, such as thiosemicarbazide [207] or thiocarbohydrazide [220], may have some utility in the study of cell wall layering in certain of the pathogenic fungi. However, it is clear that no single chemical fixative or electron cytochemical technique is ideal for the demonstration of cell wall layers, owing to the differences in cell wall chemistry and organization.

It is now known that the biochemical and physical structures comprising fungal cell walls are complex. They are composed mainly of carbohydrate polymers of glucose, mannose, galactose, and glucosamine, with substantial amounts of protein, lipid, and ash. The cell wall structure of certain pathogenic fungi has been studied by various means of mechanical disruption, chemical analysis following hydrolysis, and sequential enzymatic dissection. These techniques have proved valuable in defining cell wall structure in a variety of pathogenic fungi, including certain of the dermatophytes [170,175]; the yeast-like and mycelial phases of *Histoplasma capsulatum* [1,10,39,46-48,116,119,178,188 and others]; cell wall regeneration of protoplasts of *H. capsulatum* [24]; the yeast-like and mycelial phases of *Blastomyces dermatitidis* and *Paracoccidioides brasiliensis* [46,114, 115]; *Sporothrix schenckii* [10]; and *Candida albicans* [30].

It is generally agreed that cell wall substructure and biochemical composition of the wall layers of most pathogenic filamentous and yeast-like fungi cannot be adequately resolved by electron microscopy of thin sections of whole cells fixed by conventional chemical means. Alternative techniques of analysis provide some information regarding wall structure and chemical content, but these methods are not without their own limitations. Sequential enzymatic dissection techniques coupled with subsequent fine structural analysis have been especially useful, but such techniques have been applied to the walls of only a few of the pathogenic fungi.

C. Budding and Apical Growth

Most of the comprehensive studies on the mechanics of budding have been made on nonpathogenic fungi. However, ultrastructural changes which occur in the wall during the budding process have been studied in some detail in a few pathogenic filamentous and yeast-like fungi, including *Cryptococcus neoformans* [28], *Candida albicans* [27,52,202], *H. capsulatum* [55], *P. brasiliensis* and *B. dermatitidis* [19], *Pityrosporum ovale* [122], *Torulopsis glabrata* [16], and *Phialophora (Wangiella) dermatitidis* [87.173]. The visualization of the ultrastructural subunit components of the cell wall may well be critically affected by the nature of the chemical fixative used in preparation. These often impose limitations on the subsequent interpretations of the processes involved in bud formation.

Marchant and Smith [153] and Kreger-van Rij and Veenhuis [123] proposed that the mode of bud formation could differ in yeasts of different taxonomic affinities. The structure of the cell wall and the method of bud formation in most ascomycetous yeasts appear to differ from that found in basidiomycetous yeasts. Whether or not the mode of bud formation possesses distinct taxonomic relationships remains to be clearly demonstrated. The wall of most ascomycetous yeasts fixed with permanganate or osmium tetroxide does not show dark lamellae, but only one thin, dark outer layer and a broad, lightly stained inner layer. The bud arises from under the wall of the parental cell and a lighter primary wall is initially formed in the isthmus between parent and daughter cells. In some pathogenic yeasts, including *C. albicans, Pityrosporum ovale,* and *T. glabrata,* budding apparently involves an extension of the existing cell wall.

The wall is lamellar in basidiomycetous yeasts. The inner layer of the parent cell is continuous with the wall of the bud. In the first stage of bud ontogeny only the inner wall grows out to form the bud and the first dark layer is formed later [123]. The budding mechanism characteristic of *Cryptococcus neoformans* agrees with that shown by the basidiomycetous yeasts [28]. The wall of the bud of this pathogen is formed by an innermost, dark lamellar layer and an electron-transparent outer layer. The inner layer of the bud cell wall is seen as a direct continuation of the corresponding innermost wall layer of the parent cell. The outermost layer is derived from the inside of the lamellar wall and subsequently grows out with the emerging bud through rupture of the lateral parental cell wall.

Studies by Grove et al. [87] indicate that hyphal and yeast-like bud formation in *Phialophora dermatitidis* result from essentially identical processes. Bud initiation in this pathogenic fungus is preceded by increased electron opacity of the parental cell wall at the site of subsequent bud emergence. Changes in the nature of the wall at similar sites have been observed in the yeast-like cells of *Paracoccidioides brasiliensis* [19] and *Rhodotorula glutinis* [122,153]. It was suggested that the increased electron opacity may in some manner be related to the presence of lytic enzyme proteins associated with wall digestion, or to an increased reactivity of preexisting wall components as the result of enzymatic modifications [87]. Bud initiation in *P. brasiliensis* is characterized by a blastic rupture of the outer portions of the hyphal or yeast wall and the subsequent emergence of the bud initial. The wall of the bud initial is continuous with an inner wall layer of the parent hypha or yeast, while the ruptured outer parental wall typically forms a collar around the immediate region of the developed bud structure.

In budding yeast-like cells, bud scars arise as the result of cell division and represent one-half of the septum formed between the parental and daughter cells. The number of bud scars corresponds to the number of daughter cells produced. When present, bud scars can be readily seen by the use of fluorescent light microscopy after staining with primulin or other brighteners. Bud scars can be seen by both scanning (Fig. 2) or transmission electron microscopy, but they are generally difficult to demonstrate in many of the pathogenic yeast-like fungi. In the case of yeast-like cells of *P. brasiliensis,* bud scars are identified only after separation is complete. The scar then appears as a truncated cone covered at the sides by dense wall material and at the top by a structure that resembles the inner wall of the nonbudding cell wall [21]. This bud scar soon disappears without a trace. Bud scars have been identified on yeast-like cells of *Candida albicans* and *S. schenckii,* but have been only rarely observed on yeast-like cells of *H. capsulatum* and *B. dermatitidis* [66].

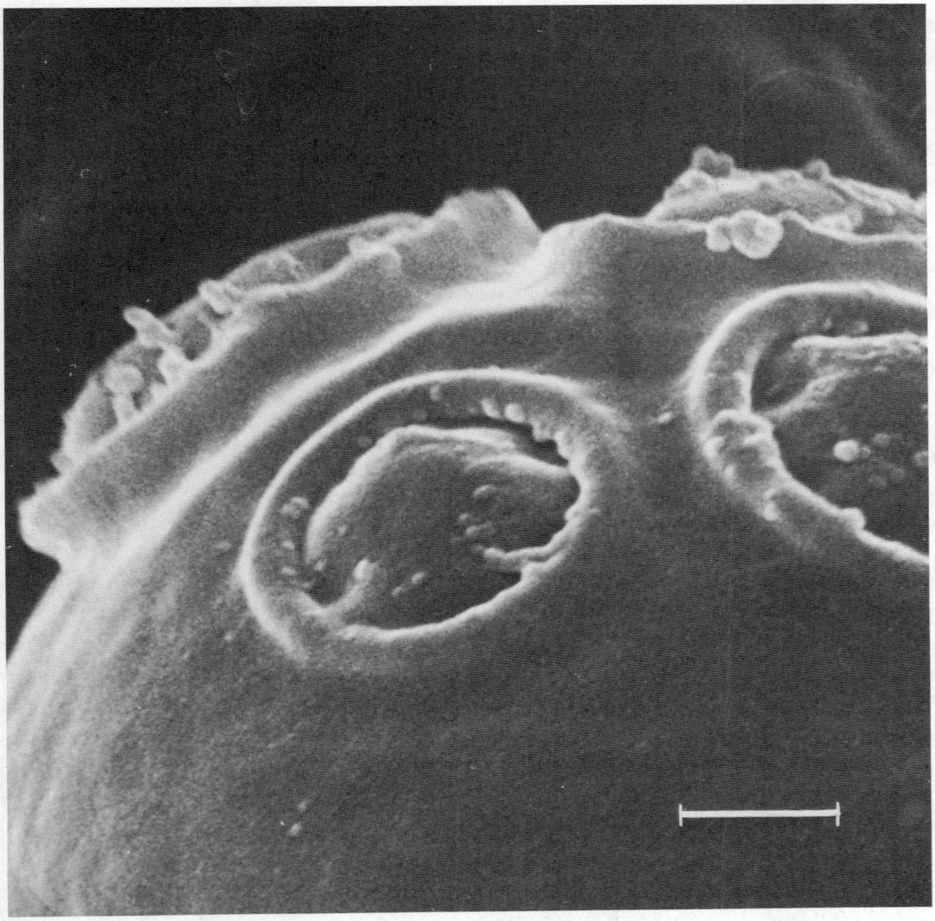

Figure 2 Bud scar formation at the cell surface of *Saccharomyces cerevisiae* as seen by scanning electron microscopy. Bar equals 0.5 μm. (From Ref. 183.)

The mechanics of cellular extension and branching was reviewed in 1965 by Robertson [193]. Since that time, cytological studies at the fine structural level have attempted to resolve some of the mechanisms involved in apical growth and extension. Apical cytoplasm can show a variable degree of vesicular differentiation, and certain subcellular structures have been described in association with hyphal tips [140].

Median longitudinal thin sections of germ tube apices of *Mucor rouxii* show the presence of an electron-opaque organelle termed the apical corpuscle [7]. This structure occurs singly, is roughly hemispherical, and is in intimate contact with the apical wall. Whether the apical corpuscle is a universal fungal organelle, or unique to *Mucor* and related fungi, remains to be clarified.

On light microscopy, an apical body termed the Spitzenkörper can be found in association with wall vesicles at the growing tips of some ascomycetous and basidiomycetous filamentous fungi, but has so far not been demonstrated in coenocytic zygomycetes.

The exact nature of the Spitzenkörper is controversial. Grove and Bracker [86] have studied the hyphal tips of a number of taxonomically diverse fungi by light and electron microscopy. These fungi included members of the Oomycetes, Zygomycetes, Ascomycotina, Basidiomycotina, and Fungi Imperfecti. In all the fungi studied, a zone was present at the hyphal apex which was rich in cytoplasmic vesicles but almost devoid of other cell components. Some of the vesicles appeared continuous with the plasma membrane at the apices. The subapical regions of the hypha contained a Spitzenkörper-like area which was associated with the growing hyphal tip. This area was not universally associated with tip growth, whereas apical vesicles were widespread in actively growing tip systems. Thus, it is believed that the Spitzenkörper may be a regional differentiation of the apical zone in intimate association with apical vesicles and with hyphal tip growth in some fungi. However, its occurrence in tip-growing systems is generally less than that of the apical vesicles. Vesicles of this type are thought to mediate cell extension by providing new materials to the plasma membrane and wall layers, or by producing substances which maintain the growing apex in a plastic condition [13]. It seems likely that such apical vesicles may be widespread throughout the septate, filamentous pathogenic fungi. Clusters of cytoplasmic vesicles can occur near the tips of buds in certain of the heterobasidiomycetous yeasts, including *R. glutinis* [141]. These vesicles appear similar to those of exocytotic vesicles seen in hyphal apices of septate mycelial fungi. In other yeast-like fungi, such vesicular clusters are either less frequent or lacking.

Trinci and Saunders [223] have applied mathematical models to the study of hyphal tip extension. If one assumes that a hypha grows at a linear rate, then the shape and length of the extension zone and the radius of the hypha at the base of the extension zone will all be constant. They concluded that hyphal extension zones are not hemispherical in shape, but rather, the shapes of rapidly growing hyphal tips approximate more closely to half ellipsoids of revolution than to hemispheres. If such mathematical models are valid, they could be of considerable value in the study of hyphal morphogenesis.

Perhaps appropriate to discussions of hyphal tip growth is the phenomenon variously referred to as self-parasitism [45], self-penetration [117], and intrahyphal hyphae [135]. These terms all refer to a unique process in which viable hyphae penetrate and grow within another hyphal element. This invasion process occurs throughout most of the major classes of fungi, and it has been observed to occur in some pathogenic species.

Electron microscopic observations of intrahyphal hyphae have been made in thin sections of the dimorphic *B. dermatitidis, P. brasiliensis* [26], and *H. capsulatum* [129]. Certain of the dermatophytic fungi, including *Trichophyton rubrum* [58] and *T. mentagrophytes* [226], have been shown to produce intrahyphal hyphae. Carbonell [20] described what was termed intrayeast hyphae in liquid cultures of yeast-like *P. brasiliensis* converting to its homologous mycelial phase. A rather high frequency of intrayeast or intrahyphal hyphae is common in liquid shaker cultures of *Blastomyces, Paracoccidioides,* and *Histoplasma.* Many other pathogenic fungi do not display this unique event when cultured under similar conditions [66]. Figure 3 illustrates what is interpreted as intrayeast hyphae in a yeast-like cell of *B. dermatitidis* grown in liquid shaker culture under cultural conditions conducive for yeast-like to mycelial phase conversion.

In the case of the dimorphic fungi, the question arises whether the high frequency of intrayeast and intrahyphal hyphae is an artifact resulting from the conditions imposed by the cultural procedure or whether the process actually possesses some advantage to the fungus. With certain plant pathogens it has been suggested that intrahyphal invasion

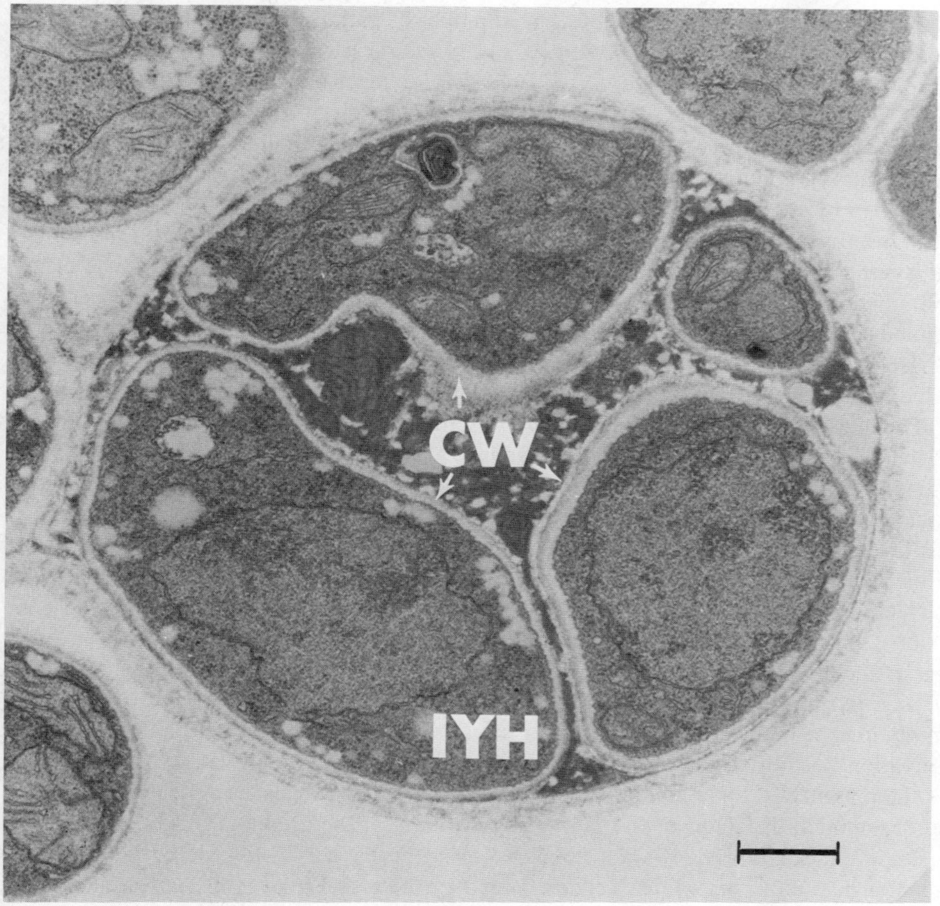

Figure 3 Intrayeast hyphae. Thin section of a yeast-like cell of *Blastomyces dermatitidis* grown in liquid shaker culture under conditions conducive for the induction of yeast-like to mycelial phase conversion. Four cross sections of intrayeast hyphae (IYH) are seen together with their cell walls (CW). Glutaraldehyde-osmium tetroxide. Bar equals 0.5 μm.

could represent a fungal defense mechanism to some toxic substance elaborated by the invaded plant host [117]. The existence of intrahyphal or intrayeast hyphae within parasitized animal tissue has not been reported. It is conceivable that this unique phenomenon possesses some physiological or environmental advantage yet to be determined.

D. Spore Cell Walls

A variety of asexual spores are produced by many of the filamentous pathogenic fungi. The external morphological attributes of these spores can often serve as a valuable diagnostic criterion for the identification of a fungal isolate. When observed by light microscopy, conidiospores may appear smooth-walled or they may possess conspicuous sculpturing of the outermost wall surface. Scanning electron microscopy has been used to visualize surface features of the conidial cell wall in a number of pathogenic fungi, including *Conidiobolus coronata* [125], *H. capsulatum* [72], and several selected strains of dermatophytic fungi of the genera *Microsporum, Epidermophyton,* and *Trichophyton* [108]. Although scanning electron microscopy provides a great depth of focus and high resolving power in the study of surface features, its usefulness is largely limited to outermost surfaces. Figures 4 and 5 illustrate the appearance of a variety of vegetative and conidial cells as seen by the scanning electron microscope.

Transmission electron microscopy of thin sections has provided some information on wall layering of mature conidiospores of certain pathogenic fungi. At times the age of the conidium and status of the medium can affect wall integrity. Highly aberrant, hypertrophied cell walls have been observed in thin sections of the conidia of *C. coronata* [77] and *Aspergillus parasiticus* [68] following loss of regulatory function of cell wall synthesis due to the depletion of an essential metabolite from the supporting medium (Fig. 1d).

The micromorphology of the tuberculate conidia of *H. capsulatum* has been studied in detail by Edwards et al. [56]. The tubercle is a part of the cell wall and is probably formed by expansion and buckling of the outermost wall layer and subsequent invasion by secreted radial filaments. The wall proper varies in its electron opacity from the innermost layers, which are almost electron translucent, through a gradually increasing number of filaments, to an organized, densely packed, and radially oriented peripheral layer. Mild oxidation with periodic acid followed by treatment with thiosemicarbazide-osmium tetroxide indicates variations in electron opacity of the spore wall and suggests biochemical differences in polysaccharide content [66]. The conidial wall of *B. dermatitidis* has been reported to consist of from two [73] to four layers [70]. On thin section the outermost surface is often seen to possess short tubercles, although this observation was not confirmed by use of the scanning electron microscope. It is possible that the presence of tuberculations on the conidium of this fungus is a reflection of the strain examined, the status of the supporting medium, or the degree of hydration.

On initial isolation many strains of *S. schenckii* produce a thick-walled, triangular-shaped, pigmented spore which is characterized by the presence of numerous, irregular aggregates of small electron-opaque particles located over the entirety of the outermost cell wall. Deposition of the particles ends abruptly at the spore-sterigmata septum. The electron-opaque particles do not occur at or along the wall of the supporting hypha [148]. This electron-opaque material is thought to represent pigment which is produced by the spore and which contributes to its dark coloration. Electron microscopy of the dimorphic *Phialophora dermatitidis* [173] does not clearly indicate the presence of any similar electron-opaque substance at the outer walls of cells characteristic of this dematiaceous pathogen.

Techniques of freeze etching and replication have been used to study aspects of the ultrastructural cytology of certain of the pathogenic fungi. Included among these are *T. rubrum, T. mentagrophytes* [91], *T. tonsurans, T. violaceum* [92,93], *Rhizopus arrhizus* [101], *Epidermophyton floccosum* [171], the spherules of *Emmonsia crescens* [99], endospores of *Rhinosporidium seeberi* [218], *Pityrosporum orbiculare* [15], and *Cryptococcus neoformans* [214,215].

Freeze etching has been successfully applied to ultrastructural investigations of fungal spores. It has been shown that this technique can provide detailed information otherwise unobtainable by other methods. Hess and co-workers [102,103] studied the surface structure of conidia from a number of species of *Penicillium* and *Aspergillus* using the freeze-etch technique (Fig. 6). This method involves the instant freezing of the spores in liquid

Freon and liquid nitrogen, and the subsequent preparation of replicas of the fractured surfaces as described by Moor [160] and Moor and Mühlethaler [161]. Fracturing ice away from spore surfaces reveals exposed "rodlet" patterns. It is believed that the rodlets of *Penicillium* consist of cutin or sporopollenin or a similar substance [197]. An examination of the conidia of 10 species each of *Penicillium* and *Aspergillus* indicated that the outer conidial wall layer was made up of rodlets which formed distinctive patterns. Some species have quite uniform conidial surface patterns from spore to spore, while in other species the patterns are extremely variable. The patterns formed by the rodlets appear to be species-specific, and therefore it was suggested that such patterns might have some taxonomic value in the study of these genera. Further study of conidial surfaces in other species of pathogenic fungi by freeze-etch techniques seems warranted.

The layers of the conidial wall are implicated in development of the hyphal wall following germination. Bartnicki-Garcia [6] described three basically different modes of vegetative wall formation following conidial germination. The mechanics by which this occurs appear to be characteristic of certain groups of fungi when observed by the electron microscope. In type I, the wall of the vegetative hypha is derived directly as an extension of the spore wall or from one of its innermost layers. This mode of wall genesis is probably the most common among those filamentous pathogenic fungi which produce conidia. Type II is probably not found in the pathogenic fungi inasmuch as it represents the de novo formation of a cell wall on a naked protoplast such as occurs during encystment of the zoospores of aquatic phycomycetes. Type III is represented by the de novo formation of the vegetative wall under the existing spore wall. This mode of wall genesis occurs in *Rhizopus* and in other genera of the Mucorales.

Aspects of the formation of the ascospore wall and its appearance on transmission or scanning electron microscopy have been studied in a number of pathogenic ascomycetous fungi. Scanning electron microscopy of mature ascospores of seven species of *Nannizzia* as well as those of *Arthroderma flavescens* indicate them to be smooth-walled, oblate sphaeroids [174]. Except for slight differences in size, these ascospores appear identical to each other. The outer surfaces of mature ascospores of *Arthroderma benhamiae (T. mentagrophytes)* [23], *A. quadrifidum (T. terrestre)* [59], and *Nannizzia gypsea (Microsporum gypseum)* [105], as well as those of *Piedraia hortai* [31], appear smooth on thin section. In contrast, the outer ascospore walls of *Emmonsiella capsulata (H. capsulatum)* have triangular excrescences (Fig. 7a), while irregular peg-like projections occur on the mature ascospores of *Ajellomyces dermatitidis (B. dermatitidis)* [75,82] (Fig. 7b). Light microscopy suggests that mature basidiospores of *Filobasidiella neoformans (Cryptococcus neoformans)* are finely roughened at the outer wall surface [126].

Since the early ultrastructural studies on ascospore formation [35,94], fine structural studies have indicated that ascospores are delimited within the ascus cytoplasm by a pair of concentric unit membranes which move apart to accommodate it. Schrantz [199,200] and Stiers [211] suggest that lomasomes may be involved in the deposition of ascospore wall material. Small, delicate, lomasomal-like membranes appear associated with the developing ascospore wall of *A. dermatitidis* [75].

It is of interest to consider briefly the wall structure of the coils which arise radially from the ascogonium of *A. dermatitidis* and *E. capsulata.* Glick and Kwon-Chung [82] have shown that electron microscopy of thin sections of these coils indicates a significant difference in size (Fig. 7c and d). Although depending on the plane of section and the area from which sections are taken, the ratio of the inner to outer wall in *Ajellomyces*

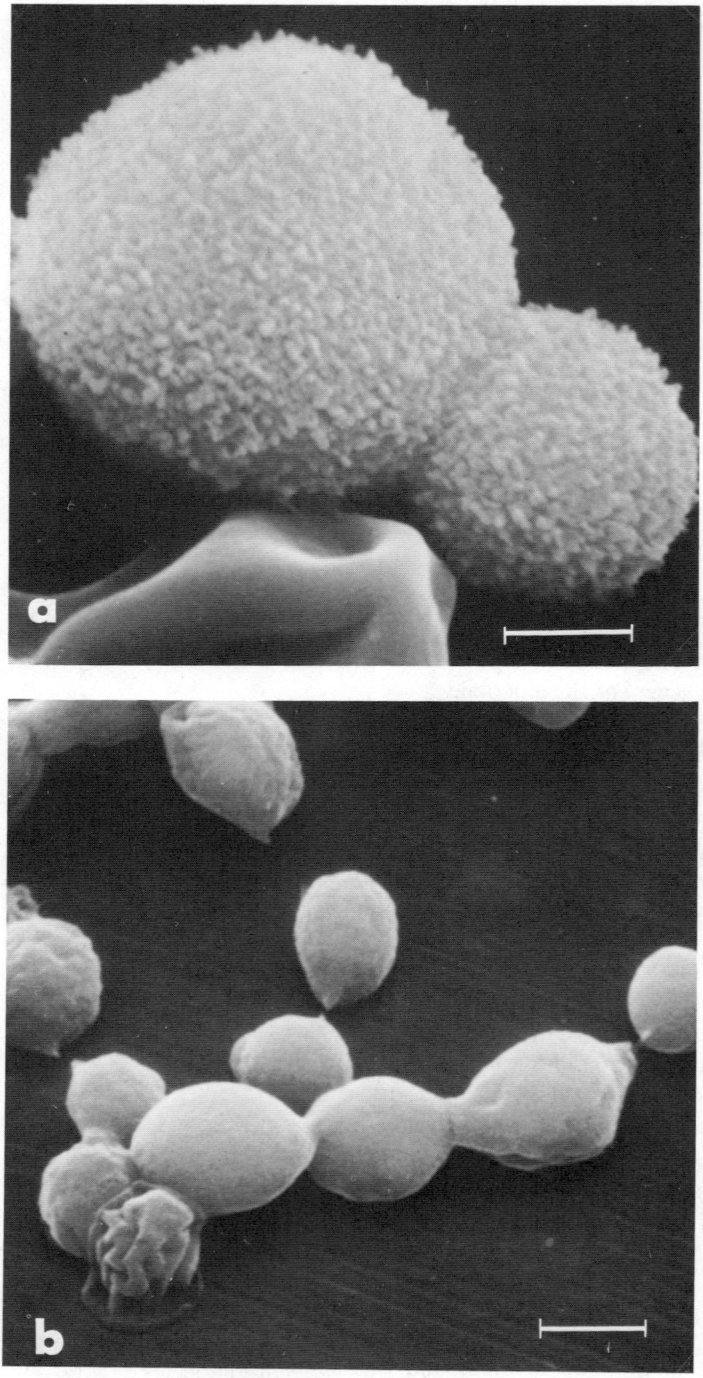

Figure 4 Scanning electron microscopy of yeast-like fungi. (a) Roughened surface features of *Candida albicans*. Bar equals 2.0 μm. (b) Surface features of yeast-like phase cells of *H. capsulatum*. Bar equals 2.0 μm.

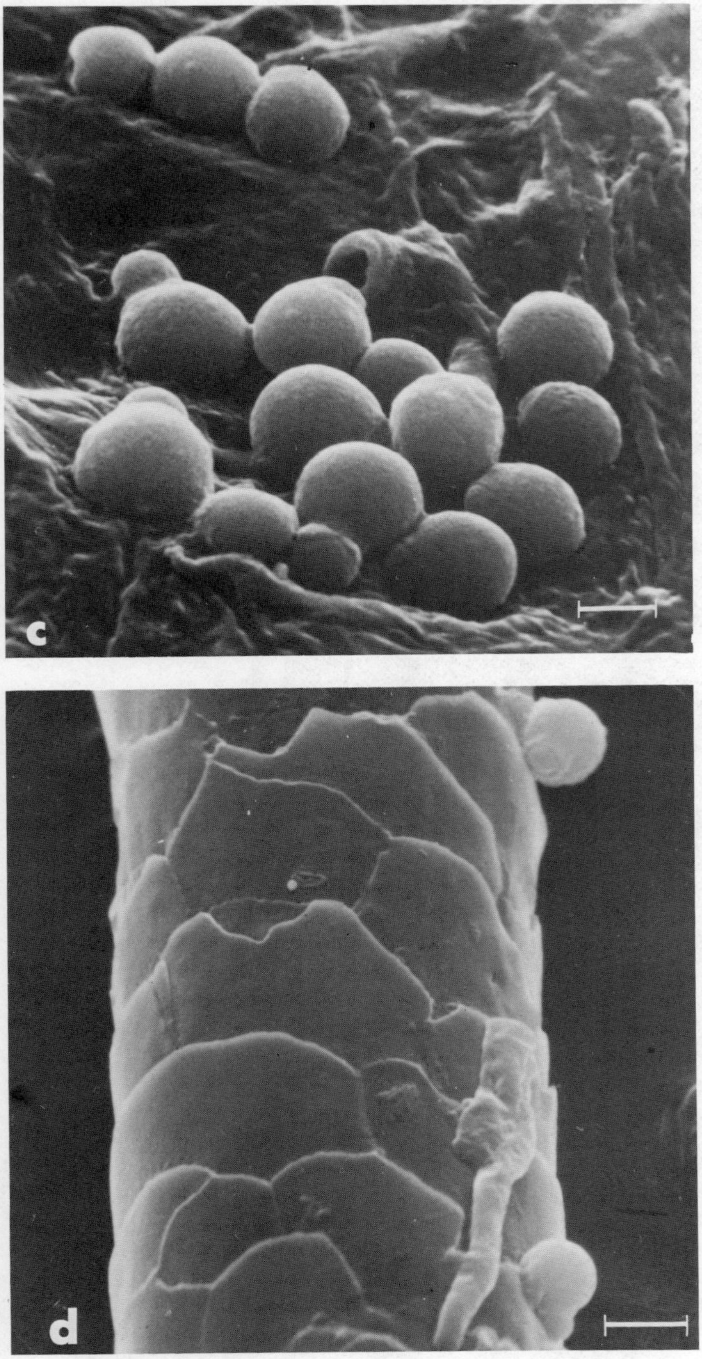

Figure 4 (c) Yeast-like cell forms of *Malassezia furfur* attached to cellophane tape showing characteristic cluster formation, budding cells, and scar formation. Bar equals 3.0 μm. (d) Attachment of hyphal and yeast-like forms of *M. furfur* to hair shaft. Bar equals 4.0 μm [Parts (a), (c), and (d) from Ref. 4., *Arch. Dermatol. 107*: 392-394 © 1973 American Medical Assoc.]

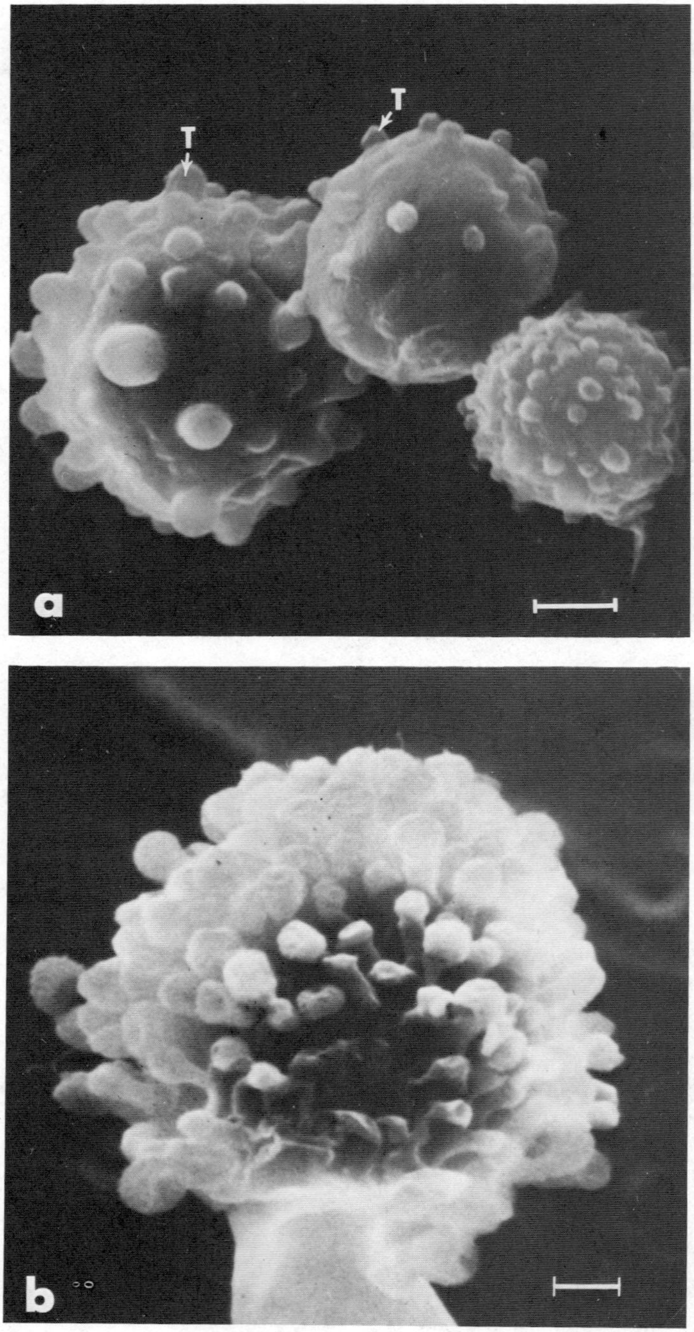

Figure 5 Scanning electron microscopy of conidia. (a) Tuberculate (T) conidia of *Histoplasma capsulatum*. Bar equals 3.0 μm. (From Ref. 72;) (b) Conidium of *Aspergillus fumigatus*. Bar equals 2.0 μm.

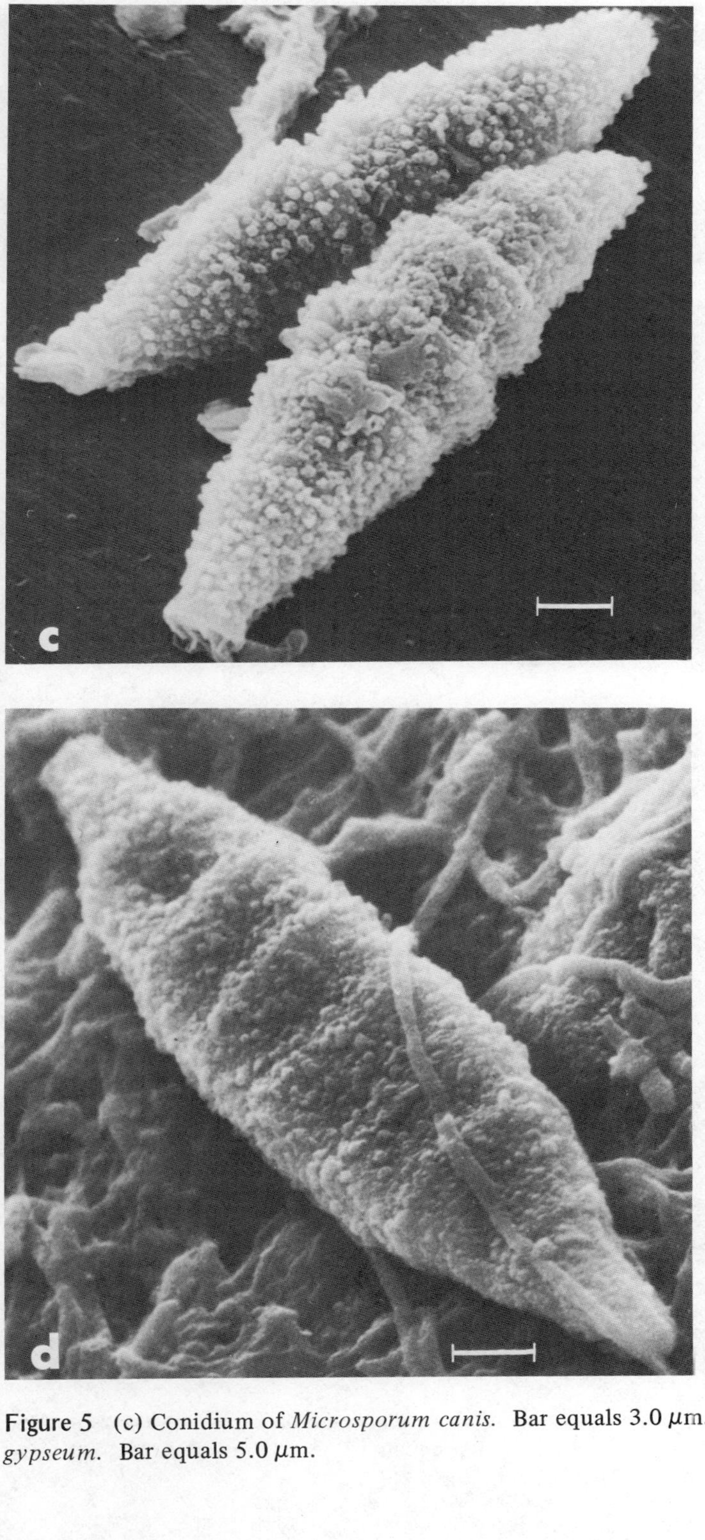

Figure 5 (c) Conidium of *Microsporum canis.* Bar equals 3.0 μm. (d) Conidium of *M. gypseum.* Bar equals 5.0 μm.

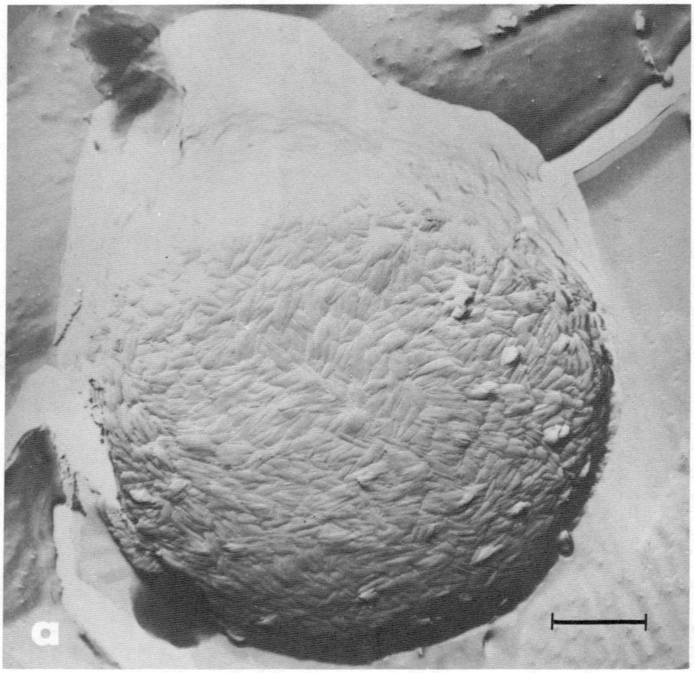

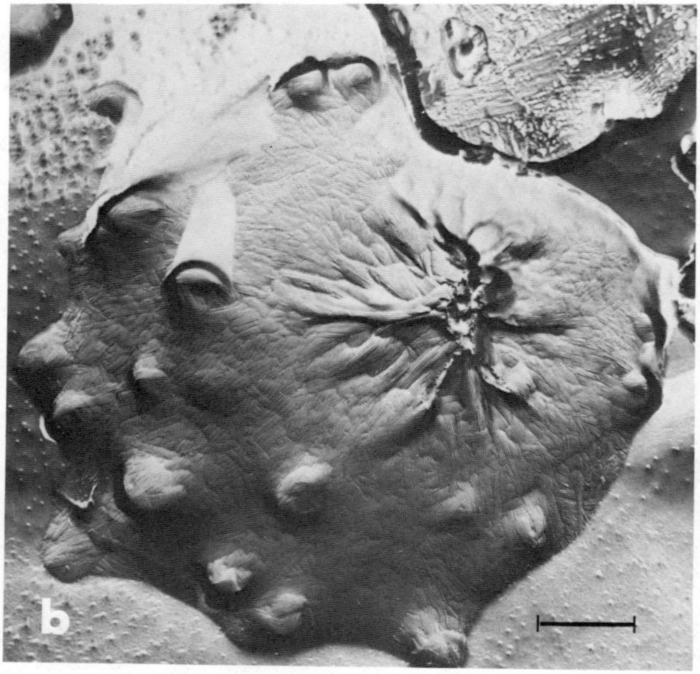

Figure 6 Surface features of *Penicillium* and *Aspergillus* conidia as seen by freeze etching. (a) Surface of a conidium of *P. camemberti*. Bar equals 1.0 μm. (From Ref. 102.) (b) Surface of a conidium of *A. niger*. Bar equals 0.5 μm.

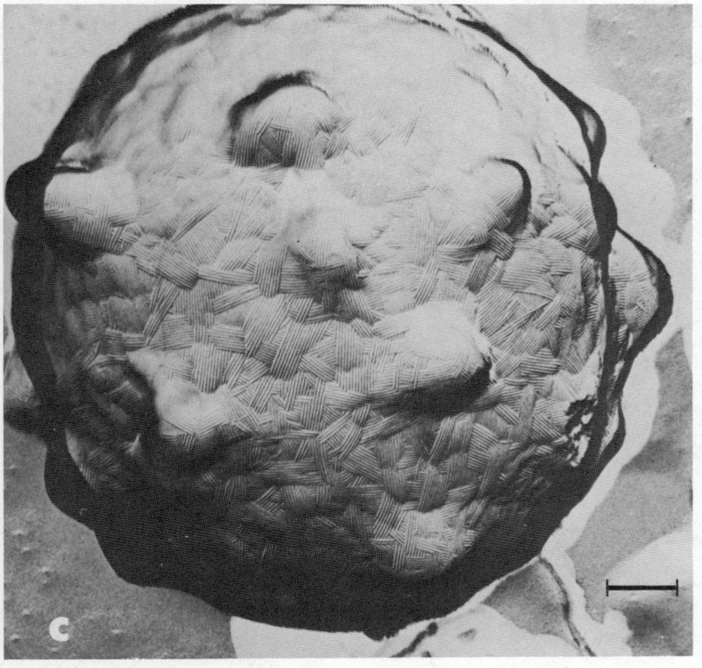

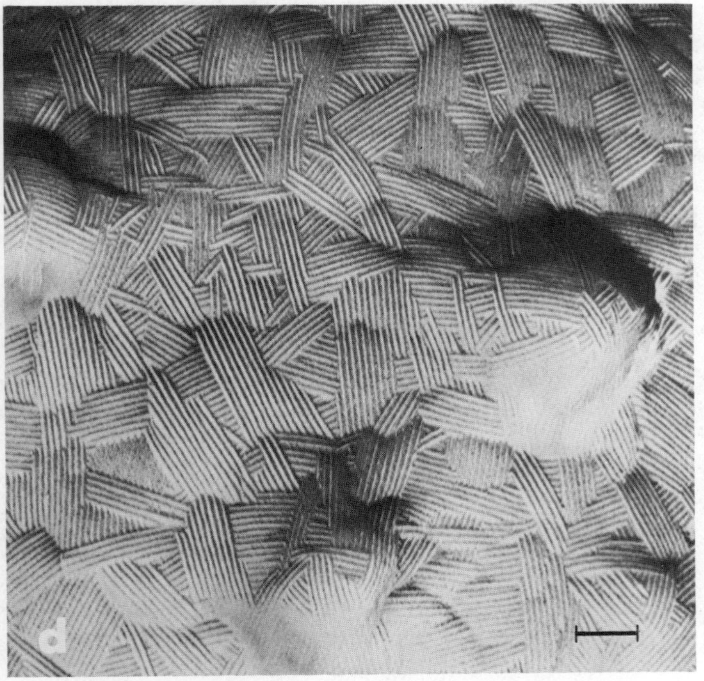

Figure 6 (c) Surface of a conidium of *A. Fumigatus*. Bar equals 0.25 μm. (d) Surface of a conidium of *A. fumigatus* showing the interlacing pattern of rodlets with no underlying spore surface showing. The surface is characterized by short narrow groups of rodlets oriented in all directions. Bar equals 0.1 μm. [Parts (b)-(d) from Ref. 103.]

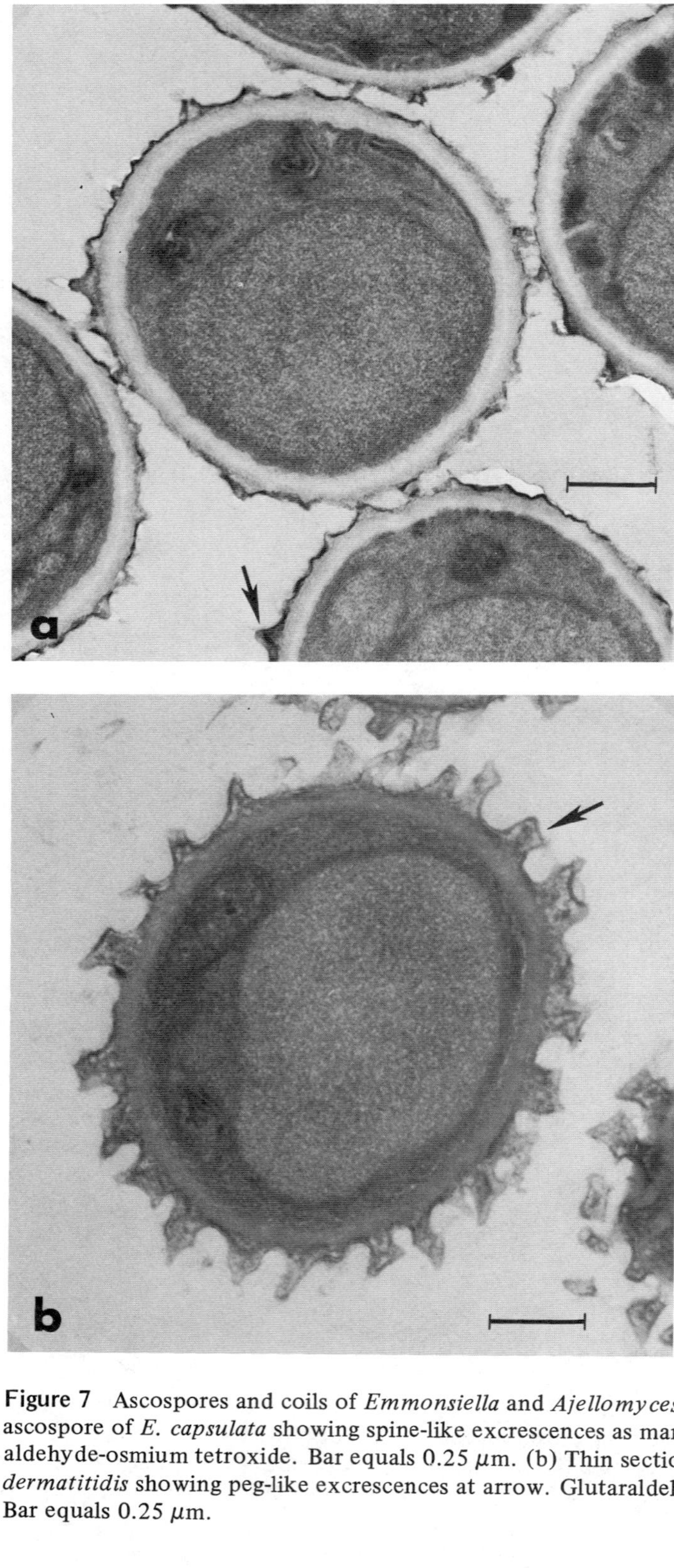

Figure 7 Ascospores and coils of *Emmonsiella* and *Ajellomyces*. (a) Thin section of an ascospore of *E. capsulata* showing spine-like excrescences as marked by the arrow. Glutaraldehyde-osmium tetroxide. Bar equals 0.25 μm. (b) Thin section of an ascospore of *A. dermatitidis* showing peg-like excrescences at arrow. Glutaraldehyde-osmium tetroxide. Bar equals 0.25 μm.

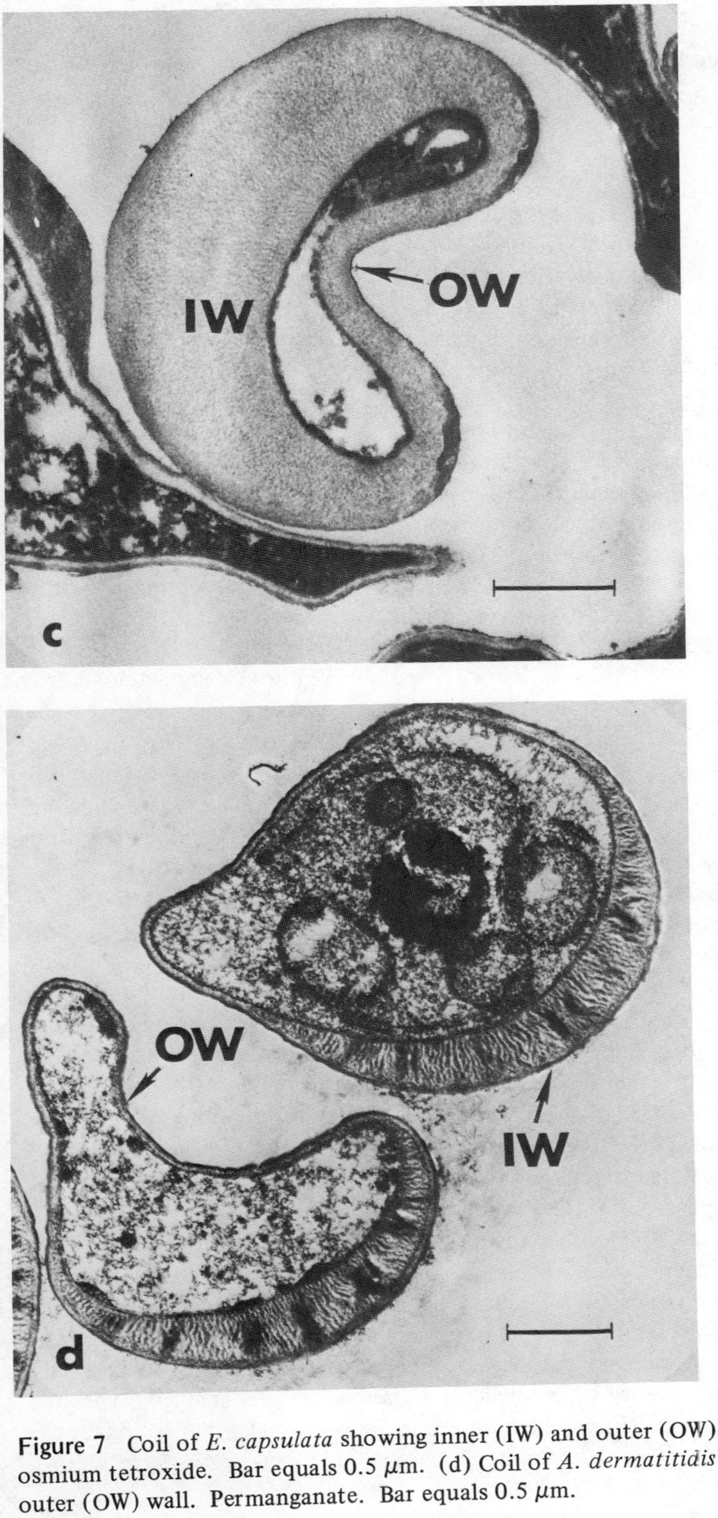

Figure 7 Coil of *E. capsulata* showing inner (IW) and outer (OW) wall. Glutaraldehyde-osmium tetroxide. Bar equals 0.5 μm. (d) Coil of *A. dermatitidis* showing inner (IW) and outer (OW) wall. Permanganate. Bar equals 0.5 μm.

varies from 1:1 to 1:5; in *Emmonsiella* this ratio is 4:1 to 15:1. Similar sections of the coils of *Arthroderma benhamiae* do not show this unusual variation in wall thickness [23].

III. THE SEPTUM

The structure of fungal septa is of considerable interest since the ultrastructural characteristics of septal types closely parallel the taxonomic position of a great many fungi. Thus, the examination of the septal area plays an important role in supporting and indicating taxonomic affinities and distinctions [164]. The fungal septum, particularly in Basidiomycotina, is not an inanimate stable structure, but can change ultrastructurally depending on its position within the mycelium and the organism's life cycle [88]. Electron microscopic study of any fungal organism should include a detailed examination of its septal organization.

In general, the fungi can be divided into the lower and higher groups by the absence or presence of regular septation. Certain of the pathogenic Zygomycetes are coenocytic. However, the Entomophthorales are unusual inasmuch as their vegetative hyphae are regularly septate, although septal organization in such fungi has received little attention. The hyphal wall of *Basidiobolus ranarum* [89] and probably *B. meristosporus* [66] appears to be composed of at least two layers. The septum is seen to be continuous with the innermost layer. A central pore or micropores are not present and the septum may therefore be termed as complete. Medial sections of *B. ranarum* often show the presence of a small central structure which appears as a depression from one side and a dome from the other. Complete septa without septal pores are characteristic of the vegetative hyphae of *Conidiobolus coronata* [77] (Fig. 8a).

The majority of pathogenic, filamentous fungi have an "ascomycetous" type of septum. These include species of the dermatophytes [110,181,230], *Phialophora* and *Cladosporium* [34], and the dimorphic fungal pathogens of the genera *Histoplasma, Blastomyces, Paracoccidioides, Sporothrix,* and *Coccidioides* [26,69,130,213], among others. The typical ascomycetous septum is relatively simple. It is generally electron translucent, wedge-shaped, with a central pore and associated Woronin (interseptal) bodies (Fig. 8b).

Woronin bodies are a class of round to oblong intracytoplasmic organelles which are frequently associated with septal pores of ascomycetous fungi (Fig. 8c). In the pathogenic ascomycetous fungi, they may vary in number and size and are generally seen on medial section through the septal pore area. They are bound by a single unit membrane, but the membrane is observed only when the organelle is sectioned medially. Although the presence of Woronin bodies has been known for some time, little information is available on their origin and development. In some nonpathogenic ascomycetes, the Woronin body appears ontogenetically related to microbodies [229]. The development of a Woronin body begins with the appearance of electron-opaque material within the microbody. This material then aggregates adjacent to the membrane of the microbody and condenses into a single paracrystalline inclusion. The inclusion is then gradually extruded and is eventually separated from the parent microbody by an exocytotic process. During this process the dense inclusions tend to become spherical in shape, but cross sections of six-sided forms may occur. In some nonpathogenic ascomycetous fungi, hexagonal crystalline structures have been noted which are believed to be functionally analogous with the spherical Woronin body [222] (Fig. 9). Hexagonal inclusions which are functionally equivalent to Woronin bodies are common in all species of the *Aspergillus glacus* group, but are not present in species of the other major groups of aspergilli [149]. Such hexagonal inclusions have not been reported from any of the major ascomycetous fungal pathogens.

Woronin bodies are generally believed to set as plugs whose presence or absence in the pore regulates cytoplasmic flow between adjacent hyphal cells of the mycelium. Their possible ontogenetic relation with microbodies might suggest a physiological function as well. At this time there is no cytochemical evidence to indicate the specific localization of any enzyme in Woronin bodies, and any physiological function which they might possibly possess remains to be described.

Septal organization of *G. candidum* should be mentioned since it is unusual among the potentially pathogenic fungi. The mature septum of this fungal opportunist is complete without a defined central pore, but rather is multiperforate with micropores [95]. These micropores have been considered as analogous with the plasmodesmata of higher plants [118,234]. Micropores are a consistent feature of the mature septum of *G. candidum* and are seen to be cylindrical connections 8.5-9.0 nm in diameter which pierce the septa of adjacent hyphal compartments, thereby creating a continuity between the plasma membranes of each compartment [209]. Although these micropores lack a desmotubule or other internal endoplasmic reticulum derivative, the lumen of the micropore is filled by an electron-opaque substance.

The most complex septal type is found in the basidiomycetous fungi. It has been stated that the micromorphology of the basidiomycete septum appears to be equal to or even exceed the basidium in characterizing this group of fungi [162]. These septa are characterized by a pronounced swelling around the central pore (the dolipore) and a perforate membranous cap (the parenthesome) located at each side of the dolipore (Fig. 8d). For these septa, there appears to be a further distinction between the homobasidomycetes, with multiperforate parenthesomes, and the jelly fungi, which possess imperforate parenthesomes. These septa have at some time a dense granule in each dolipore opening [163]. A third type of basidiomycetous septum is found in the Filobasidiaceae in which the dolipore granule is replaced by banded material and, when present, the parenthesomes are vesicular [165]. The monocaryons of the Filobasidiaceae are yeasts which are particularly assignable to the genus *Cryptococcus.* Recent studies have shown that *Filobasidiella neoformans* (*C. neoformans,* serotypes A and D) has a dolipore septum between hyphal cells and also between clamp connections and adjacent cells. The pore-occluding material is a heterogenous, flattened plate with dark margins and a lighter center. Representative basidiomycete parenthesomes are lacking and endoplasmic reticulum is seen in the dolipore region [127]. The septa of *Filobasidiella bacillispora* (*C. neoformans,* serotypes B and C), the other species of the genus, have not yet been analyzed.

Yeasts of the anascosporogenous genus *Trichosporon* are characterized by the formation of budding cells and a septate mycelium which falls apart into arthroconidia. *Trichosporon cutaneum,* the etiological agent of white piedra, produces a basidiomycetous septum [124] in which the septal structure and septal pore resembles that of *Rhizoctonia solani* [14]. Although the sexaul cycle for *T. cutaneum* is not yet known, the observation of a septal pore of the basidiomycetous type makes it probable that this fungus is related to the Basidiomycotina. The structure of the cell wall and the method of bud formation of *T. cutaneum* resembles those of yeasts of the genera *Rhodosporium* and *Leucosporidium* [123].

A unique approach to a better understanding of the dolipore-parenthesome structure has been made by Patton and Marchant [176]. In their study a total of nine different methods of measurement of dolipore-parenthesome parameters were considered, and sub-

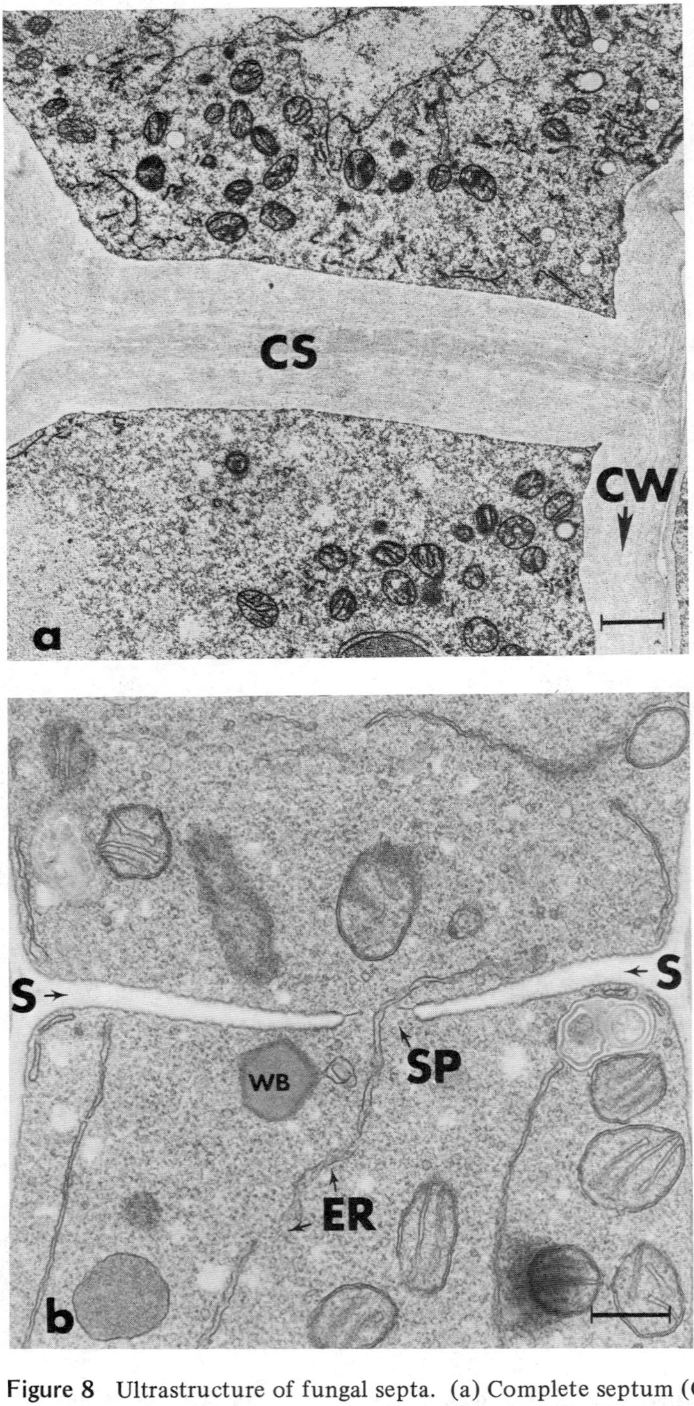

Figure 8 Ultrastructure of fungal septa. (a) Complete septum (CS) and cell wall (CW) as seen in a longitudinal thin section of a hypha of *Conidiobolus coronata*. Permanganate. Bar equals 1.0 μm. (b) Longitudinal thin section through the septal area of *Chaetomium thermophile* showing septal plates (S), septal pore (SP), endoplasmic reticulum (ER), and a Woronin body (WB). Glutaraldehyde-osmium tetroxide. Bar equals 0.25 μm.

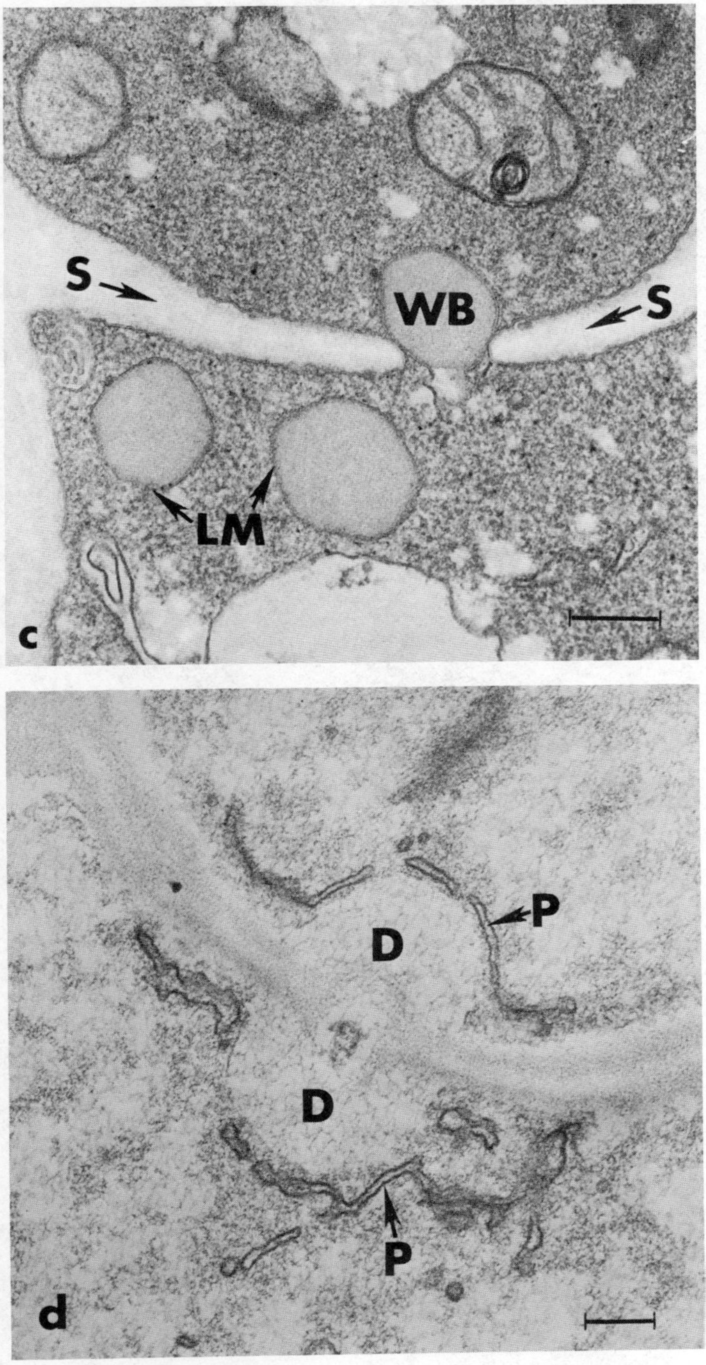

Figure 8 Ascomycetous septum (S) showing the occlusion of the septal pore by a Woronin body (WB). Note the limiting membranes (LM). Glutaraldehyde-osmium tetroxide. Bar equals 0.1 μm. (d) Basidiomycetous-type septum with a perforate parenthesome (P) and dolipore swellings (D). Permanganate. Bar equals 0.25 μm.

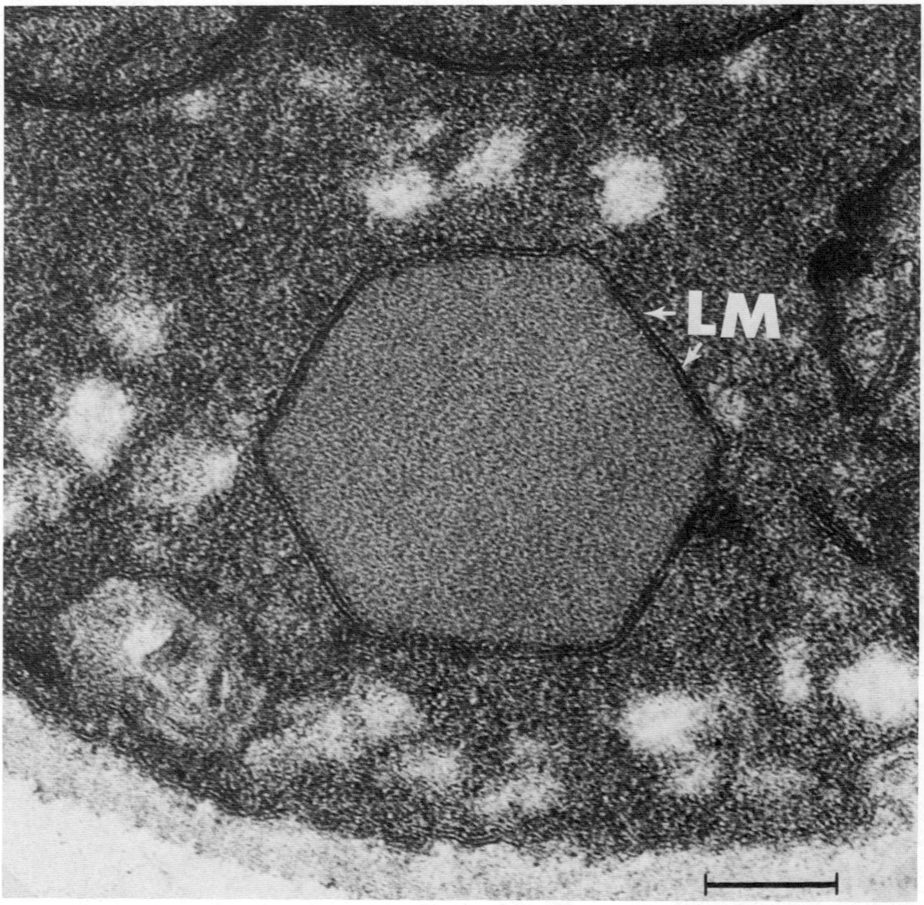

Figure 9 Portion of a thin section of a hypha of *Chaetomium thermophile* showing the presence of a hexagonal Woronin body-like inclusion. Note the limiting unit membrane (LM). Glutaraldehyde-osmium tetroxide. Bar equals 0.1 μm.

sequent mathematical analyses of these parameters indicated that quantitative differences occur between septal structures in different taxonomic groups. The results of the study suggest that the range of septal ultrastructure is a continuum and that possible evolutionary trends may be assigned to the ultrastructure of the dolipore-parenthesome complex produced by a given basidiomycetous fungus.

IV. CELL ORGANELLES AND CYTOPLASMIC INCLUSIONS

A. The Cytoplasmic Membrane and Its Ramifications

The protoplasm of all cells is surrounded by a membrane. In both procaryotic and eucaryotic cells, plasma membranes with a thickness of about 100 Å separate the cell contents from the external environment. In the eucaryotic cell, membranes act further in the compartmentalization of the cytoplasm by delimiting various organelles. Membranes provide boundaries, but they are also sites of many complex biochemical processes. Such processes include the selection or exclusion of certain ions at the cell surface, active uptake and transport of a great variety of substrates, oxidative phosphorylation, enzyme localization, cytokinesis, and others. In general, fungal membranes show a dissimilarity which is expressed in dimensions, symmetry, and lamellar versus globular or micellar substructure [12].

The plasma membrane or plasmalemma occurs as a biological entity in all living things and has been designated by Robertson [192] as the "unit membrane." The general structure of the unit membrane was predicted in 1935 by Danielli and Davson [38] to explain the physiological properties of membranes. The unit membrane consists of a bimolecular lipid layer covered on each side by a layer of protein. The water-soluble or polar ends of the lipid bilayer face outward and contact the protein surfaces. A review of the evidence for the unit membrane can be found in the work by Robertson [192].

Demonstration of the fungal plasma membrane by transmission electron microscopy of thin sections is generally achieved with glutaraldehyde-osmium tetroxide or permanganate fixation, although its integrity may be affected by the conditions used for fixation. Techniques of freeze etching and freeze fracturing may be of value in the study of the fungal plasma membrane. Such techniques have been used for the study of plasma membranes of certain of the dermatophytic fungi [18]. Plasma membranes of the pathogenic fungi are characteristically tripartite unit membranes. They are usually of increased electron opacity and vary somewhat in thickness. They may appear regular, but often they are rugose or serrate and often show vesiculation. Plasma membranes of protoplasts of pathogenic fungi may not always be clearly defined. Often, other membranes may occur parallel to the plasma membrane of the protoplast [22].

Various enzymes have been reported to be located solely in the plasma membrane fractions of certain yeasts and filamentous fungi. These enzymes include Mg^{2+}-dependent adenosine triphosphatase [156], 5′-nucleotidase and chitin synthetase [113], phospholipase 172], and phosphatidyl inositol kinase [231]. Marriott [154] obtained relatively pure preparations of plasma membranes from the yeast and mycelial forms of *Candida albicans,* and subsequently described Na^+/K^+-stimulated Mg^{2+}-dependent adenosine triphosphatase and mannan synthetase-enriched activities in the plasma membrane of both morphological forms of the fungus. Such localization of enzymes in the plasma membranes of *C. albicans* suggests that multifunctional biochemical activities may be associated with the plasma membranes of other pathogenic fungi. Of particular importance is the sensitivity of the plasma membrane to certain of the antifungal agents employed in chemotherapy [11,43, 65,111,112,120,131,171].

The plasma membrane may have important functions in the synthesis of new cell wall materials. It has been shown that the chitin synthetase of *Saccharomyces cerevisiae* occurs mainly as a zymogen, and that this zymogen is associated with many sites along the plasma membrane [51]. It is believed that during the initiation of budding in this yeast, an activator of the chitin synthetase zymogen is directed from vesicles to a specific site by an as yet unknown apparatus. There the zymogen is activated to produce a chitin septum. The chitin synthetase zymogen appears uniformly distributed on the plasma membrane and is activated only in a restricted area by the zymogen activator. This hypothesis is attractive in view of the chitin content of the walls of many of the yeast-like, pathogenic fungi.

Vesicular or tubular membranous structures located between the plasma membrane (the plasmalemma) and the cell wall were first observed in the fungi by Gibardt [84]. These structures have been termed "lomasomes" by Moore and McAlear [167] on the basis of their location. Lomasomes or similar structures have been reported in many fungi, algae, and higher plants. They appear throughout the pathogenic fungi and have been frequently mentioned in descriptions of the ultrastructural cytology of these organisms. Some workers regard lamellar bodies found after glutaraldehyde-osmium tetroxide fixation schedules as possible artifacts [12,37,224] and not true cellular components. Electron microscopic observations of fungal hyphae and yeast-like cells by conventional methods indicate that these widespread, membranous structures are probably not fixation artifacts [150].

Marchant and Robards [152] proposed that plasmalemmal structures (paramural bodies, lomasomes) were of two types with differing ontogeny. They adopted a nomenclature restricting the term "lomasome" to wall structures produced by the passage of vesicular material across the plasma membrane. They limited the term "plasmalemmasome" to structures produced exclusively from the plasma membrane by infolding. Marchant and Moore [150] described plasmalemmasomes as primarily excess plasma membrane that has accumulated in an endocellular manner. In contrast, lomasomes appear to be accumulated, ejected materials sequestered between the cell wall and plasma membrane by the deposition of wall material. Thus, lomasomes contain some wall material as part of their organization and tend to occur extracellularly in contrast to plasmalemmasomes, which appear entirely endocellularly and solely as elaborations of the plasma membrane.

Although the present evidence supports the concept that lomasomes and similar plasmalemmal structures are perhaps true entities of the living fungal cell, it is difficult to ascribe specific or even generalized functions to these structures. Bracker [12] lists some of the proposed functions attributed to lomasomes. It has been suggested that lomasome-like vesicles may perform secretory functions in the formation or incorporation of cell wall materials [121,151,233]. In most fungi an organized Golgi apparatus is lacking. Thus, it is possible that lomasome-like structures in association with the endoplasmic reticulum might in some manner be used by these organisms as a functional equivalent for the Golgi apparatus in wall formation. However, there is no convincing cytochemical evidence at this time to support these suggestions.

B. The Nucleus and Endoplasmic Reticulum

Most pathogenic fungi are typically mononucleate. Some species, e.g., yeast-like cells of *P. brasiliensis* [21] and *B. dermatitidis* [53], and certain of the filamentous, dermatophytic fungi [230] are multinucleate. Cells of the pathogenic Zygomycetes obviously contain more than one nucleus (Fig. 10a).

The nucleus is surrounded by a double membrane which is frequently seen to be continuous with profiles of endoplasmic reticulum (Fig. 10b). It is possible that this continuity results from previous nuclear division without complete separation of the envelopes in the case of multinucleate organisms. However, the nuclear membrane and endoplasmic reticulum may indeed constitute one continuous membrane system throughout the cell. A similar continuity of the plasma membrane and the nuclear membrane has been reported by others [139]. Pores are common in and over the entirety of the nuclear membrane and appear to connect the nuclear interior with the surrounding cytoplasm, thus permitting free exchange between the nucleus and its surroundings.

On fixation with glutaraldehyde-osmium tetroxide, the nucleoplasm is slightly osmiophilic and generally homogenous in appearance. It may be seen to contain a more electron opaque, usually eccentric nucleolar region. Fixation in permanganate results in both dark and lightly stained areas; the lighter areas may represent chromatin material [225] (Fig. 10c). In general, fungal chromosomes are difficult to discern, although at times synaptonemal complexes which indicate meiosis are visualized in nuclei during formation of ascospores.

Studies of nuclear division in the fungal organism are complicated by problems attendant with synchrony of the dividing cells, small nuclei, and almost insignificant chromosomes. Methods described by Barnicot and Huxley [5], Robbins and Gonatas [191], and Zickler [239] appear to be useful in the study of mitosis of fungal nuclei. Occasionally, random scanning of thin sections of fungal cells may reveal certain general features of nuclear division. In general, the fungal nucleus constricts at the midpoint between the daughter nuclei, and the nuclear membrane most often remains intact (Fig. 10d). In higher organisms the nuclear membrane dissociates and then re-forms.

Ultrastructural studies of somatic nuclear division have been made in only a few taxonomically diverse species. Little information concerning the mechanics of nuclear division is available for the pathogenic fungi. The zygomycetous *Basidiobolus ranarum* is perhaps the best characterized of the fungal pathogens in regard to nuclear behavior during division. Nuclear division in this fungus has been carefully studied by light microscopy [142,194,195] and electron microscopy [212]. The latter study is a comprehensive evaluation of somatic nuclear division in this fungus and should be consulted by those interested in this subject. Nuclear division in *B. ranarum* is mitotic, although some atypical features have been noted. The nuclear envelope partially fragments in prophase but, except for areas at the equator, it regains much of its integrity by metaphase. The spindle axis forms along the long axis of the hypha in metaphase. It is normal to the hyphal axis in metaphase. In anaphase it again lies along the hyphal axis. Chromatin could not be identified until prometaphase, when it was estimated that at least 900 minute chromosomes appear [212]. This number is surprisingly high, but appears to be in general agreement with the number of spindle microtubules observed in prometaphase. In this fungus there is only one microtubule associated with each divided kinetochore. Anaphase movement is brought about by both chromosome movement toward the poles and separation of the poles. Perinuclear centrioles then migrate to the poles in early prophase and there replicate in metaphase. In prophase the nucleolus is dispersed into granules which then collect in the polar regions during meta- and anaphase. These granules aggregate in telophase, re-forming the nucleolus without an association with loci of chromosomes. It is interesting to note that there are apparently fundamental differences between the chromosomes of *B. ranarum* and those of most eucaryotes. The lack of identifiable chromatin in sections of interphase and prophase nuclei, the conspicuous but separate kinetochores during metaphase, and the association of each kinetochore with a single spindle microtubule can be regarded as unique characters.

It is apparent that the specific mechanics of nuclear division as it occurs in *B. ranarum* cannot be regarded as typical of that found in other pathogenic fungi. As pointed out by Bracker [12], the diverse and incomplete nature of information regarding nuclear division in fungi does not permit a generalized hypothesis, and it is quite possible that several mechanisms may be operative. It is obvious that the fungal nucleus and it behavior in division need careful study. At present, the literature on this subject, especially that con-

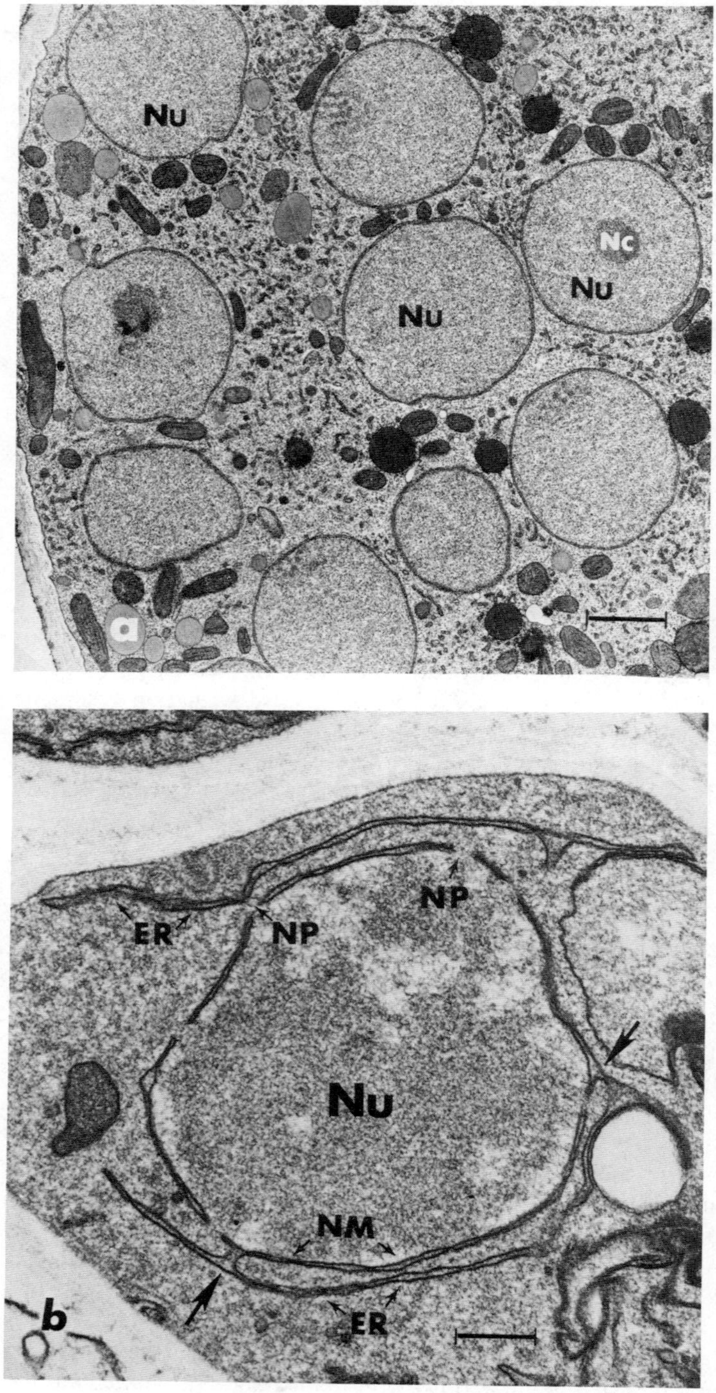

Figure 10 Fungal nuclei. (a) Thin cross section of a hypha of *Conidiobolus coronata* showing multiple nuclei (Nu) with nucleoli (Nc). Permanganate. Bar equals 1.0 μm. (b) Portion of a thin section of yeast-like *Histoplasma capsulatum* showing a nucleus (Nu), nuclear membrane (NM), nuclear pores (NP), and profiles of endoplasmic reticulum (ER). Note the continuity of the endoplasmic reticulum with the nuclear membrane as marked by the arrows. Permanganate. Bar equals 0.25 μm.

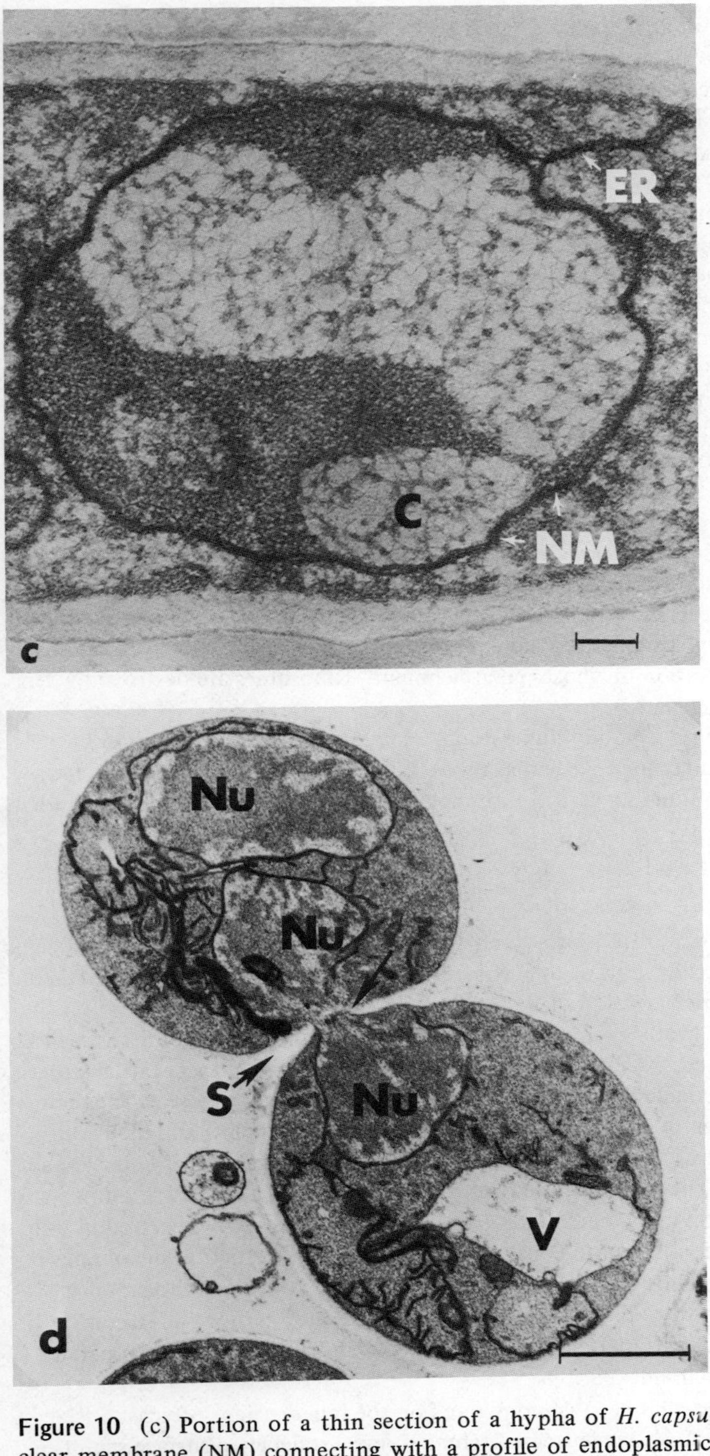

Figure 10 (c) Portion of a thin section of a hypha of *H. capsulatum* showing the nuclear membrane (NM) connecting with a profile of endoplasmic reticulum (ER), and the appearance of what is thought to represent chromatin material (C). Permanganate. Bar equals 0.1 μm. (d) Portions of a thin section of a budding cell of *H. capsulatum* showing separation of the nuclei (Nu) at the septum (S). Bar equals 1.0 μm.

cerning the pathogenic fungi, is sparse. It might well be expected that great variations exist in both nuclear morphological organization and nuclear behavior during division. More comprehensive accounts of the behavior of the fungal nucleus during mitosis can be found in the reviews by Robinow and Bakerspiegel [195] and Matile et al. [157]. The latter review contains extensive information regarding the caryology of the nonpathogenic yeasts.

The endoplasmic reticulum occurs throughout the pathogenic fungi. The typical profile of endoplasmic reticulum is represented as paired parallel membranes separated by a narrow lumen. It is generally seen as a scattered to sometimes continuous cisternal system throughout the cytoplasm; it may occur singly or in groups of lamellae. The membranes of endoplasmic reticulum appear structurally different from those of the plasma membrane. They appear thinner and more symmetrical, while the tripartite organization is somewhat less well defined than that seen in the case of the plasma membrane. Profiles of endoplasmic reticulum may be seen to interconnect or they may end abruptly. They often can be observed to associate specifically with the membranes of other structures, such as those of the nucleus and perhaps mitochondria. This association suggests a functional continuity, which is as yet not well defined.

Endoplasmic reticulum may be referred to as rough or smooth depending on the presence or absence of ribosomes in association with the membranes. Although rough endoplasmic reticulum has been shown in some fungi, most of the ribosomes occur free in the cytoplasm. Fixation in glutaraldehyde-osmium tetroxide or permanganate is often unsatisfactory for the demonstration of ribosomal attachment. Ribosomes are destroyed when cells are fixed in permanganate.

Endoplasmic reticulum is poorly understood. However, it seems to function in some manner as a transport and storage system in the cell. The major characteristics of endoplasmic reticulum in the fungi suggest its morphological identity with endoplasmic reticulum of higher plant cells.

C. Mitochondria

Mirochondria appear as ubiquitous intracytoplasmic organelles throughout the pathogenic fungi, although many descriptions of the ultrastructural cytology of pathogens often contain little more than a brief mention of their presence.

Moore and McAlear [166] examined the mitochondrial morphology of more than 50 genera of the Eumycotina. Although they found wide variation of mitochondrial structure in different fungi, they concluded that fungal mitochondria showed no special characteristics that might distinguish them from the mitochondria of higher plants and metazoa.

Mitochondria of the pathogenic fungi, as is the case in all fungi, can vary considerably in their size and shape. They may be spherical to rod-shaped or thread-like in outline and measure about 0.3-1.0 μm in diameter and up to 3.0 μm in length. When viewed in thin section, the mitochondrion is seen to consist of two membranes: an outer membrane and an inner membrane which invaginates into the interior to form cristae. Mitochondrial cristae may be tubular or plate-like. Generally, mitochondria of pathogenic Ascomycotina, Basidiomycotina, and Fungi Imperfecti have plate-like cristae which most often parallel the long axis of the mitochondrion.

A particulate structure has been demonstrated on the inner surface of yeast cristae by use of negative staining [204], while freeze etching shows a roughened surface sculptured by warts and depressions [161]. In mature spores of *B. ranarum,* freeze etching indicates a regular pattern on both sides of the outer mitochondrial membrane. Bands consisting of 5-10 parallel ridges, each about 20 nm in width, envelop the outer as well as the inner side of this membrane. Hexagonal, crystalline structures which probably represent stored protein are enclosed in special bodies with a smooth surrounding unit membrane. These crystals are formed by parallel rods consisting of globular subunits about 6.5-7.0 nm wide [8]. It is possible that similar crystalline structures may occur in the mitochondria of other pathogenic Zygomycetes. However, their occurrence in mitochondria of other pathogens remains to be determined.

The mitochondrial membranes are generally well preserved by conventional glutaraldehyde-osmium tetroxide fixation. The addition of acrolein to the primary fixative may be useful in obtaining better fixation. With aldehyde-osmium tetroxide fixation, the perimitochondrial space is electron transparent. Ribosomes and fibrils of DNA are often seen within the matrices of well-fixed, mitochrondial membranes. Mitochondrial membranes of most pathogenic fungi are usually preserved in permanganate-fixed cells as is true with other membrane structures. The mitochondrial membranes appear in sharp relief against a rather vacant background. On the other hand, potassium permanganate destroys ribosomes and can bring about changes in the cytoplasm.

The relative numbers and morphology of fungal mitochondria can vary with the age of the fungus cell or with the stage of its life cycle. In general, they are most abundant in cells with high metabolic activity. They may occur at random throughout the cytoplasm. In some of the zygomycetous fungi, mitochondria are numerous in apical regions of the hyphae and often occur in areas along the periphery of the cell wall-cytoplasm interface. In yeast-like cells of *Blastomyces dermatitidis* and *P. brasiliensis,* they are often seen at and along the plasma membrane [66]. Mitochondria have been described in some detail for a few pathogenic fungi, including *Torulopsis glabrata* [16], *B. dermatitidis* [53], and *Cryptococcus neoformans* [54].

Fungal mitochondria are functionally equivalent to the mitochondria of other eucaryotic cells. Although certain differences in the structure and components of the electron transport system exist, they are essentially similar to those of higher plants and metazoa. The biochemical properties of mitochondria from *Candida albicans* have been studied by Yamaguchi et al. [237]. These workers described a method by which relatively intact mitochondria can be prepared and which might well be applicable to those of other pathogenic fungal species.

It appears that the number of mitochondria per cell remains rather constant following cell division. This observation suggests a means for the generation of new mitochondria. Three mechanisms have been proposed: (1) division of preexisting mitochondria, (2) de novo synthesis; and (3) the transformation of other cell components into mitochondria. The studies by Luck [136-138] on mitochondria from *Neurospora crassa* are consistent with the hypothesis that new mitochondria arise by steady growth and the division of preexisting mitochondria. A comprehensive review of the structure, biochemistry, biogenesis, and degradation of mitochondria can be found in the monograph by Tandler and Hoppel [216].

D. Membrane Configurations

A variety of concentric whorls of lamellar membranes, tubules, combinations of tubules and lamellae, or similar membrane configurations have been observed frequently in thin sections of various pathogenic fungi. The micromorphology of these configurations is

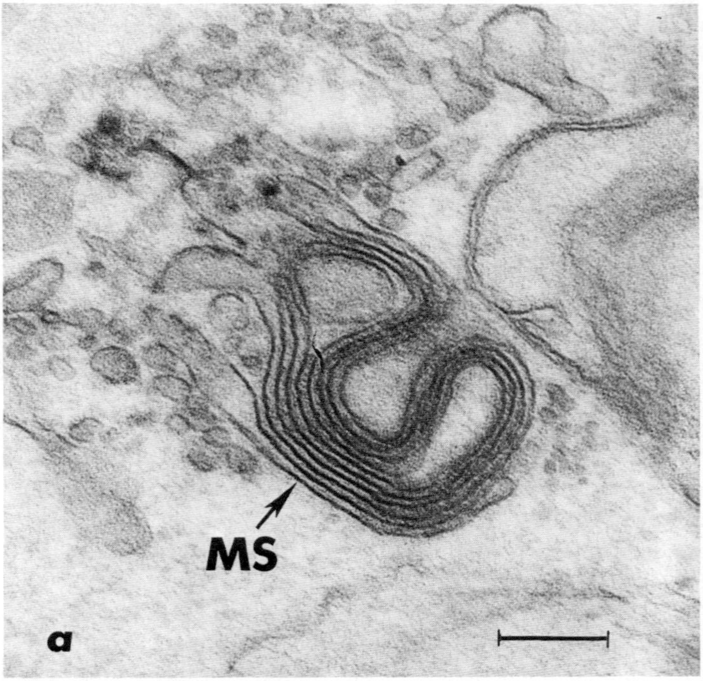

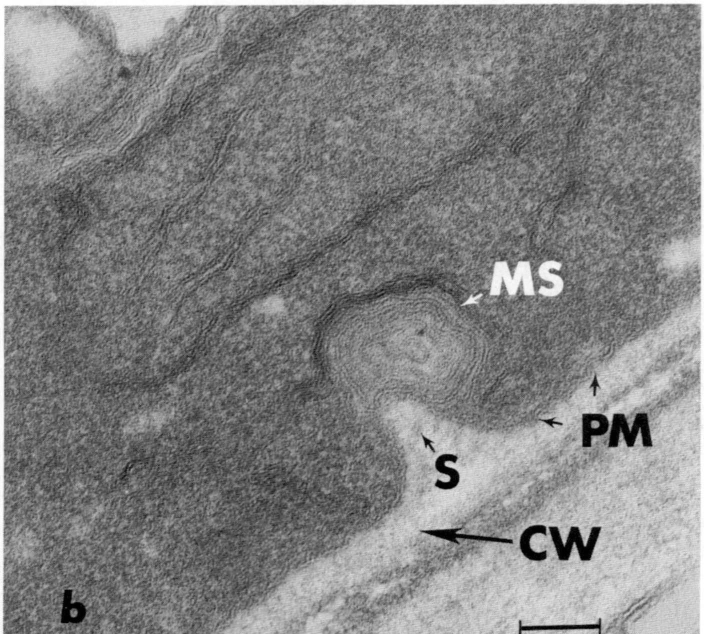

Figure 11 Membrane systems in pathogenic fungi. (a) Mesosomal-like membrane system (MS) in the cytoplasm of a hypha of *Basidiobolus ranarum*. Permanganate. Bar equals 0.25 μ. (b) Membrane system (MS) in a hypha of *Histoplasma capsulatum*. Note its relation to the plasma membrane (PM), developing septum (S), and cell wall (CW). Glutaraldehyde-osmium tetroxide. Bar equals 0.1 μm.

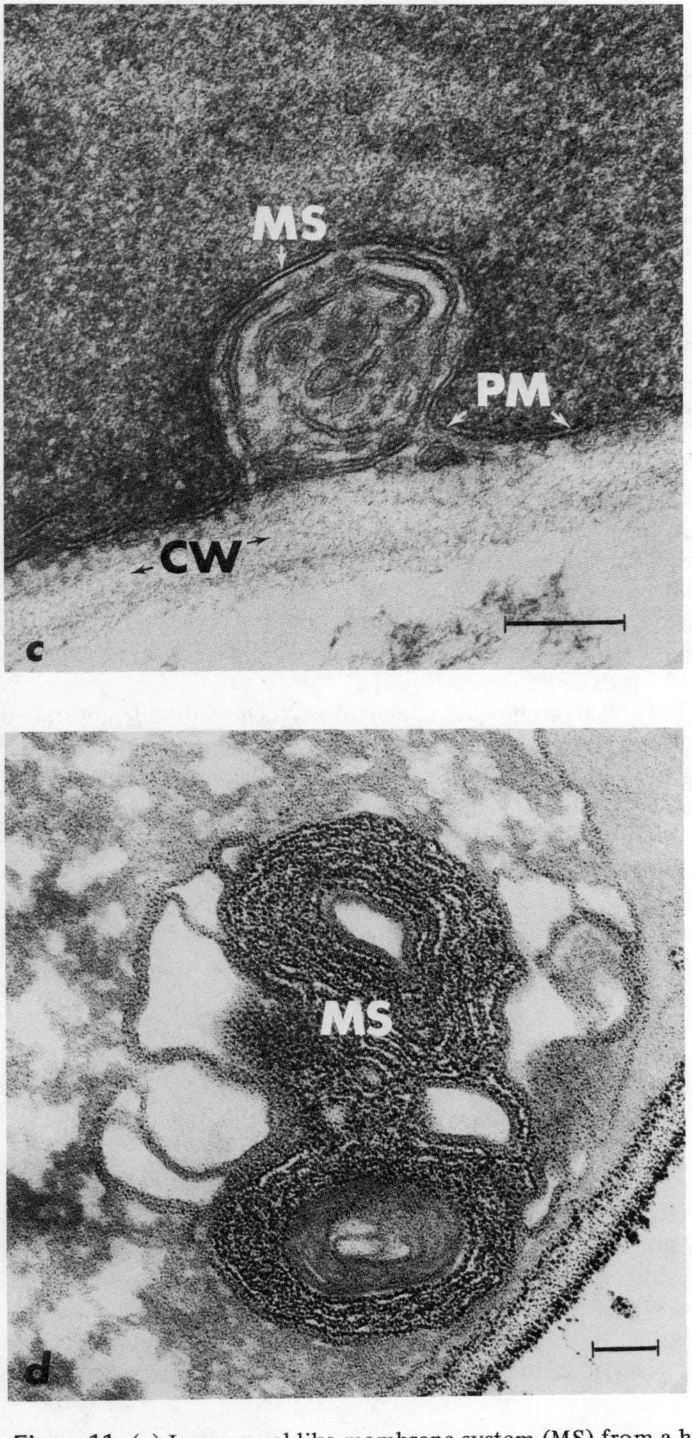

Figure 11 (c) Lomasomal-like membrane system (MS) from a hypha of *Blastomyces dermatitidis*. Note its relation to the plasma membrane (PM) and the well wall (CW). Glutaraldehyde-osmium tetroxide. Bar equals 0.1 μm. (d) Thin section of a hypha of *Sporothrix schenckii* following staining by the PATAg method of Thiéry [220]. Note the deposition of silver granules along the membranes (MS). Glutaraldehyde-osmium tetroxide-PATAg. Bar equals 0.1 μm.

generally similar, but a certain amount of control of the morphological appearance of these structures is apparently exerted by the fungal species [168]. Possibly for this reason, there is a lack of any general agreement on their nomenclature, origin, and possible functions. In the pathogenic fungi, such configurations have been variously termed myelin figures [25,76], mesosomes [21,25,26,53,57,74,76,83,130], or simply intracytoplasmic membrane systems (Fig.11a and d). Those membranous structures shown in Fig. 11b and c and designated as lomasomes or plasmalemmasomes are discussed in Sec. IV.A.

In many respects configurations with concentric whorls of lamellar membranes resemble the concentric membranes found in the myelin sheaths formed by peripheral nerves of vertebrates [79]. In general, membranous or lamellar configurations have been noted following fixation in glutaraldehyde-osmium tetroxide, and are not usually noted in fungal tissues following fixation with potassium permanganate. Some workers consider them artifacts of glutaraldehyde fixation [12] or as configurations formed from water and lipid following loss of cellular proteins. In senescent or degenerate cells this is most likely the case.

It is possible that certain membrane configurations seen in fungal cells are real entities and perhaps have some functional role in the cell. Cole and Aldrich [33] demonstrated myelin-like figures in unfixed, glycerinated, freeze-etched conidia of *Scopulariopsis brevicaulis* and suggested that these structures were indeed not artifacts. Mesosome-like figures were similarly demonstrated by freeze etching of cells of *Cryptococcus neoformans* [57], while mesosomal-like membranes have been noted in the cytoplasm (Fig. 11a) of permanganate-fixed hyphae cells of *Basidiobolus ranarum* [66].

Leiva and Carbonell [133], using techniques of electron cytochemistry, demonstrated the presence of succinic dehydrogenase activity in mesosomes, the plasma membrane, and mitochondria of the hyphae of *P. brasiliensis.* The highest activity was localized in the mesosomes. It was observed as highly electron-opaque zones with concentric ring-like patterns. In other pathogenic fungi, intracytoplasmic membrane systems or mesosomal-like structures have been noted which are strongly reactive when thin sections of the fungus are treated with the PATAg technique of Thiéry [220]. A thin section of a hypha of *Sporothrix schenckii* after treatment by this technique is shown in Fig. 11d. The deposition of silver granules along the membranes suggests the presence of either glycolipid or glycoprotein. Thus, the demonstration of membranous configurations by techniques of freeze etching, by fixation in permanganate, and by methods of cytochemistry would argue that many of these membranous structures are not simply artifacts produced by glutaraldehyde-osmium tetroxide fixation.

Functions for the mesosomal-like membrane systems in the pathogenic fungi remain to be determined, but it is possible that they are involved in septal formation [26,74], in conidial wall synthesis [80], DNA replication, or respiratory processes [57,133].

Important but as yet unresolved questions are whether all membrane configurations in fungi are fundamentally similar in their significance and possible function in the cell, and to what extent they might correspond with somewhat similar configurations in other organisms [168]. It is possible that future studies employing such techniques as megavolt electron microscopy or the use of specific electron cytochemical localization may be of some value in clarifying the status of the membrane system as a functional subcellular organelle.

E. Vacuoles and Vacuolar Inclusions

Vacuoles are intracytoplasmic structures found in all pathogenic fungi. They are membrane-bound areas within the cytoplasm which contain aqueous, nonprotoplasmic liquids or storage products. The membrane which forms the boundary between the cytoplasm and the vacuole proper is termed the tonoplast. Vacuoles can vary in size and shape and are common in the subapical regions of most of the pathogenic, filamentous fungi. They are especially numerous in older cells. In certain yeasts such as *Candida albicans,* there is generally one large vacuolar apparatus, although several smaller units can occur. This central vacuole is regarded as the most conspicuous organelle in the cytoplasm of this pathogenic yeast, and it is associated with a variety of enzymatic activities [44].

The presence of osmiophilic inclusions of varying sizes and shapes but of more or less homogenous texture has been described in a number of pathogenic yeast-like and filamentous fungi. Many of these electron-opaque structures may be located within vacuolated areas which they either partially or completely fill (Fig. 12). These vacuoles usually possess a single limiting membrane or tonoplast which often is not well defined. In some pathogenic fungi, the electron-opaque body or vacuolated area containing electron-opaque materials appear in close approximation to lipid bodies. Among the pathogenic fungi, electron-opaque materials contained within vacuoles are especially evident in *Phialophora dermatitidis* [87], *Penicillium marneffei* [67], *Trichophyton mentagrophytes* [147], *S. schenckii* [76], and *Blastomyces dermatitidis* [70]. In *S. schenckii* such electron-opaque material contained within vacuoles is a common observation in thin sections of young cells. This material is markedly osmiophilic, reacts strongly with the lipophilic reagent thiocarbohydrazide, and can be partially solubilized with the lipid solvent sodium methoxide. In older cells of *S. schenckii,* lipid bodies are commonly seen in close approximation with the tonoplast of the vacuole [76]. Osmiophilic inclusions in *Ceratocystis ulmi* appear to contain DNA as determined by autoradiographic examination after labeling with tritiated thymidine [29]. Since fungi have not previously been reported to have thymidine kinase, this latter observation may require reexamination.

Published micrographs of a variety of pathogenic fungi indicate that electron-opaque material contained within vacuoles is widespread among the fungi. It should be pointed out, however, that this does not imply that all electron-opaque inclusions seen within vacuoles are either equivalent cytological inclusions or that they are chemically similar in composition. It is possible that these electron-opaque materials represent various types of storage substances, products of their degradation, or in some cases, even pigments. A more precise understanding of electron-opaque bodies and similar materials contained within vacuoles must await additional studies on their isolation, purification, and biochemical characterization in cell-free fractions.

F. Microbodies and Microbody-like Organelles

Numerous ultrastructural studies of many different species of fungi have established the widespread occurrence of microbodies in fungal cells [62,158,227]. Such structures seem to have multifunctional purposes in metabolism, and the term "microbody" can be redefined biochemically as peroxisomes, glyoxysomes, or lysosomes.

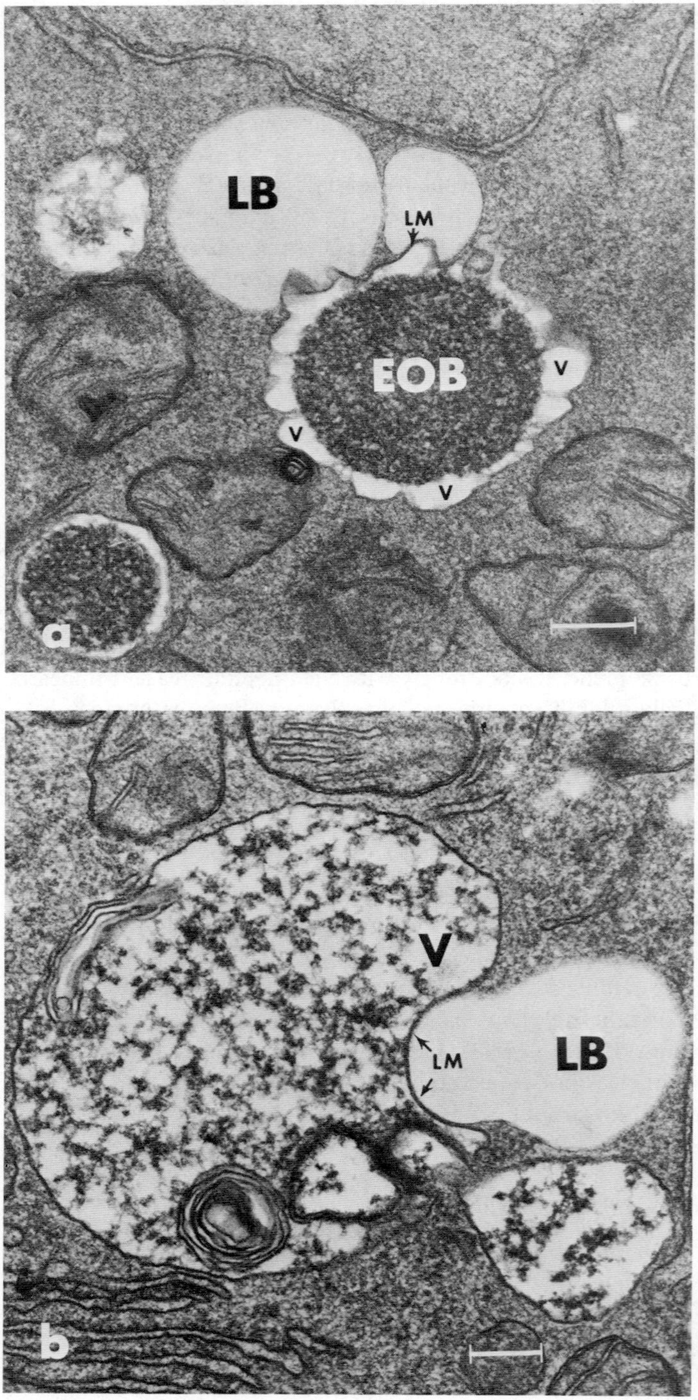
LB
LM
EOB
V
V
V
a
V
LM
LB
b

In general, microbodies are bound by a single membrane and are approximately 0.5-1.2 μm in the greatest diameter. The limiting membrane of the microbody may often be in contact with the endoplasmic reticulum. Microbodies may show spherical, elongate, or irregularly shaped profiles in thin section, and have a granular to fibrillar matrix of varying electron opacity. Dense amorphous to crystalline inclusions or nucleoids may occur in their matrix, but these inclusions are not necessarily a typical feature (Fig. 13a) and b).

Little information is available on the occurrence and distribution in the pathogenic fungi of subcellular organelles having the morphological characteristics and specific marker enzymes of peroxisomes or glyoxysomes. In well-fixed specimens of fungal tissue, these structures are usually evident on electron microscopy of thin sections, and the presence of unidentified organelles fitting the general description of microbodies can be established by appropriate cytochemical means (Fig. 13c). Only a few descriptions of the possible occurrence of peroxisomes in the pathogenic fungi have been reported. *Candida albicans* [44] and yeast-like forms of *Paracoccidioides brasiliensis* [21] are reported to contain round to oval bodies which morphologically resemble peroxisomes. However, these bodies were devoid of demonstrable catalase activity. Such enzymatic activity in *C. albicans* appears on the mitochondrial cristae.

A class of microbody-like organelles was reported from the vegetative hyphae and conidia of *Conidiobolus coronata* [77]. The cytoplasm of ungerminated conidia of this pathogenic fungus contains large numbers of microbody-like organelles in intimate association with lipid storage bodies (Fig. 13d). These organelles are round to oval to slightly reniform structures which are bound by a single membrane and composed of an electron-opaque, slightly granular matrix without evidence of nucleoid formation. The limiting membranes of lipid storage bodies are intimately associated with that of one or more of the microbody-like organelles (Fig. 14a). This intimate association of the two classes of cytoplasmic organelles suggests that the microbody-like organelle of *C. coronata* may be involved in some manner with lipid metabolism during the life cycle of the fungus. Cautious interpretation of electron cytochemical localization studies suggests that lipase, nonspecific esterase, and possibly aryl sulfatase are associated with the microbody-like organelles. Neither peroxidase nor acid phosphatase could be demonstrated. Lambiase and Yendol [128] have since described similar microbody-like organelles in the entomogenous fungus *Entomophthora apiculata*. They postulated that these organelles may contain enzymes involved with the penetration of the insect cuticle. Electron microscopy of the hyphae and conidia of *Basidiobolus meristosporus* indicates the presence of structures which closely resemble the classic descriptions of microbodies (Fig. 14b). These structures are limited by a single membrane and are occasionally seen to contain nucleoid-like inclusions within the

---

Figure 12 Vacuoles and vacuolar inclusions. (a) Thin section of yeast-like *Blastomyces dermatitidis* showing vacuoles (V) containing electron-opaque bodies (EOB), and the relation of the limiting membrane (LM) to lipid bodies. Glutaraldehyde-osmium. Bar equals 0.25 μm. (b) Yeast-like *B. dermatitidis* showing a vacuolated area (V) containing electron-opaque material. Note the relation of the limiting membrane (LM) to the lipid body (LB). Glutaraldehyde-osmium tetroxide. Bar equals 0.25 μm. (From Ref. 70.)

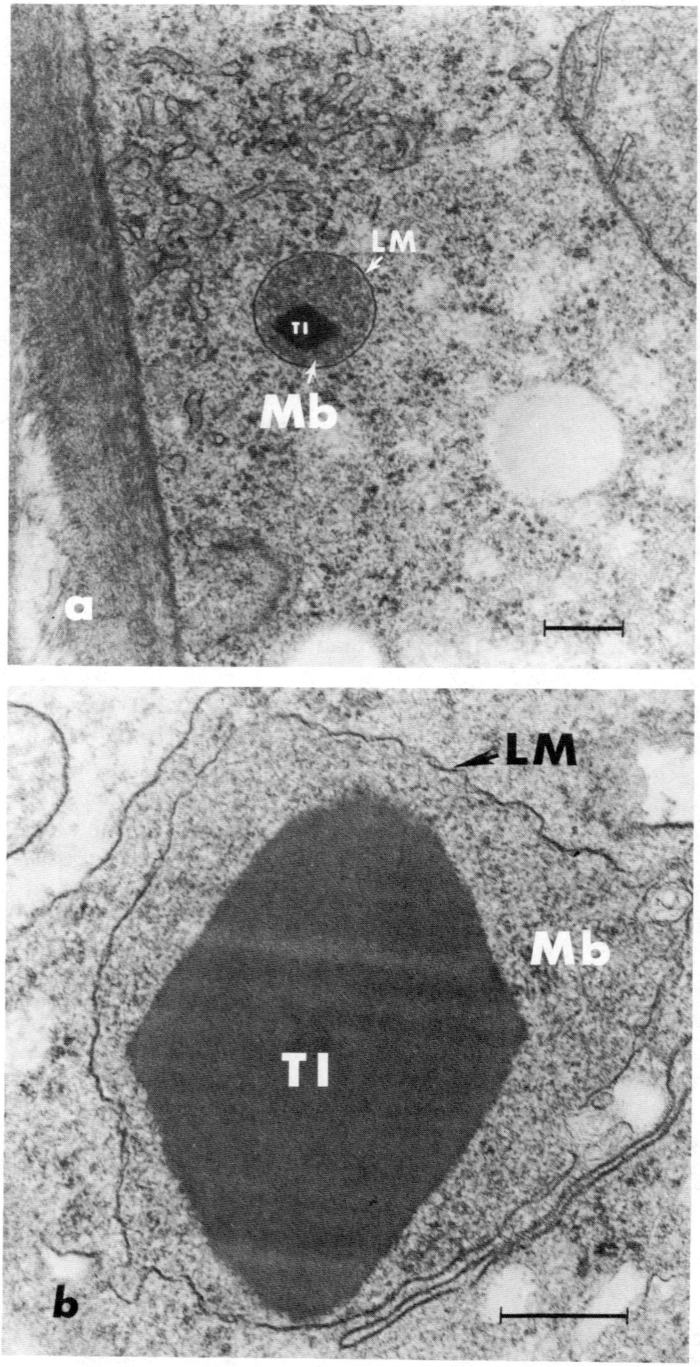

Figure 13 Microbodies and microbody-like organelles. (a) Microbody (Mb) in the hyphal cytoplasm of *Basidiobolus meristosporus.* Note the single limiting membrane (LM) and the electron-opaque tetragonal inclusion (TI). Glutaraldehyde-osmium tetroxide. Bar equals 0.25 μm. (b) Electron micrograph showing a microbody (Mb), its single limiting membrane (LM), and an electron-opaque tetragonal inclusion at high magnification. Glutaraldehyde-osmium tetroxide. Bar equals 0.25 μm.

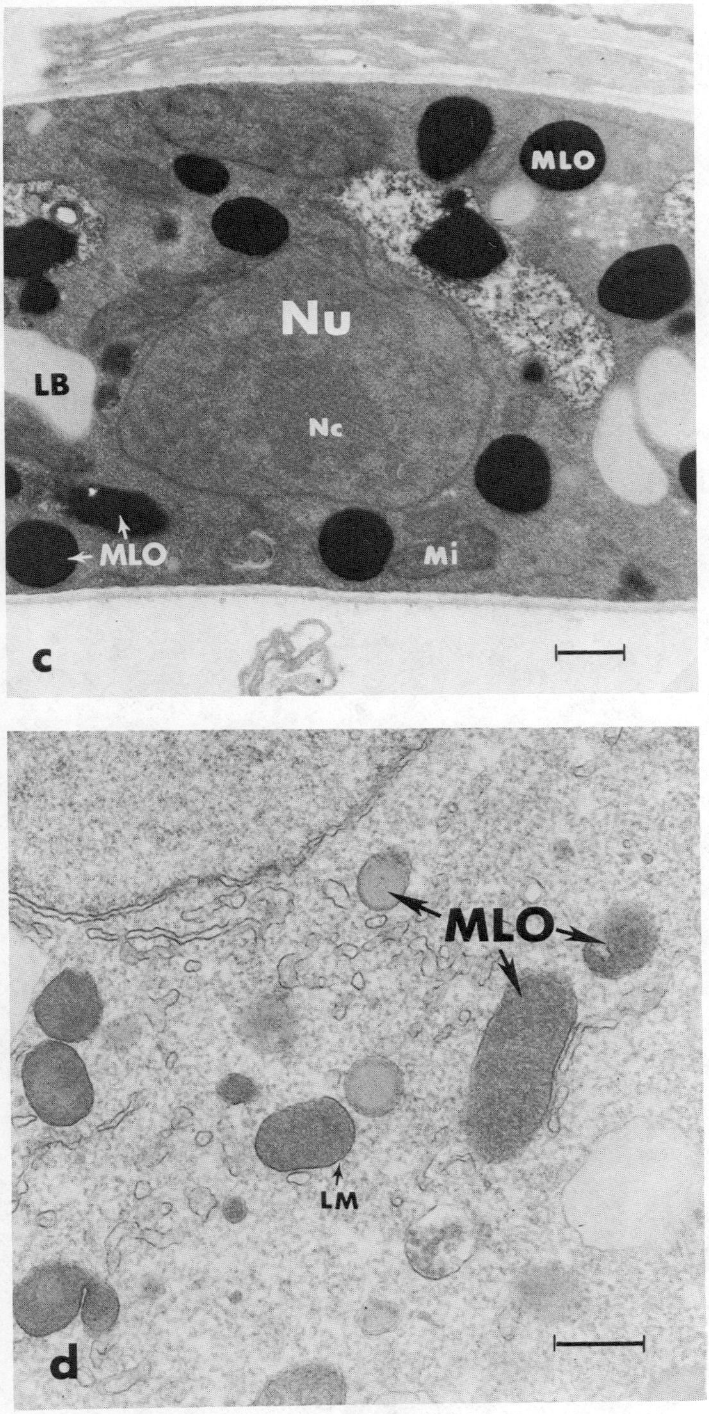

Figure 13 (c) Longitudinal thin section of a hypha of *Conidiobolus coronata* following electron cytochemical determination of catalase. Note the nucleus (Nu) with nucleolus (Nc), mitochondria (Mi), and the increased electron opacity of the microbody-like organelles (MLO). Glutaraldehyde-osmium tetroxide. Bar equals 0.5 μm. (d) Thin section of a hypha of *C. coronata* showing microbody-like organelles (MLO). Note the single limiting membrane (LM). Glutaraldehyde-osmium tetroxide. Bar equals 0.25 μm.

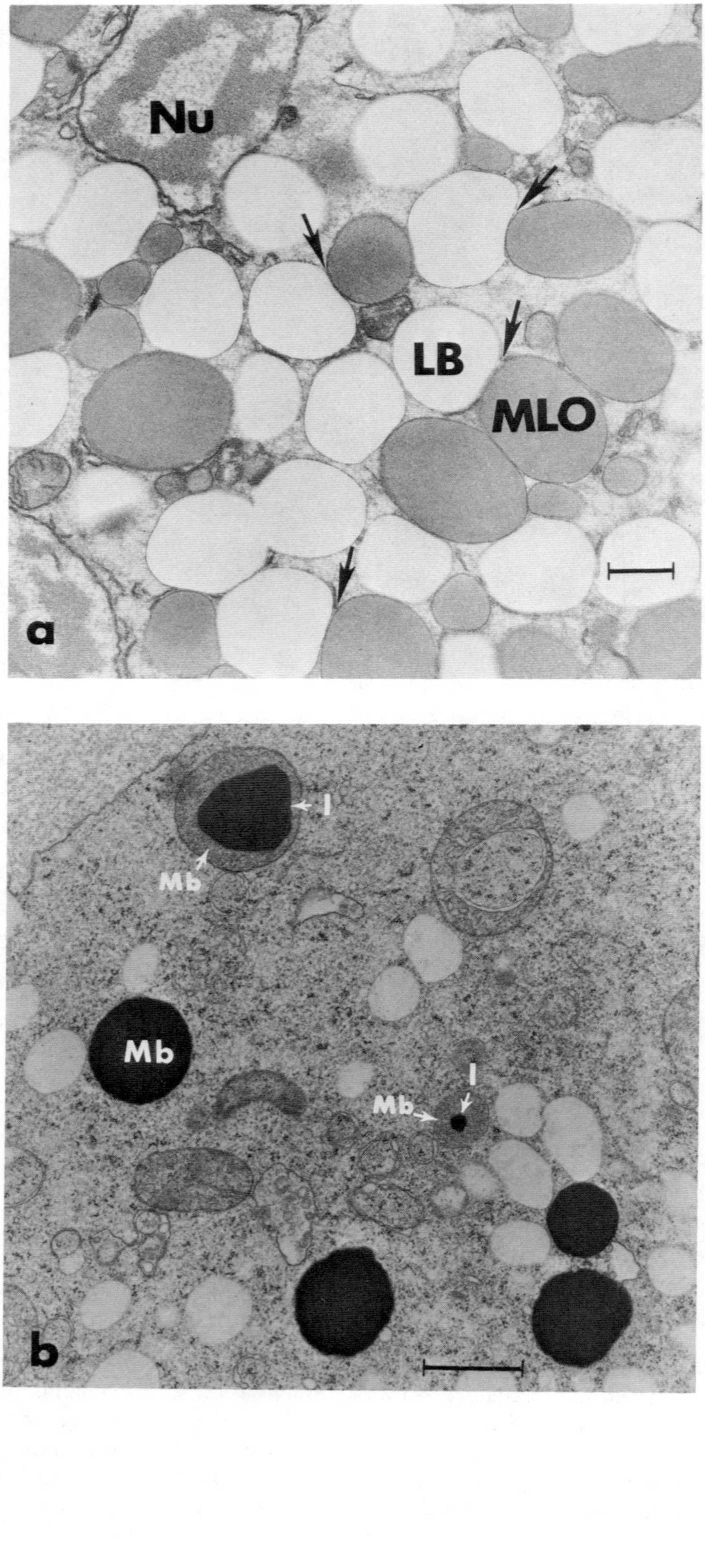
Nu
LB
MLO
a
I
Mb
Mb
I
Mb
b

matrix. These microbodies are not associated specifically with lipid bodies, and their enzymatic reactivity remains to be determined [66]. It is clear that much additional work will be necessary to clarify the status of both peroxisomes and glyoxysomes as functional subcellular components of the pathogenic fungi.

Novikoff [169] and De Duve and Wattiaux [42] first recognized the nature and importance of lysosomes as cellular organelles. These subcellular structures are widely distributed in eucaryotic organisms and play multiple roles in various biosynthetic processes of the cell. As yet, a precise definition of lysosomes is difficult since many conflicting interpretations have been made concerning their function(s).

Lysosomes are cell particles bound by a single limiting membrane. They can be localized by a variety of cytochemical procedures since they characteristically contain different acid hydrolases. Gahan [64] defines a lysosome as a cell organelle with the following diagnostic criteria: (1) a single limiting membrane, (2) liberation of hydrolases on destruction of the limiting membrane, (3) the presence of more than one demonstrable acid hydrolase, and (4) reaction with vital dyes. It has been emphasized that more than one acid hydrolase should be detected before a subcellular body is called a lysosome.

That the occurrence of lysosomal-like structures in fungi can be regarded as the fungal equivalent of the lysosomes of animal tissue seems well established. Certain cellular inclusions in fungi have been noted which give a dense staining on preparation for electron microscopy, and the presence or absence of a single limiting membrane is difficult to discern. Thus, there is a problem in interpreting the relationship of such bodies to both microbodies and lysosomal-like particles.

Early concepts of the existence of fungal lysosomes are found in the studies by Matile and Wiemken [155], Pitt and Walker [180], Pitt [179], and Armentrout and Wilson [2]. These studies describe certain lysosomal-like bodies or particles in *Botrytis cinerea* in which acid phosphatase and certain other acid hydrolases could be localized by appropriate cytochemical means. Lysosomal-like bodies which react positively to the Gomori reaction for acid phosphatase, aryl sulfatase, and deoxyribonuclease have since been described in a number of fungi, including *Basidiobolus ranarum* [235]. It was reported that lysosomal like particles appeared most abundant in hyphal apices, and that their dimensions and movement were similar among those fungi studied. Lysosomal-like particles containing acid phosphatases have been reported from cells of *G. candidum* [186]. These particles stained with the vital dyes brilliant cresyl blue and neutral red. Pretreatment of the cells with procedures that injure the lysosomal membrane resulted in a uniform staining due to the release of the enzyme into the cytoplasm.

De Duve and Wattiaux [42] expanded the lysosomal concept to include most vacuolar components of the cell. They consider the vacuolar system, which contains hydrolytic enzymes, to constitute an intracellular digestive system comparable to the digestive tract

Figure 14 Microbodies and microbody-like organelles. (a) Thin section of a conidium of *Conidiobolus coronata* showing a nucleus (Nu), and the intimate association between microbody-like organelles (MLO) and lipid bodies (LB). Permanganate. Bar equals 0.5 μm. (b) Microbodies (Mb) and electron-opaque inclusions (I) in a hypha of *Basidiobolus meristosporus*. Glutaraldehyde-osmium tetroxide. Bar equals 1.0 μm.

of higher organisms. The electron cytochemical localization of high intracellular acid hydrolase activities known to occur in many filamentous fungi has not been adequately defined. In many fungi, vacuolar elements located in the cytoplasm are believed to represent a part of a complex lysosomal system which may involve the endoplasmic reticulum and chains of small interconnected vacuoles. It is becoming increasingly evident that such systems may be common among the filamentous, septate fungi, and perhaps the yeast-like fungi as well (see Sec. V).

Fungal lysosomes or their equivalents may function in a variety of cellular processes. If they are indeed similar to those in animals, the primary function of fungal lysosomes is the compartmentalization of acid hydrolytic enzymes involved in catabolism and mobilization of storage reserves [98,145]. It is also possible that they secrete extracellular enzymes [106]. In some plant pathogens, lysosomal-like particles or their equivalents may be involved in the penetration of plant host cells as they are in animals [185], and in fungal fructifications through lysis of nonspore materials with subsequent release of both sexual and asexual spores [235].

There seems little doubt that fungal lysosomes or their equivalents occur throughout widely separated taxa, and it is reasonable to assume that they may occur in many of those fungi pathogenic for the mammalian host. However, lysosomal-like bodies, particles, or complex vacuolar systems having lysosomal-like functions have not yet been adequately studied in the pathogens. Until such studies are available, it would be premature to speculate on possible relationships among their occurrence, their functions, and their possible significance in the invasion and subsequent parasitism of animal host tissue.

G. Lipid Bodies and Other Storage Inclusions

Lipids are an integral component of the cell and contribute much to the physiological characteristics of membranes. They are also found as storage material and occur in the fat deposits of multicellular organisms as well as in the form of small droplets in lower forms of life.

Lipid bodies are ubiquitous in the pathogenic, filamentous, and yeast-like fungi. Their size and number vary with the type of cell, but they are found mainly in conidia, mature or senescent vegetative cells, and function chiefly as storage organelles. They are surrounded by a unit membrane and their content is generally osmiophilic. The degree of blackening due to reduction of osmium tetroxide varies with the composition of the stored lipid [60]. When potassium permanganate is used as the primary or secondary fixative, the limiting membrane is sharply outlined. Because of the insolubility of permanganate in lipid, the lipid body content is electron transparent (Fig. 15a). It has been mentioned that certain cellular inclusions which have been termed lipid bodies or fat bodies may indeed be organelles such as lysosomal-like bodies, and that a clearer distinction between lipid inclusions in the cell and lysosomes is needed. Further, the role of lysosomal-like bodies in lipid

---

Figure 15 Lipid bodies and vacuoles. (a) Portion of a thin section of *Basidiobolus meristosporus* showing lipid bodies (LB) and their limiting membranes (LM). Permanganate. Bar equals 0.25 μm. (b) Portion of a thin section of a hypha of *Sporothrix schenckii* showing increased electron opacity of lipid bodies (LB) after treatment by the OTO method of Seligman et al. [201]. Bar equals 0.1 μm.

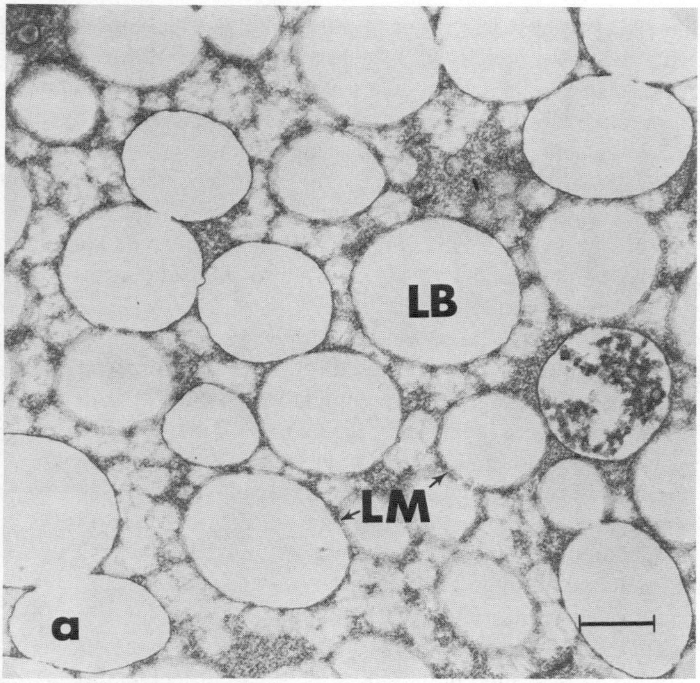
LB
LM
a

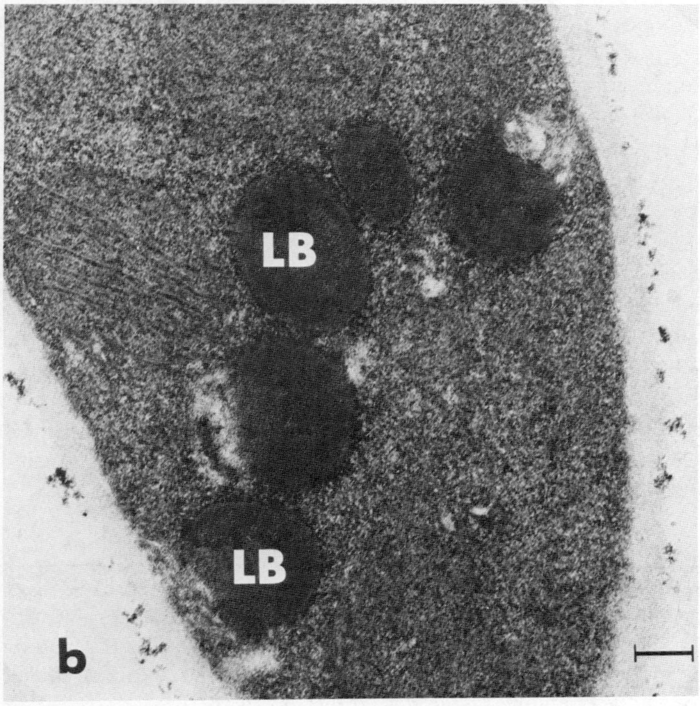
LB
LB
b

synthesis needs clarification [235]. The electron opacity of lipid inclusion bodies can be enhanced markedly following treatment of cells with the lipophilic reagent thiocarbohydrazide (Fig. 15b). Use of this osmiophilic reagent enhances the contrast of all components of the cell stained by osmium tetroxide during the initial fixation [201]. On treatment with an excess of thiocarbohydrazide, one end of the molecule attaches to the osmium bound in the tissue and, when this is followed by further exposure to osmium, more osmium is bound at these sites. Although all osmiophilic cellular components are increased in contrast, the greatest enhancement of contrast occurs in those cellular components holding the most osmium, i.e., lipid. This technique can be used effectively to delineate high lipid-containing inclusions in a variety of fungal tissue.

Little is known of the ontogeny of lipid bodies in the pathogenic fungi. However, at least two different modes of ontogeny have been proposed for other fungal species. In *Fusarium sulphureum* lipid inclusions result from an accumulation of oils intracisternally within the endoplasmic reticulum. Mature lipid bodies then detach and are surrounded by a unit membrane derived from the endoplasmic reticulum [198]. In *Claviceps purpurea* structures are observed which have been described as "lipid-forming" and "lipid-absorbing" bodies [228]. These structures appear in the chlamydospores of this fungus, and in some respects resemble osmiophilic structures reported in the hyphae and yeast-like cells of *S. schenckii* [76]. The latter observations are interesting inasmuch as lipid bodies of a number of pathogenic fungi occur in the near proximity of vacuolar areas containing electron-opaque material (see Sec. IV.E). In higher plants, structures termed "spherosomes" have been proposed as precursor organelles of lipid bodies in which the limiting membrane of the spherosome becomes that of the lipid body [63]. There is no convincing evidence to suggest that lipid bodies of any of the pathogenic fungi possess similar ontological relationships.

Little is known of the physiology of glycogen synthesis and storage systems of the pathogenic fungi, and the occurrence of glycogen storage materials in these organisms has received little attention [17]. Most of the information regarding glycogen storage and glycogen-membrane complexes comes from studies of such systems in mammalian cells [189,190,210]. Glycogen is a common storage product which can frequently be demonstrated in thin sections of various yeast-like and filamentous, pathogenic fungi. It is commonly encountered in nonapical regions of hyphae and in reproductive cells.

It is possible that more than one type of glycogen or glycogen-like materials can be stored by many of the pathogenic fungi. In some fungal cells, glycogen appears as small aggregates of increased electron opacity which measure about 15-30 nm in diameter (Fig. 16a). These small aggregates may form complexes with cytoplasmic membranes, and they are to be distinguished from rather similar ribonucleoprotein particles. In other cells, electron-translucent aggregates are often present in the cytoplasm [9]. These aggregates can measure up to 200 nm or more in diameter and are often seen along and immediately

Figure 16 Glycogen storage inclusions. (a) Thin section of a hypha of *Sporothrix schenckii* stained by the PATAg method of Thiéry [220]. Note the electron-opaque aggregates of glycogen-like material (G). Bar equals 0.25 μm. (b) Thin section of yeast-like *Histoplasma capsulatum* stained by the method of Stastná and Trávnik [207]. Note the increased electron opacity of the presumed glycogen-like material (G). Bar equals 0.5 μm.

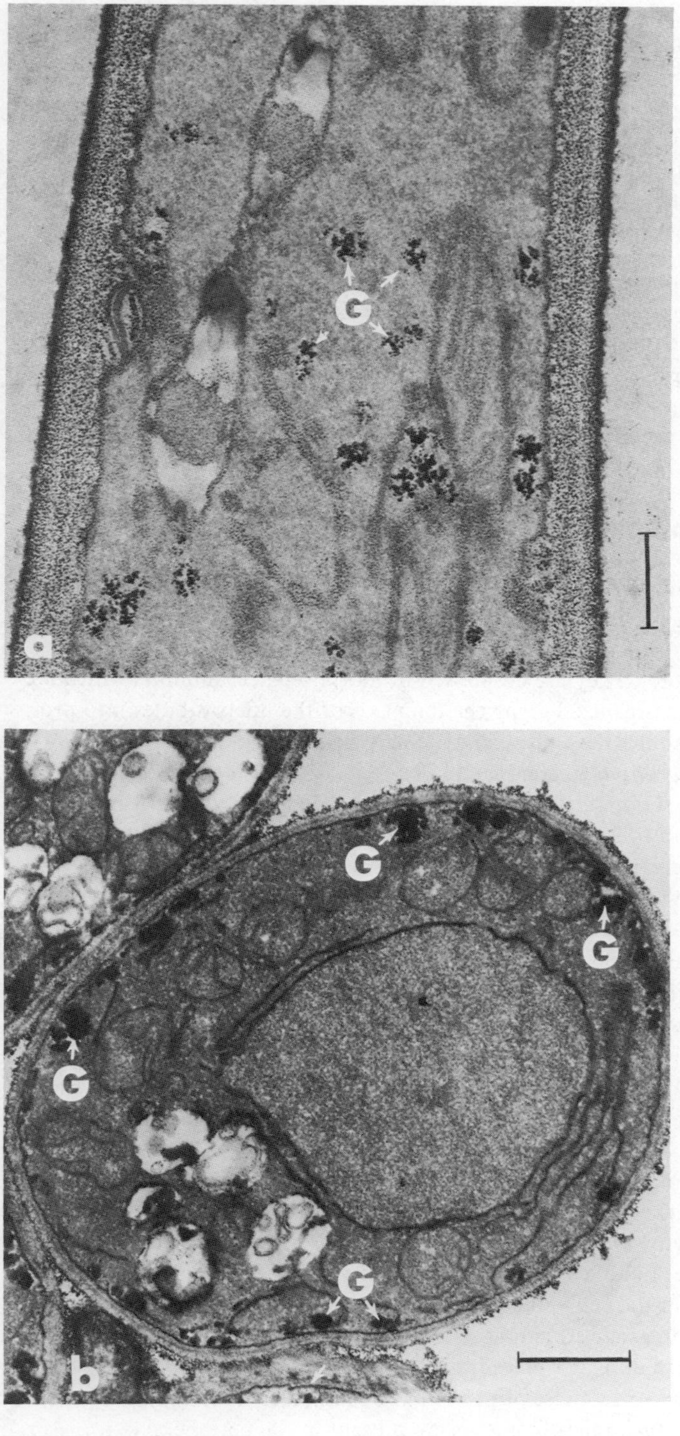
G
a
G
G
G
G
b

adjacent to the plasma membrane (Fig. 16b). In both cases, these aggregates of presumed glycogen or glycogen-like substances react strongly with electron cytochemical techniques which employ mild oxidation in dilute periodic acid and its subsequent reaction on vicinal glycol groups characteristic of polysaccharide macromolecules. The PATAg technique described by Thiéry [220] is especially useful in the identification of such periodic acid-reactive polysaccharides in thin sections of a variety of fungal organisms.

H. Microtubules and Microfilaments

Microtubules and microfilaments are recognized as components of virtually all eucaryotic cells. Their absence from the cytoplasm of any eucaryotic cell would be remarkable. Yet little information is available on the occurrence, characteristics, and distribution of these structures among the pathogenic fungi. Griseofulvin, which is used to treat a variety of dermatophytic infections, appears to interact with microtubules of certain of the pathogenic fungi, and it has been reported to inhibit specifically the mitotic spindle in various species of fungal organisms [120]. Microtubules and microfilaments have received wide attention in cells of higher plants and animals. Considerable work has been done on aspects of their structure, composition, organization, and function. A comprehensive review of the subject has been made by Hepler and Palevitz [100].

Cytoplasmic microtubules are demonstrated by using highly purified glutaraldehyde as the primary fixative. They are commonly recognized in electron micrographs as rigid, long structures having an outside diameter of about 200-250 Å with a densely stained cortex and lightly stained core. They are composed of morphological subunits which are about 40-50 Å wide, and are organized along the long axis of the microtubules into protofilaments [100]. Microtubules have been convincingly demonstrated in the nuclei of certain yeasts such as *Saccharomyces cerevisiae* [159,177], *Schizosaccharomyces pombe* [104], and certain heterobasidiomycetous yeasts, including *Rhodotorula glutinis* [143,144]. Cytoplasmic microtubules up to 3.0 μm in length have been seen within the cytoplasm of glutaraldehyde-fixed, regenerating protoplasts of *S. pombe,* and are located close to either the nuclear membrane or to mitochondria [104]. In certain fungi of the genus *Uromyces,* microtubules are found to be associated with mitochondria more frequently than expected by chance, and it is tempting to suggest that microtubules are involved in the positioning and slow movement of this organelle [97]. Cytoplasmic microtubules do not appear to be associated with any other organelle in yeast nor do they lie under the plasma membrane as is often the case in higher plants [132]. Microtubules appear to have various roles in the fungi. They have been implicated in the movement of vesicles or other protoplasmic movements, intracellular transport, maintainance of cell shape, wall formation, and nuclear division [12].

Microfilaments occur commonly throughout eucaryotic cells, including lower and higher plants. At least in higher plants, they measure about 50-70 Å in diameter and occur in bundles about 0.1 μm wide and at least several micrometers in length (Fig. 17). The number of filaments per bundle varies from only a few to perhaps more than 20. Several bundles may occur in the cell and are usually oriented parallel to the longitudinal axis or the direction of cytoplasmic streaming. In the fungi, microfilaments may have a larger diameter than that seen in higher plants and may at times reach 200 Å. Zachariah [238] first described bundles or fascicles of parallel 200-Å fibrillar structures in filamentous fungi. In *Ascobolus stercorarius,* microfilaments are present in the peripheral cytoplasm of

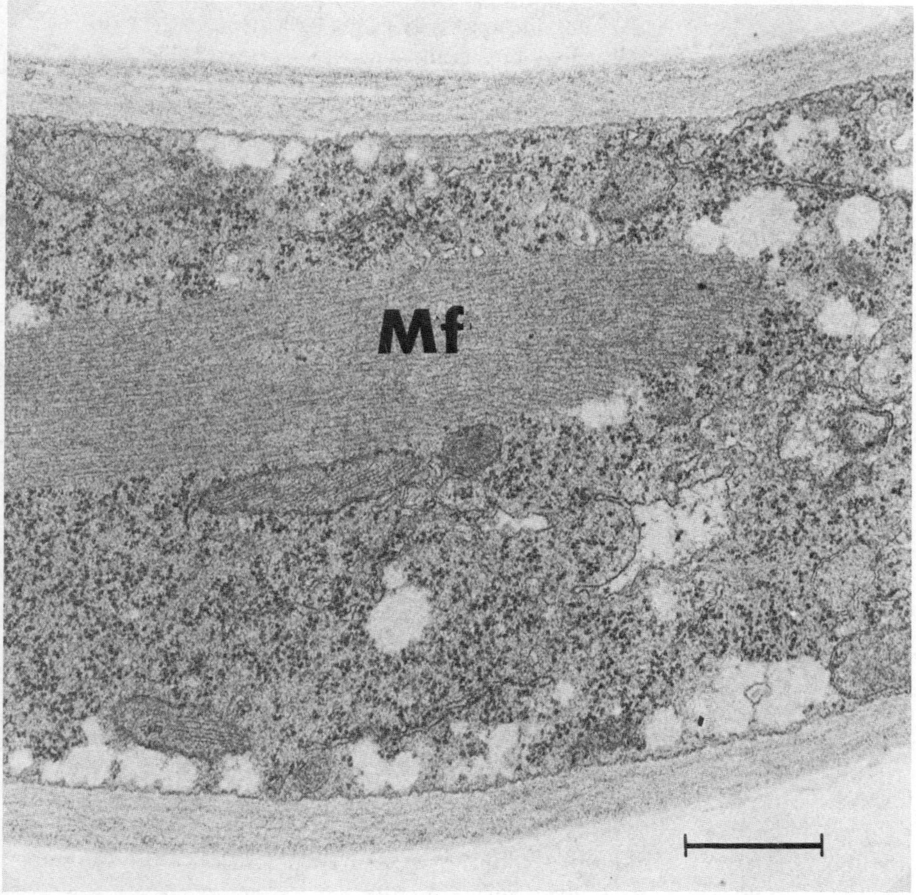

Figure 17 Thin section of a hypha of an unidentified basidiomycete showing a fascicle of microfilaments (Mf). Glutaraldehyde-osmium tetroxide. Bar equals 0.5 μm.

nonsporogenous, apothecial cells. These fascicles may reach 3-6 μm in length and appear in a more or less parallel manner. In *Drechslera sorokiniana,* microfilaments are present in conidiophores and appear to be involved in active cyclosis [32]. In the fungi, microfilaments seem to be associated with movements related to cytoplasmic streaming and in the direction of organellar migration within the cytoplasm.

Although fascicles of microfilaments can be distinguished in thin sections of fungi fixed in glutaraldehyde, their common occurrence in pathogenic fungi does not appear to have been described or otherwise noted. It is difficult to account for this fact, but it is reasonable to accept their probably occurrence within the cytoplasm of these fungi at least during some stage of their growth.

I. Fungal Ribosomes

Ribosomes are 15-25 nm masses of nucleoprotein which are found in the cytoplasm or associated with endoplasmic reticulum and the nuclear envelope. In all higher eucaryotic organisms, the ribosomes are contained not only in the basic mass of the cytoplasm and nucleus, but may be found in mitochondria and chloroplasts. After aldehyde-osmium tetroxide fixation schedules, ribosomes of the pathogenic fungi are seen under the electron microscope as electron-opaque, roundish granules. As mentioned previously, fixation of cells with permanganate results in the destruction of ribosomes.

Study of isolated ribosomal preparations permits a more accurate determination of the size, shape, and function [66,78,236]. The average value of the sedimentation coefficient for the ribosomes from 26 species of fungi was found to be 81S, with a range of variation from 79 to 85S [217].

Unfixed cytoplasmic ribosomes [77S) of *G. candidum* have been isolated and studied by electron microscopy. Negatively stained ribosomal monomers could be resolved into three images. Two of these are believed to be identical with the reported images of rat liver ribosomes. The dimensions of *G. candidum* monomer ribosomes were reported to be 290 × 270 Å [203]. In a study of the immunogenicity of isolated ribosomes from yeast-like cells of *H. capsulatum*, Feit and Tewari [61] reported a sedimentation coefficient of 77S from studies in which the ribosomes of *Saccharomyces cerevisiae* were used as an 80S marker.

Purified cytoplasmic ribosomal monomers, isolated from each of the two growth phases of the dimorphic *H. capsulatum* and *Blastomyces dermatitidis*, have been examined by bright- and dark-field electron microscopy [66]. Examples of such monomers are shown in Fig. 18. Cytoplasmic ribosomal fractions from the yeast-like and mycelial phases of these dimorphic fungi contain 760 μg RNA per milligram of proteins. The maximum dimensions of projections of glutaraldehyde-fixed, unstained monomers in dark-field electron microscopy for each of the four preparations varied over the same range, 241-370 Å, with means of 310-320 Å and a standard deviation of the means in the range 35-50 Å. A t-test applied to these data reveals no significant differences in maximum dimensions of monosomes from the four sources at the 0.5 significance level. The subunit configuration of all monosomes, as visualized in positive and negative stain, was similar quantitatively to that reported for other species of fungi. Suggestive interspecies and interphasic size differences could be observed among positively and negatively stained preparations visualized in bright-field electron microscopy. However, since the exact nature of structural alterations induced by heavy metal is not understood, any differences seen in stained monosomes must be interpreted cautiously.

J. Golgi Apparatus

The Golgi apparatus, consisting of interassociated dictyosomes, is a widespread component of eucaryotic cells. The dictyosome appears as a stack of smooth cisternal membranes which often possess tubular peripheries and exhibit polarity (Fig. 19). Vesicles may be attached or adjacent and the dictyosome may be associated with endoplasmic reticulum or nuclear envelopes.

Recognizable dictyosomes have been demonstrated in only a few fungal organisms, and comparable structures are not encountered as characteristic subcellular components of the fungal pathogen. The yeast-like cell of *B. dermatitidis* is one of the most cytologically complex of the pathogenic fungi yet studied by electron microscopy [53]. The Golgi apparatus in this fungus is believed to be represented by a small multivesicular component. This structure consists of numerous, small membrane-limited vesicles contained

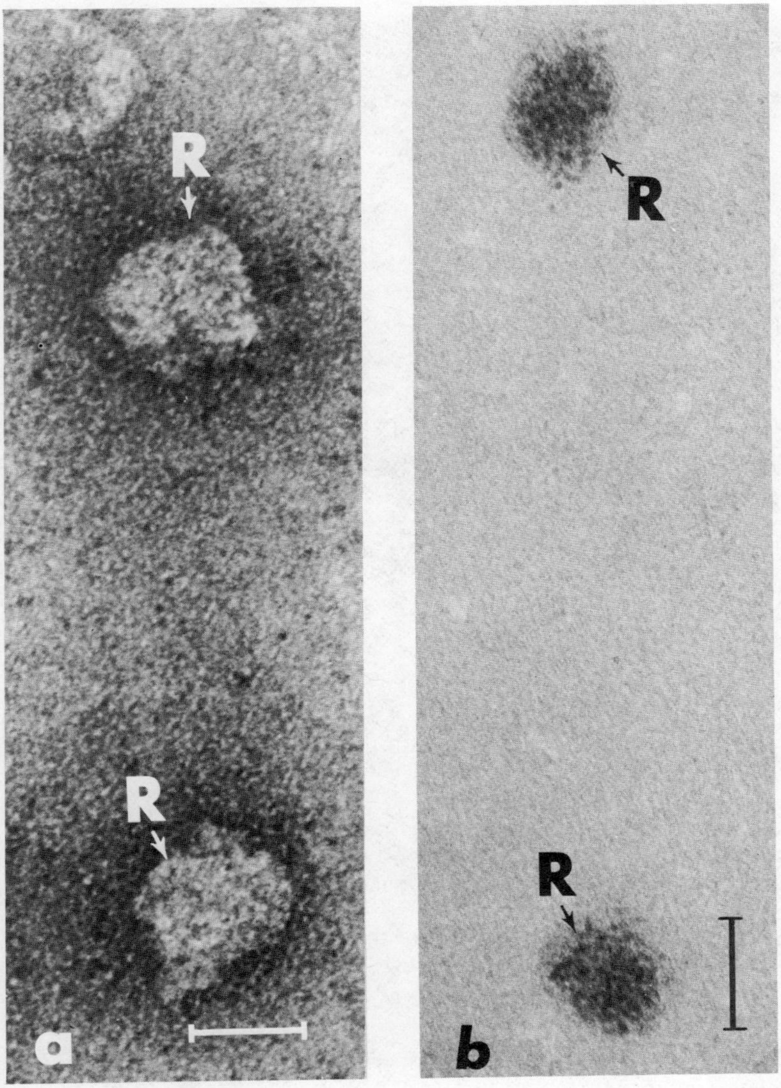

Figure 18 Fungal ribosomes. (a) Negatively stained ribosomes (R) from the mycelial phase of *Histoplasma capsulatum*. Bar equals 20 nm. (b) Positively stained ribosomes (R) from the mycelial phase of *H. capsulatum*. Bar equals 20 nm.

within a large vesicle limited by a single membrane. Clusters of smooth cisternae which may functionally constitute the Golgi apparatus have been identified in relatively ribosome-free regions of the cytoplasm of *Phialophora dermatitidis* [87]. Thus, vesicular bodies, specialized tubules, or cisternae may be either functional equivalents or otherwise primitive forms of the more highly structured Golgi apparatus. Similar structures may well be present in other fungal pathogens, but have not been recognized or identified as possible Golgi apparatus equivalents. Careful study of thin sections of well-fixed fungal cells may

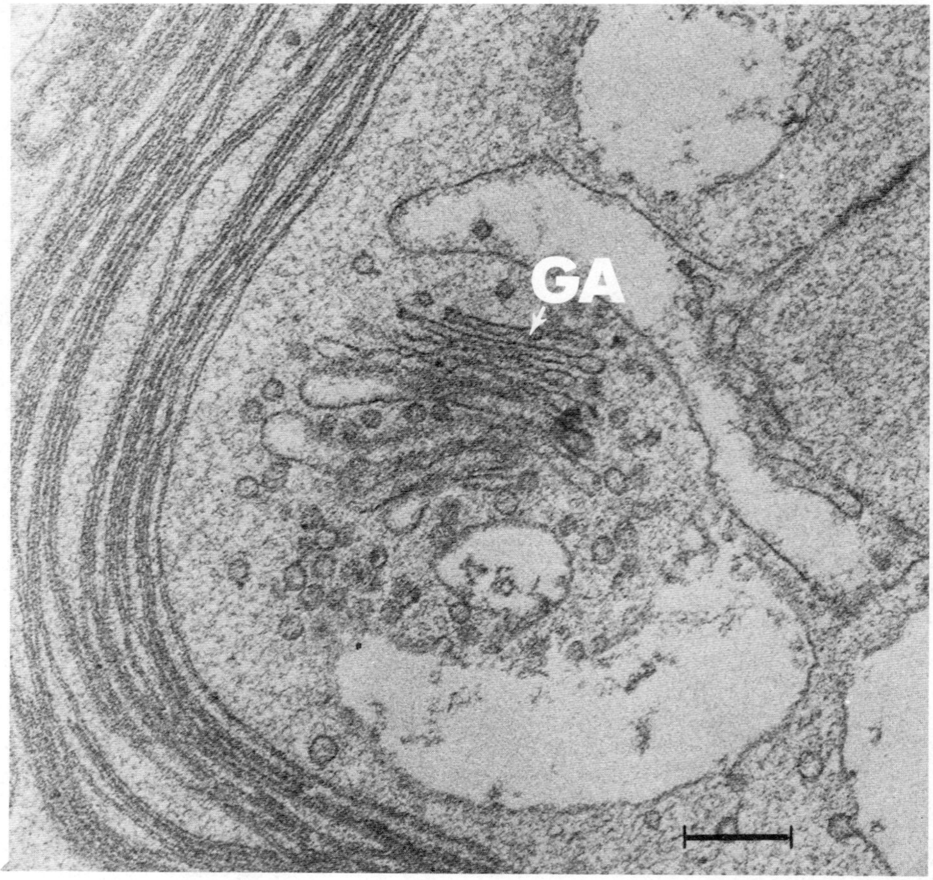

Figure 19 Portion of a thin section of an algal cell of the genus *Chlamydomonas* showing a Golgi apparatus (GA). Permanganate. Bar equals 0.25 μm.

ultimately show the widespread occurrence of such specialized subcellular components throughout the pathogenic fungi.

V. ELECTRON CYTOCHEMISTRY AND FUNGAL CYTOLOGY

Although mycologists can learn much from purely morphological descriptions, knowledge of structure-function relationships is of great potential value. An increasing body of knowledge is accumulating on such relationships in the fungi by use of electron cytochemical localization of enzymes and cytological identification of structural components. A wide range of biological applications is already possible and rapid advances are being made in existing methods and development of new techniques. Enzyme histochemistry has become much more scientifically acceptable to biologists in general.

Reliable staining methods are currently available for the localization of at least two important classes of enzyme activities: phosphatases and haemoproteins. In the latter case, such enzymes as peroxidase and catalase can be demonstrated by using 3,3′-diaminobenzidine as substrate. Methods based on the use of 3,3′-diaminobenzidine have found wide application and yield localizations of high cytological resolution. Phosphatase activities have proven to be useful markers for certain cell organelles. Acid phosphatase, for example, is a useful marker for lysosomes and for specialized regions of Golgi-associated smooth endoplasmic reticulum that is involved in the formation of lysosomes. Alkaline phosphatase can serve as a marker for membranes that contain high levels of this enzyme. Nucleoside phosphatase activities can be demonstrated in cytomembranes with various nucleoside mono-, di-, and triphosphates as substrate and manganese or magnesium as the activating ion. Other phosphatase activities that are useful as markers for endoplasmic reticulum and the nuclear envelope are nucleoside diphosphatase and glucose-6-phosphatase. General discussions of the electron microscopical localization of enzymes can be found in the reviews by Lewis [134] and Snitka and Seligman [206]. Reiss [187] has reviewed the applicability of certain cytochemical techniques as well as the distribution of the enzymes as established by these techniques in cytological components of fungi.

The results of many studies indicate that cytochemical methods, which have been developed for the study of mammalian histology and cytology, can be applied with certain modifications to investigations of the pathogenic fungi. However, cytochemical investigations of whole cell preparations may be complicated by the problem of permeation of chemical fixatives through the cell wall and plasma membrane. In other cases, substrates or cytochemical reactants may be excluded. To this end, Borgers and De Nollin [11] introduced a technique for the preservation of subcellular organelles of *Candida albicans* by sectioning solidified pellets of the cells in the presence of cold glutaraldehyde. This method permitted the demonstration of a number of enzymatic activities in various substructures typically found in this pathogenic yeast [44]. The distribution of most specific and nonspecific phosphatases and of peroxidases were at variance with that found in mammalian cells. Activities toward a variety of substrates, including β-glycerophosphate, p-nitrophenylphosphate, adenosine monophosphate, adenosine triphosphate, thiamin pyrophosphate, and glucose-6-phosphate were almost exclusively confined to the central vacuolar apparatus. Oxidative and peroxidative activities were demonstrated only in mitochondria. The specific marker enzymes for endoplasmic reticulum, the plasma membrane, Golgi apparatus, and peroxisomes were not found. In another study, Pugh and Cawson [184] described the cytochemical localization of phospholipase A and lysophospholipase as being present in developing buds and in the cell membrane and walls of nondividing cells of *C. albicans.* These studies serve to illustrate the utility of electron cytochemical localization of a variety of enzymatic activities in a single fungal organism.

Electron cytochemical localization of enzymes has a direct application to the study of fungal cytology and may ultimately provide a means for the identification of certain cell organelles or specialized regions of the cytoplasm unattainable by other methods. Cytochemical localization of certain acid hydrolases indicates that the yeast-like and mycelial phase cells of *H. capsulatum* possess acid phosphatase-reactive areas within the cytoplasm which might be equated with lysosomal-like vesicles [66]. The Gomori method with β-glycerophosphate as substrate has been used to display opaque deposits, corresponding to the acid phosphatase reaction product, in the nucleoplasm, along the nuclear membrane, and in discrete, somewhat irregular vesicular areas within the cytoplasm of the yeast-like cell (Fig. 20a). No acid phosphatase activity is associated with mitochondria, endoplasmic reticulum, the plasma membrane, or cell wall. Similar vesicles having acid

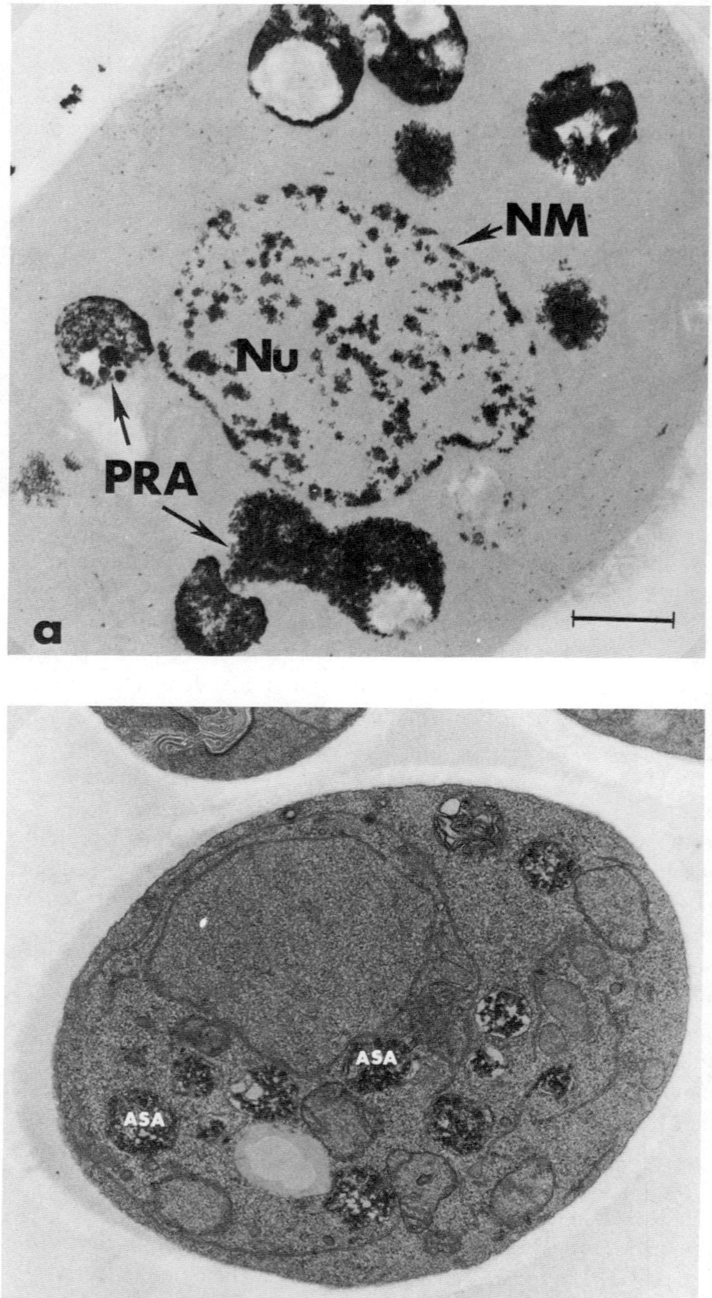
NM
Nu
PRA
a
ASA
ASA
b

phosphatase reactivity are present in the cytoplasm of the hyphae of *H. capsulatum* (Fig. 21a). These acid phosphatase-reactive areas could be demonstrated when di(dicyclohexyl)-ammonium-2-naphthyl phosphate (a synthetic substrate) was employed. Vesicles of similar size, shape, and distribution within the yeast-like and hyphal cell cytoplasm contain aryl sulfatase when p-nitrocatechol is used as substrate (Fig. 20b). The cytochemical localization of reaction product of at least two acid hydrolases within these regions suggests their possible identity as fungal lysosomal equivalents. Figure 21b illustrates the use of 3,3′-diaminobenzidine in the localization of cytochrome oxidase on the mitochondrial membranes of yeast-like *H. capsulatum.* These observations serve to illustrate the specific application of electron cytochemical procedures to the identification of function or specialized regions in the fungal cell.

The application of x-ray microanalysis to fungal cytochemistry and cytology is still in its infancy. Electron probe microanalysis provides a unique alternative method of cytochemical localization as determined by conventional histological or cytochemical studies of fungal tissue. High-resolution electron microscopy combined with energy-dispersive x-ray from either scanning or scanning transmission electron microscopes can detect in the same specimen all elements having atomic weights from 5 (boron) to 92 (uranium). Thus, the technique provides an approach to the study of structure-function relationships heretofore unattainable by any other means. Thibaut et al. [219] have applied x-ray microanalysis in conjunction with scanning electron microscopy to detect certain elements in preparations of *Coccidioides immitis, Paracoccidioides,* and *Trichosporon cutaneum.* Their results indicated the presence of calcium, chlorine, sulfur, and phosphorus. The common characteristic of these three species was their richness in both calcium and phosphorus.

There is little doubt that future investigations of the cytological components of the pathogenic fungi will profit greatly by application of available electron cytochemical techniques to the study of structure-function relationships. Indeed, the biochemical pathways which regulate growth and differentiation in the fungal cell may not be synchronous in a given cell population. Thus, observation at the ultrastructural level of cytochemically demonstrable events associated with morphogenesis may well provide insights not otherwise obtainable.

APPENDIX

The following series of transmission electron micrographs (Figs. 22-69) illustrate a variety of conidial and vegetative cells of the pathogenic fungi as seen by techniques of electron microscopy. The electron micrographs include examples of thin sections of representative pathogenic fungi or fungal opportunists from the Eumycotina.

Figure 20 Electron cytochemical localization. (a) Thin section of yeast-like *Histoplasma capsulatum* showing localization of acid phosphatase using β-glycerophosphate as substrate. Note the presence of Gomori reaction product in the nucleoplasm (Nu), along the nuclear membrane (NM), and in phosphatase-reactive areas (PRA). These vacuolated areas may represent fungal lysosomal equivalents. Bar equals 0.25 μm. (b) Thin section of yeast-like *H. capsulatum* showing the localization of aryl sulfatase using p-nitrocatechol as substrate. Note the localization of reaction product (ASA) in areas similar to that seen for acid phosphatase. Bar equals 0.25 μm.

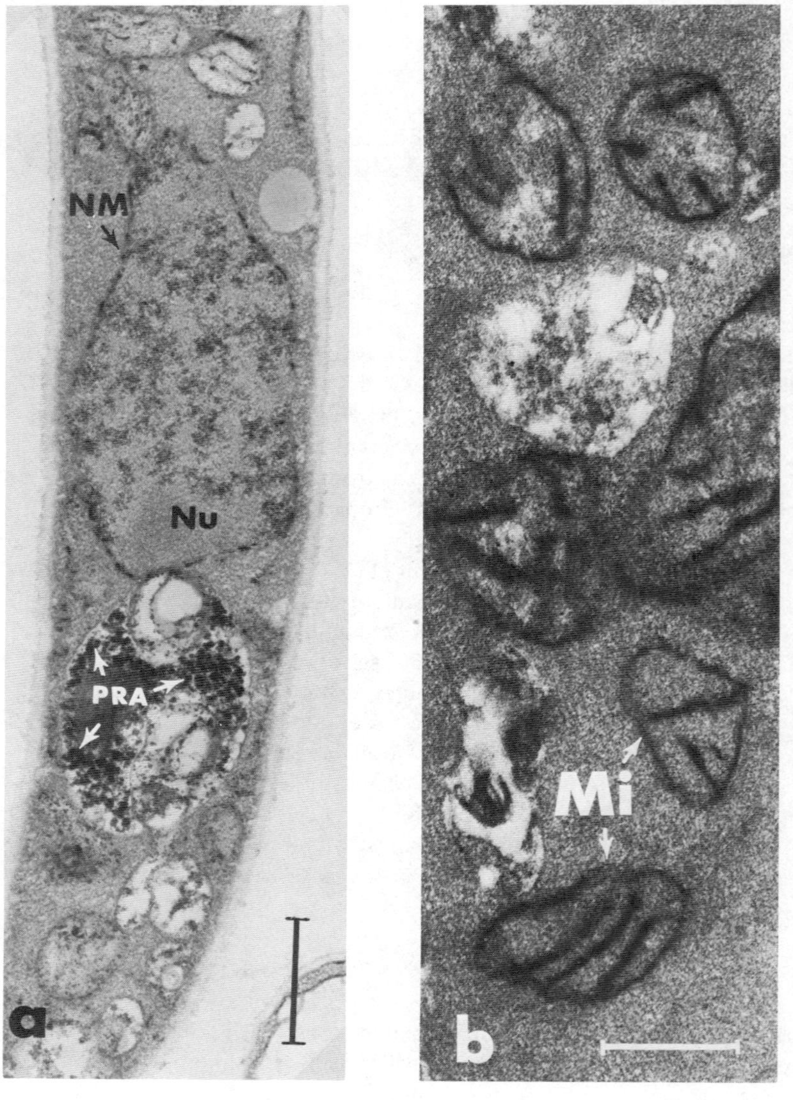

Figure 21 Electron cytochemical localization. (a) Longitudinal thin section of a hypha of *Histoplasma capsulatum* showing localization of acid phosphatase using β-glycerophosphate as substrate. Gomori reaction product is present in the nucleoplasm (Nu), on the nuclear membrane (NM), and in vacuolated areas (PRA). Bar equals 0.5 μm. (b) Thin section of a portion of yeast-like *H. capsulatum* showing the localization of cytochrome oxidase using the 3,3′-diaminobenzidine technique. Note the distribution of reaction product along the mitochondrial membranes (Mi). Bar equals 0.25 μm.

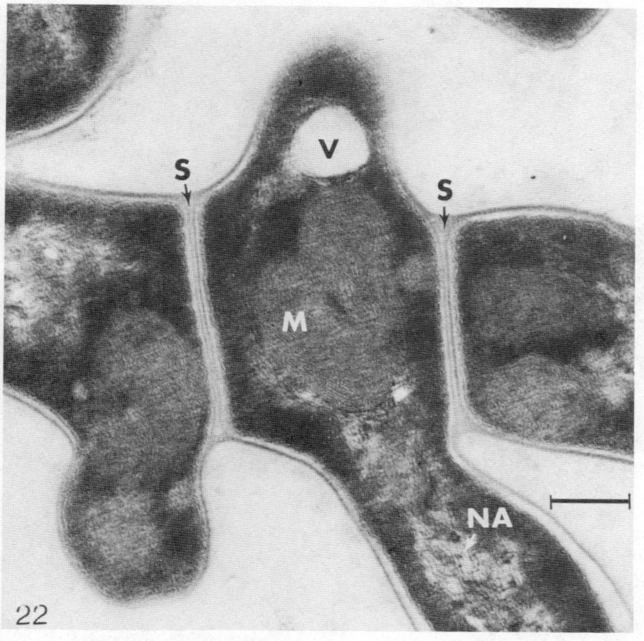

Figure 22 *Nocardia asteroides.* Longitudinal thin section showing branching cells, mesosomes (M), nuclear area (NA), vacuole (V), and septa (S). Glutaraldehyde-osmium tetroxide. Bar equals 0.25 μm.

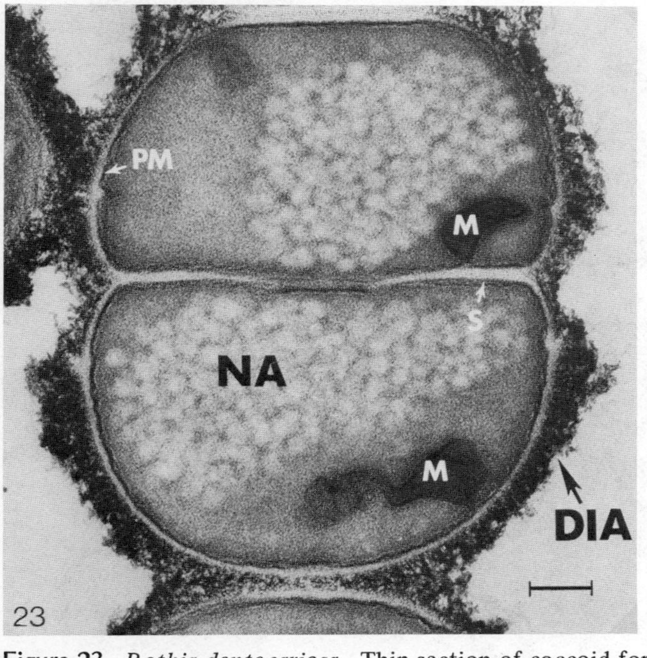

Figure 23 *Rothia dentocariosa.* Thin section of coccoid forms showing the cytological organization of a procaryote. Note the nuclear area (NA), mesosomes (M), septum (S), plasma membrane (PM), and dialyzed iron activity (DIA) at the outer wall surface (binding of dialyzed iron by acid mucosubstances). Glutaraldehyde-dialyzed iron-osmium tetroxide. Bar equals 0.1 μm.

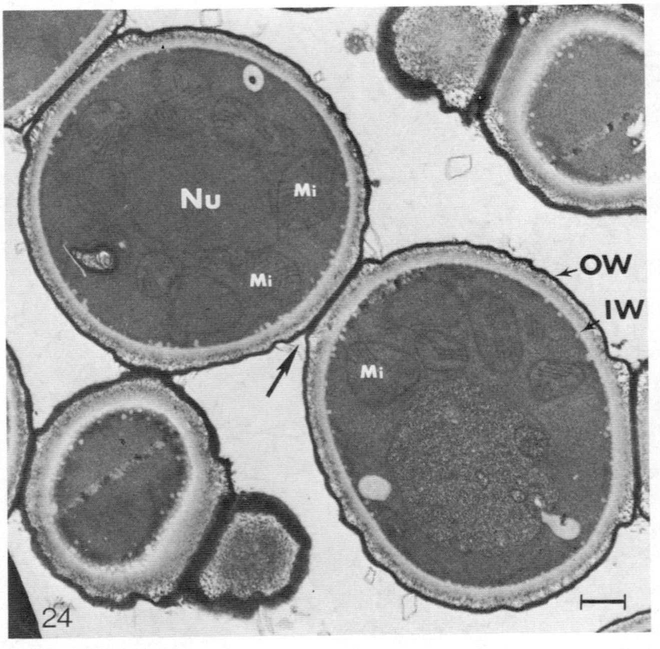

Figure 24 *Mucor pusillus.* Thin sections of sporangiospores showing scattered mitochondria (Mi), nucleus (Nu), inner cell wall (IW), and an electron-opaque outer cell wall (OW). Note the wall connections as marked by the arrow. Glutaraldehyde-osmium tetroxide. Bar equals 0.20 μm.

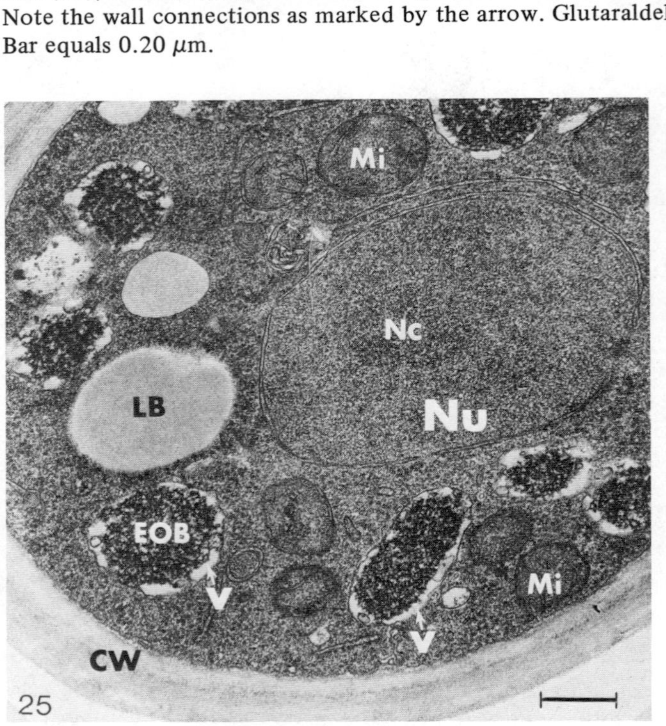

Figure 25 *Basidiobolus meristosporus.* Thin section of portions of a conidium showing a nucleus (Nu) with nucleolus (Nc), mitochondria (Mi), lipid body (LB), and vacuoles (V) containing electron-opaque bodies (EOB). Glutaraldehyde-osmium tetroxide. Bar equals 0.3 μm.

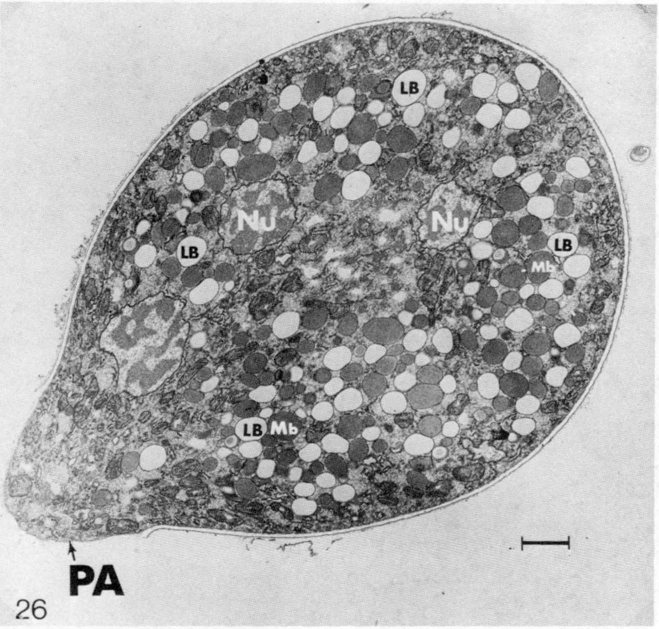

Figure 26 *Conidiobolus coronata.* Thin section of a conidium showing multiple nuclei (Nu), lipid bodies (LB), microbody-like organelles (Mb), and papillar area (PA). Permanganate. Bar equals 1.0 μm.

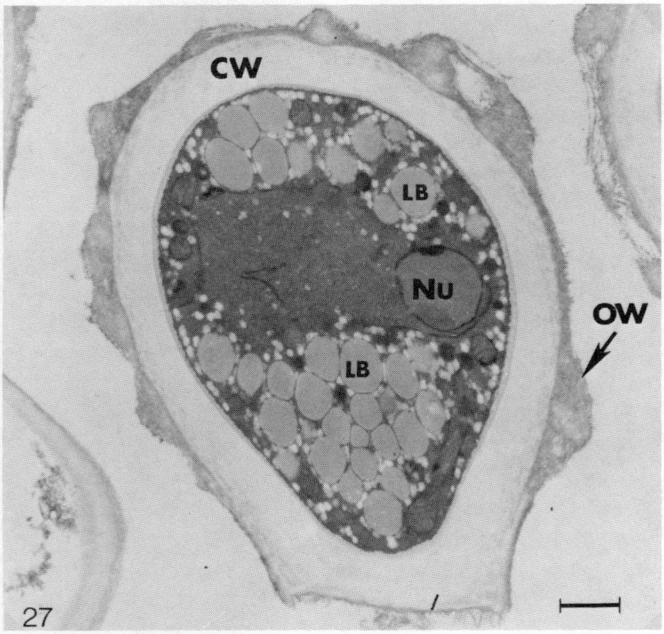

Figure 27 *Scopulariopsis brevicaulis.* Thin section of a conidium showing lipid bodies (LB), nucleus (Nu), thickened cell wall (CW), and sculpturing of the outer wall (OW). Glutaraldehyde-osmium tetroxide. Bar equals 0.5 μm.

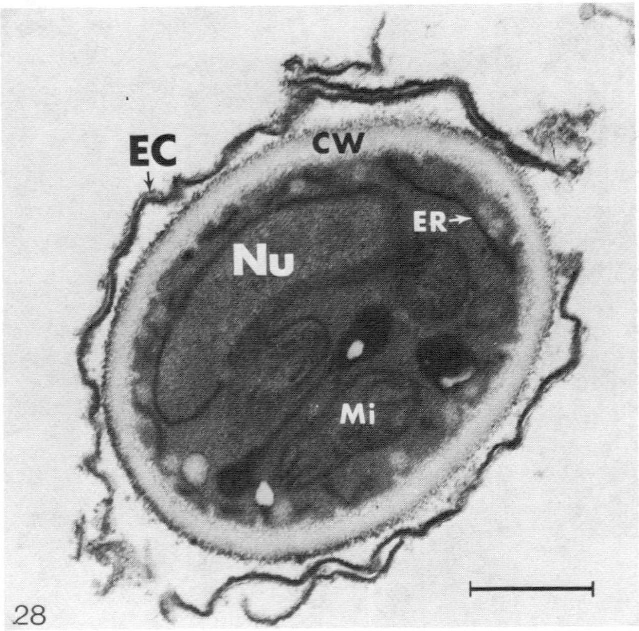

Figure 28 *Penicillium* sp. Thin section of a conidium showing cell wall (CW) and exocoat (EC), nucleus (Nu), endoplasmic reticulum (ER), and mitochondria (Mi). Glutaraldehyde-permanganate. Bar equals 0.5 μm.

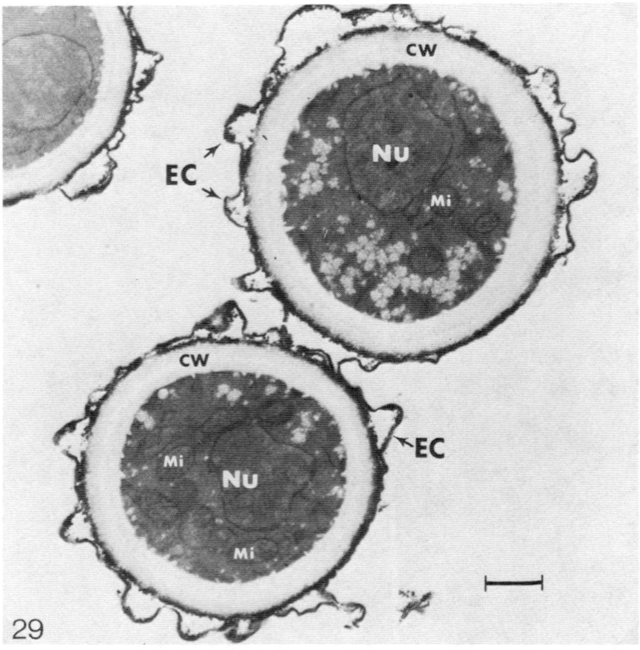

Figure 29 *Aspergillus* sp. Thin section of conidia showing nuclei (Nu), mitochondria (Mi), cell wall (CW), and exocoat (EC). Glutaraldehyde-osmium tetroxide. Bar equals 0.5 μm.

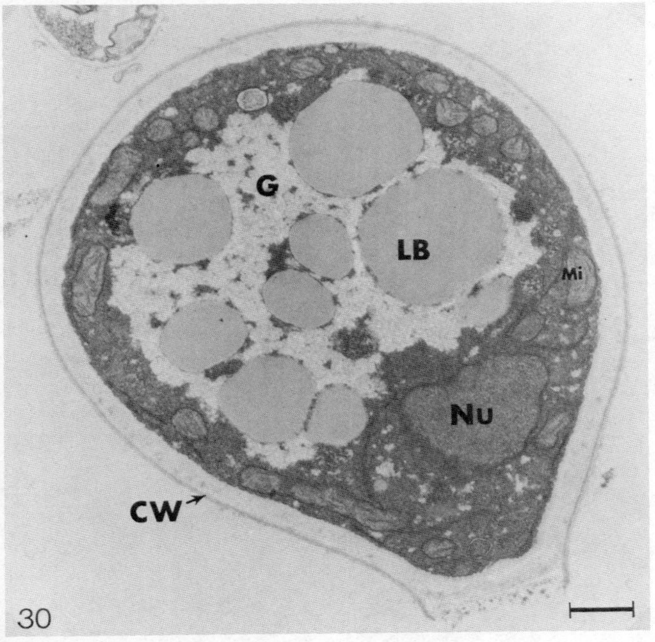

Figure 30 *Scedosporium apiospermum (Petriellidium boydii).* Thin section of a conidium showing lipid bodies (LB), glycogen-like material (G), mitochondria (MI), nucleus (Nu), and cell wall (CW). Glutaraldehyde-osmium tetroxide. Bar equals 0.5 μm.

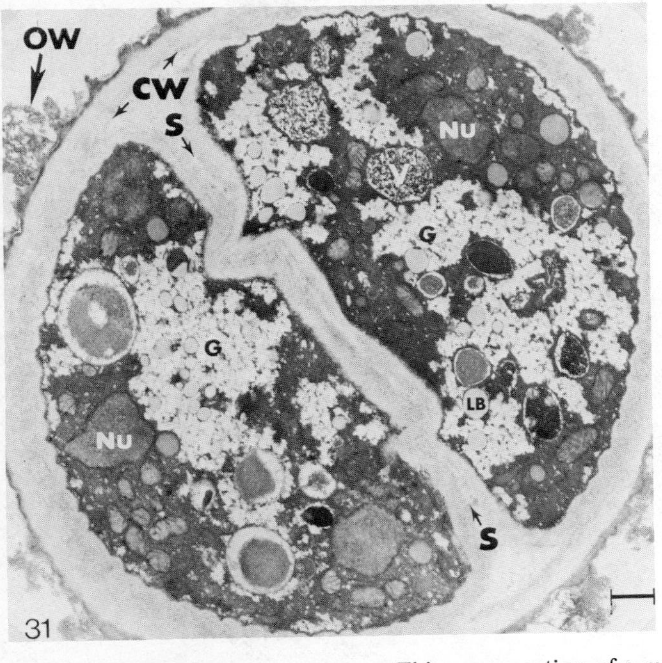

Figure 31 *Microsporum gypseum.* Thin cross section of a macroconidium showing cell wall (CW), electron-opaque, irregular outer wall (OW), septum (S), nuclei (Nu), vacuoles (V), lipid bodies (LB), and glycogen-like material (G). Glutaraldehyde-osmium tetroxide. Bar equals 0.5 μm.

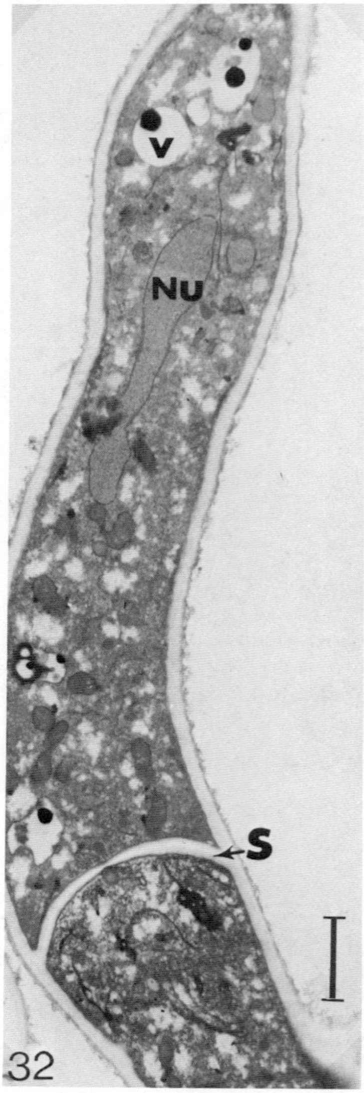

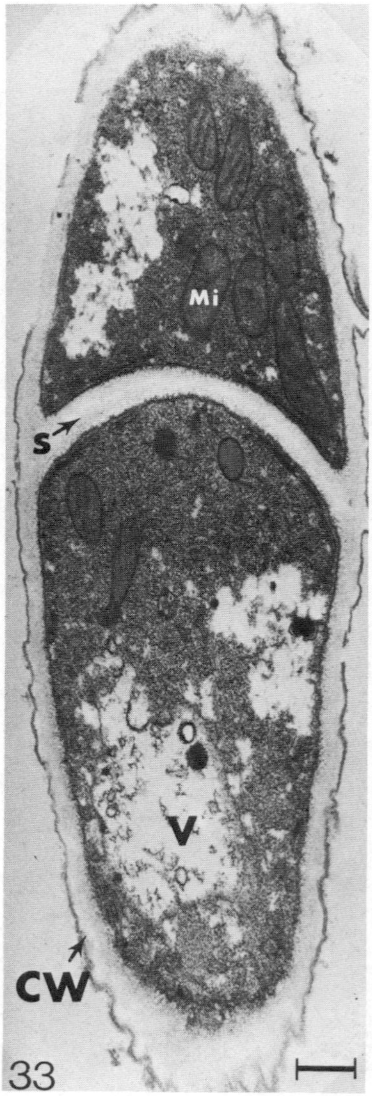

Figure 32 *Trichophyton rubrum.* Longitudinal thin section of a macroconidium showing a septum (S), nucleus (Nu), and vacuoles (V). Glutaraldehyde-osmium tetroxide. Bar equals 1.0 μm.

Figure 33 *Epidermophyton floccosum.* Longitudinal thin section of a macroconidium showing a roughened cell wall (CW), mitochondria (Mi), septum (S), and vacuole (V). Glutaraldehyde-osmium tetroxide. Bar equals 0.25 μm.

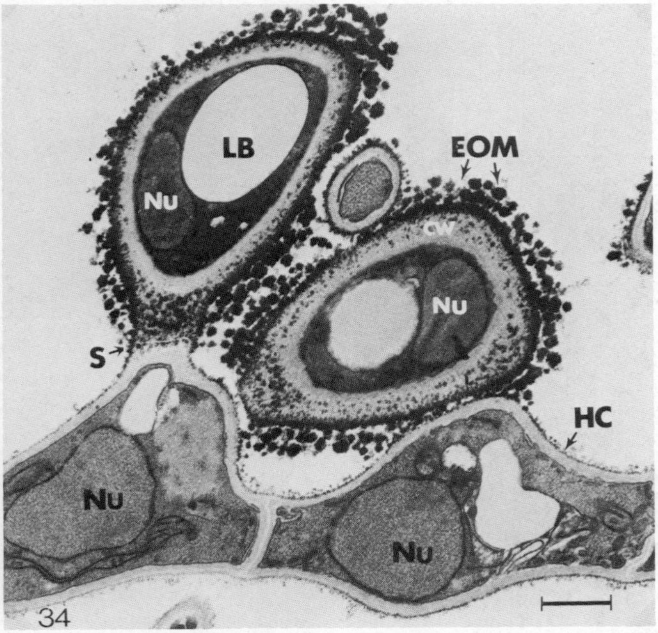

Figure 34 *Sporothrix schenckii.* Thin section of triangular, pigmented macroconidia showing lipid bodies (LB), nuclei (Nu), and portions of the supporting hypha (HC). Note the presence of electron-opaque material (EOM) at the outermost wall surface (CW) of the conidia and its absence immediately at the septum (S) of the sterigmata. Glutaraldehyde-permanganate. Bar equals 0.5 μm. (From Ref. 148.)

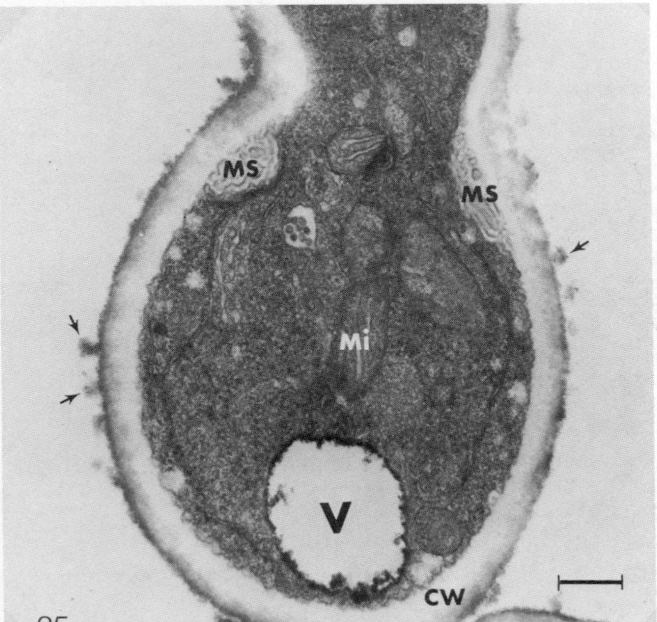

Figure 35 *Fonsecaea pedrosoi.* Thin section of a germinating conidium showing mitochondria (Mi), a large vacuole (V), and membrane systems (MS) near the origin of the germ tube. Note the electron-opaque granules (arrows) at the outermost cell wall (CW) surface. Glutaraldehyde-osmium tetroxide. Bar equals 0.25 μm.

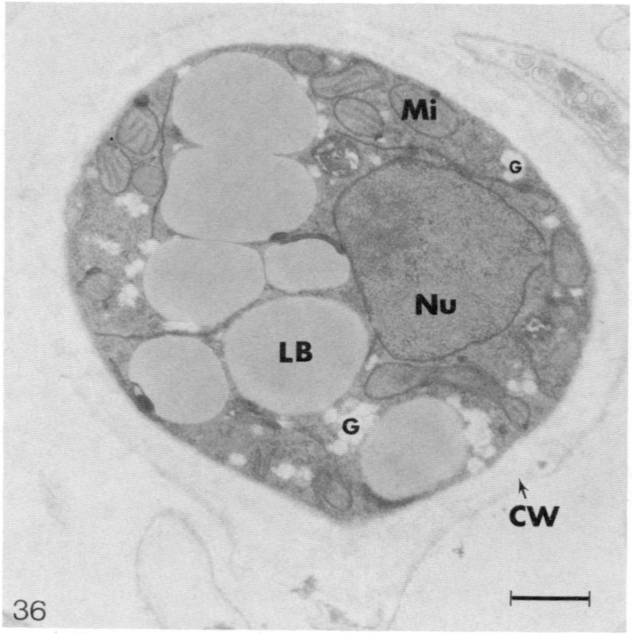

Figure 36 *Blastomyces dermatitidis.* Thin section of a conidium showing lipid bodies (LB), a single nucleus (Nu), mitochondria (Mi), glycogen-like material (G), and cell wall (CW). Glutaraldehyde-osmium tetroxide. Bar equals 0.5 μm.

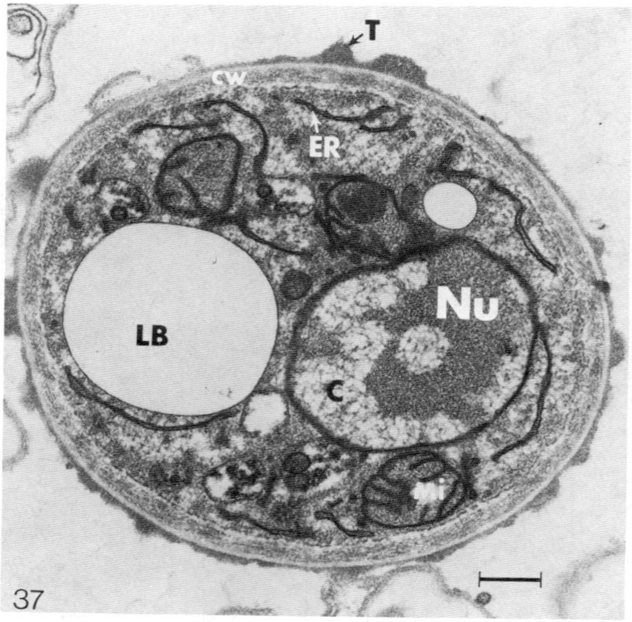

Figure 37 *Histoplasma capsulatum.* Thin section of a microconidium showing lipid bodies (LB), mitochondria (Mi), endoplasmic reticulum (ER), nucleus (Nu) with chromatin material (C), and tubercles (T) along the cell wall (CW). Permanganate. Bar equals 0.25 μm. (From Ref. 69.)

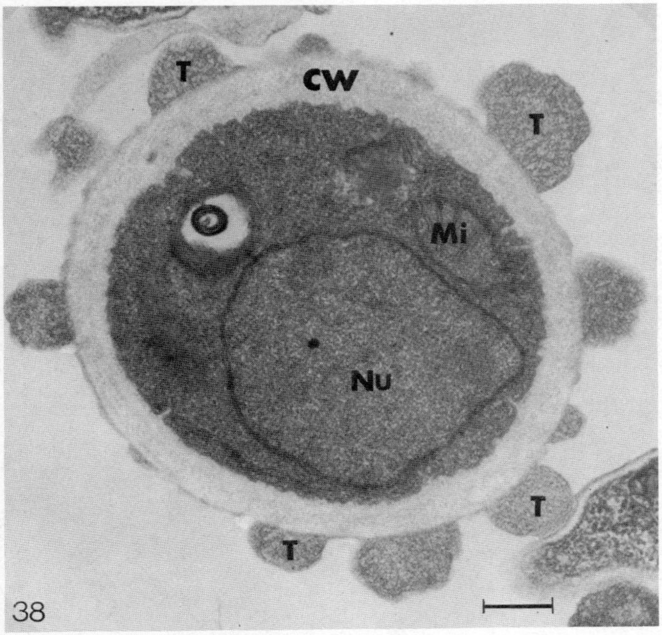

Figure 38 *Histoplasma capsulatum.* Thin section of a microconidium showing a mitochondrion (Mi), nucleus (Nu), and tubercles (T) at the outer portion of the cell wall (CW). Glutaraldehyde-osmium tetroxide. Bar equals 0.5 μm.

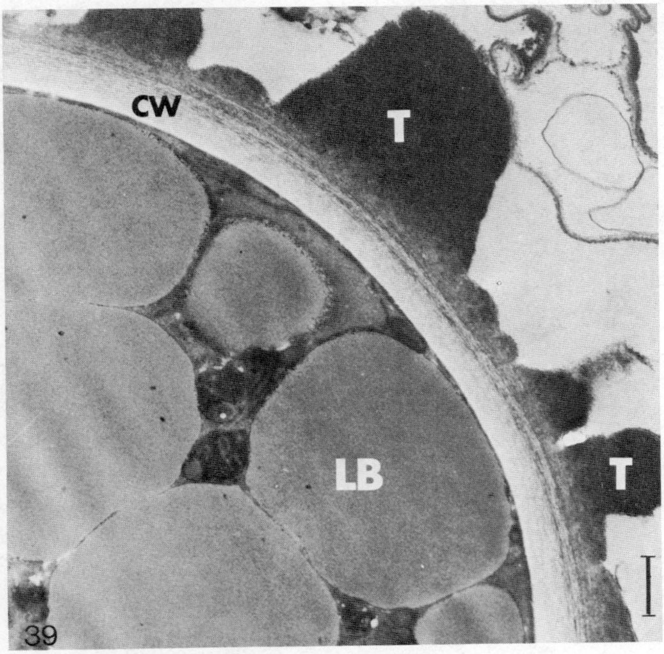

Figure 39 *Histoplasma capsulatum.* Portion of a thin section of a macroconidium showing multiple lipid bodies (LB) and tubercles (T) at the outermost surface of the laminated cell wall (CW). Glutaraldehyde-osmium tetroxide. Bar equals 0.5 μm.

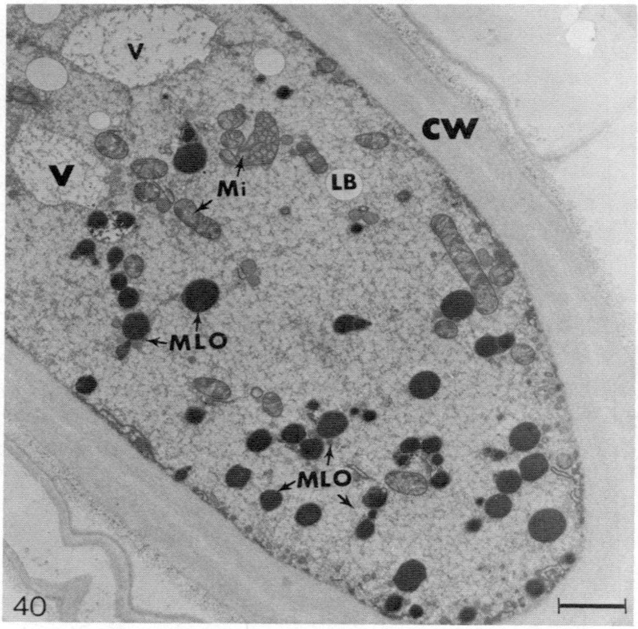

Figure 40 *Basidiobolus meristosporus.* Longitudinal thin section of a hypha showing lipid bodies (LB), vacuoles (V), mitochondria (Mi), microbody-like organelles (MLO), and a thickened cell wall (CW). Glutaraldehyde-osmium tetroxide. Bar equals 1.0 μm.

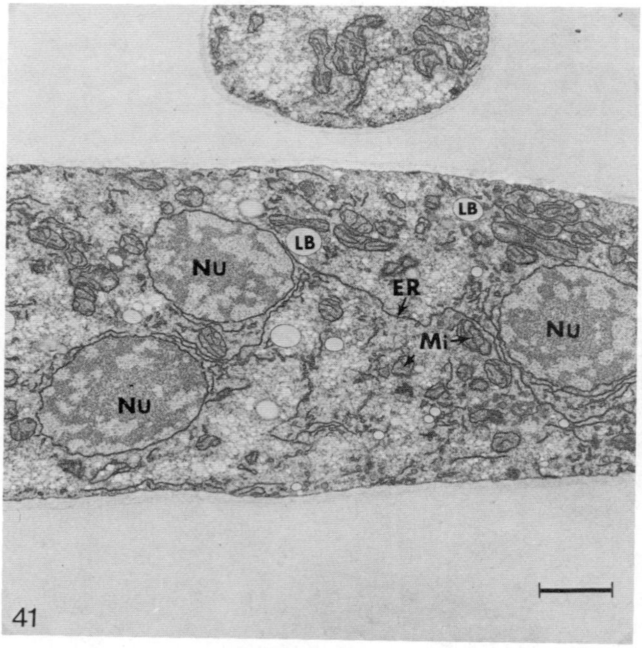

Figure 41 *Conidiobolus coronata.* Longitudinal thin section of a hypha showing lipid bodies (LB), mitochondria (Mi), endoplasmic reticulum (ER), and multiple nuclei (Nu). Permanganate. Bar equals 1.0 μm.

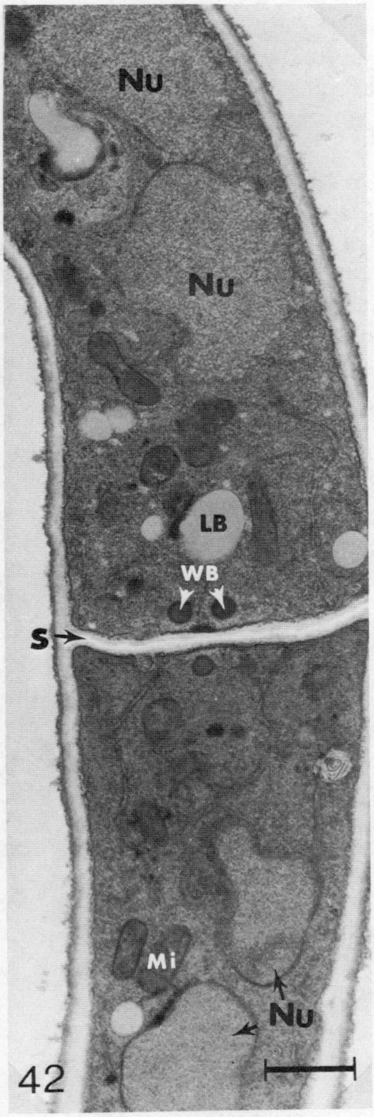

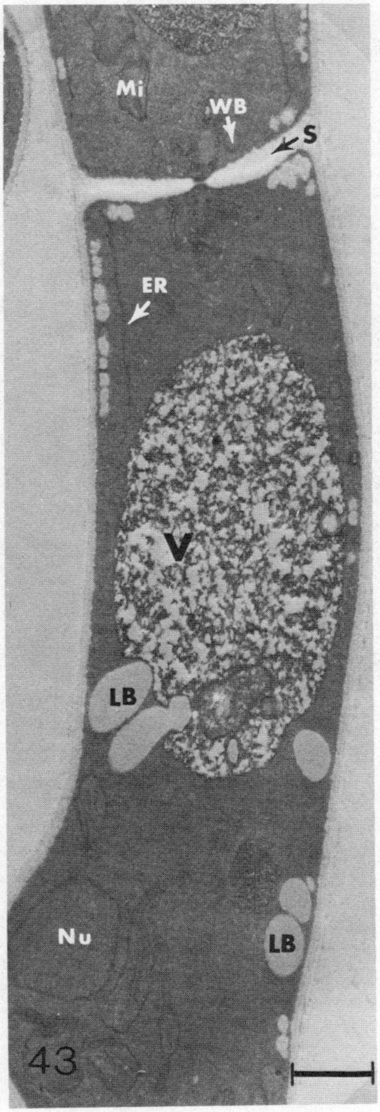

Figure 42 *Scopulariopsis brevicaulis.* Longitudinal thin section of a hypha showing lipid bodies (LB), mitochondria (Mi), multiple nuclei (Nu), and a septum (S) with associated Woronin bodies (WB). Glutaraldehyde-osmium tetroxide. Bar equals 0.5 μm.

Figure 43 *Penicillium* sp. Longitudinal thin section of a hypha showing lipid bodies (LB), vacuole (V), nucleus (Nu), mitochondria (Mi), endoplasmic reticulum (ER), and a septum (S) with Woronin bodies (WB). Glutaraldehyde-osmium tetroxide. Bar equals 0.5 μm.

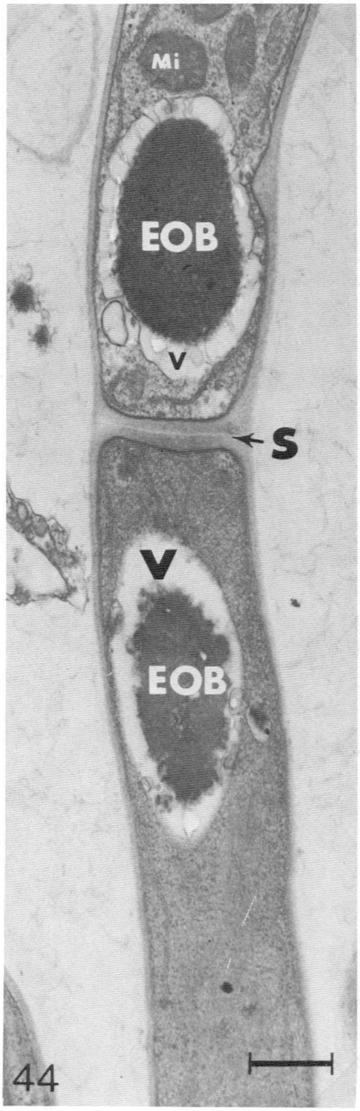

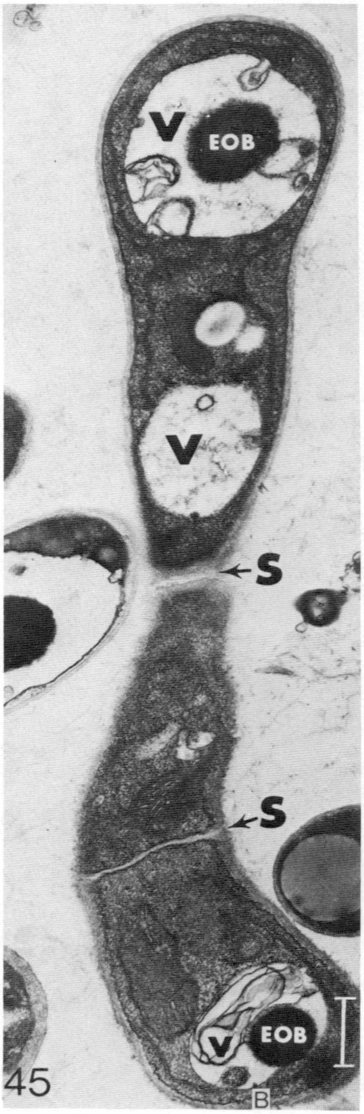

Figure 44 *Penicillium marneffei.* Longitudinal thin section of a hypha showing mitochondria (Mi), vacuoles (V) containing electron-opaque bodies (EOB), and a septum (S). Glutaraldehyde-osmium. Bar equals 0.5 μm. (From Ref. 67.)

Figure 45 *Penicillium marneffei.* Longitudinal thin section of an arthroconidium showing septa (S), and vacuolated areas (V) containing electron-opaque bodies (EOB). Glutaraldehyde-osmium tetroxide. Bar equals 0.5 μm. (From Ref. 67.)

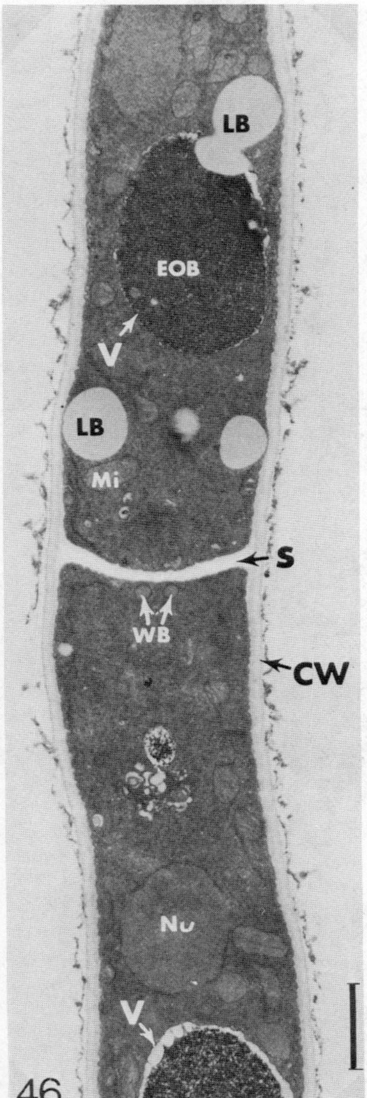

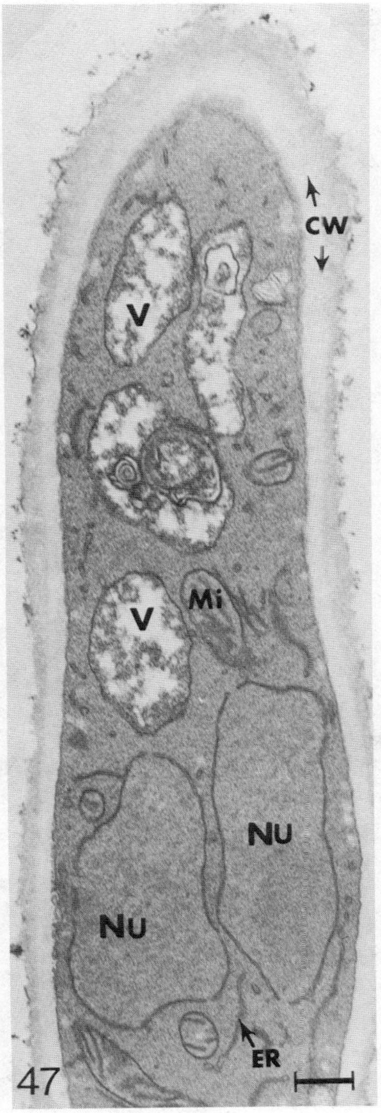

Figure 46 *Aspergillus* sp. Longitudinal thin section of a hypha showing lipid bodies (LB), mitochondria (Mi), nucleus (Nu), vacuolated areas containing electron-opaque bodies (EOB), cell wall (CW), and a septum (S). Glutaraldehyde-osmium tetroxide. Bar equals 1.0 μm.

Figure 47 *Epidermophyton floccosum*. Longitudinal thin section of a hypha showing multiple nuclei (Nu), mitochondria (Mi), endoplasmic reticulum (ER), vacuoles (V), and cell wall (CW). Glutaraldehyde-osmium tetroxide. Bar equals 0.25 μm.

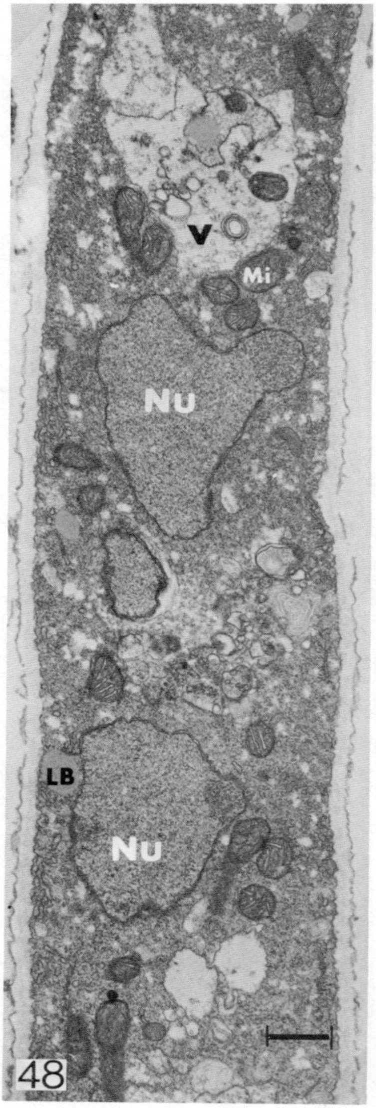

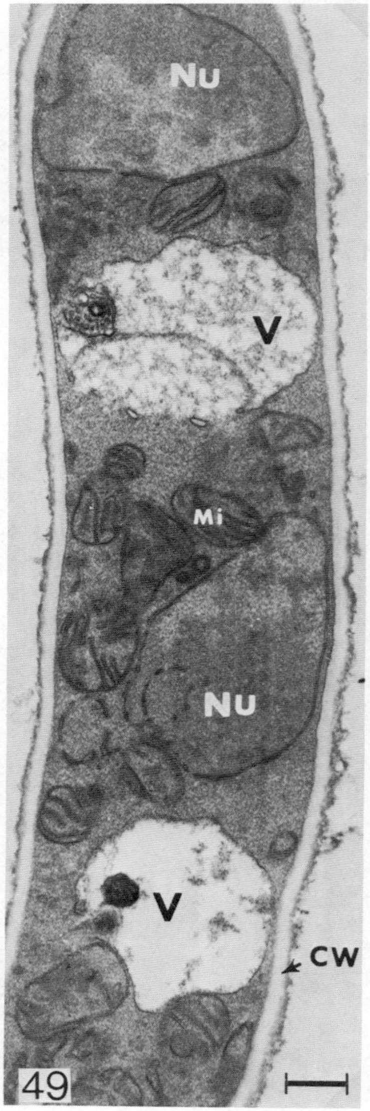

Figure 48 *Microsporum gypseum.* Longitudinal thin section of a hypha showing lipid bodies (LB), vacuole (V), mitochondria (Mi), and multiple nuclei (Nu). Glutaraldehyde-osmium tetroxide. Bar equals 0.5 μm.

Figure 49 *Microsporum distortum.* Longitudinal thin section of a hypha showing mitochondria (Mi), vacuoles (V), multiple nuclei (Nu), and cell wall (CW). Glutaraldehyde-osmium tetroxide. Bar equals 0.25 μm.

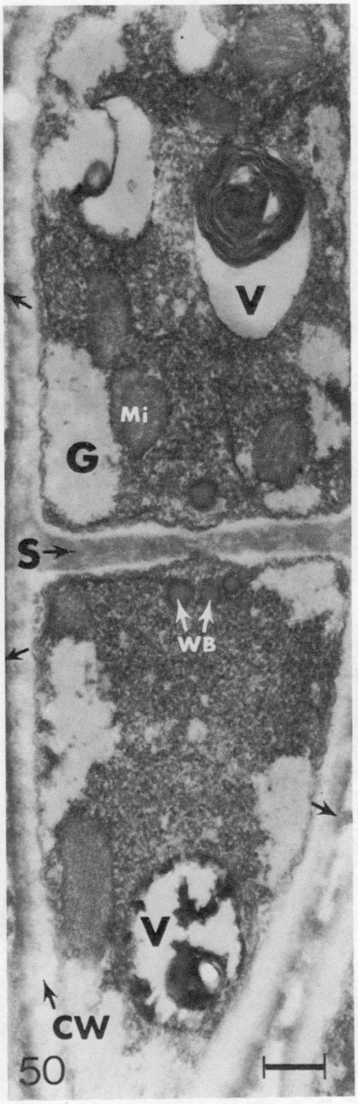

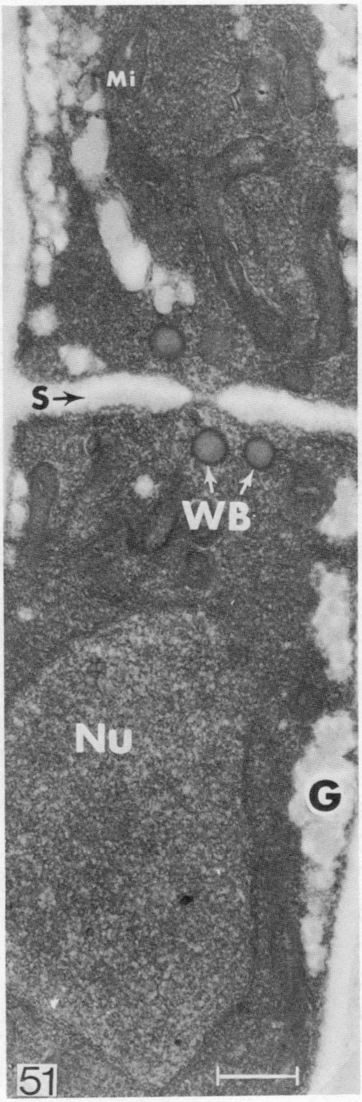

Figure 50 *Fonsecaea pedrosoi.* Longitudinal thin section of a hypha showing mitochondria (Mi), glycogen-like material (G), vacuoles (V), septum (S) with Woronin bodies (WB), and cell wall (CW). Note the electron-opaque granules (at arrows) along the cell wall. Glutaraldehyde-osmium tetroxide. Bar equals 0.25 μm.

Figure 51 *Sporothrix schenckii.* Longitudinal thin section of a hypha showing a nucleus (Nu), glycogen-like material (G), mitochondria (Mi), and a septum (S) with Woronin bodies (WB). Glutaraldehyde-osmium tetroxide. Bar equals 0.25 μm.

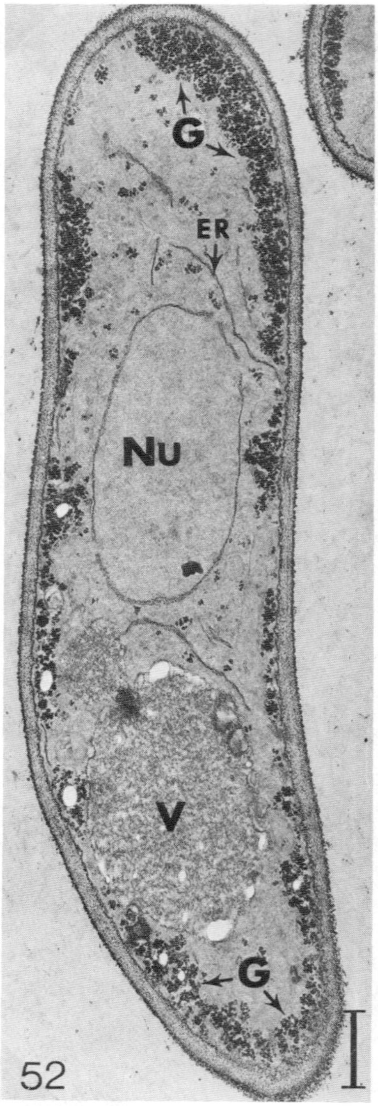

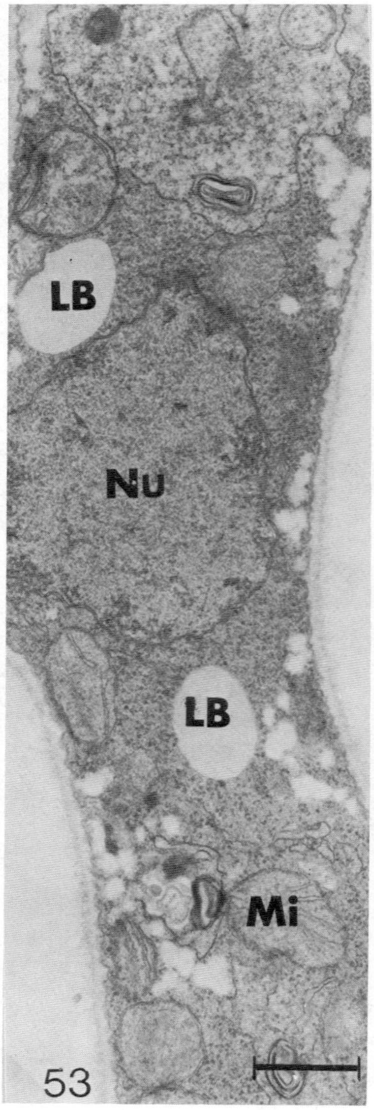

Figure 52 *Sporothrix schenckii.* Longitudinal thin section of a hypha stained by the PATAg method of Thiéry [220]. Note the nucleus (Nu), vacuole (V), endoplasmic reticulum (ER), and electron-opaque granules of glycogen-like material (G). Glutaraldehyde-osmium tetroxide PATAg. Bar equals 0.5 μm.

Figure 53 *Blastomyces dermatitidis.* Longitudinal thin section of a hypha showing lipid bodies (LB), mitochondria (Mi), and nucleus (Nu). Glutaraldehyde-osmium tetroxide. Bar equals 0.5 μm.

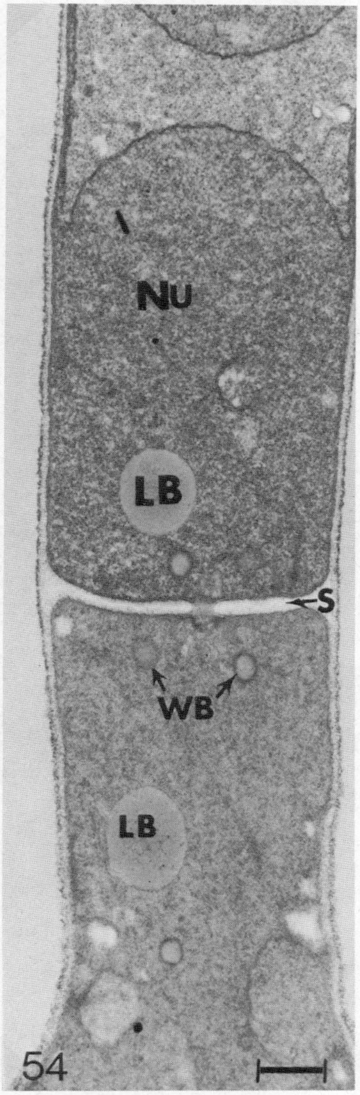

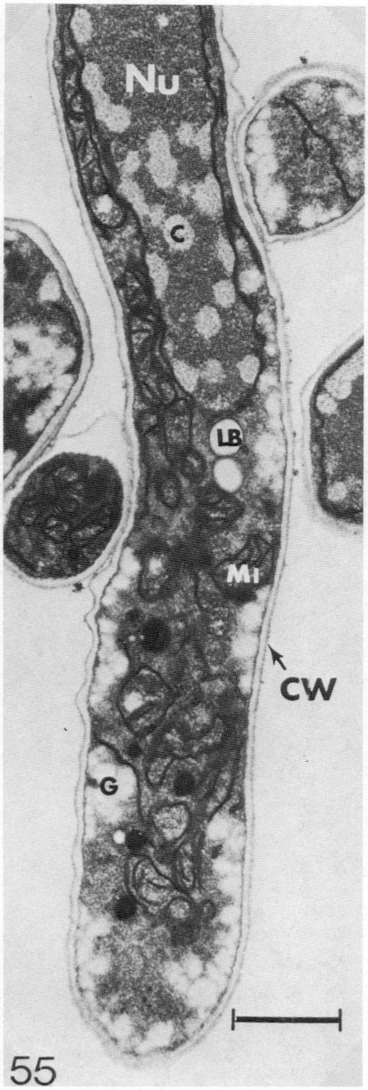

Figure 54 *Histoplasma farcinimosum.* Longitudinal thin section of a hypha showing lipid bodies (LB), nucleus (Nu), and a septum (S) with Woronin bodies (WB). Glutaraldehyde-osmium tetroxide. Bar equals 0.25 μm.

Figure 55 *Histoplasma capsulatum.* Longitudinal thin section of a hypha showing a nucleus (Nu) with chromatin material (C), lipid bodies (LB), mitochondria (Mi), glycogen-like material (G), and cell wall (CW). Permanganate. Bar equals 0.3 μm.

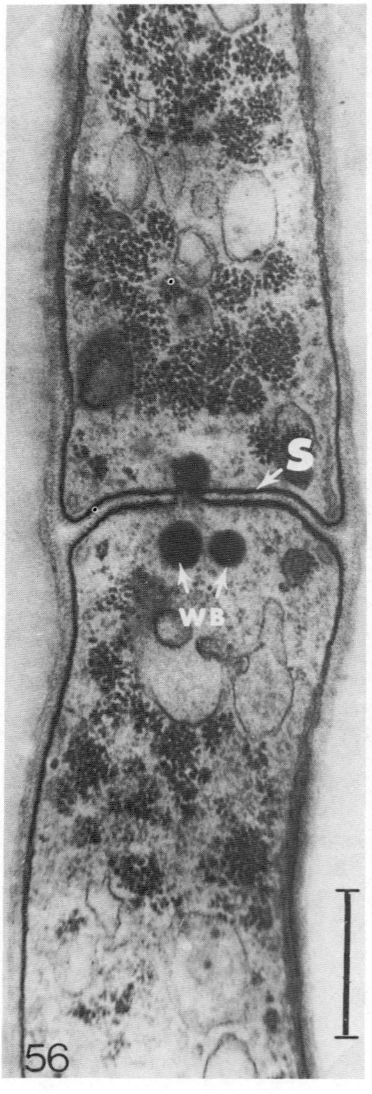

Figure 56 *Coccidioides immitis.* Longitudinal thin section of a hypha showing a septum (S) with Woronin bodies (WB). Glutaraldehyde-osmium. Bar equals 0.5 μm. (From Ref. 213.)

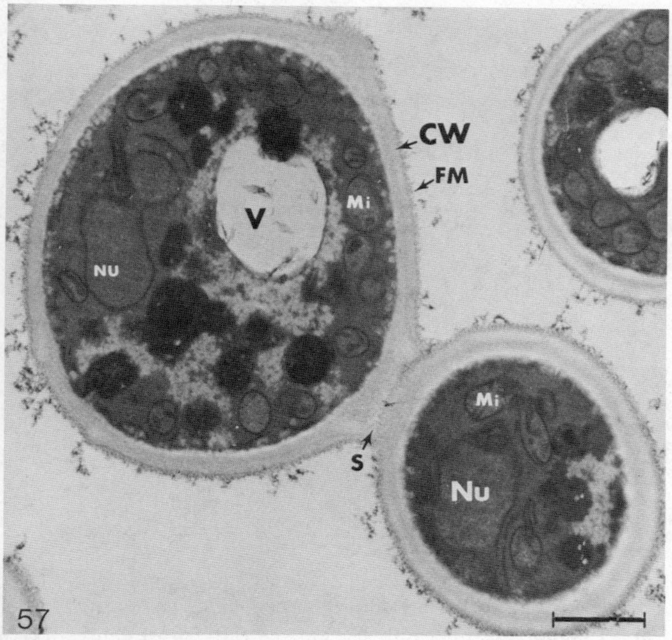

Figure 57 *Candida albicans.* Thin section of a budding cell showing nuclei (Nu), vacuole (V), mitochondria (Mi), and filamentous material (FM) at and along the outermost portion of the cell wall (CW). Glutaraldehyde-osmium tetroxide. Bar equals 0.5 μm.

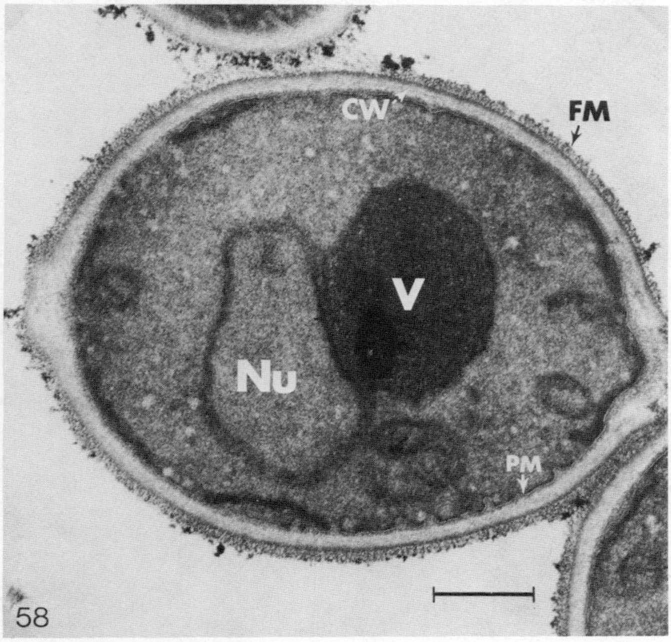

Figure 58 *Torulopsis glabrata.* Thin section showing a nucleus (Nu), vacuole (V), plasma membrane (PM), and filamentous material (FM) at the outermost cell wall (CW) surface. Glutaraldehyde-osmium tetroxide. Bar equals 0.25 μm.

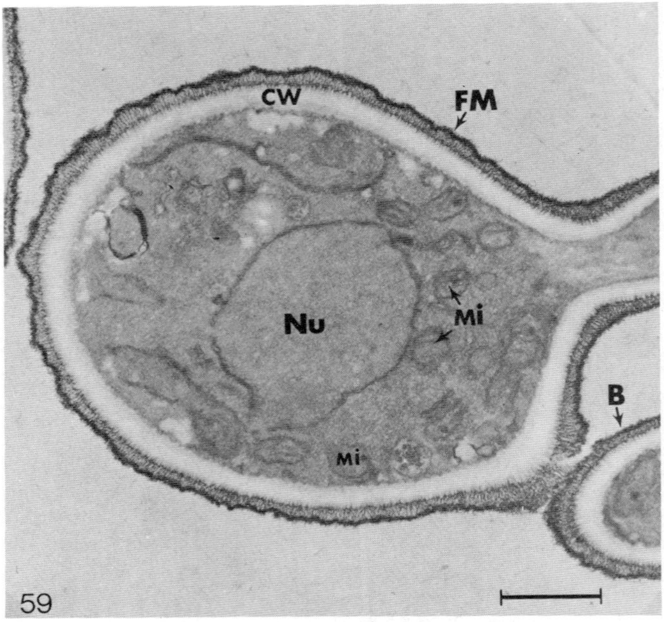

Figure 59 *Sporothrix schenckii.* Thin section of a budding yeast-like cell showing mitochondria (Mi), nucleus (Nu), and filamentous material (FM) at the outermost cell wall (CW) surfaces of the parental and bud (B) cells. Cells were stained with acidified dialyzed iron to enhance the electron opacity of the filamentous material. Glutaraldehyde-dialyzed iron-osmium tetroxide. Bar equals 0.5 μm. (From Ref. 71.)

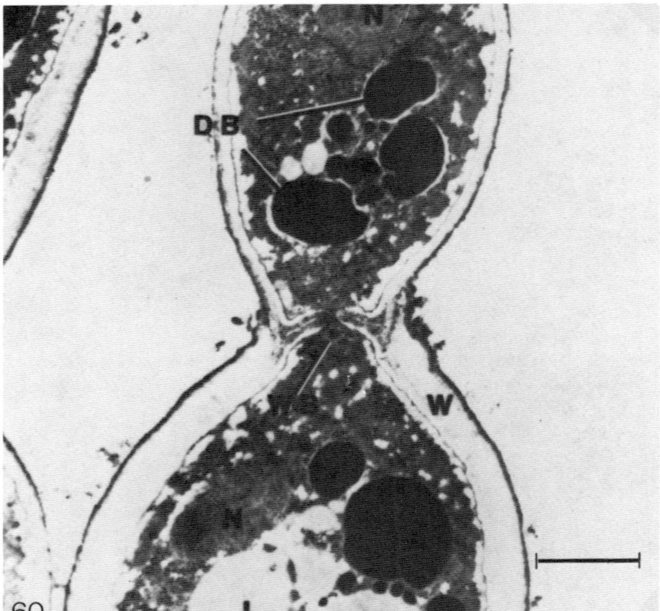

Figure 60 *Phialophora dermatitidis.* Thin section of thick-walled yeast after conversion. note the Woronin bodies (WB), cell wall (W), lipid body (L), and dense bodies (DB). Glutaraldehyde-osmium tetroxide. Bar equals 0.5 μm. (From Ref. 173.)

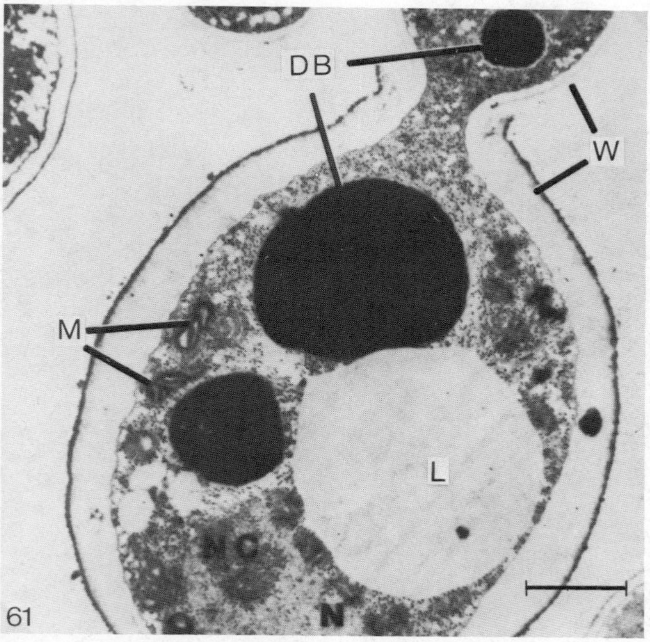

Figure 61 *Phialophora dermatitidis.* Thin section of thick-walled yeast-like cell during conversion. Note the mitochondria (M), lipid body (L), nucleus (N) with nucleolus (Nc), cell wall (W), and dense bodies (DB). Glutaraldehyde-osmium tetroxide. Bar equals 0.5 μm. (From Ref. 173.)

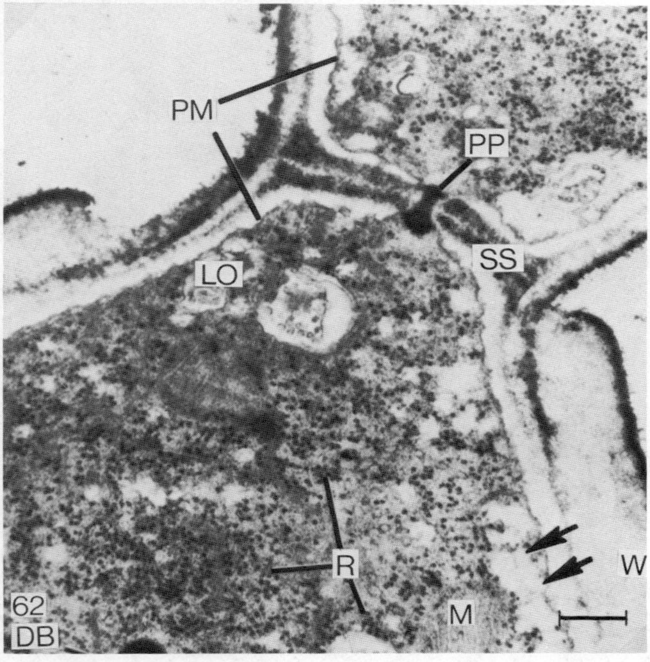

Figure 62 *Phialophora dermatitidis.* Thin section of a thick-walled yeast-like cell and portions of the first moniliform cell. Note the cell wall (W), mitochondria (M), ribosomes (R), pore plug (PP), simple septum (SS), plasmalemmosome (LO), polysaccharide storage areas at the arrows, and portions of a dense body (DB). Glutaraldehyde-osmium. Bar equals 0.25 μm. (From Ref. 173.)

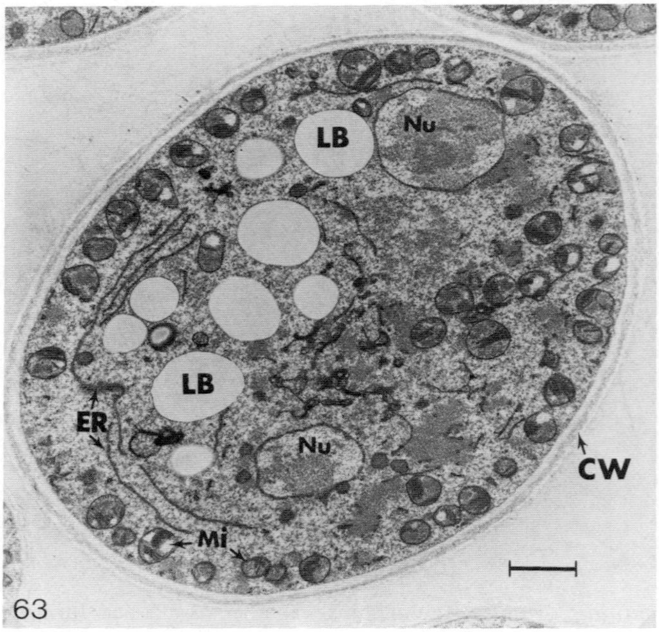

Figure 63 *Paracoccidioides brasiliensis.* Thin section of a yeast-like cell showing mitochondria (Mi), lipid bodies (LB), endoplasmic reticulum (ER), multiple nuclei (Nu), and cell wall (CW). Permanganate. Bar equals 1.0 μm.

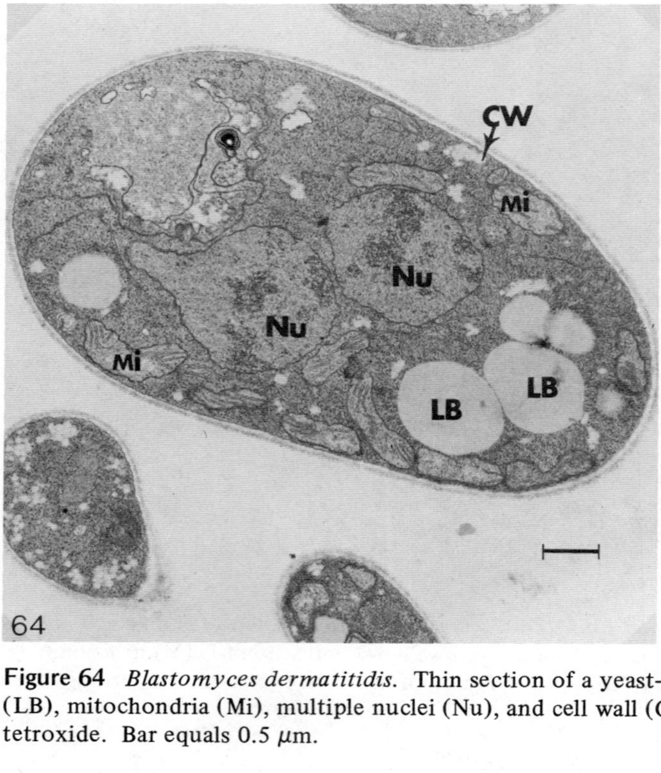

Figure 64 *Blastomyces dermatitidis.* Thin section of a yeast-like cell showing lipid bodies (LB), mitochondria (Mi), multiple nuclei (Nu), and cell wall (CW). Glutaraldehyde-osmium tetroxide. Bar equals 0.5 μm.

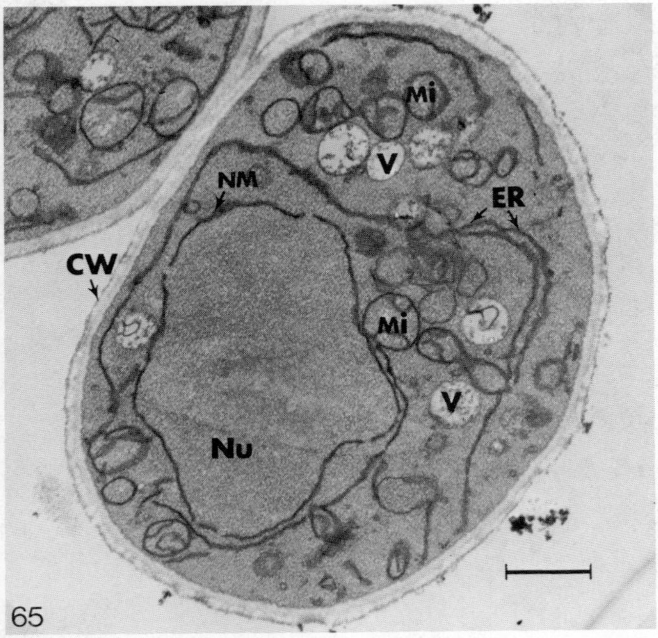

Figure 65 *Histoplasma capsulatum.* Thin section of a yeast-like cell showing mitochondria (Mi), nucleus (Nu), nuclear membrane (NM), vacuoles (V), endoplasmic reticulum (ER), and cell wall (CW). Glutaraldehyde-permanganate. Bar equals 0.5 μm.

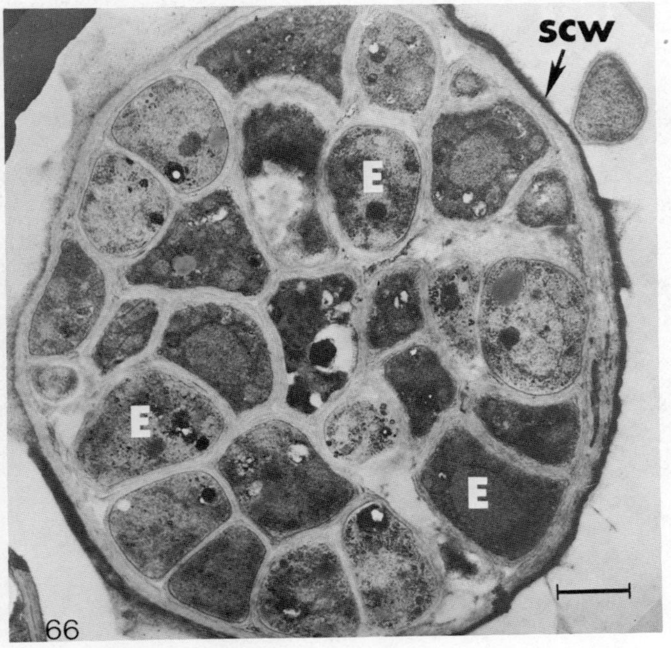

Figure 66 *Coccidioides immitis.* Thin section of a maturing spherule showing a late stage in cleavage with the presence of numerous young endospores (E) bounded by the spherule cell wall (SCW). Glutaraldehyde-osmium tetroxide. Bar equals 1.0 μm. (From Ref. 213.)

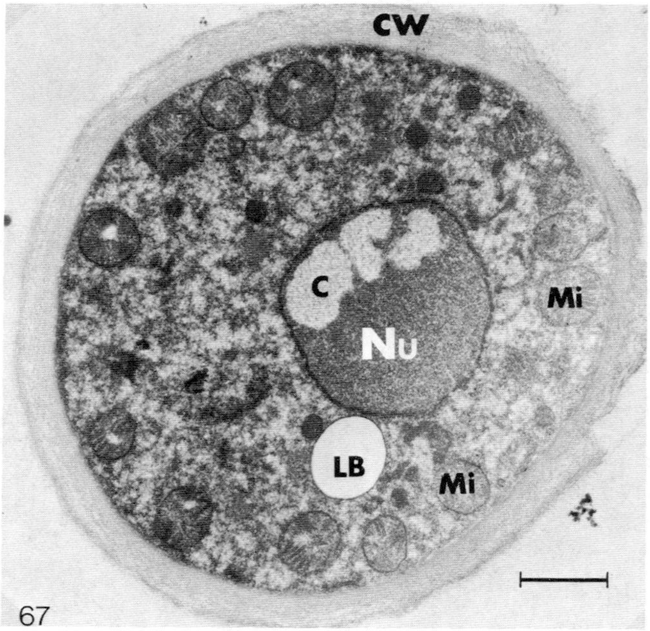

Figure 67 *Trichosporon cutaneum.* Thin section of a yeast-like form showing a thickened cell wall (CW), mitochondria (Mi), lipid body (LB), and a nucleus (Nu) with chromatin material (C). Glutaraldehyde-permanganate. Bar equals 0.5 μm.

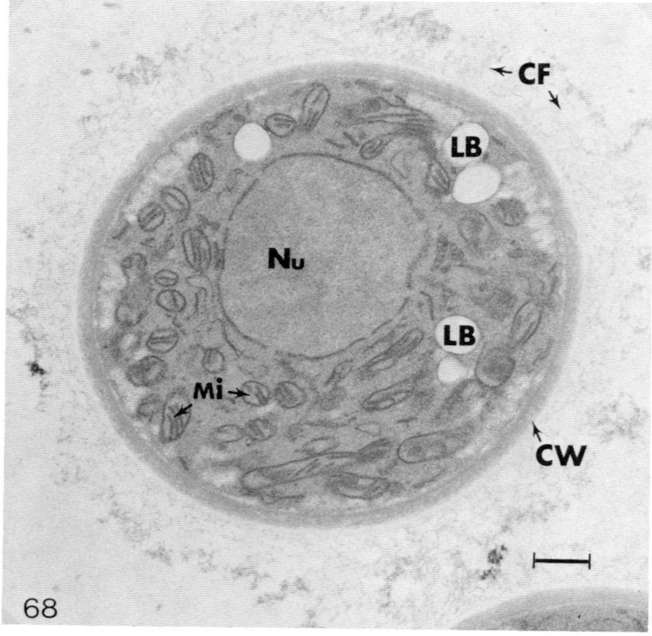

Figure 68 *Cryptococcus neoformans.* Thin section showing lipid bodies (LB), mitochondria (Mi), nucleus (Nu), and capsular filaments (CF) exterior to the cell wall (CW). Glutaraldehyde-permanganate. Bar equals 0.25 μm.

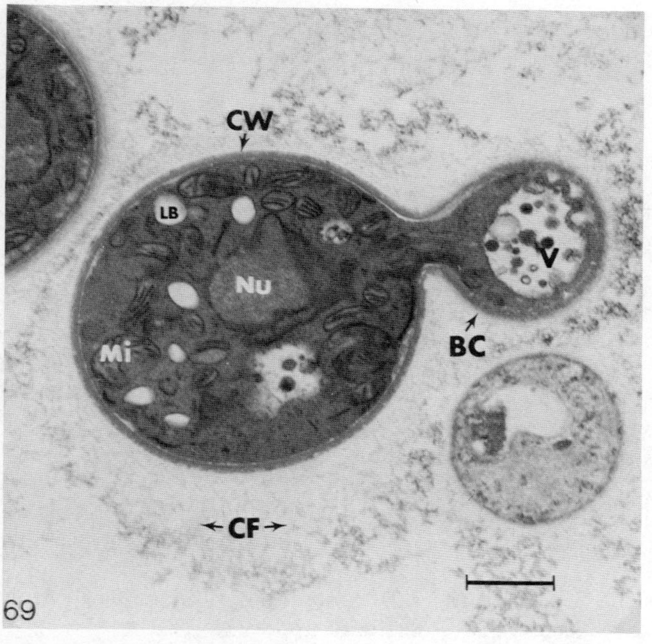

Figure 69 *Cryptococcus neoformans.* Thin section of a budding cell (BC) showing lipid bodies (LB), vacuoles (V), nucleus (NU), mitochondria (Mi), and capsular filaments (CF) exterior to the cell wall (CW). Glutaraldehyde-permanganate. Bar equals 0.5 μm.

Most of the electron micrographs in this appendix as well as those used elsewhere were obtained from thin sections of cultured materials prepared in the author's laboratory. For these micrographs, the author must assume responsibility for the quality of the preparation and its interpretation. The ultrastructural appearance of the fungal cell as seen by a variety of fixation schedules is illustrated. These procedures include standard methods of fixation using either glutaraldehyde-osmium tetroxide, glutaraldehyde-potassium permanganate, or aqueous potassium permanganate alone. Each technique possesses certain advantages in the study of the cytological organization of the fungal organism.

Pathogenic species of the Actinomycetales (*Nocardia* and *Streptomyces* species) and those of the genus *Actinomyces* of the gram-positive, nonsporeforming, anaerobic bacilli are typically procaryotic cells. Inasmuch as the ultrastructural attributes of these procaryotic organisms have not been discussed herein, it seems appropriate that examples of cells having a typical procaryotic cytological organization should be illustrated. For purposes of comparison of their fine structural appearance with that of the Eumycotic pathogenic organisms, we have included two transmission electron micrographs which illustrate the cytology of procaryotic cells of the Actinomycetales. These electron micrographs show thin sections of *Nocardia asteroides* (Fig. 22) and the coccoid forms of the highly pleomorphic *Rothia dentocariosa* (Fig. 23). The latter organism is a resident of the human oral cavity, and its fine structure has been described elsewhere [196]. In general, the typical procaryotic cell lacks a nuclear envelope, mitochondria, and a reticulum with ribosomes. A class of membranous structures termed bacterial mesosomes are commonly seen in thin sections of the procaryotic cell. These structures have been found in a wide variety of the Actinomycetales and a close association between these membranes and developing cross walls has been noted [232]. A comprehensive review of the bacterial mesosome has been made by Ghose [81].

ACKNOWLEDGMENTS

The author wishes to thank those who so generously contributed both unpublished and published micrographs used in the illustrations. Special acknowledgments are made to Mrs. Karen Boyd Grantham for her contributions and excellent technical assistance. This work was supported in part by the Veterans Administration Research Service.

REFERENCES

1. Anderson K. L., R. W. Wheat, N. R. Conant, and W. Clingenpeel. 1974. Composition of the cell wall and other fractions of the autolyzed yeast form of *Histoplasma capsulatum*. Mycopathol. Mycol. Appl. *54*: 439-451.
2. Armentrout, V. N., and C. L. Wilson. 1969. Haustorial-host interaction during mycoparasitism of *Piptocephalis virginiana* on *Mycotypha microspora*. Phytopathology *59*: 897-905.
3. Barnes, W. G., A. Flesher, A. E. Berger, and J. D. Arnold. 1971. Scanning electron microscopic studies of *Candida albicans*. J. Bacteriol. *106*: 276-280.
4. Barnes, W. G., G. C. Sauer, and J. D. Arnold. 1973. Scanning electron microscopy of tinea versicolor organisms (*Malassezia furfur-Pityrosporon orbiculare*?) Arch. Dermatol. *107*: 392-394.
5. Barnicot, N. A., and H. E. Huxley. 1965. Electron microscopic observations on mitotic chromosomes. Q. J. Microsc. Sci. *106*: 197-214.
6. Bartnicki-Garcia, S. 1968. Cell wall chemistry, morphogenesis, and taxonomy of fungi. Annu. Rev. Microbiol. *22*: 87-108.
7. Bartnicki-Garcia, S., N. Nelson, and E. Cota-Robles. 1968. A novel apical corpuscle in hyphae of *Mucor rouxii*. J. Bacteriol. *95*: 2399-2402.
8. Bauer, H., and K. Tanaka. 1968. Ultrastructure of mitochondria and crystal-containing bodies in mature ballistospores of the fungus *Basidiobolus ranarum* as revealed by freeze-etching. J. Bacteriol. *96*: 2132-2137.
9. Beckett, A., I. B. Heath, and D. J. McLaughlin. 1974. *An Atlas of Fungal Ultrastructure*. Longman, London.
10. Blank, F. 1954. On the cell walls of dimorphic fungi causing systemic infections. Can. J. Microbiol. *1*: 1-5.
11. Borgers, M., and S. De Nollin. 1974. The preservation of subcellular organelles of *Candida albicans* using conventional fixatives. J. Cell Biol. *62*: 574-581.
12. Bracker, C. E. 1967. Ultrastructure of fungi. Annu. Rev. Phytopathol. *5*: 343-374.
13. Bracker, C. E. 1971. Cytoplasmic vesicles in germinating spores of *Gilbertella persicaria*. Protoplasma *72*: 381-397.
14. Bracker, C. E., and E. E. Butler. 1963. The ultrastructure and development of septa in hyphae of *Rhizoctonia solani*. Mycologia *55*: 35-38.
15. Breathnach, A. S., M. Gross, and B. Martin. 1976. Freeze-fracture replication of cultural *Pityrosporum orbiculare*. Sabouraudia *14*: 105-113.
16. Brown, J. P., and M. R. Edwards. 1971. Observations on budding cells of *Torulopsis glabrata*. Can. J. Microbiol. *17*: 609-616.
17. Calonge, F. D. 1969. The occurrence of glycogen-membrane complexes in fungi. An electron microscopic study. Protoplasma *67*: 79-85.

18. Caputo, R., M. Innocenti, and M. Shimono. 1977. Cell wall and plasma membranes of dermatophytes: a freeze-fracture study. J. Ultrastruct. Res. *59*: 149-157.
19. Carbonell, L. M. 1967. Cell wall changes during the budding process of *Paracoccidioides brasiliensis* and *Blastomyces dermatitidis*. J. Bacteriol. *94*: 213-223.
20. Carbonell, L. M. 1969. Intrayeast hyphae in *Paracoccidioides brasiliensis*. J. Bacteriol. *98*: 1395-1396.
21. Carbonell, L. M. 1972. Ultrastructure of *Paracoccidioides brasiliensis*. In *Paracoccidioidomycosis*, Proceedings of the First Pan American Symposium, Pan Am. Health Organ. Sci. Publ. No. 254, Washington, D.C., pp. 21-28.
22. Carbonell, L. M., M. D. Berliner, and F. Gil. 1972. Kinetic and morphological observations on the yeast phase of *Histoplasma capsulatum* during protoplast formation. J. Bacteriol. *110*: 731-738.
23. Carbonell, L. M., and R. Gill. 1971. Ultrastructure of sexual reproduction in *Arthroderma benhamiae*. In A. Perez-Miravete and D. Pelaez (Eds.), *Recent Advances in Microbiology*, Associación Mexicana de la Microbiologia, Mexico, D.F., pp. 481-492.
24. Carbonell, L. M., F. Gill, and F. Yegres. 1973. Cell wall regeneration of protoplasts of *Histoplasma capsulatum*. In R. J. Villaneuva, I. Garcia-Acha, S. Gascon, and F. Uruburu (Eds.), *Yeast, Mould, and Plant Protoplasts*, Academic, New York, pp. 93-103.
25. Carbonell, L. M., and L. Pollak. 1962. "Myelin figures" in yeast cultures of *Paracoccidioides brasiliensis*. J. Bacteriol. *83*: 1356-1357.
26. Carbonell, L. M., and J. Rodriguez. 1968. Mycelial phase of *Paracoccidioides brasiliensis* and *Blastomyces dermatitidis*: an electron microscopic study. J. Bacteriol. *96*: 533-543.
27. Cassone, A., N. Simonetti, and V. Strippoli. 1973. Ultrastructural changes in the wall during germ-tube formation from blastospores of *Candida albicans*. J. Gen. Microbiol. *77*: 417-428.
28. Cassone, A., N. Simonetti, and V. Strippoli. 1974. Wall structure and bud formation in *Cryptococcus neoformans*. Arch. Microbiol. *95*: 205-212.
29. Chamberland, H., and G. B. Ouellette. 1977. Formes d'inclusions osmiophiles dans les cellules de *Ceratocystis ulmi*. Can. J. Bot. *55*: 695-710.
30. Chattaway, F., M. R. Holmes, and A. J. E. Barlow. 1968. Cell wall composition of the mycelial and blastospore forms of *Candida albicans*. J. Gen. Microbiol. *51*: 367-376.
31. Chong, K. C., B. A. Adam, and T. S. Soo-Hoo. 1975. Morphology of *Piedra hortai*. Sabouraudia *13*: 157-160.
32. Cole, G. T. 1972. Microfibrils in the cytoplasm of fertile hyphae of the imperfect fungus, *Drechslera sorokiniana*. J. Ultrastruct. Res. *41*: 563-571.
33. Cole, G. T., and H. C. Aldrich. 1971. Demonstration of myelin figures in unfixed, freeze-etched fungus spores. J. Cell Biol. *51*: 873-874.
34. Cooper, B. H., S. Grove, C. Mims, and P. J. Szaniszlo. 1973. Septal ultrastructure in *Phialophora pedrosoi, Phialophora verrucosa* and *Cladosporium carrionii*. Sabouraudia *11*: 127-130.
35. Conti, S. F., and H. B. Naylor. 1960. Electron microscopy of ultrathin sections of *Schizosaccharomyces octosporus*. III. Ascosporogenesis, ascospore structure, and germination. J. Bacteriol. *79*: 331-340.
36. Crang, R. E., and D. G. Pechak. 1978. Serial section reconstruction of black yeast, *Aureobasidium pullularis,* by means of high voltage electron microscopy. Protoplasma *96*: 225-234.
37. Curgy, J. J. 1968. Influence du mode de fixation sur la possibilité d'observer des

structures myéliniques dan les hépatocytes d'embryons de poulet. J. Microsc. (Paris) *7*: 63-80.

38. Danielli, J. F., and H. Davson. 1935. A contribution to the theory of permeability of thin films. J. Cell. Comp. Physiol. *5*: 495-508.
39. Davis, T. E., Jr., J. E. Domer, and Y. T. Li. 1977. Cell wall studies of *Histoplasma capsulatum* and *Blastomyces dermatitidis* using autologous and heterologous enzymes. Infect. Immun. *15*: 978-987.
40. Day, A. W., and N. H. Poon. 1975. Fungal fimbriae. II. Their role in conjugation in *Ustilago violacea*. Can. J. Microbiol. *21*: 547-557.
41. Day, A. W., N. H. Poon, and G. G. Stewart. 1975. Fungal fimbriae. III. The effect on flocculation in *Saccharomyces*. Can. J. Microbiol. *21*: 558-564.
42. De Duve, C., and R. Wattiaux. 1966. Function of lysosomes. Annu. Rev. Physiol. *28*: 435-492.
43. De Nollin, S., and M. Borgers. 1974. The ultrastructure of *Candida albicans* after in vitro treatment with miconazole. Sabouraudia *12*: 341-351.
44. De Nollin, S., F. Throne, and M. Borgers. 1975. Enzyme cytochemistry of *Candida albicans*. J. Histochem. Cytochem. *23*: 758-765.
45. Dodge, B. O. 1920. The life history of *Ascobolus magnificus*: origin of the ascocarp from two strains. Mycologia *12*: 115-133.
46. Domer, J. E. 1971. Monosaccharide and chitin content of cell walls of *Histoplasma capsulatum* and *Blastomyces dermatitidis*. J. Bacteriol. *107*: 870-877.
47. Domer, J. E., and J. G. Hamilton. 1971. The readily extracted lipids of *Histoplasma capsulatum* and *Blastomyces dermatitidis*. Biochim. Biophys. Acta *231*: 465-478.
48. Domer, J. E., J. G. Hamilton, and J. C. Harkin. 1967. Comparative study of the cell walls of the yeastlike and mycelial phases of *Histoplasma capsulatum*. J. Bacteriol. *94*: 466-474.
49. Dupouy, G. 1977. Advantages of megavolt electron microscopy in biological research. Ultramicroscopy *2*: 199-203.
50. Dupouy, G., F. Perrier, and L. Durrieu. 1960. L'observation de matière vivante au moyen d'un microscope électronique fonctionnant sous haute tension. C. R. Acad. Sci. (Paris) *251*: 2836-2841.
51. Duran, A., E. Cabib, and B. Bowers. 1979. Chitin synthetase distribution of the yeast plasma membrane. Science *203*: 363-365.
52. Dzaczenko, W., and A. Cassone. 1971. Visualization of new ultrastructural components in the cell wall of *Candida albicans* with fixatives containing TAPO. J. Cell Biol. *52*: 186-190.
53. Edwards, G. A., and M. R. Edwards. 1960. The intracellular membranes of *Blastomyces dermatitidis*. Am. J. Bot. *47*: 622-632.
54. Edwards, M. R., M. A. Gordon, E. W. Lapa, and W. C. Ghiorse. 1967. Micromorphology of *Cryptococcus neoformans*. J. Bacteriol. *94*: 766-777.
55. Edwards, M. R., E. L. Hazen, and G. A. Edwards. 1959. The fine structure of the yeast-like cells of *Histoplasma* in culture. J. Gen. Microbiol. *20*: 496-503.
56. Edwards, M. R., E. L. Hazen, and G. A. Edwards. 1960. The micromorphology of the tuberculate spores of *Histoplasma capsulatum*. Can. J. Microbiol. *6*: 65-70.
57. Edwards, M. R., and S. C. Holt. 1968. Mesosome-like structures in *Cryptococcus neoformans* studied by chemical fixation and by freeze-etching technique. Proc. Electron Microsc. Soc. Am. (Abstr.), p. 80.
58. Farley, J. F., R. A. Jersild, and D. J. Niederpruem. 1975. Origin and ultrastructure of intra-hyphal hyphae in *Trichophyton terrestre* and *T. rubrum*. Arch. Microbiol. *106*: 195-200.

59. Farley, J. F., R. A. Jersild, and D. J. Niederpruem. 1976. Ultrastructural aspects of ascosporulation in *Arthroderma quadrifidum* (= *Trichophyton terrestre*). Sabouraudia *14*: 337-341.

60. Fawcett, D. W. 1966. *The Cell: Its Organelles and Inclusions. An Atlas of Fine Structure.* Saunders, Philadelphia.

61. Feit, C., and R. P. Tewari. 1974. Immunogenicity of ribosomal preparations from yeast cells of *Histoplasma capsulatum.* Infect. Immun. *10*: 1091-1097.

62. Frederick, S. E., P. J. Gruber, and E. Newcomb. 1975. Plant microbodies. Protoplasma *84*: 1-19.

63. Frey-Wyssling, A., E. Grieshaber, and K. Mühlethaler. 1963. Origin of spherosomes in plant cells. J. Ultrastruct. Res. *8*: 506-516.

64. Gahan, P. B. 1967. Histochemistry of lysosomes. Int. Rev. Cytol. *21*: 1-63.

65. Gale, G. R. 1963. Cytology of *Candida albicans* as influenced by drugs acting on the cytoplasmic membrane. J. Bacteriol. *86*: 151-157.

66. Garrison, R. G. Unpublished observations.

67. Garrison, R. G., and K. S. Boyd. 1973. Dimorphism of *Penicillium marneffei* as observed by electron microscopy. Can. J. Microbiol. *19*: 1305-1309.

68. Garrison, R. G., and K. S. Boyd. 1974. Ultrastructural studies of induced morphogenesis by *Aspergillus parasiticus.* Sabouraudia *12*: 179-187.

69. Garrison, R. G., and K. S. Boyd. 1977. The fine structure of microconidial germination and vegetative cells of *Histoplasma capsulatum.* Ann. Microbiol. (Paris) *128A*: 135-149.

70. Garrison, R. G., and K. S. Boyd. 1978. Role of the conidium in dimorphism of *Blastomyces dermatitidis.* Mycopathologia *64*: 29-33.

71. Garrison, R. G., K. S. Boyd, and F. Mariat. 1975. Ultrastructural studies of the mycelium-to-yeast transformation of *Sporothrix schenckii.* J. Bacteriol. *124*: 959-968.

72. Garrison, R. G., and J. W. Lane. 1973. Scanning-beam electron microscopy of the conidia of the brown and albino filamentous varieties of *Histoplasma capsulatum.* Mycopathol. Mycol. Appl. *49*: 185-191.

73. Garrison, R. G., and J. W. Lane. 1974. The fine structure of the microconidium of *Blastomyces dermatitidis.* Mycopathol. Mycol. Appl. *52*: 93-100.

74. Garrison, R. G., J. W. Lane, and M. F. Field. 1970. Ultrastructural changes during the yeastlike to mycelial phase conversion of *Blastomyces dermatitidis* and *Histoplasma capsulatum.* J. Bacteriol. *101*: 628-635.

75. Garrison, R. G., J. W. Lane, and D. R. Johnson. 1973. Ultrastructural studies on the cleistothecium of *Ajellomyces dermatitidis.* Sabouraudia *11*: 131-136.

76. Garrison, R. G., F. Mariat, K. S. Boyd, and H. Fromentin. 1977. Ultrastructural observations of an unusual osmiophilic body in the hyphae of *Sporothrix schenckii* and *Ceratocystis stenoceras.* Ann. Microbiol. (Paris) *128B*: 319-337.

77. Garrison, R. G., F. Mariat, K. S. Boyd, and J. F. Tally. 1975. Ultrastructural and electron cytochemical studies of *Entomophthora coronata.* Ann. Microbiol. (Paris) *126B*: 149-173.

78. Garvey, M. C., and R. P. Tewari. 1972. A cell-free amino acid incorporating system from yeast-phase cells of *Histoplasma capsulatum.* Sabouraudia *10*: 113-121.

79. Geren, B. B. 1956. Structural studies on the formation of myelin sheath in peripheral nerve fibers. In R. Rudnick (Ed.), *Cellular Mechanisms in Differentiation and Growth,* Proceedings of the Fourteenth Symposium of the Society for the Study of Development and Growth, Princeton University Press, Princeton, N.J., pp. 213-220.

80. Ghiorse, W. C., and M. R. Edwards. 1973. Ultrastructure of *Aspergillus fumigatus* conidia. Development and maturation. Protoplasma *76*: 49-59.

81. Ghose, B. K. 1974. The mesosome—a clue to the evolution of the plasma membrane. Subcell. Biochem. *3*: 311-367.

82. Glick, A. D., and K. J. Kwon-Chung. 1973. Ultrastructural comparison of coils and ascospores of *Emmonsiella capsulata* and *Ajellomyces dermatitidis*. Mycologia *65*: 216-220.

83. Gil, F. 1973. Mesosomes: their role in the delimitation of the ascospore. Mycopathol. Mycol. Appl. *49*: 243-247.

84. Gibardt, M. 1961. Licht- und Electronenmikrokopische Untersuchungen an *Polystictus versicolor*. II. Die Feinstruktur von Grundplassum und Mitochondrien. Arch. Mikrobiol. *39*: 351-359.

85. Greenhalgh, G. N., and L. V. Evans. 1971. Electron microscopy. In C. Booth (Ed.), *Methods in Microbiology*, Vol. 4, Academic, New York, pp. 517-566.

86. Grove, S. N., and C. E. Bracker. 1970. Protoplasmic organization of hyphal tips among fungi: vesicles and Spitzenkörper. J. Bacteriol. *104*: 989-1009.

87. Grove, S. N., K. B. Oujezdsky, and P. J. Szaniszlo. 1973. Budding in the dimorphic fungus *Phialophora dermatitidis*. J. Bacteriol. *115*: 323-329.

88. Gull, K. 1976. Differentiation of septal ultrastructure according to cell type in the basidiomycete, *Agrocybe praecox*. J. Ultrastruct. Res. *54*: 89-94.

89. Gull, K., and A. P. J. Trinci. 1975. Septal ultrastructure in *Basidiobolus ranarum*. Sabouraudia *13*: 49-51.

90. Harris, J. L. 1975. Some three-dimensional aspects of *Ceratocystis ulmi* as observed by high-voltage electron microscopy. Mycologia *67*: 332-341.

91. Hasegawa, T. 1975. *Trichophyton rubrum* and *T. mentagrophytes* studied by freeze-etching. Sabouraudia *13*: 241-243.

92. Hasegawa, T., E. Nakai, and V. S. Rajan. 1975. Freeze-etching observations of *Trichophyton tonsurans*. Proceedings of the VIth Congress of the International Society for Human and Animal Mycology, Tokyo, pp. 72-73 (Abstr.).

93. Hasegawa, T., E. Nakai, and V. S. Rajan. 1977. Freeze-etching observations of *Trichophyton violaceum*. Sabouraudia *15*: 95-98.

94. Hashimoto, T., P. Gerhardt, S. F. Conti, and H. B. Naylor. 1960. Studies on the fine structure of microorganisms. V. Morphogenesis of nuclear and membrane structure during ascospore formation in yeast. J. Biophys. Biochem. Cytol. *7*: 305-310.

95. Hashimoto, T., T. Kishi, and N. Yoshida. 1964. Demonstration of micropores in fungal cross walls. Nature (Lond.) *202*: 1353.

96. Hawker, L. E. 1965. Fine structure of fungi as revealed by electron microscopy. Biol. Rev. Camb. Philos. Soc. *40*: 52-92.

97. Heath, I. B., and M. C. Heath. 1978. Microtubules and organelle movements in the rust fungus *Uromyces phaseoli*. Cytobiologie *16*: 393-411.

98. Heiniger, U., and P. Matile. 1974. Protease secretion in *Neurospora crassa*. Biochem. Biophys. Res. Commun. *60*: 1425-1432.

99. Hejtmánek, M., O. Nečas, and M. Havelková. The ultrastructure of the spherules of *Emmonsia crescens* studied by the technique of freeze-etching. Mycopathol. Mycol. Appl. *54*: 79-83.

100. Hepler, P. K., and B. A. Palevitz. 1974. Microtubules and microfilaments. Annu. Rev. Plant Physiol. *25*: 309-362.

101. Hess, W. M. 1973. Ultrastructure of dormant and germinating sporangiospores of *Rhizopus arrhizus*. Protoplasma *77*: 15-33.

102. Hess, W. M., M. M. A. Sassen, and C. C. Remsen. 1968. Surface characteristics of *Penicillium* conidia. Mycologia *60*: 290-303.

103. Hess, W. M., and D. L. Stocks. 1969. Surface characteristics of *Aspergillus* conidia. Mycologia *61*: 560-571.

104. Hereward, F. V. 1974. Cytoplasmic microtubules in a yeast. Planta *117*: 355-360.
105. Hill, T. W. 1975. Ultrastructure of ascosporogenesis in *Nannizzia gypsea.* J. Bacteriol. *122*:743-748.
106. Hislop, E. C., V. M. Barnaby, C. Shellis, and F. Labarda. 1974. Localization of α-L-arabinofuranosidase and acid phosphatase in mycelium of *Sclerotina fructigena.* J. Gen. Microbiol. *81*: 79-99.
107. Hoch, H. C. 1977. Use of permanganate to increase electron opacity of fungal walls. Mycologia *69*: 1209-1213.
108. Ito, Y., Y. Nozawa, H. Suzuki, and T. Setoguti. 1970. Surface structure of dermatophytes as seen by the scanning electron microscope. Sabouraudia *7*: 270-272.
109. Ito, Y., T. Setoguti, Y. Nozawa, and S. Sakurai. 1967. An electron microscopic observation of *Trichophyton violaceum.* J. Invest. Dermatol. *48*: 121-126.
110. Ito, Y., T. Setoguti, Y. Nozawa, S. Sakurai, and T. Yoneya. 1969. Further observations on the fine structure of *Trichophyton violaceum*: non-pigmented, mutant strains. Sabouraudia *7*: 138-141.
111. Iwata, K., Y. Kanda, H. Yamaguchi, and M. Osumi. 1973. Electron microscopic studies on the mechanism of action of clotrimazole on *Candida albicans.* Sabouraudia *11*: 205-209.
112. Iwata, K., H. Yamaguchi, Y. Kanda, and M. Osumi. 1972. Mechanism of action of clotrimazole on *Candida albicans*: an electron microscopic study. In K. Iwata (Ed.), *Yeasts and Yeast-like Microorganisms in Medical Science,* Proceedings of the Second International Specialized Symposium on Yeasts, University of Tokyo Press, Tokyo, pp. 335-338.
113. Jan, Y. N. 1974. Properties and cellular localization of chitin synthetase in *Phycomyces blakesleeanus.* J. Biol. Chem. *249*: 1973-1979.
114. Kanetsuna, F., and L. M. Carbonell. 1971. Cell wall composition of yeast-like and mycelial forms of *Blastomyces dermatitidis.* J. Bacteriol. *106*: 946-948.
115. Kanetsuna, F., L. M. Carbonell, I. Azuma, and Y. Yamamura. 1972. Biochemical studies on the thermal dimorphism of *Paracoccidioides brasiliensis.* J. Bacteriol. *110*: 208-218.
116. Kanetsuna, F., L. M. Carbonell, F. Gil, and I. Azuma. 1974. Chemical and ultrastructural studies on the cell walls of *Histoplasma capsulatum.* Mycopathologia *54*: 1-13.
117. Kendrick, W. B., and A. Molnar. 1965. A new *Ceratocystis* and its *Verticicladiella* imperfect stage associated with the bark beetle *Dryocoetes confusus* on *Abies lasiocarpa.* Can. J. Bot. *43*: 39-43.
118. Kirk, B. T., and J. G. Sinclair. 1966. Plasmodesmata between hyphal cells of *Geotrichum candidum.* Science *153*: 1646.
119. Kobayashi, G. S., and P. L. Guiliacci. 1967. Cell wall studies of *Histoplasma capsulatum.* Sabouraudia *5*: 180-188.
120. Kobayashi, G. S., and G. Medoff. 1977. Antifungal agents: recent developments. Annu. Rev. Microbiol. *31*: 291-308.
121. Kozar, F., and J. Weijer. 1969. Electron-dense structures in *Neurospora crassa.* II. Lomasome-like structures. Can. J. Genet. Cytol. *11*: 617-621.
122. Kreger-van Rij, N. J. W., and M. Veenhuis. 1970. An electron microscopic study of the yeast *Pityrosporum ovale.* Arch. Mikrobiol. *71*: 123-131.
123. Kreger-van Rij, N. J. W., and M. Veenhuis. 1971. A comparative study of the cell wall structure of basidiomycetous and related yeasts. J. Gen. Microbiol. *68*: 87-95.
124. Kreger-van Rij, N. J. W., and M. Veenhuis. 1971. Septal pores in *Trichosporon cutaneum.* Sabouraudia *9*: 36-38.

125. Krejzova, R. 1977. Morphology and surface structures of *Conidiobolus coronata* (Cost.) Batko. Ceska Mykol. *31*: 28-30.

126. Kwon-Chung, K. J. 1975. A new genus, *Filobasidiella,* the perfect state of *Cryptococcus neoformans.* Mycologia *67*: 1197-1199.

127. Kwon-Chung, K. J., and T. J. Popkin. 1976. Ultrastructure of septal complex in *Filobasidiella neoformans (Cryptococcus neoformans).* J. Bacteriol. *125*: 524-528.

128. Lambiase, J. T., and W. G. Yendol. 1977. The fine structure of *Entomophthora apiculata* and its penetration of *Trichopulsia ni.* Can. J. Microbiol. *23*: 452-464.

129. Lane, J. W., and R. G. Garrison. 1970. Electron microscopy of self-parasitism by *Histoplasma capsulatum* and *Blastomyces dermatitidis.* Mycopathol. Mycol. Appl. *40*: 271-276.

130. Lane, J. W., R. G. Garrison, and M. F. Field, 1969. Ultrastructural studies on the yeastlike and mycelial phases of *Sporotrichum schenckii.* J. Bacteriol. *100*: 1010-1019.

131. Lane, J. W., R. G. Garrison, and D. R. Johnson. 1972. Drug-induced alterations in the ultrastructural organization of *Histoplasma capsulatum* and *Blastomyces dermatitidis.* Mycopathol. Mycol. Appl. *48*: 289-296.

132. Ledbetter, M. C., and K. R. Porter. 1963. A "microtubule" in plant fine structure. J. Cell Biol. *19*: 239-250.

133. Leiva, S., and L. M. Carbonell. 1970. Demonstration of succinic dehydrogenase in the mesosomes of the mycelial phase of *Paracoccidioides brasiliensis.* J. Gen. Microbiol. *62*: 43-47.

134. Lewis, P. R. 1975. Electron microscopical localization of enzymes. In R. Barer and V. E. Casslett (Eds.), *Advances in Optical and Electron Microscopy,* Vol. 6, Academic, New York, pp. 171-226.

135. Lowry, R. J., and A. S. Sussman. 1966. Intrahyphal hyphae in "clock" mutants of *Neurospora.* Mycologia *58*: 541-548.

136. Luck, D. J. L. 1963. Formation of mitochondria in *Neurospora crassa.* A quantitative radioautographic study. J. Cell Biol. *16*: 483-499.

137. Luck, D. J. L. 1965. The influence of precursor pool size on mitochondrial composition in *Neurospora crassa.* J. Cell Biol. *24*: 445-460.

138. Luck, D. J. L. 1965. Formation of mitochondria in *Neurospora crassa*: a study based on mitochondrial density changes. J. Cell Biol. *24*: 461-470.

139. McAlear, J. H., and G. A. Edwards. 1959. Continuity of plasma membrane and nuclear membrane. Exp. Cell Res. *16*: 689-692.

140. McClure, W. K., D. Park, and P. M. Robinson. 1968. Apical organization in the somatic hyphae of fungi. J. Gen. Microbiol. *50*: 177-182.

141. McCully, E. K., and C. E. Bracker. 1972. Apical vesicles in growing bud cells of heterobasidiomycetous yeasts. J. Bacteriol. *109*: 922-926.

142. McCully, E. K., and C. F. Robinow. 1971. Mitosis in the fission yeast *Schizosaccharomyces pombe.* A comparative study with light and electron microscopy. J. Cell Sci. *9*: 475-507.

143. McCully, E. K., and C. F. Robinow. 1972. Mitosis in heterobasidiomycetous yeasts. I. *Leucosporidium scotti (Candida scotti).* J. Cell Sci. *10*: 857-881.

144. McCully, E. K., and C. F. Robinow. 1972. Mitosis in heterobasidiomycetous yeasts. II. *Rhodosporium* sp., *Rhodotorula glutinis* and *Aesiosporum salmonicolor (Sporobolomyces salmonicolor).* J. Cell Sci. *11*: 1-31.

145. Maggi, V. 1973. Lysosomes. In E. E. Bittar (Ed.), *Cell Biology in Medicine,* Wiley, New York, pp. 215-263.

146. Mahvi, T. A., S. S. Spicer, and N. J. Wright. 1974. Cytochemistry of acid muco-

substances and acid phosphatase in *Cryptococcus neoformans*. Can. J. Microbiol. *20*: 833-838.

147. Mares, D., G. L. Vannini, M. P. Fasulo, and A. Bruni. 1977. Submicroscopic morphology of *Trichophyton mentagrophytes* grown at different temperatures. Mycopathologia *61*: 43-48.

148. Mariat, F., R. G. Garrison, K. S. Boyd, M. A. Rouffaud, and H. Fromentin. 1978. Premières observations sur les macrospores pigmentées de *Sporothrix schenckii*. C. R. Acad. Sci., Ser. D *286*: 1429-1432.

149. Mason, P. J., and R. Crosse. 1975. Crystalline inclusions in hyphae of the glacus group of aspergilli. Trans. Br. Mycol. Soc. *65*: 129-162.

150. Marchant, R., and R. T. Moore. 1973. Lomasomes and plasmalemmasomes in fungi. Protoplasma *76*: 235-247.

151. Marchant, R., A. Peat, and G. H. Banbury. 1967. The ultrastructural basis of hyphal growth. New Phytol. *66*: 623-629.

152. Marchant, R., and A. W. Robards. 1968. Membrane systems associated with the plasmalemma of plant cells. Ann. Bot. (Lond.) *32*: 457-471.

153. Marchant, R., and G. Smith. 1968. Bud formation in *Saccharomyces cerevisiae* and a comparison with the mechanism of cell division in other yeasts. J. Gen. Microbiol. *53*: 163-169.

154. Marriott, M. S. 1975. Enzymatic activity of purified plasma membranes from the yeast and mycelial forms of *Candida albicans*. J. Gen. Microbiol. *89*: 345-352.

155. Matile, P., and A. Wiemken. 1967. The vacuole as the lysosome of the yeast cell. Arch. Microbiol. *56*: 148-155.

156. Matile, P., H. Moor, and K. Mühlethaler. 1967. Isolation and properties of plasmalemma in yeast. Arch. Microbiol. *58*: 201-211.

157. Matile, P., H. Moor, and C. F. Robinow. 1969. Yeast cytology. In A. H. Rose and J. S. Harrison (Eds.), *The Yeasts,* Vol. 1, Academic, New York, pp. 219-297.

158. Maxwell, D. P., M. D. Maxwell, G. Haussler, V. N. Armentrout, G. M. Murray, and H. C. Hoch. 1975. Microbodies and glyoxylate-cycle enzyme activities in filamentous fungi. Planta *124*: 109-123.

159. Moens, P. B., and E. Rapport. 1971. Spindles, spindle plaques, and meiosis in the yeast *Saccharomyces cerevisiae* (Hansen). J. Cell Biol. *50*: 344-361.

160. Moor, H. 1964. Die Gefrien-fixation lebender Zellen und ihre Anwendung in der Elektronmikroskopie. Z. Zellforsch. Mikrosk. Anat. *62*: 546-580.

161. Moor, H., and K. Mühlethaler. 1963. Fine structure in frozen-etched yeast cells. J. Cell Biol. *17*: 609-628.

162. Moore, R. T. 1965. The ultrastructure of fungal cells. In G. C. Ainsworth and A. S. Sussman (Eds.), *The Fungi: An Advanced Treatise,* Vol. 1, Academic, New York, pp. 95-118.

163. Moore, R. T. 1975. Early ontogenic stages in dolipore/parenthesome formation in *Polysporus biennis*. J. Gen. Microbiol. *87*: 251-259.

164. Moore, R. T. 1978. Taxonomic significance of septal ultrastructure with particular reference to the jelly fungi. Mycologia *70*: 1007-1024.

165. Moore, R. T., and N. J. W. Kreger-van Rij. 1972. Ultrastructure of *Filobasidium olive*. Can. J. Bot. *18*: 1949-1951.

166. Moore, R. T., and J. H. McAlear. 1963. Fine structure of mycota. 9. Fungal mitochondria. J. Ultrastruct. Res. *8*: 144-153.

167. Moore, R. T., and J. H. McAlear. 1968. Fine structure of mycota. 5. Lomasomes—previously uncharacterized hyphal structures. Mycologia *53*: 194-200.

168. Moore-Landeker, E. 1971. Ultrastructural observations on lamellar and tubular membrane configurations in fungi. Cytologia *36*: 563-574.

169. Novikoff, A. B. 1961. Lysosomes and related particles. In J. Bracket and A. Mirsky (Eds.), *The Cell,* Vol. 2, Academic, New York, pp. 423-488.
170. Noguchi, T., Y. Kitajima, Y. Nozawa, and Y. Ito. 1971. Isolation, composition, and structure of cell walls of *Trichophyton mentagrophytes.* Arch. Biochem. Biophys. *146*: 506-512.
171. Nozawa, Y., Y. Kitajima, T. Sekiya, and Y. Ito. 1974. Ultrastructural alterations induced by amphotericin B in the plasma membrane of *Epidermophyton floccosum* as revealed by freeze-etch electron microscopy. Biochim. Biophys. Acta *367*: 32-38.
172. Nurminen, T., E. Oura, and H. Suomalainen. 1970. The enzymatic composition of the isolated cell wall and plasma membrane of baker's yeast. Biochem. J. *116*: 61-69.
173. Oujezdsky, K. B., S. N. Grove, and P. J. Szaniszlo. 1973. Morphological and structural changes during the yeast-to-mold conversion of *Phialophora dermatitidis.* J. Bacteriol. *113*: 468-477.
174. Padhye, A. A., and J. W. Carmichael. 1972. Ascospore morphology of *Nannizzia* and *Arthroderma* species by scanning electron microscopy. Sabouraudia *10*: 313-314.
175. Page, W. J., and J. J. Stock. 1974. Sequential action of cell wall hydrolases in the germination and outgrowth of *Microsporon gypseum* macroconidia. Can. J. Microbiol. *20*: 483-489.
176. Patton, A. M., and R. Marchant. 1978. A mathematical analysis of dolipore/parenthesome structure in basidiomycetes. J. Gen. Microbiol. *109*: 335-349.
177. Peterson, J. B., R. H. Gray, and H. Ris. 1972. Meiotic spindle plaques in *Saccharomyces cerevisiae.* J. Cell Biol. *53*: 837-841.
178. Pine, L., and C. J. Boone. 1968. Cell wall composition and serological reactivity of *Histoplasma capsulatum* serotypes and related species. J. Bacteriol. *96*: 789-798.
179. Pitt, D. 1968. Histochemical demonstration of certain hydrolytic enzymes within cytoplasmic particles of *Botyritis cinerea.* J. Gen. Microbiol. *52*: 67-75.
180. Pitt, D., and P. J. Walker. 1967. Particulate localization of acid phosphatase in fungi. Nature (Lond.) *215*: 783-784.
181. Pock-Steen, B., and T. Kobayasi. 1970. Ultrastructure of the hyphal wall and septum of *Trichophyton mentagrophytes.* J. Invest. Dermatol. *55*: 404-409.
182. Poon, N. H., and A. W. Day. 1975. Fungal fimbriae. I. Structure, origin, and synthesis. Can. J. Microbiol. *21*: 537-546.
183. Pringle, A., J. R. Forsdyke, and A. H. Rose. 1979. Scanning electron microscopic study of sphaeroplast formation in *Saccharomyces cerevisiae.* J. Bacteriol. *140*: 289-293.
184. Pugh, D., and R. A. Cawson. 1975. The cytochemical localization of phospholipase A and lysophospholipase in *Candida albicans.* Sabouraudia *13*: 110-115.
185. Ragner, N., L. Linkhalm, and E. Zyche. 1968. Lysosomes of *Toxoplasma gondii* and their possible relation to the host-cell penetration of toxoplasma parasites. J. Bacteriol. *96*: 916-919.
186. Reiss, J. 1971. Lysosome-like particles in *Geotrichum candidum*: a cytochemical study. Z. Allg. Mikrobiol. *11*: 319-323.
187. Reiss, J. 1973. *Enzyme Cytochemistry of Fungi,* Gustav Fischer Verlag, Stuttgart, pp. 1-39.
188. Reiss, E., S. E. Miller, W. Kaplan, and L. Kaufman. 1977. Antigenic, chemical, and structural properties of cell walls of *Histoplasma capsulatum* yeast-form chemotypes 1 and 2 after serial enzymatic hydrolysis. Infect. Immun. *16*: 690-700.

189. Revel, J. P., L. Napolitano, and D. W. Fawcett. 1960. Identification of glycogen in electron micrographs of thin tissue sections. J. Biophys. Biochem. Cytol. *8*: 575-589.
190. Revel, J. P., L. Napolitano, and D. W. Fawcett. 1964. Electron microscopy of glycogen. J. Histochem. Cytochem. *12*: 104-114.
191. Robbins, E., and N. K. Gonatas. 1964. The ultrastructure of a mammalian cell during the mitotic cycle. J. Cell Biol. *21*: 429-463.
192. Robertson, J. D. 1960. The molecular structure and contact relationships of cell membranes. Prog. Biophys. Mol. Biol. *10*: 343-418.
193. Robertson, N. F. 1965. The mechanism of cellular extension and branching. In G. C. Ainsworth and A. S. Sussman (Eds.), *The Fungi*: *An Advanced Treatise,* Vol. 1, Academic, New York, pp. 613-623.
194. Robinow, C. F. 1963. Observations on cell growth, mitosis, and division in the fungus *Basidiobolus ranarum.* J. Cell Biol. *17*: 123-152.
195. Robinow, C. F., and A. Bakerspiegel. 1965. Somatic nuclei and forms of mitosis in fungi. In G. C. Ainsworth and A. S. Sussman (Eds.), *The Fungi*: *An Advanced Treatise,* Vol. 1, Academic, New York, pp. 119-142.
196. Roth, G. D., R. G. Garrison, and K. S. Boyd. 1976. Ultrastructure of *Rothia dentocariosa.* J. Gen. Microbiol. *95*: 45-53.
197. Sassen, M. M. A., C. C. Remsen, and W. M. Hess. 1967. Fine structure of *Penicillium megasporum* conidiospores. Protoplasma *64*: 75-88.
198. Schneider, E. F., and W. L. Seamon. 1977. Ontogeny of lipid bodies in the endoplasmic reticulum of *Fusarium sulphureum.* Can. J. Microbiol. *23*: 190-196.
199. Schrantz, J. P. 1970. Etude cytologique en microscopie optique et électronique de quelques ascomycètes. I. Le noyau. Rev. Cytol. Biol. Veg. *33*: 1-67.
200. Schrantz, J. P. 1970. Etude cytologique en microscopie optique et électronique de quelques ascomycètes. II. La paroi. Rev. Cytol. Biol. Veg. *33*: 111-168.
201. Seligman, A. M., H. L. Wasserkrug, and J. A. Hanker. 1966. A new staining method (OTO) for enhancing contrast of lipid-containing membranes and droplets in osmium tetroxide fixed tissue with osmiophilic thiocarbohydrazide (TCH). J. Cell Biol. *30*: 424-432.
202. Shannon, J. L., and A. H. Rothman. 1971. Transverse septum formation in budding cells of the yeastlike fungus *Candida albicans.* J. Bacteriol. *106*: 1026-1028.
203. Shimizu, K., J. Hosoi, M. Ueda, and Y. Miyazawa. 1977. Electron microscopic study of *Geotrichum candidum* monomer ribosomes. J. Electron Microsc. *26*: 41-46.
204. Shinagawa, Y., A. Inouye, T. Ohnishi, and B. Hagihara. 1966. Electron microscopic studies of isolated yeast mitochondria with negative staining and thin sectioning methods. Exp. Cell Res. *43*: 301-310.
205. Slack, J. M., and M. A. Gerencser. 1975. *Actinomyces, Filamentous Bacteria*: *Biology and Pathogenicity,* Burgess, Minneapolis.
206. Snitka, T. K., and A. M. Seligman. 1971. Ultrastructural localization of enzymes. Annu. Rev. Biochem. *40*: 375-396.
207. Stastná, J., and P. Trávnik. 1971. Electron microscopic detection of PAS-positive substances with thiosemicarbazide. Histochemie *27*: 63-68.
208. Steele, S. D., and T. W. Fraser. 1973. Ultrastructural changes during germination of *Geotrichum candidum* arthrospores. Can. J. Microbiol. *19*: 1031-1034.
209. Steele, S. D., and T. W. Fraser. 1973. The ultrastructure of *Geotrichum candidum.* Can. J. Microbiol. *19*: 1507-1512.
210. Steiner, J. W., K. Miyai, and M. J. Philips. 1964. Electron microscopy of mem-

brane-particle arrays in liver cells of ethionine-intoxicated rats. Am. J. Pathol. *44*: 169-214.

211. Stiers, D. L. 1974. Fine structure of ascospore formation in *Poronia punctata*. Can. J. Bot. *52*: 999-1003.

212. Sun, N. C., and C. C. Bowen. 1972. Ultrastructural studies of nuclear division in *Basidiobolus ranarum* Eidam. Caryologia *25*: 471-494.

213. Sun, S. H., S. S. Sekhon, and M. Huppert. 1979. Electron microscopic studies of saprobic and parasitic forms of *Coccidioides immitis*. Sabouraudia *17*: 265-273.

214. Takeo, K., I. Uesaka, K. Uehira, and M. Nishimura. 1973. Fine structure of *Cryptococcus neoformans* grown in vitro as observed by freeze-etching. J. Bacteriol. *113*: 1442-1448.

215. Takeo, K., I. Uesaka, K. Uehira, and M. Nishimura. 1973. Fine structure of *Cryptococcus neoformans* grown in vivo as observed by freeze-etching. J. Bacteriol. *113*: 1449-1454.

216. Tandler, B., and C. L. Hoppel. 1972. *Mitochondria,* Academic, New York.

217. Taylor, M. M., and R. Storck. 1964. Uniqueness of bacterial ribosomes. Proc. Natl. Acad. Sci. U.S.A. *52*: 958-965.

218. Teh, E. C. 1971. Surface structures of the endospores of *Rhinosporidium seeberi*. Sabouraudia *12*: 263-265.

219. Thibaut, M., M. Ansel, and J. De Azevedo Carneiro. 1977. X-ray microanalysis as applied to fungal tissue. Br. J. Exp. Pathol. *58*: 209-214.

220. Thiéry, J. P. 1967. Mise en évidence des polysaccharides sur coupes fines en microscopie électronique. J. Microsc. (Paris) *6*: 987-1018.

221. Travassos, L. R., W. de Sousa, L. Mendonca-Previato, and K. O. Lloyd. 1977. Location and biochemical nature of surface components reacting with concanavalin A in different cell types of *Sporothrix schenckii*. Exp. Mycol. *1*: 293-305.

222. Trinci, A. P. J., and A. J. Collinge. 1974. Occlusion of the septal pore of damaged hyphae of *Neurospora crassa* by hexagonal crystals. Protoplasma *80*: 57-67.

223. Trinci, A. P. J., and P. T. Saunders. 1977. Tip growth of fungal hyphae. J. Gen. Microbiol. *103*: 243-248.

224. Trump, B. F., and J. L. E. Ericsson. 1965. The effect of the fixative solution on the ultrastructure of cells and tissues. A comparative analysis with particular attention to the proximal convoluted tubule of the rat kidney. Lab. Invest. *14*: 1245-1323.

225. Tsukahara, T., M. Yamada, and T. Itagaki. 1966. Micromorphology of conidiospores of *Aspergillus niger* by electron microscopy. Jpn. J. Microbiol. *10*: 93-107.

226. Urabe, H., and T. Izu. 1969. The ultrastructure of *Trichophyton* and a double cell wall in the hyphae. J. Invest. Dermatol. *52*: 508-513.

227. Virgil, E. L. 1973. Structure and function of plant microbodies. Subcell. Biochem. *2*: 237-285.

228. Vorisek, J., J. Ludvik, and Z. Rehácek. 1974. Morphogenesis and ultrastructure of *Claviceps purpurea* during submerged alkaloid formation. J. Bacteriol. *120*: 1401-1408.

229. Wergin, W. P. 1973. Development of Woronin bodies from microbodies in *Fusarium oxysporum* f. sp. *lycopersici*. Protoplasma *76*: 249-260.

230. Werner, H. J., H. W. Jolly, and C. L. Carpenter. 1968. Observations on the fine structure of *Microsporum audounii*. J. Invest. Dermatol. *50*: 276-279.

231. Wheeler, G., R. Michell, and A. H. Rose. 1972. Phosphatidyl inositol kinase activity in *Saccharomyces cerevisiae*. Biochem. J. *127*: 64.

232. Williams, S. T., G. P. Sharples, and R. M. Bradshaw. 1973. The fine structure of the Actinomycetales. In G. Sykes and F. A. Skinner (Eds.), *Actinomycetales: Characteristics and Practical Importance,* Academic, New York, pp. 113-130.
233. Wilsenach, R., and M. Kessel. 1965. The role of lomasomes in wall formation in *Penicillium vermiculatum.* J. Gen. Microbiol. *40*: 401-404.
234. Wilsenach, R., and M. Kessel. 1965. Micropores in the cross walls of *Geotrichum candidum.* Nature (Lond.) *207*: 545-546.
235. Wilson, C. L., D. L. Stiers, and G. G. Smith. 1970. Fungal lysosomes or spherosomes. Phytopathology *60*: 216-227.
236. Yamaguchi, H., and K. Iwata. 1970. In vitro and in vivo protein synthesis in *Candida albicans.* I. Isolation and protein synthesis of polysomes. Sabouraudia *8*: 177-189.
237. Yamaguchi, H., Y. Kanda, and K. Iwata. 1971. Biochemical properties of mitochondria from *Candida albicans.* Sabouraudia *9*: 221-230.
238. Zachariah, K. 1970. Lattice bodies and filament bundles in apothecial cells of the fungus *Ascobolus.* Can. J. Bot. *48*: 1115-1118.
239. Zickler, D. 1970. Division spindle and centrosomal plaques during mitosis and meiosis in some ascomycetes. Chromosoma *30*: 287-304.

10

Dimorphism: Morphological and Biochemical Aspects

Paul J. Szaniszlo, Charles W. Jacobs, and Philip A. Geis* / The University of Texas at Austin, Austin, Texas

I. INTRODUCTION

Many fungi which cause human diseases grow vegetatively in more than one morphology. Usually, the saprophytic morphology of these fungi consists of hyphae and sporulation structures. In vivo many of the same fungi grow vegetatively in reduced phenotypes which are yeast-like, pseudomycelial, multicellular, or sporangial. A smaller number of pathogenic fungi are predominantly yeasts as saprophytes, but exhibit filamentous or other vegetative morphologies in the host.

The ability of certain pathogenic fungi to exhibit a phenotypic duality of form in which there are distinct saprophytic and parasitic phases has been termed dimorphism [2,

**Present affiliation:* The Proctor & Gamble Company, Cincinnati, Ohio

471]. In medical mycology, dimorphism implies that infection is accompanied by a transition of the pathogenic fungus from its saprophytic to its parasitic morphology [268]. The term "dimorphism" is also used extensively in general mycology to describe environmentally controlled interconversions of yeast and mycelial forms of a fungus regardless of its medical importance [479]. Frequently, the term is used more broadly to describe any distinct vegetative transformation exhibited by a fungus [533]. In general mycology, a fungus may be dimorphic but not be pathogenic. It is obvious that the use of the term dimorphism in medical mycology does not always equate directly with the use of the term in general mycology.

The ambiguity relating to the term "dimorphism" is avoided in this chapter by considering as dimorphic all pathogenic fungi that undergo distinct phase transitions either in vivo or in vitro. This broad usage of the term allows us to consider the yeast-hyphal transitions associated with a diverse array of organisms, such as *Emmonsiella (Histoplasma) capsulatum, Exophiala werneckii,* and *Candida albicans,* the hyphal-spherule transitions of *Coccidioides immitis,* and the transitions to thick-walled and often multicellular morphologies of agents such as *Phialophora verrucosa* and *Wangiella dermatitidis.* Although the process of dimorphism is generally considered to be reversible, it is apparent that with many dimorphic fungi, cyclical life cycle processes account for the phase transitions. In some instances the transitions are associated with cell cycle phenomena and involve cells and hyphae growing logarithmically. Other transitions are associated with cells having reached certain stages of maturation. In the latter case, phase transition may be associated with stationary-phase cells. Unfortunately, the distinction between the growth phase of the cell undergoing phase transition is often ignored, even though major differences in metabolism are associated with logarithmically growing and stationary-phase cells.

Our point of view in this chapter is that the majority of pathogenic fungi represent misplaced saprophytes which are able to grow in the unfriendly environments of host tissue. In the host, the fungus undoubtedly encounters environmental pressures of sufficient magnitude to alter fungal metabolism significantly. If the fungus has the inherent ability to undergo phase transitions, the altered metabolism may lead to altered phenotypic growth. The altered phenotypes of some pathogenic fungi in host tissues are monomorphic, suggesting that for these fungi the host conditions are totally permissive for the altered phenotype and totally restrictive for the saprophytic morphology. The morphology exhibited by other pathogenic fungi in the host are mixtures of altered and saprophytic phenotypes. Apparently, the host environment does not always provide every pathogenic fungus with all the essentials for uniquely monomorphic growth. Instead, the host environment may represent an intermediate environment which allows the simultaneous growth of the fungus in two or more interconverting phenotypes. These phenotypes may be localized in the same tissue or be restricted to different tissues of the same organism. It is likely that the altered phenotypes are better able to survive host defense mechanisms, although definitive proof for this supposition awaits rigorous experimentation.

The diversity of dimorphic fungi makes it impossible to identify all the morphologies they exhibit in the host. Certainly, it is not known whether the tissue phase of *C. immitis* or *Emmonsia crescens* and *E. parva* is an integral component of the saprophytic phases of these fungi. However, it does appear that the yeast cells of many medically important fungi represent the ultimate phenotypic reduction of a hyphal fungus and, in fact, represent disarticulated conidiophores.

A case can be made that altered fungal morphologies are the end result of defensive and offensive survival battles between host tissue and fungus. When the fungus survives the battle, it produces the disease state. A case can also be made that all fungi exhibit altered phenotypes in host tissue. For this reason we have chosen to focus on those fungi exhibiting distinct phase transitions either in vivo or in vitro. A listing of the fungi considered in this chapter, together with a brief description of their saprophytic and parasitic morphologies, is presented in Table 1. We have included the mucors in our review because some members represent the most intensively studied dimorphic fungi. The study of these fungi offers important insights into the myriad of factors that influence fungal morphogenesis. Also, the molecular basis of dimorphism has been studied more thoroughly in these fungi than in any others. The wide-ranging investigations of the dimorphic mucors suggest that they may be the first to reveal how form is determined. As with only a few other dimorphic fungi, the dimorphic mucors are well suited for morphogenic study because they exhibit defined periods of apical, budding, and isotropic growth. The apical mode of growth gives rise to hyphae, the budding mode to yeasts, and the isotropic mode to enlarged spherical forms.

II. SPECIES WITH PREDOMINANTLY YEAST MORPHOLOGY IN VIVO

A. *Ajellomyces (Blastomyces) dermatitidis*

Ajellomyces (Blastomyces) dermatitidis exhibits a yeast-hyphal dimorphism which is primarily regulated by temperature. The biphasic nature of this fungus was first noted by Gilchrist and Stokes, who in 1898 observed individual yeasts forming germ tubes which subsequently gave rise to hyphae [210]. They noted that upon isolation from tissue, the fungus is exclusively in the yeast morphology, measures 10-20 μm in diameter, has thick walls, and forms buds with broad bases. In laboratory culture at room temperatures, the tissue-phase yeast gives rise to a grayish white filamentous colony which later turns pure white. Conidia arise either from short lateral hyphae or directly from the sides of hyphae. The conidia tend to be smooth-walled, hyaline, pyriform cells, 3-5 μm in diameter [107]. Both yeast cells and conidia may give rise to either hyphae or yeasts, depending on cultural manipulation [210]. When yeast colonies develop, they are heaped, crumbly structures consisting of thick-walled, globose, budding cells, and chains of chlamydospores [107].

Conant and Howell [134] and Howard and Herndon [271] observed that hyphal forms give rise to budding yeasts after a shift from 25°C to 37°C. They noted that the hyphal → yeast conversion involves the sequestering of cytoplasmic contents into hyphal segments. These segments form into round or pyriform transition cells which then produce yeasts by budding. Garrison and Boyd [201] suggest that yeasts arise primarily from conidia, and that hyphal fragments also give rise to yeast cells, but that such conversion is relatively inefficient with regard to production of large numbers of cells.

The ability of conidia to convert to the yeast morphology is perhaps more important than hyphal → yeast conversion because it is believed that inhalation of airborne conidia represents the primary mode of transmission of blastomycosis [5,6,192,301]. Although conidia incubated at 25°C germinate to form hyphae, they also have the capacity to produce secondary blastoconidia by a mechanism that resembles budding [106,268]. When germination is carried out at 37°C, a bud-like structure emerges after 24 h [201]. This structure arises from the innermost layer of the conidial wall by splitting of the outer three wall layers. However, instead of forming a germ tube, the bud-like structure becomes a conidial yeast mother cell. Acting as a transitional cell, the yeast mother cell contains

Table 1 The Dimorphic Pathogenic Fungi and Their Growth Phases

In vivo morphology group	Fungus	Saprophytic phase	Parasitic phase
Predominantly yeasts	*Ajellomyces (Blastomyces) dermatitidis*	Septate hyphae; white or beige, fluffy or glabrous colonies; microconidia, cleistothecia, ascospores	Budding yeasts, 8-20 μm; neck between mother cell and daughter bud very broad
	Emmonsiella (Histoplasma) capsulatum	Septate hyphae; white or tan, fluffy or flat colonies; tuberculate macroconidia, microconidia, cleistothecia, ascospores	Budding yeasts, 2-4 μm; neck between mother cell and daughter bud very narrow
	Paracoccidioides brasiliensis	Septate hyphae; white to beige, glabrous, leathery, flat, raised, or velvety colonies; no characteristic conidia	Budding yeasts, size variable, 2-30 μm or more; buds one to many per mother cell with neck between mother cell and daughter bud very narrow
	Sporothrix schenckii	Septate hyphae; white, black, or gray, glabrous or fuzzy, wrinkled colonies; conidiogenesis similar to some *Ceratocystis* species	Budding yeasts, size variable, 2-10 μm or more; sometimes cigar-shaped and up to 30 μm long; buds one to many per mother cell
Interconverting phenotypes	*Candida albicans*	Budding yeast forms predominate, size 4-6 μm; smooth and creamy, white colonies; true hyphae, pseudohyphae, and chlamydospores not infrequent under certain environmental conditions	Mixtures of true hyphae, pseudohyphae, and attendant budding blastoconidia (yeasts)
	Exophiala werneckii	Budding yeast forms predominate, in young cultures 6-12 μm; septate hyphae arise in older cultures, many intermediate forms; white, gray, or green-black, glabrous or fuzzy colonies	Mixtures of hyphae and attendant budding blastoconidia (yeasts)

	Wangiella dermatitidis	Budding yeast forms predominate in young cultures, 6–8 μm; septate moniliform and true hyphae arise in older cultures, phialoconidia and annelloconidia	Mixtures of hyphae and attendant budding blastoconidia (yeasts); occasionally, thick-walled and swollen cells and multicellular (sclerotic) forms
Predominantly swollen (isotropic) phenotypes	*Coccidioides immitis*	Septate hyphae which yield arthroconidia upon fragmentation; white, tan, or brown, glabrous or fluffy, smooth or wrinkled colonies; arthroconidia typically barrel-shaped with disjuncters	Spherules which produce numerous endospores; size of spherules at maturity 30–60 μm in diameter, endospores formed by progressive cleavage of cytoplasm
	Emmonsia parva and *E. crescens*	Septate hyphae; clear to white, glabrous to tufted colonies; smooth or spiny aleurioconidia	Adiaspores which represent very enlarged aleurioconidia; size varies from 10 to 400 μm in diameter; growth in tissue only by enlargement
	Phialophora verrucosa and other chromomycotic fungi	Septate hyphae; green-brown to green-black, fluffy colonies; phialoconidia, annelloconidia, etc.	Sclerotic cells which are swollen, thick-walled cells and thick-walled multicellular forms
	Mucor (pathogenic species rarely dimorphic)	Sparsely septate or aseptate hyphae; wooly white or gray colonies; sporangia, sporangiospores, and zygospores; some species form budding yeast forms	Hyphae very swollen, irregular, sparsely branched, and with few septa

numerous organelles, including mitochondria, lipid bodies, vacuoles, and extensive membrane systems. In time, the yeast mother cell gives rise to daughter yeasts.

Aside from the quite obvious differences between hyphae and yeasts, there are numerous ultrastructural differences among the yeasts, hyphae, and transitional forms, particularly with regard to septa and cell walls. Yeast cells form double, imperforate septa, which are cleaved by means of a furrow that develops between them during cell separation [204]. Conversely, hyphae form single, perforate septa with associated Woronin bodies. However, the hyphal septa are thicker than those of the yeasts [105,204]. In contrast, hyphal walls are generally thinner than yeast walls and are only about 80 nm thick, increasing to about 150 nm near the septa. The hyphal wall apparently consists of two fibrillar layers of differing textures [105]. The yeast wall ranges between 250 and 1200 nm in thickness, and also consists of two layers. In this case, however, the outer fibrillar layer is thin, only about 6 nm in cross section, and is made up of indefinitely arranged, but very obvious, fibrils. The inner layer of the yeast wall is homogeneously granular and contains no fibrils [99]. The first indication of transitional cell formation occurs within 6 h after induction of yeast → hyphal transition when an intracytoplasmic membrane system forms at the plasma membrane of the yeast cell. Between 12 and 18 h the yeast cell gives rise by blastic processes to a transition cell which has a characteristic ultrastructure unlike either the yeast or hyphal cell. The septum between the yeast and transitional cell has Woronin bodies associated with its septal pore. Mitochondria are densely packed in the transitional cell, indicating a high level of metabolic activity [204]. The wall of the transitional cell is thin, resembling that of the hyphal phase, as opposed to the thick wall of the yeast phase. Hyphae emerge from the transitional cell between 18 and 24 h after induction. At this time a complex intracytoplasmic membrane system and numerous mitochondria are located throughout the transition cell-germ tube complex. Hyphal elements which develop from the germ tubes show typical perforate septa with Woronin bodies and septum-associated membrane systems. It appears that the membrane systems arising during yeast → hyphal transition are not derived from yeast-phase organelles, but rather arise de novo during transition [204].

The morphology of *A. dermatitidis* is environmentally determined and is not dependent on whether the inoculum is derived from a yeast or hyphal culture [331,492]. The type of dimorphism exhibited by *A. dermatitidis* was termed thermal dimorphism [400] because the primary regulatory factor in determining morphology is the growth temperature. However, nutritional factors may modify the primary effect of temperature. The role of temperature in regulating dimorphism in this fungus was first studied about the turn of the century by Ricketts [469] and Hamburger [238], but the major work in demonstrating the importance of temperature was performed by Levine and Ordal [331] and Salvin [492]. These investigators determined that the phase of growth is dependent on temperature and independent of carbon or nitrogen source [492] or pH [331], and that yeast cells are formed when the growth temperature is 35°C or above, regardless of whether the medium is complex or defined. Temperature is also the sole determining factor when the organism is grown in mouse tissue culture [271].

Whereas nutrition alone cannot determine the phase of growth, several reports indicate that it can modify the response of the organism to temperature. One report indicates that elevated CO_2 levels are required for the induction of hyphal → yeast conversion, but not for the maintenance of the yeast morphology [204]. Another report [81] indicates that conversion will occur in an atmosphere of 15% CO_2, but it is not clear whether the CO_2

is obligatory for transition. However, a requirement of some strains for exogenous biotin [235], a cofactor in CO_2 fixation, suggests that CO_2 metabolism may play some role in dimorphism.

Domer [162] found that the yeast forms of a number of strains are not stable at 37°C in glucose yeast extract broth and begin a yeast → hyphal conversion after the second transfer in this medium. However, the yeast form of the same strains is stable at 37°C on brain heart infusion agar (BHI). The amino acid tyrosine may play a role in destabilizing the yeast form, since yeast-phase cells grown at 35°C yield hyphae in the presence of tyrosine, but not in its absence [207]. However, as is the case with other dimorphic fungi, it is difficult to separate differences in nutritional requirements due to morphology and those due to strain variation.

The biochemical differences associated with the two morphologies can be loosely grouped into two categories: those related to the cell wall and those related to the cytoplasm. Hyphal walls contain measurable amounts of galactomannans with a mannose/galactose ratio of 3:1. Yeast walls, on the other hand, contain no measurable galactomannans [18]. Other reports which do not lend themselves to broad generalities due to wide interstrain variation, indicate different glucose/galactose/mannose ratios in yeast and hyphal walls [297]. These reports also indicate that hyphal walls have about the same amount, or slightly more, hexose, but less amino sugar than the yeast walls [154,162, 297]. Not only does the monomeric composition of the walls exhibit differences correlated with morphological phase, but yeast and hyphal walls also exhibit differences in the structures of their polymers. For example, total yeast wall glucan is composed of 95% α-glucan and 5% β-glucan. Glucans of the hyphal wall are present in the ratio 60% α-glucan to 40% β-glucan [297]. The main glucan of both forms is apparently (1 → 4) linked [154]. The relative lack of α(1 → 3)-glucan in the yeast cell is accompanied by a high β(1 → 3)-glucanase activity. This activity is present in both hyphal and yeast forms, but the higher activity of the enzyme at temperatures required for induction of the yeast phase may result in degradation of β(1 → 3)-glucan as it is made. On the other hand, the possible compartmentation of synthetic and hydrolytic enzymes may play an undetermined role in the regulation of β(1 → 3)-glucan.

Interestingly, α-glucosidase activities are absent in both forms, but protease and α(1 → 3)-glucanase activities are present in both. This suggests that the latter enzymes are required for growth, but their exact roles are uncertain [154]. Differences in the amino acid compositions of the cell wall proteins, as well as differences in lipid composition, in the two forms are also apparent. As a percentage of total cell wall, hyphae contain about three times the amount of each amino acid, except histidine. Histidine levels are equal in the two forms [297,482,483]. As might be expected from the ultrastructural evidence, additional protein components are synthesized upon induction of the alternate morphology [483], indicating an active transition between the phases. Finally, the hyphal walls contain about twice as much lipid as the yeast walls [163,297]. The significance of the wall differences between the two forms of *A. dermatitidis* has not been rationalized in an appropriate model, such as that described for the thermally dimorphic fungus *Paracoccidioides brasiliensis.* However, the many similarities of the two species suggest that the same model might prove valid for both species and that possibility deserves evaluation.

The shift in morphology of *A. dermatitidis* results from a complex series of events which involves more than a shift in the nature of the cell wall. It involves changes in ribonucleic acid (RNA) and protein synthesis, and induction of new enzymes. Taylor [547] has shown that whereas the deoxyribonucleic acid (DNA) content per milligram of dry weight remains constant, the synthesis of RNA rises for abour 1 week or 10 days after the inoculation of the yeast phase onto a fresh agar medium at either 25°C or 37°C, then drops precipitously to a more or less steady state. The value for the RNA content of the yeast phase reaches a maximum at about 7 days and that for the mold phase at 10 days. The yeast phase reaches a higher peak value than the mold phase. The level of RNA in the hyphal form is always lower than it is in the yeast form, a fact that may be attributable to a slower and more uniform growth rate of the hyphal form. There is no significant difference in the guanine (G) + cytosine (C) content of the DNAs of the two phases, as determined by CsCl buoyant density sedimentation [44], as nuclear DNA from both yeasts and hyphae has a G + C content of 48.5%. Mitochondrial DNA is 33.7-34.9% G + C, depending on the strain. The amount of trichloroacetic acid (TCA)-precipitable nitrogen, which mostly represents protein, remains at a low level in the hyphal phase, but slowly falls from a very high level in the yeast phase. Taylor [547] interprets this as meaning that RNA and protein synthesis are involved in the dimorphic process, and claims that preliminary work indicates a causal role for RNA.

Roy and Landau have demonstrated a causal role of RNA and protein synthesis to a certain extent with their observation that yeast-hyphal interconversion involves the variation of the number and composition of cell wall proteins and malate dehydrogenase isoenzymes [481,483]. Their results show a minimum of four isoenzymes of malate dehydrogenase in the crude extract of the yeast form, and only two, which may or may not be the same as the two of the yeast form, in the hyphal form. Other studies showed that hyphae have two cytoplasmic and two mitochondrial isoenzymes, all of which are relatively slow in migrating in a disc-gel system, and that the yeast phase has four slowly migrating cytoplasmic types and two rapidly migrating mitochondrial types of enzymes. Sodium dodecyl sulfate (SDS) electrophoresis of cell wall proteins indicates a single resolvable protein band in the hyphal form and four in the yeast form. Although these results may not present an accurate picture of the populations of cell wall proteins, the bands may serve as indicators by representing the major proteins present in the two walls.

The ability of cells of the two morphologies to transport amino acids also differs. Results of uptake studies indicate that the Michaelis-Menten constants (K_m) for transport of L-leucine and L-cystine vary in the two growth forms [43]. The K_m values for cystine and leucine transport (which represent the external concentration of amino acid required to give half-maximal uptake) were about equal in the hyphal phase, whereas in the yeast phase the value for leucine transport was nine times higher than it was for cystine. The K_m value for leucine transport was four times higher for yeast-phase cells than for hyphae, but the value for cystine transport by yeast-phase cells was only half the corresponding value for the hyphae.

The rate of uptake of L-leucine and L-cystine per milligram of dry weight was faster by the hyphal phase than by the yeast phase. The uptake rate of leucine was 11 times faster and incorporation four times higher in the hyphal phase. The rate of cystine uptake was 10 times faster and incorporation was five times higher in the hyphal than it was in the yeast phase. It is not clear at this point how much of the disparity in uptake is attributable to differences in diffusion rates as opposed to the efficiency of active transport, although the relatively thick yeast cell wall probably interferes with diffusion to a great extent. The activities of the active transport systems do vary drastically between morphologies, however, and this may play an important role in the induction and maintenance of morphology.

A report that *A. dermatitidis* can survive for significant periods of time (up to 12 months) in tap water prompted Sheiban [508] to investigate the ability of the organism to undergo phase transitions in the same medium. Sheiban's findings indicate that either the yeast or hyphal morphologies may be derived in the absence of exogenous substrate. This confirms and extends a report [573] that exogenous hexoses are not required for conversion. It also raises the most interesting question as to whether a specific endogenous substrate is used during the phase change, as in the case of *Neurospora tetrasperma* and *Schizophyllum commune* in which trehalose [333] and mannitol and arabitol [4], respectively, are used as specific substrates during spore germination.

Nickerson and Edwards [400] noted that temperature affects O_2 consumption in an unusual way. In general, O_2 consumption of mycelia varies linearly with temperature until the temperature reaches about 32°C, whereupon the relation becomes curvilinear downward. They suggested that this phenomenon is the result of increasingly important reversible denaturation processes. When the temperature of incubation is shifted drastically (more than 15-20°C), the respiration rate exhibits an "overshoot" or "undershoot" phenomenon. That is, if a culture is equilibrated to 20°C and then quickly raised to 40°C, the rate of O_2 consumption is initially much higher than would be observed for a culture that had been slowly equilibrated to 40°C. The rate of O_2 consumption slowly returns to normal over a period of about 30 min. The opposite situation holds true for drastic temperature drops and the rate of O_2 consumption undershoots the expected rate. The authors suggested that the overshoot phenomenon and the postulated reversible denaturation shifts in the organism's metabolism induce the change in morphology.

Yeast-hyphal interconversion in *A. dermatitidis* involves a large number of events. Far from being a simple shifting of the mode of cell wall synthesis from construction of cylindrical to spherical walls, the transition is an active process. Transition is not direct and proceeds through specialized intermediate cells. Yeasts give rise to hyphal transition cells. All of these cell types are both morphologically and functionally transitional, often combining features of both the initial and final morphologies. However, this should not be surprising considering the differences between yeasts and hyphae. Certainly, the two morphologies differ in the compositions of their cell wall carbohydrates [18,154,162, 297], RNAs [483,547], isoenzymes of central metabolic pathways [483], and resistance to antibiotics [367]. Expression of dimorphism in this organism is a process which involves growth not only in two morphological forms, but in two very distinct physiological forms. It is these physiological distinctions which may favor the expression of one morphology over the other in establishing and propagating the disease state.

B. *Emmonsiella (Histoplasma) capsulatum*

Emmonsiella (Histoplasma) capsulatum, the causative agent of histoplasmosis, is the most studied of the dimorphic fungi that have a parasitic phase consisting predominantly of yeast cells. The nature of dimorphism in *E. capsulatum* has been the subject of continuous, intensive study by a host of investigators ever since this fungus was first cultured in vitro and discovered to have a hyphal saprophytic phase in addition to its parasitic yeast phase [157,161,240,241,462]. The results of this study provide a surfeit of information which is only beginning to yield insight into the regulatory processes responsible for phase maintenance and phase transition.

The hyphal phase of *E. capsulatum* is found in soil and can be maintained in vitro on a variety of media at temperatures below 34°C. Two types of hyphal morphologies have been described [48] and are designated type A (albino) and type B (brown). Type A hyphae produce a white colony without diffusible pigment, whereas the B type grows as a pigment-producing brown colony. The hyphal types remain morphologically constant when *E. capsulatum* is maintained in the yeast form, but the brown type tends to convert to the white type during prolonged mycelial culture. Each of the two morphological forms elaborate microconidia and macroconidia as asexual reproductive propagules [179]. Type A hyphae produce few macroconidia, but many microconidia, whereas type B hyphae mostly produce macroconidia [48].

Although it is generally recognized that *E. capsulatum* produces at least two types of conidia, which are differentiated primarily by size, it is not established whether or not the two spore types possess different physiological characteristics [200]. However, the small size of the microconidium (less than 5 μm in diameter) indicates that this spore is the primary infecting agent of *E. capsulatum.* The ability of the germinated microconidium to undergo phase transition and give rise to a yeast phase at 37°C, with or without the intervention of a hyphal phase, enhances the epidemiological importance of the microconidium [200]. It is probable that microconidia are inhaled into alveolar spaces, germinate, possibly after being phagocytized by host cells, and then give rise to the small budding yeasts characteristic of the parasitic phase.

The yeast cells of *E. capsulatum* are generally found in vivo in cells of the reticuloendothelial tissue [179]. They are small, round, oval, or pyriform cells having an average size of 2-4 μm in the parasitic form, as well as in culture at 37°C [175]. Growth of the yeast cell is by budding. The attachment between a mother cell and daughter bud is characteristically narrow and delicate. Unbudded yeast cells exhibit only a single nucleus [175,266]. Growth of the yeast phase in vitro is characteristically slow, with reported generation times at 37°C ranging from 8 h on a glucose cysteine blood agar medium [480] to more than 30 h in a defined synthetic medium supplemented with cysteine or serine [58,532]. The generation time reported for yeast growth on glucose cysteine blood agar [480] is probably near minimum, because it agrees well with the 10-h generation time estimated for the intracellular growth of *E. capsulatum* yeast cells in mouse histiocytes maintained at 37°C [270].

Ultrastructural studies document that the yeast cell wall is a thin, laminated structure having an average thickness of 90 nm [175]. The cytoplasm of the yeast cell contains the usual complement of organelles, and the nucleus occupies a large portion of an unbudded cell. Typical mesosome and lomasome profiles are observed. Lipid bodies are reported to be common, but critical evidence regarding their identification is lacking. The process of bud formation and nuclear division is not unusual and appears to be similar to that reported for other budding yeasts. The presence of the mesosome and lomasome profiles in the yeasts, their absence in yeast protoplasts, and their reappearance in regenerated yeasts indicate that these membranous structures are important to cell wall deposition [101].

Vegetative hyphal elements measure approximately 1.2-1.5 μm in diameter and exhibit thin walls which are approximately 50 nm in thickness [199]. The hyphal elements are usually mononucleate and have a moderately extensive endoplasmic reticulum which may or may not be connected to the lomasome-like structures scattered throughout the cytoplasm. Material thought to be glycogen deposits appears near the plasma membrane. Near-median sections of hyphae document that the septa are similar to those of other euascomycetes and have a central septal pore and associated Woronin bodies.

Reversible phase transitions in *E. capsulatum* are induced in the laboratory by proper manipulation of environmental conditions. Whereas temperature of incubation is of paramount importance in determining whether *E. capsulatum* grows as a budding yeast or filamentously [266,439,501], certain nutritional factors are also known to be important for phase maintenance or phase transition. Conversion of yeast cells to hyphae occurs easily in both simple and complex media upon temperature shift from 37°C to 25°C. As in yeast → hyphal transition of other dimorphic fungi, outgrowth of the yeast cell results in the formation of one or rarely two transitional cells which give rise to true hyphal elements [205].

Ultrastructural studies of yeast → hyphal transition in *E. capsulatum* suggest that significant cytological changes occur in the yeast within the first 6-8 h after induction of transition [204,205]. At this time there is a rapid appearance of numerous small intracytoplasmic membrane systems and an apparent increase in the number of mitochondria. The membrane systems do not appear to arise from preexisting organelles of the yeast cells, but develop de novo from the plasma membrane and mature rapidly during conversion [204]. The germ tube that develops into the transitional cell generally appears between 18 and 24 h [205].

Measuring about 30 nm in thickness, the cell wall of the developing transitional cell is only one-third as thick as the wall of the yeast mother cell and arises invariably from the inner portion of that wall [205]. Septum formation between the yeast mother cell and the transitional cell usually occurs between 30 and 48 h at 37°C. The exact timing of this event during conversion is not well documented, even though the appearance of the septum may be correlated with the transient increases in RNA and protein synthesis noted to occur at about the same time. Woronin bodies are in close association with the pore of the septum separating the yeast mother cell and the transitional cell [205]. The presence of a perforate septum indicates the hyphal nature of the transitional cell [205].

Transition from the hyphal phase to the yeast phase is more difficult to achieve than the yeast → hyphal transition and is induced by several factors, depending on strain, including elevated temperature [132,436], a reducing milieu [378,470], or the presence in the culture medium of cystine, cysteine, serine, or some other organic sulfur molecules [58,357,359,491,501,532]. Four morphological events are associated with hyphal → yeast transition [168,199,200,204,215,268,383,441,452]. These events are also associated with yeast cell development from germinated microconidia [199,200]. The four events involve (1) the formation of yeast cells by budding of hyphal cells, (2) the probable direct conversion of hyphal cells to the yeast phase by monilial chain formation, (3) the formation of stalked yeast cells arising in a manner reminiscent of microconidial formation, and (4) the direct conversion of microconidia to yeast forms by polar and nonpolar budding. In the latter case, the microconidium becomes a yeast mother cell directly, or forms yeast-phase cells by budding and gradually loses its identity as a microconidium. Alternatively, the microconidium germinates to form a hypha which gives rise to yeasts by methods 1, 2, and 3 [199,268,441].

Ultrastructural studies of hyphal → yeast transition are lacking, but most details are probably similar to those associated with microconidial germination at 37°C [199]. After 48 h the microconidia develop bud-like structures having a wall which is about 75 nm thick and continuous only with the inner layer(s) of the microconidial wall. Within 72 h, the bud-like structure elongates into a short septate hypha or becomes a very enlarged, nonelongated structure which appears to be a yeast mother cell. The cytoplasm of the yeast mother cell is highly organized and contains many mitochondria, lipid bodies, vacuoles, glycogen-like deposits, and scattered membrane systems [199]. Lomasome-like membranes are associated with the plasma membrane and may function in providing preformed membrane that allows for the rapid enlargement of the yeast mother cell. The yeast mother cell gives rise by blastic processes to single or multiple buds which constitute the beginnings of the yeast phase of growth. As in germ tube formation, the wall of a bud is continuous with the inner portion of the yeast mother cell wall [199]. Electron micrographs suggest that the single nucleus of the yeast mother cell migrates to the site of bud emergence and then undergoes mitosis in the isthmus between the mother cell and daughter bud.

Numerous studies have documented the physiological and biochemical differences between the two phases of *E. capsulatum.* The differences include vitamin requirements [383,491], cell wall composition [164,299,308], lipid content [163], total chemical composition [126], ability to take up various substances [203,208,209,229], ability to synthesize various amino acids [532], cyclic adenosine 3′-5′-monophosphate (cAMP) levels [359], DNA-dependent RNA polymerases [59-62,373], sensitivities to various antibiotics and inhibitors [121,123,358], and a multiplicity of other differences associated with conditions that induce phase transition in either the yeast or hyphal form.

The early observations that maintenance of the yeast phase and induction of yeast → hyphal transition requires complex media such as blood agar slopes contained in tubes sealed with paraffin [132,157], modified Francis' glucose cystine blood agar [95], and others [311,489] provided the original impetus for studying the differences between the yeast and hyphal phases of *E. capsulatum.* Salvin [491] was probably the first investigator to document specific nutritional differences between yeast and hyphae when he showed that the yeast phase of certain strains has a specific requirement for biotin, whereas this same requirement is not exhibited by the hyphal phase. The biotin requirement is expressed only in the yeast phase when cysteine is the sole nitrogen source. More important than the finding of a need for biotin was Salvin's [491] discovery that the addition of cysteine or cystine as the sole amino acid in a glucose salts medium maintains the yeast phase and inhibits yeast → hyphal transition. An apparent need for reduced sulfur probably explains why previous investigators determined that only complex media supported yeast maintenance. Salvin [491] further showed that more complex nitrogen sources, such as casamino acids, proteose peptone, and glutathione, also maintain the yeast phase, most assuredly because they contain sufficient reduced sulfur. The complexity of the system was demonstrated when Salvin [491] found that DL-methionine, although not able to support yeast growth when employed as the sole sulfur and nitrogen source, supports such growth when incorporated into a medium containing 16 nonsulfur-containing amino acids. In this medium, yeast-phase growth occurs only when compounds containing sulfide or sulfhydryl groups are used to supplement the nonsulfur-amino acid complex. No compound, organic or inorganic, that contains oxidized sulfur promotes growth of the yeast phase. However, these compounds support mycelial growth even at 37°C.

Subsequent research corroborated and extended many of the findings of Salvin, although it has yet to be established exactly how sulfur regulates *E. capsulatum* morphogenesis. In an extensive series of studies, Pine [436-441] demonstrated that *E. capsulatum* is a strict aerobe and thus not able to grow anaerobically in the yeast phase as previously reported [489]. This finding explains some of the inconsistencies evident in the earlier nutritional studies that were mostly carried out under oxygen-limiting conditions [132,

157,489,491,501]; these conditions probably tended to maintain sulfhydryl groups longer than might be maintained under standard shake culture or aerobic slant culture conditions. The importance of sulfhydryl (–SH) groups was further emphasized when Pine [436] showed that under most conditions of culture at 37°C, sulfhydryl groups must be present in the growth medium for the initiation of yeast growth. He suggested that the requirement for cysteine is not a requirement for –SH groups alone, but is necessary for the maintenance of viability of the yeast phase until growth begins. This suggestion is supported by the fact that cysteine cannot be replaced by glutathione. On the basis of other results, Pine and Peacock [440] postulated that conversion of hyphae to yeasts in *E. capsulatum* resolves itself into two problems: (1) satisfaction of the growth requirements of the yeast phase, and (2) inhibition of the growth of the hyphal phase without the inhibition of conversion to the yeast phase.

Studies by Scherr [501] of the roles of –SH groups and incubation temperature revealed that the concentration of –SH groups (or cysteine) is a more critical factor than a temperature of 37°C for maintenance of yeast growth. In these studies *E. capsulatum* was maintained in the yeast phase for long periods at 25°C with little evidence of yeast → mold transition. This suggested that the amount of hyphal growth is directly proportional to the decrease in incubation temperature and inversely proportional to the increase in the molar concentration of cysteine. Scherr [501] deduced that a molecule containing a reduced sulfur group is necessary for yeast growth, because at 37°C *E. capsulatum* must have a limited capacity to reduce sulfur-containing molecules. A number of recent studies [58,208,209,357,359,532] confirm the absence of certain reductases which are important in yeast-phase sulfur metabolism. Scherr [501] speculated that a temperature of 37°C provides optimum conditions for the activity of enzymes directing hyphal → yeast transition. However, yeast growth, once initiated, is maintained indefinitely at temperatures less than 37°C if the –SH concentration of the medium is maintained at some optimal level. Scherr's inability to bring about hyphal → yeast transition at temperatures lower than 37°C indicates that the metabolic pathways governing the maintenance of yeast phase growth and those directing mold → yeast conversion must be distinct from each other [501].

The possibility that organic sulfur is not needed just for the satisfaction of a nutritional requirement of the yeast cells, but may also be important in regulating the oxidation-reduction potential of the culture medium at levels required for phase transition, has been suggested [501] and confirmed [378]. Scherr [501] pointed out that the oxidation-reduction potential for media supporting yeast growth in tightly closed tubes is significantly different from that supporting hyphal growth. He speculated that yeast growth, once initiated, raises the oxidation-reduction potential of the medium sufficiently to permit yeast → hyphal transition in the absence of a concentration of cysteine normally necessary to maintain the yeast phase. He also suggested that a low oxidation-reduction potential, provided by free –SH groups, is required for hyphal → yeast transition. Evidence in support of this idea is provided by Rippon [470], who showed that hyphae give rise to yeast cells in brain heart infusion broth without added cysteine when the medium is maintained at a low oxidation-reduction potential. In Rippon's experiments the medium potential was regulated by electrolytic means in a continuously monitored environmental system. However, the complexity of the medium employed in these experiments did not exclude the presence of –SH groups. Other evidence provided by McVeigh and Houston [378] suggested that even in a defined medium, mycelial fragments of *E. capsulatum,* exposed to a low oxidation-reduction potential at 37°C for just 18-24 h, are induced along the pathway leading to hyphal → yeast transition. These investigators determined that the induction process occurs in the presence or absence of an organic sulfur source, either with or without gaseous oxygen. They also suggested that although transition can be induced in the absence of oxygen and an organic sulfur source, both are required for yeast growth after induction. Their suggestion was based on the finding that in media without organic sulfur, mycelial fragments exposed to short periods of low oxidation-reduction potentials give rise to yeasts much more rapidly than controls after transfer to medium containing cysteine or other organic sulfur molecules.

Although many studies suggest that certain organic sulfur compounds act by lowering the oxidation-reduction potential of the culture medium, it appears that a nutritional role may also be involved in their action, because all organic sulfur molecules are not equally effective in initiation or maintenance of the yeast phase of all strains of *E. capsulatum* [203,378,491]. The possible nutritional role of certain sulfur-containing organic compounds, particularly cystine and cysteine, has prompted investigators to examine the uptake, biosynthesis, and metabolism of these molecules [58,208,209,357,359,532].

Studies of the uptake of L-cystine demonstrate that transport of this amino acid in the yeast phase of *E. capsulatum* is mediated by an energy-requiring system [203,208]. Gilbert and Howard [208] showed that the uptake system is very specific in the strain they studied and not inhibited by other amino acids. Their results support the conclusion that transport is specific either because cystine is a dimer and thus very different from most other amino acids or because the site of transport might be associated with a reductase which only allows the sulfhydryl form to cross the cell membrane.

Garrison et al. [203] independently determined that cysteine, as well as cystine, is actively transported by yeast-phase cells. Their studies showed that transport is mediated by an adenosine triphosphate (ATP)-dependent permease. They documented that the two organic sulfur compounds stimulate yeast-phase, but not hyphal-phase, respiration. The stimulation of respiration is postulated to be by entry of cystine or cysteine into the Krebs cycle after their conversion to pyruvic acid.

Maresca et al. [359] proposed that cysteine and cystine participate in morphogenesis by providing critical sulfhydryl groups that affect the activity of cAMP-phosphodiesterase, which in turn regulates the levels of intracellular cAMP. In their model, the temperature of incubation changes membrane redox potentials, thereby activating or inactivating the cystine-cysteine permease system [359] or, alternatively, by modifying certain reductases involved in sulfur amino acid metabolism [58,357,532]. Thus, at 37°C, the permease or reductase(s) is activated and facilitates the transport of critical sulfhydryl groups necessary to activate cAMP-phosphodiesterase. As lower concentrations of intracellular cAMP develop, due to the increased activity of the cAMP-phosphodiesterase, the yeast phase is maintained or hyphal → yeast transition occurs. In contrast, at 27°C the permease or reductase(s) is not active. Thus, the transport of the appropriate sulfhydryl groups is inhibited and the intracellular cAMP concentration increases. The higher levels of cAMP maintain the hyphal phase or induce yeast → hyphal transition.

The proposal by Maresca et al. [359] is based on a number of elegant and simple experiments that confirm and greatly extend the suggestions and findings of Gilbert and Howard [209] and Garrison et al. [203]. The experiments showed that yeast cells of the Downs strain incubated at 37°C in a glucose yeast extract medium have a higher level of respiration than hyphae incubated at 25°C. The difference in the level of respiration is attributed to the higher temperature of incubation and not some inherent quality of the phases themselves. This conclusion is based on the fact that yeast cells exhibit hyphal-phase respiration rates when incubated at 25°C, whereas hyphae exhibit yeast-phase rates when incubated at 37°C [359]. In agreement with Garrison et al. [203], Maresca et al. [359] confirmed that the addition of cysteine or cystine to a glucose yeast extract medium stimulates respiration of the yeast phase, but not the hyphal phase. Shifts of stimulated yeasts from 37°C to 25°C result in the loss of stimulated respiration, whereas non-stimulated yeasts (yeasts growing in the presence of cystine or cysteine and incubated at 25°C) show rapid increases in respiration when shifted to 37°C, even in the presence of cycloheximide or chloramphenicol. However, hyphae have the potential for stimulation by cystine or cysteine because when shifted to 37°C the hyphae become almost immediately responsive to the sulfur amino acids, even if treated with cycloheximide and chloramphenicol. Apparently, new protein synthesis is required for transport and the permease(s) is present in the membranes of both phases, but is only operative at 37°C.

The possibility that this model is correct was enhanced by experiments showing that the sulfhydryl blocking agent p-chlormercurphenisulfonic acid allows hyphal growth, but inhibits hyphal → yeast transition, in the presence of cysteine at 37°C [359]. This blocking agent also accelerates the transition of yeast to hyphae at both 25°C and 37°C. The opposite effects are found when experiments are carried out with dithiothreitol, a sulfhydryl protector. With this agent, yeast cells incubated at 25°C are unable to give rise to hyphae.

The model was also supported by the finding that the level of cAMP is about five times higher in the hyphal phase than in the yeast phase [359]. On the average, hyphae contain almost five times as much cAMP per milligram of protein as yeasts. The addition of appropriate concentrations of dibutyryl cAMP or theophylline to yeast cultures induces yeast → hyphal transition at 37°C, as would be expected if the levels of cAMP were truly regulating morphogenesis.

That reductases involved in sulfur amino acid metabolism are more important to the model than are the permeases responsible for cystine or cysteine transport is suggested by the findings that cystine reductase [357] and sulfite reductase [532] activities are not uniformly expressed in the two phases. Stetler and Boguslawski [532] have divided the strains of *E. capsulatum* into two groups on the basis of yeast-phase sulfur metabolism. One group includes those strains having a stringent requirement for cysteine (or other form of reduced sulfur) for growth in a minimal, defined medium which consists of glucose, vitamins, salts, and trace elements, and which supports growth of the hyphal phase. The strains in this group lack the enzyme sulfite reductase and thus cannot reduce inorganic sulfite to the sulfide necessary for cysteine biosynthesis. The other group includes strains also having hyphae that can grow in the minimal medium, but whose yeasts require either cysteine (0.2 mM) or serine (5.0 mM). The Downs strain employed by Maresca et al. [359] is a member of this second group. This strain possesses sulfite reductase activity when grown in the absence of cysteine. However, the enzyme in yeasts is completely repressed in the presence of cysteine. Thus, hyphae of *E. capsulatum* appear to be constitutive for sulfite reductase, but yeasts have a cysteine-repressible enzyme (group 2). Apparently, at 37°C various defects in the sulfur biosynthetic pathway among the strains of *E. capsulatum* are expressed, dictating medium supplementation by cysteine or precursors of cysteine biosynthesis such as serine [532]. The presence of these compounds at 37°C may so lower cAMP levels and stimulate respiration that yeast-phase growth results. The variety of defects in sulfur metabolism among strains of *E. capsulatum* probably explains some of the inconsistent results related to the nutrition of yeast-phase maintenance and hyphal → yeast transition.

Many of the myriad of physiological conditions affecting morphogenesis in *E. capsulatum* can be rationalized as regulating the activity of adenyl cyclase and cAMP-phosphodiesterase, which control intracellular cAMP concentrations. The mechanisms by which the cAMP concentrations might ultimately generate a morphological change in *E. capsulatum* is by no means established. As pointed out by Stewart and Rogers [533], the cAMP could (1) regulate gene transcription with the gene products directly or indirectly affecting cell wall biosynthesis and morphogenesis, (2) activate cAMP-dependent protein kinases that are concerned with cell wall biosynthesis, or (3) interact with microtubules or microfilaments and thus regulate the supply of precursors to the site(s) of wall synthesis. Although evidence is unavailable for directly implicating cAMP in any of these processes in *E. capsulatum,* it is tempting to speculate that cAMP may regulate gene transcription in a manner that produces changes in the rates of RNA synthesis and the nature of the polymerases involved in RNA synthesis. Certainly, the latter could be accomplished by cAMP-mediated post-translational phosphorylation of RNA polymerase. A number of changes relating to RNA synthesis have been documented [59-62,122,310,374,375]. Among the changes, the most dramatic is associated with the transition from hyphal to yeast morphology. During this transition, RNA synthesis decreases to 26.7% of normal for at least 4 h [122]; a reduction in total net RNA synthesis also occurs. Yeast cultures incubated at 37°C for 96 h increase in RNA content 325%, whereas the RNA content of hyphal cultures incubated at 25°C for the same length of time increases only 185%. More important, hyphal cultures exhibit an actual decrease in net RNA content for the first 32 h after a shift from 25°C to 37°C. Subsequently, net RNA synthesis increases slightly above inoculum levels. The changes in RNA synthesis associated with the hyphal → yeast transition correlate well with morphological changes. In general, no morphological changes are associated with the initial 32-h period of incubation at 37°C and bud-like hyphal tip constrictions and yeast cells appear at 48 h and 72 h, respectively. Fractionation of the RNA synthesized during the transition reveals no obvious selective inhibition of the various RNA classes.

A transient increase in RNA synthesis relative to dry weight also occurs at about 36 h in exponentially growing yeast cultures shifted from 37°C to 25°C (G. S. Kobayashi, Washington University, personal communication). In these cultures, germ tubes begin appearing after 1 h, suggesting that the increase in RNA synthesis is associated with the formation of a septum between the parent yeast cell and germ tube and not with the initiation of the conversion process itself.

The RNA polymerases from both the yeast and hyphal phases of *E. capsulatum* have been isolated and characterized. Each phase possesses three RNA-polymerase activities [61,373]. However, this was not evident initially because hyphal extracts exhibit very low levels of RNA-polymerase activity which elute from chromatography columns as a single peak. The low activity is attributed to the presence in the extracts of a potent inhibitor of RNA-polymerase activity [60] and also to the fact that some of the RNA polymerases are inhibited by low concentrations of RNA molecules [374]. Further purification of the mycelial RNA-polymerase activity has resolved the three RNA-polymerase activities of the hyphal phase [373].

Although each phase contains three polymerase activities, the two sets of polymerases differ in chromatographic behavior and in sensitivities to the DNA-dependent RNA-polymerase inhibitor, α-amanitin. Two of the three enzymes from the yeast phase are sensitive to this drug, with the major species, polymerase III (polymerase designations here are based on elution profiles and do not necessarily correspond with the enzyme designations established for other eucaryotes), being most sensitive, polymerase II resistant, and polymerase I intermediate in sensitivity [373]. The yeast phase polymerases are reported to be temperature sensitive in vitro and inhibited at 37°C. The data may indicate that the same polymerases function in both the yeasts and hyphae, although the major species of polymerase is different in the two forms, or that the polymerases are modified for their role depending on conditions of growth. Alternatively, the data may indicate that two completely different sets of polymerases are present in yeast and hyphal forms and that the two sets have different temperature sensitivities. The correlation between elevated cAMP levels and yeast → hyphal transition favors the hypothesis that the polymerases are modified post-translationally via protein phosphorylation. Regardless of the mechanism, when cultures are shifted from 25°C to 37°C a decrease in RNA synthesis is expected, since no species of RNA polymerase is present that could function optimally for RNA synthesis [61]. Thus, a period of time would elapse before polymerases are converted to temperature-resistant forms or before new polymerases are made.

As indicated by this review, exciting advances have been made during attempts to elucidate the molecular events controlling dimorphism in *E. capsulatum*. Many of these advances are summarized by the sequences depicted in Fig. 1. However, the exact mechanism responsible for morphogenesis in *E. capsulatum* is still poorly understood. It is clear that temperature regulates the synthesis or function of key enzymes that ultimately determine phase morphology. The major candidates for these enzymes are those involved in sulfur metabolism leading to cysteine biosynthesis. The fact that some of these enzymes are variously present in the yeast and hyphal phases, or in the yeast phase itself, makes study of their regulation very relevant because the variation can account for strain differences. The intriguing hypothesis of Maresca et al. [359] that cysteine or other sulfur-containing molecules exert their effect by regulating intracellular cAMP levels provides a model for directing future research efforts. Possibly, cAMP is involved in RNA synthesis or modification. Alternatively, cAMP may act by modifying a hypothetical cytoskeletal system, although this possibility has yet to be addressed by investigators of dimorphism in *E. capsulatum.*

C. *Paracoccidioides brasiliensis*

Paracoccidioides brasiliensis Is the etiologic agent of South American blastomycosis. This dimorphic fungus has a filamentous, saprophytic morphology, whereas its parasitic morphology is that of a multiple budding yeast [179,314]. In culture on simple or complex media, dimorphism is temperature dependent, with the mycelial morphology developing at room temperature and the yeast at 37°C [207,400].

The slow-growing, white-to-tan, folded colony of the hyphal phase exhibits no definitive, morphological structures. However, extended incubation in glucose-deficient media allows the development of arthroconidia and blastoconidia [443,458,466,467]. The unusual blastoconidia are from 2 to 3 μm in diameter and are directly associated with hyphae or borne singly on the tips of short conidiophores [443]. Chlamydospores, ranging in size from 6.5 to 30 μm in diameter, are found in terminal, intercalary, and lateral locations on the hyphae [104,443].

Investigations of hyphae reveal cells with dimensions of 0.8-2.4 μm by 6.4-29 μm

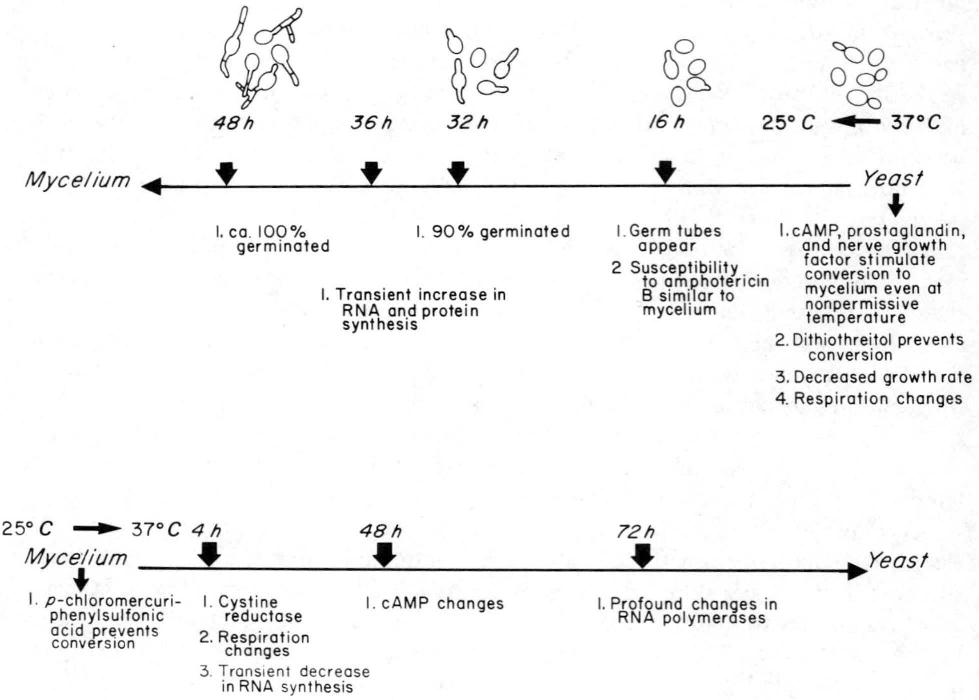

Figure 1 Model for the sequences of events associated with yeast → mycelium and mycelium → yeast transitions in the Downs strain of *Emmonsiella capsulatum*. (Courtesy of G. S. Kobayashi and G. Medoff.)

[104]. These cells are multinucleate and possess mesosomes, lomasomes, and numerous mitochondria whose long axes are parallel to those of the hyphal unit [105]. The septal region has been described as "bulged" [443]. Within this region, one encounters a septum which is penetrated by a single pore associated with Woronin bodies and a septal plug [100,105]. The cell walls of the hyphae are from 80 to 150 nm in width and are composed of from one to three discernible layers [105]. The outermost layer is electron dense and has been described as finely fibrillar [99,105].

The parasitic morphology is that of a budding yeast whose production in vitro has been accomplished classically by incubation at 35-37°C [7,314]. Two types of budding are observed. The first is the so-called "regular" method, which involves the production of narrow-necked buds, 2-10 μm in diameter [193]. These buds can be spherical or bottle-shaped and their production does not result in the distortion of the mother cell [193,465,568]. The second method of budding is associated with "cryptosporulation," a term attributed by Lacaz [314] to Ciferri and Redaelli. Here, chromatin is divided into chromidia which function as potential spores. These chromidia become associated with the cell membrane and form numerous, small buds which encircle the mother cell. Furtado et al. [193] suggested that these buds are simultaneously formed by the expansion of the central vacuole of the mother cell, and that this process can result in the death of the mother cell [193]. In both forms of budding, an electron-dense layer appears over the site shortly before bud emergence [97,99]. When the bud develops, its wall is continuous with the inner layer of the cell wall of the mother yeast cell. The bud separates from the mother cell, leaving a bud scar which soon disappears [97,99].

Examination of the yeast cell reveals it to be from 2 to 35 μm in diameter in culture and up to 60 μm in diameter in vivo [104,179]. Electron microscopic examination of the yeast cell shows that it possesses multiple nuclei, mesosomes, lomasomes, and concentric lamellae termed "myelin figures" [96,97,100,103]. The cell wall is from 200 to 600 nm thick and comprises from one to three layers [496]. The demonstration of layers is dependent on the age of the cells, the cutting angle of the section examined, and perhaps other factors [99]. Chemical treatment of cell wall samples reveals an inner, chitinous layer in the form of long, thin fibers and an outer layer of short, thick fibers composed of α-glucan [102].

Temperature of incubation is the primary determinant of the in vitro morphology of *P. brasiliensis.* The conversion of a yeast culture to a hyphal one is accomplished by shifting the temperature of incubation from 37°C to room temperature [400]. Transition is first evidenced in submerged culture at 24 h by the production of elongated buds having diameters greater than those of the hyphae [457]. These multinucleate, transitional cells have thinner cell walls than do mother yeast cells, and also have perforated septa similar to those found in hyphae [97,100]. The elongated forms are produced singly of in multiples and continue to grow in an unidirectional manner as the mother cell dies. Growth is exclusively filamentous by 102 h at room temperature [457]. Both protein and DNA synthesis are required for formation of the transitional, elongated buds just as they are for normal bud initiation [170,457].

The transition events described above were noted with the use of a single isolate, IVIC-Pb9 [97,100,457]. Microscopic studies of other strains indicate that some events in the transition may be limited to the IVIC-Pb9 strain. Vieira de Silva et al. [568] described elongated, bottle-shaped buds as a normal component of the yeast population of one strain (Fig. 2). Garrison and Boyd [197] investigated the dimorphic phenomenon of a strain of *P. brasiliensis* mistakenly identified as *Histoplasma farciminosum* [303a]. These investigators found that the yeast → hyphal conversion of this strain takes place without the production of a transitional form. They further noted that the cell wall of the hypha appears to be continuous with the inner layer of the yeast cell wall [197]. The development of a live hypha inside a dead yeast cell during yeast → hyphal transition has also been reported, but the relevance of this process to dimorphism is obscure [98].

Unlike the yeast → hyphal transition, the in vitro mold → yeast transition involves very few hyphal cells. By 18 h, after the temperature shift to 37°C, most hyphal cells are dead, and the few viable remaining cells begin to swell and lose the electron-dense outer layer of the cell wall [99,100]. These multinucleate, transitional forms, termed interseptal spaces [100,197] or chlamydospores [99,197], continue to swell and produce numerous buds. During this transformation, the cell wall thickens and attains the dimensions of a yeast cell. The transition form itself then disarticulates and becomes a yeast cell [99,100]. Experimental infections with hyphal fragments show the in vivo conversion to be slower than, but morphologically similar to, the in vitro transition [104].

The temperature-dependent, morphological transitions of *P. brasiliensis* are asynchronous and, for the most part, nonconservative [99,100]. Each yeast cell can produce the elongated forms which give rise to the hyphal cell, but cannot itself become a hyphal

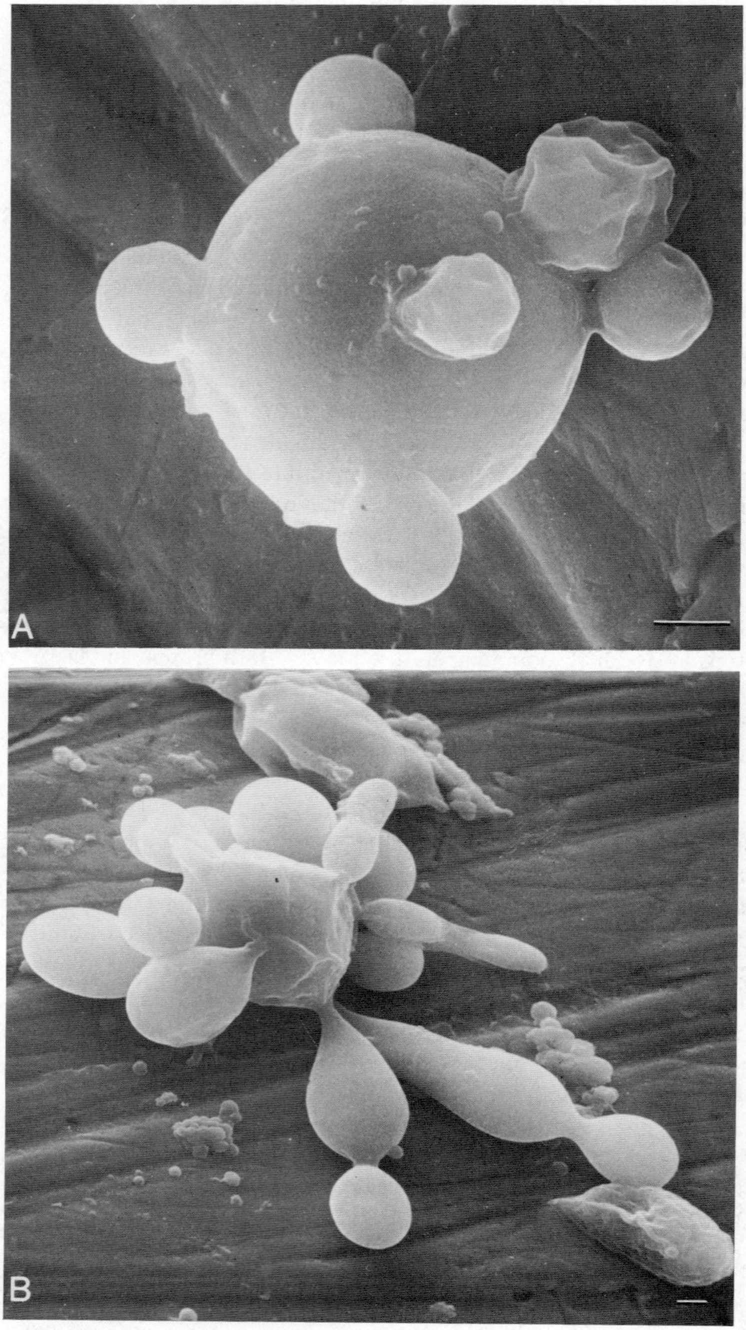

Figure 2 Scanning electron micrographs of budding yeast cells of *Paracoccidioides brasiliensis:* (A) spherical buds; (B) elongated buds. Bars represent 2 μm. (From Ref. 568.)

cell. In contrast, few of the hyphal cells give rise to the yeast population upon the temperature shift; however, those that do separate become budding yeast cells [99].

The importance of environment in the study of the dimorphic nature of *P. brasiliensis* is not limited to the temperature of incubation. Certainly, temperature has been the sole determinant classically employed for morphologic conversion [207,400]. However, other factors, both physical and chemical, can modify, if not replace the temperature effect on morphogenesis. Negroni [397] obtained hyphal → yeast transition at 28°C in shake cultures. Extended incubation on snake dung medium at 23-28°C has been shown to produce free-floating, multiple-budded cells [63]. Gilardi and Laffer [207] obtained hyphal growth at 37°C on basal medium with 0.5% tyrosine. Ramirez-Martinez produced pseudohyphal cells from yeast cells by incubation on glucose-deficient medium at 37°C [458]. She further showed the inhibitory effects of medium components commonly employed in cultural investigations of the dimorphic phenomenon [457,458], and emphasized that control of media ingredients may be as important as any other environmental variable for the study of dimorphism in *P. brasiliensis* [458].

The study of the regulation and biochemistry of dimorphism in *P. brasiliensis* has long been focused on the terminal morphologies rather than on the regulatory events associated with the transition from one morphology to another. As Ramirez-Martinez [456,458] points out, research has been concerned with the characteristics of the end products of dimorphism such as cell wall morphologies, while the mechanisms by which the differences are wrought are comparatively neglected.

The enzymatic activities of the yeast and hyphal morphologies of *P. brasiliensis* have been compared in a number of studies. Nickerson and Edwards [400], using mycelial clumps and yeast cells from broth cultures, compared the phases in respirometer studies. A high rate of endogenous respiration was found associated with both morphologies, although that of the yeast was noted to be five to six times that of the hyphal form. The yeast cells showed exogenous respiration in the presence of glucose or acetate, whereas none was noted with starved or resting hyphae with either substrate [400].

Kanetsuna and Carbonell [295] examined the two morphologies of *P. brasiliensis* for enzymes employed in oxidative glucose utilization. Cell-free extracts of broth-grown hyphal and agar-grown yeast cells possess the enzymes of the hexose monophosphate pathway, glycolysis, and the citric acid cycle. The levels of these enzymes are from 1.1 to 4.5 times greater in the extract of the yeasts for all but two enzymes. Kanetsuna [294] and Kanetsuna and Carbonell [295] later found protein disulfide reductase activity to be fivefold greater in yeast cells and β-glucanase activity to be five- to eightfold greater in hyphal cells. Cytochemical experiments showed more alkaline phosphatase and L-leucyl aminopeptidase activity in yeast cells [10].

The enzymatic activity of the yeasts is greater than that of the hyphae. However, one must not overlook the limits of the methodologies employed. Mycelial clumps may not be readily comparable to yeast cells, in that most of the yeast cells are actively respiring and synthesizing cell wall components, whereas these same activities occur primarily at the apex of hyphal filament [33,37,394,568,593]. Kanetsuna [294] pointed out that there are undoubtedly physiological differences between the apical cell and older hyphal segments. Carbonell and Rodriguez [105] showed that hyphal clumps incubated 18 days at room temperature possess viable cells only in the peripheral region, the region that is most susceptible to mechanical damage during centrifugation and other procedures [218].

In contrast, the central area of hyphal clumps is composed of empty cells. The comparisons are further complicated by a soluble, lethal metabolite present in cultures of both morphologies [15,493]. Comparison of the enzymatic capacities of the yeast cell with the apical region of the hypha would be more significant.

Although the regulation of dimorphism in *P. brasiliensis* is obscure, the final determinant of fungal morphology, and therefore of the dimorphic phenomenon, is the cell wall [223,294]. The cell wall of the two morphologies of *P. brasiliensis* has been the subject of considerable research. It was shown that the amino acid content of the yeast cell wall is approximately one-third, and the glucosamine content which is felt to represent chitin [391] is three times that of the hyphal wall [300]. In addition to glucose, mannose, galactose, and N-acetylglucosamine, both N-acetylgalactosamine and uronic acid have been cytochemically demonstrated [534]. Although the two morphologies possess similar hexose quantities in their walls, the predominant hexose wall polymer of the yeasts is an α-glucan and that of the hyphae is a β-glucan [296,300,496]. San-Blas et al. [496,497] showed that the presence of α-glucan in the yeast cell wall is critical to the establishment of a successful systemic infection. All three molecular species possess integral roles in cell wall integrity because lytic enzymes which act on each are required for protoplast formation [221]. Kanetsuna et al. [298] proposed a model for the dimorphic phenomenon at the cell wall level. In this model, the structural rigidity of the hyphal cell wall is maintained by linear β-glucan and disulfide-linked protein molecules. In contrast, the yeast cells are thought to be spherical due to the more active synthesis of α-glucan and chitin. The role of α-glucan may be the same as that of chitin, which in *Saccharomyces cerevisiae* is claimed to maintain the structural integrity of the mother-daughter pair [87]. The temperature-sensitive nature of the synthesis of the α- and β-glucans has been documented [296] and contributes to this hypothesis. The α-glucan synthetase has been isolated from both yeast and hyphal cells, but the isolation of β-glucan synthetase has not been accomplished [495]. Of the cell wall lytic enzymes, only β-glucanase has been shown in *P. brasiliensis* [187,495].

The control of dimorphism in *P. brasiliensis* remains obscure. Further investigations into the differences in RNA populations between the yeast and hyphal morphologies [456] and the use of inhibitors [498] may lead to elucidation. Relevant data may also be found in the work performed on the dimorphic phenomena of *E. capsulatum* and *A. dermatitidis.* The similarities of the latter organism and *P. brasiliensis* have prompted some to place them in the same genus [133,207]. Others have suggested that *E. capsulatum* and *A. dermatitidis* should share the same perfect genus [372]. It is conceivable that these three dimorphic organisms, all of which possess a hyphal-yeast and temperature-dependent dimorphism, are closely related and may show common mechanisms for the control of their dimorphic behavior.

D. *Sporothrix schenckii*

Sporothrix schenckii, the agent of sporotrichosis, exhibits a yeast-hyphal dimorphism which is primarily controlled by temperature and CO_2 concentration. The dimorphic nature of the fungus was first noted by Schenck [500], who observed only yeasts in tissue, but upon primary isolation obtained hyphal growth. The yeast form was first grown in culture in 1908 [343], but the environmental factors controlling dimorphism in this organism, and the morphological changes accompanying phase transition, were not determined until much later [169,267,490].

Hyphae of *S. schenckii* are 1.5-3 μm in diameter, septate, and granular in appearance [267,272]. Hyphal cell walls are relatively thin (80-140 nm thick) and appear to be bilayered when observed by electron microscopy [322]. The thicker inner layer has a low affinity for dialyzed iron, whereas the much thinner outer layer has a higher affinity. Garrison et al. [202] suggest that the high-affinity layer contains a large amount of acidic polysaccharides. Pyriform to ovoid conidia arise laterally on the hyphae, in clusters at the distal region of conidiophores, or occasionally from the proximal conidiophore regions [267,272]. The conidia often give rise to secondary conidia, apparently by a blastic process [267,272,548]. At the level of resolution of light microscopy, the conidia seem to be attached to hyphae by sterigmata [267,272]. However, at the greater resolving power of the electron microscope, thick plate-like structures are evident at the point of hyphal attachment [322]. The conidia germinate mono- or bipolarly to form hyphae [202,267,272], but direct budding to form yeasts has also been reported [575].

The yeast phase of *S. schenckii* ranges in shape from fusiform [169,267], to spherical [267], but no cytoplasmic differences correlate with the different cell shapes [322]. Yeasts usually bud at the polar regions, but may occasionally bud elsewhere [267,322, 490]. Like conidial germination, budding usually occurs at one pole at a time, but occasionally multiple budding occurs at the same or opposite poles [267,322,490]. The yeast wall is bilayered and consists of an inner, electron-transparent layer and an outer, electron-dense layer. The outer layer of the yeast wall, like that of the hyphal wall, has a high affinity for dialyzed iron [202,322], but in contrast to the hyphal wall, the outer layer of yeast wall is very thick and contains long, intertwining microfibrils [202,322]. This microfibrillar outer layer is not only a property of yeast walls, but also of the walls of those structures on hyphae destined to become yeasts [202,322]. Often, yeasts, but not hyphae, are surrounded by a slime layer [202,322].

The interconversion of yeasts and hyphae has been studied in tissue culture [258, 267] and on a number of complex [202,267,272,322] and defined [360,380] media. Although conidia have been reported to bud directly to yield the yeast phase [575], this does not seem to be a universal characteristic of hyphal → yeast transition [202,267]. More often, the hyphal morphology is established and then gives rise directly to the yeast phase. Shifting conidia to 37°C in appropriate medium induces germination and hyphal outgrowth. The tips and branches of the hyphae differentiate to form thick, multiple-branched structures which apparently have a composition different from hyphae since they stain differently with cytological stains [267]. However, no obvious cytoplasmic changes can be seen by electron microscopy [202]. Buds that form on the tips and sides of the branched transition structures elongate and form additional buds. Because cell separation between mother and daughter cells does not occur immediately, short branched chains of yeasts are transiently formed. Eventually, the multiple-branched structure separates from the hypha, but continues budding [267]. Buds may also emerge directly from the hyphae, usually near septa, but occasionally from hyphal tips in a manner similar to the emergence of conidia [202]. The cell wall of the bud is formed from the inner layer of the hyphal wall and has a thick, electron-dense, iron-binding layer, comparable to the wall of a yeast. A nonperforate septum is formed between the hypha and bud, but the appearances of this septum and the conidial plate [202,322] suggest that they are not homologous structures, although they perform similar functions.

After the hyphae have apparently exhausted the capacity to generate yeasts blastically, they may convert to yeasts by another means. This is an arthric mechanism in which cytoplasm condenses into hyphal segments to form chains of endogenous oidialike cells. These chains then fragment and the cells emerge from the original hyphal wall and begin to reproduce blastically [202,267]. The cell wall of the oidial cell is derived from the inner layer of the hyphal wall, and the cell is released by the breakdown of the hyphal wall [202].

Transition of yeasts to hyphae occurs by the outgrowth of a transitional cell from the yeast [321]. The intracytoplasmic membranes of the yeast reorganize near the point of outgrowth and the transitional cell arises blastically like a yeast bud, its cell wall being derived from the inner cell wall layer of the yeast. Unlike a bud septum, the septum between the yeast and transitional cell is perforate. The transitional cell gives rise to a hypha typical of *S. schenckii.*

There seems to be a complex interrelationship among temperature, CO_2 concentration, and nutrition in determining the morphology of *S. schenckii.* Drouhet and Mariat [169] in an elegant study of environmental factors influencing dimorphism in this organism, determined that temperature and CO_2 concentration play an important role in determining and maintaining morphology. More recently, Mendonca et al. [380] were able to eliminate temperature as a factor and obtain pure cultures of yeasts by using a defined medium in shaken, sealed flasks. Unfortunately, they did not investigate the mechanism by which the medium exerted its effects. Drouhet and Mariat [169] found that yeasts can be induced to grow even on very simple, defined media if the temperature is above 37°C and the atmosphere contains 5% CO_2. Below 37°C, hyphae develop, and in the absence of exogenous CO_2, the development of the yeast form is more difficult. The requirement for CO_2 can be replaced with casein hydrolysate or a mixture of amino acids, or with the amino acids arginine, glycine, or alanine. These may supply CO_2 by decarboxylation, especially since hyphae tend to develop if biotin, a cofactor in CO_2 metabolism, is limiting. In addition, high aeration rates which tend to remove CO_2, or inorganic nitrogen sources which do not supply carboxyl groups, favor hyphal development. However, the amino acids do not completely substitute for CO_2, since agitation is required; yeasts will not develop on solid medium or in stationary culture. In addition, pH has little or no effect on morphology, but near-neutral pH seems to be particularly favorable for yeast growth. Neither sulfhydryl groups nor sulfur salts induce the yeast morphology as they do in *E. capsulatum* and *Candida albicans* (see Secs. II.B and III.A).

Mariat and de Bievre [360] determined that the metabolisms of yeasts and hyphae differ. Hyphae seem to be the more efficient growth form, since the molar growth yield (moles of nitrogen incorporated per mole of glucose consumed) is higher for hyphae than yeasts. The ratio of RNA to DNA is approximately 10 times higher in hyphae. Thus, transition to the yeast phase may be a way of responding to adverse environmental conditions (high temperature, elevated CO_2 tension) without producing a dormant state.

Biochemically, there appear to be only a few differences in the walls of hyphae and yeasts. The differences in amounts of acidic iron-binding material (possibly acidic polysaccharide) in yeasts and hyphae have been noted [202]. Similarly, yeasts react more strongly with concanavalin A than do hyphae; the concanavalin A-binding is described as thick in yeasts and absent in hyphae [560]. The material is thought to be a peptidorhamnomannan with a composition of 50% D-mannose, 33% L-rhamnose, and 17% peptide [335]. Since the alkali-extractable rhamnomannan alone is nonreactive, the moieties that bind concanavalin A may be alkali-labile, mannose-containing oligosaccharides attached to serine or threonine residues [560]. It is tempting to think that the iron-binding and concanavalin A-binding materials are related, especially since the peptidorhamnomannan can be rich in dicarboxylic amino acids [335] or glutamic acid [560]. However, it seems more likely that the layer in question is heterogeneous and contains a number of components which are more abundant in yeasts than in hyphae. This is supported by the observation that basic amino acids are also frequent in the peptidorhamnomannan [560], and that *S. schenckii* produces both neutral monorhamnosyl rhamnomannans and acidic rhamnomannans containing 4-O-substituted glucuronic acid residues and $\alpha(1 \rightarrow 2)$-linked dirhamnosyl side chains [559].

The rhamnomannan seems to be the major antigenic determinant in *S. schenckii* and has received considerable attention. Yeasts have been shown to produce only monorhamnosyl side chains, regardless of the incubation temperature [380]. Antiserum against monorhamnosyl rhamnomannans reacts with yeasts, yeast primordia on hyphae, and conidia [336], and the structure of rhamnomannan in yeasts and conidia is similar [558]. Conversely, antiserum against dirhamnosyl rhamnomannans reacts only against hyphae [336]. The structure of the rhamnomannans seems to consist of a $\alpha(1 \rightarrow 6)$-D-mannopyranose backbone with $\alpha(1 \rightarrow 3)$-linked rhamnose side chains. The rhamnose moieties of the dirhamnosyl side chains are linked in their $\alpha(1 \rightarrow 2)$ configuration [559]. A significant amount of 4-O- and 2,4-di-O-substituted mannose polymers is also found [557]. Synthesis of both polymers seems to be inhibited at 37°C [380], although the 2,4-di-O-substituted mannan may be inhibited more than the 4-O-substituted polymer. There is also some evidence [337] that synthesis of the dirhamnosyl moieties is inhibited at 37°C. The significance of these variations in structure to the determination of morphology is unclear at the present. Conceivably, they represent a significant structural alteration which is active in determining morphology. Conversely, they could equally well be secondary effects which are the result, and not the cause, of altered morphology.

There may be considerably more neutral carbohydrate in yeast walls than in hyphal walls, and more total nitrogen in the hyphal walls. However, Previato et al. [449] note that their preparations were washed extensively and a portion of the external layer of the yeast wall may have been lost. There are nearly identical percentages of insoluble polysaccharides (24.4% ± 1.1%) [449] and chitin [448] in the walls of the two forms. Both forms have $\beta(1 \rightarrow 3)$, $\beta(1 \rightarrow 4)$, and $\beta(1 \rightarrow 6)$-linked glucopyranose polymers, and the percentages of each type of linkage are identical in the two forms ($\pm{<}1\%$) [449]. Both total nitrogen [449] and protein [448] content are lower in yeast walls than hyphal walls, although three specific amino acids, threonine, serine, and alanine, are present in higher proportions in the yeast [448]. This indicates that different populations of proteins or peptides are associated with the walls of the two forms.

The lipid content of yeast cells is lower than hyphae, and there are significant differences in two particular lipids. Palmitic acid associates with the acid-extractable lipids in hyphae and the free lipids in yeasts. A $C_{18:3}$ acid (linolenic?) is present in yeasts, but undetectable in hyphae. Finally, there is a higher proportion of phosphate in hyphal walls than in yeast walls [448].

Hyphal → yeast transition in *S. schenckii* may occur by means of an unusual transitional structure as well as arthric and blastic mechanisms similar to those seen in other fungi exhibiting yeast-hyphal dimorphism. In this instance, a differentiated, branched structure functions as a transitional cell between hypha and yeast. Yeast → hyphal transitions in *S. schenckii* fit the pattern of yeast → transitional cell → hypha which seems to be characteristic of phase transitions in this direction.

The thermal dependence of dimorphism and the apparent temperature sensitivity of synthesis of certain cell wall polymers, suggest that the stability of specific proteins at elevated temperatures plays a role in determining morphology. The dependence on CO_2 for optimum conversion indicates that some product of CO_2 fixation may be necessary for expression of the yeast morphology. A similar hypothesis was proposed for the hyphal → yeast transition of *Mucor rouxii* [32,39], after it was found that, in the presence of excess CO_2, yeasts are induced and considerable CO_2 is incorporated into aspartic acid, glutamic acid, and threonine. It would be interesting to know if incorporation of CO_2 into cellular constituents of *S. schenckii* is dependent on morphology. Certainly, the medium described by Mendonca et al. [380] will be useful in studying the CO_2 dependence of phase transitions and could be used to investigate the points noted above. The fact that medium, CO_2 concentration, and temperature all play roles in morphogenesis indicates that a number of pathways may exist to regulate morphology. The common factor in all cases of yeast development is that either CO_2 metabolism is involved or that CO_2 effects on metabolism cannot be eliminated. The study of CO_2 metabolism as it relates to *S. schenckii* dimorphism may yield invaluable information on the molecular basis of morphogenesis in this fungus.

III. SPECIES WITH MIXTURES OF INTERCONVERTING MORPHOLOGIES IN VIVO

A. *Candida albicans*

Candida albicans is a fungus of significant pathological importance. Although temperature is an important factor in the regulation of morphogenesis in *C. albicans,* it is not central as in the temperature-dependent dimorphisms of *A. dermatitidis* and *P. brasiliensis.* Instead, morphogenesis in *C. albicans* is extremely complex and modified in its expression by a number of physical and chemical factors. The complexity is accentuated because the organism is polymorphic and grows as a multipolarly budding yeast, as a pseudohypha of elongated yeast-like cells, and as a true septate hypha. Both the true hypha and pseudohypha may regenerate the yeast phase by producing clusters of yeast-like blastoconidia (Figs. 3 and 4). The polymorphic nature of *C. albicans* is further augmented by the production of thick-walled chlamydospores [279].

The yeast, hyphal, and pseudohyphal morphologies are all associated with host lesions [219,345,388,521,583,590]. The yeast morphology is considered by many authorities to be the saprophytic one which colonizes body surfaces, whereas conversion to the hyphal morphology is required for tissue invasion [263,445,477]. However, the decreased virulence of *C. albicans* upon intravenous or intraperitoneal challenge with hyphae or pseudohyphae, compared to blastoconidium challenge [182,355,521], indicates the requirement of all morphologies for the establishment, dissemination, and maintenance of the parasitic relationship.

Yeast cells of *C. albicans* are round to oval, measure 2-5 μm in diameter [179], and are capable of multipolar budding. The daughter buds often remain attached to mother cells until stationary growth conditions prevail, at which time cells accumulate as singlets [527]. The yeast cell wall is a laminated structure from 100 to 300 nm thick [446,586, 589]. Most investigators note from four to five layers by electron microscopic examination [160,502,586], but Poulain et al. [446] claim an eight-layered structure. This eight-layer hypothesis is an eclectic one synthesized from data of cells grown under different

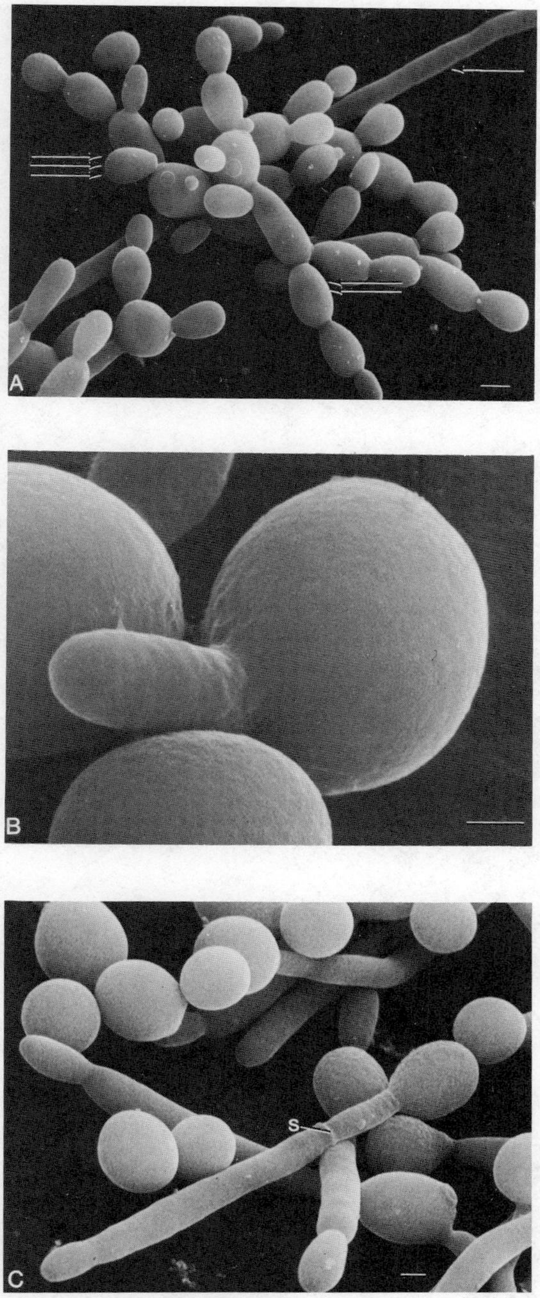

Figure 3 Scanning electron micrographs of *Candida albicans*. (A) Hyphal (arrow), pseudohyphal (double arrow), and blastoconidium (triple arrow) morphologies. Bar represents 2 μm. (B) Germ tube. Bar represents 1 μm. (C) Young hyphal element with intermediate septum (S). Bar represents 1 μm. (Courtesy of L. M. Pope and G. T. Cole.)

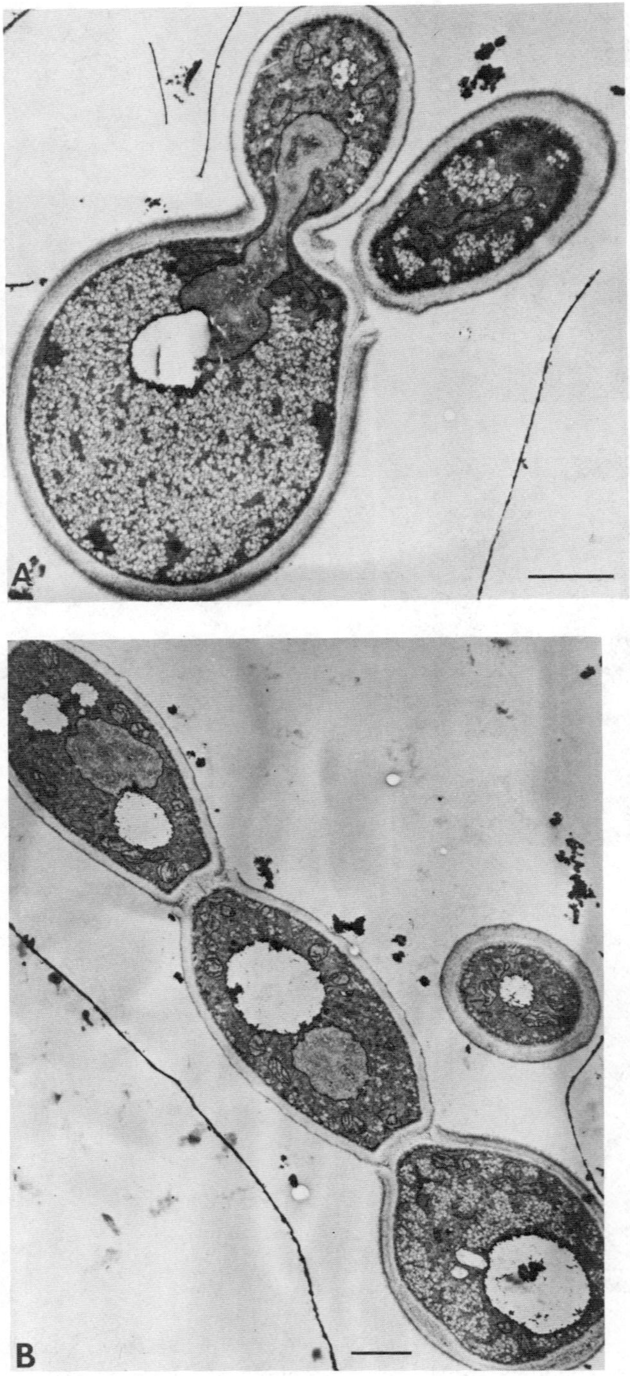

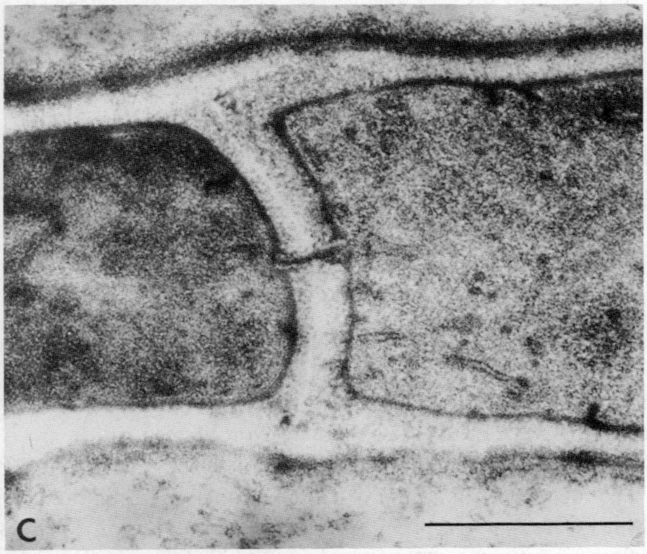

Figure 4 Transmission electron micrographs of *Candida albicans* morphologies. (A) Multiple budding yeast cell. Line represents 1 μm. (B) Pseudohyphal element with yeast-like septa. Bar represents 1 μm. (C) Perforate, hyphal septum. Line represents 0.5 μm. (From Ref. 502.)

cultural conditions. The importance of cultural conditions [98], as well as the localized occurrence of some wall layers [351] and the intermixing of wall components in stationary phase [109], suggest that this approach may be invalid.

Reproductive growth occurs by a blastic process in which a major part of the incipient, daughter cell wall is synthesized continuously with the fourth, electron-transparent layer of the mother cell wall [502]. After bud growth and cytokinesis, a bud scar remains at the site of separation [502], but the birth scar of the daughter cell is virtually undetectable [507].

Investigation of the intracellular ultrastructure of yeast cells reveals the presence of ovoid nuclei, round to elongate mitochondria located around the periphery of the cytoplasm, ribosomes, a sparse endoplasmic reticulum, polysomes, and vacuoles [447,589]. The cytoplasm is surrounded by a sinuous cell membrane which possesses tubular invaginations [589].

The pseudohyphal morphology of *C. albicans* consists of elongated blastoconidia possessing complete septa [502] (Fig. 4B). Individual cells are from 2 to 5 μm in diameter and 26 to 30 μm in length [518]. The intracellular elements of the in vivo, pseudohyphal cells are reported to include cytoplasmic vesicles, endoplasmic reticulum, and Spitzenkörper [455]. However, the Spitzenkörper described for pseudohyphal cells of *C. albicans* is, by the authors' own admission, inconsistent with published descriptions of such bodies.

True hyphal cells of *C. albicans* are usually longer than those of yeast cells and measure from 0.8 to 1.3 μm in diameter [518,589]. When induced by yeast extract-deficient medium, the true hyphal cells possess a triple-layered wall from 70 to 90 nm in width, whereas biotin-deficiency induces hyphal cells having a four-layered wall 170-200 nm in width [586]. A perforated septum 0.17-0.22 μm wide, with no evidence of Woronin bodies or septal plugs, separates the hyphal cells [502] (Fig. 4C). Within the hyphal cell are elongated, mitochondrial profiles and nuclei whose long axes are parallel to that of the cell and vacuoles [589]. The hyphal cell plasma membrane is significantly different from the yeast cell membrane and exhibits invaginations that are deeper and much more frequent [589]. In addition, an extensive development of an intracytoplasmic membrane system appearing as a whorled structure originates with the infolding of the plasma membrane [589].

Yeast cells are easily distinguished from the other morphologies of *C. albicans,* but the recognition of hyphae as true hyphae or pseudohyphae is more difficult (Figs. 3 and 4). Wickerham [579] suggested two criteria: (1) the presence of true, refractile, nonconstricted septa in hyphal cells versus the curved constricted cellular junctions without septa of the pseudohyphal cells; and (2) the increased length of the terminal hyphal cell over that of its adjacent cell as compared to the tip cell of the pseudohypha, which is shorter than or equal to the penultimate element. Despite these criteria, the accurate designation of the two elements is exceedingly difficult. Although some taxonomists choose to group pseudohyphae and hyphae under one term [316], others have reevaluated the use of these terminologies [45]. Possibly the most productive and accurate approach to the problem would be the demonstration by electron microscopy of the perforated septum of true hyphae and the bud scar and complete septa of pseudohyphal associations [502] (Fig. 4B and C).

Under some conditions the hyphae or pseudohyphae of *C. albicans* produce chlamydospores [279,563]. These are thick-walled, nondeciduous, intercalary or terminal, asexual spores formed by the rounding up of preexisting cells [3]. In *C. albicans*, chlamydospores have a double- or triple-layered cell wall which is much thicker than that of the yeast or mycelial cell and which is continuous, in part, with that of the "suspensor cell" [279,382]. Young chlamydospores possess an irregular cytoplasmic membrane, mitochondria, and lipid bodies [23,112,382]. These cells are quite fragile, extremely short lived (60-72 h) but are able initially to germinate by either budding or germ tube formation [112,279, 381,382]. The short life span suggests the incorrect use here of the term "chlamydospores," especially as reviewed and characterized by Griffiths [220].

The morphogenetic processes by which one population of *C. albicans* cells gives rise to another of an alternative morphology have been the subjects of extensive but often narrowly focused studies. Many of these studies were prompted by the observation of germ tube formation by blastoconidia (Fig. 3B and C), which was initially noted in blood and was thought to be mediated by a heat-stable, nondialyzable "filament-transforming factor" [468]. This phase transition has become a critical criterion for the rapid identification of *C. albicans* in the clinical laboratory.

Initially, germ tube formation is similar to yeast budding. The major portion of the four- or five-layered germ tube cell wall is continuous with the electron-transparent fourth layer of the blastoconidium wall [502]. The outer layers of the mother cell wall are continuous with the germ tube wall but with dimensional changes. The third, main layer of the mother cell is not noted in the germ tube [502]. Cassone et al. [111] noted that all layers of the mother cell are resynthesized at some point during germ tube elongation, but are subject to quantitative changes due to mechanical and lytic factors. However, some antigenic components of the mother cell wall are incorporated into the surface of the growing germ tube [216].

No constriction is noted at the juncture of the mother cell and the germ tube, nor is a septum usually formed at that site. However, Yamaguchi et al. [589] noted a uniperforate septum at this locus when germination was induced by yeast extract-deficient medium. At maturity, the septum has a single central pore without associated Woronin bodies or septal plugs [502,589] and delimits an average hyphal cell length of 15.6 μm [384]. The hyphal units continue to grow by apical extension, producing a branching mycelium on a suitably nutritious substrate [153,345,502]. In the absence of such a medium, growth ceases and chlamydospores may form [112,174,520,536]. A similar sequence of events leading to hyphal outgrowth appears to take place in vivo [46,94,263, 307,345,590].

In contrast to mycelial development following germ tube formation [319], pseudohyphal development from yeasts is by a blastic process [502]. The pseudohyphal outgrowth is initially similar to that of a germ tube but is distinguished by the development of typical yeast septa with constrictions and no septal pores [502]. Both hyphal elements and pseudohyphal elements form blastoconidia by lateral and terminal budding [384,413]. Initially, the blastoconidia are only capable of budding growth and cannot germinate until they reach stationary phase or are exposed to morphogenetic inducers such as serum or N-acetylglucosamine [113,181,527]. The blastoconidia and yeast cells are probably functionally equivalent, as has been suggested for their counterparts in *Wangiella dermatitidis* [228].

The cell wall compositions of the yeast and hyphal forms of *C. albicans* are qualitatively similar, but quantitatively distinct. Cell walls of the two morphologies induced by different biotin levels have similar concentrations of lipid, protein, and total carbohydrate, but the mannan content of the hyphal wall is 20% less than that of the yeast wall, whereas the quantity of alkali-insoluble glucan is greater in the hyphal wall [586,587]. Chattaway et al. [117], employing medium and temperature manipulations to obtain the two morphologies, found the chitin content of an alkali-insoluble fraction of mycelial wall to be three times that of the same yeast wall fraction. However, the protein content of this fraction is only one-third that of the yeast wall fraction. The glucan polymers of the cell walls are of $\beta(1 \rightarrow 6)$ and $\beta(1 \rightarrow 3)$ linkages, while those of the mannans are $\alpha(1 \rightarrow 2)$ and $\alpha(1 \rightarrow 6)$ with variable branching [55,515]. Variations in mannan linkage, as well as protein constitution, are responsible for the existence of two serotypes of *C. albicans* [248, 386,540].

The similarities in the compositions of the cell walls of the two morphologies have led some investigators to examine the distribution of wall polymers within and upon the cell wall. It is evident that yeast and hyphal walls share common antigenic determinants on both their inner and outer wall surfaces [567]. Both agglutination of blastoconidia with concanavalin A [110,280] and mild extraction procedures [515] indicate a mannan layer on the outer surface of the yeast wall. Lectin-mediated staining shows that this polysaccharide is distributed throughout the wall and that electron microscopic layering reflects the alkali-insoluble complexes both over and in the glucan-chitin matrix [110]. These observations and others [514] that suggest a covalent linkage of chitin and β-glucan in fungal walls indicate the complexity of wall structure.

The many similarities, rather than the differences, of yeast and hyphal cell walls are striking. Significant factors such as the relative amounts of chitin, mannan, and alkali-soluble and alkali-insoluble glucans exist [117,586], but as suggested by Chattaway et al. [116], differences in enzymatic sensitivity of the hyphal and yeast walls may indicate considerable differences in tertiary wall structure. Such an interpretation appears to be supported by the wall layering observations. It is worthy of note that recent analytical and cytological data indicate the presence of uronic acid and N-acetylgalactosamine [461, 534], in addition to the saccharide monomers previously cited in these walls (glucose, mannose, and N-acetylglucosamine) [117]. However, no morphogenetic role has been suggested for either of these minor components, although they both appear to be localized at the cell surface [534] and uronic acid is associated with the mannan polysaccharides [461].

Limited analysis of the walls of a pseudohyphal variant of *C. albicans* incapable of hyphal growth indicates differences in amino acid composition from the yeast morphology [354]. Cell wall analysis of a divisionless mutant, which some investigators suggest to be pseudohyphal rather than hyphal [413], reveals the cell wall composition to be virtually identical to that of the parental yeast strain [305].

Whole cell analysis and electron microscopic examination of chlamydospores and their suspensor cells indicate that these acid-fast cells possess a wall of two layers: an outer glucan-chitin layer and an inner electron-dense, proteinaceous layer [112,278].

A number of macromolecular and enzymatic differences exist among the various morphologies of *C. albicans*. Adenosines tri-, di-, and monophosphates differ in concentrations between the yeast and hyphal morphologies and within the same morphology during different stages in the growth cycle [116]. Yamaguchi [588] employed zinc starvation to obtain the hyphal morphology and found a decrease in the RNA content and an increase in the inorganic polyphosphate content compared to that of the yeast cells. Similar results in RNA contents are noted when biotin or yeast extract are employed as morphological inducers [585,587]. Synchronous cultures of the yeast and hyphal phases reveal exponential protein and RNA syntheses and parallel, stepwise DNA synthesis [145, 570]. Induction of germ tube formation by temperature shift of stationary-phase blastoconidia results in the synthesis of lipids similar to those noted with the induction of budding growth [27]. Analysis of plasma membranes reveals more lipid and less carbohydrate in the yeast morphology than in the hyphal morphology [363]. Sphingolipid and phosphatidyl serine are present only in the yeast membranes, which also contain more sterol and less free fatty acid than do those of the hyphae [363]. Consistent with this is the cell wall lipid content of the two morphologies, which show more free fatty acid in the hyphal walls and the presence of sphingosine phosphatides in the yeast wall [52]. The differences noted may indicate the significant distinction between lipids of germ tubes and those of mature hyphae or they may merely reflect the varied lipid composition of the whole cell versus that of the membranes or cell walls. The concentrations of reduced and oxidized pyridine nucleotides are the same in the yeast and hyphal morphologies but vary within the same morphology during the growth phase (W. L. Chaffin, Texas Tech University, personal communication).

Acid and alkaline phosphatase, peptidase, phospholipase, and β-glucosidase activities are cytochemically demonstrable in both the yeast and hyphal morphologies. The acid phosphatase is a cell wall-associated activity [118,120], with expression critically controlled by the inorganic phosphorus content of the medium [414]. Both the acid and alkaline phosphatase activities are the same in the yeast or hyphal morphologies [312]. Aminopeptidase is detectable in either the yeast or pseudohyphal cell vacuole as well as in the cell wall [387]. A number of specific aminopeptidases are liberated into the medium from the blastoconidia of *C. albicans* and other pathogenic fungi [328]. Phospholipase is an intracellular activity whose presence in vivo is suggested to be associated with pathogenicity and which, together with acid phosphatase and glucosaminidase activities, is concentrated at the bud site of yeasts and the apical tips of germ tubes and pseudohyphae, but is generalized in nondividing blastoconidia [453,454].

Marriot investigated the enzymatic activity in purified plasma membranes of the yeast and hyphal forms of *C. albicans* [362]. Enrichment of specific activity with plasma membrane purification is found with Mg^{2+}-dependent and Na^+/K^+-stimulated adenosine triphosphatases, and mannan synthase, whereas no such enrichment was noted with acid or alkaline phosphatase, NADH oxidase, or 5′-nucleotidase. Chattaway et al. [116] employed conditions in which temperature was the determinant of morphology and found that with maximum germ tube development, phosphofructokinase activity and glucose catabolism via the hexose monophosphate pathway were low, while aminotransferase activity was high. With some variation, this situation is reversed in blastoconidium budding or in production of lateral blastoconidia by mycelium. Chitin synthesis is limited by the decreased aminotransferase activity, which is consistent with the cell wall chitin contents [117]. The NADPH thus made available would presumably fulfill the conditions for yeast growth by maintaining the cell wall sulfhydryl groups in a reduced state and thereby allowing the generalized, nondirectional expansive growth of the yeast morphology as suggested by Nickerson and Falcone [401].

Braun and Calderone [69] have demonstrated that chitin synthetase activity of hyphal cells grown in a synthetic medium is greater than that of yeast cells grown in Sabouraud's medium, an observation consistent with the chitin contents of the yeast and mycelial cells

[117]. The incorporation of radiolabeled N-acetylglucosamine into the acid- and alkali-insoluble cell wall fraction was 10-fold that of the yeast cells and was localized in the apical tips of hyphae and budding sites of blastoconidia. This incorporation via chitin synthetase was stimulated by a vacuolar fraction from yeast cells, but a similar fraction from hyphal cells stimulated neither yeast nor hyphal synthetase activity [69].

Bhattacharya et al. [49,51] have shown that addition of N-acetylglucosamine to *C. albicans* results in the specific induction of N-acetylglucosamine kinase, an enzyme that phosphorylates N-acetylglucosamine. This induction is not subject to catabolite repression by glucose, glycerol, or succinate [49,523]. The relationship of this induction and the function of chitin synthetase with the N-acetylglucosamine-induced germination of *C. albicans* blastoconidia [522] is intriguing. It is doubtful that this compound serves as a primary carbon source for growth under the conditions of this germination [112,522]. However, *C. albicans* can utilize N-acetylglucosamine as the sole source of carbon as do other pathogenic *Candida* spp. [524].

The enzymatic differences between the yeast and hyphal cells are, like those of the cell walls, more quantitative than qualitative. A number of potentially productive studies have been performed such as those on the effects of cAMP on protein and RNA synthesis [50], on the existence of an alternate, cyanide-insensitive respiration [511], and on the utilization of trehalose and glycogen during release from stationary phase [16]. In general, these studies have not included morphogenetic observations. Land et al. [317] showed that a number of other factors are morphologically significant to the germination phenomenon, but their role in continued maintenance and growth of the hyphal morphology remains to be elucidated.

The phenomenon by which blastoconidia of *C. albicans* "germinate" to produce germ tubes in serum [29,274,287,290,291] provides not only a valuable clinical tool [344, 545], but also a model for studying mediators and effectors of this morphological transition. The transition occurs in complex substances such as saliva [30,306], cerebrospinal fluid [318], plasma [293], seminal plasma [29], egg white [80], and trypticase and peptone broths [290,291], as well as in defined media such as that of Lee et al. [327] and tissue culture medium 199 [318].

One factor suggested as being responsible for the yeast → hyphal transition in serum is transferrin [319]. That the saturation of transferrin with ferrous ammonium sulfate allows increased germ tube production [319], and that the use of plasma treated with ethylenediaminetetraacetic acid (EDTA) decreases germ tube production [318], indicate a possible role for a metal ion(s). Such a factor may be protein bound, as it is not removed by dialysis [318]. However, some investigators feel that iron is required for filamentous growth [259], whereas others conjecture that added iron is acting not by stimulating filamentous growth, but by binding transferrin and thereby removing its inhibition [318]. Transferrin also binds copper and zinc [319], but the possibility that a small concentration of these ions is made available for filamentous growth with iron saturation is unlikely in light of the fact that the addition of copper or zinc favors yeast growth [45, 566,588].

Examination of the serum components reveals that the most active fraction in inducing germ tube formation is 95% albumin, but comparison with commercial albumins and with a similar but more active fraction of seminal plasma indicates that albumin itself is not responsible [29]. Whatever the serum factor, it is heat stable (56°C), nondialyzable, not an agglutinin, and is not degraded during the yeast → hyphal transition process [468,

546]. Also, hyperimmune serum or purified gamma globulin from such serum prevents both germination of blastoconidia and their respiration [217]. Neither proteins of whole serum and gamma globulin nor free amino acids appear to be metabolized during germ tube formation [546]. Acid hydrolysis of a germ tube-active, chromatographically isolated fraction of seminal plasma provides a highly active mixture of transition-inducing components. A mixture of amino acids which roughly corresponds to the hydrolysate described above also induces 90% germination in a number of strains when combined with glucose and inorganic phosphorus [119]. The critical amino acids of this mixture are aspartic acid, lysine, histidine, threonine, proline, and β-alanine. Lee et al. [327] employed a mixture of amino acids, selected on the basis of the aminopeptidase activities of *C. albicans* blastoconidia [328], together with biotin, inorganic salts, and glucose, to grow either morphology with the temperature of incubation as the morphological determinant.

A number of individual medium components or combinations of components determine the morphology of *C. albicans.* Cysteine is said to limit *C. albicans* growth to that of the yeast morphology [146,402,546]. However, this effect is strain variable [589] and has been shown to be without effect in some germination systems [408,584] and to inhibit all growth in others [571]. Glycine or L-α-n-butyric acid in a salts biotin glucose medium induces filamentous growth at 37°C [23,356]. Such a glycine-containing medium was employed by Mardon and Balish [25,26,352,353] to induce filamentation and to demonstrate an increased methionine-synthetic capacity and a greater concentration of S-adenosylmethionine in the vacuoles of the mycelial phase versus that of the yeast morphology. As the sole source of carbon and nitrogen, L-proline allows germination of blastoconidia at 37°C [146]. This germination is inhibited by addition of cysteine, D-proline and proline analogs, as well as concentration ranges of glucose (0.001-0.28 M), phosphate (0.1-0.5 M), and ammonium ions (0.0002-0.2 M) [146]. Members of the glutamate, succinyl, or acetoacetyl-CoA families of amino acids allow filamentation in a phosphate-buffered medium with biotin and a low glucose concentration, whereas increased glucose ($\geqslant 10^{-1}$ M) allows only yeast growth [316]. Nishioka and Silva-Hutner [408] showed that with one strain of *C. albicans,* glutamate as the sole carbon and nitrogen source induces germination that is unaffected by cysteine or glucose concentrations. Suboptimal biotin (0.1 ng/ml) is a hyphal effector [587], and an alleviation of biotin starvation by a biotin-sparing effect is seen with glycerol-monooleate (50 ng/ml), but not aspartic acid. Observations concerning amino acids as dictators or modifiers of morphology are made more difficult by a number of factors, such as the growth phase, which greatly alters transport of specific amino acids or groups of amino acids [281,412,525].

In addition to the substrates mentioned before, a number of carbohydrates influence morphology. Galactose, xylose, lactose, or trehalose, under conditions of decreased oxygen availability, produce a hyphal morphology, whereas glucose, of 0.7% or greater concentration, results in a yeast morphology [556]. The former effect may be the result of lack of substrate utilization, since the increased galactose concentration allows no increase in cell mass. Polysaccharides such as glycogen or starch used as the sole source of carbon allow slow, filamentous growth [402]. Continuous cultivation with maltose promotes a mycelial morphology, whereas lactate promotes a yeast morphology. Fructose or glucose engenders mixtures of yeasts and pseudohyphae [512]. The glucose/phosphate ratio is said to control morphology via mediation of the intracellular magnesium level [580].

Simonetti et al. [522] described an interesting phenomenon in which N-acetylglucosamine induces germ tube production from blastoconidia. These workers showed that induction is optimized with magnesium or manganese in imidazole-HCl buffer and is markedly inhibited or absent at alkaline pH or in the presence of cysteine.

Davies and Denning [152] observed that uniformly small, single blastoconidia of *C. albicans* isolated with a linear density gradient will all develop germ tubes if placed in serum at 37°C. Chaffin and Sogin [114] revealed that this population was in stationary phase and capable of "germinating" if diluted in fresh defined medium at 37°C or of synchronously budding if in fresh medium at 25°C. Control experiments revealed that no logarithmic-phase cells could germinate. Soll and co-workers [384,527,528] extended these observations and noted that stationary-phase cells also bud when released into fresh medium with a pH of 4.5 at 37°C. Budding cells and their progeny do not form germ tubes following temperature or pH shift prior to reaching stationary phase. Outgrowth of buds or germ tubes are closely preceded by a point of "commitment." Nuclear division takes place when the germ tube length is equal to, or greater than, the diameter of the mother cell, and both outgrowth and nuclear migration take place before DNA replication. Soll and Bedell [527] believe that conversion competency of stationary-phase blastoconidia or yeasts is due to factors other than the exhaustion of nutrients. These factors may be "autoregulatory" compounds [180,250,487]. Bell and Chaffin [46a] showed that a loss of inducibility of germ tube formation follows a decrease of growth by limitation of nutrients. The loss is variable with the limiting nutrient and is most pronounced with glucose. Certainly, stationary-phase cells differ from logarithmic growth-phase cells in a number of ways, including cell wall structure [78,110], sterol content [513], mode of respiration, and amino acid transport [281], but the relationship with morphogenesis is still only empirical.

A number of factors, in addition to those mentioned, are said to modify or dictate the morphology of *C. albicans*. These include defined chemicals such as cobalt and penicillin [404] and other conditions such as yeast extract deficiency [585] or light [488]. Chattaway et al. [119] suggest that mycelial growth is the product of four critical criteria: (1) an alkaline pH, (2) a glucose concentration of 0.2% or less, (3) an incubation temperature of 37°C or higher, and (4) an inoculum of from 10^5 to 10^6 yeast cells per milliliter. To these, Odds [413] adds a liquid formulation of a polysaccharide plus non-sulfur-containing amino acids. As indicated above, research has shown that amino acids individually [146,316,317] and in mixtures [118,318,327], as well as carbon sources [512,556], especially glucose, are significant morphological effectors.

The glut of observations concerning the morphogenesis of *C. albicans* is complicated by three factors. The first is the differentiation between the pseudohyphal and the hyphal forms. They are structurally and probably functionally distinct, and the former, although it possesses a yeast-like septum [502], is usually grouped with the hyphal morphology in most studies, owing to its directional mode of growth and gross morphology [180,316]. The second is the use of many strains and media. Obscure clinical isolates or diverse mutants are often employed in investigations [e.g., 180,354,404,556]. The germination characteristics of such strains are often variable [23,142,346,354,415,488,569]. A phenomenon described for one strain is not always subject to generalization of all isolates [569]. The situation is further complicated by variation in germination-induction capabilities of commercially available, undefined media components [414], and although serum-induced germination is rarely perturbed, the ability of this constituent to provide for true hyphal growth is variable [174,520,570]. Finally, we must view germination itself as a variable phenomenon. Although the germ tube is established as an incipient hyphal entity [319,345], it is not clear if the same factors are involved with hyphal production and ultimate maintenance of the apical mode of hyphal growth. This uncertainty has major significance in the study of those factors that promote germination but do not provide for hyphal growth [174,520,536], as well as in the study of those factors that induce morphologically distinct germination phenomena, such as yeast extract-deficient media [589] and release from stationary phase in fresh medium at 37°C [114,527].

Many of the factors that induce mycelial growth in *C. albicans* also induce chlamydospore formation [249,334], although hyphal morphogenesis is not a necessary prerequisite [24,329]. Whereas glucose is a potent inhibitor of chlamydospore formation [249], N-acetylglucosamine induces massive chlamydosporogenesis following hyphal development [113]. Chlamydospores so induced can germinate initially, as do blastoconidia [522], if placed in suitable media with serum or N-acetylglucosamine [113]. The limited vitality of these cells [381,382] indicates that rather than being true chlamydospores [220], they are terminal units of hyphal differentiation produced by isotropic growth processes subsequent to "starvation" [113,174,520,536] or exposure to membrane-altering [183] or enzyme-activating agents [113].

Despite the voluminous data concerning polymorphism in *C. albicans,* few workers attempt to formulate a hypothesis of control for this phenomenon, and only two models are generally recognized. The model of Nickerson et al. [398,399,401,403,404] is based on the hypothesis that mycelial growth is the product of cell growth without accompanying cell division. The directional integrity of a hypha is maintained by the disulfide bonds of the cell wall, which preserve its cylindrical nature. The yeast cell has in its wall reduced sulfhydryls which allow for nondirectional growth. This reduction of the sulfhydryls occurs via a flavoprotein, disulfide reductase, with reduced pyridine nucleotide produced by the rapid catabolism of substrate. Nutritional controls, such as the poorly utilized polysaccharides, dictate a hyphal morphology because an insufficient amount of reduced pyridine nucleotide is produced. Chattaway et al. [116] extended Nickerson's theory to explain the chemical differences in cell wall composition [117] by correlating the differences with activities of the hexose monophosphate pathway and to aminotransferase activity.

The cytochemical presence of sulfhydryl groups is noted only in yeast cells and not in pseudohyphae [442,476]. Moreover, cysteine, apparently by reducing cell wall disulfide bonds, promotes a yeast morphology [398]. However, the absence of sulfhydryl groups from hyphae of *C. albicans* is not true for all hyphae because *Neurospora crassa* hyphal tips are very rich in sulfhydryl groups [594]. A number of suboptimal nutritional circumstances such as growth on a poorly utilized substrate [556] or an insufficient biotin concentration [587] result in mycelial growth. It is significant to note that stationary-phase blastoconidia, which some workers suggest [194] have a greater proportion of cell wall disulfide bonds, are able to germinate and form germ tubes with the appropriate stimuli [113]. The observation that the concentrations of reduced pyridine nucleotides are the same in the two morphologies (W. L. Chaffin, Texas Tech University, personal communication) indicates that wall morphology is dependent on more than the simple availability of these compounds within the cell. The description of the mycelial morphology as cell growth and metabolism without cell "division" is possibly more appropriate for pseudohyphal elongation. Certainly, the cell wall ultrastructure, chemical composition, and the existence of a perforate septum in true hyphae indicate the greater complexity of dimorphism than can be explained by a single enzyme complex.

Land et al. [316,317] recently suggested that a number of diverse observations concerning the dimorphism of *C. albicans* may be integrated into a hypothesis in which filamentation is secondary to a Crabtree-like repression of mitochondrial activity. Such observations include induction of filamentous growth by high glucose concentrations [316], the glucose-to-phosphate ratio [580], the phosphate concentration [146], and growth with certain amino acids [316]. This theory invokes some of the features of the Nickerson theory, especially the necessity of cell wall sulfhydryl maintenance via reduced pyridine nucleotides for yeast growth.

Nickerson's model of the biochemical basis of dimorphism has been used to explain morphogenesis in many systems [398]. In contrast, that of Land et al. [316,317] is focused on *C. albicans* and is not considered in the dimorphism of organisms which express no Crabtree phenomenon, such as *C. tropicalis* [155], or without functional mitochondria, such as *C. slooffii* [17]. In addition, the germination effect of amino acids, said to operate via this phenomenon, is strain variable and is not complete in its effect on a group of cells [146].

Odds and co-workers [181,413] have proposed an alternative model in which temperature is the critical criterion. Here, the wall is synthesized in two stages: a soft primary wall is initially produced and later solidified. At high temperatures, solidification occurs rapidly, forming a hyphal cylinder. At lower temperatures this process is slow, so that the primary wall may swell under cytoplasmic turgor pressure and produce the yeast cell which later hardens its wall. This simplistic theory further cites pseudohyphae as products of intermediate temperature. The great number of effectors of morphology previously cited are not integrated into the theory, nor is the stationary-phase induction phenomenon. The formation of perforate hyphal septa would, of necessity, be a temperature-dependent event, and it is doubtful that such a structure could be the product of simple alteration by solidification of previously synthesized polymers. Finally, the production of lateral blastoconidia by both true hyphae and pseudohyphae disputes this theory as the sole control of morphology, since blastoconidia and yeasts appear to be both structurally and functionally equivalent. Increased temperature is included in most systems that induce mycelial growth in *C. albicans,* but it is rarely the sole determinant and can be bypassed by employing chemical and physical factors.

The rationalization of morphogenesis in this organism as viewed with germ tube formation as the central phenomenon may be placed under one of two schemes on the basis of the number of seemingly unrelated observations involved. In the first of these, a central sequence of dependent events exists and leads to germination. Seemingly unrelated factors initiate or modify the activation of this sequence at any number of points in such a way that the intracellular control mechanisms are constantly subject to environmental controls [413]. The models proposed by Land et al. [316,317] and Nickerson [398] and Nickerson and Mankowski [402] may be viewed in this way. An alternative manner of interpreting the current literature is illustrated in Fig. 5. In such a scheme, more than one general developmental sequence may converge at some point or may independently result in germination. Certainly the germination-competent stationary-phase cells have many characteristics that may be related to other germination factors, such as altered amino acid transport [281], respiration changes [281,511], and increased cell wall sulfhydryl reduction [194], but this mechanism appears unrelated to that of serum and N-acetylglucosamine [365] or autoregulators [250]. The involvement of multiple pathways is also suggested by Bedell and Soll [45], who found that germination in stationary cultures of *C. albicans*

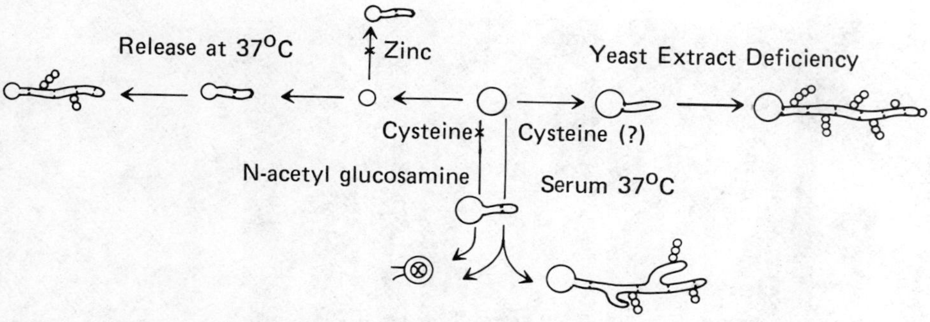

Figure 5 Proposed alternate pathways for germ tube formation and hyphal growth in *Candida albicans.* Inducing agents are serum, N-acetylglucosamine, biotin insufficiency, and release from stationary phase at 37°C.

was blocked by zinc, whereas germination in cultures released from stationary phase at 37°C was not affected.

Superimposed on the phenomenon of dimorphism is germ tube formation, which is a transitional, not a terminal, phenomenon. Certain inducing systems promote apical extension, septum formation, and hyphal growth [356,512], but others will not, probably due to lack of substrate [112,522]. Substrate-limiting conditions will inhibit further elongation [556] and, often, chlamydospore formation [112,174,520,536]. Differences between hyphal and germ tube wall and lipid compositions have been suggested [27,111] and these differences may reflect a wider morphological identity for *C. albicans.*

B. *Exophiala werneckii*

Exophiala werneckii, the causative agent of the superficial human skin disease tinea nigria, was noted as early as 1922 to be exceedingly polymorphic in culture [323]. Carrión [108] pointed out that several distinct morphologies are characteristically produced in liquid media or in aging surface colonies on agar media. Included among these morphologies are three types of true hyphae and two types of yeast growth. Undoubtedly, the vegetative variability exhibited by *E. werneckii* has contributed to the confusion in taxonomy associated with this form-species [64,261,370,371].

In general, the initial colonial growth of *E. werneckii* on a variety of agar media is black, moist, shiny, and consists of yeast cells, whereas older colonies are greenish black and velvety in texture due to hyphal development [8,370,371]. Conidiogenous cells produced by the hyphae are terminal and intercalary monoblastic annellides [130,230,370, 371]. Both hyphae and budding yeast cells are evident in vivo [471].

Ultrastructural studies of the hyphal form of *E. werneckii* are lacking, but the microscopic aspects of the yeast form (Fig. 6) and its mechanism of budding have been described in detail [230]. In general, unbudded single yeast cells are ovoid to peanut-shaped, and measure 5-8 μm in length. The budding yeast cells exhibit a definite polarity which is recognized ultrastructurally by the presence in younger cells of a birth scar distal to the budding apex. Frequently, there is a vacuole near the birth scar end. The generative apex of a nonbudding cell varies in morphology according to the budding history of the cell; those yeast cells that have never produced a daughter bud exhibit a rounded apex, whereas

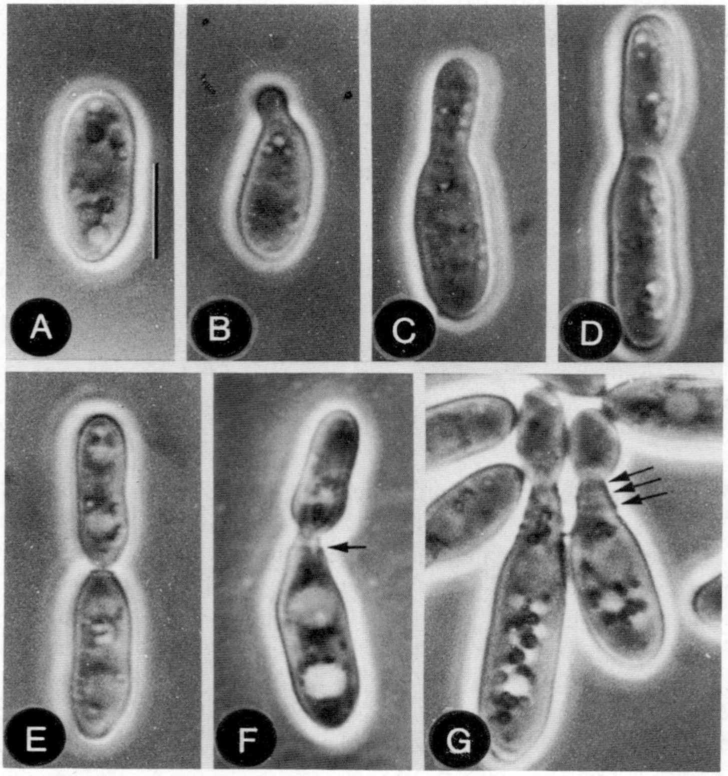

Figure 6 Phase-contrast micrographs showing typical budding sequence in *Exophiala werneckii*. (A) Nonbudding yeast cell. (B, C) Daughter cell enlargement. (D) Formation of septum between daughter and mother cell. (E) Separation of daughter cell. (F) Mother cell with third daughter cell; distinct annellation (arrow) formed from second daughter cell. (G) Mother cell (right cell) with fourth daughter cell; three annellations are evident (arrows) on elongated generation apex of mother cell; cell on left does not exhibit obvious annellations, but elongated generative apex is indicative of past production of daughter cells. Bar represents 5 μm. (From Ref. 230, reproduced with permission of the authors and the New York Botanical Garden.)

cells that have produced daughter cells have a flattened, often elongated apex. The production of successive daughter cells results in the formation of distinct annellations around the generative apex and an elongation and narrowing of the apex. The mother cells are typically uninucleate and contain numerous mitochondrial profiles, endoplasmic reticulum, and masses resembling both lipid globules and glycogen deposits. Two other structures, one resembling a stack of endoplasmic reticulum cisternae and a second resembling a rudimentary Golgi apparatus, are frequently observed in the cytoplasm of rapidly growing cells. These structures are possibly involved in the budding process. The developmental aspects of budding growth suggest that each yeast cell of *E. werneckii* represents a disarticulated annellophore and each bud cell, a conidium.

The dimorphism exhibited by *E. werneckii* is mediated by factors of nutrition, incubation temperature, cell density, and atmospheric composition [242,265,405]. Nielsen and Conant [405] were able to induce mold → yeast transition and to maintain yeast-phase growth in a cysteine starch agar medium overlaid with cysteine starch broth. Houston et al. [265], after observing growth on a variety of agar and broth media, reported that in general, yeast growth is induced by cysteine enrichment, elevated temperature, shake culture conditions, and increased CO_2 concentrations, whereas mycelial growth is favored by lower temperatures and nitrogen or oxygen atmospheres. Their work demonstrated that *E. werneckii* is very sensitive to environmental perturbations and quickly and easily changes its morphology accordingly. Hardcastle and Szaniszlo [242] investigated the effects of a number of factors on phase maintenance and phase transition in liquid medium using homogeneous populations of logarithmically growing yeast-phase cells as inocula. In agreement with Houston et al. [265], these investigators determined that yeasts can be propagated with ease at 21°C in an agitated sucrose nitrate salts medium. Either enrichment of yeast inoculum cultures with yeast extract or a shift of inoculum cultures from 21°C to 30°C induces yeast → hyphal transitions [242]. The addition of yeast extract at a final concentration of 1% to cultures containing 10^4 yeast cells per milliliter induces nearly total conversion of the yeasts to true hyphae (hyphae with parallel walls as seen in optical section). Two types of morphological changes appear to accompany this conversion, depending on the cell cycle position of the yeast involved. Cells not having entered a typical budding sequence simply convert directly to hyphae by elongation. In contrast, only the buds of budded yeasts elongate and form hyphae. Under the conditions of culture employed by Hardcastle and Szaniszlo [242], the hyphal nature of a culture was difficult to maintain because of the tendency of hyphae to produce buds (blastoconidia), which in turn produce yeast progeny.

Hardcastle and Szaniszlo [242] also showed that dimorphism in *E. werneckii* is exhibited not only by conversion of yeasts to true hyphae, but also by conversion under appropriate conditions to moniliform hyphae (hyphae made up of elongated hyphal elements which exhibit constrictions in optical section at points where septa form). Moniliform hyphae result from the use of a large number of yeast cells for inoculum, higher temperatures, or in some instances very rapid agitation. The moniliform hyphae of *E. werneckii* are not equivalent to the pseudohyphae produced by some other dimorphic fungi because when observed in thin section, moniliform hyphae exhibit perforate septa with associated Woronin bodies (R. V. Hardcastle, R. A. Gustafson, and P. J. Szaniszlo, unpublished results). In contrast, pseudohyphae of other fungi appear to represent chains of yeast cells which have not completed cell separation and have double nonperforate septa between cells.

The importance of the inoculum growth conditions to the transition potential of the yeast-phase cells is demonstrated by comparing the transition competency of yeast cells from enriched medium to those from sucrose nitrate salts medium [242]. When yeast cells from the enriched medium are reinoculated at a concentration of 10^5 cells per milliliter into enriched medium, a perpetuation of a transition-incompetent yeast population results. When the inoculum is 10^4 cells per milliliter or less, an abortive attempt at transition occurs resulting in pleomorphic budding units. Only when transition-incompetent yeast cells undergo a number of cell division cycles in minimal medium is nutritional conversion competence restored. Hardcastle and Szaniszlo [242] suggested that the yeast cell must exhaust an intracellular supply of some factor, either present in the yeast cell or induced by the presence of yeast extract, which inhibits the transition of yeast cells to hyphae.

The changes in particle number, average particle volume, and also relative volume which accompanies yeast → hyphal conversions mediated by either medium enrichment or temperature were evaluated using a Coulter Counter and Coulter Channelizer [242]. Based on relative volume changes, nutritionally mediated conversion is attended by an increase in growth rate. In contrast, the pronounced temperature-mediated conversion at 30°C is attended by a decrease in growth rate through the entire initial conversion period.

Although no explanation can be provided concerning the regulation of dimorphism in *E. werneckii,* it is evident that this fungus is well suited for further morphogenetic research. The ability of the yeast phase to grow well in a simple, nonreduced, sucrose salts medium makes it particularly amenable to nutritional investigation. Also, the ease with which the yeast → moniliform and moniliform → yeast transitions can be monitored by automated methods indicates that *E. werneckii* is an excellent candidate for basic studies of phase transitions. The nutritional-dependent and temperature-dependent responses of *E. werneckii* and the relationships of these responses to population density effects are depicted in Fig. 7. Unpublished data of Hardcastle and Szaniszlo indicate that the population effects may be related to increased CO_2 concentrations developing in cultures inoculated with very high numbers of yeast cells. This hypothesis would be in accord with the results of Houston et al. [265], who reported that incubation of *E. werneckii* under CO_2 leads to yeast growth.

C. *Wangiella dermatitidis*

The dimorphism of only one agent of phaeohyphomycosis [8], *Wangiella dermatitidis,* has been studied in sufficient detail to warrant extensive comment. This dematiaceous fungus exhibits a dimorphism in which the vegetative yeast form apparently must become thick-walled and acquire spore-like qualities before it converts to either moniliform or true hyphae [424]. Transition of hyphae to the yeast phase occurs when lateral or terminal bud cells (blastoconidia) are produced and when these buds or their progeny give rise to daughter cells. However, interest in the morphogenetic regulation of *W. dermatitidis* has been focused on its ability to convert to thick-walled yeasts and multicellular forms (Figs. 8 and 10B). The resemblance of these forms to sclerotic bodies, the most characteristic tissue phase of chromomycotic fungi, has historically prompted most investigators to include *W. dermatitidis* among the agents of chromomycosis [11,109,179,292]. However, the differences in the tissue reactions usually elicited by the agents of chromomycosis and by *W. dermatitidis* suggest that Ajello [8] may be correct in including this fungus among the agents of phaeohyphomycosis. Nevertheless, the ability of *W. dermatitidis* to grow in the yeast, hyphal, and multicellular morphologies and undergo phase transitions among these morphologies [228,273,292,338,361,424,425,473-475,543] makes this polymorphic fungus an excellent model for elucidating growth and developmental patterns among the numerous agents of both phaeohyphomycosis and chromomycosis. Certain factors have already been discovered which similarly regulate conversions to the multicellular morphology in *W. dermatitidis* and the classical agents of chromomycosis [543]. A generalized asexual life cycle illustrating each cell type of *W. dermatitidis* is presented in Fig. 9.

Strains of *W. dermatitidis* produce two types of colonies on the surface of a variety

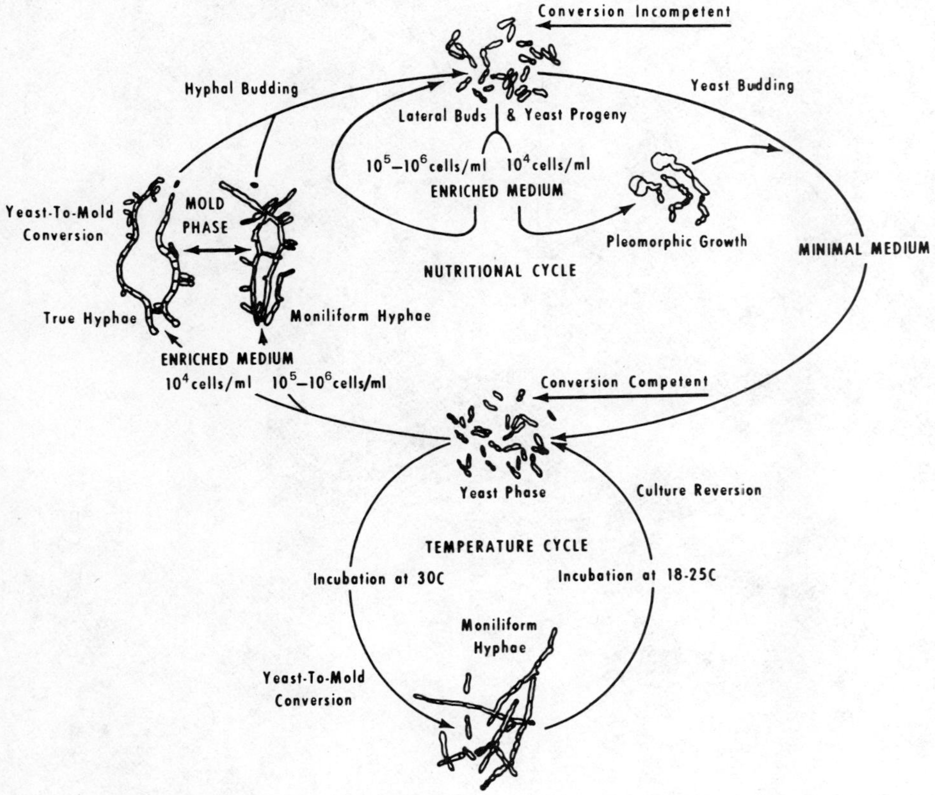

Figure 7 Proposed cyclic sequences depicting the morphological responses exhibited by *Exophiala werneckii* to in vitro changes in nutrition, cell number, and temperature. (From Ref. 242.)

of agar culture media [292]. The colonies of most strains are typically smooth, dark, and pasty, whereas a few are dry and granular. The smooth, pasty colonies are transparent at first and subsequently become brown, then black with age. After a number of days, hyphae develop at the periphery of the colonies and below the agar surface. With some strains the colony morphology does not change further, regardless of the duration of the cultivation. The colonies of other strains become mold-like in appearance following overgrowth by gray aerial hyphae. The granular colonies are slow growing, dry, and rough.

A variety of vegetative growth forms can be identified by microscopic examination of the different colony types [228,424]. Very young smooth colonies are predominantly made up of hyaline, thin-walled and thick-walled yeasts, and moniliform and true hyphae with clusters of yeast-like conidia. The conidia appear to be produced in a basipetalous fashion from the apex of relatively undifferentiated determinate conidiophores [130,368-370,425], although this matter is a point of disagreement [156]. Very old smooth colonies or granular colonies are composed of a few to many thick-walled cells and thick-walled multicellular forms [109,292,424,543].

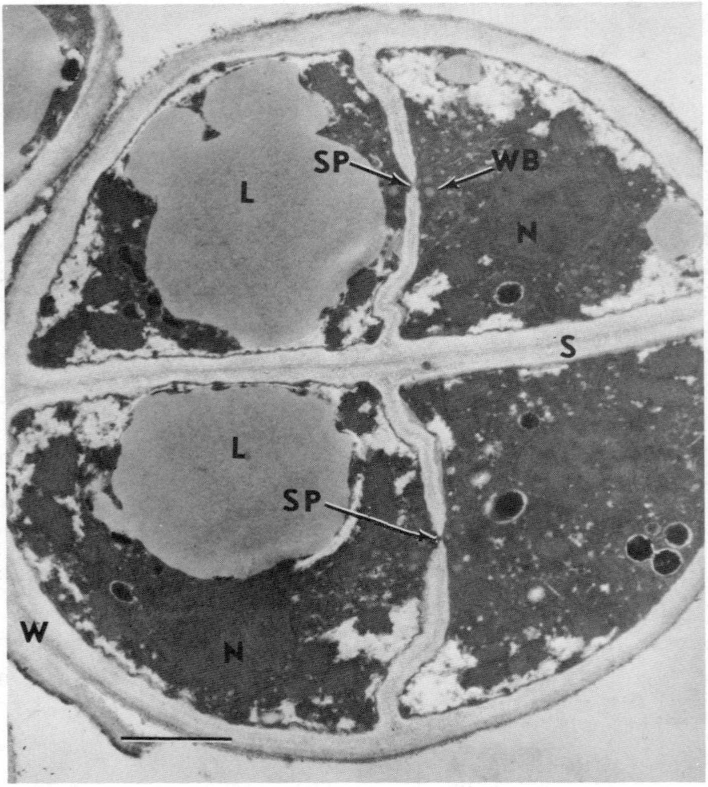

Figure 8 Transmission electron micrograph of a multicellular body of *Wangiella dermatitidis* induced from a yeast cell by 3 days of incubation in acidic (pH 2.5) medium. The multicellular body may have two types of septa. The thick variety has not been observed to have a pore. The thin variety often exhibits a septal pore (SP) with associated Woronin bodies (WB). The thickened wall (W) has a tendency to slough away as the multicellular body expands. Lipid bodies (L) and nuclei (N) are evident in most cells. Bar represents 1 μm. (From Ref. 543, reproduced with permission of the authors and the New York Botanical Garden.)

Detailed morphological and ultrastructural studies of *W. dermatitidis* have aided in the elucidation of the relationships among the cell types and the events associated with phase transitions [228,273,361,424,543]. Homogeneous populations of the cell types can be obtained in vitro by appropriate culture manipulations [424,543]. During exponential growth of *W. dermatitidis* cultured in a yeast extract-supplemented Czapek-Dox broth (MCDB), cell populations are totally homogeneous, consisting of single, unbudded yeast cells in various stages of bud formation (Fig. 10A). The yeast cells measure about 5 by 6 μm in the unbudded condition and about 6 by 12 μm in the budded condition. The wall of the yeast is about 75-125 nm thick [228]. The protoplasm of the thin-walled yeast possesses all the ultrastructural characteristics of rapidly growing fungal cells, such as numerous ribosomal and mitochondrial profiles, few vacuoles, and very little storage polysaccharide or lipid. However, as the yeasts age on agar surfaces and become thick-

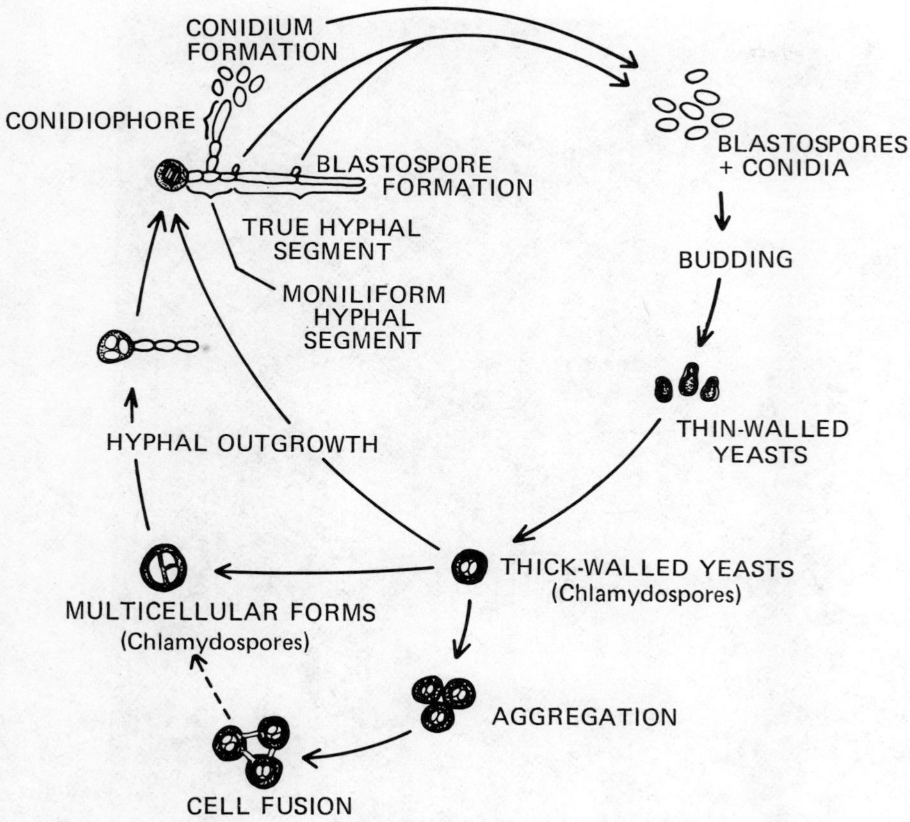

Figure 9 Proposed cyclic sequence depicting the morphological responses exhibited by *Wangiella dermatitidis* to in vitro changes of nutrition, temperature, and pH. (Courtesy of P. J. Szaniszlo and K. B. Oujezdsky.)

walled, they progress through at least two stages of development. In the first stage, the yeast wall thickens, ribosomes and mitochondria become less visible, and there is a pronounced accumulation in the cytoplasm of electron-transparent areas which probably represent glycogen storage areas. In the second stage, the wall is composed of at least two layers, and there is an obvious reduction in the cytoplasm of storage polysaccharide and a concomitant increase in lipid bodies. These bodies are very evident in electron microscopic profiles of the thick-walled yeasts and impart a refractile quality to these cells when viewed with the phase-contrast microscope [424]. Only thick-walled yeasts that have matured to the second stage of development are observed to undergo yeast → hyphal transition. Apparently, the ability of *W. dermatitidis* yeast cells to form germ tubes is dependent on the accumulation of suitable endogenous substrate and the attainment of the spore-like qualities. One candidate for the endogenous substrate is the lipid which accumulates during aging. Calderone [93] showed that young, thin-walled, log-phase cells contain approximately 5.1% fatty acid, whereas cells incubated at 37°C for very long periods of time (14-35 days) are composed of approximately 30% fatty acid. Evidence for the supposition

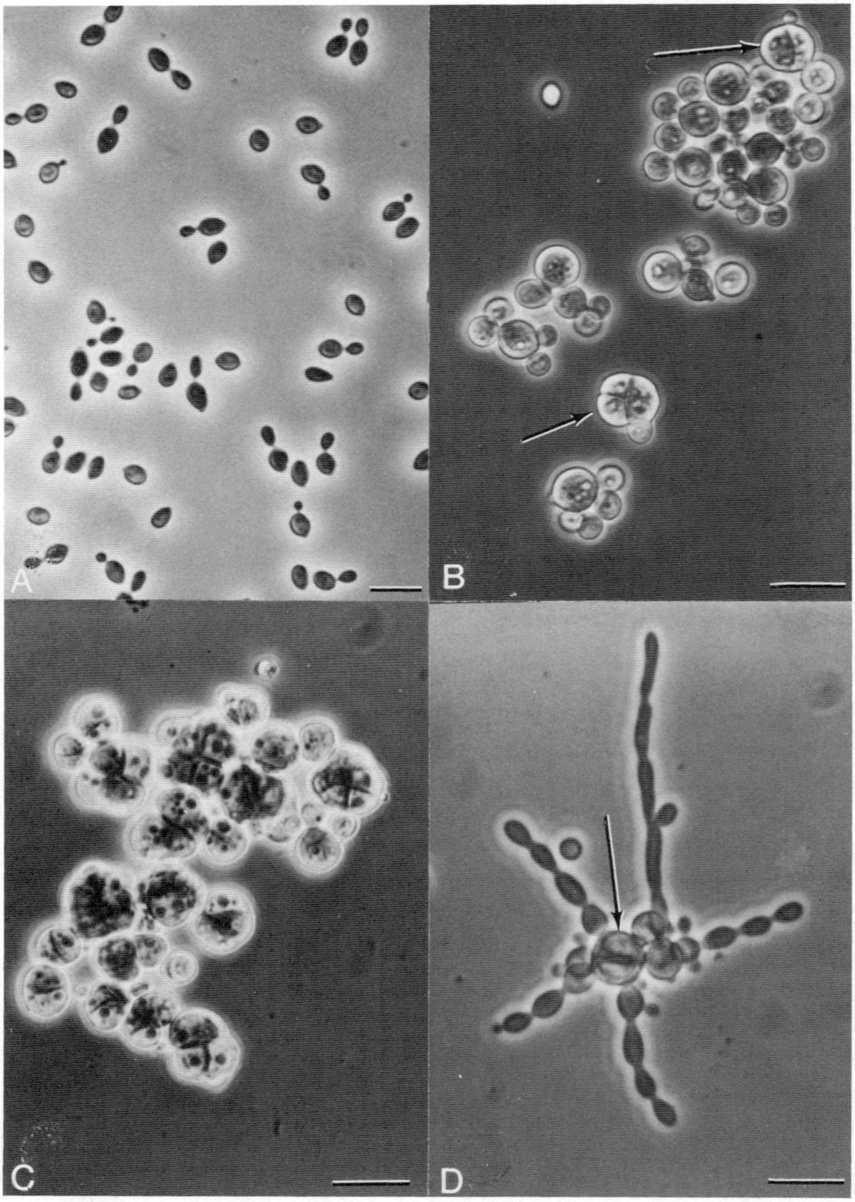

Figure 10 Phase-contrast micrographs showing conversion of *Wangiella dermatitidis* yeast cells (A) to multicellular bodies (B,C) and the hyphal outgrowth (D) of one body (arrow) by the formation of moniliform hyphae. The conversion of the yeast cells to the multicellular bodies (A-C) was induced by culture at pH 7. Bars represent 5 μm. (From Ref. 543, reproduced with permission of the authors and the New York Botanical Garden.)

that the yeast must reach the stage 2 developmental level resides in the fact that log-phase yeasts, stage 1 yeasts, or stationary-phase yeasts have not been induced to form germ tubes (P. H. Hsieh and P. J. Szaniszlo, unpublished results). A similar situation exists in *C. albicans*: exponential-phase yeast cells of this fungus are also unable to form germ tubes, although stationary-phase cells can [114].

The stage 2 thick-walled yeast cells, when inoculated into medium which supports exponential growth, may bud under certain culture conditions, but more frequently form moniliform and true hyphae [424]. Invariably, the transition process is characterized by the initial outgrowth of a moniliform hypha consisting of two to many moniliform hyphal cells. These cells act as transition cells and often give rise to true hyphae. Direct transitions of yeast cells to true hyphae are not observed. Apparently, in *W. dermatitidis* a series of blastically derived hyphal cells are produced while the major metabolic alterations necessary for true hyphal growth occur. The presence of a simple septum and Woronin bodies between the thick-walled yeast and the first moniliform hyphal cell attests to the hyphal, as opposed to pseudohyphal nature of the moniliform transition cells. The large populations of ribosomes, mitochondria, and lipid bodies in the cytoplasm of the thick-walled yeast confirm that major metabolic alterations occur in the thick-walled yeast prior to hyphal outgrowth.

In common with the hyphal outgrowth of yeast cells of most other dimorphic fungi, the wall of the first moniliform hyphal cell of *W. dermatitidis* is thin and continuous with an inner layer of the thick-walled parental yeast [424]. The wall of the hypha is about 200-250 nm thick, whereas the wall of the thick-walled cell is at least 500 nm thick. As the moniliform hypha elongates, prominent septa form between the hyphal cells. In near median section these septa are observed to have the characteristics of simple septa with a central pore, often closed by a pore plug, and associated Woronin bodies.

True hyphal cells are elongate, thin-walled, and contain numerous ribosomes, electron-dense bodies, glycogen masses, mitochondria and single nuclei [424]. Hyphal branch formation and lateral hyphal bud (blastoconidium) formation occur frequently at the younger ends of both moniliform and true hyphal cells. The young lateral hyphal buds can be distinguished from the young lateral hyphal branches because the wall of the hyphal branch is continuous with the entire wall of the main hypha, whereas the wall of the bud is continuous with only an inner layer of the hyphal wall.

Ultrastructural comparisons of yeast bud formation and hyphal bud (blastoconidium) formation reveal that both processes are characterized by blastic rupture of the outer portion of the yeast or hyphal wall and the emergence of the bud through the resulting opening. During the early stages of bud formation, the nucleus of the parent cell becomes situated opposite the emerging bud, and mitochondria localize in the cell periphery. At the bud apex a ribosome-free zone of cytoplasm containing a cluster of vesicles is observed, suggesting that the mechanism of budding in *W. dermatitidis* is regulated by mechanisms similar to those thought to regulate budding in monomorphic yeasts such as *Saccharomyces cerevisiae* [351,364]. The ultrastructural evidence suggests that the lateral hyphal buds and the yeast buds are structurally and functionally equivalent, and therefore, can either bud new progeny or convert into thick-walled cells. The latter can remain dormant or undergo transition to hyphae or conversion to multicellular forms. Young conidia apparently behave similarly to blastoconidia and thin-walled yeasts [424].

The multicellular morphology of *W. dermatitidis* can be induced in thin-walled yeasts in low numbers by long-term culture on agar medium and in large numbers by 3-5 day culture in very acidic medium (Fig. 10) or 24- to 48-h culture of temperature-sensitive multicellular (Mc) mutants [228,475,543]. The fine-structural details associated with the conversion are available only for the multicellular forms induced by low pH [543]. In acidified medium, exponential-phase yeast cells of *W. dermatitidis* exhibit few changes in gross morphology during the first 24 h. However, after 2 or 3 days, many yeast forms are enlarged, and some exhibit one or more internal septa (Fig. 10B). By 5 days, a high percentage of the yeast forms are multicellular (Fig. 10C). Culture of these multicellular forms in nonacidic medium induces hyphal outgrowth involving moniliform hyphal segments, which often give rise to true hyphal segments (Fig. 10D).

Yeast cells cultured at low pH do not exhibit the extreme cytoplasmic rearrangement associated with the thick-walled yeasts in old cultures on agar surfaces [543]. Ribosomes and mitochondria remain prominent, and the cells retain most of the characteristics associated with rapidly growing cells. However, in common with the transition-competent thick-walled cells, the cytoplasm of the multicellular forms developing in acid accumulate one or more large lipid bodies. The initial septum tends to divide the yeast cell's cytoplasm into two relatively equal portions. Subsequent septa are represented by both thick and thin varieties, although the thick septa may only represent secondarily thickened thin septa. When observed in the proper plane of section, the thin septa are perforate and Woronin bodies are associated with the septal pore. The wall of the multicellular bodies becomes very thick, commonly approaching 500 nm in thickness. Nuclei are visible in most cells, indicating that nuclear division precedes septum formation. Cell separation of multicellular body cells is infrequent, but apparently can occur. The ultrastructural aspects of hyphal outgrowth of multicellular body cells is essentially identical to the outgrowth of thick-walled cells.

The effect of pH on yeast cells of *W. dermatitidis* resembles a gradient effect [543]. At near neutral pH values (5.5-6.5), the yeast cells grow well in a budding fashion, and at the end of the growth period show little morphological change beyond becoming uniformly nonbudded. At lower pH levels (3.5-4.5), reproductive growth is equally good, but near the end of the growth period, the nonbudded yeast cells become thick walled and refractile when viewed with phase-contrast optics. Only at the lowest pH levels, where reproductive growth is inhibited, are multicellular forms produced. These results suggest that wall thickening and cross-wall formation during the yeast → multicellular form conversion are dependent on the availability of adequate substrates and the absence of conditions that promote budding growth. The results also suggest that the acidic cultural conditions specifically inhibit the event of bud emergence by yeast cells without inhibiting the events of growth, nuclear division, and cytokinesis.

The suggestion that the conversion of yeasts to multicellular forms involves the inhibition of yeast budding without the inhibition of growth, nuclear division, and cytokinesis was confirmed using temperature-sensitive Mc mutants [473-475,543]. These mutants, designated Mc 1, Mc 2, and Mc 3, represent a new and unique class of fungal morphological mutants and may be the only experimentally derived cell cycle morphological mutants available among the pathogenic dimorphic fungi. At the permissive temperature (25°C) the mutants and wild type all grow in the yeast morphology, exhibit very similar kinetics of growth and synthesis of macromolecules, and have identical nuclear staining patterns. However, at the restrictive temperature (37°C) in a glucose salts yeast extract medium the mutants express the multicellular phenotype, whereas the wild type continues to grow in the yeast morphology. The only apparent difference between the mutants and the parental strain is the temperature-dependent induction of the multicellular phenotype, one of the three alternative vegetative morphologies characteristic of the parental strain.

Each Mc mutant expresses the multicellular phenotype in a different manner [475]. Yeast cells of Mc 2 and Mc 3 incubated at 37°C cease budding rapidly and within 48 h convert almost totally to the multicellular form. Mutant Mc 2 differs phenotypically from Mc 3 in that the mutation(s) carried by Mc 2 is lethal at 37°C. Mutant Mc 1, unlike Mc 2 and Mc 3, continues budding for several generations during incubation at the restrictive temperature. It is suspected that the mutations carried by Mc 2 and Mc 3 result in the production of a temperature-sensitive protein which rapidly brings about the arrest of budding growth. Mc 1, unlike Mc 2 and Mc 3, is thought to be thermolabile for a process involved in gene product synthesis. This allows Mc 1 to continue budding until protein is degraded or diluted to an ineffectual level.

The incorporation of radiolabeled precursors into nucleic acids and proteins at the restrictive temperature reflects the manner in which the multicellular phenotype is expressed in each mutant [475]. The total incorporation into multicellular forms of Mc 2 and Mc 3, which is significantly higher than for cells of wild type, reflects the rapid cessation of budding growth, the enlargement of the cellular unit, and the increase in the number of cells in the multicellular forms. In agreement with the late cessation of budding growth and delayed expression of the multicellular morphology of Mc 1, this mutant accumulates only slightly higher amounts of precursor as RNA and DNA per colony-forming unit than wild type. However, inordinant amounts of precursor are incorporated into protein, suggesting that in Mc 1, protein degradation is defective. After 48 h, Mc 1 incorporates about twice the amount of adenine as wild type, whereas Mc 3 accumulates 8-10 times as much. This finding agrees with the results of investigations of the nuclear conditions of the mutants [474]. Staining with the fluorescent nuclear stain mithramycin documented that after 48 h at the restrictive temperature the average number of nuclei per multicellular unit of Mc 1 and Mc 2 is about 2.5 to 3, whereas for Mc 3 it is 5. However, the number of nuclei ranges from 1 to 13 in Mc 1, 1 to 22 in Mc 2, and 1 to 31 in Mc 3.

The results with the Mc mutants indicate that yeast-phase cell cycle events are inhibited in the mutants at 37°C. Thus yeast → multicellular form conversion in *W. dermatitidis* is very different from yeast → hyphal transition. In the case of the former, exponentially growing yeasts can give rise to multicellular forms, whereas with the latter only certain differentiated stationary-phase cells appear capable of hyphal outgrowth. Evidently, little rearrangement of metabolism is necessary for yeasts to convert to Mc forms. In contrast, a dormant or nearly dormant metabolic period prior to the resumption of rapid growth is required for hyphal outgrowth (germination).

The yeast-phase cell cycle of *W. dermatitidis* was analyzed because of the apparent requirement that specific cell cycle events must be inhibited prior to conversion of yeasts to multicellular forms [338,473]. By using inhibitors, particularly the DNA-synthesis inhibitor hydroxyurea (HU), partially synchronized cultures and flow-microfluorometry results were obtained which allowed the construction of a tentative cell cycle map of a dimorphic fungus [475a,544] (Fig. 11). The results suggest that the yeast-phase cell cycle of *W. dermatitidis*, like that of the hemiascomycetous yeast *S. cerevisiae* [244], appears to consist of two independent pathways of dependent events, a DNA-nuclear division sequence and a budding-growth sequence. The existence of the two independent pathways

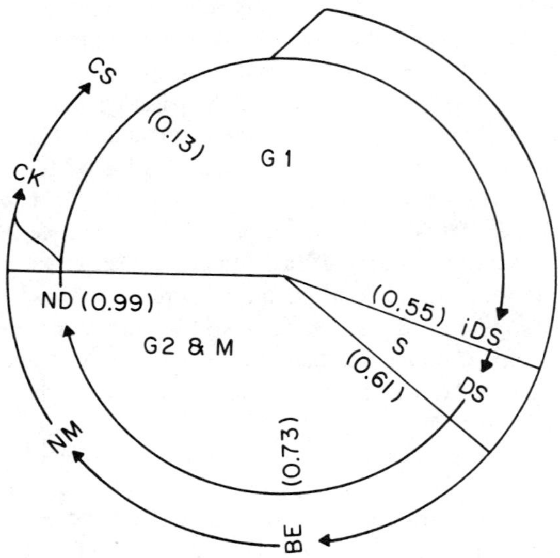

Figure 11 Diagrammatic representation of the yeast-phase cell cycle of *Wangiella dermatitidis.* iDS, initiation of DNA synthesis; DS, DNA synthesis; BE, bud emergence; NM, nuclear migration; ND, nuclear division; CK, cytokinesis; CS, cell separation. The times for these events and for the phases in the cycle are estimated as fractions of the cell cycle based on a generation time of 4.5 h. The fractions of the cycle occupied by G1, S, and G2 and M and the time of BE and CS were calculated by application of the age distribution equation [29] to the data derived by FMF analysis and cell counts of mithramycin-stained cells. (From Ref. 475a.)

is supported by two observations. First, cells arrested by HU in DNA synthesis are able to complete bud initiation, bud enlargement, and nuclear migration, whereas nuclear division, cytokinesis, and cell ceparation do not occur. Second, the temperature-sensitive multicellular mutants, like the mutant of *S. cerevisiae* blocked in bud emergence [247], continue to synthesize DNA and undergo nuclear division even in the absence of bud emergence [474,475]. Thus, there is no causal connection between the nuclear division cycle and the budding growth cycle; bud emergence can occur in the absence of DNA synthesis and mitosis or DNA synthesis and mitosis can occur in the absence of bud emergence. Because HU-treated G1 cells are able to complete bud emergence and enlargement, it is likely that the two pathways diverge in G1. The convergence of the independent sequence is suggested by the fact that yeast cells contain at least one nucleus per cell.

Although the yeast-phase cell cycle of *W. dermatitidis* resembles that of *S. cerevisiae,* it also differs in a number of respects [338,473]. First, bud emergence, which occurs early in S phase in *S. cerevisiae* [244], probably occurs during the G2 phase in *W. dermatitidis.* Second, the fraction of the cell cycle spent in G1, S, and G2 and M is estimated as 55% for G1, 6% for S, and 39% for G2 and M. For *S. cerevisiae* the fractions represent roughly one-third of the cycle for each of the three phases [244]. However, it must be pointed out that the generation time for *S. cerevisiae* is approximately 2 h, whereas for *W. dermatitidis* it is about 4.5 h under the conditions of culture employed. Since the length of the G1 interval may vary depending on extracellular and intracellular conditions, it is possible that the fractions would be more similar to those of *S. cerevisiae* if *W. derma-*

titidis could be grown at a faster rate. Generation times approaching 2 h have been observed for yeast cells growing in peritoneal macrophage (P. J. Szaniszlo, J. D. Marlowe, and S. Keen, unpublished results). Finally, the mechanism controlling bud formation and growth in *W. dermatitidis* may differ from that in *S. cerevisiae.* Cells of the latter fungus arrested in the initiation of DNA synthesis produce only single buds. This suggests that bud initiation in *W. dermatitidis* may be more stringently controlled than in *S. cerevisiae* [475a].

Cell wall analyses of yeast cells and multicellular forms of Mc 3 indicate significant, quantitative differences in protein, neutral sugars, melanin, and chitin, with the greatest differences found in chitin and melanin (P. A. Geis, unpublished data). Walls of multicellular forms contain five times more chitin and three times more melanin than walls of yeasts cultured at 37°C. The melanin appears not to be crucial to morphogenesis because white, melaninless mutants are capable of conversion under appropriate stimuli (P. A. Geis and P. J. Szaniszlo, unpublished data). However, histochemical studies suggest that chitin synthesis parallels, and may be critical to, morphogenesis [277; P. A. Geis and P. J. Szaniszlo, unpublished data].

The early events that occur during the conversion of Mc 3 from the yeast form to the multicellular form are beginning to be investigated. The approximate execution point of the mutation carried by Mc 3 was determined by use of yeast cells derived from synchronous and asynchronous cultures, by "shift-up" experiments (shifts of cells from 25°C to 37°C), and by time-lapse photography [473; C. W. Jacobs and P. J. Szaniszlo, unpublished data]. All available results localize the execution point for the mutation leading to multicellularity in G2, at or just prior to bud formation. The placement of the execution point in G2 is supported by the observations that unbudded cell populations of Mc 2 shifted to 37°C increase only slightly in cell number and do not bud, whereas budded cells complete the cycle and double the cell number before arresting as unbudded cells. Apparently, only one or a few critical proteins are temperature sensitive in Mc 3, and when the yeast cells are shifted from the permissive to the restrictive temperature, the function of this protein is impaired. Since cell wall synthesis and growth continue in Mc 3 at the restrictive temperature, it is unlikely that the temperature-sensitive protein(s) is involved directly in these processes. More likely, the temperature-sensitive proteins or protein are responsible for activating wall synthesis and degradation at a specific budding locus. Candidates for such proteins are cytoskeletal components which may direct bud initiation, nuclear division, and nuclear migration in *W. dermatitidis* in a manner similar to that thought to occur in *S. cerevisiae* [84,385]. Thus, the conversion of yeasts to multicellular forms may be regulated by the dysfunction of only one or a few site-specific proteins. In contrast, yeast → hyphal transition probably involves the rearrangement of whole metabolic pathways and consequently the synthesis of many new and different proteins, and the degradation of others. These recent results with the wild-type strain and the temperature-sensitive mutant, Mc 3, demonstrate that *W. dermatitidis* is an excellent organism in which to study one or more discrete morphological events in a polymorphic fungus.

IV. SPECIES WITH PREDOMINANTLY ISOTROPIC (SWOLLEN) MORPHOLOGY IN VIVO

A. *Coccidioides immitis*

Coccidioides immitis exhibits a unique dimorphism in the host characterized by the transition of arthroconidia to spherules. The arthroconidia are derived by disarticulation of a saprophytic mycelial phase consisting of septate hyphae. Depending on strain, the arthroconidia exhibit variable morphologies, but are usually barrel-shaped, measuring 2.5-4 by 3-6 μm. The spherules, which become as large as 80 μm or more in diameter, develop by the isotropic enlargement of the arthroconidia and give rise to a self-perpetuating spherule-endospore phase. Reproductive growth of the spherule-endospore phase in the host or in vitro is via liberation of endospores followed by their development into new endosporulating spherules.

Our understanding of the regulatory events responsible for induction or maintenance of the spherule-endospore phase in *C. immitis* is rudimentary at the physiological level and nonexistent at the biochemical and molecular levels. Morphogenetic research on this organism has been limited because of its highly infectious nature and the lack of proper containment facilities in most mycological research laboratories. However, in view of the facts that the spherule-endospore phase was first perpetuated in vitro as early as 1914 [347], that the induction of the spherule-endospore phase in vitro could provide a rapid, inexpensive, method for the presumptive identification of *C. immitis* [538], and that the spherule-endospore phase affords superior protection compared to equal doses of the mycelial-arthroconidial phase when used to vaccinate mice [330], it is surprising how little is known about the regulation of spherule development. There is little doubt that insight into the biochemistry and molecular biology of morphogenesis in *C. immitis* would significantly augment studies currently being conducted pertaining to the therapy of disseminated coccidioidomycosis.

Because *C. immitis* is not fastidious and can be grown on most artificial or natural substrates that support fungal growth, the majority of nutritional studies are concerned with the in vitro cultivation of spherules [206]. MacNeal and Taylor [347] may have been the first to cultivate spherules successfully in vitro when they observed spherule multiplication under anaerobic culture conditions in complex natural substrates inoculated with spherules in pus.

The early success with the anaerobic perpetuation of the spherule-endospore phase greatly influenced subsequent investigations, and considerable effort was directed toward determining whether conversion is enhanced by oxygen-limiting conditions or by increased carbon dioxide conditions [74,137,139,315,339,342,410]. The rationale for the earliest studies involving variable oxygen tensions was the avascularity of coccidioidal granulomas, which suggests that anoxia might be a factor in spherule induction [21]. However, later studies were generally prompted by demonstration of the favorable effect of metabolic or bubbled CO_2 on the growth of the endospore-spherule phase [206].

Numerous media are successfully employed to cultivate spherules. Most early media consisted largely or entirely of complex organic materials such as ascitic fluid or gelatinized horse serum containing kidney tissue [347], glucose broth solidified with partially coagulated egg albumin [315], coconut milk [82], Tyrode's solution containing chick embryo extract and rooster serum [342], and tissue culture medium fortified by serum and HeLa cells [31]. More recently, spherules have been cultivated in semidefined or defined media. Baker and Mrak [22] obtained spherules in a variety of solid media after 45-60 days when active mycelial growth ceased due to drying of the medium. Among their media was an acetate ammonium inorganic salts medium which was modified by Converse [136] to provide the first chemically defined medium to support the production of spherules and the perpetuation of a spherule-endosphere phase through several subcultures. Prompted by the success of others with Tween 80 [135], Converse modified his original medium by the incorporation of the anionic surface-active agent, Tamol N. The latter, like Tween 80, acts as a stimulus for spherule production, endospore formation, and endospore liberation [137-139]. Either the original basal medium of Converse or a variation is used in most contemporary studies of the in vitro cultivation of the spherule-endospore phase of *C. immitis.*

The development of a defined liquid medium for culturing spherules encouraged investigators to determine the absolute requirements for perpetuating in vitro a pure (hyphae-free) spherule-endospore phase. The studies met with varying degrees of success and produced a conflicting literature which is only partly explained by strain differences obscuring common regulatory themes. The importance of dilute media in the successful induction of the spherule-endospore phase was demonstrated by Converse [136-138] and Converse and Besemer [139]. A fivefold dilution of his liquid synthetic inoculum medium provides the proper medium conditions at 34°C for maximum conversion of viable units to spherules following a short period of mycelial growth. The undiluted inoculum medium provides copious quantities of arthroconidia after a period of extensive mycelial growth under the same incubation conditions. No other factor tested by Converse was effective in stimulating spherule development in other than diluted medium [139]. Although it was not demonstrated whether spherule induction in dilute medium is caused by changes in osmotic pressures or changes in metabolism caused by starvation conditions, it was shown that further development of nonsporulating spherules is inhibited by nutritional deficiency. The nutritional inhibition was demonstrated by supplementing cultures of spherules devoid of endospores with additional glucose and acetate. The supplement allowed resumption of spherule development, producing endospores that were liberated into the medium. In contrast, inhibited spherules which were not supplemented did not produce endospores. Evidently, starvation conditions or dilute media are necessary for induction of spherules from hyphae or arthroconidia, but enough substrate must still be available to support development leading to the release of mature endospores. Future studies with nonmetabolizable compounds which only change the medium viscosity may elucidate the diluted medium effect.

The importance of elevated temperature (between 34 and 41°C) in generating and maintaining spherules is well documented [74,136-139,330,342,347], but is not the only factor crucial for phase transition [74,139]. Converse [137] showed that an increase in temperature from 34°C to 40°C is accompanied by an increase in spherule production. At 41°C almost 100% of an arthroconidium inoculum converts to spherules. At lower temperatures arthroconidia give rise to both hyphae and spherules. Apparently, incubation at 41°C completely inhibits germination by hyphal outgrowth. At this temperature arthroconidia grow isotropically until their cytoplasm is cleaved, whereas hyphae are rendered nonviable [74]. Spherules survive the elevated temperatures with no known adverse affects [74].

The effect of Tamol N or other surface-active agents is thought to result from changes in permeability of the spherule wall or from accelerated growth [139]. Although both effects might participate in the inhibition of polar growth leading to outgrowth of arthroconidia and the promotion of isotropic growth leading to spherulation, it is difficult to envision how endosporulation is influenced. Nevertheless, Tamol N is reported to be a growth stimulant at concentrations of 0.05-1.0%, a spherulation stimulant at concentrations of 0.005-1.0%, and a dispersing agent at concentrations of 0.01-1.0%. Its effect as a dispersing agent is dramatic and greatly facilitates perpetuation of a spherule-endospore cycle by promoting endospore release [137]. However, not all investigators have demonstrated a need for surface-active agents and, in fact, the spherule-endospore phase was maintained at 40°C for years in a modified Converse medium devoid of Tamol N, but pre-bubbled with a mixture of 20% CO_2 and 80% air.

The beneficial effect of CO_2 continues to be implied by investigators using it in the routine production and maintenance of spherules. Lubarsky and Plunkett [342] were among the first to show the favorable effect of increased carbon dioxide on production of the parasitic phase of *C. immitis.* Converse [137] noted that reduced aeration results in increased spherule formation. He speculated that the effect is due to an accumulation of CO_2. Sodium bicarbonate substitutes for reduced aeration and prompted the routine addition of bicarbonate to Converse's basal spherule medium (BSM) [72]. When gaseous carbon dioxide is used with the BSM with added bicarbonate, the stimulatory effect of carbon dioxide is not demonstrated. Instead, optimal sporulation occurs in atmospheres containing 20% O_2 in the absence of CO_2. Concentrations of 10% gaseous CO_2 have no effect, but concentrations greater than 10% inhibit spherulation [139]. Apparently, the stimulatory effect of gaseous carbon dioxide is masked by the identical stimulatory effect of supplemental bicarbonate.

Other experiments showed that large arthroconidial inocula cultured in Converse's BSM convert almost quantitatively to endosporulating spherules under shake culture conditions at 34°C, whereas low arthroconidial inocula in the same medium do not. At the low inoculum level, the *C. immitis* arthroconidia germinate and produce mycelial cultures. A gaseous mixture of 90% air and 10% CO_2 reverses the effect of low inoculum levels and inhibits the germination of arthroconidia while inducing nearly quantitative conversion to spherules. Evidently, the favorable effect of the larger inoculum is due to the greater concentration of metabolic CO_2 in the cultures [339]. In another similar set of experiments [74] spherules could not be grown at 40°C without bubbled CO_2. Arthroconidia incubated at 40°C in the absence of gaseous CO_2 germinated, yielding hyphal cultures, whereas arthroconidia cultured under increased CO_2 tension transformed into spherules. The CO_2 requirement for spherulation was confirmed when spherules, subcultured into fresh medium without bubbled CO_2, germinated; whereas germination was eliminated when the arthroconidia were used to inoculate media that had been bubbled with CO_2 prior to inoculation [74]. The recent physiological studies of Brosbe [76], Lones et al. [340], and Sun et al. [538] and the excellent cytological studies of Sun and Huppert [537] and Sun et al. [539] involving spherule culture under increased CO_2 tension tend to support the utility of using increased CO_2 tensions for the induction and maintenance of the spherule-endospore phase. The elucidation of the role of CO_2 on morphogenesis in *C. immitis* awaits rigorous investigations such as those applied to the CO_2 effects among the Mucorales.

The cell wall composition of the morphological forms of *C. immitis* indicates the same qualitative composition for spherules as for mycelial walls. In addition to the major neutral sugars glucose and mannose, minor quantities of galactose and 3-methoxymannose are found in both forms [577,578]. Chitin has been identified in and is critical to the structural integrity of the spherule wall [57,71,131]. It is also presumed to be the source of at least part of the substantial amount of glucosamine in mycelial wall hydrolysates [36]. Some differences in tertiary wall structure are evidenced by the different susceptibilities of wall components to extraction by surfactants and by digestion with lytic enzymes [131,578].

The cytological details of the mycelial-arthroconidial phase and the endospore-spherule phase of *C. immitis* have been described by numerous investigators [1,71,72,77,144,165,

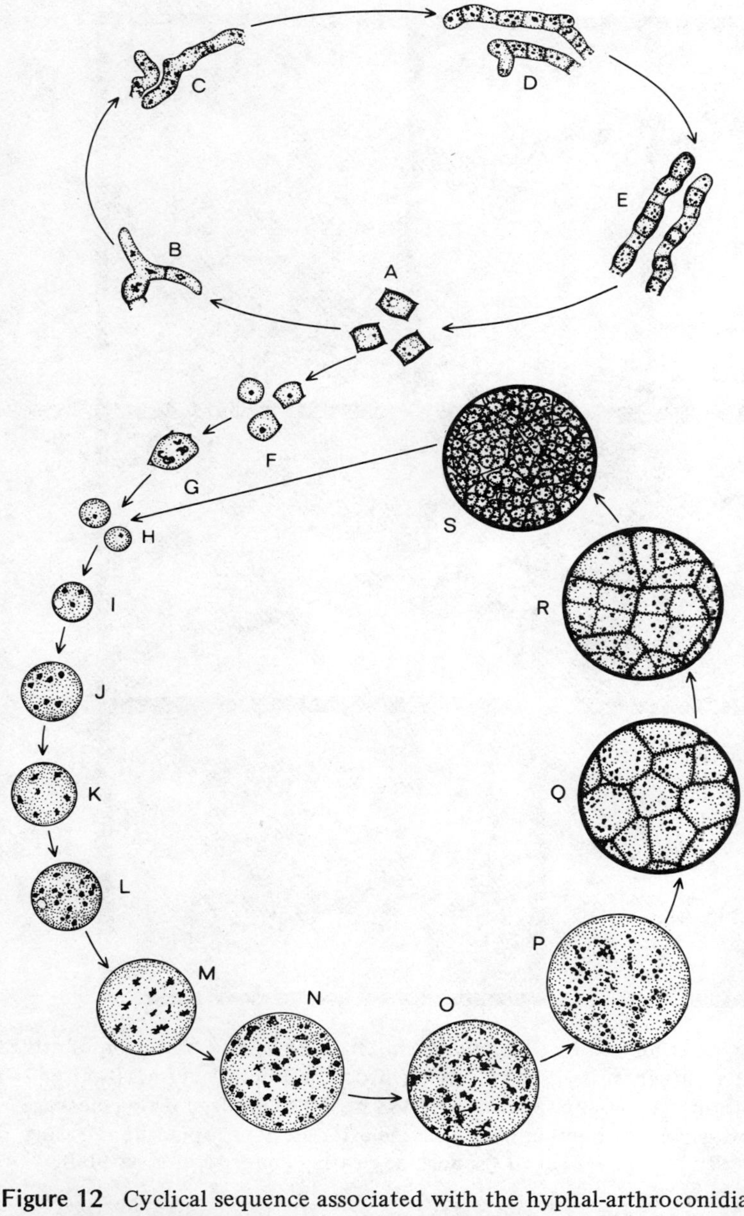

Figure 12 Cyclical sequence associated with the hyphal-arthroconidial cycle (A–E), the endospore-spherule cycle (H-S), and the transitions of one cycle to the other (H-B and A-H) in *Coccidioides immitis.* (From Ref. 537.)

313,416,537-539]. Only the most recent studies, however, have described the cytological and ultrastructural aspects of conversion of one phase to another [537,539]. Successive cytological stages in the mycelial-arthroconidium cycle (A through E) and the endospore-spherule cycle (H through S), as well as their transition forms (A to H and H to B), are diagrammatically illustrated in Fig. 12. Using acetoorcein staining procedures, Sun and

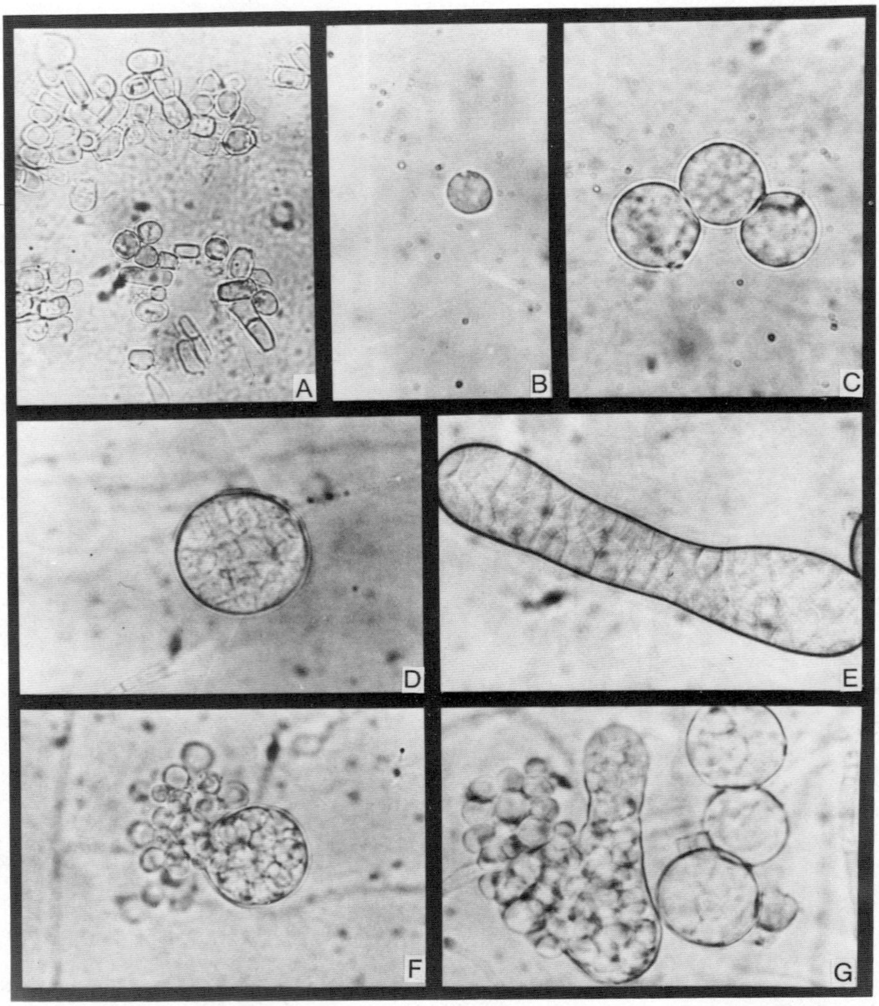

Figure 13 Morphologies exhibited by *Coccidioides immitis* during the transition of arthroconidia to endosporulating spherules (500X): (A) arthroconidium inoculum; (B) after 24 h under inducing conditions; (C) young spherule after 48 h; (D) spherule showing cleavage after 72 h; (E) "peanut-shaped" spherule with cleavage after 72 h; (F) spherule releasing endospores after 96 h; (G) "peanut-shaped" spherule releasing endospores after 96 h. (From Ref. 537.)

Huppert showed [537] that most arthroconidia contain two to three nuclei, but may have as few as one or as many as eight nuclei (Fig. 12A). Conversion of the arthroconidia to spherules in a modified Converse agar medium at 40°C in an atmosphere of 20% CO_2 and 80% air is first indicated at the light microscope level by the formation of a single, large nucleus in each arthroconidium (Fig. 12F). It is not known whether the enlarged nucleus develops from nuclear fusion or from enlargement of one nucleus with the degeneration of the others. Under the conversion conditions employed by Sun and Huppert [537], most arthroconidia by 24 h are rounded, enlarged, and have varying numbers of nuclei, depending on cell size (Figs. 12G-K and 13B). Synchronous nuclear divisions continue for about another 24-48 h (Fig. 12L-P), after which time cleavage becomes apparent (Figs. 12Q-S and 13C and D). At first the cells of the spherule are multinucleate, but as cleavage continues the number of nuclei per cell becomes diminished until each endospore has one or rarely two or more nuclei. In vitro some spherules are elongate or peanut-shaped, although most are spherical (Fig. 13E-G). By 96 h, spherule walls begin to rupture, releasing endospores (Figs. 12S-H and 13F and G). The general development of endosporulating spherules in vivo appears to differ very little from the development in vitro [537].

Electron microscope studies demonstrate that the spherule wall is approximately 500 nm or more in thickness and is composed of at least two layers [73,165,539]. Upon spherule → hyphal transition, the outer electron-dense layer of the wall ruptures, and the inner, less electron-dense portion is continuous with the wall of the germ tube [539]. The new germ tube wall is thin and measures approximately 150 nm in thickness. The first septum which forms between a cell of the spherule and the germ tube appears to be a simple septum with a septal pore and associated Woronin bodies, although the micrograph presented as evidence for this supposition does not conclusively demonstrate this fact [539]. However, simple septa, septal pores, Woronin bodies, and septal pore plugs are characteristic of *C. immitis* hyphae, and there is no reason to doubt that the first septum formed upon spherule transition is any different [144,539].

Some interesting similarities and differences are associated with the ultrastructural aspects of phase transition in *C. immitis* compared to other pathogenic dimorphic fungi having a growth phase derived predominantly by isotropic growth. For instance, it appears that *C. immitis, Rhinosporidium seeberi, Emmonsia parva,* and *E. crescens,* the chromomycotic fungi, and *W. dermatitidis* all have the ability to grow isotropically without subsequent obligatory bud or hyphal formation [14]. Hyphal outgrowth by the isotropically derived structures of these fungi, where it is known to occur, is always by extension of an inner wall layer with the formation of a simple septum between the mother cell and the first germ tube cell. Septum formation in both the spherule and the sclerotic cell appears to result from centripetal wall deposition similar to that described for *W. dermatitidis* [539, 543] and not from the coalescence of numerous cytoplasmic vesicles leading to free cell formation, as in the case of phycomycetous fungi of the Zygomycota [66]. However, the completed interior walls appear to be very different in *C. immitis* and *W. dermatitidis* [539,543]. In the former, nonperforate, double septa form between cytoplasmic units. This type of septum allows for the ultimate liberation of the endospores as independent cells. In contrast, the septa that cleave the cytoplasm of *W. dermatitidis* may all be perforate septa. Such septa are not usually associated with the fragmentation of fungi or the cleavage of cytoplasmic units prior to spore formation. It is evident that the endospore cycles of *C. immitis* and *R. seeberi* confer a distinct advantage in vivo to these fungi compared to *E. crescens* and *E. parva,* the chromomycotic fungi, and *W. dermatitidis* by providing a method for reproductive growth functionally equivalent to the budding-reproductive growth cycles of some other dimorphic pathogenic fungi.

B. *Emmonsia parva* and *E. crescens*

The etiologic agents of adiaspiromycosis, *Emmonsia parva* and *E. crescens,* express a temperature-dependent dimorphism. The saprophytic morphology of these organisms is a hyaline, septate mycelium bearing sessile, single-celled conidia and is produced in culture at room temperature [107,125]. The two species are differentiated on the basis of the temperature effects and sizes of their parasitic phase [423]. This parasitic morphology or adiaspore is produced in vitro by cultivation at elevated temperatures in broth or on agar [143,255,282] and is a spherical cell from 25 to 700 μm in diameter with no internal division or endosporulation [143,178]. Growth takes place by generalized expansion [572] and the suggestion that these adiaspores or their developmental forms, trophocytes, may bud [256,257] is perhaps due to variation in temperature of incubation [237]. The adiaspore is surrounded by a thick, chitin-containing wall [56] and possesses an endoplasmic reticulum, mitochondria, glycogen deposits, and fatty droplets [256]. The adiaspore of *E. parva* is produced in vitro by incubation at temperatures above 41°C, is thermotolerant and uninucleate, and may reach a diameter of 30 μm. In comparison, that of *E. crescens* is produced at 37°C, is less thermotolerant, and is multinucleate [143,423]. Adiaspores of *E. crescens* may reach a diameter of 700 μm and may be visible upon gross examination [167,178,561].

Upon temperature-induced transformation to the parasitic morphology, the conidium swells and produces an intermediate trophocyte phase [256,257]. The conidium often continues to swell and may increase in volume by a factor of 10^6 compared with the mature adiaspore of *E. crescens* [178]. The adiaspore → hyphal transition takes place in vitro at decreased incubation temperatures. The adiaspores germinate asynchronously by germ tube formation: the uninucleate *E. parva* by a single hyphal outgrowth and the multinucleate *E. crescens* by numerous germ tubes [254,423].

The relationship of these form-species with other dimorphic fungi is intriguing. The adiaspore has been termed a "spherule" [256,282], and thus some authorities relate the form-species of *Emmonsia* to the conidial Zygomycetes by viewing the adiaspore as a zygomycetous sporangium with a single sporangiospore [106,167,561]. However, the saprophytic morphology of the form-species of *Emmonsia* is virtually identical to that of other medically important fungi whose morphology is critically controlled by temperature of incubation, such as *Ajellomyces dermatitidis, Emmonsiella capsulatum* (if one does not consider the tuberculate macroconidia), and *Paracoccidioides brasiliensis* [106,197].

C. *Phialophora verrucosa* and Other Chromomycotic Fungi

Chromomycosis is a disease caused by at least four distinct imperfect dematiaceous fungi [11,109,179]. The most commonly identified agents of the disease are *Phialophora verrucosa, Rhinocladiella pedrosoi, R. compacta,* and *Cladosporium carrionii.* All of these agents have very similar colony morphologies in the saprophytic phase which consist of extensive, filamentous, black, dark green or gray, conidia-producing hyphae. However, the distinctiveness of the conidiogenous apparatus and mode of conidium production allows the identification of each form-species. In contrast to the well-developed saprophytic phase, the parasitic phase is often a much reduced thallus consisting of single, spherical, thick-walled cells which do not bud, but instead may grow by isotropic enlargement and by the formation of internal septa [11,109,320,379,517].

Medlar [379] and Lane [320] designated individual, thick-walled, nonseptate cells as sclerotic cells and the multicellular forms derived from the sclerotic cells as sclerotia. Although mycologically inappropriate, the use of the term "sclerotic cell" persisted and is well known to the medical mycologist as representing the parasitic morphology of a chromomycotic fungus. However, the term "sclerotium" is not used for the multicellular form and tends to be replaced by sclerotic body [11,426]. In this review, sclerotic body refers to both the thick-walled cells and the thick-walled multicellular forms which develop in tissue or in vitro. Whether the bodies which develop in vitro are homologous to similar structures which develop in vivo is not known, although there is no a priori reason to assume that they are not. It is recognized that the structures formed in vitro are similar to the multicellular forms of *W. dermatitidis* [543] and have many of the characteristics of chlamydospores [3,220].

Currently, it is not possible to explain why a number of different fungi cause clinically indistinguishable diseases. However, it appears that the explanation resides in a common aspect of metabolism that allows each agent to infect tissue, reside in that tissue for long periods in the sclerotic-body morphology, and grow predominantly by isotropic enlargement with frequent formation of septa and occasionally by hyphal development. Possibly the thick-walled cells and the thick-walled multicellular forms are induced in vitro by the continuous pressures of cellular events combined with the nonspecific factors associated with granulomatous lesions. These may include acids, anaerobiosis, caseation, fibrosis, and calcification. Howard suggests that microbes inhibited by these factors may eventually succumb [269]. The chronic nature of chromomycosis suggests that this is not always the case with a chromomycotic fungus. Nevertheless, while the cellular and nonspecific factors brought to bear on a chromomycotic fungus are not always fungicidal, they appear to be sufficient to inhibit significantly normal phenotypic fungal development. Consequently, the thick-walled sclerotic bodies are induced. It is assumed that these forms are much more resistant to host pressures than are the conidia, which probably represent the infecting agents, or hyphae. However, proof for this assumption is lacking.

Although most investigations of tissue infected with a chromomycotic fungus reveal the presence of sclerotic bodies, the physiology of these forms and the control of their development is obscure, largely because this phase of the fungus has not been elicited homogeneously in vitro. Also, sclerotic bodies cannot be obtained in sufficient quantity from infected tissue to guarantee an adequate source of experimental material. Cultural and experimental studies of the saprophytic phase have also been limited. This latter situation results, in part, from the difficulties associated with the critical evaluation of mold growth, the characteristic vegetative phase of a chromomycotic fungus. The review of Carrion and Silva [109] reflects the paucity of knowledge prior to 1947 about the growth and development of chromomycotic fungi.

As early as 1915, Medlar [379] and Lane [320] described the in vitro production of sclerotic bodies by a chromomycotic fungus. Hydrocele agar and Loeffler's blood serum were used to induce the morphology in *P. verrucosa.* According to these investigators sclerotic bodies are produced predominantly from the end cells of short, lateral, hyphal branches, although they are also formed by any cell in a branch and, in some cases, by cells near the tip of a main hypha. The development involves the gradual enlargement of a hyphal cell into a round or ovoid form 8-15 μm in diameter. If development involves a nonterminal cell, the enlargement is usually toward one side. The wall is thin at first, but as the cell ages, it becomes thickened and brown or golden in color. Subsequently, the enlarged cells divide into two, three, or four or even more cells by a process of septation in more than one plane. The resulting sclerotic bodies produce hyphae and aerial conidia when cultured using proper conditions. Medlar [379] and Lane [320] postulated that the thick-walled structures are produced as a result of unfavorable or abnormal conditions and that subsequent septation does not lead to endogenous spore formation. These early observations and interpretations were confirmed in essence by most subsequent investigators [273,361,516].

Silva [516] recognized that studies of the induction of sclerotic bodies in vitro might aid in understanding the adaptation of chromomycotic fungi to parasitism and provide knowledge for development of rational therapies. Working with several strains of *R. pedrosoi, R. compactum,* and *P. verrucosum,* she obtained inconsistent production of sclerotic bodies with *R. pedrosoi* and *R. compactum* in a "synthetic lymph" which was actually a mineral-base blood plasma. However, another medium, Francis' glucose cystine blood agar, induces strains of *R. pedrosoi* and *P. verrucosa* to produce numerous sclerotic bodies.

During a much less comprehensive study, West [576] was unable to induce sclerotic bodies in a strain of *P. verrucosa* cultured in the medium of Francis. However, abundant sclerotic bodies formed in a glutamic acid salts sucrose medium and a glucose salts ammonium chloride medium. Lesser numbers of sclerotic bodies were formed in the latter medium when proline or guanine served as the nitrogen source. Neither the studies of Silva [516] nor West [576] provide insight into how the various media might regulate hyphal → sclerotic body dimorphism.

The more recent study of Reiss and Nickerson [464] represents the most definitive effort to date to explain the regulation of morphogenesis in a hyphal chromomycotic fungus. These investigators, working with a number of strains of *P. verrucosa,* systematically studied the effects of divalent cations on growth, sporulation, and the development of sclerotic bodies. Experiments were carried out at 25°C using a glucose potassium nitrate salts medium rendered deficient in Zn^{2+}, Cu^{2+}, Fe^{2+}, and Mn^{2+} by extraction with 8-hydroxyquinoline. The extracted medium was subsequently supplemented with all, none, or all-but-one of the four aforementioned cations. When all the cations were present in the basal medium, significant filamentous growth and conidial production occurred. However, when Mn^{2+} was omitted from the same medium, conidiogenesis ceased, and a dispersed growth consisting of short hyphae with swollen terminal structures developed. The swollen structures resembled tissue-phase sclerotic bodies, although none of the structures became septate. Identical structures were also produced when the basal medium was supplemented with 0.02% yeast extract or when the KNO_3 was replaced with 1% peptone, urea, glycine, or ammonium phosphate. When ammonium sulfate was used as the nitrogen source instead of KNO_3, abundant chlamydospores formed even in the presence of Mn^{2+}. This effect was attributed to the low pH (2.6) which developed in the culture medium, not to the ammonium sulfate nitrogen source. Evidence that the low pH was affecting morphogenesis was provided by the fact that in the presence of $(NH_4)_2HPO_4$ and Mn^{2+}, the pH after growth remained high (7.0) and no sclerotic bodies developed. Sclerotic bodies also formed in an unextracted yeast extract peptone glucose medium (YPG). However, when this medium was supplemented with Mn^{2+}, sclerotic body formation was inhibited and normal hyphal growth was restored. Atomic absorption spectroscopy confirmed that the hydroxyquinoline-treated YPG was deficient in Mn^{2+}. It is interesting to note that strains of *R. pedrosoi* and *P. richardsiae* were not induced to form sclerotic bodies in Mn^{2+}-deficient media. This would be expected for *P. richardsiae* because it is not known to form sclerotic bodies in tissue. However, *R. pedrosoi* is a classical agent of chromomycosis and could be expected to form sclerotic bodies by the same regulatory mechanisms that induce *P. verrucosa.*

In an attempt to determine how Mn^{2+} regulates dimorphism in *P. verrucosa,* Reiss and Nickerson [464] carried out radiolabeling studies with $^{54}MnCl_2$. They determined that 23% of the radioactivity localized in phialoconidia, whereas 77% localized in hyphae.

Mycelial and conidial fractionation studies further localized 52% of the radioactivity in the soluble fraction of the cytoplasm, 6% in subcellular particles, and 19% in the cell wall. The radioactivity in the soluble cytoplasmic fraction represented either $^{54}MnCl_2$ or possibly some low molecular weight (mol wt 700) organometallic compound. The radioactivity in the cell wall fraction was efficiently extracted with EDTA, but was only slightly vulnerable to extraction by sodium dodecyl sulfate (SDS). Lipid extractions using an alcohol ether solution followed by chloroform did not release the $^{54}Mn^{2+}$ bound in the walls. In contrast, when melanin pigment was extracted using hot alkali and peroxide, the radioactivity was removed. However, the radioactivity did not co-precipitate with the pigment. Because this procedure would have solubilized considerable protein and polysaccharide, it is possible that the Mn^{2+} was bound with a wall enzyme or a polysaccharide fraction, although it appears that this possibility was not investigated. Reiss and Nickerson hypothesized that their results are consistent with a model involving the storage of Mn^{2+} in the form of a low molecular weight organometallic chelate with the cation eventually functioning in the cell wall, perhaps as a cofactor in wall biosynthesis.

Structures similar to those produced by *P. verrucosa* were also induced in *Aspergillus parasiticus* under Mn^{2+}-deprivation conditions [158,196]. Mn^{2+} concentrations of 7.3×10^{-5} mM induced the development of swollen structures, whereas higher concentrations allowed normal hyphal growth. The response of *A. parasiticus* spores to the Mn^{2+} starvation conditions was identical to that of *P. verrucosa.* In *A. parasiticus,* as in *P. verrucosa,* spores germinated, producing a modified hyphal phase, prior to converting to sclerotic-like bodies. Apparently, endogenous Mn^{2+} became limiting. Evidence for this suggestion was provided by experiments in which the swollen bodies of *A. parasiticus* were transferred to either Mn^{2+}-deficient or Mn^{2+}-supplemented media [158]. In the deficient medium, growth was arrested as evidenced by the absence of protein synthesis. However, normal growth recommenced when the swollen bodies were cultured in the Mn^{2+}-supplemented medium.

During attempts to elucidate the regulation of sclerotic body development in strains of *W. dermatitidis,* Hsieh discovered that culture of this organism at 25 or 37°C in medium of very low pH induced the quantitative conversion of yeast cells to sclerotic bodies [273]. The induction was brought about by a pH of 2.5, but normal phenotypic growth occurred at higher pH levels [273,543]. Using these conditions, both thick-walled cells and thick-walled multicellular forms were induced in shaken acidified Czapek-Dox broth supplemented with 1% yeast extract or with an amino acid mixture and thiamine [543]. The utility of *W. dermatitidis* as a model for elucidating growth and developmental patterns among chromomycotic fungi was demonstrated when identical cultural conditions were found to induce the sclerotic morphology in a number of strains of *P. verrucosa, R. pedrosoi,* and *C. carrionii* [273,361,543]. Other dematiaceous zoopathogenic fungi which are not noted for developing sclerotic bodies in vivo, including strains of *C. trichoides, Exophiala jeanselmei, E. spinifera,* and *P. gougerotii,* did not form the thick-walled cells or thick-walled multicellular forms under identical conditions [273, 361,543].

Strains of *R. pedrosoi, P. verrucosa,* and *C. carrionii* produce sclerotic bodies in variable numbers after about 1 week of culture at 25°C (Figs. 14-16). In general, the sclerotic bodies are produced in a similar manner by all the strains and are formed in both terminal and intercalary positions. Ultrastructural studies [361] indicate that these structures, although having a variety of shapes and being composed of different numbers of cells, are

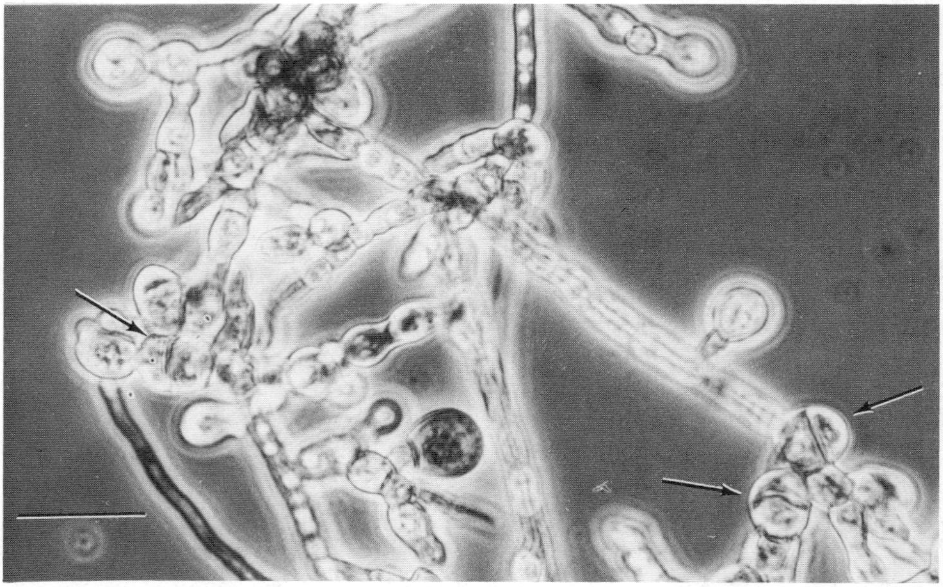

Figure 14 Phase-contrast micrograph of *Rhinocladiella pedrosoi* with numerous nonseptate and septate (arrows) terminal and intercalary sclerotic bodies after culture in acidic medium (pH 2.5) for 1 week. Bar represents 10 μm. (Courtesy of P. J. Szaniszlo and J. D. Marlowe.)

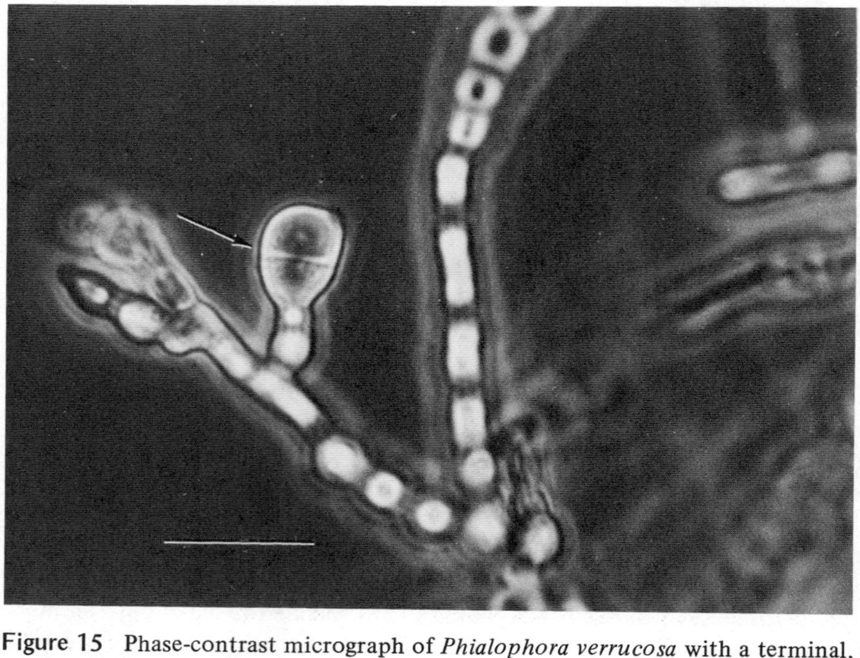

Figure 15 Phase-contrast micrograph of *Phialophora verrucosa* with a terminal, septate, sclerotic body (arrow) after culture in acidic medium (pH 2.5) for 1 week. Bar represents 10 μm. (Courtest of P. J. Szaniszlo and J. D. Marlowe.)

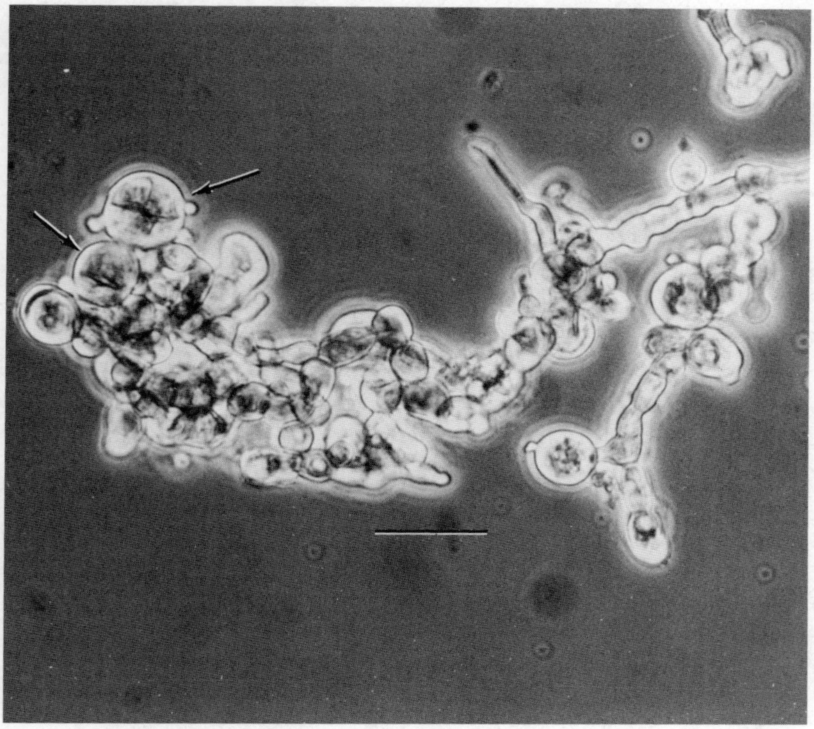

Figure 16 Phase-contrast micrograph of *Cladosporium carrionii* with numerous septate (arrows) and nonseptate, terminal and intercalary, sclerotic bodies after culture in acidic medium (pH 2.5) for 1 week. Bar represents 10 μm. (Courtesy of P. J. Szaniszlo and J. D. Marlowe.)

structurally identical (Figs. 17-19). Young sclerotic bodies have cytoplasm characteristic of actively growing cells, whereas very old bodies have more spore-like cytoplasm containing glycogen deposits and numerous lipid bodies. The thickened cell walls are usually multilayered. Septa frequently exhibit a septal pore with accompanying Woronin bodies (Fig. 17), as would be expected of the septa of a modified hyphal structure. Similar septa have been observed in the normal hyphae of chromomycotic fungi [140] and in the sclerotic bodies of *W. dermatitidis* [543]. Electron micrographic profiles of hyphae are sometimes observed, which suggest that thick-walled, intercalary sclerotic bodies can break away or out of the "parental" hyphal cell (Fig. 20). In general, the sclerotic bodies formed in vitro by acid induction of chromomycotic fungi and the multicellular bodies of *W. dermatitidis* have the ultrastructural characteristics of sclerotic bodies formed in human tissue [406,407]. The acid-induced bodies are also ultrastructurally very similar to structures formed by *A. parasiticus* in Mn^{2+}- deficient medium [196].

Although sclerotic bodies are formed in vitro by all the agents of chromomycosis cultured in acidified media, conversion is not as complete as with *W. dermatitidis* [273, 543]. In general, the acid induction of dimorphism in the chromomycotic fungi is virtually identical with the results obtained by Reiss and Nickerson [464]. Very possibly the

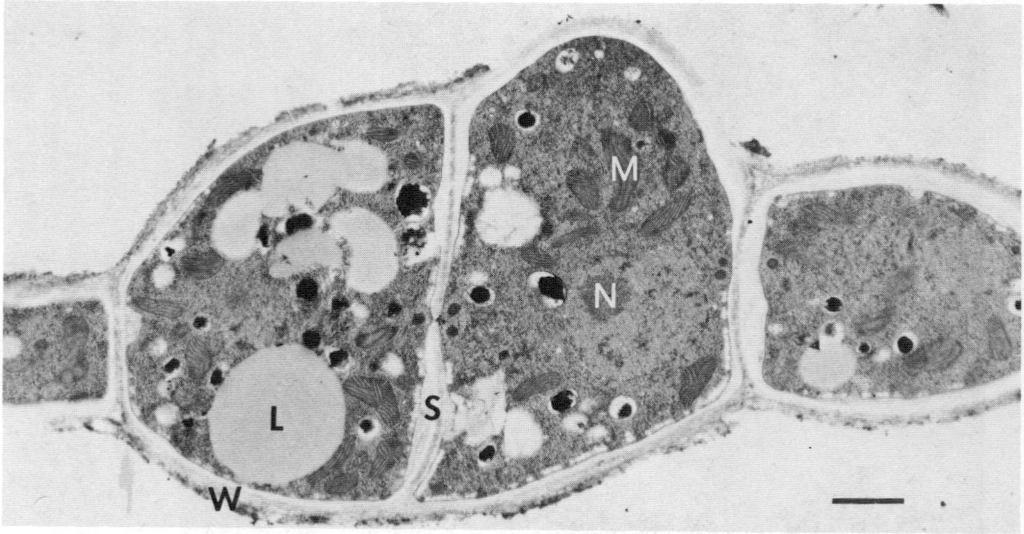

Figure 17 Transmission electron micrograph of intercalary sclerotic body of *Rhinocladiella pedrosoi* formed in acidic (pH 2.5) culture medium. Note septum (S) with septal pore and associated Woronin bodies. Other structures labeled are wall (W), lipid body (L), mitochondrion (M), and nucleus (N). Bar represents 1 μm. (Courtesy of P. J. Szaniszlo and J. D. Marlowe.)

transport of Mn^{2+} is inhibited by extremely acid pH, and thus induction in *W. dermatitidis* and the chromomycotic fungi may also be dependent on Mn^{2+} deprivation. In fact, Reiss and Nickerson [464] suggested that *P. verrucosa* can form sclerotic bodies in vitro in the presence of Mn^{2+} if the pH of the culture medium is 2.6.

The absence of septa in the sclerotic bodies formed by *P. verrucosa* in the study of Reiss and Nickerson [464] is possibly attributable to the lack of sufficient nutrients in the medium near the end of the growth cycle when septation usually occurs. Alternatively, the Mn^{2+} may be directly involved in septation. Hsieh [273] observed that the formation of multicellular bodies in *W. dermatitidis* is nearly quantitative in acidified medium containing yeast extract or a complex amino acid mixture and thiamine. In contrast, only mixtures of a few multicellular forms and many nonbudding thick-walled and thin-walled, nonseptate yeasts develop when NH_4Cl, or single amino acids, serve as the sole source of nitrogen. No multicellular bodies are formed in the absence of thiamine, which is a known requirement for many chromomycotic fungi [517] and for *W. dermatitidis* [543]. The addition of thiamine to the defined medium employed for culturing *P. verrucosa* might have allowed the sclerotic bodies of that organism to become septate. The apparent lack of a need for thiamine by the *P. verrucosa* strain employed by Reiss and Nickerson [464] may have resulted from carryover in inoculum grown on a nondefined medium, although it is not clear what medium was employed for the raising of the phialoconidium inoculum. At least two 48-h time courses of growth in unsupplemented medium are necessary to demonstrate a thiamine requirement in *W. dermatitidis* [273].

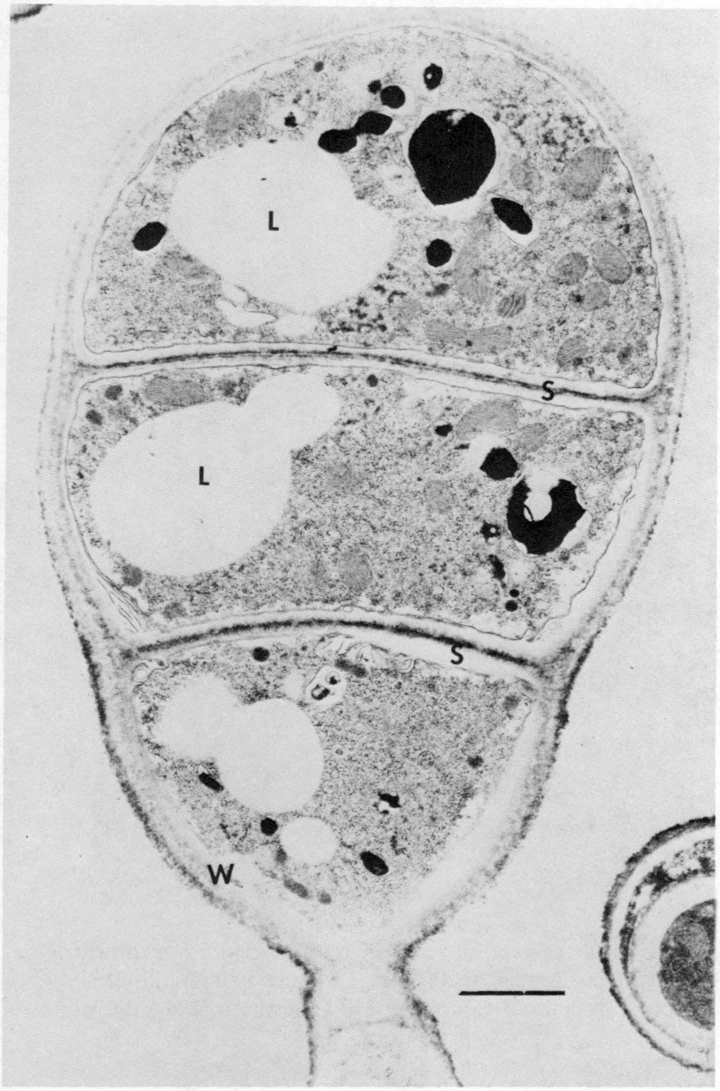

Figure 18 Transmission electron micrograph of a terminal sclerotic body of *Rhinocladiella pedrosoi* formed in acidic (pH 2.5) culture medium. Each cell of the sclerotic body exhibits abundant ribosomes and mitochondria, as well as one or more lipid bodies (L). The wall (W) is multilayered and the septa (S) in this section do not exhibit septal pores. Bar represents 1 μm. (Courtesy of P. J. Szansizlo and J. D. Marlowe.)

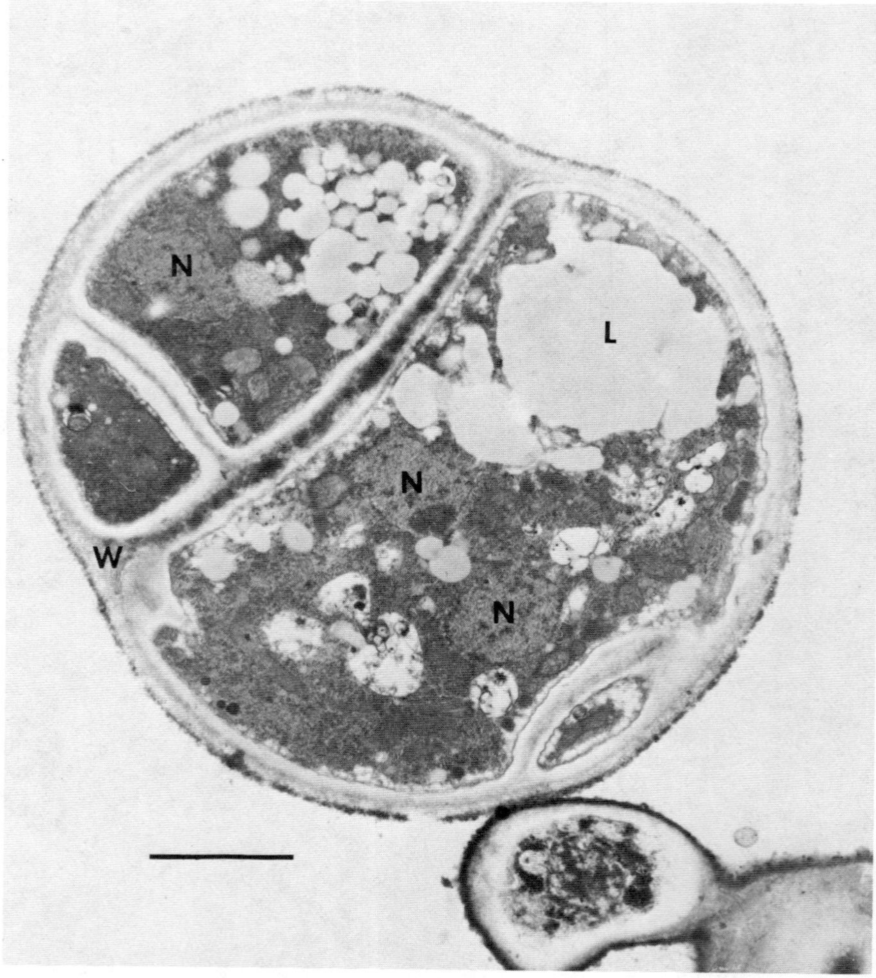

Figure 19 Transmission electron micrograph of sclerotic body of *Cladosporium carrionii* formed in acidic (pH 2.5) culture medium. Labeled structures are wall (W), lipid body (L), and nuclei (N). Bar represents 1 μm. (Courtesy of P. J. Szaniszlo and J. D. Marlowe.)

It is unclear from the experimental results collected thus far how the Mn^{2+} deprivation or the acidic cultural conditions, whether they are essentially the same effect or not, induce sclerotic body formation among the chromomycotic fungi. Certainly, Mn^{2+} affects a plethora of metabolic processes, including the transport of other ions and the synthesis of cell walls, DNA, RNA, and fatty acids [519]. However, in general, the large number of enzymes for which Mn^{2+} serves as a cofactor are also effectively activated by Mg^{2+}. An exception to this generalized rule may be the DNA-dependent RNA polymerases. It is hypothesized that the change in properties of the RNA polymerase early in sporulation of *Bacillus subtilis* [191], coupled with a change in the intracellular free Mg^{2+}/Mn^{2+} ratio [505],

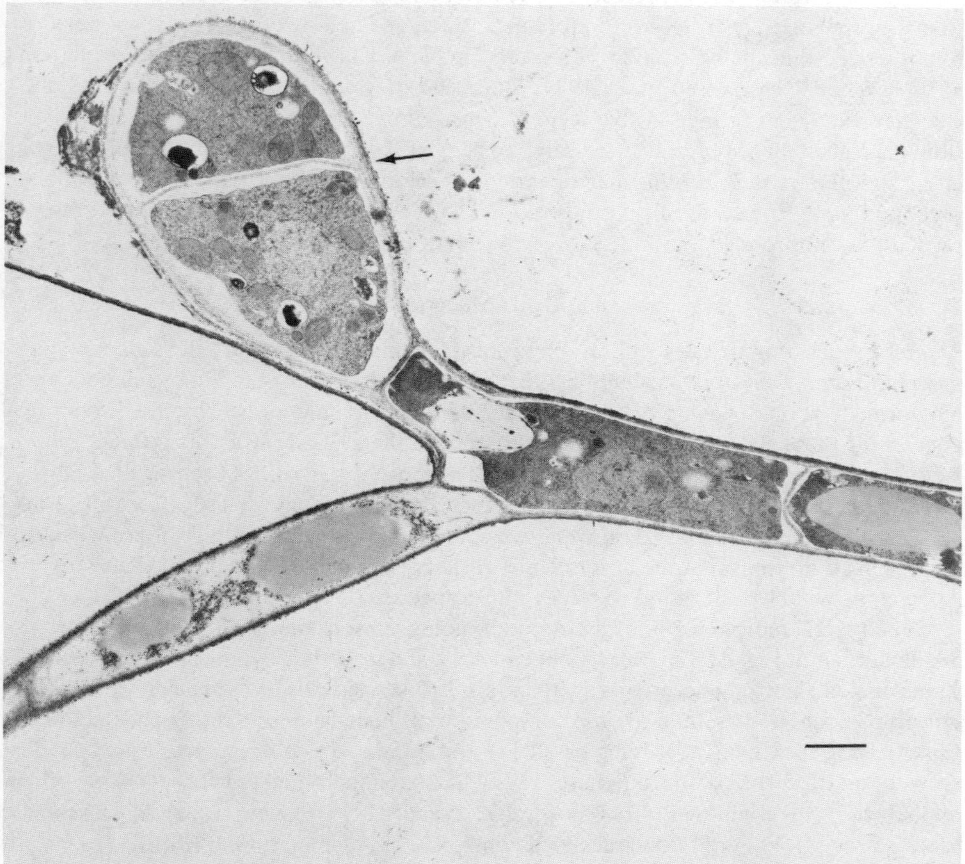

Figure 20 Transmission electron micrograph of *Rhinocladiella pedrosoi* showing an intercalary sclerotic body (arrow) that may be disarticulating from the parental hypha. Bar represents 1 μm. (Courtesy of P. J. Szaniszlo and J. D. Marlowe.)

may lead to differences in polymerase specificity and thus to new and different "sporulation-specific" messenger RNA and protein synthesis [433]. Alternatively, it has been hypothesized that the key role for Mn^{2+} early in bacterial sporulation involves regulation at the level of RNA transcription [191]. Since the sclerotic bodies formed in vitro and the sclerotic bodies formed in vivo seem to be vegetative forms in the process, albeit slowly, of becoming spores, it is tempting to suggest that processes similar to those regulating sporulation in *B. subtilis* may operate in developmental processes in the chromomycotic fungi. However, only future research will reveal the similarities, if any, in the regulation of morphogenesis in these very different microorganisms.

D. *Mucor rouxii, M. racemosus,* and Other Mucorales

Several species of mucorales exhibit dimorphism characterized by growth as a coenocytic mycelium or, alternatively, as budding yeasts. Although none of the dimorphic mucors are normally zoopathogenic [9], they are included in this chapter because the literature concerning these fungi offers important insight into the myriad of factors influencing fungal morphogenesis, and the molecular basis of morphogenesis has been studied more intensely in the mucors than in any of the other dimorphic fungi. This review will center on *Mucor rouxii* and *M. racemosus,* the two species that provide the bulk of knowledge concerning dimorphism in the order. Unless otherwise specified, the use of the genus name alone will be understood to refer to both species.

The typical morphologies of dimorphic zygomycetes, exemplified by *M. racemosus,* are illustrated in Fig. 21. Hyphae are coenocytic, and growth is by apical extension [37]. Transmission electron microscopy [40] indicates that the walls of hyphae are thin, of uniform thickness (50-100 nm), and composed of a single layer of densely packed microfibrils having their main axes lying parallel to the surface. In contrast, yeasts are spherical, grow by multipolar budding, and have walls which are much thicker (500-1000 nm) than hyphal walls and composed of two resolvable layers [324]. The outer, dense, yeast wall layer is only 90-200 nm in thickness and contains 6-nm-diameter microfibrils. The remainder of the thickness of the wall consists of a less dense, homogeneous inner layer. To further compare the thickness, yeast walls contribute 30-40% of the cell volume, whereas hyphal walls contribute only 5-10% of hyphal volume [40]. Yeast → hyphal transition occurs by outgrowth of germ tubes from yeast cells. At first, the germ tube resembles a bud initial, but its hyphal nature soon becomes evident. Like buds, germ tubes appear multipolarly and converting yeasts with several hyphal outgrowths are not uncommon [32,39; J. Peters, University of California at Irvine, personal communication].

Among the dimorphic fungi, species of *Mucor* were the first in which the phenomenon of dimorphism was recognized for what it is: a single organism capable of vegetative growth in two different morphologies. Berkeley [47] and Bail [20] noticed both morphologies in their cultures, but misinterpreted the phenomenon as a transmutation of species. Reess [463], Fitz [186], and Pasteur [428], on the other hand, all recognized that the yeasts and hyphae found in cultures of *Mucor* are different morphological expressions of the same organism. Pasteur noted that yeasts develop in anaerobic conditions at the bottom of standing cultures, whereas hyphae develop in aerobic conditions at the top of a culture. These and similar observations led to the early idea that the balance between fermentation and respiration determines cellular morphology. Although other hypotheses were advanced [70], this concept persisted until recently.

It is interesting that the fermentation-respiration theory prevailed for so long, since the

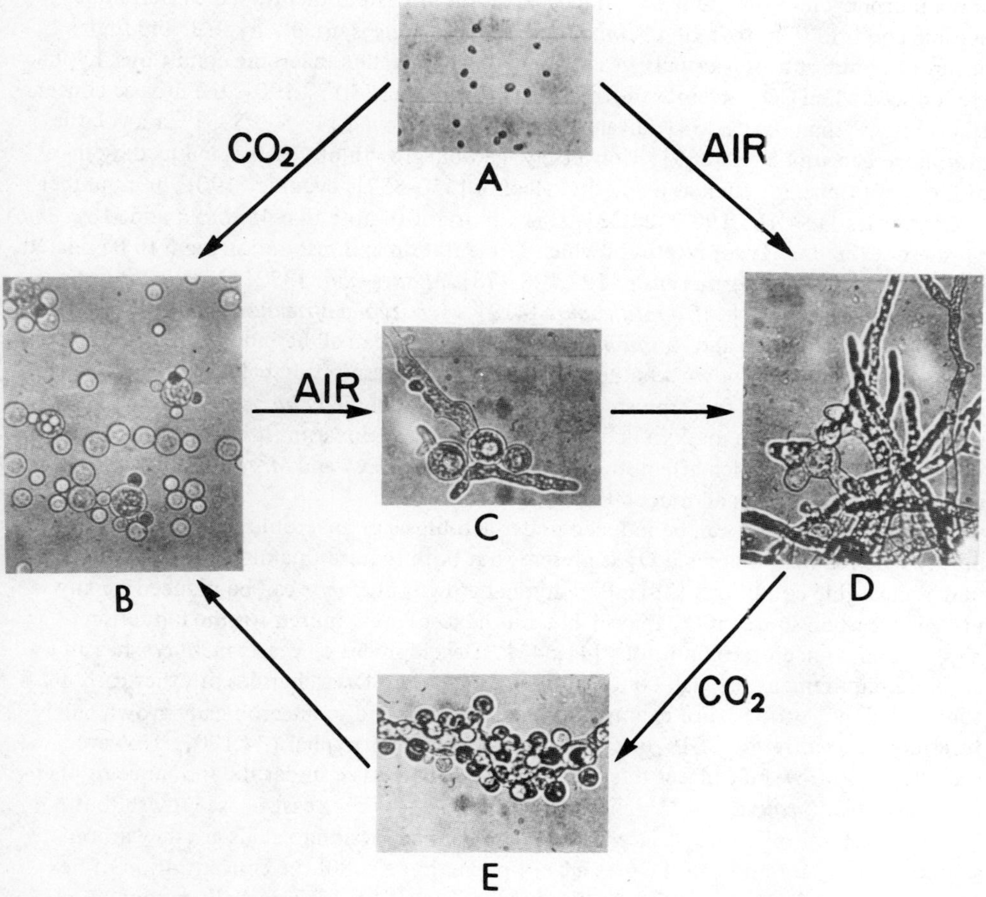

Figure 21 Morphologies of *Mucor racemosus* induced by aerobic (air) and anaerobic (CO_2) conditions: (A) sporangiospores; (B) budding yeast; (C) yeast → hyphal conversion; (D) hyphae; (E) hyphal → yeast conversion. Other factors may also induce these morphologies, as described in the text. [From Ref. 541, with permission from P. S. Sypherd, P. T. Borgia, and J. L. Paznokas. 1978. Biochemistry of dimorphism in the fungus *Mucor*. *Adv. Microb. Physiol. 18*: 67-104. © by Academic Press Inc. (London) Ltd.]

induction of anaerobic hyphae was first noted by Brefeld over 100 years ago [70]. In addition, Wehmer [574] noted at the beginning of this century that in *M. racemosus* and *M. javanicus,* morphology and fermentative metabolism are not closely linked; yeasts and hyphae both carry out extensive ethanolic fermentation. Nevertheless, relatively recent evidence [552] was interpreted to mean that dimorphism in *Mucor* is a morphological manifestation of the Pasteur effect; that is, respiratory metabolism is required for expression of the morphologically more complex hyphal form, whereas fermentative metabolism supports only the simpler yeast morphology. In actuality, in the mucors, as in many of the other dimorphic fungi already discussed, regulation of morphology is mediated by a complex interplay of carbon and nitrogen metabolism, atmospheric conditions, divalent cations, and a morphogen. Consistent with the observation of Pasteur mentioned earlier, under aerobic conditions, growth of *M. rouxii* and *M. racemosus* is usually hyphal, and under anaerobic conditions it is usually yeast-like. However, under anaerobic conditions, hyphae are induced when CO_2 is removed from the culture [39,176,177,190], the hexose concentration is less than 0.01% [34], divalent cations are removed [41,529,595], or a volatile morphogen accumulates [389]. Conversely, aerobic growth of yeasts is induced by inhibitors of respiration such as phenethyl alcohol [550-552], cyanide [190], or a number of antibiotics [184,185,190,232,233]. It is important to note that *M. rouxii* and *M. racemosus* are not the only zygomycetes in which fermentation and respiration seem to be related to morphology. *Mucor genevensis* [127,128,478], *M. alternans* [332], *M. bacilliformis* [535], *M. pusillis* [184], *M. subtillisimus* [418], *Mycotypha africana* and *Mycotypha microspora* [234,503], and *Actinomucor elegans* [184] can all be induced to grow in the yeast phase. It is not known whether the fermentation-respiration balance in these species is the key to control of dimorphism or, as in the better studied *Mucor* species, the fermentation-respiration balance masks a deeper level of control. However, it is likely that the environmental factors which affect morphogenesis in *M. rouxii* and *M. racemosus* similarly affect morphogenesis in all mucorales.

Either morphology can be induced under aerobiosis or anaerobiosis. Neither yeasts nor hyphae require vitamins if O_2 is present, but both require thiamine and nicotinic acid under anaerobic conditions [38]. Pure hyphal growth in *Mucor* can be induced on any utilizable carbon source if O_2 is available, but hexoses are required for the induction of hyphae under anaerobic conditions [41,541]. Development of yeasts in *Mucor* has an absolute requirement for high concentrations of hexose. Disaccharides or other carbon sources will not suffice. For example, if hexose is supplied to anaerobically grown cultures at a concentration of 1% hexose or below, growth is hyphal [34,190]. However, the rate of gas flow may affect this observation [389]. Also, under the influence of phenethyl alcohol, *M. rouxii* and *M. genevensis* require hexose for yeast-phase growth, but hyphae develop if pentoses, disaccharides, glucosamine, or amino acids are the carbon source [551]. Moreover, in the presence of phenethyl alcohol the concentration of hexose is critical, as hyphal growth is induced if the level of hexose is initially below 2%. *Mucor genevensis* also grows in the yeast phase under high hexose concentration [127, 128,478], even under aerobic conditions, but under glucose-limiting conditions in a chemostat, this species reverts to the hyphal phase [478]. Sypherd et al. [541] cite unpublished observations that progressively higher glucose concentrations delay the onset of germ tube formation. Dibutyryl cAMP and cAMP can induce mucors to grow as yeasts under aerobic conditions [325,429], but these agents have no effect if the carbon source is not a hexose.

The hyphal-phase nitrogen requirements of *M. rouxii* and *M. racemosus,* like the carbon requirements of aerobic hyphae, are less stringent than for yeasts. It has been shown that yeast growth of these fungi [434] is dependent on an organic nitrogen source. Hyphal growth, on the other hand, does not require organic nitrogen and can occur in the presence of inorganic nitrogen salts. In *M. racemosus* the requirement is absolute, but in *M. rouxii* some limited yeast growth can occur in the presence of inorganic nitrogen.

The role played by atmosphere of incubation in determining the morphology of *M. rouxii* and *M. racemosus* is of critical importance, as indicated by the earlier reference to the roles of aerobiosis and anaerobiosis in the effect of carbon source on morphology. Even the composition of the anaerobic atmosphere has a morphogenetic effect on these fungi. However, this effect is not absolute and may be overridden by carbon source. Both yeasts and hyphae grow under anaerobic conditions, in conjunction with the proper carbon source. In 2% glucose medium, *M. rouxii* requires 30% CO_2/70% N_2 atmosphere for total yeast-phase development [39]. Higher partial pressures of CO_2 will offset the hypha-inducing effect of low concentrations of hexose, just as CO_2 is unnecessary if glucose concentration is sufficiently high [34]. Either yeasts or hyphae can be derived even under 100% N_2 if all conditions are carefully controlled [389]. Bartnicki-Garcia [32] indicated that hyphae grown under 100% N_2 and yeasts grown under 100% CO_2 metabolize glucose in the same manner with equal efficiency, and that CO_2 evolution is independent of the partial pressure of CO_2 within a range of zero to 1 atm. Inderlied and Sypherd [275] reported similar findings for *M. racemosus.*

It appears that CO_2 affects morphology neither by inhibition of decarboxylation reactions nor by lowering of pH. Also, it is not likely that the complementary effects of glucose and CO_2 are a result of endogenous CO_2 produced by fermentative metabolism, since CO_2 evolution by anaerobic yeasts and hyphae are identical [34], although the N_2 flow rate through the hyphal cultures was not rigidly controlled, and lowering of CO_2 concentration by flushing cannot be discounted. Bartnicki-Garcia and Nickerson [39] determined that significant CO_2 is incorporated into cellular constituents. Two-thirds of the CO_2 incorporated is found in protein with the preponderance located in aspartic acid residues and smaller amounts in glutamic acid and threonine. Under the conditions used, CO_2 assimilation is a function of the partial pressure of CO_2, as is induction of the yeast morphology. It is hypothesized that CO_2 exerts its morphogenetic effect by increasing the levels of aspartic acid, which in turn stimulates the deposition of aspartic acid-rich mannan-protein into the cell wall. The mannan-proteins are postulated to cause a shift from tubular to spherical cell wall growth, resulting in the induction of the yeast phase [32].

As with a number of other dimorphic fungi, chelating agents affect morphogenesis in the mucors. Yeast → hyphal conversion is induced by the addition of chelating agents such as EDTA [41]. N-acetic acid-type chelators reverse the effect of CO_2, and yeasts grown in high CO_2 tension convert to hyphae. The EDTA appears to work by chelation of divalent cations. This effect is reversed by the addition of zinc and other transition group metal ions. Zorzopoulos et al. [595] claimed that the effect of EDTA is due to a property other than that of a chelating agent. However, as has been noted [541], the medium of Zorzopoulos et al. [595] contained commercial casein hydrolysate, which contains large amounts of metal salts. Thus, the purported metal-deficient medium probably contained large quantities of metal ions and the EDTA was actually exerting its effect by chelating divalent cations.

The discovery by Mooney and Sypherd [389] that a volatile morphogen is produced by *M. rouxii* and *M. racemosus* added a new dimension to dimorphism research. In a series of elegant experiments, these investigators found that under anaerobic conditions, morphology is dependent on the rate at which N_2 or argon is bubbled through the medium. High flow rates induce the yeast phase and low flow rates induce hyphae. The effect is independent of glucose concentration, but is affected by temperature. The compound is volatile, can be passed from one culture to another culture connected by an airstream, and is not a component of the culture medium. Since the effect runs counter to the CO_2 effect (high concentrations of the morphogen stimulate hyphal growth), the factor is not CO_2.

Many aspects of macromolecular synthesis have been extensively studied in the yeast and hyphal phases of *Mucor* spp. and during their transitions. Synthesis of RNA appears to be tightly regulated during morphological conversion. During the first 5 min of aeration the RNA produced becomes quickly associated with ribosomes and is probably ribosomal RNA (rRNA) [233]. The G + C content of RNA produced after 20 min, however, resembles that of DNA (41% versus 39%), and is thus probably messenger RNA (mRNA). Later during conversion, the G + C content of the RNA being synthesized rises to about 47%, a figure identical to that found in established yeast and hyphal phase organisms. Haidle and Storck [232] showed that total RNA synthesis in yeast cells growing vegetatively increases linearly, whereas during conversion synthesis is discontinuous and increases at a much faster rate.

Young and Whiteley [591,592] found that there are three species of DNA-dependent RNA polymerase in *M. rouxii.* These species are in both yeasts and mycelia, although no differences in any properties are noted for the yeasts or hyphae [592]. The three species resemble those isolated from other fungi in approximate molecular weights, subunit makeup, stimulation by Mg^{2+} or Mn^{2+}, and α-amanitin sensitivity [236,592]. The specific activities of the three species vary during morphogenesis [591]. Young and Whitely noted two bursts in the synthesis of total RNA and in the specific activity of the three enzymes [591]. The first coincides with swelling of yeasts after aeration, the second with the emergence of germ tubes. According to evidence gathered with *M. racemosus,* the bursts of RNA synthesis are composed of a number of independently regulated syntheses of rRNA, pre-rRNA, mRNA, and transfer RNA (tRNA) [422]. All RNA species reach a maximum specific rate of synthesis 2 h after aeration and by 6 h have returned to the level of yeasts grown at high CO_2 tensions [422]. The second burst of synthesis seen in *M. rouxii* was not seen in *M. racemosus,* but this dichotomy in synthetic pattern may be due to different methods of measuring specific activity. In the first case [591] label incorporation was referenced to protein, whereas in the second reference was to RNA. The specific activities and ratios of the active RNA polymerases responsible for transcribing the various classes of RNA vary during these bursts. For example, prior to aeration 31% of the enzyme activity is due to enzyme I, thought to transcribe rRNA; by 2 h afterwards, the amount is 64%; and by 6 h, it declines to 25%. The activity of enzyme II, thought to transcribe mRNA, does not change until 6 h after aeration, when it accounts for 65% of the activity. Activity of enzyme III, thought to synthesize tRNA, does not change during morphogenesis [591].

Like RNA synthesis, protein synthesis is also affected during morphogenesis. Protein synthesis increases at a linear rate in anaerobic yeast-phase cultures, but increases exponentially soon after aeration and continues in this manner for at least 12 h [232]. The rate of protein synthesis, as measured by incorporation of radioactively labeled leucine into protein, increases during air-induced conversion, reflecting the increased growth rate associated with morphogenesis. When the specific rate of leucine incorporation is calculated, however, there is a pronounced increase in the rate of protein synthesis at the time of germ tube emergence, followed by a decline as hyphal elongation continues [419]. Autoradiographic evidence indicates that there are no preferred sites for protein synthesis during conversion [420].

The increase in protein synthesis appears to be correlated with an increase in RNA synthesis. However, since the percentage of polyadenylated RNA (putative mRNA) is constant with respect to the total RNA pool [422], the increase in protein synthesis is apparently due to an increase in the rate of translation during the conversion process. Yeast cells, transition forms, and hyphae have different specific rates of translation [421]. Yeast cells translate at a rate of 2.3 amino acids per second per ribosome (aa/sec/rib). Upon aeration, the rate increases to 5.2 aa/sec/rib during germ tube emergence and later rises to 8.2 aa/sec/rib. There is no change in the total RNA/protein ratio, percentage of rRNA to total RNA, or the number of ribosomes per milligram of cellular protein. In fact, there is a decrease in the number of active ribosomes and the number of mRNA molecules being translated during the most active period of protein synthesis. This increase in the rate of protein synthesis is not associated with the shift from anaerobiosis to aerobiosis, but the velocity of ribosome movement along mRNA molecules is regulated during morphogenesis. Since a specific ribosomal protein is phosphorylated during conversion [421], phosphorylation of selected proteins may be a mechanism of regulation of morphology at the level of translation.

Cell wall compositions in the yeast and hyphal phases are virtually identical, with the main significant differences being that yeasts contain about four times as much mannose as do hyphae and about half again as much protein [33]. Mannose is present in *M. rouxii* as a heteropolysaccharide called mucoran [36], whereas in *M. racemosus* it is present in a homopolysaccharide resembling the mannan of *Saccharomyces cerevisiae* [541]. Strongly acidic polysaccharides of hyphae have a greater proportion of D-galactose and L-fucose than do those of yeasts [166]. Both yeast and hyphal walls [40,150] contain chitin and chitosan, but cell wall synthesis is regulated differently in the two forms [37]. The synthesis of chitin and chitosan occurs preferentially in the apex of hyphal cells, but evenly over the surface of yeast buds. Similar delocalization of chitin synthesis has been observed in the *cdc 24* mutants of *S. cerevisiae,* where delocalization accompanies inhibition of bud emergence [526]. Delocalization of chitin also occurs during yeast → multicellular conversion in *W. dermatitidis* [277].

Bartnicki-Garcia and co-workers extensively investigated the mechanism and regulation of chitin synthesis in *M. rouxii.* They found that about 85% of the chitin synthase activity is located in the cell wall fraction and most of the remainder in the microsomal fraction. Chitin synthase is found in the apices of disrupted hyphal cells, and in situ chitin synthesis by hyphal walls resembles that seen in intact cells [377]. Yeast and mycelial chitin synthase are very similar, if not identical. Both require divalent cations, with Mn^{2+} being more effective than Mg^{2+}, and both use uridine diphosphate-N-acetylglucosamine as substrate, both are activated by N-acetylglucosamine, and both are competitively inhibited by uridine diphosphate and polyoxin. The enzymes are activated by endogenous proteases [376,485], and their stability in crude extracts differs greatly. Hyphal enzyme rapidly loses activity upon incubation at 28°C, but yeast enzyme increases in specific activity over the course of several hours before activity declines [485]. These results are interpreted as meaning that the hyphal enzyme is present in the active form and is rapidly degraded by proteases. Conversely, yeast enzyme would be present as a zymogen which is activated by protease before extensive proteolysis causes inactivation of chitin synthesis. Activation can be blocked by a soluble inhibitory protein [376,485]. This fact suggests that chitin synthase in *Mucor* may be regulated in a manner different from that of *S. cerevisiae.* In the latter instance chitin synthase is activated by protease B, but inhibited by protease B inhibitor [564,565]. The inhibition in *Mucor* apparently blocks the action of activated enzyme [341]. The concentration of the inhibitor in yeast cells of *M. rouxii* is 2.5 times that of hyphae [485], but the high level of zymogen in yeast cells suggests that a saccharomyces-type regulatory mechanism may exist undetected. How this regulates morphogenesis is not clear; however, spatial activation of chitin synthase correlates with morphology whether or not it plays a regulatory role.

Cyclic AMP has a profound effect on the morphology of *Mucor.* In vivo, cAMP levels are five times higher in yeasts than in hyphae. This is not due to an increase in synthesis of cAMP, but to a reduction in phosphodiesterase activity [429]. Cyclic AMP dramatically increases the activity of the glycolytic regulator phosphofructokinase in vitro, and in vivo phosphofructokinase activity is 30-40 times higher in the yeast than in the aerobically grown hyphal phase [430,549]. However, when yeast or hyphae are grown under N_2, there is no difference in intracellular cAMP levels [431]. Exogenously added cAMP (10 mM) or dibutyryl cAMP (2 mM) prevents induction of the hyphal morphology upon aeration of anaerobically grown yeast cells [325,429].

Passeron and co-workers [392] found in *M. rouxii* a cAMP-dependent histone kinase which binds to DNA. Thus, there is a possibility that cAMP plays a role in the determination of morphology in *Mucor.* Cyclic AMP-dependent nuclear protein kinases are not novel and, in fact, one has been found which specifically phosphorylates the 25,000-dalton subunit of rat RNA polymerase II [231]. The nucleotide stimulates the activity of purified *Escherichia coli* RNA polymerase [283] and acts with S-adenosylmethionine as a regulator of RNA synthesis in *S. cerevisiae* [326]. Since enzymatic modification of histone occurs more often in template-active regions of the chromosome [212] and is believed to have a general effect in promoting RNA synthesis by relaxing cellular chromatin [276], it is possible that cAMP mediates transcriptional control by directly or indirectly modifying specific chromatin proteins or RNA polymerases. This hypothesis has the added attractiveness that in vitro transcripts of chromatin closely resemble in vivo transcripts, but bear little resemblance to in vitro transcripts of DNA [115,151,159,212,348]. It is possible that selective morphology-dependent mRNA transcription could be mediated by modification of specific DNA-binding proteins. Cyclic AMP could play a key regulatory role by activating protein kinases, which then phosphorylate positive and negative modifiers of transcription, or it could stimulate the yeast phase of growth by acting directly on the DNA-binding proteins. Modifications of RNA polymerases could change the affinity of the polymerase for positive regulators of the operator sites for specific genes. In contrast to cAMP, cGMP does not seem to play a role in yeast → hyphal conversion but may be involved in regulating the growth of germinating spores [418].

Sypherd and co-workers (P. S. Sypherd, University of California at Irvine, personal communication) determined that methylation of an abundant protein is correlated with morphogenesis and may play a regulatory role in yeast → hyphal conversion. The resulting major methylated polypeptide (MMP) is apparently a structural protein and is rather common in fungi, since it is said to be found in *Candida, Saccharomyces,* and *Neurospora,* in addition to *M. racemosus.* During CO_2 to air-induced or N_2-induced yeast → hyphal conversion, methylation occurs mostly at lysine residues, and MMP in hyphae is methylated three times more heavily than in yeasts.

Using *coy,* a nonauxotrophic mutant which requires exogenous methionine for transition from yeast → hyphae, it was determined that methionine (probably in the form of S-adenosylmethionine) is the methyl donor. In the absence of methionine, and under CO_2 or in the presence of exogenous methionine in air, *coy* methylates MMP exactly as does wild type. Under aerobic conditions in the absence of methionine, *coy* hypermethylates MMP, but at different amino acid residues. Improper methylation of MMP correlates well with inhibition of morphogenesis. If MMP is a structural, cytoskeletal protein, as Sypherd hypothesizes, it may play a role in determination of morphology and its methylation may serve a regulatory role in morphogenesis.

Although hyphae with fermentative metabolism occur, respiration has not been demonstrated in yeast cells. The uncoupling or poisoning of oxidative phosphorylation is sufficient to induce aerobic, fermentative yeast growth without impairing other mitochondrial functions [39,41,275,551]. Yeast cells with full respiratory capacity can be induced, but measurement of respiratory capacity in vitro does not necessarily imply that the full capacity is being used in vivo. This point is supported by the observation that cytochrome oxidase and mitochondrial number both increase dramatically during yeast → hyphal conversion, even when it is carried out under anaerobic conditions [595]. Fermentation-respiration seems to play a negative role in morphogenesis (respiration will not allow yeast cell formation). A possible exception to this rule may be the *coy* mutant mentioned previously, but as of this writing O_2 metabolism had not been investigated in relation to morphology in this mutant.

The requirement for hexose in producing the yeast phase may be partially explained by the effect of cAMP and hexoses on α- and β-glucosidase synthesis. β-Glucosidase in *M. racemosus* does not require an inducer for synthesis, but is subject to hexose repression and inactivation in the presence of cAMP [65]. α-Glucosidase is induced by disaccharides [188,529], but is not synthesized in significant amounts by yeast cells [529]. Perhaps a combination of these two regulatory mechanisms prevents yeast cells from using disaccharides as a carbon source, thus inhibiting hyphal → yeast transition. However, neither the failure of yeasts to synthesize α-glucosidase, nor the inactivation of β-glucosidase by cAMP, explains why anaerobic growth on fermentable nonhexoses yields hyphae under conditions where both morphologies contain equal amounts of intracellular cAMP.

Glucose catabolism in anaerobically grown yeasts and hyphae follows almost identical patterns, with most of the glucose being catabolized by the Embden-Myerhoff-Parnas (EMP) pathway and the remainder being diverted through the pentose phosphate (PP) pathway [275]. Ethanol and CO_2 are the primary end products of the fermentation. Aerobically grown hyphae have a respiratory metabolism. Therefore, in *Mucor*, morphology can be altered without radical changes in carbon metabolism. Since there is little difference in the carbon flux between anaerobically grown yeasts and hyphae, morphogenesis in *Mucor* is not closely regulated by catabolic enzymes and metabolites. Inderlied and Sypherd [275] suggested that since anaerobic mycelia in their experiments were grown in the presence of a CO_2 trap, the high CO_2 levels produced by fermentation are required for yeast cell development. Nonhexose carbon sources, they suggested, are not metabolized at a rate sufficient to maintain high CO_2 levels.

Specific activity of pyruvate kinase is three to five times higher in yeasts than hyphae [427]. There are two isozymes of pyruvate kinase, but synthesis is unrelated to morphology and controlled by glucose [432]. Aerobically grown hyphae and aerobically grown dibutyryl cAMP-induced yeasts have identical respiratory capacities [431]. Activity of NAD-dependent glutamate dehydrogenase (NAD-GDH) is correlated with morphology [435]. NAD-GDH-specific activity is about 10 times higher in hyphae than in yeasts and its induction always precedes hyphal morphogenesis. Suppression of hyphal morphogenesis by dibutyryl cAMP also suppresses the increase in NAD-GDH activity. Peters and Sypherd [435] outline one path toward yeast → hyphal morphogenesis in *Mucor*. Intracellular cAMP levels drop within 30 min after CO_2 is removed, budding ceases after 2 h, NAD-GDH and MMP methylation levels increase after 3 h, and hyphae appear by about 4 h after induction.

Ornithine decarboxylase (ODC) activity increases 30- to 50-fold during morphogenesis.

The increase in activity always precedes hyphal development, and requires protein synthesis from preformed mRNA (P. S. Sypherd, University of California at Irvine, personal communication). Since ODC is a major enzyme in the pathway of polyamine biosynthesis, there is a possible role for putrescine or spermidine in the induction of hyphal growth. Sypherd suggests that such a role might be effective at the level of transcription, RNA processing, or translation.

The experimental data related to morphogenesis in the mucors present a number of interpretive problems. Perhaps the greatest problem is related to the inocula used to initiate experiments. Because the morphological change is so well defined, published reports deal with transition from the yeast to the hyphal phase. However, the inoculum used to begin the experiments varies from laboratory to laboratory. Early experiments by Bartnicki-Garcia and Nickerson used yeast cells, germinated spores, or spores [39-41]. Both investigators subsequently used spore inocula [34,176,177], as did Sypherd and his co-workers. On the other hand, Passeron's, Terenzi's, Storck's, and Whiteley's laboratories routinely use yeast cell inocula obtained by germinating spores and growing them for periods ranging from 15 [595] to 48 h [190] prior to onset of the experiment. Since there seems to be no difference in the environmental factors inducing hyphal germination from spores and yeast → hyphal conversion, it is important to consider that spore germination and vegetative morphogenesis are not equivalent phenomena. Upon induction of hyphae from yeast cells, preferably in the logarithmic phase of growth, one can observe physiological phenomena related to vegetative morphogenesis. Upon induction of yeasts or hyphae from spores, one must necessarily observe the physiology of germination prior to morphogenesis. It is then necessary to separate physiological events related to spore germination from those related to morphogenesis, or to ignore the problem of comparing results from experiments in which different sorts of starting inocula were used.

Using spores as the experimental inoculum can also introduce problems in interpreting results. For example, Haidle and Storck [232] reported that *M. rouxii* NRRL 1894 developed as yeast cells under N_2 anaerobiosis. Bartnicki-Garcia [34] and Elmer and Nickerson [176] criticized the results on the basis that no growth data were shown, and that photographs of the yeast cells did not show budding yeasts, but rather swollen spores. Since Elmer and Nickerson were able to obtain hyphal growth of NRRL 1894 under N_2, the criticism appears valid. It may also be that the state of the inoculum as much as the difference in organisms caused the differing results of Orlowski and Sypherd's [422] and Young and Whiteley's [591] investigations of specific activities of RNA synthesis. It can be seen that unless care is taken and populations monitored closely, factors may be introduced by studying spore outgrowth and calling it yeast → hyphal conversion which can greatly confuse the issue. The greatest errors would be introduced in short-term experiments in which most of the events studied would be spore germination events. Ideally, all yeast → hyphal conversions should be carried out using established, budding yeasts as the starting inoculum.

The literature pertaining to the dimorphism of the mucors is difficult to summarize and rationalize into a model for morphogenesis. However, it is apparent that carbon and nitrogen metabolism play important roles in the morphogenesis of *M. rouxii* and *M. racemosus.* Hexoses are required for yeast growth and are metabolized primarily through the EMP and PP pathways. Nonhexoses, including disaccharides, are not used. If the hexoses are metabolized through the respiratory chain, hyphae develop. Thus, it is probable that some metabolite of the EMP or PP pathways is required in high concentration for induction of yeast phase growth.

Glutamate may be the most critical amino acid for determining morphology (J. Peters, University of California at Irvine, personal communication), although this fact needs to be confirmed by subsequent observations. Since NAD-glutamate dehydrogenase and ornithine decarboxylase activities are highly correlated with morphology, it is probable that their reactions provide some necessary intermediate in the morphogenetic process. NAD-GDH catalyzes the reaction

$$\text{Glutamate} \rightarrow NH_4^+ + \alpha\text{-ketoglutarate}$$

and ODC catalyzes the reaction

$$\text{Ornithine} \rightarrow CO_2 + \text{putrescine}$$

Aside from the role of polyamines as regulators of DNA, RNA, and protein and lipid biosynthesis, this does not seem too promising at first glance. However, ornithine may be derived from arginine by the reaction

$$\text{Arginine} \rightarrow \text{urea} + \text{ornithine}$$

J. Peters (unpublished data) demonstrated that urea induces hyphal development under conditions which normally yield only yeasts. If arginine is degraded to urea and putrescine, putrescine may condense with the propylamine moiety of S-adenosylmethionine to form spermine and spermidine, thus explaining, in part, the methionine requirements of the *coy* mutant [542].

It seems that the generation of ammonia by GDH, of putrescine by ODC, and the formation of S-adenosylmethionine for use as (1) methyl donor, (2) propylamine donor, and (3) transcription regulator are all interrelated in control of morphogenesis. Just how they interact is unclear, but ultimately they must direct cell wall growth toward either highly polar or isotropic growth.

V. CONCLUSIONS

An understanding of dimorphism is dependent on integrated experimental approaches involving cytology, physiology, biochemistry, and genetics. Major advances have been made in recent years toward elucidating the determinants of morphogenesis in the two monomorphic fungi *Saccharomyces cerevisiae* and *Neurospora crassa* (see reviews in Refs. 75, 86, 244, and 504). The extensive genetic literature associated with these two fungi has greatly accelerated and facilitated these advances. By contrast, the absence of well-defined and characterized genetic systems among the dimorphic fungi has significantly hampered morphogenetic studies in thes organisms. This aspect of the study of dimorphism among the medically important fungi will continue to be troublesome until the sexual cycle for each fungus is discovered and extensively exploited for genetic purposes.

Many of the lessons learned with *N. crassa* and *S. cerevisiae* are beginning to be applied to the study of other fungi. The studies of Hartwell and his co-workers [244,246, 451] have set a pattern for developmental research which will be invaluable in elucidating the nature of the yeast-phase cell cycle of certain dimorphic fungi. By subjecting developmental mutants of *S. cerevisiae* to genetic analysis, Hartwell and co-workers [243-247, 260] established the order of expression of certain genes necessary for completion of the normal sequences constituting the yeast cell cycle. The position of a cell in the cycle was predicted from the size of the bud relative to the mother cell; in logarithmically growing populations unbudded cells are new cells, whereas older cells have buds which become progressively larger as the cell progresses toward cytokinesis and cell separation. Thus, cell wall growth in the bud of *S. cerevisiae* is intimately related to the cell division cycle. The same is true for other morphogenetic events in *S. cerevisiae,* such as bud emergence, nuclear migration, mitosis, and septation. From the fragmentary results obtained in studies of the yeast phase of a few dimorphic fungi, it can be hypothesized that *S. cerevisiae* is a useful model for elucidating yeast-phase development. The basic regulation and control of the budding growth cycle and its coordination with the DNA synthesis-nuclear division cycle will probably be very similar among all fungi, regardless of their taxonomy, although interesting variations on this theme should be expected. The elegant studies of Cabib and his co-workers [86,88,89,92,171-173,564,565] on septum biogenesis in *S. cerevisiae* set a standard for dissecting a yeast cell cycle event at the biochemical and enzymatic level.

Based on an extensive literature concerning the cytology and ultrastructure of hyphal tips, Bartnicki-Garcia [35] formulated a model for hyphal tip growth in particular and cell wall growth in general. The model is based on the assumption that wall expansion results from the cumulative action of minute hypothetical units of wall growth. Thus, apical growth, budding growth, and isotropic growth are treated in terms of patterns of distribution of these units. The model incorporates the idea that wall precursors, wall synthetic enzymes, and wall lytic enzymes are all present at the sites of wall growth and that some or all of these components arrive at the sites compartmentalized in vesicles. These vesicles fuse with the plasma membrane, deposit their contents external to the plasma membrane (exocytosis), and thereby provide the essential ingredients for wall growth. The major events considered essential for the growth of one wall unit (consisting for simplicity of only two components, a microfibrillar skeleton and amorphous embedding material) are as follows:

1. Vesicles containing lytic enzymes fuse with the plasma membrane and secrete the enzymes into the existing wall (exocytosis).
2. The lytic enzymes weaken the existing wall by splitting inter- or intramolecular bonds of the microfibrillar skeleton.
3. The weakened and partly dissociated microfibrillar complex expands, because of high turgor pressure, and allows an increase in surface area.
4. The expanded microfibrillar complex is rebuilt by microfibrillar synthesizing enzymes located in the wall or on the outer surface of the new plasma membrane. The synthetases involved in the microfibrillar rebuilding are also envisioned to be secreted via vesicles, whereas the soluble precursors for the synthesis of the microfibrils are thought to be transported directly across the plasma membrane by active or passive membrane transport mechanisms.
5. Other vesicles containing the amorphous wall components in largely or entirely preformed states deposit their contents into the periplasmic space against the inner wall surface. This material is then forced into or through the microfibrillar skeletal network by turgor pressure.

The combined action of all these events reconstitutes the expanded wall unit without loss of structural properties, including the coaxial arrangement (differential layering) of wall polymers.

Although the model for cell wall growth outlined by Bartnicki-Garcia is described in terms of a limited number of discrete events, it is probable that normal phenotypic wall expansion involves the simultaneous action of all the events [35]. However, an ultimate model of dimorphism must also embody mechanisms that account for altered cell shapes. As pointed out by Stewart and Rogers [533] in their perceptive review of fungal dimorphism, it is necessary to think in terms of differential rates of growth of different cell wall regions. Also, whether wall expansion is accompanying apical growth, budding growth, or isotropic growth, Bartnicki-Garcia's model predicts that vesicles should be evident at sites of rapid wall synthesis. Such vesicles are easily demonstrated [223] and were well documented in hyphal tips by electron microscopy prior to the formulation of the model [224, 226,252]. Additionally, vesicles are observed in developing buds of various yeast forms [83,141,228,350,366,390,411,506], and prior to the initiation of growth in spores, cysts, and sporangia [67,222,225,262]. Vesicles are also evident at sites of developing clamp connections, septa, and blastoconidia [211,228,459]. In hyphal tips that are growing until prepared for electron microscopy, the vesicles are present in the apical zone to the exclusion of most other cytoplasmic organelles and inclusions [211,224,227]. Grove et al. [227] suggested that many components for wall growth are derived in the subapical hyphal regions where they are transferred to dictyosomes (Golgi-like structures) from the endoplasmic reticulum. The transfer is accomplished by blebbing of endoplasmic reticulum and refusion of the resulting vesicles into a cisterna at the dictyosome's proximal pole. The cisternal contents and membranes are then transformed as the cisterna is displaced to the distal pole (toward the hyphal apex) by the continued formation of new cisternae by the endoplasmic reticulum. Evidence for the transformation of the cisternal contents is based on the differential staining characteristics of proximal and distal cisternae [147,148,226,227]. Ultimately, the distal cisternae vesiculate and transport their contents to the sites of wall growth. Cytochemical methods for the localization of polysaccharides and enzymes have confirmed that dictyosomal cisternae and vesicles contain some constituents necessary for wall synthesis [147,148,226,227,409]. Other techniques involving autoradiography and fluorescent stains have been used to document that hyphae grow by deposition of cell wall material predominantly at the hyphal tip [35,37,67,213, 214,302,303,309].

In contrast to hyphal wall growth, which usually generates a cylinder with parallel side walls when observed in optical section, budding yeasts exhibit wall growth that generates ellipsoidal shapes. This implies that yeast wall growth involves processes that are somewhat different from apical growth processes. It appears that in addition to polar extension of the bud tip, some equatorial wall enlargement is required. Cabib [86] rules out apical growth followed by simple stretching of the wall at the bud equator as a possible valid explanation for this nonpolar enlargement because the wall of the bud is of uniform thickness. Bartnicki-Garcia and Lippman [37] and Braun and Calderone [69] state that in *Mucor rouxii* and *Candida albicans*, respectively, wall growth, as indicated by the incorporation of labeled N-acetylglucosamine, is uniform over the surface of a developing bud. Evidence is compelling, however, that some aspects of wall growth in the developing buds of other yeasts may involve the apical insertion of molecules. Autoradiography after growth of *S. cerevisiae* and *Schizosaccharomyces pombe* in tritiated glucose suggests the polar incorporation of wall glucans [284,286]. Also, the inhibition of glucan synthesis with 2-deoxyglucose causes lysis of the bud apex of *Saccharomyces cerevisiae*, which suggests that the bud apex has been weakened preferentially due to the absence of glucans essential for wall stability [285]. Furthermore, no fluorescence is observed at the apices of buds on cells exposed to fluorescein-conjugated concanavalin A and then allowed to grow in the absence of the dye. The concanavalin A is thought to bind specifically with mannan polymers [19,553]. However, when tritiated mannose [553,554] is used to label mannan prior to high-resolution autoradiography of isolated cell walls of *S. cerevisiae*, incorporation of mannose is found to occur uniformly over the surface of very small buds, chiefly at the tips of intermediate-size buds, and at both the tips and sides of large buds. Possibly, these results indicate the relocation of wall polymers after their synthesis. In yeast and hyphal morphologies of *C. albicans*, and in the hyphae of *Penicillium chrysogenum*, previously synthesized wall polymers are not all fixed and may be found in areas of new wall synthesis [149,216]. Similar results with Calcofluor-staining materials in the walls of *Wangiella dermatitidis* (C. W. Jacobs and P. J. Szaniszlo, unpublished results) indicate that septum material may become incorporated into the walls of developing buds. Such reports for the growth of *S. cerevisiae* [149] have been disputed [553]. Other results dealing with the incorporation of glucose into the cell wall of *S. cerevisiae* indicate that the rate of synthesis of glucan and mannan vary through the cell cycle with major changes in close correlation to the point of bud emergence [53,54]. Possibly, polysaccharide polymers are differentially deposited in the new bud during development.

To reconcile the conflicting and confusing fluorescent and autoradiographic data related to wall deposition in the multipolar budding yeast, *S. cerevisiae*, Cabib [86] suggested that at least the insertion of new mannan can be rationalized in terms of a beam of wall-synthesizing particles (which could be vesicles containing wall components and enzymes) acting upon the site of bud formation. At the time of bud emergence, all of the bud is labeled. At later times the polarly directed beam of particles reaches only the bud tip, giving rise to localized labeling. Finally, as the bud becomes even larger, it is impossible to maintain the direction of all the particles in a straight path. At this time some of the particles reach the bud side walls, allowing for interstitial wall growth or secondary deposition. The interstitial wall growth presumably contributes to the equatorial expansion of the bud.

Hyphal apical extension and most aspects of yeast budding, as outlined, represent highly directed, polar growth processes. Even if yeast budding is predominantly interstitial at later stages, early bud emergence and growth must be highly polarized to inhibit wall softening and synthesis at sites remote from the primary bud locus. The formation of true hyphae by dimorphic fungi may constitute the purest form of sustained and directed wall synthesis. Perfect continuous regulation and coordination of wall lytic and synthetic processes would appear to be required for the generation of a cylindrical hypha with parallel side walls as seen in optical section. An interruption of these processes could lead to cessation of apical growth and the solidification of the wall into a ridged stabilized structure. Little alteration in wall ultrastructure would be expected upon the resumption of apical growth unless long-term, major solidification of the wall had occurred. In this case, new wall growth would probably be characterized by the blastic emergence of an inner layer of cell wall through overlying wall material. Most, if not all, dimorphic fungi which give rise to lateral hyphal buds (blastoconidia) exhibit wall ultrastructures indicative of the yeast cell wall being contiguous with only the inner portion of the hyphal wall [198,201,228,424]. Yeast budding in some dimorphic fungi is also often characterized by the new bud wall being contiguous with an inner wall portion of the yeast mother cell [97,98,228,502].

The similarity in the processes involved in the formation of buds from yeasts or hyphae in *W. dermatitidis* led Grove et al. [228] to suggest that both are not only functionally equivalent but also structurally equivalent. Apparently, environment determines whether yeast or hyphal buds become reproductive budding yeasts or dormant blastoconidia. In vivo, in liquid medium, and on some solid media, the yeast and hyphal bud cells usually form daughter buds. However, because a nonperforate septum isolates daughter buds from mother cells, hyphal buds produced on aerial hyphae may not come in direct contact with substrate and consequently become spores. Possibly, other dimorphic fungi which produce blastoconidia in a manner similar to *W. dermatitidis* also have structurally and functionally equivalent yeast and hyphal buds.

Polarly directed yeast wall growth, or for that matter blastoconidia wall growth, appears to differ from apical wall extension by involving periodic processes. In polarized budding growth, periods of highly directed, dynamic lytic activity leading to bud emergence alternate with periods of less directed, more quiescent lytic activity associated with maintaining wall growth. In yeasts, once a critical volume is reached [288,289], isotropic wall growth ceases and only the periodic polarized growth associated with each new bud occurs. The sometimes conflicting reports of Cabib and co-workers [88,90,91,171-173,304,564, 565] and Bartnicki-Garcia and co-workers [68,341,376,484-486] on the regulation of chitin synthesis suggest a possible explanation for this phenomenon. Cabib believes that synthase zymogen is located on the plasma membrane in *S. cerevisiae.* Bartnicki-Garcia believes that chitin synthase zymogen is located in chitosomes in *M. rouxii,* although he has reported that some zymogen is associated with the plasma membrane. Schekman and Brawley [499] found both soluble (chitosome-like) and membrane-bound (plasma membrane-like) zymogen in *S. cerevisiae.* Chitin synthase zymogen must be activated by a protease and is thus subject to proteolysis. Therefore, if two different populations of chitin synthase (and by analogy, other cell wall synthesizing enzymes) exist in yeast cells, one may be limited in concentration, plasma membrane bound, subject to random proteolysis, and participate in isotropic growth. The other could be synthesized throughout the life of the cell, found in vesicles, periodically activated, and participate in budding growth. Thus, the plasma membrane-bound enzyme, being continuously degraded at low levels, would eventually reach negligible levels of active enzyme and isotropic growth would cease. The vesicular enzyme could always be available for budding growth in sufficient quantities to assure propagation.

Superimposed on the polar-isotropic mechanism of wall growth would be the septum-building cycle which partitions the mother cell and daughter bud. After septum formation, controlled lysis of certain cell wall components near the septum results in cell separation. Such lytic action appears to be very specific because yeast cells of a glucosamine auxotroph of *S. cerevisiae,* grown in the absence of glucosamine, form defective septa and do not separate [28]. The lytic activity may conceivably be limited to a specific wall region by fungal lectins, such as those found associated with cell wall chitin in *N. crassa* and which inhibit chitinase activity [450]. If this controlled lytic activity does not occur at the appropriate stage of the cell cycle, pseudohyphae might be expected to form.

Variations on the theme of pure budding growth and pure apical growth may lead to transitional cells and moniliform hyphae. These blastically derived hyphal forms are usually associated with yeast → hyphal transitions and probably reflect the ability of a hyphal fungus to grow reproductively in the yeast phase. A case can be made that germ tube formation in some dimorphic fungi requires that large concentrations of lytic enzymes become polarly directed at a germ tube emergence locus, much as they are prior to yeast bud emergence. The blastic eruption that follows leads to a transitional cell which may have the morphological appearance of a yeast bud cell. However, the transitional cell is identified ultrastructurally as a hyphal cell by the formation of a perforate septum between the yeast mother cell and the transition cell. The inability of the yeast mother cell to direct accurately the lytic and synthetic enzymes of wall growth, or their activators, only toward a single wall locus probably results in the elliptical shape characteristic of most transition cells. Whereas most dimorphic fungi have only one or two transitional cells, a few form long transitional segments which constitute moniliform hyphae. Among the latter, *W. dermatitidis* is interesting because microscopic study suggests that the length of the moniliform hyphal segment reflects the amount of time necessary for the hyphae to eliminate all vestiges of discontinuous and less than perfectly directed wall synthesis [424]. With this fungus, under some growth conditions, only a few cells make up the moniliform portion of the hypha, whereas under other growth conditions numerous moniliform cells develop. However, in general, a gradual transition in shape can be observed among the cells of the moniliform segment, whereby the proximal cells are more spherical and the distal cells progressively more oblong [424].

The complete elimination of polarity during wall expansion probably leads to isotropic growth, which may result from a uniform distribution of wall growth units around the cell periphery [35]. The spherical growth associated with spore swelling prior to germ tube formation, with yeast cells prior to budding, and with the formation of sclerotic bodies, spherules, and adiospores are examples of isotropic growth [14,35,37,537,543,572] and probably results from the loss of polarity in wall synthesis due to adverse environmental conditions. In conidiogenesis, this delocalization to isotropic growth is noted ultrastructurally by Spitzenkörper disintegration, dispersion of wall-associated vesicles, and migration of mitochondria into apical hyphal regions [395]. The more complete the loss of polar growth, the more spherical the resulting cell morphology.

Evidence that at least some wall units are distributed over the entire surface of a cell during isotropic growth was provided by Sloat and Pringle [526]. They examined the distribution of chitin (as revealed by the fluorescence of Calcofluor) on the surface of a conditional *S. cerevisiae* cell division mutant (*cdc 24;* see Ref. 246) that fails to bud at a nonpermissive temperature, but continues nuclear division and cellular growth. They found that this mutant deposits chitin (or at least Calcofluor-staining material) nearly uniformly over the enlarging cell surface, instead of only at sites of septal rings as characteristic of normal budding growth. Similar results have been obtained with conditional multicellular mutants of *W. dermatitidis* [277] which show very localized staining during budding growth, but uniform staining beginning at very early times during yeast → multicellular form transition. These studies, together with the earlier studies of Bartnicki-Garcia and Lippman [37], tend to substantiate that at least chitin microfibrillar synthesis is largely, if not entirely, randomized over the entire cell surface during isotropic growth.

Although not all aspects of the literature pertaining to cell wall morphogenesis can be reconciled by the discussion related above, it seems clear that polar growth (budding and apical) requires site-specific wall synthesis and lysis, whereas isotropic growth involves less stringent random regulation of wall growth. Certainly, any type of asymmetry implies the nonrandom distribution of wall components. Perhaps the most intriguing feature of dimorphism is not the question of how the fungal wall can come to possess different shapes, but how the fungal cell brings about site-specific synthesis. In the case of the vesicular models for transport of materials to sites of wall synthesis, it is necessary to comprehend the factors controlling vesicle movement. Therefore, it may be necessary to postulate in fungi the presence of a polarization director which is able to direct site-specific wall synthesis and lysis in addition to numerous other functions. In hyphae, general wall expansion is polarly directed and site-specific wall synthesis occurs during apical growth, branch formation, lateral hyphal bud formation, and septum formation. In yeast forms, the events of bud emergence and septum formation are site-specific [86,277,499]. Sexual mating in *S. cerevisiae* is also preceded by the hormone-controlled, localized deposition of wall polymers which have been histochemically identified as chitin by primulin staining and as mannan by concanavalin A associations [499,555]. Additionally, nuclear migration and mitosis and other organellar movements in fungi are often polarly directed in line with the long axis of growth [253,396,444,460].

The roles ascribed to the hypothetical polarization director can be filled by the known functions exhibited by cytoskeletal components, microtubules (diameter most commonly 25 nm), microfilaments (diameter 6 nm), and possibly intermediate filaments (diameter 10 nm) [189,417,472,531]. The cytoskeletal components may be capable of segregating portions of the fungal cell surface by being attracted to membrane recognition sites for vesicles, while not affecting overall membrane fluidity. The possibility of cytoskeletal involvement in fungal dimorphism suggests directions for new morphogenetic research and has been provocatively addressed by Stewart and Rogers [533]. As they pointed out, however, the identification of microtubules and microfilaments in dimorphic fungi is not common and *W. dermatitidis* may represent the only medically important dimorphic fungus documented to have microtubules [424]. Nevertheless, microtubules and microfilaments are well documented ultrastructurally in a wide variety of vegetatively monomorphic fungi [12,83,84,251-253,396,444,459,460] and recently have been identified in *Aspergillus nidulans, Neurospora crassa,* and *S. cerevisiae* by chemical techniques [13,42,129,393, 509,510]. The discovery by Byers and Goetsch [83,84] that microtubules normally extend from the spindle plaque in the nuclear envelope to the site of the growing bud has provided some of the impetus for the study of microtubule function in *S. cerevisiae.* The implication is that in *S. cerevisiae* microtubules may be involved in directing vesicles containing wall material and appropriate enzymes to the site of bud emergence, as well as being involved in directing nuclear migrations. Recent studies of the green filamentous alga, *Mougeotia* sp., suggest that microtubules are intimately involved in the cell wall deposition associated with regenerating protoplasts [349]. Microtubules are always present and free in the spherical protoplasts of this alga, but become progressively cross-bridged to the plasma membrane during the first 3 h of elongation that accompanies cell wall regeneration. Treatment of protoplasts with drugs that interfere with microtubule polymerization inhibits elongation and the regenerating protoplasts remain spherical. Cell wall microfibrils in untreated regenerating protoplasts are highly oriented, whereas they are randomly oriented in the treated regenerating protoplasts. The involvement of intermediate filaments (diameter 10 nm) in *S. cerevisiae* budding was documented when a ring of these filaments was found in the cytoplasm underlying the mother-bud junction [85]. Brown's [79] supposition that colchicine depolymerizes microtubules and prevents the uncovering of receptors necessary for fusion of vesicles with the plasma membrane in the unicellular alga *Pleurochrysis scherfelii* similarly suggests a polarization role for microtubules. Thus, it is tempting to postulate that extracellular factors may act on fungal membranes, much as they do in animal cells, and there activate intracellular regulators which control specific enzyme functions and processes that, in part, alter the fungal cytoskeletal system.

With some of dimorphic fungi, cAMP and the enzymes involved in its metabolism, adenylate cyclase and 3′5′-cAMP phosphodiesterase, are suggested to be important for phase transition [325,359,429,431]. It is well known that cAMP acts in animals as an intracellular regulator. Willingham [581] postulates that in animal cells cAMP may interact directly with microtubules and microfilaments to regulate shape and motility. Also, agents that interfere with microtubular function prevent many of the morphological changes in animal cells induced by cAMP [581]. Willingham and Pastan [582] propose that cAMP stimulates microtubule assembly, and inhibits the contraction of microfilaments. Both Garland [195] and Steiner [530] demonstrated that cAMP affects microtubular protein in vitro, although the mechanism of the different effects attributed to cAMP were not elucidated. The possible involvement of similar processes in fungi is suggested by the studies of two *slime* varients of *N. crassa* [562]. These varients are essentially devoid of walls and grow by pinching-off beaded spheroplasts from migratory flattened spheroplasts. The pinching off of the beaded spheroplasts is enhanced by a number of drugs, including cytochalasins A and B, which are known to cause disaggregation of microfilaments, and also reagents that affect cyclic nucleotides, such as dibutyryl cAMP, caffein, and sulfhydryl reagents. As pointed out by Stewart and Rogers [533], the significance of such observations to dimorphism is not that shape is directly altered, since the shape of the fungal protoplast is determined largely by the external wall of the intact cell, but that microtubules and microfilaments may control delivery of precursor material through cytoplasm to the sites of wall synthesis. It has been shown that in the absence of cell walls, such as in fungal zoospores, microtubules probably act as the major morphogenetic agents [264]. However, the possibility of interaction of cAMP with microtubules and microfilaments should not be emphasized to the exclusion of other roles of cAMP, such as cAMP regulation of gene transcription and cAMP activation of cAMP-dependent protein kinases, which are both concerned with wall biosynthesis [533].

The influenced of divalent cations on the dimorphism of many fungi may be partly reconciled with their known effects on adenyl cyclase, cAMP-phosphodiesterase, microtubule disassembly and microtubule polymerization. The adenyl cyclase from *Mucor rouxii* requires Mn^{2+} for activity and EDTA inhibits cAMP-phosphodiesterase activity, presumably by withholding a cation [429]. The effects of Zn^{2+}, Mn^{2+}, and chelators such as EDTA on *Phialophora verrucosa, Emmonsiella capsulatum, A. parasiticus, M. rouxii,* and *C. albicans* [39,158,196,440,464,588,595] could conceivably be affecting cytoskeletal structures directly by interfering with Ca^{2+} and Mg^{2+} involvement in microtubule disassembly and assembly [531] or may be acting indirectly by interfering with the activities of adenyl cyclase or cAMP-phosphodiesterase.

Alternatively, divalent cations and their chelating agents could affect morphogenesis by a direct effect on the cell membrane. By changing the membrane charge density in the neighborhood of wall growth, the regulation of cell wall synthesis could be disrupted by altering the microenvironment of relevant proteins. In this case, an intact, functional cytoskeletal system could exist, yet be unable to direct wall synthesis because the target site for vesicle attachment to the plasma membrane is unrecognizable. Such a hypothesis could explain some of the effects of acid medium on fungal morphology, if the excess protons change charge density in a manner similar to the addition and removal of divalent cations. It could also explain the continuation of site-specific cell wall growth in the presence of agents which are believed to disrupt the cytoskeletal system [239; C. W. Jacobs and P. J. Szaniszlo, unpublished results]. Additionally, this hypothesis may explain the recent results of Novick and Schekman [411], who found that both cellular growth and cell surface growth are inhibited while vesicles accumulate in a temperature-sensitive mutant of *S. cerevisiae* which synthesizes, but does not secrete, acid phosphatase and invertase. This mutant, *sec 1,* exhibits a fivefold increase in intracellular phosphatase during a 2.5-h incubation period at 37°C compared to the parental strain from which it was derived. Similarly, the intracellular pool of invertase increased sixfold during a 1-h period at 37°C. Although the function of the *sec 1* gene product is not known, Novick and Schekman [411] speculate that the vesicles containing acid phosphatase and invertase are unable to fuse with the plasmalemma because of defective transport, membrane recognition, or exocytosis. Return of the mutant to the permissive temperature allows resumption of normal growth and development [411]. Because the vesicles accumulate instead of fusing randomly with the plasma membrane, it appears that membrane recognition sites which allow exocytosis are affected in *sec* mutants.

The divalent cation Ca^{2+}, in addition to affecting microtubule disassembly, is also known to exert profound influence on many other biological processes, such as cell motility, muscle cell contraction, axonal flow, cytoplasmic streaming, chromosome movement, neurotransmitter release, endocytosis, and most important, to the model of wall biogenesis, exocytosis [124]. Possibly all of these particular processes are indirectly affected by microtubule and microfilament function or recognition. Enzymes other than adenyl cyclase and cAMP-phosphodiesterase are also affected by Ca^{2+}, including myosin light-chain kinase, Ca^{2+}-dependent protein kinase, NAD kinase, guanylate kinase, and phospholipase A_2 [124]. The plethora of effects mediated by Ca^{2+} and its Ca^{2+}-binding protein calmodulin has led to its identification as a second major intracellular regulator, which, together with the nucleotide monophosphates, regulates the majority of extracellularly stimulated processes in animal cells [124]. It is unfortunate indeed that the effects of Ca^{2+} are not usually monitored in studies of the influence of divalent cations on fungal dimorphism. Such studies might implicate the direct involvement of the cytoskeletal system in the polarity exhibited by fungi during normal growth and the modulation of polarity during phase transition.

Regardless of whether or not the cytoskeletal system is the ultimate polarization director in fungi, it is clear that future studies of dimorphism must take this hypothesis into account. The prediction of Stewart and Rogers [533] that only a microscopic approach closely coupled with integrated chemical, immunochemical, and radiographic procedures will (1) permit the specific identification of the macromolecules constituting intracellular precursors of cell wall fibrils and matrix components, (2) reveal how and when and which vesicles fuse with each other and with the plasma membrane so that exocytotic and membrane circulation patterns can be identified, and (3) permit the identification of microtubules and microfilaments and subunits of these structures as they may exist in pools in the cells (presumably directing site-specific cell wall expansion) still seem appropriate to the subject of dimorphism. The identification of the hypothetical polarization director and the determination of its regulation will do much to explain the mechanism by which medically important fungi alter their morphology and possibly, as a result, resist being killed by specific and nonspecific host pressures.

NOTE ADDED IN PROOF

Recent experiments using *M. racemosus* indicate that if oxygen is rigorously excluded in cultures sparged with nitrogen, growth is in the yeast phase. Under these conditions, morphology is independent of nitrogen flow rate. Thus the "volatile factor" reported by Mooney and Sypherd [389] is likely to be molecular oxygen (P. Borgia, University of Southern Illinois, personal communication).

REFERENCES

1. Ahlfeldt, F. E. 1929. Special observations on the morphology of *Coccidioides immitis.* J. Infect. Dis. *44*: 277-281.
2. Ainsworth, G. C. 1955. Pathogenicity of fungi in man and animals. In J. W. Howie and A. J. O'Hea (Eds.), *Mechanisms of Microbial Pathogenicity,* Cambridge University Press, Cambridge, pp. 242-262.
3. Ainsworth, G. C. 1971. *Ainsworth and Bisby's Dictionary of the Fungi,* 7th ed., Commonwealth Mycological Institute, Kew, Surrey, England.
4. Aitken, W. B., and D. J. Niederpruem. 1970. Ultrastructural changes and biochemical events in basidiospore germination of *Schizophylum commune.* J. Bacteriol. *104*: 981-988.
5. Ajello, L. 1967. Comparative ecology of respiratory mycotic disease agents. Bacteriol. Rev. *31*: 6-24.
6. Ajello, L. 1970. The medical mycological iceberg. In *Proceedings of the First International Symposium on Mycoses,* Pan Am. Health Organ. Sci. Publ. No. 205, Washington, D.C., pp. 3-10.
7. Ajello, L. 1972. In *Paracoccidioidomycosis,* Proceedings of the First Pan American Symposium, Pan Am. Health Organ. Sci. Publ. No. 254, Washington, D. C.
8. Ajello, L. 1978. The black yeasts as disease agents: historical perspective. In *The Black and White Yeasts,* Proceedings of the Fourth International Conference on the Mycoses, Pan Am. Health Organ. Sci. Publ. No. 356, Washington, D.C., pp. 9-16.
9. Ajello, L., D. F. Dean, and R. S. Irwin. 1976. The zygomycete *Saksenaea vasiformis* as a pathogen of humans with a critical review of the etiology of zygomycosis. Mycologia *68*: 52-62.
10. Albornz, M. B. de, and I. C. Aasen. 1971. Etudis citoquinico de las principales enzimas hidrolyticas del *Paracoccidioides brasiliensis.* Sabouraudia *9*: 139-143.
11. Al-Doory, Y. 1972. *Chromomycosis.* Mountain Press, Missoula, Mont.
12. Allen, E. D., R. Lowry, and A. S. Sussman. 1974. Accumulation of microfilaments in a colonial mutant of *Neurospora crassa.* J. Ultrastruct. Res. *48*: 455-464.
13. Allen, E. D., and A. S. Sussman. 1978. Presence of an actin-like protein in the mycelium of *Neurospora crassa.* J. Bacteriol. *135*: 713-716.
14. Anderson, J. G. 1978. Temperature-induced fungal development. In J. E. Smith and D. R. Berry (Eds.), *The filamentous Fungi,* Vol. 3, Wiley, New York, pp. 358-375.
15. Aranjo, M., and A. Restrepo. 1976. Determination of the growth curves of the mycelial and yeast forms of *Paracoccidioides brasiliensis.* Mycopathologia *59*: 163-169.
16. Arnold, W. N., and M. N. McLellan. 1975. Trehalose and glycogen levels during the initial stages of growth of *Candida albicans.* Physiol. Chem. Phys. *7*: 369-380.
17. Arthur, H., K. Watson, C. R. McArthur, and G. D. Clark-Walker. 1978. Naturally occurring respiratory deficient *Candida slooffi* strains resemble petite mutants. Nature (Lond.) *271*: 750-752.
18. Azuma, I., F. Kanetsuna, Y. Tanaka, Y. Yamamura, and L. M. Carbonell. 1974. Chemical and immunological properties of galactomannans obtained from *Histoplasma duboisii, Histoplasma capsulatum, Paracoccidioides brasiliensis,* and *Blastomyces dermatitidis.* Mycopathol. Mycol. Appl. *54*: 111-121.
19. Baenziger, J. V., and D. Fiete. 1979. Structural determinants of concanavalin A specificity for oligosaccharides. J. Biol. Chem. *254*: 2400-2407.
20. Bail, T. 1857. Über Hefe. Flora (Jena) *40*: 417-444.
21. Baker, O., and A. I. Braude. 1956. A study of stimuli leading to the production of spherules in coccidioidomycosis. J. Lab. Clin. Med. *47*: 169-181.
22. Baker, E. E., and E. M. Mrak. 1941. Spherule formation in culture by *Coccidioides immitis* (Rixford and Gilchrist, 1896). Am. J. Trop. Med. Hyg. *21*: 589-596.

23. Bakerspigel, A. 1964. Some observations on the cytology of *Candida albicans*. J. Bacteriol. *87*: 228-230.
24. Bakerspigel, A., and S. Burke. 1974. A possible function of the chlamydospores of *Candida albicans*. Mycopathol. Mycol. Appl. *54*: 147-152.
25. Balish, E. 1973. Methionine biosynthesis and S-adenosylmethionine degradation during an induced morphogenesis of *Candida albicans*. Can. J. Microbiol. *19*: 847-853.
26. Balish, E., and G. Svihla. 1966. Ultraviolet microscopy of *Candida albicans*. J. Bacteriol. *92*: 1812-1820.
27. Ballmann, G. E., and L. Chaffin. 1979. Lipid synthesis during reinitiation of growth from stationary phase cultures of *Candida albicans*. Mycopathologia *67*: 39-43.
28. Ballou, C. F., S. K. Maitia, J. W. Walker, and W. L. Whelan. 1977. Developmental defects associated with glucosamine auxotrophy in *Saccharomyces cerevisiae*. Proc. Natl. Acad. Sci. U.S.A. *74*: 4351-4355.
29. Barlow, A. J. E., T. Aldersly, and F. W. Chattaway. 1974. Factors present in serum and seminal plasma which promote germ-tube formation and mycelial growth in *Candida albicans*. J. Gen. Microbiol. *82*:261-272.
30. Bartels, H. A. 1969. Germ-tube formation in *Candida albicans*. J. Dent. Res. *48*: 230-235.
31. Bartels, P. A., N. Wagoner, and H. W. Larsh. 1968. Conversion of *Coccidioides immitis* in tissue culture. Mycopathol. Mycol. Appl. *35*: 37-41.
32. Bartnicki-Garcia, S. 1963. Symposium on biochemical bases of morphogenesis in fungi. III. Mold-yeast dimorphism of *Mucor*. Bacteriol. Rev. *27*:293-303.
33. Bartnicki-Garcia, S. 1968. Cell wall chemistry, morphogenesis and taxonomy of fungi. Annu. Rev. Microbiol. *22*: 87-108.
34. Bartnicki-Garcia, S. 1968. Control of dimorphism in *Mucor* by hexoses: inhibition of hyphal morphogenesis. J. Bacteriol. *96*: 1586-1594.
35. Bartnicki-Garcia, S. 1973. Fundamental aspects of hyphal morphogenesis. In J. M. Ashworth and J. E. Smith (Eds.), *Microbial Differentiation*, Vol. 23, Cambridge University Press, Cambridge, pp. 245-267.
36. Bartnicki-Garcia, S., and B. Lindberg. 1972. Partial characterization of mucoran: the glucuronomannan component. Carbohydr. Res. *23*: 75-85.
37. Bartnicki-Garcia, S., and E. Lippman. 1969. Fungal morphogenesis: cell wall construction in *Mucor rouxii*. Science *165*: 302-304.
38. Bartnicki-Garcia, S., and W. J. Nickerson. 1961. Thiamine and nicotinic acid: anaerobic growth factors for *Mucor rouxii*. J. Bacteriol. *82*: 142-148.
39. Bartnicki-Garcia, S., and W. J. Nickerson. 1962. Induction of yeast-like development in *Mucor* by carbon dioxide. J. Bacteriol. *84*: 820-840.
40. Bartnicki-Garcia, S., and W. J. Nickerson. 1962. Isolation, composition, and structure of cell walls of filamentous and yeast-like forms of *Mucor rouxii*. Biochim. Biophys. Acta *58*: 102-119.
41. Bartnicki-Garcia, S., and W. J. Nickerson. 1962. Nutrition, growth, and morphogenesis of *Mucor rouxii*. J. Bacteriol. *84*: 841-858.
42. Baum, P., J. Thorner, and L. Honig. 1978. Identification of tubulin from the yeast *Saccharomyces cerevisiae*. Proc. Natl. Acad. Sci. U.S.A. *75*:4962-4966.
43. Bawdon, R. E., and R. G. Garrison. 1974. The uptake and incorporation of leucine and cystine by the mycelial and yeastlike phases of *Blastomyces dermatitidis*. Mycopathol. Mycol. Appl. *54*: 421-433.
44. Bawdon, R. E., R. G. Garrison, and L. R. Fina. 1972. Deoxyribonucleic acid base composition of the yeastlike and mycelial phases of *Histoplasma capsulatum* and *Blastomyces dermatitidis*. J. Bacteriol. *111*: 593-596.
45. Bedell, G. W., and D. R. Soll. 1979. Effects of low concentrations of zinc on the growth and dimorphism of *Candida albicans*: evidence for zinc-resistant and -sensitive pathways for mycelium formation. Infect. Immun. *26*: 348-354.

46. Belcher, R. W., J. F. Carney, and F. G. Monahan. 1973. An electron microscopic study of phagocytosis of *Candida albicans* by polymorphonuclear leukocytes. Lab. Invest. *29*: 620-627.

46a. Bell, W. M., and W. L. Chaffin. 1980. Nutrient-limited yeast growth in *Candida albicans:* effect on yeast-mycelial transition. Can. J. Microbiol. *26*: 102-105.

47. Berkeley, M. J. 1838. On a conferoid state of *Mucor clavatus* Link. Mag. Zool. Bot. *2*: 340-343.

48. Berliner, M. D. 1968. Primary subcultures of *Histoplasma capsulatum*. I. Macro- and micromorphology of the mycelial phase. Sabouraudia *6*: 111-118.

49. Bhattacharya, A., S. Banerjee, and A. Datta. 1974. Regulation of N-acetylglucosamine kinase in yeast. Biochim. Biophys. Acta *374*: 384-391.

50. Bhattacharya, A., and A. Datta. 1977. Effect of cyclic AMP on RNA and protein synthesis. Biochem. Biophys. Res. Commun. *77*: 1438-1444.

51. Bhattacharya, A., M. Puri, and A. Datta. 1974. Induction of N-acetylglucosamine kinase in yeast. Biochem. J. *141*: 593-595.

52. Bianchi, D. E. 1967. The lipid content of cell walls obtained from juvenile, yeast-like and filamentous cells of *Candida albicans*. Antonie Leeuwenhoek J. Microbiol. Serol. *33*: 324-332.

53. Biely, P. 1978. Changes in the rate of synthesis of wall polysaccharides during the cell cycle of yeast. Arch. Microbiol. *119*: 213-214.

54. Biely, P., J. Kovarik, and S. Bauer. 1973. Cell wall formation in yeast: an electron microscopic study. Arch. Microbiol. *94*: 365-371.

55. Bishop, C. T., F. Blank, and P. E. Gardner. 1960. The cell wall polysaccharides of *Candida albicans*: glucan, mannan and chitin. Can. J. Chem. *38*: 869-881.

56. Blank, F. 1957. Note on the chemical composition of the cell wall of *Haplosporangium parvum*. J. Histochem. Cytochem. *5*: 500-502.

57. Blank, F., and R. C. Burke. 1954. Chemical composition of the cell wall of *Coccidioides immitis*. Nature (Lond.) *173*: 829.

58. Boguslawski, G., J. M. Akagi, and L. G. Ward. 1976. Possible role for cysteine biosynthesis in conversion from mycelial to yeast form of *Histoplasma capsulatum*. Nature (Lond.) *261*: 336-338.

59. Boguslawski, G., G. S. Kobayashi, D. Schlessinger, and G. Medoff. 1975. Characterization of an inhibitor of RNA polymerase from the mycelial phase of *Histoplasma capsulatum*. J. Bacteriol. *122*: 532-537.

60. Boguslawski, G., G. Medoff, D. Schlessinger, and G. S. Kobayashi. 1975. Histin, an RNA polymerase inhibitor isolated from *Histoplasma capsulatum*. Biochem. Biophys. Res. Commun. *64*: 625-632.

61. Boguslawski, G., D. Schlessinger, G. Medoff, and G. Kobayashi. 1974. RNA polymerase of the yeast phase of *Histoplasma capsulatum*. J. Bacteriol. *118*: 480-485.

62. Boguslawski, G., W. A. Zehring, R. R. Meyer, and J. S. Parr. 1977. Characterization of a novel ribonucleotide polymerizing enzyme from a fungus *Histoplasma capsulatum*. J. Biol. Chem. *252*: 4337-4346.

63. Borelli, D. 1972. Some ecological aspects of *Paracoccidioides brasiliensis*. In *Paracoccidioidomycosis,* Proceedings of the First Pan American Symposium, Pan Am. Health Organ. Sci. Publ. No. 254, Washington, D.C., pp. 59-64.

64. Borelli, D. 1978. *Aureobasidium werneckii*: acervuli, sporodochia, and pycnidia. In *The Black and White Yeasts,* Proceedings of the Fourth International Conference on the Mycoses, Pan Am. Health Organ. Sci. Publ. No. 356, Washington, D.C., pp. 79-90.

65. Borgia, P., and P. S. Sypherd. 1977. Control of β-glucosidase synthesis in *Mucor racemosus*. J. Bacteriol. *130*: 812-817.

66. Bracker, C. E. 1968. The ultrastructure and development of sporangia in *Gilbertella persicaria.* Mycologia *60*: 1016-1017.
67. Bracker, C. E. 1971. Cytoplasmic vesicles in germinating spores of *Gilbertella persicaria.* Protoplasma *72*: 381-397.
68. Bracker, C. E., J. Ruiz-Herrera, and S. Bartnicki-Garcia. 1976. Structure and transformation of chitin synthetase particles (chitosomes) during microfibril synthesis in vitro. Proc. Natl. Acad. Sci. U.S.A. 4570-4574.
69. Braun, P. C., and R. A. Calderone. 1978. Chitin synthesis in *Candida albicans*: comparison of yeast and hyphal forms. J. Bacteriol. *133*: 1472-1477.
70. Brefeld, O. 1873. *Mucor racemosus* und Hefe. Flora (Jena) *56*: 385-399.
71. Breslau, A. M. 1955. Comparative histochemical studies on *Coccidioides immitis* and *Haplosporangium parvum.* J. Histochem. Cytol. *3*: 141-147.
72. Breslau, A. M., and J. O. Erickson. 1963. Correlation histochemistry and fine-structure in the development of the "in vitro" spherules of *Coccidioides immitis.* Ann. Histochim. *8*: 165-172.
73. Breslau, A. M., T. J. Hensely, and J. O. Erickson. 1961. Electron microscopy of cultured spherules of *Coccidioides immitis.* J. Biophys. Biochem. Cytol. *9*: 627-637.
74. Breslau, A. M., and M. Y. Kubota. 1964. Continuous in vitro cultivation of spherules of *Coccidioides immitis.* J. Bacteriol. *87*: 468-472.
75. Brody, S. 1973. Metabolism, cell walls and morphogenesis. In S. J. Coward (Ed.), *Developmental Regulation*: *Aspects of Cell Differentiation,* Academic, New York, pp. 107-154.
76. Brosbe, E. A. 1967. Use of refined agar for the in vitro propagation of the spherule phase of *Coccidioides immitis.* J. Bacteriol. *93*: 497-498.
77. Brosbe, E. A., J. N. Kietzman, and J. E. Snellen. 1967. Time lapse cinemicrographic studies on behavior of *Coccidioides immitis* in vitro. In L. Ajello (Ed.), *Coccidioidomycosis,* Papers from the Second Symposium on Coccidioidomycosis, University of Arizona Press, Tucson, pp. 409-414.
78. Brown, J. B. 1971. Susceptibility of the cell walls of some yeasts to lysis by enzymes of *Helix pomatia.* Can. J. Microbiol. *17*: 205-208.
79. Brown, R. M. 1974. New observations on membrane flow and product transport in the Golgi apparatus of *Pleurochrysis.* J. Cell Biol. *63*: 39.
80. Buckley, H. R., and N. van Uden. 1963. The identification of *Candida albicans* within two hours by the use of an egg white slide preparation. Sabouraudia *2*: 205-208.
81. Bullen, J. J. 1949. The yeastlike form of *Cryptococcus farciminosus* (Rivolta): *(Histoplasma farciminosum).* J. Pathol. Bacteriol. *61*: 117-120.
82. Burke, R. C. 1951. In vitro cultivation of the parasitic phase of *Coccidioides immitis.* Proc. Soc. Exp. Biol. Med. *76*: 332-335.
83. Byers, B., and L. Goetsch. 1973. Duplication of spindle plaques and integration of the yeast cell cycle. Cold Spring Harbor Symp. Quant. Biol. *38*: 123-131.
84. Byers, B., and L. Goetsch. 1975. Behavior of spindles and spindle plaques in the cell cycle and conjugation of *Saccharomyces cerevisiae.* J. Bacteriol. *124*: 511-523.
85. Byers, B., and L. Goetsch. 1976. A highly ordered ring of membrane associated filaments in budding yeasts. J. Cell Biol. *69*: 717-721.
86. Cabib, E. 1975. Molecular aspects of yeast morphogenesis. Annu. Rev. Microbiol. *29*: 191-214.
87. Cabib, E., and B. Bowers. 1971. Chitin and yeast budding. J. Biol. Chem. *246*: 152-159.
88. Cabib, E., and V. Farkas. 1971. The control of morphogenesis: an enzymatic

mechanism for the initiation of septum formation in yeast. Proc. Natl. Acad. Sci. U.S.A. *68*: 2052-2056.

89. Cabib, E., V. Farkas, R. E. Ulane, and B. Bowers. 1973. Yeast septum formation as a model for morphogenesis. In J. R. Villaneuva, I. Garcia-Acha, S. Gascon and F. Uruburu (Eds.), *Yeast, Mould and Plant Protoplasts,* Academic, New York, pp. 105-116.
90. Cabib, E., and F. A. Keller. 1971. Chitin and yeast budding: allosteric inhibition of chitin synthetase by a heat-stable protein from yeast. J. Biol. Chem. *246*: 167-173.
91. Cabib, E., R. Ulane, and B. Bowers. 1973. Yeast chitin synthetase: separation of the zymogen from its activating factor and recovery of the latter from the vacuole fraction. J. Biol. Chem. *248*: 1451-1458.
92. Cabib, E., R. Ulane, and B. Bowers. 1974. A molecular model for morphogenesis: the primary septum in yeast. In B. L. Horecker and E. R. Stadtman (Eds.), *Current Topics in Cellular Regulations,* Vol. 8, Academic, New York, pp. 1-32.
93. Calderone, R. A. 1976. Endogenous respiration and fatty acids of *Phialophora dermatitidis*. Mycologia *68*: 99-107.
94. Calderone, R. A., M. F. Rotondo, and M. A. Sande. 1978. *Candida albicans* endocarditis: ultrastructural studies of vegetation formation. Infect. Immun. *20*: 279-289.
95. Campbell, C. C. 1947. Reverting *Histoplasma capsulatum* to the yeast phase. J. Bacteriol. *54*: 263-264.
96. Carbonell, L. M. 1963. Ultrastructura del *Paracoccidioides brasiliensis* en cultivos de la fase leva duriforme. Mycopathol. Mycol. Appl. *19*: 184-204.
97. Carbonell, L. M. 1967. Cell wall changes during the budding process of *Paracoccidioides brasiliensis* and *Blastomyces dermatitidis*. J. Bacteriol. *94*: 213-223.
98. Carbonell, L. M. 1969. Intrayeast hyphae in *Paracoccidioides brasiliensis.* J. Bacteriol. *98*: 1395-1396.
99. Carbonell, L. M. 1972. Ultrastructure of *Paracoccidioides brasiliensis.* In *Paracoccidioidomycosis,* Proceedings of the First Pan American Symposium, Pan Am. Health Organ. Sci. Publ. No. 254, Washington, D.C., pp. 21-28.
100. Carbonell, L. M. 1969. Ultrastructure of dimorphic transformation in *Paracoccidioides brasiliensis.* J. Bacteriol. *100*: 1076-1082.
101. Carbonell, L. M., M. D. Berliner, and F. Gil. 1972. Kinetic and morphological observations on the yeast phase of *Histoplasma capsulatum* during protoplast formation. J. Bacteriol. *110*: 731-738.
102. Carbonell, L. M., F. Kanetsuna, and F. Gil. 1970. Chemical morphology of glucan and chitin in the cell wall of the yeast phase of *Paracoccidioides brasiliensis.* J. Bacteriol. *101*: 636-642.
103. Carbonell, L. M., and L. Pollak. 1962. "Myelin figures" in yeast cultures of *Paracoccidioides brasiliensis.* J. Bacteriol. *83*: 1356-1357.
104. Carbonell, L. M., and J. Rodriguez. 1965. Transformation of mycelial and yeast forms of *Paracoccidioides brasiliensis* in cultures and in experimental inoculations. J. Bacteriol. *90*: 504-510.
105. Carbonell, L. M., and J. Rodriguez. 1968. Mycelial phase of *Paracoccidioides brasiliensis* and *Blastomyces dermatitidis:* an electron microscopic study. J. Bacteriol. *96*: 533-543.
106. Carmichael, J. W. 1951. The pulmonary fungus *Haplosporangium parvum.* II. Strain and generic relations. Mycologia *43*: 605-624.
107. Carmichael, J. W. 1962. *Chrysosporium* and some other aleuriosporic hyphomycetes. Can. J. Bot. *40*: 1137-1173.

108. Carrión, A. L. 1950. Yeast-like dematiaceous fungi infecting the human skin. Arch. Dermatol. Syphilol. *61*: 966-1009.

109. Carrión, A. L., and M. Silva. 1947. Chromoblastomycosis and its etiologic agents. Ann. Cryptogam. Phytopathol. *6*: 20-62.

110. Cassone, A., D. Kerridge, and E. F. Gale. 1979. Ultrastructural changes in the cell wall of *Candida albicans* following cessation of growth and their possible relationship to the development of polyene resistance. J. Gen. Microbiol. *110*: 339-349.

111. Cassone, A., E. Mattia, and L. Boldrini. 1978. Agglutination of blastospores of *Candida albicans* by concanavalin A and its relationship with the distribution of mannan polymers and the ultrastructure of the cell wall. J. Gen. Microbiol. *105*: 263-273.

112. Cassone, A., N. Simonetti, and V. Strippoli. 1973. Ultrastructural changes in the wall during germ-tube formation from blastospores of *Candida albicans*. J. Gen. Microbiol. *77*: 417-426.

113. Cassone, A., N. Simonetti, and V. Strippoli. 1974. Production, ultrastructure, and germination of *Candida albicans* chlamydospores. In P. Gerhardt, R. N. Costilow, and H. L. Sadoff (Eds.), *Spores VI,* American Society for Microbiology, Washington, D.C., pp. 172-178.

114. Chaffin, W. L., and S. J. Sogin. 1976. Germ tube formation from zonal rotor fractions of *Candida albicans*. J. Bacteriol. *126*: 771-776.

115. Chan, J. Y. H., R. Loor, and T. Y. Wang. 1976. Transcription of Ehrlich ascites tumor DNA and chromatin by purified homologous RNA polymerase II (or B). Arch. Biochem. Biophys. *173*: 564-576.

116. Chattaway, F. W., R. Bishop, M. R. Holmes, and F. C. Odds. 1973. Enzyme activities associated with carbohydrate synthesis and breakdown in the yeast and mycelial forms of *Candida albicans*. J. Gen. Microbiol. *75*: 97-109.

117. Chattaway, F. W., and M. R. Holmes. 1968. Cell wall composition of the mycelial and blastospore forms of *Candida albicans*. J. Gen. Microbiol. *51*: 367-376.

118. Chattaway, F. W., F. C. Odds, and A. J. E. Barlow. 1971. An examination of the production of hydrolytic enzymes and toxins by pathogenic strains of *Candida albicans*. J. Gen. Microbiol. *67*: 255-263.

119. Chattaway, F. W., J. O'Reilly, A. J. E. Barlow, and T. Aldersly. 1976. Induction of the mycelial form of *Candida albicans* by hydrolysates of peptides from seminal plasma. J. Gen. Microbiol. *96*: 317-322.

120. Chattaway, F. W., S. Shenolikar, J. O'Reilly, and A. J. E. Barlow. 1976. Changes in the cell surface of the dimorphic forms of *Candida albicans* by treatment with hydrolytic enzymes. J. Gen. Microbiol. *95*: 335-347.

121. Cheung, S. C., G. S. Kobayashi, and G. Medoff. 1976. Antibiotic sensitivity of the yeast and mycelial phases of *Histoplasma capsulatum*. Sabouraudia *14*: 99-103.

122. Cheung, S. C., G. S. Kobayashi, D. Schlessinger, and G. Medoff. 1974. RNA metabolism during morphogenesis in *Histoplasma capsulatum*. J. Gen. Microbiol. *82*: 301-307.

123. Cheung, S. C., G. Medoff, D. Schlessinger, and G. S. Kobayashi. 1975. Response of yeast and mycelial phases of *H. capsulatum* to amphotericin B and actinomycin D. Antimicrob. Agents Chemother. *8*: 498-503.

124. Cheung, W. Y. 1980. Calmodulin plays a pivotal role in cellular regulation. Science *207*: 19-27.

125. Ciferri, R., and A. Montemartini. 1958. Taxonomy of *Haplosporangium parvum*. Mycopathol. Mycol. Appl. *10*: 303-316.

126. Cino, P. M., and R. P. Tewari. 1972. Chemical composition of *Histoplasma capsulatum*. Mycopathol. Mycol. Appl. *47*: 285-294.

127. Clark-Walker, G. D. 1972. Development of respiration and mitochondria in *Mucor genevensis* after anaerobic growth: absence of glucose repression. J. Bacteriol. *109*: 399-408.

128. Clark-Walker, G. D. 1973. Relationship between dimorphology and respiration in *Mucor genevensis* studied with chloramphenicol. J. Bacteriol. *116*: 972-980.

129. Clayton, L., C. I. Pogsan, and K. Gull. 1979. Microtubule proteins in the yeast, *Saccharomyces cerevisiae*. FEBS Lett. *106*: 67-70.

130. Cole, G. T. 1978. Conidiogenesis in the black yeasts. In *The Black and White Yeasts,* Proceedings of the Fourth International Conference on the Mycoses, Pan Am. Health Organ. Sci. Publ. No. 356, Washington, D.C., pp. 66-78.

131. Collins, M., and D. Pappagianis. 1973. Effects of lysozyme and chitinase on the spherules of *Coccidioides immitis* in vitro. Infect. Immun. *7*: 817-822.

132. Conant, N. F. 1941. A cultural study of the life cycle of *Histoplasma capsulatum* Darling, 1906. J. Bacteriol. *41*: 563-579.

133. Conant, N. F., and A. Howell. 1941. Etiological agents of North and South American blastomycosis. Proc. Soc. Exp. Biol. Med. *46*: 426-428.

134. Conant, N. F., and A. Howell. 1942. Similarity of the fungi causing South American blastomycosis (paracoccidioidal granuloma) and North American blastomycosis (Gilchrist's disease). J. Invest. Dermatol. *5*: 353-370.

135. Conant, N. F., and R. A. Vogel. 1954. The parasitic growth phase of *Coccidioides immitis* in culture. Mycologia *46*: 157-160.

136. Converse, J. L. 1955. Growth of spherules of *Coccidioides immitis* in a chemically defined medium. Proc. Soc. Exp. Biol. Med. *90*: 709-711.

137. Converse, J. L. 1956. Effect of physicochemical environment on spherulation of *Coccidioides immitis* in a chemically defined medium. J. Bacteriol. *72*: 784-792.

138. Converse, J. L. 1957. Effect of surface active agents on endosporulation of *Coccidioides immitis* in a chemically defined medium. J. Bacteriol. *74*: 106-107.

139. Converse, J. L., and A. R. Besemer. 1959. Nutrition of the parasitic phase of *Coccidioides immitis* in a chemically defined liquid medium. J. Bacteriol. *78*: 231-239.

140. Cooper, B. H., S. Grove, C. Mims, and P. J. Szaniszlo. 1973. Septal ultrastructure in *Phialophora pedrosoi, Phialophora verrucosa* and *Cladosporium carrionii.* Sabouraudia *11*: 127-130.

141. Cortat, M., P. Matile, and A. Wiemkin. 1972. Isolation of glucanase-containing vesicles from budding yeast. Arch. Microbiol. *82*: 189-205.

142. Crow, S. A., P. I. Bowman, and D. G. Ahearn. 1977. Isolation of atypical *Candida albicans* from the North Sea. Appl. Environ. Microbiol. *33*: 738-739.

143. Cueva, J. A., and M. D. Little. 1971. *Emmonsia crescens* infection (adiaspiromycosis) in man in Honduras. Am. J. Trop. Med. Hyg. *20*: 282-287.

144. Cutler, J. E., and K. H. Erke. 1971. Ultrastructural characteristics of *Coccidioides immitis,* a morphological variant of *Cryptococcus neoformans* and *Podosypha ravenelii.* J. Bacteriol. *105*: 438-444.

145. Dabrowa, N., D. H. Howard, J. W. Landau, and Y. Schechter. 1970. Synthesis of nucleic acids and proteins in the dimorphic forms of *Candida albicans.* Sabouraudia *8*: 163-169.

146. Dabrowa, N., S. S. S. Taxer, and D. H. Howard. 1976. Germination of *Candida albicans* induced by proline. Infect. Immun. *13*: 830-835.

147. Dargent, R. 1975. Sur l'ultrastructure des hyphes en croissance de l'*Achlya bisexualis* Coker. Mise en évidence d'une secrétion polysaccharidique et d'une activitié, phosphatasique alkaline dans l'appareil de Golgi et au niveau des vesicles cytoplasmiques apicales. C. R. Acad. Sci., Ser. D. *280*: 1445-1448.

148. Dargent, R., and J. Denisse. 1976. Sur l'ultrastructure apicale de l'*Achlya bisexualis* Coker. Mise en évidence d'une activité phosphatasique acide. C. R. Acad. Sci., Ser. D *282*: 1601-1604.
149. Darken, M. 1962. Absorption and transport of fluorescent brighteners by microorganisms. Appl. Microbiol. *10*: 387-394.
150. Datema, R., J. G. H. Wessels, and H. Van Den Ende. 1977. The hyphal wall of *Mucor mucedo*. Eur. J. Biochem. *80*: 621-626.
151. Daubert, S., D. Peters, and M. E. Dahmus. 1977. Selective transcription of ribosomal sequences in vitro by RNA polymerase I. Arch. Biochem. Biophys. *178*: 381-386.
152. Davies, R. R., and T. J. V. Denning. 1972. Growth and form in *Candida albicans.* Sabouraudia *10*: 180-188.
153. Davies, R. R., and M. A. Savage. 1974. Observations on 5-fluorocytosine and *Candida albicans.* Sabouraudia *12*: 302-308.
154. Davies, T. E., J. Domer, and Y. T. Li. 1977. Cell wall studies of *Histoplasma capsulatum* and *Blastomyces dermatitidis.* J. Bacteriol. *107*: 870-877.
155. De Deken, R. H. 1966. The Crabtree effect: a regulatory system in yeast. J. Gen. Microbiol. *44*: 149-156.
156. de Hoog, G. S. 1977: *Rhinocladiella* and allied genera. In *Studies in Mycology,* Vol. 15, Centraalbureau voor Schimmelcultures, Baarn, p. 118.
157. DeMonbreun, W. A. 1939. The dog as a natural host for *Histoplasma capsulatum.* Am. J. Trop. Med. *19*: 565-588.
158. Detroy, R. W., and A. Ciegler. 1971. Induction of yeastlike development in *Aspergillus parasiticus.* J. Gen. Microbiol. *65*: 259-264.
159. Diecks-Ventling, C., and F. Bieri-Bonniot. 1976. DNA-dependent RNA polymerase I and II activation by 17-β-estradiol-receptor complex on liver chromatin from immature chicks. In D. P. Nierlich, W. J. Rutter, and C. F. Fox (Eds.), *Molecular Mechanism in the Control of Gene Expression,* Academic, New York, pp. 337-347.
160. Djaczenko, W., and A. Cassone. 1971. Visualization of new ultrastructural components in the cell wall of *Candida albicans* with fixatives containing TAPO. J. Cell Biol. *52*: 186-190.
161. Dodd, K., and E. H. Tompkins. 1934. A case of histoplasmosis of Darling in an infant. Am. J. Trop. Med. *14*: 127-137.
162. Domer, J. E. 1971. Monosaccharide and chitin content of cell walls of *Histoplasma capsulatum* and *Blastomyces dermatitidis.* J. Bacteriol. *107*: 870-877.
163. Domer, J. E., and J. G. Hamilton. 1971. The readily extracted lipids of *Histoplasma capsulatum* and *Blastomyces dermatitidis.* Biochim. Biophys. Acta *231*: 465-478.
164. Domer, J. E., J. G. Hamilton, and J. C. Harkin. 1967. Comparative study of the cell walls of the yeastlike and mycelial phases of *Histoplasma capsulatum.* J. Bacteriol. *94*: 466-474.
165. Donnely, W. H., and E. J. Yunis. 1974. The ultrastructure of *Coccidioides immitis.* Arch. Pathol. *98*: 227-232.
166. Dow, J. M., and P. H. Rubery. 1977. Chemical fractionation of the cell walls of mycelial and yeast-like forms of *Mucor rouxii:* a comparative study of the polysaccharide and glycoprotein components. J. Gen. Microbiol. *99*: 29-41.
167. Dowding, E. S. 1947. The pulmonary fungus. *Haplosporangium parvum* and its relationship with some human pathogens. Can. J. Res. *25*: 195-206.
168. Dowding, E. S. 1948. The spores of *Histoplasma capsulatum.* Can. J. Res. *26*: 265-273.
169. Drouhet, E., and F. Mariat. 1952. Etude des facteurs determinant le développe-

ment de la phase levure de *Sporotrichum schenckii.* Ann. Inst. Pasteur (Paris) *87*: 506-515.
170. Drouhet, E., and R. C. Zapater. 1954. Phase levure et filamenteuse de *Paracoccidioides brasiliensis;* et de des noyaux. Ann. Inst. Pasteur (Paris) *87*: 396-403.
171. Duran, A., B. Bowers, and E. Cabib. 1975. Chitin synthetase zymogen is attached to the yeast plasma membrane. Proc. Natl. Acad. Sci. U.S.A. *72*: 3952-3955.
172. Duran, A., and E. Cabib. 1978. Solubilization and partial purification of yeast chitin synthetase: confirmation of the zymogenic nature of the enzyme. J. Biol. Chem. *253*: 4419-4425.
173. Duran, A., E. Cabib, and B. Bowers. 1979. Chitin synthetase distribution on the yeast plasma membrane. Science. *203*: 363-365.
174. Dutt-Choudhuri, R., and R. Dutt. 1961. A simple medium for the demonstration of chlamydospores of *Candida albicans.* Nature (Lond.) *189*: 418.
175. Edwards, M. R., E. L. Hazen, and G. A. Edwards. 1959. The fine structures of the yeastlike cells of *Histoplasma capsulatum* in culture. J. Gen. Microbiol. *20*: 496-503.
176. Elmer, G. W., and W. J. Nickerson. 1970. Filamentous growth of *Mucor rouxii* under nitrogen. J. Bacteriol. *101*: 592-594.
177. Elmer, G. W., and W. J. Nickerson. 1970. Nutritional requirements for growth and yeast-like development of *Mucor rouxii* under carbon dioxide. J. Bacteriol. *101*: 595-602.
178. Emmons, C. W. 1964. Budding in *Emmonsia crescens.* Mycologia *56*: 415-419.
179. Emmons, C. W., C. H. Binford, J. P. Utz, and K. J. Kwon-Chung. 1977. *Medical Mycology,* Lea & Febiger, Philadelphia.
180. Evans, E. G. V., F. C. Odds, M. D. Richardson, and K. T. Holland. 1974. The effect of growth medium on filament production in *Candida albicans.* Sabouraudia *12*: 112-119.
181. Evans, E. G. V., F. C. Odds, and K. T. Holland. 1975. Resistance of the *Candida albicans* filamentous cycle to environmental change. Sabouraudia *13*: 231-238.
182. Evans, Z. A., and D. N. Mardon. 1977. Organ localization in mice challenged with a typical *Candida albicans* strain and a pseudohyphal variant. Proc. Soc. Exp. Biol. Med. *155*: 234-238.
183. Evron, R., and S. Ganor. 1968. The use of sodium taurocholate medium for identifying *Candida albicans.* J. Invest. Dermatol. *51*: 108-110.
184. Fischer, D. J. 1977. Induction of yeast-like growth in mucorales by systemic fungicides and other compounds. Trans. Br. Mycol. Soc. *68*: 397-402.
185. Fischer, D. J., R. J. Pring, D. V. Richomond, and A. Michael. 1973. Production of yeastlike forms of *Mucor* by antibiotics and fungicides. J. Gen. Microbiol. *75*: xiv.
186. Fitz, A. 1873. Über alkoholische Gahrung durch *Mucor mucedo.* Ber. Dtsch. Chem. Ges. *2*: 48-58.
187. Flores-Carreon, A., A. Gomez-Villaneuva, and G. San-Blas. 1979. β-1,3-Glucanase and dimorphism in *Paracoccidioides brasiliensis.* Antonie Leeuwenhoek J. Microbiol. Serol. *45*: 265-274.
188. Flores-Carreon, A., E. Reyes, and J. Ruiz-Herrera. 1970. Inducible cell wall-bound α-glucosidase in *Mucor rouxii.* Biochim. Biophys. Acta *222*: 354-360.
189. Forer, A. 1974. Possible roles of microtubules and actin-like filaments during cell division. In G. M. Padilla, I. L. Cameron, and A. Zimmerman (Eds.), *Cell Cycle Controls,* Academic, New York, pp. 319-336.
190. Friedenthal, M., A. Epstein, and S. Passeron. 1974. Effect of potassium cyanide, glucose and anaerobiosis on morphogenesis of *Mucor rouxii.* J. Gen. Microbiol. *84*: 15-24.

191. Fukuda, R., G. Keilman, E. McVey, and R. H. Doi. 1975. Ribonucleic acid polymerase pattern of sporulating *Bacillus subtilis* cells. In P. Gerhardt, R. N. Costilow and H. L. Sadoff (Eds.), *Spores IV,* American Society for Microbiology, Washington, D.C., pp. 213-220.
192. Furcolow, M. L., and C. D. Smith. 1973. A new hypothesis on the epidemiology of blastomycosis and the ecology of *Blastomyces dermatitidis.* Trans. N.Y. Acad. Sci. II *35*: 421-430.
193. Furtado, J. S., T. de Brito, and E. Freymuller. 1967. The structure and reproduction of *Paracoccidioides brasiliensis* in human tissue. Sabouraudia *5*: 226-229.
194. Gale, E. F., A. M. Johnson, D. Kerridge, and T. Y. Kon. 1975. Factors affecting the changes in amphotericin sensitivity of *Candida albicans* during growth. J. Gen. Microbiol. *87*: 20-36.
195. Garland, D. L. 1979. cAMP inhibits the in vitro assembly of microtubules. Arch. Biochem. Biophys. *198*: 335-337.
196. Garrison, R. G., and K. S. Boyd. 1971. Ultrastructural studies of induced morphogenesis by *Aspergillus parasiticus.* Sabouraudia *12*: 179-187.
197. Garrison, R. G., and K. S. Boyd. 1975. Aspects of the dimorphism of *Histoplasma farciminosum:* a light and electron microscopic study. Sabouraudia *13*: 174-184.
198. Garrison, R. G., and K. S. Boyd. 1977. Electron microscopy of yeast cell development from the microconidium of *Histoplasma capsulatum.* Abstr. Annu. Meet. Am. Soc. Microbiol., 1977, F18, p. 121.
199. Garrison, R. G., and K. S. Boyd. 1977. The fine structure of microconidial germination and vegetative cells of *Histoplasma capsulatum.* Ann. Microbiol. (Paris) *128A*: 137-149.
200. Garrison, R. G., and K. S. Boyd. 1978. Electron microscopy of yeast-like cell development from the microconidium of *Histoplasma capsulatum.* J. Bacteriol. *133*: 345-353.
201. Garrison, R. G., and K. S. Boyd. 1978. Role of the conidium in dimorphism of *Blastomyces dermatitidis.* Mycopathologia *64*: 29-33.
202. Garrison, R. G., K. S. Boyd, and F. Mariat. 1975. Ultrastructural studies of the mycelium-to-yeast transformation of *Sporothrix schenckii.* J. Bacteriol. *124*: 959-968.
203. Garrison, R. G., H. T. Dodd, and J. W. Hamilton. 1970. Uptake of low molecular weight sulfur-containing compounds by *Histoplasma capsulatum* and related dimorphic fungi. Mycopathol. Mycol. Appl. *40*: 171-180.
204. Garrison, R. G., J. W. Lane, and M. F. Field. 1970. Ultrastructural changes during the yeastlike to mycelial phase conversion of *Blastomyces dermatitidis* and *Histoplasma capsulatum.* J. Bacteriol. *101*: 628-635.
205. Garrison, R. G., J. W. Lane, and D. R. Johnson. 1971. Electron microscopy of the transitional conversion cell of *Histoplasma capsulatum.* Mycopathol. Mycol. Appl. *44*: 121-129.
206. Gilardi, G. L. 1965. Nutrition of systemic and subcutaneous pathogenic fungi. Bacteriol. Rev. *29*: 406-424.
207. Gilardi, G. L., and N. C. Laffer. 1962. Nutritional studies on the yeast phase of *Blastomyces dermatitidis* and *B. brasiliensis.* J. Bacteriol. *83*: 219-227.
208. Gilbert, B. E., and D. H. Howard. 1970. Incorporation of cystine by yeast cells of *Histoplasma capsulatum.* Infect. Immun. *2*: 169-174.
209. Gilbert, B. E., and D. H. Howard. 1970. Uptake of cystine by the yeast phase of *Histoplasma capsulatum.* Infect. Immun. *2*: 139-144.
210. Gilchrist, T. C., and W. R. Stokes. 1898. A case of pseudo-lupus vulgaris caused by a blastomyces. J. Exp. Med. *3*: 53-78.

211. Girbart, M. 1969. Die Ultrastruktur der Apekalregion von Pilzhyphen. Protoplasma *67*: 413-441.
212. Gjerset, R. A., H. Biessman, B. Levy, and B. J. McCarthy. 1976. Relationship between chromatin structure and transcription. In D. P. Neirlich, W. J. Rutter, and C. F. Fox (Eds.), *Molecular Mechanisms in the Control of Gene Expression,* Academic, New York, pp. 279-307.
213. Gooday, G. W. 1971. An autoradiographic study of hyphal growth of some fungi. J. Gen. Microbiol. *67*: 123-133.
214. Gooday, G. W. 1978. The enzymology of hyphal growth. In J. E. Smith and D. R. Berry (Eds.), *The Filamentous Fungi,* Vol. 3, Wiley, New York, pp. 51-77.
215. Goos, R. D. 1964. Germination of the macroconidium of *Histoplasma capsulatum.* Mycologia *56*: 662-671.
216. Goos, R. D., and D. F. Summers. 1964. Use of fluorescent antibody techniques in observations on morphogenesis of fungi. Mycologia *56*: 701-707.
217. Grappel, S. F., and R. A. Calderone. 1976. Effect of antibodies on the respiration and morphology of *Candida albicans.* Sabouraudia *14*: 51-60.
218. Greenaway, W., and S. Ward. 1978. Use of blue dextran to measure interhyphal spaces in pellets of fungal mycelium. Trans. Br. Mycol. Soc. *70*: 309-311.
219. Gresham, G. A., and C. H. Whittle. 1961. Studies on the invasive, mycelial form of *Candida albicans.* Sabouraudia *1*: 30-33.
220. Griffiths, D. A. 1974. The origin, structure and function of chlamydospores in fungi. Nova Hedwigia *25*: 503-546.
221. Grimaldi de Jimenez, G., F. Yegres, and F. San-Blas. 1979. Liberation of protoplasts from the yeast-like form of *Paracoccidioides brasiliensis* by enzymatic treatment. Sabouraudia *17*: 279-285.
222. Grove, S. N. 1972. Apical vesicles in germinating conidia of *Aspergillus parasiticus.* Mycologia *64*: 638-641.
223. Grove, S. N. 1978. The cytology of hyphal tip growth. In J. E. Smith and D. R. Berry (Eds.), *The Filamentous Fungi,* Vol. 3, Wiley, New York, pp. 28-50.
224. Grove, S. N., and C. E. Bracker. 1970. Protoplasmic organization of hyphal tips among fungi: vesicles and Spitzenkörper. J. Bacteriol. *104*: 989-1009.
225. Grove, S. N., and C. E. Bracker. 1979. Protoplasmic changes during zoospore encystment and cyst germination in *Pythium aphanidermatum.* Exp. Mycol. *2*: 51-98.
226. Grove, S. N., C. E. Bracker, and D. J. Morre. 1968. Cytomembrane differentiation in the endoplasmic reticulum-Golgi apparatus-vesicle complex. Science *161*: 171-173.
227. Grove, S. N., C. E. Bracker, and D. J. Morre. 1970. An ultrastructural basis for hyphal tip growth in *Pythium ultimum.* Am. J. Bot. *59*: 245-266.
228. Grove, S. N., K. B. Oujezdsky, and P. J. Szaniszlo. 1973. Budding in the dimorphic fungus *Phialophora dermatitidis.* J. Bacteriol. *115*: 323-329.
229. Gupta, R. K., and D. H. Howard. 1971. Comparative physiological studies of the yeast and mycelial forms of *H. capsulatum:* uptake and incorporation of L-leucine. J. Bacteriol. *105*: 690-700.
230. Gustafson, R. A., R. V. Hardcastle, and P. J. Szaniszlo. 1975. Budding in the dimorphic fungus *Cladosporium werneckii.* Mycologia *67*: 942-951.
231. Guzik, G., and K. M. Anderson. 1976. Cyclic nucleotides and other factors affecting the apparent activity of RNA polymerases solubilized from rat mammary gland nuclei. Int. J. Biochem. *7*: 607-614.
232. Haidle, C. W., and R. Storck. 1966. Control of dimorphism in *Mucor rouxii.* J. Bacteriol. *92*: 1236-1244.
233. Haidle, C. W., and R. Storck. 1966. Inhibition by cycloheximide of protein and RNA synthesis in *Mucor rouxii.* Biochem. Biophys. Res. Commun. *22*: 175-180.

234. Hall, M. J., and N. Kolankaya. 1974. The physiology of mould-yeast dimorphism in the genus *Mycotypha* (Mucorales). J. Gen. Microbiol. *82*: 25-34.
235. Halliday, W. J., and E. McCoy. 1955. Biotin as a growth requirement of *Blastomyces dermatitidis*. J. Bacteriol. *70*: 464-468.
236. Halvorson, H. O., D. Kabak, S. Sogin, E. M. Sajdel-Sulkowska, and I. Takano. 1974. Control of RNA synthesis in yeast. Can. J. Genet. Cytol. *16*: 751-764.
237. Hamacek, F., F. Kyntera, J. Dvorak, and M. Otcenasek. 1971. Influence of some factors on the growth and development of adiaspores of the fungus *Emmonsia crescens* Emmons et Jellison 1960. Mycopathol. Mycol. Appl. *43*: 201-209.
238. Hamburger, W. W. 1907. A comparative study of four strains of organisms isolated from four cases of generalized blastomycosis. J. Infect. Dis. *1*: 201-209.
239. Hammerschlag, R. S., and H. D. Sisler. 1973. Benomyl and methyl-2-benzimidazole carbamate (MBC): biochemical, cytological and chemical aspects of toxicity to *Ustilago maydis* and *Saccharomyces cerevisiae*. Pestic. Biochem. Physiol. *3*: 42-54.
240. Hansmann, G. H., and J. R. Schenken. 1933. New disease caused by a yeastlike organism. Science *77*: 460.
241. Hansmann, G. H., and J. R. Schenken. 1934. A unique infection in man caused by a new yeastlike organism, a pathogenic member of the genus *Sepedonium*. Am. J. Pathol. *10*: 731-738.
242. Hardcastle, R. V., and P. J. Szaniszlo. 1974. Characterization of dimorphism in *Cladosporium werneckii*. J. Bacteriol. *119*: 294-302.
243. Hartwell, L. H. 1971. Genetic control of the cell division cycle in yeast. II. Genes controlling DNA replication and its initiation. J. Mol. Biol. *59*: 183-194.
244. Hartwell, L. H. 1974. *Saccharomyces cerevisiae* cell cycle. Bacteriol. Rev. *38*: 164-198.
245. Hartwell, L. H. 1976. Sequential function of gene products relative to DNA synthesis in the yeast cell cycle. J. Mol. Biol. *104*: 803-817.
246. Hartwell, L. H., J. Culotti, J. R. Pringle, and B. J. Reid. 1974. Genetic control of the cell division cycle in yeast. Science *183*: 46-51.
247. Hartwell, L. H., R. K. Mortimer, J. Culotti, and M. Culotti. 1973. Genetic control of the cell division cycle in yeast. V. Genetic analysis of *cdc* mutants. Genetics *74*: 267-286.
248. Hasenclever, H. F., and W. O. Mitchell. 1961. Antigenic studies of *Candida*. I. Observations of two antigenic groups in *Candida albicans*. J. Bacteriol. *82*: 570-573.
249. Hayes, A. B. 1966. Chlamydospore production in *Candida albicans*. Mycopathol. Mycol. Appl. *29*: 87-96.
250. Hazen, K. C., and J. E. Cutler. 1979. Autoregulation of germ tube formation by *Candida albicans*. Infect. Immun. *24*: 661-666.
251. Heath, I. B. 1975. The effect of antimicrotubule agents on the growth and ultrastructure of the fungus *Saprolegnia fecax* and their ineffectiveness in disrupting hyphal microtubules. Protoplasma *85*: 147-176.
252. Heath, I. B., J. L. Gay, and A. D. Greenwood. 1971. Cell wall formation in the Saprolegniales: cytoplasmic vesicles underlying developing walls. J. Gen. Microbiol. *65*: 225-232.
253. Heath, B., and M. C. Heath. 1978. Microtubules and organelle movements in the rust fungus *Uromyces phaseoli* var. *orgnae*. Cytobiology *16*: 393-411.
254. Hejtmanek, M. 1976. Scanning electron microscopy of experimental adiaspiromycosis. Mycopathologia *58*: 91-95.
255. Hejtmanek, M., and J. Bartek. 1976. Mutants of *Emmonsia crescens*: their pathogenicity and size of adiaspores in vivo. Folia Microbiol. *21*: 297-300.

256. Hejtmanek, M., O. Necas, and M. Havelkova. 1974. The ultrastructure of the spherules of *Emmonsia crescens* studied by the technique of freeze-etching. Mycopathol. Mycol. Appl. *54*: 79-83.
257. Hejtmanek, M., O. Necas, and M. Havelkova. 1976. Ultrastructure of *Emmonsia crescens* studied by the technique of freeze-etching. Acta Univ. Palackl. Olomuc. Fac. Med. *97*: 73-75.
258. Hempel, H., and N. L. Goodman. 1975. Rapid conversion of *Histoplasma capsulatum, Blastomyces dermatitidis* and *Sporothrix schenckii* in tissue culture. J. Clin. Microbiol. *1*: 420-424.
259. Hendry, A. T. 1973. Inhibition of catalase activity of *Candida albicans* by serum. Sabouraudia *10*: 193-204.
260. Hereford, L. M., and L. H. Hartwell. 1974. Sequential gene function in the initiation of *Saccharomyces cerevisiae* DNA synthesis. J. Mol. Biol. *84*: 445-461.
261. Hermanides-Nijhof, E. J. 1977. *Aureobasidium* and allied genera. In *Studies in Mycology,* Vol. 15, Centraalbureau voor Schimmelcultures, Baarn, p. 174.
262. Hess, S. L., P. J. Allen, D. Nelson, and H. Lester. 1975. Mode of action of methyl cis-ferulate, the self-inhibitor of stem rust uredospore germination. Physiol. Plant Pathol. *5*: 107-112.
263. Hill, D. W., and L. P. Gebhardt. 1956. Morphological transformation of *Candida albicans* in tissues of mice. Proc. Soc. Exp. Biol. Med. *96*: 640-644.
264. Holloway, S. A., and I. B. Heath. 1979. Morphogenesis and the role of microtubules in synchronous populations of *Saprolegnia* zoospores. Exp. Mycol. *1*: 9-29.
265. Houston, M. R., K. H. Meyer, N. Thomas, and F. T. Wolf. 1969. Dimorphism in *Cladosporium werneckii.* Sabouraudia *3*: 195-198.
266. Howard, D. H. 1959. Observations on tissue cultures of murine peritonial exudates inoculated with *Histoplasma capsulatum.* J. Bacteriol. *78*: 69-78.
267. Howard, D. H. 1961. Dimorphism of *Sporotrichum schenckii.* J. Bacteriol. *81*: 464-469.
268. Howard, D. H. 1962. The morphogenesis of the parasitic form of the dimorphic fungi. Mycopathol. Mycol. Appl. *18*: 127-139.
269. Howard, D. H. 1973. Further studies on the inhibition of *Histoplasma capsulatum* within macrophages from immunized animals. Infect. Immun. *8*: 577-581.
270. Howard, D. H. 1974. Intracellular behavior of *Histoplasma capsulatum.* J. Bacteriol. *87*: 33-38.
271. Howard, D. H., and R. L. Herndon. 1960. Tissue cultures of mouse peritoneal exudates inoculated with *Blastomyces dermatitidis.* J. Bacteriol. *80*: 522-527.
272. Howard, D. H., and G. F. Orr. 1963. Comparison of strains of *Sporothrix schenckii* isolated from nature. J. Bacteriol. *85*: 816-821.
273. Hsieh, P. H. 1974. Growth and developmental characteristics in *Phialophora dermatitidis* and other chromoblastomycotic fungi. M.A. thesis, University of Texas.
274. Hu, F., C. S. Livingood, P. Johnson, and C. M. Pomerat. 1954. Tissue culture studies on human skin. Arch. Dermatol. Syphilol. *70*: 1-15.
275. Inderlied, C. B., and P. S. Sypherd. 1978. Glucose metabolism and dimorphism in *Mucor.* J. Bacteriol. *133*: 1282-1286.
276. Jackson, V., A. Shires, R. Chalkley, and D. K. Granner. 1975. Studies on highly metabolically active acetylation and phosphorylation of histones. J. Biol. Chem. *250*: 4856-4863.
277. Jacobs, C. W., and P. J. Szaniszlo. 1979. Cell wall associated changes during early yeastlike to multicellular conversion in the polymorphic fungus *Wangiella dermatitidis.* Abstr. Annu. Meet. Am. Soc. Microbiol., 1979, F10, p. 364.

278. Jansons, V. K., and W. J. Nickerson. 1970. Chemical composition of chlamydospores of *Candida albicans.* J. Bacteriol. *104*: 922-932.
279. Jansons, V. K., and W. J. Nickerson. 1970. Induction, morphogenesis, and germination of the chlamydospore of *Candida albicans.* J. Bacteriol. *104*: 910-921.
280. Jansons, V. K., and J. A. Paktor. 1977. The effect of temperature on concanavalin A-mediated agglutination of cells with rigid receptors. Biochim. Biophys. Acta *467*: 321-326.
281. Jayakumar, S., M. Singh, and R. Prasad. 1978. Characteristics of proline transport in normal and starved cells of *Candida albicans.* Biochim. Biophys. Acta *514*: 348-355.
282. Jellison, L. W., L. Glesne, and C. Owen. 1962. The passive cutaneous anaphylaxis (PCA) test in adiaspiromycosis. Mycologia *54*: 466-475.
283. Jiresova, M., and J. Janacek. 1977. The effect of *Escherichia coli* DNA-dependent RNA polymerases on DNA of different origin: effect of cyclic AMP. Folia Microbiol. *22*: 168-172.
284. Johnson, B. F. 1965. Autoradiographic analysis of regional cell wall growth of yeasts. *Schizosaccharomyces pombe.* Exp. Cell Res. *39*: 613-624.
285. Johnson, B. F. 1968. Lysis of yeast cell walls induced by 2-deoxyglucose at their sites of glucan synthesis. J. Bacteriol. *95*: 1169-1172.
286. Johnson, B. F., and E. J. Gibson. 1966. Autoradiographic analysis of regional cell wall growth of yeasts. III. *Saccharomyces cerevisiae.* Exp. Cell Res. *41*: 580-591.
287. Johnson, S. A. M. 1954. *Candida (Monilia) albicans.* Effect of amino acids, glucose, pH, chlorotetracycline (aureomycin), dibasic sodium and calcium phosphates, and anaerobic and aerobic conditions on its growth. Arch. Dermatol. Syphilol. *70*: 49-60.
288. Johnston, G. C., J. R. Pringle, and L. H. Hartwell. 1977. Coordination of growth with cell division in the yeast *Saccharomyces cerevisiae.* Exp. Cell Res. *105*: 79-98.
289. Johnston, G. C., R. A. Singer, and E. S. McFarlane. 1977. Growth and cell division during nitrogen starvation of the yeast *Saccharomyces cerevisiae.* J. Bacteriol. *132*: 723-730.
290. Joshi, K. R., J. B. Gavin, and D. A. Bremner. 1973. The formation of germ tubes by *Candida albicans* in various peptone media. Sabouraudia *11*: 259-262.
291. Joshi, K. R., D. A. Bremner, J. B. Gavin, P. B. Herdson, and D. N. Parr. 1973. The formation of germ tubes by *Candida albicans* in sheep serum and trypticase soya broth. Am. J. Clin. *60*: 839-842.
292. Jotisankasa, V., H. S. Nielsen, and N. F. Conant. 1970. *Phialophora dermatitidis*: its morphology and biology. Sabouraudia *8*: 98-107.
293. Kaminsky, D., and L. Quinlan. 1963. A simple, inexpensive method for rapid identification of *Candida albicans.* Am. J. Clin. Pathol. *40*: 143-145.
294. Kanetsuna, F. 1972. Biochemical characteristics of *Paracoccidioides brasiliensis.* In *Paracoccidioidomycosis,* Proceedings of the First Pan American Symposium, Pan Am. Health Organ. Sci. Publ. No. 254, Washington, D.C., pp. 31-37.
295. Kanetsuna, F., and L. M. Carbonell. 1966. Enzymes in glycolysis and the citric acid cycle in the yeast and mycelial forms of *Paracoccidioides brasiliensis.* J. Bacteriol. *92*: 1315-1320.
296. Kanetsuna, F., and L. M. Carbonell. 1970. Cell wall glucans of the yeast and mycelial forms of *Paracoccidioides brasiliensis.* J. Bacteriol. *101*: 675-680.
297. Kanetsuna, F., and L. M. Carbonell. 1971. Cell wall composition of the yeastlike form of *Blastomyces dermatitidis.* J. Bacteriol. *106*: 946-948.
298. Kanetsuna, F., L. M. Carbonell, I. Azuma, and Y. Yamamura. 1972. Biochemical studies on the thermal dimorphism of *Paracoccidioides brasiliensis.* J. Bacteriol. *110*: 208-218.

299. Kanetsuna, F., L. M. Carbonell, F. Gil, and I. Azuma. 1974. Chemical and ultrastructural studies on the cell walls of the yeastlike and mycelial forms of *Histoplasma capsulatum*. Mycopathol. Mycol. Appl. *54*: 1-13.
300. Kanetsuna, F., L. M. Carbonell, R. E. Moreno, and J. Rodriguez. 1969. Cell wall composition of the yeast and mycelial forms of *Paracoccidioides brasiliensis*. J. Bacteriol. *97*: 1036-1041.
301. Kaplan, W. 1973. Epidemiology of the principal systemic mycoses of man and lower animals and the ecology of their etiologic agents. Vet. Med. Assoc. *163*: 1043-1047.
302. Katz, D., and R. F. Rosenberger. 1970. The utilization of galactose by an *Aspergillus nidulans* mutant lacking galactose phosphate-UDP glucose transferase and its relation to cell wall synthesis. Arch. Mikrobiol. *74*: 41-51.
303. Katz, D., and R. F. Rosenberger. 1971. Hyphal wall synthesis in *Aspergillus nidulans*: effect of protein synthesis inhibition and osmotic shock on chitin insertion and morphogenesis. J. Bacteriol. *108*: 184-190.
303a. Kaufman, L. 1976. Letter to the editor. Sabouraudia *14*: 114.
304. Keller, F. A., and E. Cabib. 1971. Chitin and yeast budding: properties of chitin synthetase from *Saccharomyces carlsbergensis*. J. Biol. Chem. *246*: 160-166.
305. Kessler, G., and W. J. Nickerson. Glucomannan-protein complexes from cell walls of yeasts. J. Biol. Chem. *234*: 2281-2285.
306. Knight, L., and J. Fletcher. 1971. Growth of *Candida albicans* in saliva: stimulation by glucose associated antibiotics, corticosteroids and diabetes mellitus. J. Infect. Dis. *123*: 371-377.
307. Kobayashi, G. S., L. Friedman, and J. F. Kofroth. 1964. Some cytological and pathogenic properties of spheroplasts of *Candida albicans*. J. Bacteriol. *88*: 795-801.
308. Kobayashi, G. S., and P. L. Guiliacci. 1967. Cell wall studies of *Histoplasma capsulatum*. Sabouraudia *5*: 180-188.
309. Kritzman, G., I. Chet, Y. Henis, and A. Hütterman. 1978. The use of the brightener "calcofluor white M2R new" in the study of fungal growth. Isr. J. Bot. *27*: 138-146.
310. Kumar, B. V., R. McMillian, G. Medoff, D. Schlessinger, and G. S. Kobayashi. 1977. Mechanism of the inhibition by RNA of the RNA polymerases of *Histoplasma capsulatum*. Biochim. Biophys. Acta *478*: 192-200.
311. Kurong, J. M., and D. Yegian. 1954. Medium for the maintenance and conversion of *Histoplasma capsulatum* to yeastlike phase. Am. J. Clin. Pathol. *24*: 505-508.
312. Kurup, P. V. 1963. Histochemical demonstration of alkaline phosphatase in *Candida*. Curr. Sci. *32*: 408-409.
313. Kwon-Chung, K. J. 1969. *Coccidioides immitis*: cytological study on the formation of arthrospores. Can. J. Genet. Cytol. *11*: 436-449.
314. Lacaz, C. S. 1953. South American blastomycosis. A review. Mycopathol. Mycol. Appl. *6*: 241-259.
315. Lack, A. R. 1938. Spherule formation and endosporulation of the fungus *Coccidioides* in vitro. Proc. Soc. Exp. Biol. Med. *38*: 907-909.
316. Land, G. A., W. C. McDonald, R. J. Stjernholm, and L. Friedman. 1975. Factors affecting filamentation in *Candida albicans*: relationship of the uptake and distribution of proline to morphogenesis. Infect. Immun. *11*: 1014-1023.
317. Land, G. A., W. C. McDonald, R. L. Stjernholm, and L. Friedman. 1975. Factors affecting filamentation in *Candida albicans*: changes in respiratory activities of *Candida albicans* during filamentation. Infect. Immun. *12*: 119-127.
318. Landau, J. W., N. Dabrowa, and V. D. Newcomer. 1965. The rapid formation in serum of filaments by *Candida albicans*. J. Invest. Dermatol. *44*: 171-179.

319. Landau, J. W., N. Dabrowa, V. D. Newcomer, and J. R. Rowe. 1964. The relationship of serum transferrin and iron to the rapid formation of germ tubes by *Candida albicans.* J. Invest. Dermatol. *43*: 473-482.

320. Lane, C. G. 1915. A cutaneous lesion caused by a new fungus *(Phialophora verrucosa).* J. Cutaneous Dis. *33*: 840-846.

321. Lane, J. W., and R. G. Garrison. 1970. Electron microscopy of the yeast to mycelial phase conversion of *Sporotrichum schenckii.* Can. J. Microbiol. *16*: 747-749.

322. Lane, J. W., R. G. Garrison, and M. F. Field. 1969. Ultrastructural studies on the yeastlike and mycelial phase of *Sporothrix schenckii.* J. Bacteriol. *100*: 1010-1019.

323. Langeron, M., and P. P. Horta. 1922. Note complémentaire sur le *Cladosporium werneckii* Horta 1921. Bull. Soc. Pathol. Exot. *15*: 381-383.

324. Lara, S. L., and S. Bartnicki-Garcia. 1974. Cytology of budding in *Mucor rouxii*: wall ontogeny. Arch. Microbiol. *97*: 1-16.

325. Larsen, A. D., and P. S. Sypherd. 1974. Cyclic adenosine 3′,5′-monophosphate and morphogenesis in *Mucor racemosus.* J. Bacteriol. *117*: 432-438.

326. Law, R. E., and A. J. Ferro. 1977. Effect of S-adenosylmethionine and cyclic adenosine-3′,5′-monophosphate on RNA synthesis during glucose repression in *Saccharomyces cerevisiae.* FEBS Lett. *80*: 153-156.

327. Lee, K. L., H. R. Buckley, and C. C. Campbell. 1975. An amino acid liquid synthetic medium for the development of mycelial and yeast forms of *Candida albicans.* Sabouraudia *13*: 148-153.

328. Lee, K. L., M. E. Reca, R. R. Watson, and C. C. Campbell. 1975. Identification of yeast phase of pathogenic fungi by the specificity of their amino-peptidase(s). Sabouraudia *13*: 132-141.

329. Lee, W. S. 1973. Chlamydospore-like cells in *Candida albicans* by 5-fluorocytosine. Can. J. Microbiol. *19*: 1449-1480.

330. Levine, H. B., J. M. Cobb, and C. E. Smith. 1960. Immunity to coccidioidomycosis induced in mice by purified spherule, arthrospore and mycelial vaccines. Trans. N.Y. Acad. Sci. *22*: 436-449.

331. Levine, S., and Z. J. Ordal. 1949. Factors influencing the morphology of *Blastomyces dermatitidis.* J. Bacteriol. *52*: 687-694.

332. Leyritz, M., and L. Kapica. 1975. Yeast-like growth of *Mucor alternans* (van Tieghem) in tissue-culture medium 199. Can. J. Microbiol. *21*: 75-78.

333. Lingappa, B. T., and A. S. Sussman. 1959. Endogenous substrates of dormant, activated and germinating ascospores of *Neurospora tetrasperma.* Plant Physiol. *34*: 466-472.

334. Liu, P., and A. Newton. 1955. Rapid chlamydospore formation by *Candida albicans* in a buffered alkaline medium. Am. J. Clin. Pathol. *25*: 93-97.

335. Lloyd, K. O., and M. A. Bitoon. 1971. Isolation and purification of a peptidorhamnomannan from the yeast form of *Sporothrix schenckii*: structural and immunochemical studies. J. Immunol. *107*: 663-671.

336. Lloyd, K. O., L. Mendonca-Previato, and L. R. Travassos. 1978. Distribution of antigenic polysaccharides in different cell types of *Sporothrix schenckii* as studied by immunofluorescent staining with rabbit antisera. Exp. Mycol. *2*: 130-137.

337. Lloyd, K. O., and L. R. Travassos. 1975. Immunochemical studies on L-rhamno-D-mannans of *Sporothrix schenckii* and related fungi by use of rabbit and human antisera. Carbohydr. Res. *40*: 89-97.

338. Lo, R. J. 1978. A preliminary study of the yeastlike cell division cycle of *Wangiella dermatitidis* 8656. M.A. thesis, University of Texas.

339. Lones, G. W., and C. L. Peacock. 1960. Role of carbon dioxide in the dimorphism of *Coccidioides immitis.* J. Bacteriol. *79*: 308-309.

340. Lones, G. W., C. L. Peacock, and F. A. McNey. 1971. Factors affecting the reversion of *Coccidioidies immitis* spherules to mycelium. Sabouraudia *9*: 287-296.

341. Lopez-Romero, E., J. Ruiz-Herrera, and S. Bartnicki-Garcia. 1978. Purification and properties of an inhibitory protein of chitin synthetase from *Mucor rouxii.* Biochim. Biophys. Acta *525*: 338-345.

342. Lubarsky, R., and O. A. Plunkett. 1955. In vitro production of the spherule phase of *Coccidioides immitis.* J. Bacteriol. *70*: 182-186.

343. Lutz, A., and A. Splendore. 1908. Über eine bei Menschen und Ratten beobachtete Mykose. Zentralbl. Bakteriol., Parasitenkd., Infektionskr. Hyg., Abt. I: Orig., *45*: 631-637.

344. Mackenzie, D. W. R. 1962. Serum tube identification of *Candida albicans.* J. Clin. Pathol. *15*: 563-565.

345. Mackenzie, D. W. R. 1963. Morphogenesis of *Candida albicans* in vivo. Sabouraudia *3*: 225-232.

346. MacKinnon, J. E. 1940. Dissociation of *Candida albicans.* J. Infect. Dis. *66*: 59-77.

347. MacNeal, W. J., and R. M. Taylor. 1914. *Coccidioides immitis* and coccidioidal granuloma. J. Med. Res. *30*: 261-274.

348. Mahadevan, P. R., and A. S. Bhagwat. 1973. Control of transcription in *Neurospora crassa.* In B. B. Biswas, R. K. Mandel, A. Stevens, and W. E. Cohn (Eds.), *Control of Transcription,* Plenum Press, New York, pp. 223-239.

349. Marchant, H. J., and E. R. Hines. 1979. The role of microtubules and cell-wall deposition in elongation of regenerating protoplasts of *Mougeotia.* Planta *146*: 41-48.

350. Marchant, R., and D. G. Smith. 1967. Wall structure and bud formation in *Rhodotorula glutinis.* Arch. Mikrobiol. *58*: 248-256.

351. Marchant, R., and D. G. Smith. 1968. Bud formation in *Saccharomyces cerevisiae* and a comparison with the mechanism of cell division in other yeasts. J. Gen. Microbiol. *53*: 163-169.

352. Mardon, D. N. 1973. In vivo synthesis of sulfur containing amino acids in *Candida albicans.* Can. J. Microbiol. *19*: 155-161.

353. Mardon, D. N., and E. Balish. 1971. Methionine biosynthesis in *Candida albicans.* I. S-Adenosyl-L-methionine (or S-methyl-L-methionine): homocysteine methyltransferase in cell-free extracts from yeast-like cells. Can. J. Microbiol. *17*: 795-805.

354. Mardon, D., E. Balish, and A. W. Phillips. 1969. Control of dimorphism in a biochemical variant of *Candida albicans.* J. Bacteriol. *100*: 701-707.

355. Mardon, D. N., J. L. Gunn, and E. Robinette. 1975. Variation in the lethal response in mice to yeast-like and pseudohyphal forms of *Candida albicans.* Can. J. Microbiol. *21*: 1681-1687.

356. Mardon, D. N., S. K. Hurst, and E. Balish. 1971. Germ-tube production by *Candida albicans* in minimal liquid culture media. Can. J. Microbiol. *17*: 851-856.

357. Maresca, B., E. Jacobson, G. Medoff, and G. Kobayashi. 1978. Cystine reductase in the dimorphic fungus *Histoplasma capsulatum.* J. Bacteriol. *135*: 987-992.

358. Maresca, B., A. M. Lambowitz, G. Kobayashi, and G. Medoff. 1979. Respiration in the yeast and mycelial phases of *Histoplasma capsulatum.* J. Bacteriol. *138*: 647-649.

359. Maresca, B., D. Schessinger, G. Medoff, and G. S. Kobayashi. 1977. Regulation of dimorphism in the pathogenic fungus *Histoplasma capsulatum.* Nature (Lond.) *266*: 447-448.

360. Mariat, F., and C. de Bievre. 1968. Caractères de la croissance et évolution de la teneur en constituants cellulaires des phases levure et filamenteuse de *Sporotrichum schenckii,* champignon dimorphique pathogène pour l'homme. Ann. Inst. Pasteur (Paris) *115*: 1082-1098.

361. Marlowe, J. D. 1977. The development of the sclerotic cell in agents of chromoblastomycosis: an ultrastructural study. M.A. thesis, University of Texas.
362. Marriott, M. S. 1975. Enzymic activity of purified plasma membranes from the yeast and mycelial forms of *Candida albicans.* J. Gen. Microbiol. *89*: 345-352.
363. Marriott, M. S. 1975. Isolation and chemical characterization of plasma membranes from the yeast and mycelial forms of *Candida albicans.* J. Gen. Microbiol. *86*: 115-132.
364. Matile, P., M. Cortat, A. Wiemken, and A. Frey-Wyssling. 1971. Isolation of glucanase-containing particles from budding *Saccharomyces cerevisiae.* Proc. Natl. Acad. Sci. U.S.A. *68*: 636-640.
365. Mattia, E., and A. Cassone. 1979. Inducibility of germ-tube formation in *Candida albicans* at different phases of yeast growth. J. Gen. Microbiol. *113*: 439-442.
366. McCully, E. K., and C. E. Bracker. 1972. Apical vesicles in growing bud cells of heterobasidiomycetous yeasts. J. Bacteriol. *109*: 922-926.
367. McDonough, E. S., L. K. Georg, L. Ajello, and S. Brinkman. 1960. Growth of human dimorphic fungi on media containing chloramphenicol and cycloheximide. Mycopathol. Mycol. Appl. *13*: 113-120.
368. McGinnis, M. R. 1977. *Wangiella,* a new genus to accommodate *Hormiscium dermatitidis.* Mycotaxon *5*: 353-363.
369. McGinnis, M. R. 1977. *Wangiella dermatitidis,* a correction. Mycotaxon *6*: 367-369.
370. McGinnis, M. R. 1978. Human pathogenic species of *Exophiala, Philalophora* and *Wangiella.* In *The Black and White Yeasts,* Proceedings of the Fourth International Conference on the Mycoses, Pan Am. Health Organ. Sci. Publ. No. 356, Washington, D.C., pp. 37-59.
371. McGinnis, M. R. 1979. Taxonomy of *Exophiala werneckii* and its relationship to *Microsporum mansonii.* Sabouraudia *17*: 145-154.
372. McGinnis, M. R., and B. Katz. 1979. *Ajellomyces* and its synonym *Emmonsiella.* Mycotaxon *8*: 157-164.
373. McMillian, R., B. V. Kumar, G. Medoff, and G. S. Kobayashi. 1977. RNA polymerases of the mycelial phase of *Histoplasma capsulatum.* Abstr. Annu. Meet. Am. Soc. Microbiol., 1977, F60, p. 128.
374. McMillian, R., B. V. Kumar, G. Medoff, D. Schlessinger, and G. S. Kobayashi. 1977. Inhibition of *Histoplasma capsulatum* RNA polymerases by homologous and heterologous RNA. J. Bacteriol. *130*: 1387-1389.
375. McMillian, R., G. Medoff, D. Schlessinger, and G. S. Kobayashi. 1977. Inhibition of isolated yeast and mycelial phase RNA polymerases of *Histoplasma capsulatum* by rifamycin derivatives. Sabouraudia *15*: 257-261.
376. McMurrough, I., and S. Bartnicki-Garcia. 1973. Inhibition and activation of chitin synthesis by *Mucor rouxii* cell extracts. Arch. Biochem. Biophys. *158*: 812-816.
377. McMurrough, I., A. Flores-Carreon, and S. Bartnicki-Garcia. 1971. Pathway of chitin synthesis and cellular localization of chitin synthetase in *Mucor rouxii.* J. Biol. Chem. *246*: 3999-4007.
378. McVeigh, I., and W. E. Houston. 1972. Factors affecting mycelial to yeast phase conversion and growth of the yeast phase of *Histoplasma capsulatum.* Mycopathol. Mycol. Appl. *47*: 135-151.
379. Medlar, E. M. 1915. A cutaneous infection caused by a new fungus, *Phialophora verrucosa,* with a study of the fungus. J. Med. Res. *32*: 507-521.
380. Mendonca, L., P. A. J. Gorin, K. O. Lloyd, and L. R. Travassos. 1976. Polymorphism of *Sporothrix schenckii* surface polysaccharides as a function of morphological differentiation. Biochemistry *15*: 2423-2431.
381. Miller, S. E., and W. R. Finnerty. 1979. Age-related physiological studies com-

paring *Candida albicans* chlamydospores to yeasts. Can. J. Microbiol. *25*: 765-772.

382. Miller, S. E., B. O. Spurlock, and G. E. Michaels. 1974. Electron microscopy of young *Candida albicans* chlamydospores. J. Bacteriol. *119*: 992-999.

383. Milne, H. A. 1957. The morphology and cytochemistry of *Histoplasma capsulatum*. J. Med. Lab. Technol. *14*: 142-163.

384. Mitchell, L. H., and D. R. Soll. 1979. Commitment to germ tube or bud formation during release from stationary phase in *Candida albicans.* Exp. Cell Res. *120*: 167-179.

385. Moen, S. P. B., and E. Rapport. 1971. Spindles, spindle plaques, and meiosis in the yeast *Saccharomyces cerevisiae* (Hansen). J. Cell Biol. *50*: 344-361.

386. Montplaisir, S., B. Nabarra, and E. Drouhet. 1976. Susceptibility and resistance of *Candida* to 5-fluorocytosine in relation to the cell wall ultrastructure. Antimicrob. Agents Chemother. *9*: 1028-1032.

387. Montes, L. F., and V. S. Constantine. 1968. Cytochemical demonstration of aminopeptidase in *Candida albicans.* J. Invest. Dermatol. *51*: 1-3.

388. Montes, L. F., and W. H. Wilborn. 1968. Ultrastructural features of host-parasite relationship in oral candidiasis. J. Bacteriol. *96*: 1349-1356.

389. Mooney, D. T., and P. S. Sypherd. 1976. Volatile factor involved in the dimorphism of *Mucor racemosus.* J. Bacteriol. *126*: 1266-1270.

390. Moor, H. 1967. Endoplasmic reticulum as the initiator of bud formation in yeast *(Saccharomyces cerevisiae).* Arch. Mikrobiol. *57*: 135-146.

391. Moreno, R. E., F. Kanetsuna, and L. M. Carbonell. 1969. Isolation of chitin and glucan from the cell wall of the yeast form of *Paracoccidioides brasiliensis.* Arch. Biochem. Biophys. *130*: 212-217.

392. Moreno, S., C. Paveto, and S. Passeron. 1977. Multiple protein kinase activities in the dimorphic fungus *Mucor rouxii.* Arch. Biochem. Biophys. *180*: 225-231.

393. Morris, N. R., M. H. Lai, and E. E. Oakley. 1979. Identification of a gene for α tubulin in *Aspergillus nidulans.* Cell *16*: 437-442.

394. Nagasaki, S. 1968. Physiological activities of various enzyme activities in relation to the culture age of *Aspergillus niger* mycelia. J. Gen. Appl. Microbiol. *14*: 147-161.

395. Najim, L. 1979. Conidiogenous loss of structuro-functional polarity in the hyphal tips of *Sclerotinia fructigena.* Eur. J. Cell Biol. *20*: 24-27.

396. Nakai, Y., and R. Ushiyama. 1978. Fine structure of shiitake, *Lentinus edodes.* VI. Cytoplasmic microtubules in relation to nuclear movement. Can. J. Bot. *56*: 1206-1211.

397. Negroni, P. 1968. Factores determinantes de la fase levaduriforme del *Paracoccidioides brasiliensis.* Mycopathol. Mycol. Appl. *34*: 280-284.

398. Nickerson, W. J. 1948. Enzymatic control of cell division in microorganisms. Nature (Lond.) *162*: 241-245.

399. Nickerson, W. J. 1954. An enzymatic locus participating in cellular division of a yeast. J. Gen. Physiol. *37*: 483-494.

400. Nickerson, W. J., and G. A. Edwards. 1949. Studies on the physiological bases of morphogenesis in fungi. I. The respiratory metabolism of dimorphic pathogenic fungi. J. Gen. Physiol. *33*: 41-55.

401. Nickerson, W. J., and G. Falcone. 1956. Identification of protein disulfide reductase as a cellular division enzyme in yeasts. Science *124*: 722-723.

402. Nickerson, W. J., and Z. Mankowski. 1953. Role of nutrition in the maintenance of the yeast shape in *Candida.* Am. J. Bot. *40*: 584-592.

403. Nickerson, W. J., W. A. Taber, and G. Falcone. 1956. Physiological bases of mor-

phogenesis in fungi 5. Effect of selenite and tellurite on cellular division of yeast-like fungi. Can. J. Microbiol. *2*: 575-584.
404. Nickerson, W. J., and N. J. W. Van Rij. 1949. The effect of sulfhydryl compounds, penicillin, and cobalt on the cell division mechanism of yeasts. Biochim. Biophys. Acta *3*: 461-475.
405. Nielsen, H. S., and N. F. Conant. 1967. Practical evaluation of antigenic relationships of yeast-like dematiacious fungi. Sabouraudia *5*: 283-294.
406. Nishimoto, K. 1970. The study of chromomycosis. Jpn. J. Dermatol. Ser. B *80*: 36-42.
407. Nishimoto, K. 1971. Electron microscopic study of chromomycosis (II) skin lesions. Jpn. J. Dermatol. Ser. B *81*: 33-37.
408. Nishioka, Y., and M. Silva-Hutner. 1974. Dimorphism, sensitivity to nystatin and acriflavin uptake in a strain of *Candida albicans* grown with glutamate as sole nitrogen and carbon source. Sabouraudia *12*: 295-301.
409. Nolan, R. A., and A. K. Bol. 1974. Cellulase localization in hyphae of *Achlya ambisexualis.* J. Bacteriol. *117*: 840-843.
410. Northey, W. T., and L. D. Brooks. 1962. Studies on *Coccidioides immitis.* I. A simple medium for in vitro spherulation. J. Bacteriol. *84*: 742-746.
411. Novick, P., and R. Schekman. 1979. Secretion and cell surface growth are blocked in a temperature-sensitive mutant of *Saccharomyces cerevisiae.* Proc. Natl. Acad. Sci. U.S.A. *76*: 1858-1862.
412. Obukhovskaya, A. S., E. B. Lishnevskaya, and I. M. Tereshin. 1979. Mechanism of the effect of levorin on transport of amino acids in *Candida albicans.* Curr. Microbiol. *2*: 219-222.
413. Odds, F. C. 1979. *Candida and Candidosis.* University Park Press, Baltimore.
414. Odds, F. C., C. A. Hall, and A. B. Abbott. 1978. Peptones and mycological reproducibility. Sabouraudia *16*: 237-246.
415. Ogletree, F. F., A. T. Abdelal, and D. G. Ahearn. 1978. Germ-tube formation by atypical strains of *Candida albicans.* Antonie Leeuwenhoek J. Microbiol. Serol. *44*: 15-24.
416. O'Hern, E. M., and B. S. Henry. 1956. A cytological study of *Coccidioides immitis* by electron microscopy. J. Bacteriol. *72*: 632-645.
417. Olmsted, J. B., and G. G. Borisy. 1973. Microtubules. Annu. Rev. Biochem. *42*: 507-540.
418. Orlowski, M., and P. S. Sypherd. 1976. Cyclic guanosine 3′,5′-monophosphate in the dimorphic fungus *Mucor racemosus.* J. Bacteriol. *125*: 1226-1228.
419. Orlowski, M., and P. S. Sypherd. 1977. Protein synthesis during morphogenesis of *Mucor racemosus.* J. Bacteriol. *132*: 209-218.
420. Orlowski, M., and P. S. Sypherd. 1978. Location of protein synthesis during morphogenesis of *Mucor racemosus.* J. Bacteriol. *133*: 399-400.
421. Orlowski, M., and P. S. Sypherd. 1978. Regulation of translation rate during morphogenesis in the fungus *Mucor.* Biochemistry *17*: 569-575.
422. Orlowski, M., and P. S. Sypherd. 1978. RNA synthesis during morphogenesis of the fungus *Mucor racemosus.* Arch. Microbiol. *119*: 145-152.
423. Otcenasek, M., K. Krivanec, and J. Slais. 1975. *Emmonsia parva* as a causal agent of adiaspiromycosis in a fox. Sabouraudia *13*: 52-57.
424. Oujezdsky, K. B., S. N. Grove, and P. J. Szaniszlo. 1973. Morphological and structural changes during the yeast-to-mold conversion of *Phialophora dermatitidis.* J. Bacteriol. *113*: 468-477.
425. Oujezdsky, K. B., and P. J. Szansizlo. 1974. Conidial ontogeny in *Phialophora dermatitidis.* Mycologia *66*: 537-542.

426. Padhye, A. A., M. R. McGinnis, and L. Ajello. 1978. Thermotolerance of *Wangiella dermatitidis*. J. Clin. Microbiol. *8*: 424–426.
427. Passeron, S., and H. Terenzi. 1970. Activation of pyruvate kinase of *Mucor rouxii* by manganese ions. FEBS Lett. *6*: 213-216.
428. Pasteur, L. 1876. Etudes sur la bière. Gauthier-Villars, Paris.
429. Paveto, C., A. Epstein, and S. Passeron. 1975. Studies on cyclic adenosine 3′,5′-monophosphate levels, adenylate cyclase and phosphodiesterase activities in the dimorphic fungus *Mucor rouxii*. Arch. Biochem. Biophys. *169*: 449–457.
430. Paveto, C., and S. Passeron. 1977. Some kinetic effects of *Mucor rouxii* phosphofructokinase: effect of cyclic adenosine-3′,5′-monophosphate. Arch. Biochem. Biophys. *178*: 1-7.
431. Paznokas, J. L., and P. S. Sypherd. 1975. Respiratory capacity, 3′,5′-cyclic adenosine monophosphate and morphogenesis of *Mucor racemosus*. J. Bacteriol. *124*: 134-139.
432. Paznokas, J. L., and P. S. Sypherd. 1977. Pyruvate kinase isozymes of *Mucor racemosus*: control of synthesis by glucose. J. Bacteriol. *130*: 661-666.
433. Pero, J., J. Nelson, and R. Losick. 1975. In vitro and in vivo transcription by vegetative and sporulating *Bacillus subtilis*. In P. Gerhardt, R. N. Costilow, and H. L. Sadoff (Eds.), *Spores IV*, American Society for Microbiology, Washington, D.C., pp. 202-212.
434. Peters, J., and P. S. Sypherd. 1978. Enrichment of mutants of *Mucor racemosus* by differential freeze-killing. J. Gen. Microbiol. *105*: 77-81.
435. Peters, J., and P. S. Sypherd. 1979. Morphology-associated expression of nicotinamide adenine dinucleotide dependent glutamate dehydrogenase in *Mucor racemosus*. J. Bacteriol. *137*: 1134-1139.
436. Pine, L. 1954. Studies on the growth of *Histoplasma capsulatum*. I. Growth of the yeast phase in liquid media. J. Bacteriol. *68*: 671-679.
437. Pine, L. 1955. Studies on the growth of *Histoplasma capsulatum*. II. Growth of the yeast phase on agar media. J. Bacteriol. *70*: 375-381.
438. Pine. L. 1957. Studies on the growth of *Histoplasma capsulatum*. III. Effect of thiamin and other vitamins on the growth of the yeast and mycelial phase of *Histoplasma capsulatum*. J. Bacteriol. *74*: 239-245.
439. Pine, L. 1970. Growth of *Histoplasma capsulatum*. VI. Maintenance of the mycelial phase of *Histoplasma capsulatum*. Appl. Microbiol. *19*: 413–420.
440. Pine, L., and C. L. Peacock. 1957. Studies on the growth of *Histoplasma capsulatum*. IV. Factors influencing conversion of the mycelial phase to the yeast phase. J. Bacteriol. *75*: 167-174.
441. Pine, L., and R. E. Webster. 1962. Conversion in strains of *Histoplasma capsulatum*. J. Bacteriol. *83*: 149-157.
442. Pitt, D. 1969. Cytochemical observation on the localization of sulphydryl groups in budding yeast cells and in the phialides of *Penicillium notatum* Westling during conidiation. J. Gen. Microbiol. *59*: 257-262.
443. Pollak, L. 1971. Aleuriospores of *Paracoccidioides brasiliensis*. Mycopathol. Mycol. Appl. *45*: 217-219.
444. Poon, N. H., and A. W. Day. 1976. Somatic nuclear division in the sporidia of *Ustilago violaceae*. IV. Microtubules and spindle pole body. Can. J. Microbiol. *22*: 507-522.
445. Pope, L. M., G. T. Cole, M. N. Guentzel, and L. J. Berry. 1979. Systemic and gastrointestinal candidiasis of infant mice after intragastric challenge. Infect. Immun. *25*: 702-707.
446. Poulain, D., G. Tronchin, J. F. Dubremetz, and J. Biguet. 1978. Ultrastructure of

the cell wall of *Candida albicans* blastospores: study of its constitutive layers by use of a cytochemical technique revealing polysaccharides. Ann. Microbiol. (Paris) *129*: 141-153.

447. von Preusser, H. J., H. Rostek, M. Klein, H. Becker, and W. P. Ness. 1979. Elektronenmikroskopische Untersuchungen an *Candida albicans* aus vaginalem Abstrichmaterial wahrend einer 3-Tage-Therapie mit Econazol. Arzneim-Forsch. Drug Res. *29*: 1432-1437.

448. Previato, J. O., P. A. J. Gorin, and L. R. Travassos. 1979. Cell-wall composition in different cell-types of the dimorphic species *Sporothrix schenckii.* Exp. Mycol. *3*: 83-91.

449. Previato, J. O., P. A. J. Gorin, R. H. Haskins, and L. R. Travassos. 1979. Soluble and insoluble glucans from different cell-types of the human pathogen *Sporothrix schenckii.* Exp. Mycol. *3*: 92-105.

450. Prick, G., and H. Diekman. 1979. A chitin-binding lectin in *Neurospora crassa.* FEMS Microbiol. Lett. *6*: 427-429.

451. Pringle, J. R. 1978. The use of conditional lethal cell cycle mutants for temporal and functional sequence mapping of cell cycle events. J. Cell. Physiol. *95*: 393-406.

452. Procknow, J. J., M. I. Page, and C. G. Loosli. 1960. Early pathogenesis of experimental histoplasmosis. Arch. Pathol. *69*: 413-426.

453. Pugh, D., and R. A. Cawson. 1977. The cytochemical localization of phospholipase in *Candida albicans* infecting the chick chorioallantoic membrane. Sabouraudia *15*: 29-35.

454. Pugh, D., and R. A. Cawson. 1978. The surface layer of *Candida albicans.* Microbios *23*: 19-23.

455. Rajasingham, K. C., and R. A. Cawson. 1978. "Spitzenkörper" in the invasive pseudohyphae of *Candida albicans.* Cell Struct. Funct. *3*: 265-267.

456. Ramirez-Martinez, J. R. 1971. A comparative study of ribonucleic acid species in yeast-like and mycelial forms of *Paracoccidioides brasiliensis.* Sabouraudia *9*: 157-163.

457. Ramirez-Martinez, J. R. 1971. *Paracoccidioides brasiliensis*: conversion of yeast-like forms into mycelia in submerged culture. J. Bacteriol. *105*: 523-526.

458. Ramirez-Martinez, J. R. 1972. Growth and physiological characteristics of *Paracoccidioides brasiliensis.* In *Paracoccidioidomycosis,* Proceedings of the First Pan American Symposium, Pan Am. Health Organ. Sci. Publ. No. 254, Washington, D.C., pp. 13-20.

459. Raudaskoski, M. 1970. Occurrence of microtubules and microfilaments, and origin of septa in dikaryotic hyphae of *Schizophyllum commune.* Protoplasma *70*: 415-422.

460. Raudaskoski, M. 1972. Occurrence of microtubules in the hyphae of *Schizophyllum commune* during intercellular nuclear migration. Arch. Mikrobiol. *86*: 91-100.

461. Ray, T. L., A. Hanson, L. F. Ray, and K. D. Wuepper. 1979. Purification of a mannan from *Candida albicans* which activates serum complement. J. Invest. Dermatol. *73*: 269-274.

462. Redaelli, P., and R. Ciferri. 1937. Ulteriori ricerche micologishe e sperimentali sulla istoplasmosi de Darling. Identita tia "*Histoplasma capsulatum*" e "*Posadasia pyriformis*" genessi della forma animalizgata de feinga. Arch. Ital. Med. Sper. *1*: 1-20.

463. Reess, M. 1870. *Botanische Untersuchungen über die Alkoholgärungspilze.* Arthur Felix, Leipzig.

464. Reiss, E., and W. J. Nickerson. 1971. Control of dimorphism in *Phialophora verrucosa.* Sabouraudia *12*: 202-213.

465. Restrepo, A., D. L. Greer, and M. Vasconcellos. 1973. Paracoccidioidomycosis: a review. Rev. Med. Vet. Mycol. *8*: 97-123.

466. Restrepo, A., L. H. Moncado, and M. Quinters. 1969. Effect of hydrogen ion concentration and of temperature on the growth of *Paracoccidioides brasiliensis* in soil extracts. Sabouraudia 7: 207-215.

467. Restrepo-Moreno, A. 1970. A reappriasal of the microscopic appearance of the mycelial phase of *Paracoccidioides brasiliensis*. Sabouraudia *8*: 141-144.

468. Reynolds, R., and A. Braude. 1956. The filament-inducing property of blood for *Candida albicans*: its nature and significance. Clin. Res. Proc. *4*: 40.

469. Ricketts, H. T. 1901. Oidiomycosis (blastomycosis) of the skin and its fungi. J. Med. Res. *6*: 373-546.

470. Rippon, J. W. 1968. Monitored environmental system to control cell growth, morphology and metabolic states in fungi by oxidation-reduction potential. Appl. Microbiol. *16*: 114-121.

471. Rippon, J. W. 1974. *Medical Mycology: The Pathogenic Fungi and the Pathogenic Actinomycetes.* Saunders, Philadelphia.

472. Roberts, K. 1974. Cytoplasmic microtubules and their functions. Prog. Biophys. Mol. Biol. *28*: 371-420.

473. Roberts, R. L. 1979. Characterization of temperature-sensitive multicellular mutants of *Wangiella dermatitidis*. Ph.D. dissertation, University of Texas.

474. Roberts, R. L., R. J. Lo, and P. J. Szaniszlo. 1979. Nuclear division in temperature-sensitive multicellular mutants of *Wangiella dermatitidis*. J. Bacteriol. *137*: 1456-1458.

475. Roberts, R. L., and P. J. Szaniszlo. 1978. Temperature-sensitive multicellular mutants of *Wangiella dermatitidis*. J. Bacteriol. *135*: 622-632.

475a. Roberts, R. L., and P. J. Szaniszlo. 1980. Yeast-phase cell cycle of the polymorphic fungus *Wangiella dermatitidis*. J. Bacteriol. *144*: 721-731.

476. Robson, J. E., and H. E. Stockley. 1962. Sulphydryl metabolism of fungi grown in submerged culture. J. Gen. Microbiol. *28*: 57-68.

477. Rogers, K. B. 1957. An aid to the differentiation between cases of moniliasis and monilia carriers. J. Clin. Pathol. *10*: 406-407.

478. Rogers, P. J., G. D. Clark-Walker, and P. R. Stewart. 1974. Effects of oxygen and glucose on energy metabolism and dimorphism of *Mucor genevensis* grown in continuous culture: reversibility of yeast-mycelium conversion. J. Bacteriol. *119*: 282-293.

479. Romano, A. H. 1966. Dimorphism. In G. C. Ainsworth and A. S. Sussman, (Eds.), *The Fungi: An Advanced Treatise,* Vol. 2, Academic, New York, pp. 181-209.

480. Rowley, D. A., and M. Huber. 1955. Pathogenesis of experimental histoplasmosis in mice. I. Measurement of infecting dosages of the yeast phase of *Histoplasma capsulatum*. J. Infect. Dis. *96*: 174-183.

481. Roy, I., and J. W. Landau. 1971. Polymorphism of malate dehydrogenase in cytoplasmic and mitochondrial fractions of *Blastomyces dermatitidis*. Sabouraudia *9*: 39-42.

482. Roy, I., and J. W. Landau. 1972. Composition of the alkali resistant cell wall material of dimorphic *Blastomyces dermatitidis*. Sabouraudia *10*: 107-112.

483. Roy, I., and J. W. Landau. 1972. Protein constituents of the cell walls of the dimorphic phases of *Blastomyces dermatitidis*. Can. J. Microbiol. *18*: 473-478.

484. Ruiz-Herrera, J., and S. Bartnicki-Garcia. 1974. Synthesis of a cell wall microfibrils in vitro by a "soluble" chitin synthetase from *Mucor rouxii*. Science *186*: 357-359.

485. Ruiz-Herrera, J., and S. Bartnicki-Garcia. 1976. Proteolytic activation and inactivation of chitin synthetase from *Mucor rouxii*. J. Gen. Microbiol. *97*: 241-249.

486. Ruiz-Herrera, J., E. Lopez-Romero, and S. Bartnicki-Garcia. 1977. Properties of chitin synthetase in isolated chitosomes from yeast cells of *Mucor rouxii.* J. Biol. Chem. *252*: 3338-3343.
487. Saltarelli, C. G. 1973. Growth stimulation and inhibition of *Candida albicans* by metabolic by-products. Mycopathol. Mycol. Appl. *51*: 53-63.
488. Saltarelli, C. G., and C. P. Coppola. 1979. Effect of light on growth and metabolic synthesis in *Candida albicans.* Mycologia *71*: 773-785.
489. Salvin, S. B. 1947. Cultural studies on the yeastlike phase of *Histoplasma capsulatum* Darling. J. Bacteriol. *54*: 655-660.
490. Salvin, S. B. 1947. Multiple budding in *Sporotrichum schenckii* Maltruchot. J. Invest. Dermatol. *9*: 315-320.
491. Salvin, S. B. 1949. Cysteine and related compounds in the growth of the yeast-like phase of *Histoplasma capsulatum.* J. Infect. Dis. *84*: 275-283.
492. Salvin, S. B. 1949. Phase determining factors in *Blastomyces dermatitidis.* Mycologia *41*: 311-319.
493. San-Blas, F., and L. J. Cova. 1975. Growth curves of the yeast-like form of *Paracoccidioides brasiliensis.* Sabouraudia *13*: 22-29.
494. San-Blas, F., G. San-Blas, and L. J. Cova. 1976. A morphological mutant of *Paracoccidioides brasiliensis* strain IVIC Pb9. Isolation and wall characterization. J. Gen. Microbiol. *93*: 209-218.
495. San-Blas, G. 1979. Biosynthesis of glucans by subcellular fractions of *Paracoccidioides brasiliensis.* Exp. Mycol. *3*: 249-258.
496. San-Blas, G., and F. San-Blas. 1977. *Paracoccidioides brasiliensis;* cell wall structure and virulence. Mycopathologia *62*: 77-86.
497. San-Blas, G., F. San-Blas, and L. E. Serrano. 1977. Host-parasite relationships in the yeast-like form of *Paracoccidioides brasiliensis* strain IVIC-Pb9. Infect. Immun. *15*: 343-346.
498. San-Blas, G., and F. J. Yegres. 1978. Effects of cyanein and ramihyphin A on the dimorphism of *Paracoccidioides brasiliensis.* Folia Microbiol. *23*: 198-201.
499. Schekman, R., and V. Brawley. 1979. Localized deposition of chitin on the yeast cell surface in response to mating pheromone. Proc. Natl. Acad. Sci. U.S.A. *76*: 645-649.
500. Schenck, B. R. 1898. On refractory subcutaneous abscesses caused by a fungus possibly related to the sporotricha. Bull. Johns Hopkins Hosp. *9*: 286-290.
501. Scherr, G. H. 1957. Studies on the dimorphism of *Histoplasma capsulatum.* I. The roles of –SH groups and incubation temperature. Exp. Cell Res. *12*: 92-107.
502. Scherwitz, C., R. Martin, and H. Ueberberg. 1978. Ultrastructural investigations of the formation of *Candida albicans* germ tubes and septa. Sabouraudia *16*: 115-124.
503. Schulz, B., G. Kraepelin, and W. Hinkelmann. 1974. Factors affecting dimorphism in Mycotypha (Mucorales): a correlation with the fermentation/respiration equilibrium. J. Gen. Microbiol. *82*: 1-13.
504. Scott, E. A. 1976. Biochemical genetics of morphogenesis in *Neurospora.* Annu. Rev. Microbiol. *30*: 85-104.
505. Scribner, H. E., J. Mogelson, E. Eisenstadt, and S. Silver. 1975. Regulation of cation transport during bacterial sporulation. In P. Gerhardt, R. N. Costilow, and H. L. Sadoff (Eds.), *Spores IV,* American Society for Microbiology, Washington, D.C., pp. 346-355.
506. Sentandreu, R., and D. H. Northcote. 1969. The formation of buds in yeasts. J. Gen. Microbiol. *55*: 393-398.
507. Shannon, J. L., and A. H. Rothman. 1971. Transverse septum formation in budding cells of the yeastlike fungus *Candida albicans.* J. Bacteriol. *106*: 1026-1028.

508. Sheiban, Z. B. 1974. Morphological conversion of *Blastomyces dermatitidis* in tap water. Bull. W.H.O. *51*: 423–427.
509. Sheir-Neiss, G., R. V. Nardi, M. A. Gealt, and N. R. Morris. 1976. Tubulin-like protein from *Aspergillus nidulans*. Biochem. Biophys. Res. Commun. *69*: 285-290.
510. Sheir-Neiss, G., M. H. Lai, and N. R. Morris. 1978. Identification of a gene for β tubulin in *Aspergillus nidulans*. Cell *15*: 639-647.
511. Shepherd, M. G., C. M. Chin, and P. A. Sullivan. 1978. The alternate respiratory pathway of *Candida albicans*. Arch. Microbiol. *116*: 61-67.
512. Shepherd, M. G., and P. A. Sullivan. 1976. The production and growth characteristics of yeast and mycelial forms of *Candida albicans* in continuous culture. J. Gen. Microbiol. *93*: 361-370.
513. Shimizu, M. T., and F. Alterthum. 1979. Estudo comparativo entre o conteudo de esterosis e resistencia a polienos em diferentes especies de leveduras. Rev. Microbiol. *10*: 14-18.
514. Sietsma, J. H., and J. G. H. Wessels. 1979. Evidence for covalent linkages between chitin and β-glucan in a fungal wall. J. Gen. Microbiol. *114*: 99-108.
515. Sikl, D., L. Masler, and S. Bauer. 1964. Mannan from the extracellular surface of *Candida albicans* Berkhout. Experientia *20*: 456.
516. Silva, M. 1957. The parasitic phase of the fungi of chromoblastomycosis: development of the sclerotic cell in vivo and in vitro. Mycologia *49*: 318-331.
517. Silva, M. 1960. Growth characteristics of the fungi of chromoblastomycosis. Ann. N.Y. Acad. Sci. *89*: 17-29.
518. Silva-Hutner, M. 1970. Yeasts. In J. Blair, E. H. Lennette, and J. P. Truant (Eds.), *Manual of Clinical Microbiology*, Williams & Wilkins, Baltimore, pp. 352-363.
519. Silver, S., and P. Jasper. 1977. Manganese transport in microorganisms. In E. D. Weinberg (Ed.), *Microorganisms and Minerals*, Dekker, New York, pp. 105-149.
520. Simonetti, N., and V. Strippoli. 1971. A simple method for rapid acquisition of chlamydospores in *Candida albicans*. Experientia *27*: 985.
521. Simonetti, N., and V. Strippoli. 1973. Pathogenicity of the Y form as compared to the M form in experimentally induced *Candida albicans*. Mycopathol. Mycol. Appl. *51*: 19-28.
522. Simonetti, N., V. Strippoli, and A. Cassone. 1974. Yeast-mycelial conversion induced by N-acetyl-D-glucosamine in *Candida albicans*. Nature (Lond.) *250*: 344-346.
523. Singh, B. R., and A. Datta. 1978. Glucose repression of the inducible catabolic pathway for N-acetylglucosamine in yeast. Biochem. Biophys. Res. Commun. *84*: 58-64.
524. Singh, B., and A. Datta. 1979. Regulation of N-acetylglucosamine uptake in yeast. Biochim. Biophys. Acta *557*: 248-258.
525. Singh, M., A. Jayakumar, and R. Prasad. 1978. The effect of altered lipid composition on the transport of various amino acids in *Candida albicans*. Arch. Biochem. Biophys. *191*: 680-686.
526. Sloat, B. F., and J. R. Pringle. 1978. A mutant of yeast defective in cellular morphogenesis. Science *200*: 1171-1173.
527. Soll, D. R., and G. W. Bedell. 1978. Bud formation and the inducibility of pseudomycelium outgrowth during release from stationary phase in *Candida albicans*. J. Gen. Microbiol. *108*: 173-180.
528. Soll, D. R., M. Stasi, and G. Bedell. 1978. The regulation of nuclear migration and division during pseudo-mycelium outgrowth in the dimorphic yeast *Candida albicans*. Exp. Cell Res. *116*: 207-215.
529. Sorrentino, A. P., J. Zorzopoulos, and H. F. Terenzi. 1977. Inducible alpha-gluco-

sidase of *Mucor rouxii*: effects of dimorphism on the development of wall-bound activity. Arch. Biochem. Biophys. *180*: 232-238.
530. Steiner, M. 1978. 3′,5′-cyclic AMP binds to and promotes polymerisation of platelet tubules. Nature (Lond.) *272*: 834-835.
531. Stephens, R. E., and K. T. Edds. 1976. Microtubules: structure, chemistry and function. Physiol. Rev. *56*: 709-776.
532. Stetler, D. A., and G. Boguslawski. 1979. Cysteine biosynthesis in a fungus, *Histoplasma capsulatum*. Sabouraudia *17*: 23-34.
533. Stewart, P. R., and P. J. Rogers. 1978. Fungal dimorphism: a particular expression of cell wall morphogenesis. In J. E. Smith and D. R. Berry (Eds.), *The Filamentous Fungi*, Wiley, New York, pp. 164-196.
534. Stoddart, R. W., and B. M. Hebertson. 1978. The use of fluorescein-labelled lectins in the detection and identification of fungi pathogenic for man: a preliminary study. J. Med. Microbiol. *11*: 315-324.
535. Storck, R., and R. C. Morrill. 1971. Respiratory-deficient, yeastlike mutant of *Mucor*. Biochem. Genet. *5*: 467-479.
536. Strippoli, V., and N. Simonetti. 1975. Specific induction of chlamydospore formation in *Candida albicans* by N-acetyl-D-glucosamine. Experientia *31*: 130-131.
537. Sun, S. H., and M. Huppert. 1976. A cytological study of morphogenesis in *Coccidioides immitis*. Sabouraudia *14*: 185-198.
538. Sun, S. H., M. Huppert, and K. R. Vukovich. 1976. Rapid in vitro conversion and identification of *Coccidioides immitis*. J. Lab. Clin. Med. *3*: 186-190.
539. Sun, S. H., S. S. Sekhon, and M. Huppert. 1979. Electron microscopic studies of saprobic and parasitic forms of *Coccidioides immitis*. Sabouraudia *17*: 265-273.
540. Sunayama, H. 1970. Studies on the antigenic activities of yeasts. IV. Analysis of the antigenic determinant groups of the mannan of *Candida albicans* serotype A. Jpn. J. Microbiol. *14*: 27-39.
541. Sypherd, P. S., P. T. Borgia, and J. L. Paznokas. 1978. Biochemistry of dimorphism in the fungus *Mucor*. Adv. Microb. Physiol. *18*: 67-104.
542. Sypherd, P. S., M. Orlowski, and J. Peters. 1979. Models of fungal dimorphism: control of dimorphism in *Mucor racemosus*. Microbiology *1979*: 224-227.
543. Szaniszlo, P. J., P. H. Hsieh, and J. D. Marlowe. 1976. Induction and ultrastructure of the multicellular (sclerotic) morphology in *Phialophora dermatitidis*. Mycologia *68*: 117-130.
544. Szaniszlo, P. J., R. J. Lo, and R. L. Roberts. 1978. Effects of hydroxyurea on the cell cycle of *Phialophora (Wangiella) dermatitidis*. Abstr. XII Int. Congr. Microbiol. *12*: 82.
545. Taschdjian, C. L., J. J. Burchall, and P. J. Kozinn. 1960. Rapid identification of *Candida albicans* by filamentation on serum and serum substitutes. Am. Med. Assoc. J. Dis. Child. *99*: 212-215.
546. Taschdjian, C. L., and P. J. Kozinn. 1961. Metabolic studies of the tissue phase of *Candida albicans* induced in vitro. Sabouraudia *1*: 73-82.
547. Taylor, J. J. 1961. Nucleic acids and dimorphism in *Blastomyces*. Exp. Cell Res. *24*: 155-158.
548. Taylor, J. J. 1970. The nature of the secondary conidia of *Sporothrix schenckii*. Mycopathol. Mycol. Appl. *41*: 379-382.
549. Terenzi, H. F., E. Rosolino, and S. Passeron. 1971. Two types of pyruvate kinase in *Mucor rouxii*: changes in the enzymatic pattern related to aerobic development. Eur. J. Biochem. *18*: 342-350.
550. Terenzi, H. F., and R. Storck. 1968. Stimulation by phenethyl alcohol of aerobic fermentation in *Mucor rouxii*. Biochem. Biophys. Res. Commun. *30*: 447-452.

551. Terenzi, H. F., and R. Storck. 1969. The effect of phenethyl alcohol on the morphology, growth and metabolism of some mucorales. Mycologia *61*: 894-901.

552. Terenzi, H. F., and R. Storck. 1969. Stimulation of fermentation and yeast-like morphogenesis in *Mucor rouxii* by phenethyl alcohol. J. Bacteriol. *97*: 1248-1261.

553. Tkacz, J. S., E. B. Cybulska, and J. O. Lampen. 1971. Specific staining of wall mannan in yeast cells with fluorescein-conjugated concanavalin A. J. Bacteriol. *105*: 1-5.

554. Tkacz, J. S., and J. O. Lampen. 1972. Wall replication in *Saccharomyces* species: use of fluorescein conjugated concanavalin A to reveal the site of mannan insertion. J. Gen. Microbiol. *72*: 243-247.

555. Tkacz, J. S., and V. L. MacKay. 1979. Sexual conjugation in yeast: cell surface changes in response to the action of mating hormones. J. Cell Biol. *80*: 326-333.

556. Ton, S. H., and M. C. Karunairatnum. 1976. Some factors affecting the transformation of the yeast-like to the mycelial-like forms of *Candida albicans*. Southeast Asian J. Trop. Med. Public Health *7*: 72-76.

557. Travassos, L. R., P. A. J. Gorin, and K. O. Lloyd. 1973. Comparison of the rhamnomannans from the human pathogen *Sporothrix schenckii* with those from the *Ceratocystis* species. Infect. Immun. *8*: 685-693.

558. Travassos, L. R., and L. Mendonca-Previato. 1978. Synthesis of monorhamnosyl L-rhamno-D-mannans by conidia of *Sporothrix schenckii*. Infect. Immun. *19*: 1-4.

559. Travassos, L. R., L. Mendonca-Previato, and P. A. J. Gorin. 1978. Heterogeneity of rhamnomannans from one strain of the human pathogen *Sporothrix schenckii* determined by ^{13}C nuclear magnetic resonance spectroscopy. Infect. Immun. *19*: 1107-1109.

560. Travassos, L. R., W. de Sousa, L. Mendonca-Previato, and K. O. Lloyd. 1977. Location and biochemical nature of surface components reacting with concanavalin A in different cell types of *Sporothrix schenckii*. Exp. Mycol. *1*: 293-305.

561. Trevis, L. 1956. Additional records of *Haplosporangium parvum* in mammals in Britain. Nature (Lond.) *177*: 437.

562. Trevithick, J. R., and P. R. Galsworthy. 1977. Morphology of slime variants of *Neurospora crassa* growing on a glass surface in liquid medium. Arch. Microbiol. *115*: 109-118.

563. Uden, N. van, and H. Buckley. 1974. *Candida* Berkhout. In J. Lodder (Ed.), *The Yeasts. A Taxonomic Study*. North-Holland, Amsterdam, pp. 893-1087.

564. Ulane, R. E., and E. Cabib. 1974. The activating system of chitin synthetase from *Saccharomyces cerevisiae*: purification and properties of an inhibitor of the activating factor. J. Biol. Chem. *249*: 3418-3422.

565. Ulane, R. E., and E. Cabib. 1976. The activating system of chitin synthetase from *Saccharomyces cerevisiae*. Purification and properties of activating factor. J. Biol. Chem. *251*: 3367-3374.

566. Vaughn, V. J., and E. D. Weinberg. 1978. *Candida albicans* dimorphism and virulence: role of copper. Mycopathologia *64*: 39-42.

567. Venezia, R. A., and R. C. Lachapelle. 1973. The use of ferritin-conjugated antibodies in the study of cell wall components of *Candida albicans*. Can. J. Microbiol. *19*: 1445-1448.

568. Vieira de Silva, C. R., M. C. F. Iwama de Mattos, and K. Fujimore. 1974. Scanning electron microscopy of *Paracoccidioides brasiliensis*. Mycopathol. Mycol. Appl. *54*: 235-251.

569. Wain, W. H., A. R. Brayton, and R. A. Cawson. 1974. Variations in the response to N-acetyl-D-glucosamine by isolates of *Candida albicans*. Mycopathologia *58*: 27-29.

570. Wain, W. H., M. F. Price, A. R. Brayton, and R. A. Cawson. 1976. Macromolecular synthesis during the cell cycles of yeast and hyphal phases of *Candida albicans.* J. Gen. Microbiol. *97*: 211-217.
571. Wain, W. H., M. F. Price, and R. A. Cawson. 1975. A re-evaluation of the effect of cysteine on *Candida albicans.* Sabouraudia *13*: 74-82.
572. Watts, J. C., and C. S. Callaway. 1975. Human pulmonary adiaspiromycosis. Arch. Pathol. *99*: 11-15.
573. Weeks, R. J. 1964. A rapid simplified medium for converting the mycelial phase of *Blastomyces dermatitidis* to the yeast phase. Mycopathol. Mycol. Appl. *22*: 153-156.
574. Wehmer, C. 1905. Unabhängigkeit der Mucorineengärung von Sauerstoffabschluss und Dugelhefe. Ber. Dtsch. Bot. Ges. *23*: 122-125.
575. Weise, E. C. 1931. Sporotrichosis in Connecticut. N. Engl. J. Med. *205*: 951-955.
576. West, B. 1967. Nutrition of *Phialophora verrucosa* A126. Mycopathol. Mycol. Appl. *31*: 12-16.
577. Wheat, R., and E. Scheer. 1977. Cell walls of *Coccidioides immitis.* Neutral sugars by aqueous alkaline polymers. Infect. Immun. *15*: 340-341.
578. Wheat, R. W., C. Tritschler, N. F. Conant, and E. P. Lowe. 1977. Comparison of *Coccidioides immitis* arthrospore, mycelium and spherule cell walls, and influence of growth medium on mycelial cell wall composition. Infect. Immun. *17*: 91-97.
579. Wickerham, L. J. 1951. *Taxonomy of Yeasts.* U.S. Dept. Agric. Tech. Bull. No. 1029, pp. 1-55.
580. Widra, A. 1964. Phosphate directed Y-M variation in *Candida albicans.* Mycopathol. Mycol. Appl. *23*: 197-202.
581. Willingham, M. C. 1976. Cyclic AMP and cell behavior in cultured cells. Int. Rev. Cytol. *44*: 319-357.
582. Willingham, M. C., and I. Pastan. 1975. Cyclic AMP and cell morphology in cultured fibroblasts. Effects on cell shape, microfilaments and microtubule distribution and orientation to substratum. J. Cell Biol. *67*: 146-159.
583. Winblad, B. 1975. Experimental renal candidiasis in mice and guinea pigs. Acta Pathol. Microbiol. Scand. Sect. A *83*: 406-414.
584. Winsten, S., and T. J. Murray. 1956. Virulence enhancement of a filamentous strain of *Candida albicans* after growth on media containing cysteine. J. Bacteriol. *71*: 738.
585. Yamaguchi, H. 1974. Dimorphism in *Candida albicans.* I. Morphology-dependent changes in cellular content of macromolecules and respiratory activity. J. Gen. Appl. Microbiol. *20*: 87-99.
586. Yamaguchi, H. 1974. Effect of biotin insufficiency on composition and ultrastructure of cell wall of *Candida albicans* in relation to its mycelial morphology. J. Gen. Appl. Microbiol. *20*: 217-228.
587. Yamaguchi, H. 1974. Mycelial development and chemical alteration of *Candida albicans* from biotin insufficiency. Sabouraudia *12*: 320-328.
588. Yamaguchi, H. 1975. Control of dimorphism in *Candida albicans* by zinc: effect on cell morphology and composition. J. Gen. Microbiol. *86*: 370-372.
589. Yamaguchi, H., Y. Kanda, and M. Osmi. 1974. Dimorphism in *Candida albicans.* II. Comparison of fine structure of yeast-like and filamentous phase growth. J. Gen. Appl. Microbiol. *20*: 101-110.
590. Young, G. 1958. The process of invasion and the persistence of *Candida albicans* injected intraperitoneally into mice. J. Infect. Dis. *102*: 114-120.
591. Young, H. A., and H. R. Whiteley. 1975. Changes in the levels of DNA-dependent

RNA polymerases during the transition of the dimorphic fungus *Mucor rouxii* from yeast-like to mycelial growth. Exp. Cell Res. *91*: 216-222.

592. Young, H. A., and H. R. Whiteley. 1975. Deoxyribonucleic acid-dependent ribonucleic acid polymerases in the dimorphic fungus *Mucor rouxii.* J. Biol. Chem. *250*: 479-487.

593. Zalokar, M. 1959. Enzyme activity and cell differentiation in *Neurospora.* Am. J. Bot. *46*: 555-559.

594. Zalokar, M. 1959. Growth and differentiation of *Neurospora* hyphae. Am. J. Bot. *46*: 602-610.

595. Zorzopoulos, J., A. J. Jobbagy, and H. F. Terenzi. 1973. Effects of ethylenediaminetetracetate and chloramphenicol on mitochondrial activity and morphogenesis in *Mucor rouxii.* J. Bacteriol. *115*: 1198-1204.

11

Conidium and Sporangiospore Formation in Pathogenic Microfungi

Garry T. Cole / The University of Texas at Austin, Austin, Texas

Robert A. Samson / Centraalbureau voor Schimmelcultures, Baarn, The Netherlands

I. INTRODUCTION

The application of developmental characters to the taxonomy of conidial fungi has been reviewed by several authors [85,88,193,305]. However, few workers have directed their attention to ontogenetic studies of pathogenic microfungi [232,233]. In this chapter we have chosen examples of imperfect, or anamorphic fungi that belong to the subdivisions Fungi Imperfecti and Zygomycotina and are causative agents of a wide range of mycoses, some commonly encountered in the clinical laboratory.

In addition to its taxonomic significance, knowledge of the mechanisms of conidiogenesis and sporangiosporogenesis has pertinent "spin-offs" for the epidemiologist. Many microfungi that are recognized pathogens are commonly isolated as soil saprobes [8,117, 262]. Information on modes of sporulation and physical factors that promote the formation of disseminating propagules are, therefore, of considerable interest to those workers who investigate epidemiological aspects of both plant and animal mycoses. Ajello [8] has pointed out that a significant void exists in our knowledge of such factors for those fungi that cause pulmonary mycoses in humans. The author stated that studies are required which would "precisely define the various physical and biochemical conditions" for growth and reproduction of these naturally occurring soil saprobes as well as the conditions necessary for these fungi to survive "in the face of the competitive forces exerted against them by the myriads of micro-organisms existing in the soil." Efforts to develop new antifungal antibiotics may also benefit from the availability of data on conidium and sporangiospore ontogeny. Investigators considering future experimentation with such compounds as the polyoxins [120,140,173] and nikkomycin [47,93], potent nucleoside antimycotic agents capable of blocking chitin synthetase activity, should take into account the effects of these antibiotics on conidium and sporangiospore formation as well as yeast budding [43]. For the cytologist and morphogeneticist, examination of the kinds of sporulation in microfungi provides data on novel processes of eucaryotic cell differentiation and may reveal experimental systems that could be useful for investigating mechanisms of exogenous and endogenous control of cell development. As in other scientific disciplines, basic research on apparently esoteric topics such as conidiogenesis and sporangiosporogenesis, may yield information that is relevant to applied aspects of even distantly related fields of study.

II. CONIDIOGENESIS

A. Blastic and Thallic Development

Two basic methods of conidium ontogeny are recognized and referred to as blastic and thallic development. Blastic conidia differentiate from the "blown-out" apex or lateral protrusions of a fertile hypha. The conidium forms by de novo growth, is delimited from its "mother" cell (i.e., conidiogenous cell) by a basal septum and usually secedes as a result of the splitting of that same septum. The latter process has been called schizolysis [88,92, 180,181]. Thallic development, on the other hand, involves conversion of an entire segment of the fertile hypha into a single terminal or intercalary conidium (holothallic ontogeny), or conversion and disarticulation of a hyphal segment into several conidia (thallic-arthric ontogeny). Terminal holothallic conidia usually secede by autolysis of the cell distal to the basal septum of the newly formed conidium. This kind of conidial secession has been termed rhexolysis [92,181]. Thallic-arthric conidia may secede either by schizolysis (holoarthric development) or rhexolysis (enteroarthric development). A diagrammatic summary of these features of conidium ontogeny is presented in Fig. 1.

1. Hyphal Tip Growth and Blastic Development

Since a blastic conidium can differentiate directly from a swollen hyphal apex, it is logical that investigations of the cytological events of hyphal tip growth should provide information on the morphogenesis of this propagule. Ultrastructural studies have revealed morphologically similar cytoplasmic vesicles at the hyphal tips of septate fungi and buds of yeast cells [143]. Careful examination of electron micrographs of conventionally prepared thin sections of these cells demonstrate two populations of vesicles at the cell apices large membrane-bound macrovesicles (approximately 0.1 μm in diameter) which usually contain amorphous or fibrous material, and microvesicles (approximately 0.05 μm in diameter) which are surrounded by a membrane-like shell and contain no visible material [78,88,138]. In septate fungi, the microvesicles are a dense cluster at the hyphal tip which appears as a dark central body or "Spitzenkörper" [48,136,137,144]. Macro- and micro-

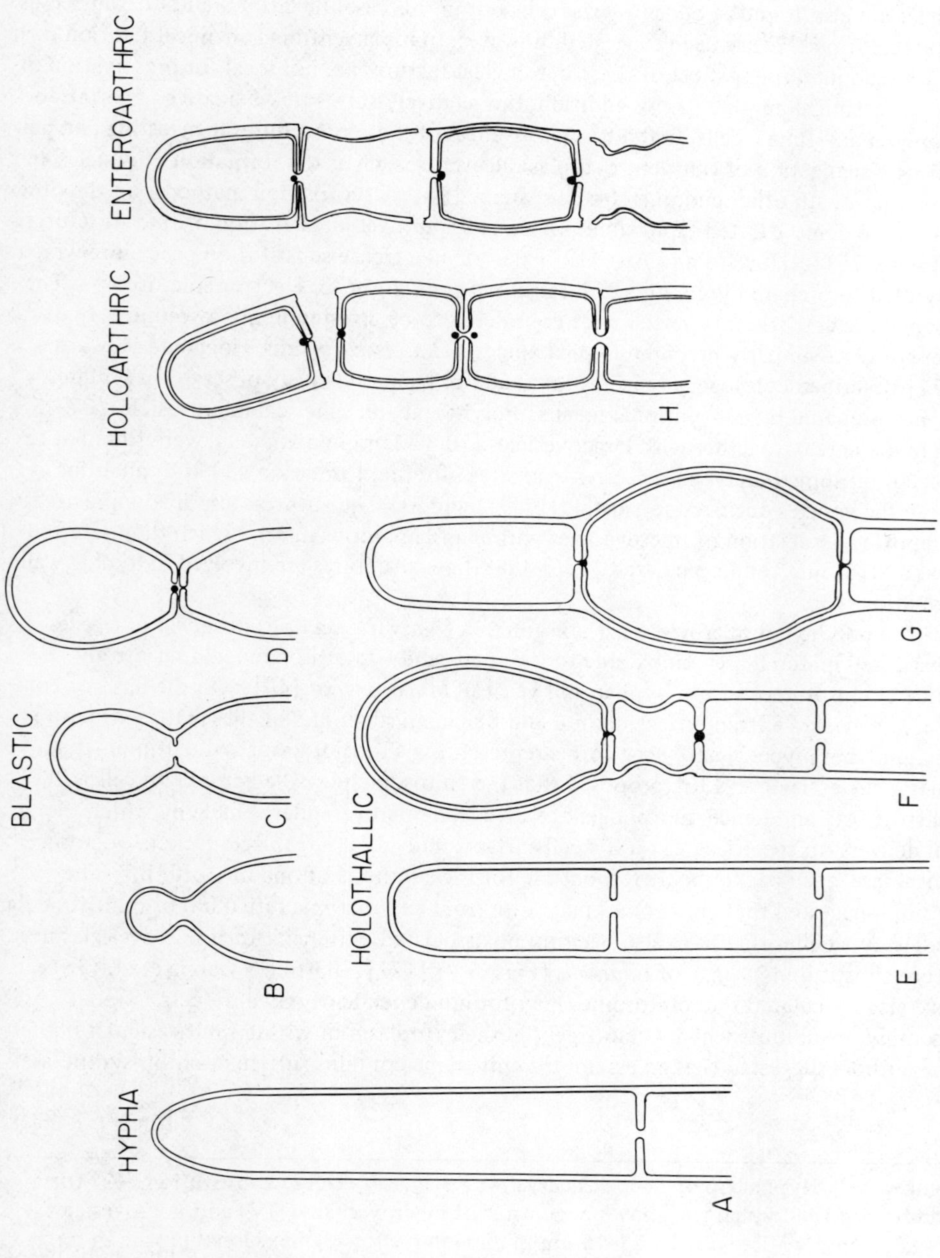

Figure 1 Diagrammatic interpretation of blastic (B-D), terminal holothallic (E,F), intercalary holothallic (G), holoarthric (H), and enteroarthric (I) conidium formation from a fertile hypha (A).

vesicles are also found in buds of yeast cells but in this case the latter are not organized as a Spitzenkörper. It has been suggested, however, that conventional chemical fixation techniques used in defining most of these cytological features are not ideal for preservation of all cytoplasmic components. In addition, the relatively slow rate of fixative penetration through many fungal cells limits the effectiveness of these procedures in arresting and permitting visualization of transient cytoplasmic events, such as the formation of vesicles and their fusion with other endomembrane systems [94]. Cryofixation methods, on the other hand, have demonstrated rapid stabilization and improved preservation of fine structures [126,169,251]. Howard and Aist [177] described a freeze-substitution procedure which they used to prepare hyphal tips of *Fusarium acuminatum* for electron microscopy. The cytoplasmic activities of the cell were considered to be arrested in microseconds. In a later study of similarly prepared hyphal apices of *Laetisaria arvalis,* Hoch and Howard [171] described cytological features that had not been previously observed in septate hyphae prepared by conventional means. Polyhedral and spherical microvesicles (27-45 nm in diameter), together with larger vesicles (70-120 nm in diameter), were found at regions of presumed dictyosome activity, at sites of septal formation and at hyphal tips where the vesicles occurred in clusters (Figs. 2 and 3). The authors described a previously unreported association of microvesicles within a filamentous matrix which they referred to as a "filasome" and speculated [176] that these structures are involved in hyphal wall formation (Fig. 4).

The polyhedral microvesicles (hexagonal in transverse section) observed in freeze-substituted material are comparable to the shape and size of isolated and negatively stained "chitosomes" from hyphae and yeast of *Mucor rouxii* [46]. On the basis of considerable evidence from ultrastructural and biochemical studies of these latter structures isolated from mycelial and yeast forms representing all major groups of chitinous fungi, Bartnicki-Garcia et al. [36] proposed that the chitosome plays "a key role in cell wall construction, and hence, morphogenesis: it is the microvesicular vehicle by which the cell delivers, to specific sites on the cell surface, individual, organized packets of chitin synthetase, each packet being responsible for the synthesis of one microfibril." The authors suggested that chitosomes may arise from endoplasmic reticulum or multivesicular bodies, organelles that have also been found associated with hexagonal microvesicles in freeze-substituted hyphae of *L. arvalis* (Figs. 5-7) [171]. Bartnicki-Garcia et al. [36] have also speculated that chitosomes may form independently by a process of self-assembly of subunits which occur freely in the cytoplasm or within multivesicular bodies. The authors suggested that an assembled chitosome contains sufficient chitin synthetase

---

Figures 2-4 Hyphal tip of *Laetisaria arvalis* preserved by freeze substitution. (2) Ultrastructure of the hyphal tip, showing a cluster of macrovesicles (MV) and microvesicles (mv). (3, inset) Microvesicle, 27-45 nm in diameter, showing hexagonal profile in transverse section. (4) Filasomes (F) composed of filaments 5.2-13.0 nm in diameter. 79,800X, 210,000X, and 139,900X, respectively. (From Ref. 171.)

Figures 5-7 Multivesicular bodies (Fig. 5) and dictyosomes, or cisternal rings (Figs. 6 and 7) of freeze-substituted hyphae of *Laetisaria arvalis.* Arrows in Fig. 6 indicate points of apparent vesiculation. mv, microvesicle; MV, macrovesicle. 63,800X, 57,500X, and 61,400X, respectively. (From Ref. 171.)

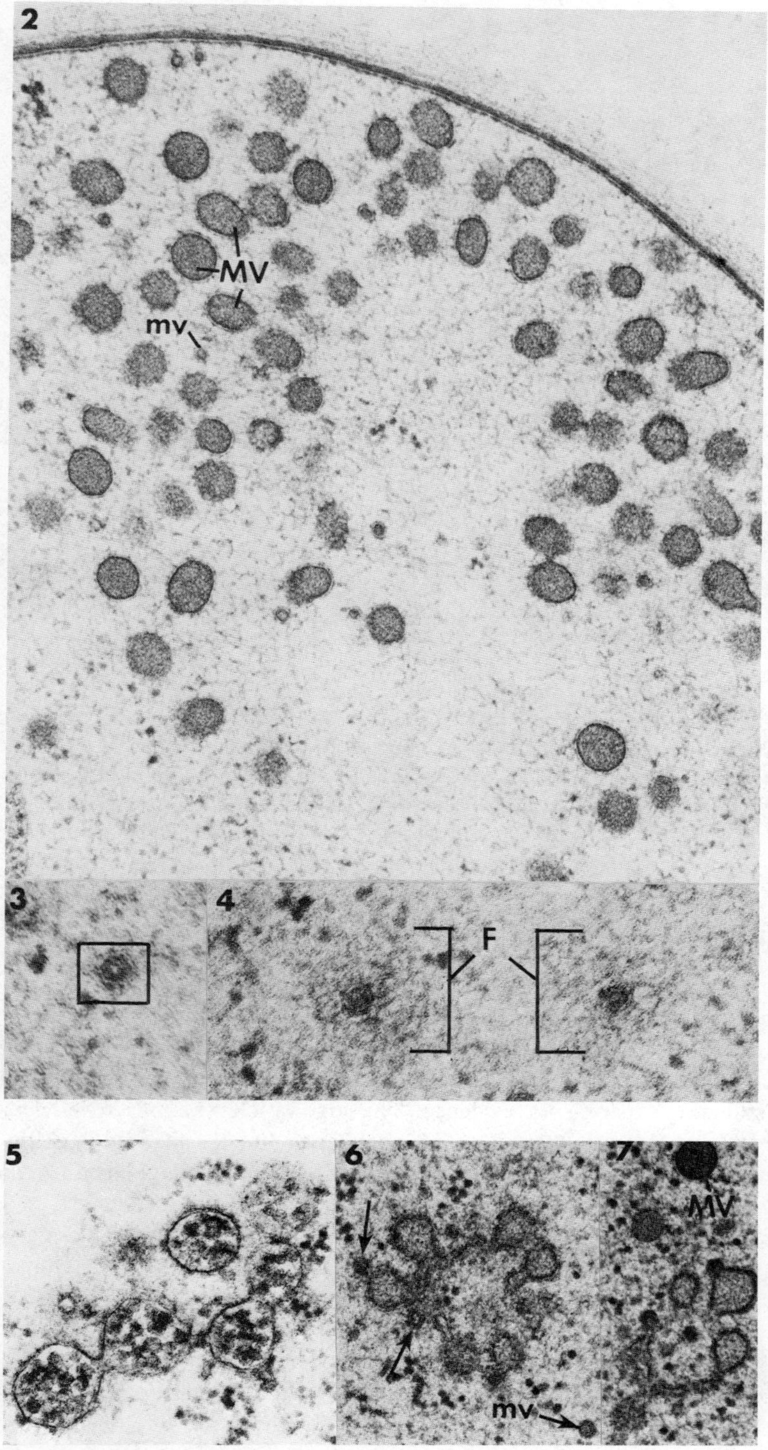
2
MV
mv
3
4
F
5
6
mv
7
MV

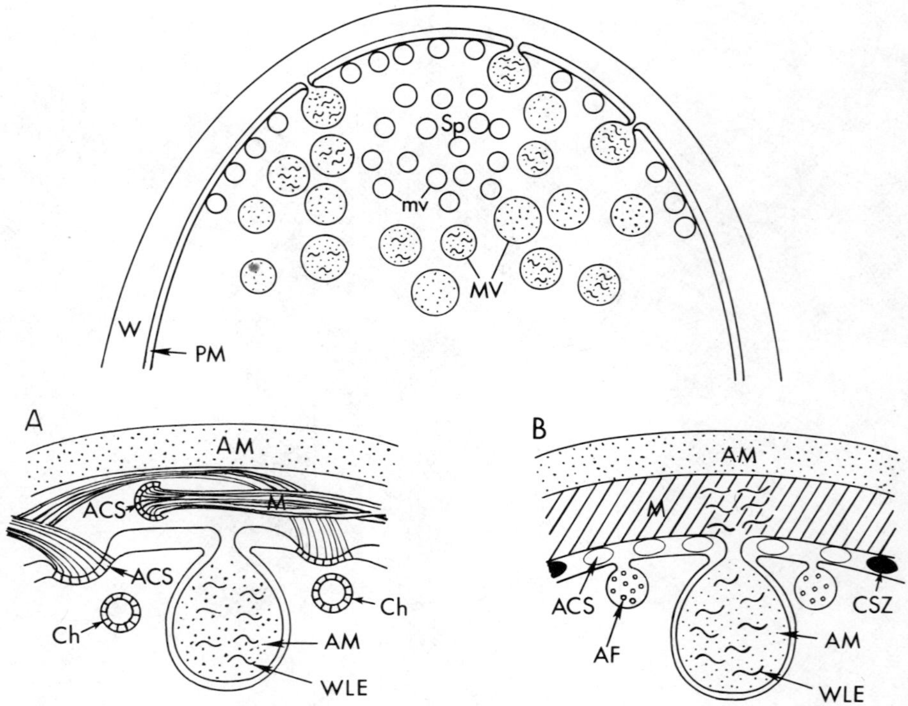

Figure 8 Diagrammatic interpretation of cytoplasmic and wall structures associated with chitin fibrillogenesis: (A) *sensu* Bartnicki-Garcia et al. [36]; (B) *sensu* Cabib et al. [59]. ACS, activated chitin synthetase; AF, activating factor (e.g., protease for chitin synthetase); AM, amorphous component of cell wall; Ch, chitosome; CSZ, chitin synthetase zymogen; M, microfibrils; mv, microvesicle; MV, vacrovesicle; PM, plasma membrane; Sp, Spitzenkörper, W, wall; WLE, wall lytic enzymes.

molecules to produce a single chitin microfibril and must subsequently fuse with the plasmalemma or pass through the membrane (exocytosis?) at specific sites of the fungal cell (e.g., hyphal tip) before fibrillogenesis is initiated. The synonymy of hexagonal microvesicles revealed by freeze substitution and chitosomes isolated from various fungi is tempting to suggest, but still lacks experimental evidence. The large vesicles (i.e., macrovesicles) observed at hyphal tips and buds of yeast cells are probably not directly involved in chitin biosynthesis, but may instead contain glycoproteins and lytic (wall softening) enzymes which are delivered to sites of new wall formation when the vesicles fuse with the cell membrane [46,122]. This fusion process also contributes new membrane to the plasmalemma which is expanding as wall intussusception proceeds and the cell grows. A summation of these concepts of cytological control of hyphal tip growth is presented in Fig. 8A.

Of primary significance in this scheme is that chitosomes provide both a spatial and temporal coordination of hyphal tip growth, an essential aspect of any hypothesis that attempts to explain polarized growth in fungi. An alternative proposal has been presented by Cabib and co-workers, however, which also satisfies this condition [57-59,108,122, 317]. The results presented by these investigators suggest that chitin synthetase zymogen (inactive form) is uniformly distributed on the plasma membrane at multiple sites and that chitin synthesis is initiated by localized activation of the enzyme. The latter phenomenon was proposed to occur by delivery of an activating factor (protease) contained within vesicles (microvesicles) to specific sites on the plasmalemma (hyphal tip, yeast bud, septum). Also, part of this hypothesis is the presence of an inhibitor which functions as a "safety valve" neutralizing any activating factor released into the cytoplasm and thus preventing random activation of the zymogen at sites other than those where fusion of the microvesicles with the plasmalemma have occurred (Fig. 8B). Several of the same fungi have been examined by proponents of the two hypotheses for chitin fibrillogenesis presented in Fig. 8A and B. While in the first case (Fig. 8A) it was suggested that chitin synthetase in its activated state is delivered to the cell membrane by chitosomes, the second scheme (Fig. 8B) suggests that most of the enzyme is associated with the plasmalemma in its zymogenic state. Cabib et al. [59] concluded that their "results are not consistent with the hypothesis that localized formation of chitin is achieved by directing particles containing the synthetase [chitosomes] to a specific site. If this were the case, the enzyme would presumably be found only in one restricted area on the plasma membrane, and most of it would be in the active rather than in the zymogen state. The reason for finding most of the synthetase in a precursor form is," in the authors' view, "that only the small portion of it which is found at the location of the growing" region of the cell "is functioning at any given time. . . ."

On the basis of current evidence both concepts of chitin formation seem plausible. However, it is not our purpose to discuss the detailed arguments for or against each proposal. Instead, we use these schemes simply as a basis for examining concepts of blastic and thallic conidium development. Chitin is a major component of the conidial wall of all filamentous fungi so far examined [21]. If one considers the cytological events associated with formation of a single holoblastic conidium at the apex of a fertile hypha, certain extrapolations can be made from the foregoing discussion of chitin synthesis and hyphal growth. As the conical apex of the hypha "blows out" to form the spherical conidium initial, wall intussusception is no longer polarized but probably involves the entire hemisphere of the growing cell (Fig. 9). Evidence from several ultrastructural studies of hyphal tips and conidial initials support this conclusion. The macro- and microvesicles present in the vegetative hyphal tip are at least partly dependent on cytoplasmic microtubules for their apical movement and accumulation [177a]. In the swollen apex of the fertile hypha (conidium initial), the cluster of vesicles has dispersed [239,313,316], but vesicles similar in morphology to those in hyphal tips are still visible adjacent ot the plasmalemma encompassing the young conidium [88]. The disposition of microtubules during conidiogenesis is unknown. Reiss and Herth [257b] have presented evidence that a Ca^{2+} gradient is essential for polarized growth of pollen tubes of *Lilium longiflorum* and if this gradient is disturbed by use of a cationophore (X-537A), tip elongation is arrested, the apex of the tube swells, and the apical wall thickens. The authors point out that other ions and relative ion balances may also be disturbed by the cationophore which may be critical for tip growth, but they conclude that "oriented exocytosis is more dependent on the specific Ca^{2+} distribution, whereas vesicle transport and formation is also dependent on other ions." A tip-to-base Ca^{2+} gradient has been demonstrated in fungal hyphae [257a] and further investigations of the significance of this phenomenon in hyphal tip growth and conidial differentiation are required. Dictyosomes, which are identified as cisternal rings in conidial fungi [80,88] and suggested to be a source of the secretory vesicles found at

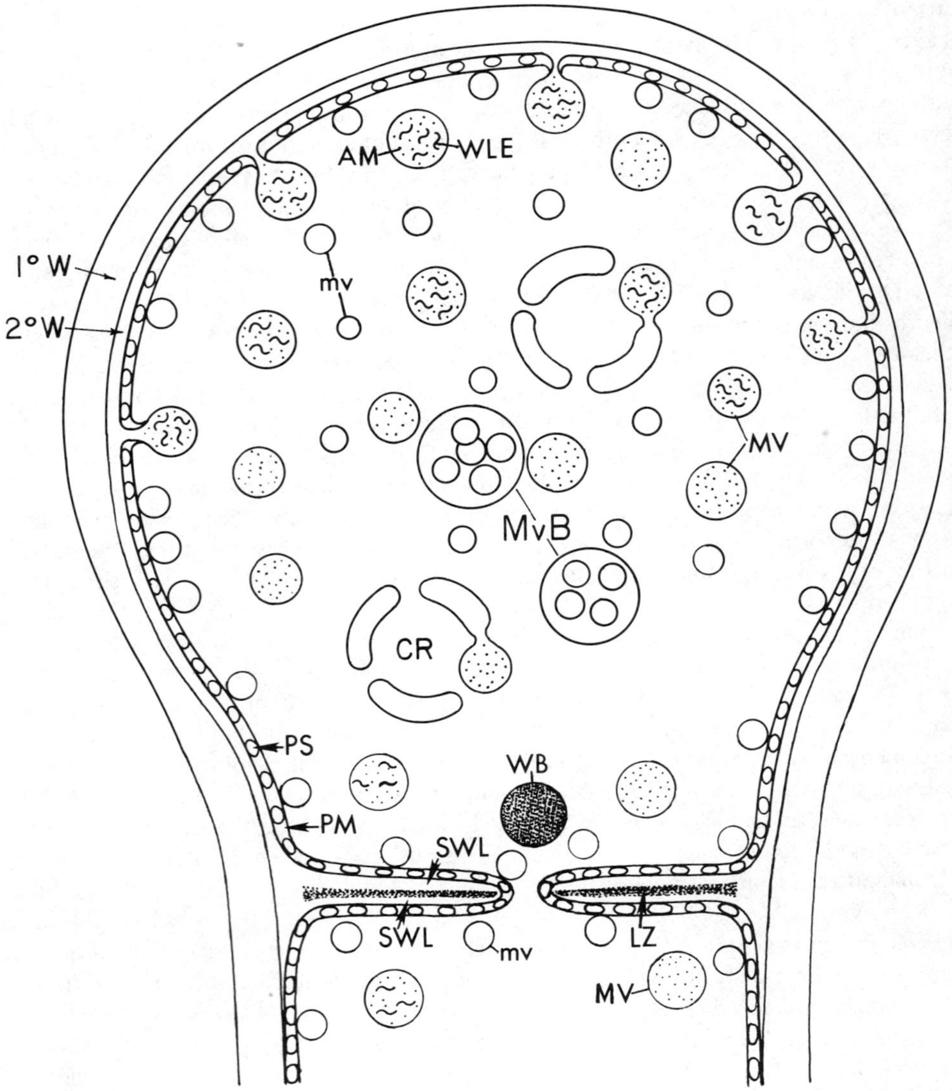

Figure 9 Diagrammatic interpretation of cytoplasmic structures associated with blastic conidium formation. 1°W, primary conidial wall; 2°W, secondary conidial wall; AM, amorphous material; CR, cisternal ring (dictyosomal equivalent); LZ, lytic zone (electron-transparent region) between septal wall layers; mv, microvesicles; MV, macrovesicles; MvB, multivesicular body; PM, plasma membrane; PS, polysaccharide synthetases; SWL, septal wall layer; WB, Woronin body; WLE, wall lytic enzymes.

the hyphal tip (Figs. 6 and 7) [133,143,144,171], are located in the subapical region of vegetative hyphae but accumulate in the cytoplasm of the conidium initial [88]. Numerous multivesicular bodies are also found in the young conidium [88]. A thin, inner electron-transparent wall layer adjacent to the outer wall encompasses the swollen apex of the fertile hypha [80,88]. The diagrammatic interpretation of blastic conidium formation presented in Fig. 9 has incorporated all these observations of cytodifferentiation. The newly formed conidial wall (secondary wall) results from the activity of chitin synthetase along with other polysaccharide synthetases on the plasmalemma. The microvesicles, which appraently fuse with the plasma membrane, may be the vehicles for transport of chitin synthetase (*sensu* Bartnicki-Garcia et al. [36]) or the activating factor for the zymogenic form of this enzyme (*sensu* Cabib et al. [59]). The macrovesicles, which also apparently fuse with the plasmalemma, may transport amorphous wall components (e.g., glycoproteins) as well as lytic enzymes to the site of conidium differentiation. Such wall-softening enzymes are suggested to play a pivotal role in altering the balance between wall rigidification and wall plasticization in the region of new wall growth [34,88,193]. An abrupt shift in favor of plasticization allows the outer fertile hyphal wall to stretch rapidly during the initial blowing-out stage of blastic conidium formation.

Initiation of septum development at the base of the conidium usually occurs while the cell is still growing. Thin sections have shown that early septation is signaled by circumferential invagination of the plasmalemma and concomitant deposition of wall material between the fold of membrane. This results in formation of a disk-like septum characteristic of most ascomycetous fungi. On the basis of ultrastructural investigations of septation in yeast and vegetative hyphae, it is suggested that microfilaments are associated with the initiation of septum development (Figs. 10 and 11). Microfilaments are firmly attached to the plasmalemma and arranged in a ring [54] or belt [139] at the site of septation. Girbardt [139] has reported that the microfilamentous septal belt (FSB) in hyphae of *Trametes versicolor* appears 2-3 min before the first signs of cross-wall formation. This same author has pointed out that microfilaments can attach to the membranes of various organelles [236,248], including mitochondria, microvesicles, and multivesicular bodies in fungal hyphae, which "may partially fulfill the demand for sequestration of factors needed for septum formation. It is possible . . . that the FSB works mainly as a mechanical device for entrapment of vesicles (chitosomes?. . .) bearing chitin synthetase and . . . other vesicles with proteases for activation of the zymogen." Girbardt [139] also reported that cytochalasin B had no effect on the FSB in *T. versicolor.* On this basis he proposed that centripetal movement of the plasmalemma during septation is the result of ingrowth of the septal wall rather than contraction of microfilaments and the latter, therefore, function as cytoskeletal rather than contractile elements. However, Allen et al. [13] have pointed out that fungi are less sensitive to cytochalasin B than to cytochalasin A. The latter has been shown to decrease the distance between hyphal branches, cause periodic constrictions along the hyphae, and induce formation of swollen or spatulate tips. The authors have also indicated that cytochalasin A affects intracellular organization associated with wall deposition (i.e., arrangement of vesicles at the hyphal tip). Such studies may prove useful in determining the interrelationships of microfilaments, microvesicles, wall deposition, and growth [311].

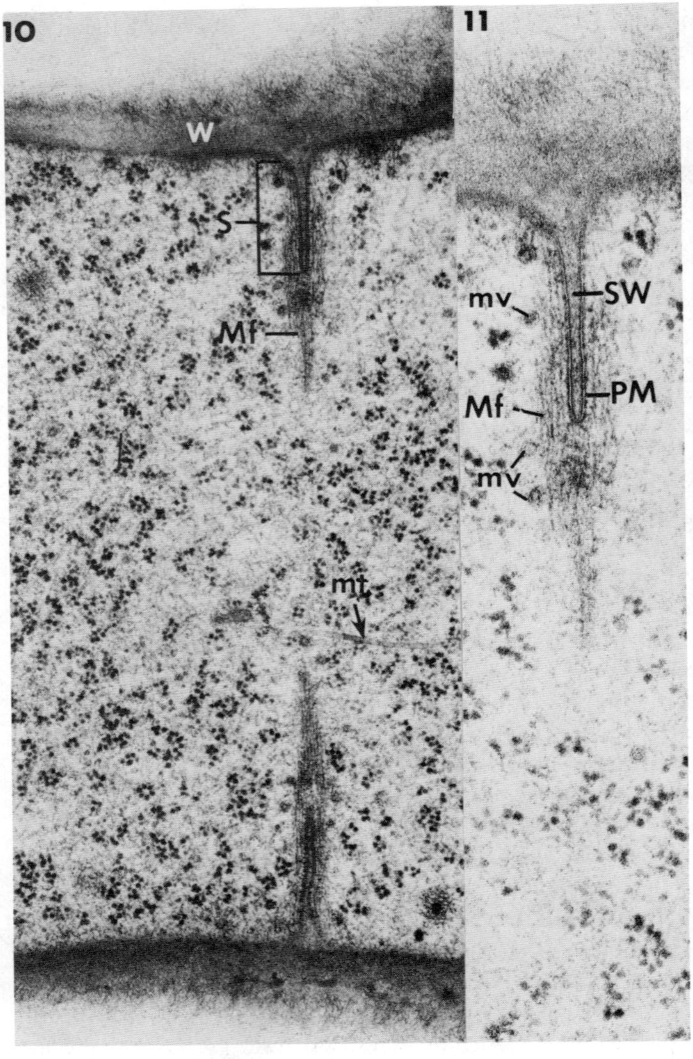

Figures 10 and 11 Septum of *Laetisaria arvalis* preserved by freeze substitution. Microfilaments (Mf), 5.2-5.4 nm wide, encompass the developing septum (S). Microvesicles (mv) are found adjacent to the developing septum and intimately associated with the microfilaments. mt, microtubule; PM, plasma membrane; SW, septal wall; W. hyphal wall. 37,800X and 79,600X, respectively. (From Ref. 171.)

Thickening of the conidial and septal wall occurs while cytoplasmic continuity is still maintained between the blastic conidium and fertile hypha, or conidiogenous cell (Fig. 9). The septal wall subsequently undergoes differentiation to form two unequally thickened layers, each presumably containing chitin as the principal wall component [141], and an intervening electron-transparent layer suggested to be a zone of lytic enzyme activity (Fig. 13) [88]. The two septal wall layers eventually separate along this electron-transparent zone, resulting in a process of conidial secession referred to earlier as schizolysis. The septal pore at the base of the conidium is plugged, usually with a membrane-bound organelle called a Woronin body [44,68,80,151]. The pore through the thin lower septal layer is commonly sealed by centripetal growth of the wall. In many forms, this lower septal layer becomes the new growing apex of the proliferating conidiogenous cell, which is destined to give rise to a second blastic conidium (e.g., phialidic and annellidic development, Figs. 28 and 29, respectively).

2. Thallic Development

The morphogenetic differences between blastic and thallic development are reflections of different sequences of cytological events. Initiation of thallic conidium formation is signaled by the arrest of tip growth of the fertile hypha [74,76,84]. It is suspected that this event is associated with disappearance of the Spitzenkörper and dispersion of the apical vesicles [88]. The ontogenetic processes that follow govern whether the determinate fertile hypha differentiates into a single, terminal, or intercalary, holothallic conidium (Figs. 1E-G and 12) or multiple arthric conidia (Fig. 1H and I).

a. Holothallic Conidia In *Microsporum gypseum,* which exemplifies holothallic conidium formation, the apical segment of the fertile hypha first begins to swell and is then delimited from the supporting cylindrical hypha by a transverse septum [76,88,129]. This initial swelling process is probably the result of plasticization of the preexisting cell wall. A basic difference between blastic and thallic development is recognized at this point: the blastic conidium initial involves blowing out of the apical dome of the fertile hypha due to polarized wall synthesis and plasticization (Fig. 9), while the holothallic conidium arises initially from a large segment of the fertile hypha and its development involves depolarized activity of lytic and synthesizing enzymes in the preexisting cell wall (Fig. 12A-C). On this basis, it would seem that at least during early conidium development the amount of surface area of the plasmalemma involved in processes of vesicle fusion and enzyme activation associated with wall differentiation determines whether a blastic or thallic conidium will be formed. As the wall encompassing the conidium initial of *M. gypseum* begins to stretch, new wall synthesis is accelerated, presumably due to increased activity of polysaccharide synthetases (e.g., chitin synthetase) and an inner, secondary wall appears (2°W, Fig. 12C) [129]. In addition to the involvement of vesicles in this process of wall synthesis, it has been suggested that lomasomes (Lo) which appear adjacent to the newly forming wall of the young holothallic conidium may also contribute to the synthesis of wall components [88,152,153,164,221,225-227,331,332]. Soon after the septum at the base of the initial has formed, a second septum appears distal to the young conidium, delimiting a short cylindrical cell. The processes of septal initiation and development are considered to be similar to those described earlier for blastic conidium formation. As the wall surrounding the initial continues to thicken, additional septation occurs, which results in formation of the mature phragmoconidium. The lateral walls of the basal cell, on the other hand, do not thicken or undergo any apparent differentiation until this cell suddenly collapses. The thin, lateral walls fold inward, the cell degenerates, and the terminal, holothallic conidium (macroconidium) of *M. gypseum* is released from its supporting hypha [76,88]. This is an example of the rhexolytic method of conidial secession. The cytological events associated with this autolytic process are unexplored. These same features of holothallic conidium ontogeny are demonstrated during macroconidium formation in *Epidermophyton* as well as macro- and microconidium formation

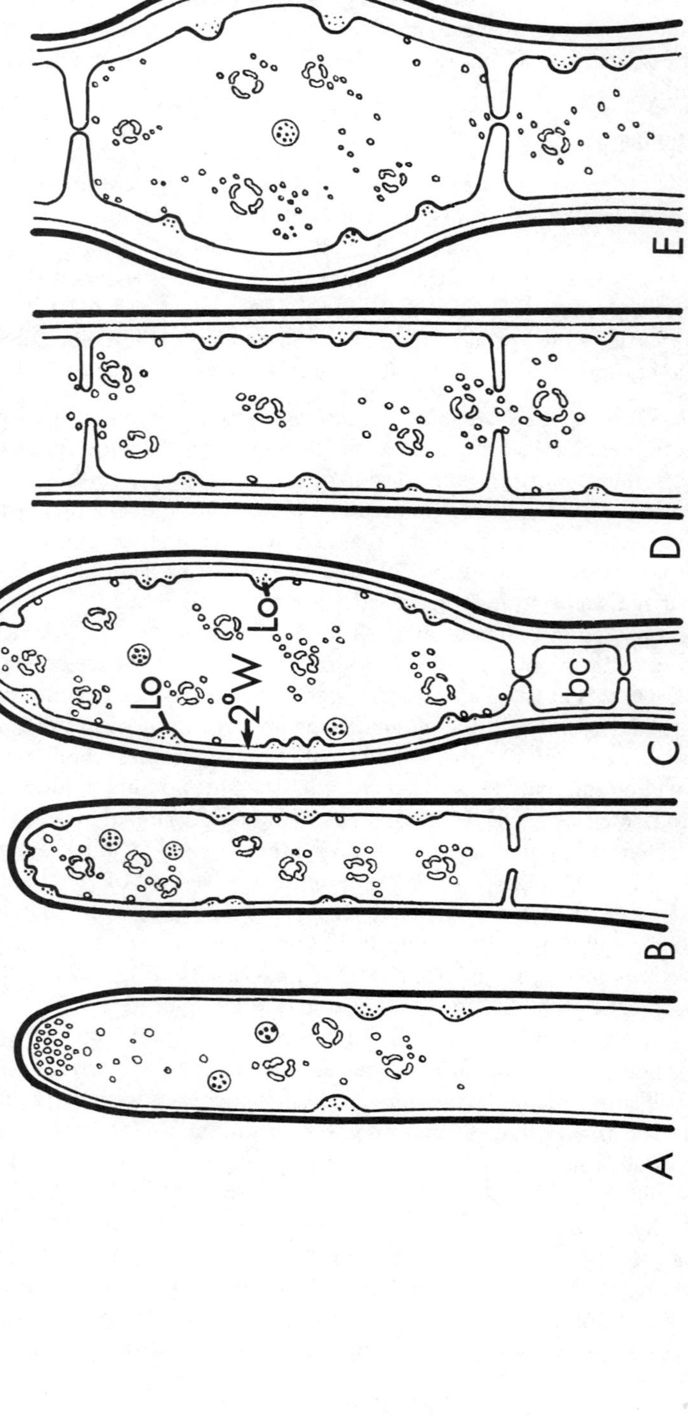

Figure 12 Diagrammatic interpretation of protoplasmic and wall differentiation during formation of holothallic terminal conidia (A-C) and intercalary holothallic conidia (D,E). bc, basal cell which later autolyses; Lo, lomasome; 2°W, secondary conidial wall. (From Ref. 88.)

in *Trichophyton* and *Microsporum.* Intercalary, holothallic conidium formation occurs by a similar process of secondary thickening of the cell, which is delimited from the rest of the hypha by cross walls (Fig. 12D and E). No secession mechanism is recognized for this kind of conidium.

b. Thallic-Arthric Conidia Polarized growth of a vegetative hypha is arrested, cross walls form at first in the subapical region of the hypha and then closer to the apex, the septa split centripetally, and the hypha disarticulates; these are successive stages of holoarthric conidium formation as demonstrated by *Geotrichum candidum* (Fig. 34) [74,84, 88]. Karyological examinations of this mode of conidiogenesis suggest that an association may exist between the positioning of nuclei within the fertile hypha and sites of septum formation. Kendrick and Chang [194] have shown in *G. candidum* that prior to septation of the determinate hypha, nuclei are fairly regularly spaced. Septa then form so that large segments of the fertile hypha are delimited, each containing two nuclei. These large segments subsequently divide by septation and each cell, destined to become an "arthroconidium," contains a single nucleus. Sometimes this nucleus subsequently undergoes mitosis and a binucleate arthroconidium is produced. Arita [20] has reported similar results based on his examination of nuclear behavior during formation of holoarthric conidia in *Pholiota adiposa.* Wall thickening and some swelling occurs concomitantly with the delimitation of the uninucleate conidia of *G. candidum.* Thin sections of the septal wall between adjacent conidia examined under the electron microscope reveal two electron-translucent layers separated by an electron-transparent zone [74]. Disarticulation of the chain of holoarthric conidia occurs by schizolysis as described earlier.

A second kind of thallic-arthric development also involves progressive septation of a determinate, fertile hypha but subsequent differentiation of conidia occurs within the hyphal wall. This process is referred to as enteroarthric conidiogenesis (Figs. 1I and 36-42). Certain hyphal segments delimited by cross walls differentiate into conidia while adjacent segments collapse and degenerate, resulting in rhexolytic conidial secession. Thin sections of differentiating fertile hyphae reveal that a secondary wall is laid down within the hyphal wall in those segments destined to become conidia [74]. In contrast, no such secondary wall encompasses the segments that later undergo autolysis. The lateral and septal walls of these cells gradually become thinner while the cell contents degenerate (Figs. 36-38). As in the case of holoarthric development, nuclei are regularly spaced within the fertile hypha prior to septation, and each conidium finally contains one to four nuclei [74]. When cell degeneration occurs, nuclei are trapped within the autolysing segments of the fertile hypha. This process of conidium formation, which is wasteful from the standpoint of energy conservation, requires further investigation.

B. Wall Chemistry and Conidiogenesis

A final point of consideration in this discussion of conidiogenesis concerns the interrelationships of wall chemistry and morphogenesis. Bartnicki-Garcia [35] has concluded that the final shape of a fungal cell is determined by the way the wall is constructed. By comparing the wall composition of conidia and vegetative hyphae from the same fungus, certain trends become evident which may reflect relationships between alterations in wall chemistry and morphogenesis. An interpretation of the wall composition and arrangement of wall components in hyphae of *Neurospora crassa* is presented in Fig. 13. The different wall layers actually occur as gradations rather than distinct zonations, so that

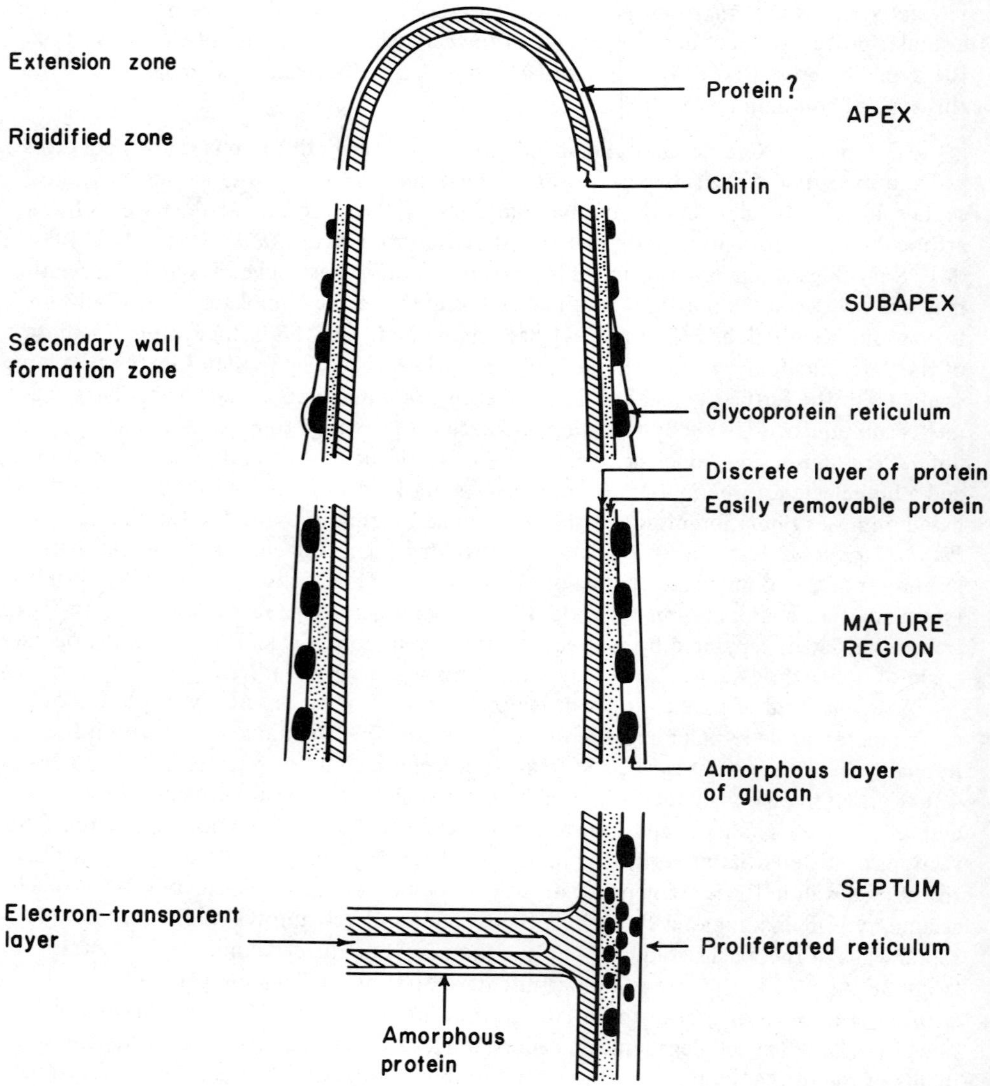

Figure 13 Diagrammatic representation of the hyphal and septal wall structure of *Neurospora crassa* based on interpretations of Burnett [51], Gooday and Trinci [141], Hunsley and Gooday [184], Hunsley and Kay [185], and Trinci [314].

this model is somewhat oversimplified. However, using this scheme for reference, let us examine the chemical composition of hyphal walls reported for several conidial fungi (Table 1). Note the relatively high concentration of neutral carbohydrates with glucose as the predominant monosaccharide in all representative species. The protein and lipid content, on the other hand, are comparatively low. The percentage of hexosamine in *Fusarium sulphureum, Epidermophyton floccosum,* and *Trichophyton mentagrophytes* is high in comparison to other fungi in Table 1. However, hexosamine estimates vary con-

Table 1 Composition of Hyphal Walls of Conidial Fungi

Component	Fungus							
	Aspergillus niger	*Aspergillus nidulans*	*Fusarium sulphureum*	*Helminthosporium sativum*	*Ceratocystis ulmi*[a]	*Epidermophyton floccosum*	*Trichophyton mentagrophytes*	*Geotrichum candidum*[b]
Total neutral carbohydrates[c] (%)	64.5	55.1	22.0	41.0	56.0	46.3	47.9	60.3 [56.8]
Glucose	(70.6)[d]	(28.9)	(12.0)	(37.0)	+++	(35.2)	(36.2)	(86.0)
Galactose	(22.0)	(3.9)	(3.0)	(2.0)	++	tr[e]	tr[e]	(4.0)
Mannose	(6.5)	(2.8)	(7.0)	(2.0)	+	(5.3)	(11.7)	(5.0)
Hexosamine (%)	14.9	14.3[f]	39.0[f]	8.6[f]	3.4[g]	30.9[h]	30.4	9.7
Protein (%)	6.6	10.6	7.3	18.0	7.9	10.3	7.8	8.8 [5.0]
Lipid (%)	13.6	9.0	–	11.4	4.5	4.6	6.6	5.3 [1.0]

[a]Vegetative hyphae of anamorph.
[b]Two sets of data reported for *G. candidum* adapted from Ebina et al. [110] (left column) and Kier et al. [200] (data in square brackets).
[c]Calculated from original sources.
[d]Monosaccharide content expressed as percentage of total carbohydrate.
[e]Trace.
[f]Expressed as chitin in original source.
[g]Expressed as N-acetylglucosamine in original source.
[h]Expressed as glucosamine in original source.

Source: Data for *A. niger, A. nidulans, F. sulphureum, H. sativum, C. ulmi, E. floccosum,* and *T. mentagrophytes,* respectively, are adapted from Cole et al. [89], Bull [50], Barran et al. [31], Applegarth and Bozoian [19], Harris and Taber [154], Nozawa et al. [242], and Noguchi et al. [241].

Table 2 Composition of Conidial Walls

Component	Blastic conidia					Thallic conidia		
	Aspergillus niger		*Aspergillus oryzae*	*Sporothrix schenckii*	*Fusarium sulphureum*	*Trichophyton mentagrophytes* (holothallic)	*Trichophyton mentagrophytes* (holoarthric)	*Geotrichum candidum* (holoarthric)[a]
	outer wall	inner wall						
Total neutral carbohydrates[b] (%)	2.5	31.2	27.0	47.0	37.6	56.1	53.0	69.7 [50.6/51.2]
Glucose[b] (%)	(37.5)[c]	(36.4)	+	(57.9)	(24.7)	+++	+++	(23.7)
Galactose	(24.2)	(30.7)	+	tr[d]	(6.7)	tr[d]	+	(21.6)
Mannose	(31.5)	(29.5)	+	(25.1)	(6.2)	+	+	(54.7)
Hexosamine (%)	2.2	6.5	20.0[e]	8.3	12.5[f]	16.0[f]	30.0[f]	12.2
Protein (%)	63.4	20.0	19.0	24.2	21.6	22.6	3.0	8.3 [6.7/7.5]
Lipid (%)	24.8	34.5	–	19.6	–	6.5	1.0	1.3 [1.0/1.0]

[a]Two sets of data reported for *G. candidum* adapted from unpublished results of Nozawa and Cole (left column) and reported results of Kier et al. [200] (data in square brackets indicate composition of two kinds of holoarthric conidia produced under different growth conditions; cylindrical conidia/ellipsoidal conidia).
[b]Calculated from original sources.
[c]Monosaccharide content expressed as percentage of total carbohydrate.
[d]Trace.
[e]Expressed as D-glucosamine in original source.
[f]Expressed as chitin in original source.
Source: Data for *A. niger, A. oryzae, S. Schenckii, F. sulphureum, T. mentagrophytes* (holothallic microconidia), and *T. mentagrophytes* (holoarthric conidia), respectively, are adapted from Cole et al. [89], Horikoshi and Iida [174, Previato et al. [253], Schneider et al [276], Wu-Yuan and Hashimoto [334], and Hashimoto et al. [161].

siderably depending on the analytical procedure employed. Also, total hexosamine is not necessarily equivalent to the chitin content of hyphal walls [89]. Table 2 provides data on the chemical composition of conidial walls of representative fungi which undergo both blastic and thallic development. The most striking contrast between Tables 1 and 2 is the difference in ratios of total neutral carbohydrate to protein; in the case of hyphal and holoarthric conidial walls (i.e., *T. mentagrophytes* and *G. candidum;* Table 1) the ratios are high, but they are significantly lower for the walls of blastic conidia. In addition, the total lipid content of hyphal and thallic conidial walls is generally lower than that of blastic conidial walls. Insufficient information is available at present to determine the precise effects of these variations in relative amounts of major wall components on morphogenesis. We do suggest, however, that these data provide additional support for our distinction between blastic, holothallic, and holoarthric development. Since a blastic conidium develops by de novo growth and usually arises from the apex of a fertile hypha, wall composition of the conidium initial would be expected at first to be comparable to that of the hyphal tip (Fig. 13). However, as the hyphal apex swells during conidium differentiation, the composition of the newly synthesized wall encompassing the young conidium would be predictably different from the composition of the wall which maintained the conical shape of the hyphal tip. During holothallic development both incorporation of preexisting hyphal wall material as well as new wall synthesis are involved in differentiation of the conidial wall. Holoarthric conidia are formed by simple hyphal fragmentation and involve little new wall synthesis. It is suggested that the data in Tables 1 and 2 are reflections of these morphogenetic differences. Whereas the composition of holoarthric conidia is very similar to that of the hyphal wall, the composition of blastic conidia is markedly different. The degree of variation between the wall composition of holothallic conidia and associated hyphae is less dramatic.

It is proposed that conidiogenesis in general is associated with an increase in deposition of glycoproteins and lipoproteins on the surface of conidial walls. Such wall components may contribute to the hydrophobicity demonstrated by many conidia [37,87]. Perhaps of particular significance in this regard is the distinct fibrillar layer formed on the outer surface of conidia revealed by freeze fracturing (Figs. 14 and 15) and identified as a crystallized protein [38,162,334] or lipoprotein complex [87]. This wall component, referred to as the rodlet layer [72,167], has been found on the wall surface of both blastic and thallic conidia [73,74,89] but not on the surface of vegetative hyphae produced by conidial fungi. Synthesis of these and other proteinaceous surface wall components may replace some of the amorphous glucans found in the outer wall layer of hyphae (see Fig. 13).

Future studies of wall chemistry and morphogenesis in conidial fungi should take advantage of the technical advances made in fermenter systems for growing fungal cultures [294]. It is now possible "to examine conidiation under quite exacting controlled experimental conditions, thus permitting some understanding of how individual environmental parameters like aeration, temperature, pH, nutrient type and concentration affect the overall process" [295]. In addition, it is necessary for cultures to be developmentally synchronized in order to obtain reproducible results from biochemical analyses. Microcycle conidiation, which has been defined as the recapitulation of conidiation following conidial germination without an intervening phase of mycelial growth [296], provides the ideal model system for correlated investigations of wall chemistry and morphogenesis [15,16,96,97,107, 281-283].

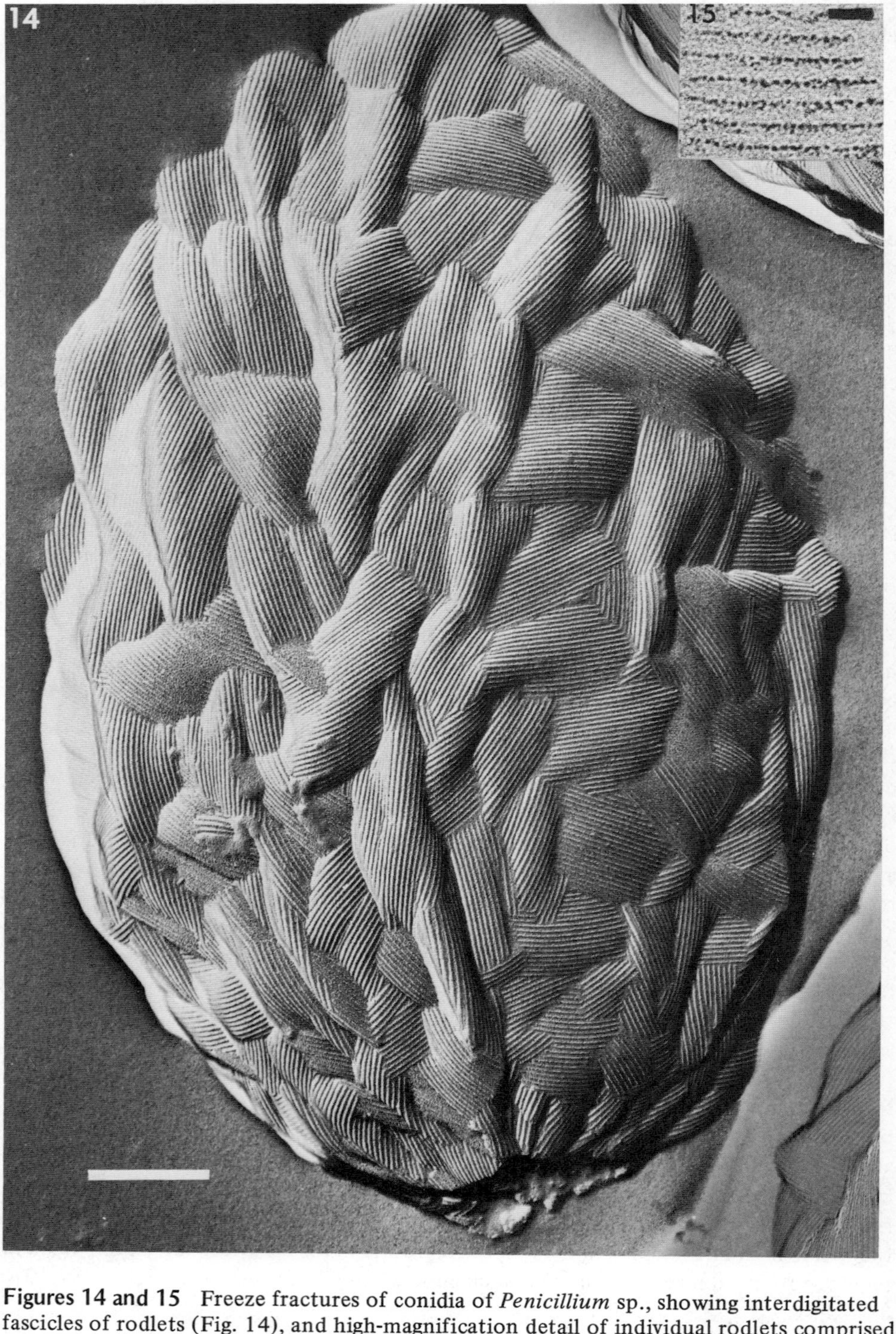

Figures 14 and 15 Freeze fractures of conidia of *Penicillium* sp., showing interdigitated fascicles of rodlets (Fig. 14), and high-magnification detail of individual rodlets comprised of chains of subunits (Fig. 15). Bar in Fig. 14 = 1 μm; bar in Fig. 15 = 20 nm. (From Ref. 89.)

C. Categories of Conidium and Conidiogenous Cell Development

A summary of the different kinds of conidium and conidiogenous cell ontogeny is presented in Fig. 16. In addition to the basic modes of blastic and thallic development, several subcategories of conidiogenesis are recognized. These are based on differences in (1) processes of wall differentiation during conidium formation (e.g., holoblastic versus enteroblastic), (2) order of conidium production (e.g., acropetal versus basipetal), and (3) kinds of conidiogenous cell development (e.g., phialidic versus annellidic). Not all mechanisms of conidiogenesis in Fig. 16 are demonstrated by fungi recognized as pathogens of animals and humans. For a more general treatment of conidium ontogeny in the Fungi Imperfecti, the reader is referred to other sources [85,88,179,193].

1. Holoblastic (Solitary) Conidia; Determinate Conidiogenous Cells

Histoplasma capsulatum is a well-known fungal pathogen which occurs naturally as a saprobe in soil or associated with decaying plant material enriched with animal feces (commonly bat guano) or other nitrogenous waste materials [95,223,256,327]. When the fungus is grown in pure culture on Sabouraud's dextrose agar at 25°C, both macroconidia (Ma) and microconidia (Mi) are produced (Figs. 17 and 18). These terminal solitary conidia are referred to as "aleuriospores" by some authors. However, the term "aleuriospore" has been used in many different contexts in mycological literature, which has generated considerable confusion. We have, therefore, rejected the use of this term [88,193]. Both macro- and microconidia commonly form at the tips of short, apically swollen, fertile branches (conidiogenous cells) which grow out from vegetative hyphae at right angles. The fertile cell ceases elongation at the time of conidium initiation. The conidiogenous cell may also be reduced to a short, lateral protrusion of the hyphal wall (CC, Fig. 17). During macro- and microconidium formation, the outer wall of the fertile hypha remains intact and blows out (holoblastic development) to encompass the smooth-walled conidium initial (CI, Fig. 18). The surface of the macroconidial wall soon develops many small protuberances (arrow in Fig. 18) which later form the tuberculate appendages that characterize the mature propagule (Fig. 17). These outgrowths react positively with PAS stain and have been suggested to consist of mucopolysaccharide. The SEM also reveals tiny wall protuberances on the surface of microconidia (Mi, Fig. 17). A cross wall has formed at the base of the mature macroconidium in Fig. 17 (arrow), delimiting it from the fertile hypha (CC). The mechanism of secession in *Histoplasma* appears to be by rhexolysis, although ultrastructural evidence is still lacking. The slightly collapsed conidiogenous cell (CC) in Fig. 17 is suggestive of this process. Van Oorschot [322] has demonstrated that secession of conidia in related species (e.g., *Chrysosporium*) occurs by rhexolysis, involving fracture of the fertile cell below the basal septum of the conidium. The similarity in habitat, morphology, and growth conditions between certain species of *Chrysosporium* and *Histoplasma capsulatum,* both anamorphs of the Gymnoascaceae, has been discussed by several authors [64,134,291]. However, *Chrysosporium* spp. differ from *H. capsulatum* in that the former are not pathogenic and lack the ability to convert to a yeast phase at 37°C. Hyphal-yeast conversion, or dimorphism, in pathogenic microfungi is discussed briefly later in this chapter and in detail by other authors of this volume (see Chap. 10).

Solitary holoblastic conidia are also produced by *Emmonsia parva,* a soil hyphomycete which is the causative agent of a pulmonary mycosis (adiaspiromycosis) of humans, rodents, and certain soil burrowing animals [117]. Considerable morphogenetic plasticity is demonstrated by this fungus when grown in pure culture (Fig. 19). On Sabouraud's

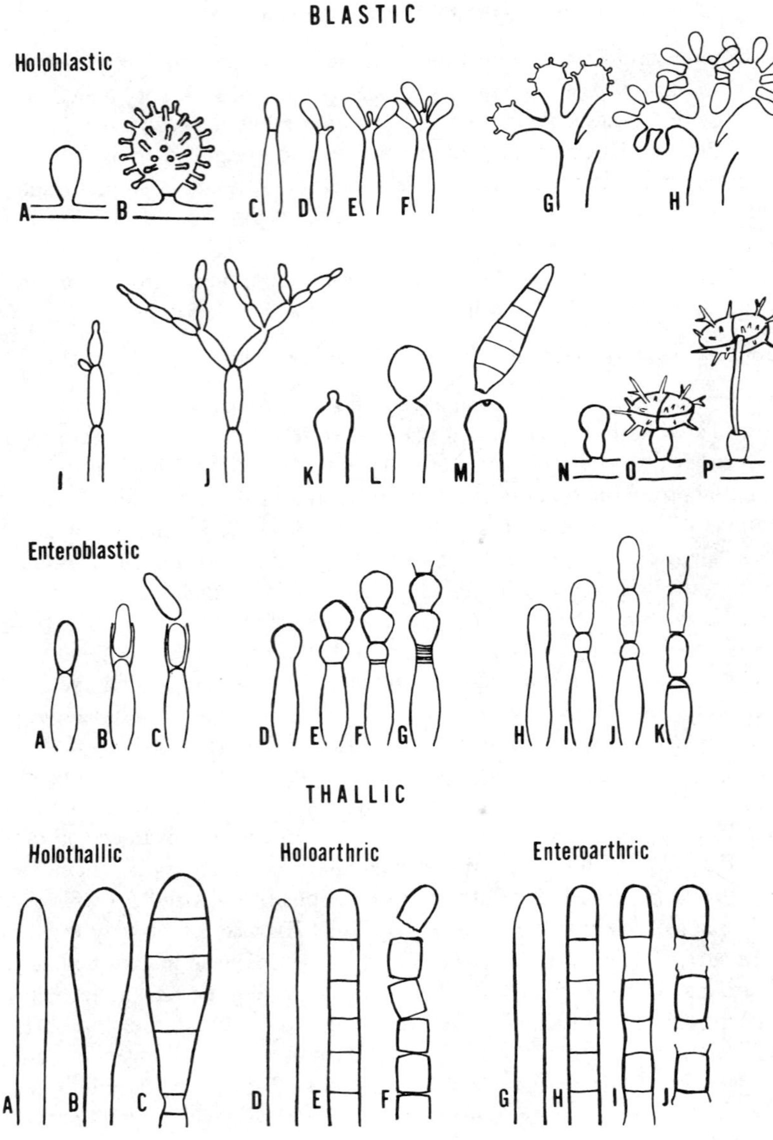

Figure 16 Diagrammatic summary of modes of conidial and conidiogenous cell development examined in the text. The following symbols are used to denote the three major criteria that distinguish the developmental categories: CW, conidial wall differentiation; OA, order of production and arrangement of conidia on the conidiogenous cell; CC, process of conidiogenous cell proliferation. Classification of ontogenetic processes based on modes of conidial wall differentiation (CW) under blastic and thallic development is indicated on the diagram. Additional subdivisions and a list of the representative species illustrated in Fig. 16 are presented below.

CW: Holoblastic. (A,B) *Histoplasma capsulatum.* OA: terminal, solitary; CC: determinate. (C-F) *Tritirachium oryzae.* OA: Terminal, solitary; CC: proliferous (sympodial). (G,H) *Botrytis cinerea.* OA: synchronous, botryose; CC: determinate. (I,J) *Cladosporium* agar at 24°C, both terminal and lateral, solitary, holoblastic conidia may be produced at the apices of cylindrical fertile branches, or on short protrusions of the hyphal wall. Conidia are also borne on ampulliform swellings or in chains of two to three [322]. The conidia may be smooth-walled or finely echinulate.

When conidia of *E. parva* are inhaled by the host, they enlarge in the alveoli or bronchioles where they become calcified and cause pulmonary dysfunction. This enlarged holoblastic conidium may also develop in vitro at 37-40°C and has been called an adiaspore [116]. Van Oorschot [322] has distinguished between two varieties of *E. parva* on the basis of whether the adiaspores are uninucleate (*E. parva* var. *parva*; Fig. 19B) or multinucleate (*E. parva* var. *crescens;* Fig. 19A). The latter, whose adiaspores have been examined by electron microscopy [166,207,328], may later be induced in vitro to undergo endosporulation [206]. This suggests that the multinucleate adiaspore functions as a disseminating agent within host tissue. *Emmonsia parva* var. *crescens* was recently implicated for the first time in a human skin infection [189].

Blastomycosis is a disease that originates as a pulmonary infection but later involves osseous and cutaneous tissues [117,279]. At first it was thought that this mycosis was restricted to the North American continent, but legitimate cases have now been observed in Africa [118] and the Middle East [212]. The causative agent, *Blastomyces dermatitidis,* has been accommodated in the Endomycetales [1] but produces holoblastic conidia in culture and is morphologically similar to *Emmonsia* (Fig. 20). At 37°C, the fungus forms enlarged, budding cells both in vivo and in vitro (Fig. 20B and C). These are apparently conidia that have converted to budding yeast and, in host tissue, may function in dissemination of the pathogen [131].

2. Holoblastic (Botryose) Conidia; Proliferous Conidiogenous Cells

Conidiogenesis in *Sporothrix schenckii* (Fig. 21), causative agent of a chronic, subcutaneous, lymphatic mycosis (sporotrichosis), exemplifies this developmental category. The natural habitat of this fungus appears to be wood from living or dead trees. Dixon et al. [102] have described a useful procedure for isolating pathogens, including *S. schenckii,* from natural substrates. Wood and soil samples were collected, washed in a mixture of sterile water and mineral oil, and the suspended material isolated from the oil-water interface was used for intraperitoneal inoculation of hamsters. Animals were killed 7 weeks post inoculation, and the organs were removed and homogenized; the homogenates were plated on media supplemented with chloramphenicol and then screened for the presence of fungal pathogens.

During conidiogenesis in *S. schenckii,* the primary holoblastic conidium is produced

---

Figure 16 (continued)
bantianum. OA: acropetal (catenulate); CC: proliferous. (K-M) *Deightoniella torulosa.* OA: terminal, solitary; CC: porogenous. (N-P) *Spegazzinia tessarthra.* OA: terminal solitary; CC: basauxic. *CW: Enteroblastic.* (A-C) *Phialophora lagerbergii.* OA: basipetal, solitary; CC: phialidic. (D-G) *Scopulariopsis brevicaulis.* OA: basipetal, solitary; CC: proliferous (annellidic). (H-K) *Cladobotryum varium.* OA: basipetal, solitary; CC: retrogressive. *CW: Holothallic.* (A-C) *Microsporum gypseum.* OA: terminal, solitary; CC: determinate. *CW: Holoarthric.* (D-F) *Geotrichum candidum.* OA: random, chains; CC: determinate. *CW: Enteroarthric.* (G-J) *Coccidioides immitis.* OA: random, chains; CC: determinate. (From Ref. 88.)

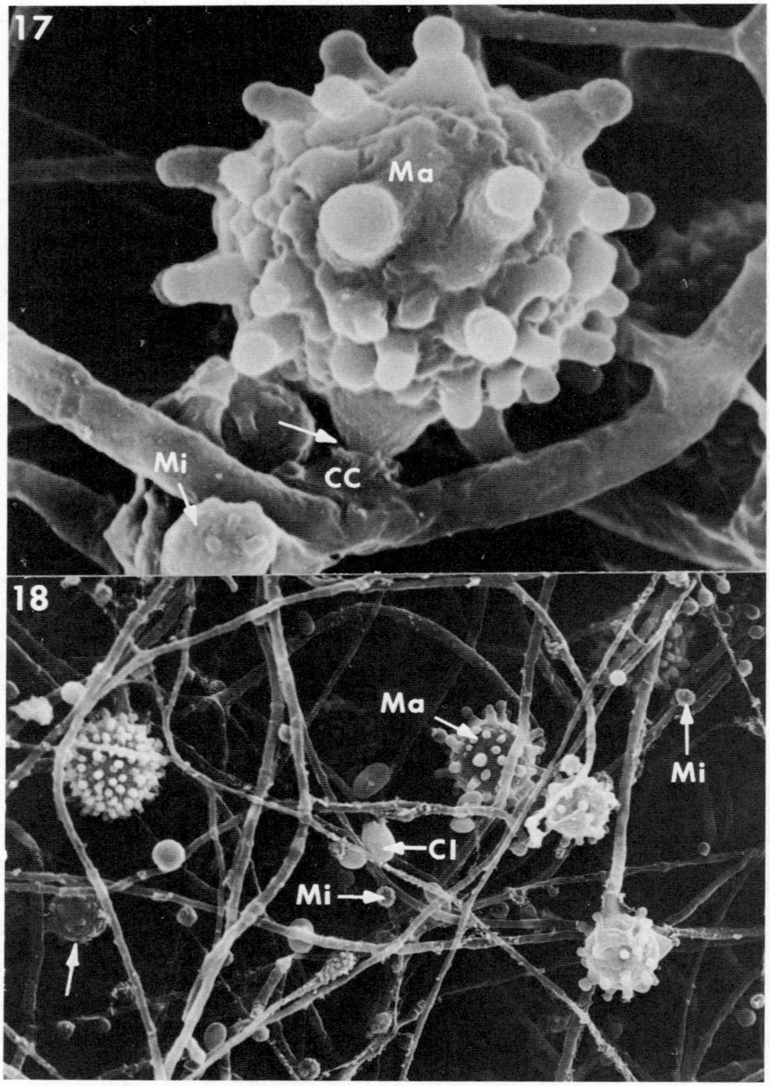

Figures 17 and 18 *Histoplasma capsulatum.* Macroconidia (Ma) and microconidia (Mi) produced at the apices of determinate conidiogenous cells (CC). Arrow in Fig. 17 locates cross wall at junction of macroconidium and conidiogenous cell. Arrow in Fig. 18 indicates young macroconidium. CI, conidium initial. 5500X and 1100X, respectively.

at the apex of the fertile cell. The latter subsequently proliferates from the base and to one side of the terminal conidium, elongates slightly and gives rise to a second, terminal, holoblastic conidium. The same sequence of events is repeated several times, which results in formation of an acropetal succession of conidia arranged in a sympodial pattern at the conidiogenous cell apex. This same developmental process has been illustrated by

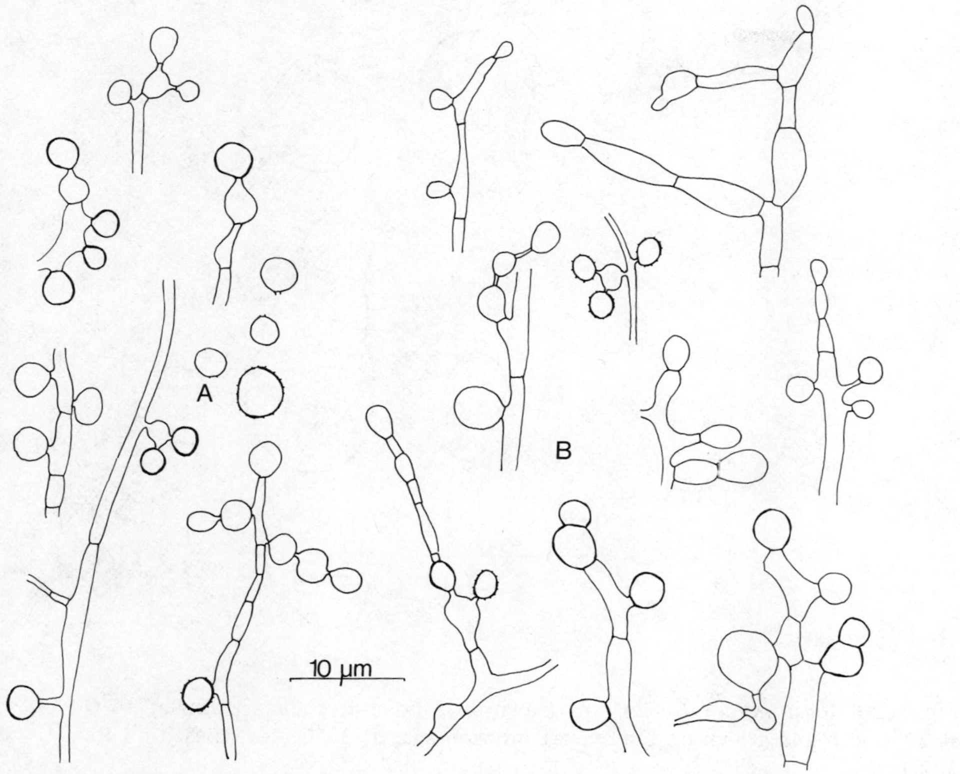

Figure 19 (A) *Emmonsia parva* var. *crescens*; (B) *E. parva* var. *parva* (Based on Ref. 322.)

time-lapse, photomicrographic sequences of several other hyphomycetous fungi [71,88, 195]. In *S. schenckii,* only slight elongation of the fertile cells usually occurs between formation of successive propagules (Fig. 21), resulting in a dense cluster or "botryose" arrangement of conidia when examined under a light microscope. However, considerable developmental variability during conidium formation in this fungus occurs, which results in pleomorphism [98]. This phenomenon, which is quite common among pathogenic hyphomycetes grown in pure culture, is illustrated and discussed later in this chapter. An account of the kind of conidiogenesis in *S. schenckii* outlined above was presented in a previous report [75]. Two kinds of propagules are produced: thin-walled, hyaline, ellipsoidal conidia (Fig. 21) produced as described above, and thick-walled pigmented conidia which arise as holoblastic, lateral outgrowths of the fertile hyphae. The latter have been referred to as "chlamydospores" [103], but their formation is typical of holoblastic conidia [88]. These pigmented conidia are produced in large numbers, forming "sleeves" around the hypha [117]. Both kinds of conidia may convert to yeast forms at room temperature.

A high degree of pleomorphism is also demonstrated by *Fonsecaea pedrosoi* [61,277]. However, the most common morphology of the conidia and conidiogenous cells of both *S. schenckii* and *F. pedrosoi* are illustrated in Figs. 21 and 22, respectively. The apex of

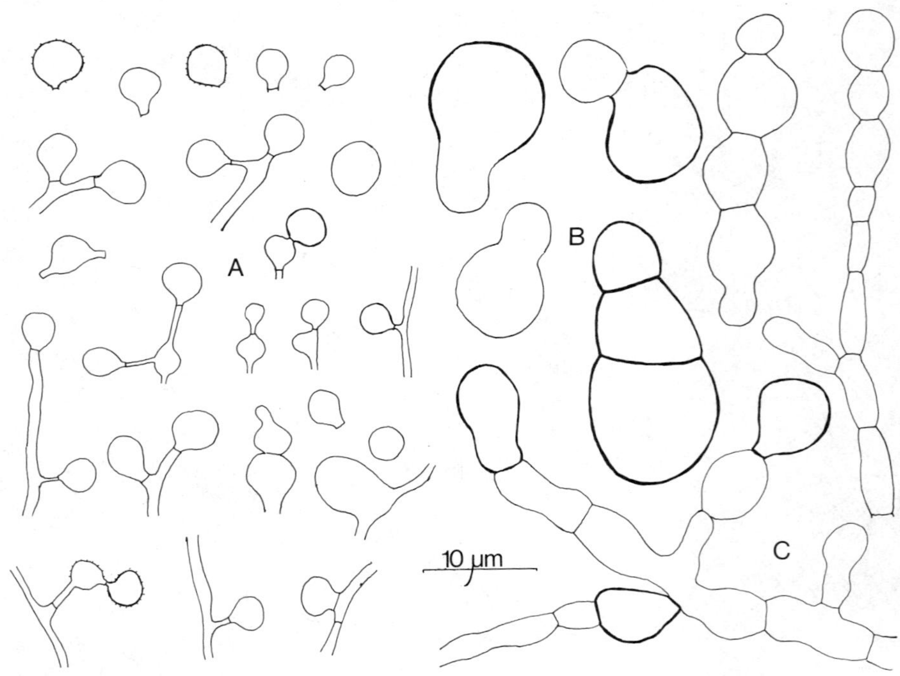

Figure 20 *Blastomyces dermatitidis* grown on Sabouraud's glucose agar at 24°C (A), at 37°C (B), and grown on brain-heart infusion agar at 37°C (C). (Based on Ref. 322.)

the conidiogenous cell of *F. pedrosoi* (Fig. 22) reveals a cluster of denticles (D), or scars of seceded conidia, and a single conidium (C) which is still attached. The conidia are formed holoblastically and the conidiogenous cell undergoes the same succession of sympodial proliferations as described for *S. schenckii.* The dense, botryose arrangement of conidia in *F. pedrosoi* makes it even more difficult, under the light microscope, to observe the alternate arrangement of denticles. This is not a problem, however, with *Dactylaria gallopava,* whose conidia are widely separated at the apex of sympodially proliferating conidiogenous cells (Fig. 23). Similar modes of conidium and conidiogenous cell development are demonstrated by *Scolecobasidium humicola,* a pathogen of coho salmon and rainbow trout [12].

3. Holoblastic (Catenulate) Conidia

A common mode of production of aerial conidia is by formation of acropetal chains of cells [88]. The conidiogenous cells are often morphologically undifferentiated from vegetative hyphae. A conidiogenous cell gives rise to a chain of conidia (Fig. 24A) which are cytoplasmically connected, at least in early development, and the youngest cell is always at the apex of the chain (i.e., acropetal arrangement). Conidia at or near the base of the chain are often larger and capable of branching to form secondary conidial chains (Fig. 24A). These ontogenetically specialized basal cells have been appropriately called "ramoconidia" [113,179]. Members of *Cladosporium* (e.g., *C. bantianum*; [170]) best

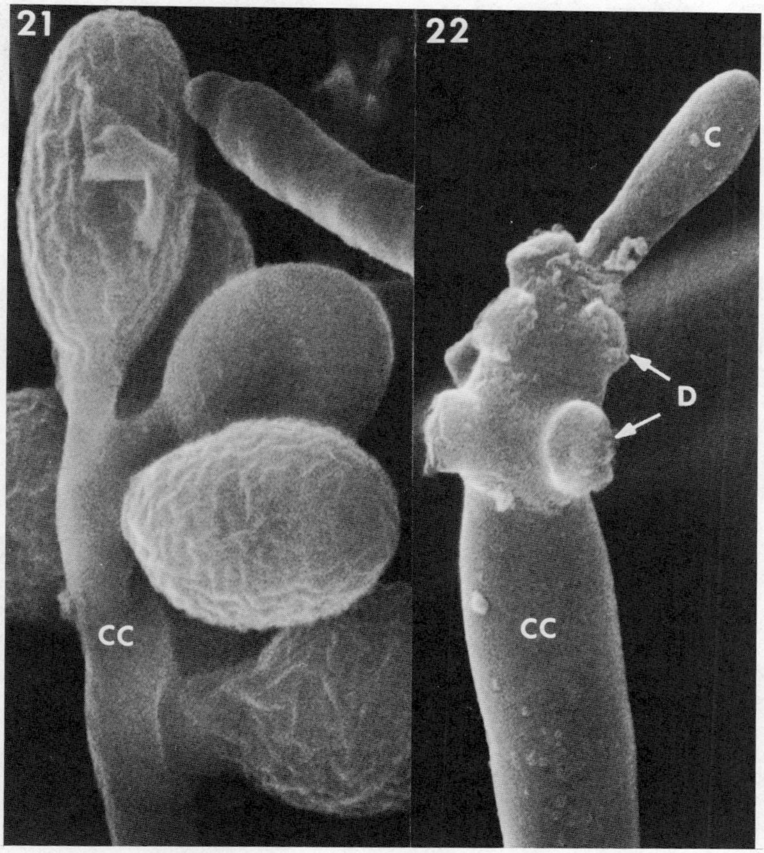

Figure 21 *Sporothrix schenckii.* Cluster of holoblastic conidia produced by sympodially proliferated conidiogenous cell (CC). 13,000X. (From Ref. 88.)

Figure 22 *Fonsecaea pedrosoi.* Cluster of denticles (D) arranged in sympodial pattern and young conidium (C) still attached to the conidiogenous cell (CC) apex. 11,000X. (From Ref. 88.)

exemplify this mode of conidiogenesis. Proliferous and well-differentiated conidiogenous cells are also produced by certain species in this genus [113].

Acropetal chains of conidia are also produced by the insidious fungal pathogen *Loboa loboi,* causative agent of lobomycosis (Fig. 24B). This fungus is known only in its parasitic or tissue form; it has never been grown in pure culture [117]. Lacaz et al. [217] have accommodated this pathogen in *Paracoccidioides* (as *P. loboi*). However, additional comparative studies of the two fungi are required. Chains of elliptical- to lemon-shaped conidia are produced by *L. loboi,* with adjacent cells joined by short, tubular connectives. Some cells are capable of multiple proliferation (Fig. 24B).

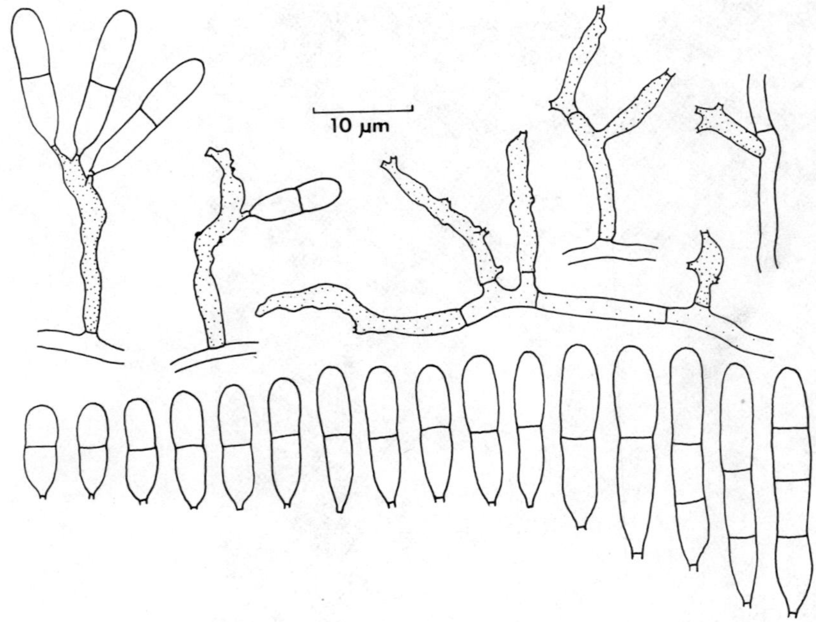

Figure 23 *Dactylaria gallopava.*

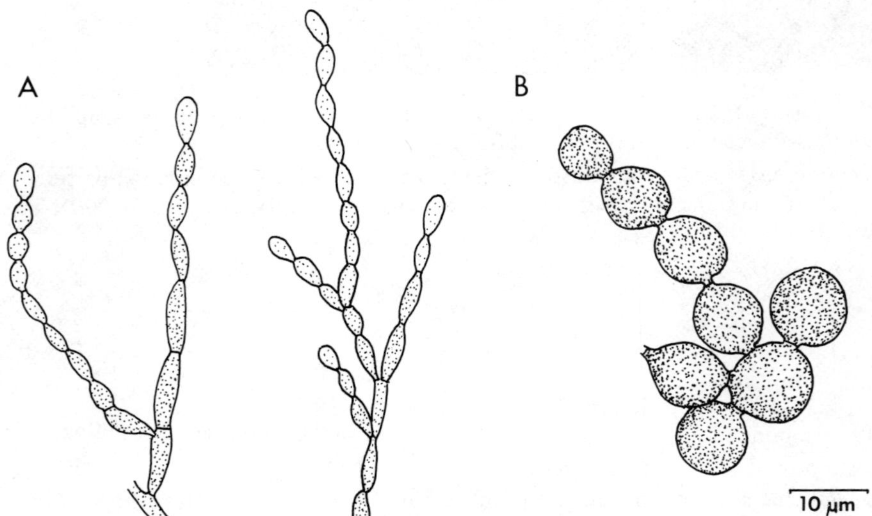

Figure 24 Catenulate conidial forms: (A) *Cladosporium bantianum*; (B) *Loboa loboi* (based on Ref. 117).

4. Holoblastic Conidia; Porogenous Conidiogenous Cells

Hughes [179] recognized a special mode of conidium and conidiogenous cell development in which the conidium initial arises from a restricted region of the fertile cell apex and appears to be growing out through a pore or channel in its wall [195]. However, thin sections of representative fungi have indicated that the conidium initial develops holoblastically, presumably involving increased activity of lytic enzymes in the isolated region of the thickened conidiogenous cell wall which permits it to blow out (CI, Fig. 25A). However, as the initial enlarges, the outer wall of the conidiogenous cell becomes very thin and, in most cases, soon begins to break down [72,88]. The distinction between holoblastic de-

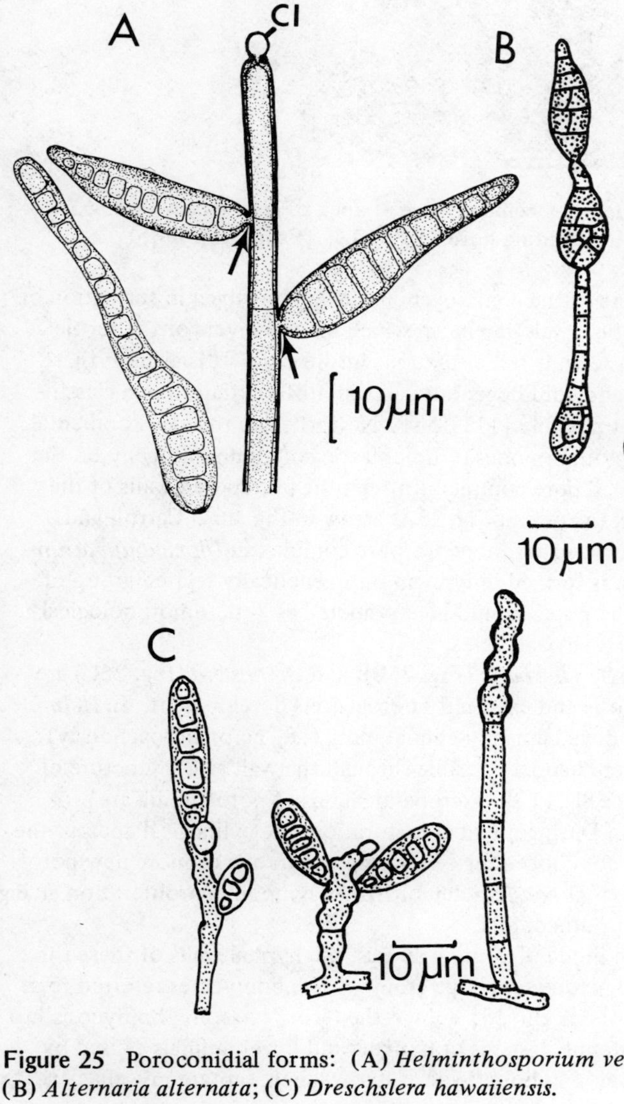

Figure 25 Poroconidial forms: (A) *Helminthosporium velutinum* (CI, conidium initial); (B) *Alternaria alternata*; (C) *Dreschslera hawaiiensis*.

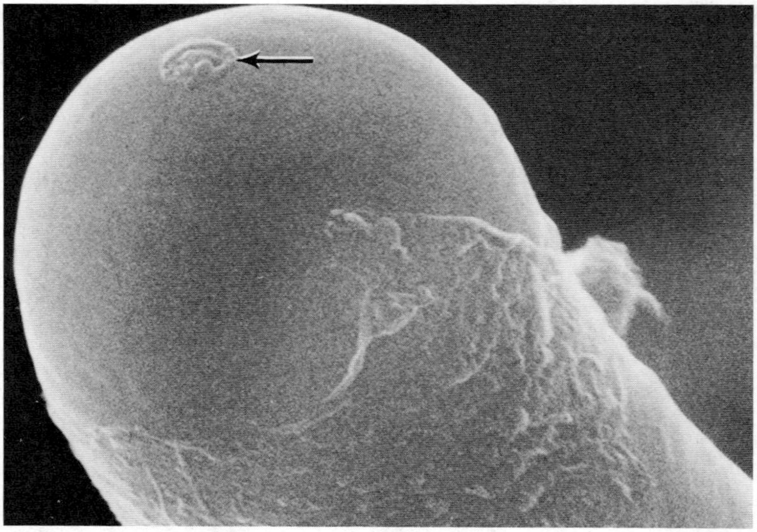

Figure 26 *Helminthosporium solani.* Occluded pore at apex of fertile cell. Arrow indicates thickened ring or annulus surrounding pore. 10,800X. (From Ref. 88.)

velopment, in which all wall layers of the conidiogenous cell are involved in formation of the conidial wall, and enteroblastic development, in which all wall layers of the fertile cell are not involved in conidiogenesis, is rather tenuous in this case. It is evident that strict use of features of conidial and conidiogenous cell wall differentiation in a classification of anamorphic fungi is not possible [111,158]. Nevertheless, this developmental process is still distinguished from other modes of holoblastic conidium ontogeny on the basis of formation of a well-defined pore complex through the juxtaposed walls of the conidial base and fertile cell apex (arrows in Fig. 25A; arrow in Fig. 26). Carroll and Carroll [67] who described the ultrastructure of the pore complex in *Ulocladium atrum,* pointed out that such a structure is formed only in an ontogenetically related group of anamorphs, and suggested that the pores should be considered as "micromorphological indicators of a natural taxon of Hyphyomycetes."

Helminthosporium (Fig. 25A), *Alternaria* (Fig. 25B), and *Drechslera* (Fig. 25C) are three agents of phaeohyphomycosis and exemplify porogenous development. In *Helminthosporium,* poroconidia are produced apically and laterally (i.e., acropleurogenously) from the fertile hypha. A distinct channel is visible through the wall at the juncture of the conidium and conidiogenous cell [112]. Acropetal chains of poroconidia are produced by members of *Alternaria.* During conidial maturation, the cylindrical apex of the cell elongates and functions as a conidiogenous cell, giving rise to one or more new poroconidia. The conidiogenous cell of *Drechslera* demonstrates sympodial proliferation giving rise to an acropetal succession of poroconidia.

In addition to their common mode of conidiogenesis, the hyphal walls of these three genera are darkly pigmented and belong to a large group of Hyphomycetes referred to as the dematiaceous fungi [113,114]. Ajello [5] coined the term "phaeohyphomycosis" to include all subcutaneous and systemic diseases of humans and lower animals caused by hyphomycetous fungi which develop in host tissue as dark-walled, septate mycelia. In spite of the etymological problems, the author [5] pointed out that the causative agents of phaeohyphomycoses include both hyphomycetous and coelomycetous fungi, as well as ascomycetous teleomorphs. Various etiologic agents embraced by this collective term demonstrate several different modes of conidiogenesis (Table 3). The nature of the diseases they cause are also heterogeneous, including encephalitis of broiler chickens associated with *Dactylaria gallopava* [135,325], cerebral infections of humans caused by the soil saprobe *Cladosporium trichoides* [117,243,259], as well as nasal and skin lesions in cattle and meningoencephalitis attributed to *Drechslera* spp. [117,128,237]. However, Ajello [5] has pointed out that the "common thread connecting all these diverse fungi to a single disease category is their tissue form. Basically all of them occur in the tissues of their hosts, be they fish or fowl or mammals, as dematiaceous mycelium." Ajello [9] distinguished phaeohyphomycosis from another fungal disease caused by dematiaceous fungi commonly referred to as chromoblastomycosis. The latter, which is known to occur with certainty only in humans, is caused by *Acrotheca aquaspersa, Cladosporium carrionii, Fonsecaea compacta, F. pedrosoi,* and *Phialophora verrucosa* [9,292]. The "diagnostic hallmark of chromoblastomycosis is the large, dark, thick-walled muriform cells produced in tissue" which is not a characteristic of phaeohyphomycosis [9]. The term "chromomycosis" has been used interchangeably with chromoblastomycosis [319] and for encompassing diseases listed as phaeohyphomycotic [117,329]. Ajello [9] has proposed to reject the term "chromomycosis" on the basis that it is illegitimate and superfluous, and has argued that chromoblastomycosis and phaeohyphomycosis should be recognized as "two completely distinct and separate disease entities." We have followed this convention in distinguishing between mycoses caused by dematiaceous fungi (Table 3).

5. Holoblastic Conidia; Basauxic Conidiogenous Cells

An unusual developmental process is demonstrated by *Wallemia sebi,* another causative agent of phaeohyphomycosis. The morphology of the tapered conidiogenous cell that gives rise to a chain of conidia at its apex (Fig. 27) is reminiscent of a phialide [224]. However, electron-microscopic examinations of the fertile cell have indicated that successive conidial initials are formed holoblastically. In fact, the swollen, rough-walled cell, which differentiates above the constriction of the conidiogenous cell apex in Fig. 27A and B (arrows), is an extension of the fertile hypha. The latter continues to elongate from a basal meristem while simultaneously undergoing septation (Fig. 27C and D). The fertile hypha first separates into long segments (Fig. 27C) which later subdivide into four conidia (Fig. 27E). These subsequently swell as the septa between them separate schizolytically (Fig. 27F-H) and the chain disarticulates (Fig. 27I). Such fragmentation of a fertile hypha is comparable to holoarthric development [22,32,113,163]. This somewhat bipartisan mode of conidiogenesis has been accommodated in a separate developmental category characterized by the formation of holoblastic conidia from basauxically developing conidiogenous cells (Fig. 16N-P) [88].

6. Enteroblastic Conidia; Phialidic-Annellidic Conidiogenous Cell Development

Phialophora verrucosa produces a specialized conidiogenous cell called a phialide, which gives rise to a basipetal succession of conidia (phialoconidia) through the ruptured apex of the fertile cell [82,88]. The first-formed conidium is produced holoblastically at the apex of the phialide. However, as the second phialoconidium develops at the base of the first, it pushes the latter upward, rupturing the outer phialide wall and leaving a cup-

Table 3 Modes of Conidiogenesis in Pathogenic Filamentous Fungi and Corresponding Mycoses[a]

Mode of conidiogenesis	Pathogen	Mycosis
Blastic		
Holoblastic		
Solitary	*Histoplasma capsulatum*	Histoplasmosis
	Emmonsia parva	Adiaspiromycosis
	Blastomyces dermatitidis	Blastomycosis
Botryose	*Sporothrix schenckii*	Sporotrichosis
	Fonsecaea pedrosoi	Chromoblastomycosis
	Fonsecaea compacta	Chromoblastomycosis
	Acrotheca aquaspersa	Chromoblastomycosis
	Dactylaria gallopava	Phaeohyphomycosis
	Scolecobasidium humicola	Phaeohyphomycosis
	Aureobasidium pullulans	Phaeohyphomycosis
Catenulate	*Cladosporium* spp.	Phaehyphomycosis
	Loboa loboi	Lobomycosis
Porogenous	*Helminthosporium* spp.	Phaeohyphomycosis
	Alternaria spp.	Phaeohyphomycosis
	Drechslera spp.	Phaeohyphomycosis
Basauxic	*Wallemia sebi*	Phaeohyphomycosis
Enteroblastic		
Phialidic-annellidic	*Phialophora verrucosa*	Chromoblastomycosis
	Phialophora richardsiae	Phaeohyphomycosis (phaeomycotic cyst)
	Phialophora parasitica	Phaeohyphomycosis (phaeomycotic cyst)
	Acremonium sp.	Phaeohyphomycosis
	Paecilomyces lilacinus	Phaeohyphomycosis
	Penicillium marneffei	Phaeohyphomycosis
	Aspergillus fumigatus	Aspergillosis
	Exophiala jeanselmei	Mycetoma
	Exophiala spinifera	Phaeohyphomycosis
	Exophiala werneckii	Tinea nigra
	Wangiella dermatitidis	Phaeohyphomycosis
	Scopulariopsis brevicaulis	Phaeohyphomycosis

[a]See Fig. 16 (p. 456) for diagrammatic illustrations of modes of conidiogenesis.

Table 3 (continued)

Mode of conidiogenesis	Pathogen	Mycosis
Thallic		
Holothallic	*Microsporum* spp.	Tinea
	Trichophyton spp.	Tinea
	Epidermophyton spp.	Tinea
Thallic-arthric		
Holoarthric	*Trichophyton mentagrophytes*	Tinea
	Geotrichum candidum	Geotrichosis
	Trichosporon beigelii	White piedra
Enteroarthric	*Coccidioides immitis*	Coccidioidomycosis

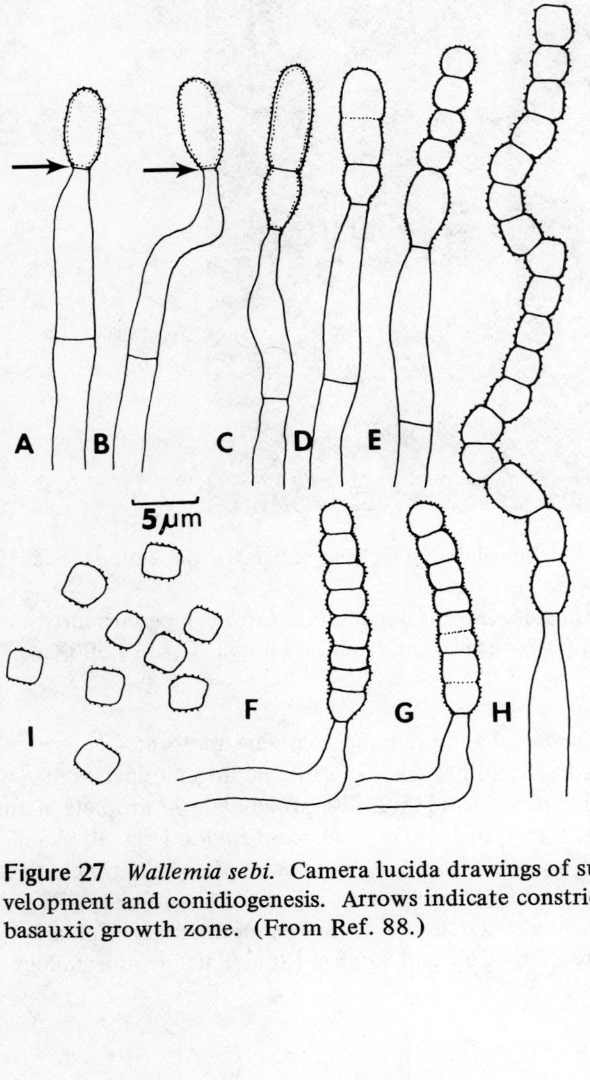

Figure 27 *Wallemia sebi.* Camera lucida drawings of successive stages of basauxic development and conidiogenesis. Arrows indicate constriction at fertile cell apex just below basauxic growth zone. (From Ref. 88.)

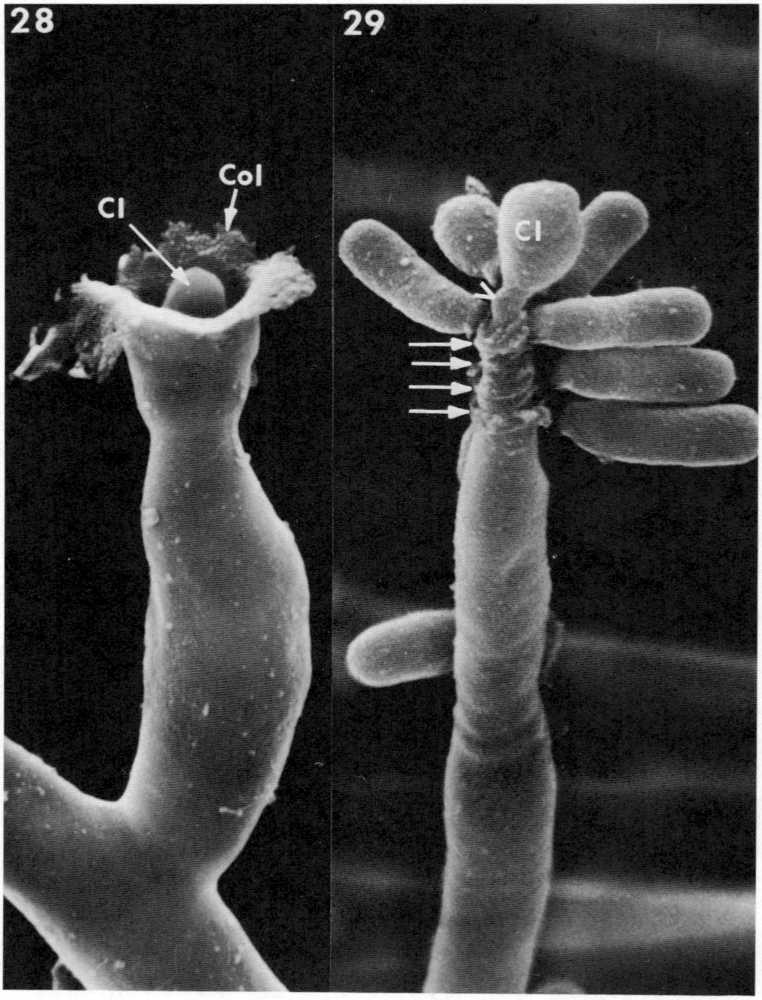

Figure 28 *Phialophora verrucosa.* Phialide showing distinct collarette and endogenous conidium initial (CI). 9500X.

Figure 29 *Exophiala spinifera.* Annellide showing annellations (arrows), percurrently proliferated region of the fertile cell (arrowhead), and conidium initial (CI). 10,500X.

shaped collarette (Col, Fig. 28). The second and subsequent conidia are formed enteroblastically. The conidia of *P. verrucosa,* as well as those of other pathogenic species of *Phialophora* (e.g., *P. richardsiae* and *P. parasitica*; [11,278]), accumulate in droplets at the apex of the phialide. Some phialidic forms, such as *Paecilomyces lilacinus* (Fig. 30A) and *Penicillium marneffei* (Fig. 30B), on the other hand, produce chains of conidia which are held together by thick wall deposits or connectives between adjacent cells [82,245,273, 274]. Differences in wall development associated with these two mechanisms of phialoconidium formation have been discussed by Cole and Samson [88]. Phialide morphology

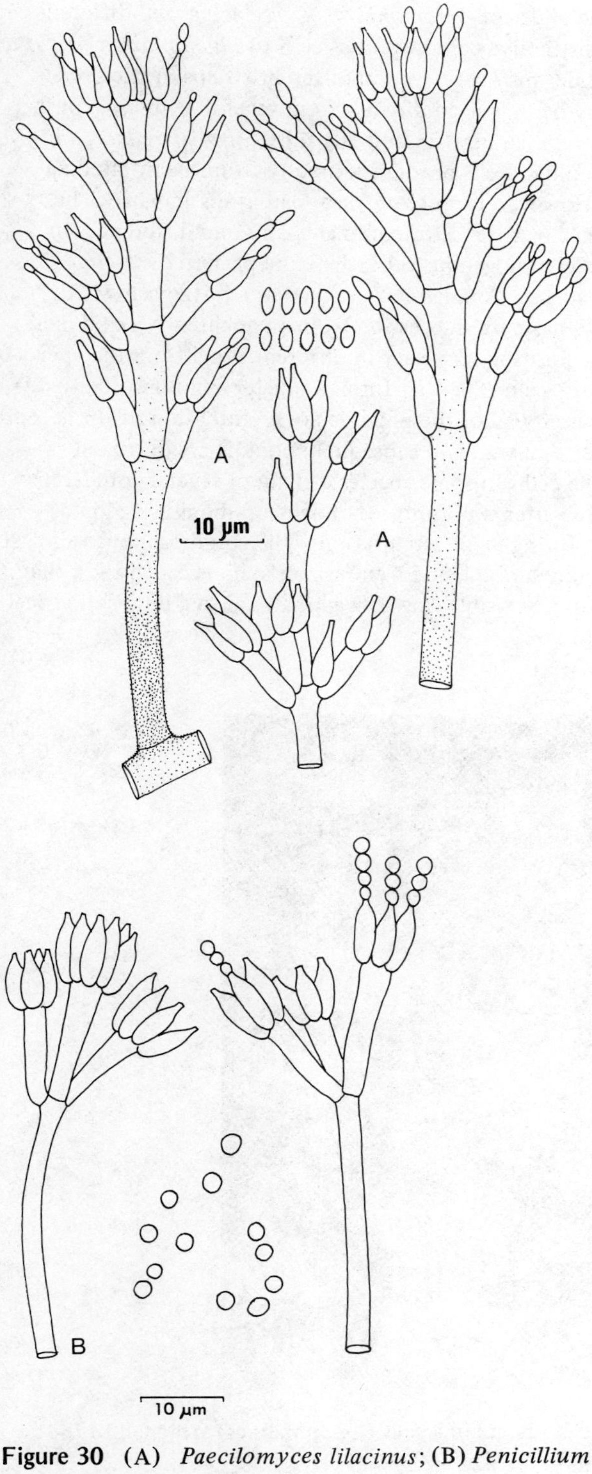

Figure 30 (A) *Paecilomyces lilacinus*; (B) *Penicillium marneffei*.

is also variable. The long, tapered phialides of *Acremonium,* which lack a well-differentiated collarette, contrast with the distinctive conidiogenous cells of *Phialophora verrucosa* (Fig. 28). The arrangement of phialides may vary even in different strains of the same fungus. The conidiogenous cells may be produced directly from vegetative hyphae or in diverse patterns on erect conidiophores (see Figs. 28 and 30) [88]. In *Aspergillus fumigatus* a dense cluster of flask-shaped phialides is produced directly from the inflated apex of the conidiophore (Cd, Fig. 31). However, in spite of the apparent dissimilarities between phialides, based on the presence or absence of conidial chains and differences in the morphology and arrangement of conidiogenous cells, the same pattern of conidiogenesis is demonstrated by all these forms. A phialide is defined as a fertile cell which produces conidia (phialoconidia) in a basipetal succession from an apparently fixed meristem, or conidiogenous locus, whose position may vary in different fungi from the apex of the cell to deep within the body of the cell. The first-formed phialoconidium usually differentiates within the unruptured, outer wall of the phialide apex. Initiation of the second conidium often causes the rupture of that wall and subsequent conidia are extruded through the resulting open end of the cell. In some species a chain of several conidia may develop within the phialide before its outer wall ruptures. Each conidium is clad in a *secondary wall* especially laid down during its differentiation. The conidiogenous cell wall remaining between the point of rupture and the endogenous meristem is called the collarette. The collarette varies from a scarcely visible ring of wall material to a long-cylindrical structure enclosing several conidia [82,88].

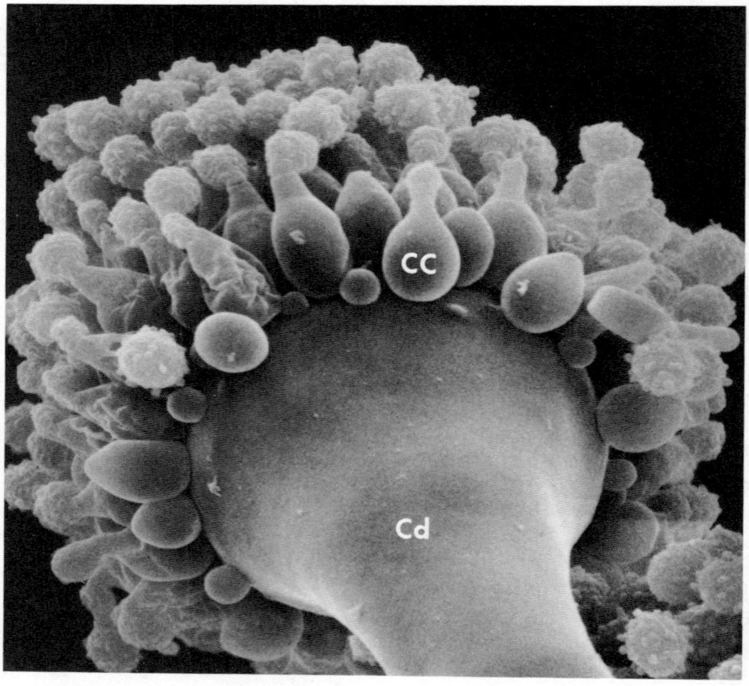

Figure 31 *Aspergillus fumigatus.* Conidiogenous cells (CC; phialides) attached to the swollen, fertile apex of the conidiophore (Cd). 3400X.

At first glance, the conidiogenous cell in Fig. 29 bears little resemblance to the phialide in Fig. 28. However, on closer scrutiny, especially after examining thin sections and photomicrographic sequences of the development of these two kinds of conidiogenous cells, it is evident that they are ontogenetically related [80,88,147-149]. The fertile cell of *Exophiala spinifera* shown in Fig. 29 is called an annellide [233]. This term was coined because of the recognition of circumscissile wall scars, or annellations (arrows), which represent sites where the conidiogenous cell is ruptured during release of successive conidia [179]. The distance separating adjacent annellations represents the zone of percurrent proliferation of the fertile cell. A conidium initial (CI) is attached to the apex of the annellide in Fig. 29. The short-cylindrical extension at the base of the initial (arrowhead) is the percurrently proliferated segment of the conidiogenous cell. *Scopulariopsis brevicaulis* exemplifies annellidic development [80,83,149a] and has been reported as the causative agent of onychomycosis in humans [125a,245a]. The presence of annellations is sometimes difficult to detect using conventional light microscopy. *Exophiala jeanselmei,* for example, had been accommodated in *Phialophora* and considered to produce phialides [115,117]. However, careful examination of the fertile apices of conidiogenous cells in this and other species of *Exophiala,* employing Nomarski interference light microscopy and the SEM, have revealed characteristic annellations [77,230,232,233].

The distinction between most phialides and annellides can be made with the light microscope; the former produce a collarette at each fertile locus, whereas the latter have two or more, usually closely spaced annellations at their fertile apex. However, a number of forms are recognized which cannot be easily identified as one or the other kind of conidogenous cell, and it is these that demonstrate convergence of the phialidic and annellidic developmental concepts [88]. For example, the fertile cells of *Cryptosporiopsis* [306], *Conioscypha* [287,288,320], and *Ascochyta* [42] produce a collarette characteristic of phialidic development, but thin sections of the fertile apex reveal distinct annellations hidden within the collarette which can be construed as a feature of annellide formation. On the other hand, thin sections of typical phialides have also shown endogenous annellations at the conidiogenous locus, although these are rather indistinct [149]. The point is that both annellides and certain phialides demonstrate a succession of percurrent proliferations at their fertile apices during conidiogenesis, a feature that bridges between the two developmental concepts [88]. Although the external morphological characters used to distinguish annellides and phialides (e.g., annellations and collarettes) are reliable in most cases, they should be interpreted with caution.

7. Holothallic Conidia; Determinate Conidiogenous Cells

The most common dermatophytes capable of superficial infection of keratinized tissue include members of *Microsporum, Trichophyton,* and *Epidermophyton.* These naturally occurring soil saprobes can be easily isolated by "baiting" the soil with hair, thus taking advantage of their selective keratinophilic feature. Holothallic conidium formation in *Microsporum gypseum* has already been described (Figs. 1E and F and 12A-C). Both macro- and microconidia are produced by *M. gypseum* and at maturity, the septate macroconidia are characteristically rough-walled (Fig. 32). *Trichophyton rubrum* produces macro- and microconidia, but the former, in this case, are septate and smooth-walled (Fig. 33). *Epidermophyton* does not form microconidia; only smooth-walled, septate macroconidia are produced. In all members of the three genera, however, the conidia (macro- or micro-) develop and secede by the same mechanisms–holothallic development

Figure 32 *Microsporum gypseum.* Macroconidia (Ma) in various stages of development. 1400X.

and rhexolytic secession [88,129,130] (see Table 3). The relatedness of these three genera based on morphological, developmental, and cultural studies is corroborated by their biochemical similarities [6]. Swenson and Ulrich [307], for example, have shown that dermatophytic genera, species, and ecological groups have very similar fatty acid composition. Initiation of experimental taxonomic techniques, such as pyrolysis-gas-liquid chromatography [66] and comparative keratinase activity [260], have been initiated in an attempt to supplement existing morphological data and formulate a more functional classification of dermatophytic fungi. These investigations are continuing.

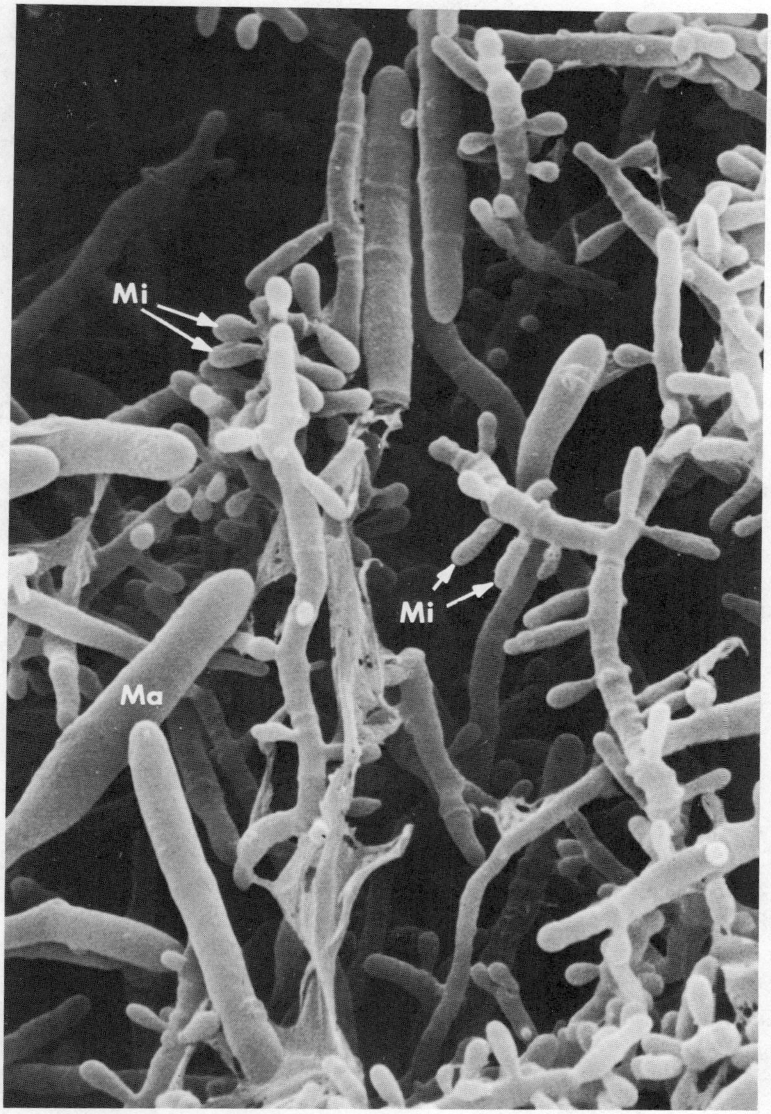

Figure 33 *Trichophyton rubrum.* Smooth-walled macroconidia (Ma) and microconidia (Mi). 1700X.

The term "chlamydospore" has been used to identify those "thick-walled, resistant resting spores" produced by "many fungi . . . particularly in old cultures or under unfavorable conditions of temperature, nutrition, or dryness" [117]. On the other hand, most chlamydospores produced by conidial fungi appear to undergo typical holothallic development. The term "chlamydospore" is so entrenched in mycological literature that it cannot be easily rejected and replaced with alternative terminology (e.g., holothallic conidium). However, we suggest that at least the definition of the chlamydospore should be expanded to include thick-walled, holothallic, terminal or intercalary spores (conidia) which do not readily separate from their parent hyphae but secede only after lysis of the adjacent cell(s), as in rhexolytic secession in *M. gypseum* [88,193].

8. Holoarthric Conidia; Determinate Conidiogenous Cells

Under certain defined growth conditions *Trichophyton mentagrophytes* is capable of forming holoarthric conidia (= arthroconidia) by a process of septation and disarticulation of fertile hyphae [41,119]. Of the three kinds of conidia produced by this fungus, only holoarthric conidia are found in active ringworm lesions [260]. Conditions that "inhibit apical extension of the hyphae but do not affect general metabolic processes" in *T. mentagrophytes* tend to stimulate holoarthric conidium formation [119]. Such conditions include incubation at 32-37°C or under limited oxygen tension, and growth on minimal media or in the presence of sublethal concentrations of certain antifungal agents. The authors have suggested that these observations provide an explanation for the occurrence of this conidial type when the dermatophyte grows in host skin. These findings also support the statement made earlier in this chapter that holoarthric development is initiated only after hyphal tip growth is arrested. Hashimoto and Blumenthal [160] have pointed out that holoarthric conidia of *T. mentagrophytes* show greater survival in the presence of certain deleterious physical and chemical agents than the hyphae from which they were formed. This suggests that although the overall chemical composition of the hypnal and holoarthric conidial walls are similar (Tables 1 and 2), subtle differences in relative amounts and arrangement of certain wall components (e.g., polysaccharides; see Refs. 321 and 334) may be highly significant in conidial wall differentiation.

Hyphal and holoarthric conidial development in *Geotrichum candidum* has been extensively investigated [74,75,249,269,270,315]. This fungus is ubiquitous in nature [63]. It is commonly isolated from soil, dairy products, many different plants, and is frequently found as part of the indigenous microflora of the gastrointestinal tract of animals (including humans). It is also the causative agent of oral, intestinal, and bronchopulmonary infections in humans [3,117], as well as cutaneous and renal infections of other animals [268,293,298]. Development of geotrichosis probably occurs under conditions that have compromised the host's immunological defenses, allowing the previously innocuous resident microbe to become a pathogen. However, *G. candidum* generally demonstrates a low level of pathogenicity for animal tissues [268].

When *G. candidum* is grown in a thin, glass culture chamber [81] and conidium differentiation is examined under the light microscope, the following sequence of events are recognized:

1. Septation is at first acropetal, but is later random.
2. Hyphal tip growth ceases before septa are formed in the apical region.
3. Continued septation yields progressively shorter segments.
4. Septation ceases and the conidia enlarge slightly, their end walls visibly become convex.
5. The conidial chain disarticulates as centripetal separation of the septa, or schizolysis occurs (Fig. 34).

Park and Robinson [249] have proposed that conidiogenesis in *G. candidum* is initiated by a metabolic change ("sporulation factor") which leads to cessation of apical growth and is "correlated with a change in form and disposition of the mitochondria" in the fertile region of the hypha and vacuolation of that part of the hypha which subtends

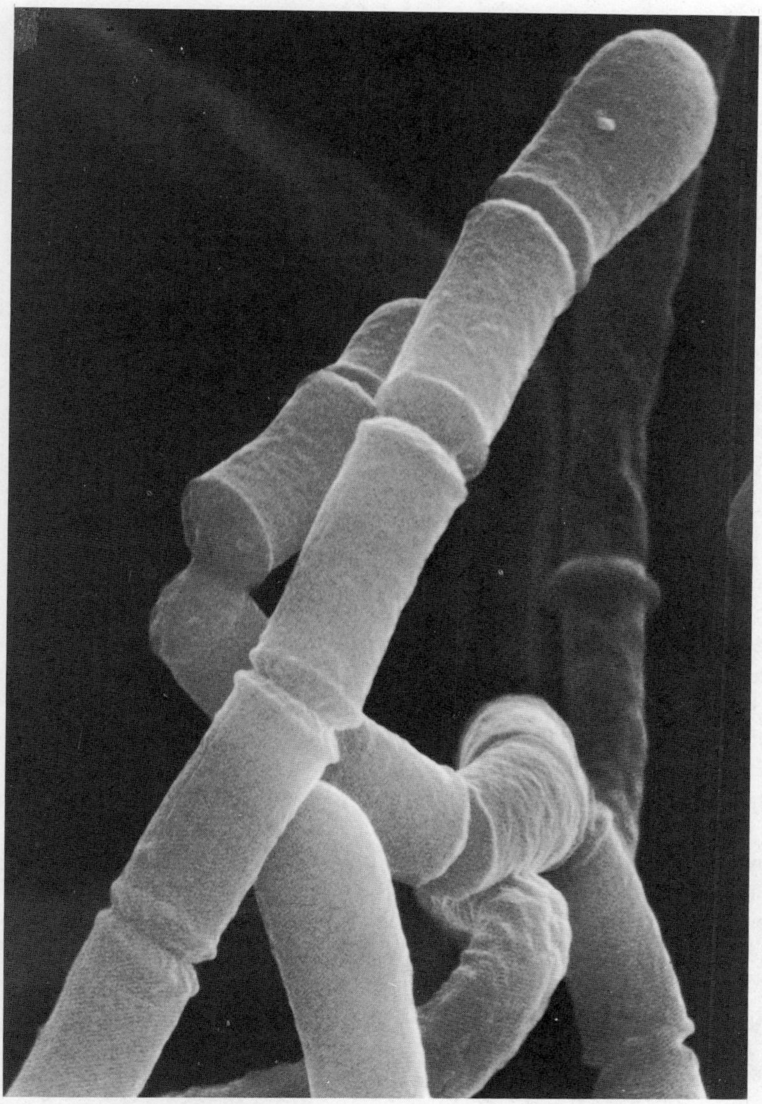

Figure 34 *Geotrichum candidum.* Holoarthric conidia. 7000X.

the conidial chain. These protoplasmic alterations in somatic hyphae are influenced by composition of the growth medium (e.g., concentration of carbon and nitrogen sources) as well as certain physical factors (e.g., low pH stimulates conidium formation in *G. candidium*) [109]. The authors pointed out that developmental events leading to conidium differentiation are reversible by transfer to noninducing conditions which permit a rapid renewal of somatic growth. Kier et al. [199] have speculated that such a sporulation factor could initiate changes in respiration rate and glucose metabolism which lead to conidium formation, but only if the initial glucose concentration of the growth medium is sufficiently high (e.g., 0.5%). Two exponential growth phases are demonstrated under these conditions; the first involves an increase in mycelial growth, while the second, after only 15% of the glucose in the medium has been utilized, is a phase of conidiogenesis. At lower initial concentrations of glucose in the growth medium (e.g., 0.1%), the fungus first uses up all the exogenous glucose and then conidia are formed in stationary phase. The holoarthric conidia produced under these two different growth conditions are morphologically and biochemically distinct (Table 2) [14,200]. *Geotrichum candidum* has been shown to be particularly well suited for examination of cell development in chemostat culture [266, 297] and future studies of factors influencing conidiogenesis should take advantage of this technique.

9. Enteroarthric Conidia; Determinate Conidiogenous Cells

Saëz [269,270] has described an unusual method of conidiogenesis in *G. candidum* and *G. loubieri* in which chains of conidia are formed endogenously within the fertile hypha, disarticulate, and are then released by breakdown of the outer hyphal wall. The author [269] observed that when his strain of *G. candidum* was grown on Sabouraud's glucose (2%) agar at 25-30°C for 2-3 weeks, it yielded abundant enteroarthric conidia, especially near the outer perimeter of the colony.

The saprophytic phase of *Coccidioides immitis* demonstrates typical enteroarthric conidium development (Figs. 35-42) [213,304]. This soil saprobe readily disperses its dry conidia into the air with the slightest disturbance of its natural habitat. Infection occurs via the respiratory tract and transformation of the conidia into multinucleate spherules initiates the parasitic cycle in the host (Fig. 35) [303]. During conidium formation, the determinate fertile hypha becomes septate in a manner comparable to holoarthric ontogeny. However, a distinguishing feature of enteroarthric development is that certain hyphal compartments are converted into conidia, whereas others undergo autolysis (Figs. 36-38). The wall of the degenerating cells eventually breaks down releasing the endogenously formed conidia (Fig. 39). The ruptured outer wall of the fertile hypha (OW) is clearly differentiated from the inner conidial wall (IW) and between them is an amorphous layer (A). Part of the septal walls of the adjacent autolyzed cells are also visible (arrows). The conidia may subsequently germinate, giving rise to vegetative hyphae (Fig. 40) which again undergo septation and enteroarthric development (i.e., saprobic cycle, Fig. 35). If a suitable animal host is infected, the conidia swell (Fig. 41), their nucleus (N) undergoes multiple mitoses (Fig. 35), and cleavage is initiated (Fig. 42) which leads to the formation of endospores. Thin-section, freeze-etch, and karyological examinations of enteroarthric development in *Sporendonema purpurascens* [74] have demonstrated that early autolysis is recognized by the presence of swollen mitochondria and nuclei surrounded by envelopes with irregular morphology, as revealed by freeze fractures. In comparison to the conidium, the cytoplasm of the autolysing cell is much less dense. As the inner conidial wall thickens, the wall of the degenerate cell becomes thinner. The nuclei trapped within these cells, however, continue to react positively to aceto-orecein stain [264], even at an advanced stage of degeneration [74]. When the wall of the fertile hypha finally ruptures, the partially decomposed cytoplasmic material is lost. The same karyological events apparently occur during holoarthric development in *C. immitis* [303] (see Table 3).

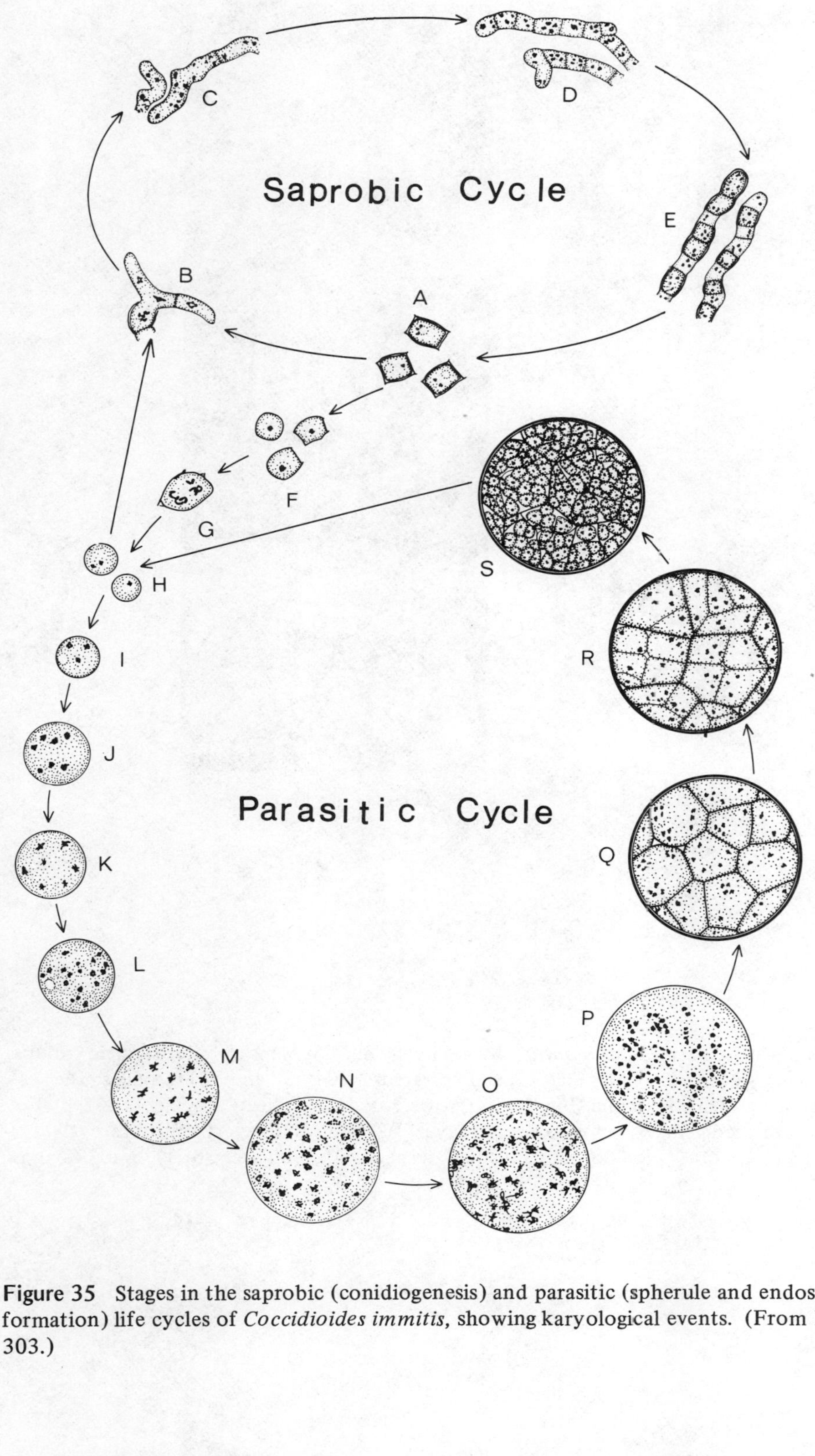

Figure 35 Stages in the saprobic (conidiogenesis) and parasitic (spherule and endospore formation) life cycles of *Coccidioides immitis*, showing karyological events. (From Ref. 303.)

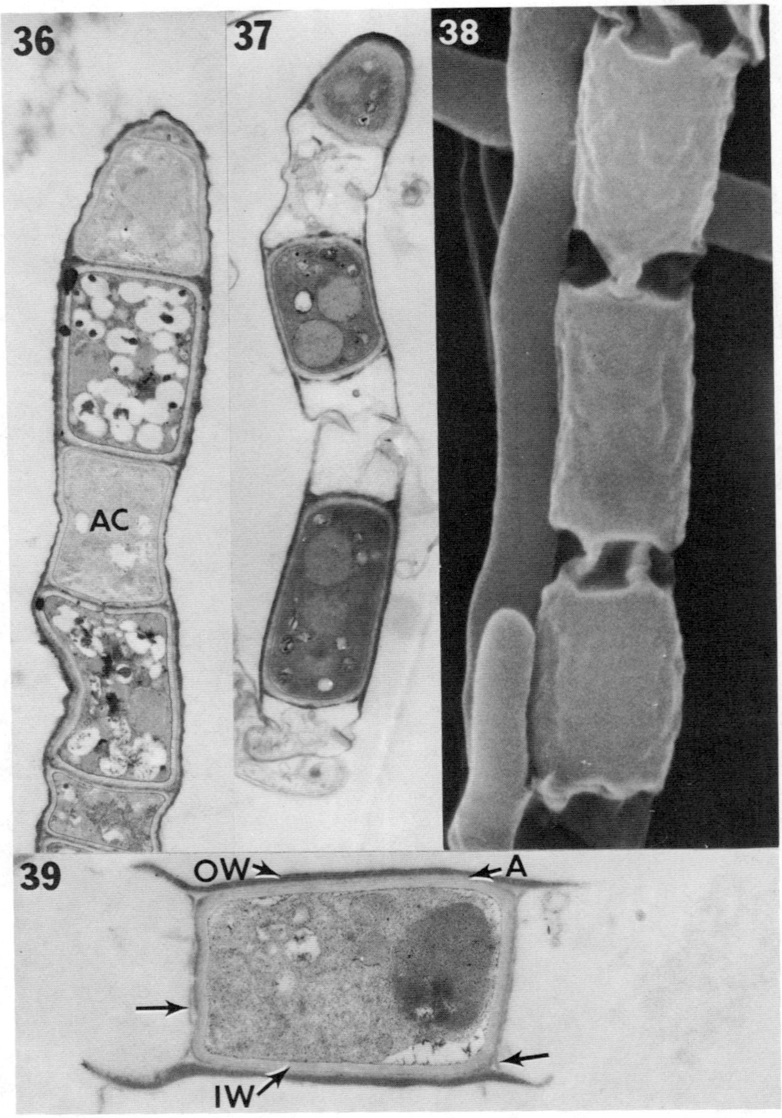

Figures 36-39 *Coccidioides immitis.* Stages in the development of enteroarthric conidia, showing autolysing cells (AC, Fig. 36) which fracture (Figs. 37 and 38) to release the endogenously formed conidia (Fig. 39). Arrows in Fig. 39 indicate fragments of septal wall of autolysed cell. A, amorphous wall layer; IW, inner wall layer; OW, outer wall layer. 6400X, 6400X, 6000X, and 11,200X, respectively. (Figures 36, 37, and 39 from Ref. 304; Fig. 38 courtesy of Dr. S. Sun.)

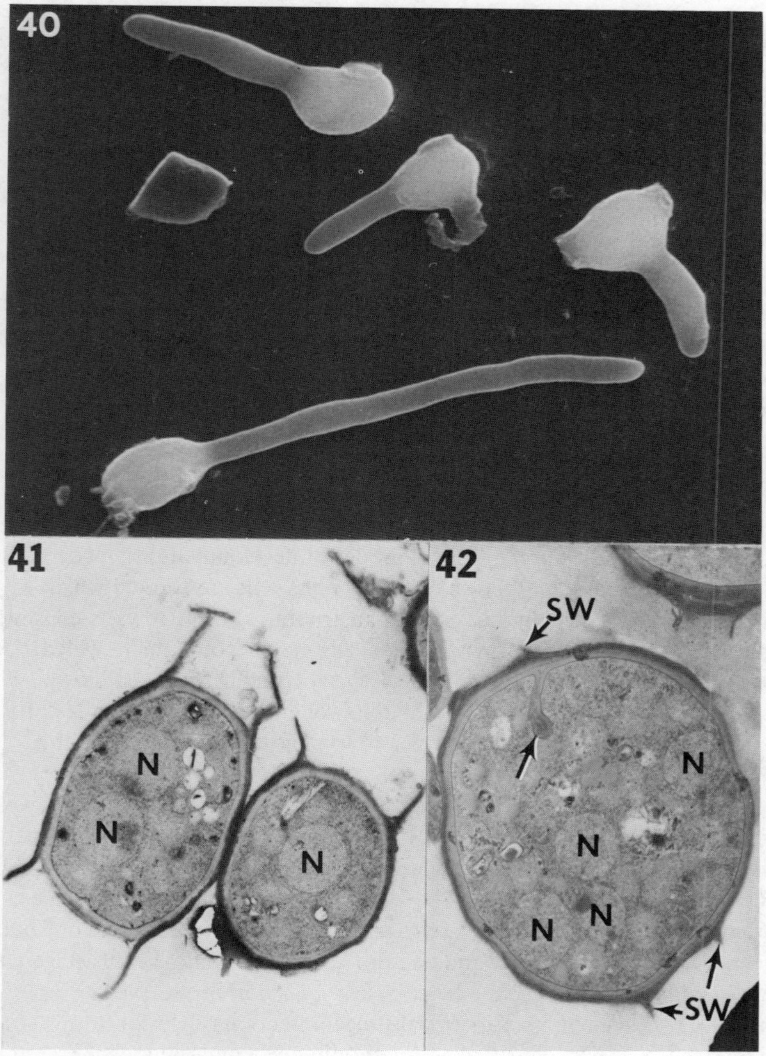

Figure 40 *Coccidioides immitis.* Germinating conidia grown on glucose (1%)–yeast extract (0.5%) agar at room temperature for 24 h. 2400X. (Courtesy of Dr. S. Sun.)

Figures 41 and 42 *Coccidioides immitis.* Swollen enteroarthric conidia after incubation on modified Converse agar at 40°C in 20% CO_2–80% air for 24 h (Fig. 41) and 48 h (Fig. 42). The developmental stage shown in Fig. 41 corresponds with I-J in Fig. 35, and Fig. 42 corresponds with P-Q in Fig. 35. Arrow in Fig. 42 locates ingrowing wall associated with "cleavage." N, nucleus; SW, septal wall of enteroarthric conidium. 11,200X and 11,200X, respectively. (From Ref. 304.)

D. Pleomorphism

Emmons et al. [117] have defined pleomorphism as "the occurrence of different growth forms in the life history of a fungus." This same term, however, has been used improperly to refer to mutational changes in dermatophytes which lead to formation of atypical hyphae, partial or complete loss of sporulation, and in some cases, general degeneration of the culture [1,117]. In this chapter we interpret pleomorphism as a condition in which more than one mode of conidiogenesis or sporangiosporogenesis occurs on the same mycelium [88,326]. A striking example of this phenomenon is demonstrated by *Wangiella dermatitidis* (Figs. 43-43) [77,229,230,309], frequently associated with chronic skin lesions of patients in Japan and other areas of the Orient [117]. The most common mode of conidiogenesis which occurs when this fungus is grown on potato dextrose agar at 22°C is formation of phialoconidia (Fig. 43; Table 3) [246]. The fertile hyphae are unusual, however, in that a single cell may have several, simultaneously active conidiogenous loci. Such conidiogenous cells with more than one open end (collarette) from each of which a basipetal succession of phialoconidia is produced are called polyphialides [88,178,182]. The polyphialides of *W. dermatitidis* are also capable of percurrent proliferation (Fig. 43), a process in which each new axis grows through the previous fertile apex (as in annellides).

Under these same growth conditions, additional modes of development are recognized in *W. dermatitidis* (Figs. 44 and 45). Polyblastic conidiogenous cells are formed which give rise to a succession of holoblastic conidia. Several abstriction scars (arrows) are visible at the apices of fertile cells in Fig. 44. Similarly shaped cells may become thick-walled, secede, and then proliferate to produce single propagules at their apices in yeast-like fashion, or germinate to produce moniliform (irregularly swollen) and typical hyphae. In Fig. 45, yeast-like cells (Y) are attached to the surface of hyphae and adjacent to a phialide (P).

Wangiella dermatitidis is a member of the "black yeasts" [99] which encompass several other pleomorphic, conidial fungi of medical interest, including *Aureobasidium pullulans, Exophiala jeanselmei* (Fig. 29), *E. spinifera,* and *E. werneckii* (Figs. 46 and 47; Table 3). Ajello [7] has defined the black yeasts as "dematiaceous, filamentous fungi which, in certain stages of their development or under certain environmental conditions, have a unicellular phase during which multiplication is by a budding process. The colonies at this stage are pasty with some shade of black." *Exophiala werneckii* (= *Cladosporium werneckii*) was first recognized as a cause of tinea nigra, a superficial fungal infection of the stratum corneum, in 1921 [175] but was not identified as a naturally occurring soil saprobe until almost 50 years later [7,203]. The taxonomy of this fungus has been critically reviewed by McGinnis [231], who has confirmed that formation of conidia from aerial hyphae as well as yeast budding involve annellidic development (Figs. 46 and 47) [22,88,146,230]. Annellations (arrows) are visible at the fertile apex of the conidiogenous cell (CC) in Fig. 46 and at budding loci of the yeast in Fig. 47.

Pleomorphic anamorphs which reproduce both by budding and formation of conidia from aerial hyphae are not restricted to the "black yeast." In pure culture, *Trichosporon beigelii* produces conidia by holoarthric development and yeast by a typical budding mechanism (Fig. 48) [201]. Holoarthric conidia may also swell and subsequently produce acropetal chains of holoblastic conidia, which further complicates the pleomorphism of this fungus [88]. The ability to alternate between mycelial and budding states under certain environmental conditions is characteristic of many conidial fungi, expecially the pathogenic forms [86,132,133,219,247].

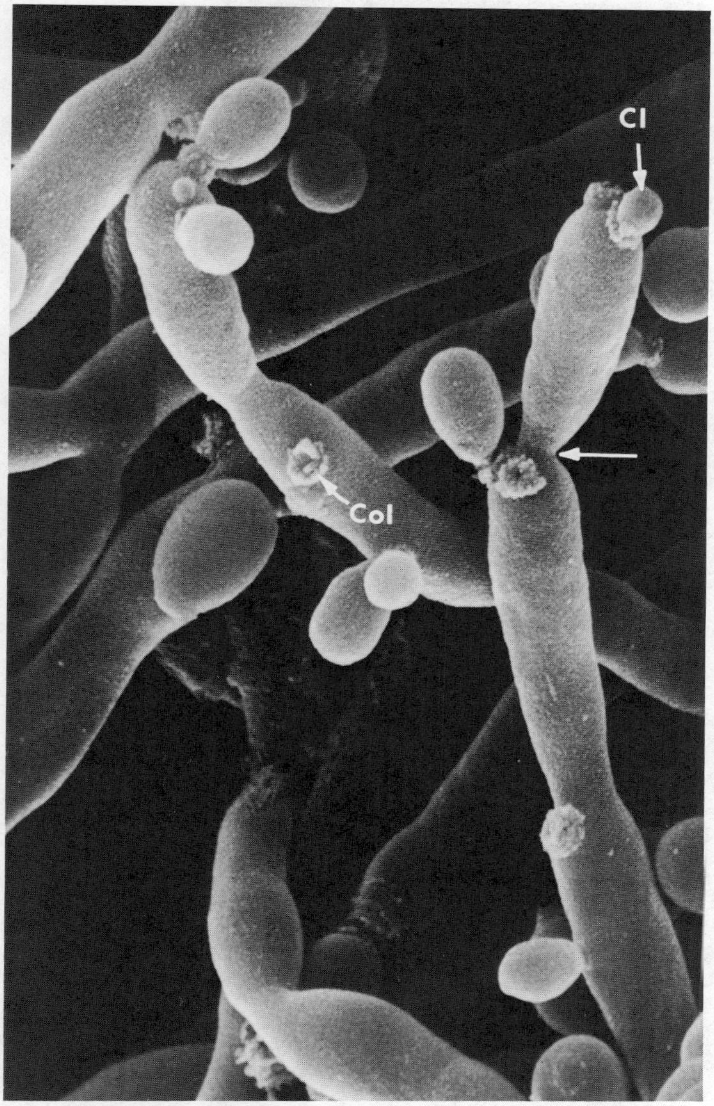

Figure 43 *Wangiella dermatitidis.* Phialides showing several collarettes (Col) on the same fertile cell (polyphialides). Arrow indicates constriction between percurrently proliferated phialides. CI, conidium initial. 7400X.

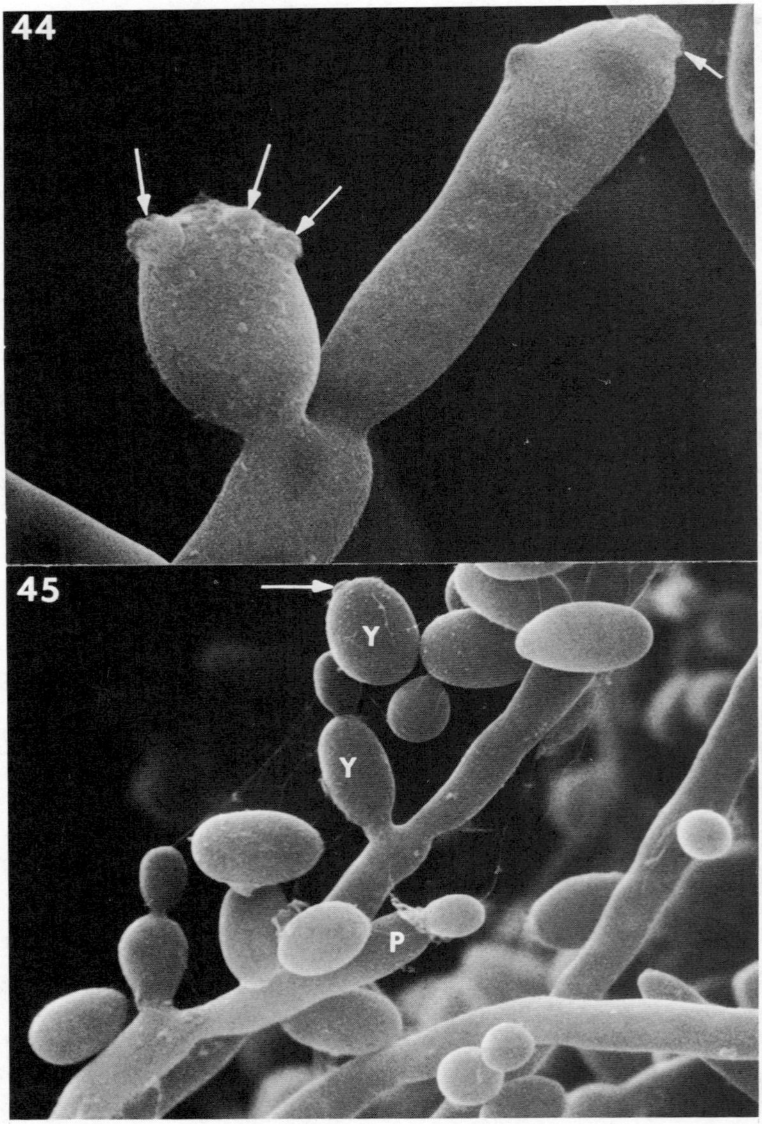

Figures 44 and 45 Pleomorphic nature of *Wangiella dermatitidis* showing polyblastic fertile cells (Fig. 44; arrows locate denticles of seceded conidia) and budding yeast (Y, Fig. 45). Arrow in Fig. 45 locates bud scar. P, phialide. 7700X and 3600X, respectively.

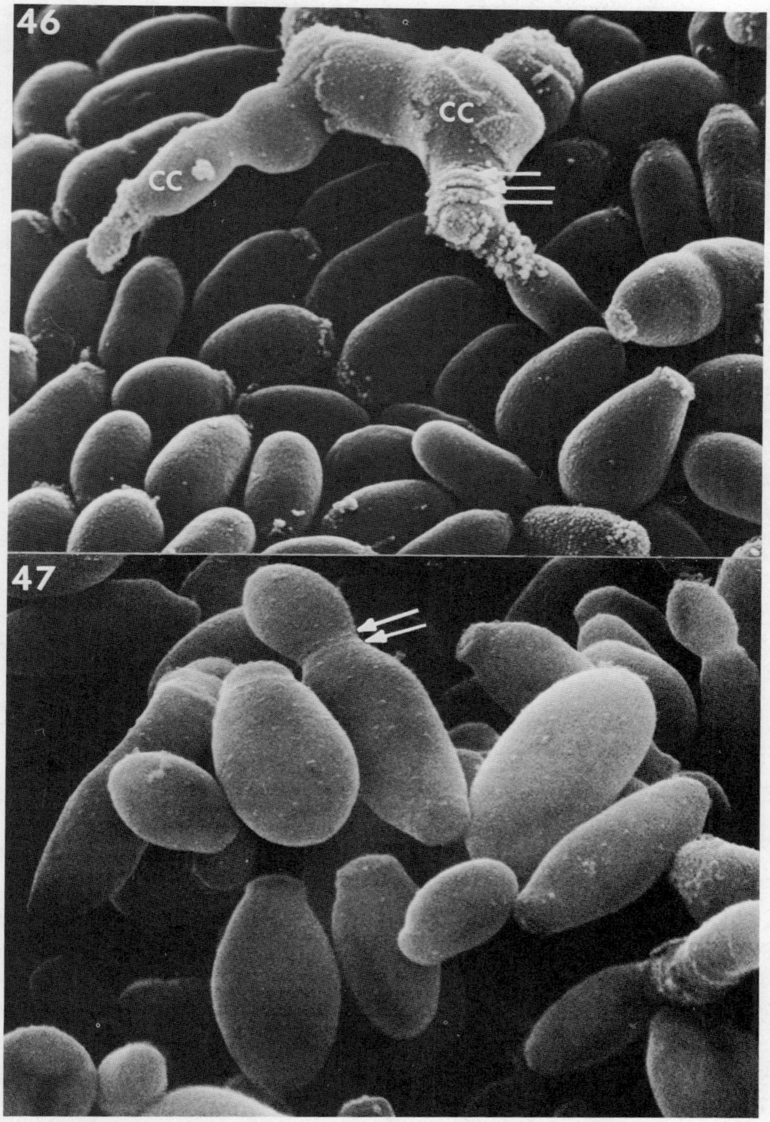

Figures 46 and 47 *Exophiala werneckii.* Annellate conidiogenous cells (CC, Fig. 46) and yeast (Fig. 47). Arrows locate annellations. 4000X and 5400X, respectively.

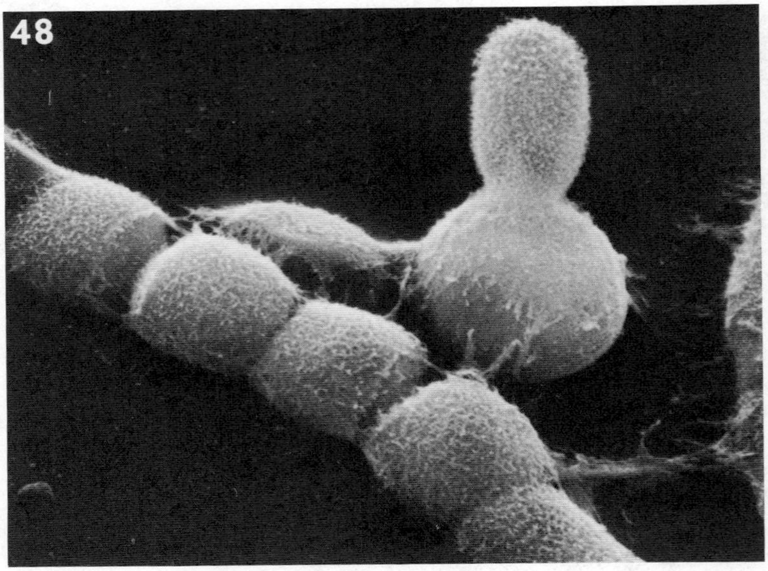

Figure 48 *Trichosporon beigelii.* Holoblastic and holoarthric conidium formation. 5500X.

The broad definition of pleomorphism embraces the phenomenon of dimorphism, which in fungi refers to yeast-hypha conversion. Thus, many conidial fungi discussed above are also referred to as dimorphic forms, such as *Histoplasma capsulatum, Blastomyces dermatitidis, Sporothrix schenckii, Coccidioides immitis,* and *Trichosporon beigelii.* Carmichael [65] has pointed out that the yeast, which is both assimilative and propagative, is better adapted to spread through a liquid medium. The filamentous form, on the other hand, is better adapted for rapid colonization of solid substrates. Rippon [261] has observed that in the case of pathogenic, dimorphic fungi, the ability to invade animal tissue and spread hematogenously is associated with their capacity to form yeast or some other "tissue" phase. The author has concluded that "morphogenesis and disease production" for these fungi are intimately related but "this pathogenic ability is an 'accidental' phenomenon for the fungus and not necessary for the maintenance and dissemination of the species." Each is capable of conidiogenesis or sporangiosporogenesis and may exist as a saprobe, free of the more restrictive growth conditions of host tissue. For example, *Trichosporon beigelii* is a keratinophilic soil fungus in nature which may cause superficial infections of human hair and nail beds. Under these conditions it produces hyphae, holoarthric and holoblastic conidia, as well as white to light brown nodules within and on the diseased hair (i.e., white piedra). On rare occasions, however, the fungus causes systemic infection, usually in patients who are immunosuppressed by their primary disease or its therapy [202]. In this case the yeast is the propagative agent and is probably disseminated in the blood stream.

It is not our purpose to review aspects of fungal dimorphism in this chapter. Extensive discussions of this important morphogenetic phenomenon can be found elsewhere [33,86,261,267,300; Chap. 10]. However, as a parallel to the general remarks on conidium formation presented in the introduction to this chapter, the following is a summary of some of the significant concepts of yeast morphogenesis based primarily on investigations of *Saccharomyces cerevisiae.*

E. Yeast Morphogenesis

Figure 49 illustrates three stages of the cell cycle of *Saccharomyces cerevisiae*; bud emergence (A, B, B′), bud growth (C-E), and yeast cell secession (F, F′). Byers and Goetsch [52] have proposed that duplication and subsequent separation of the spindle pole plaques

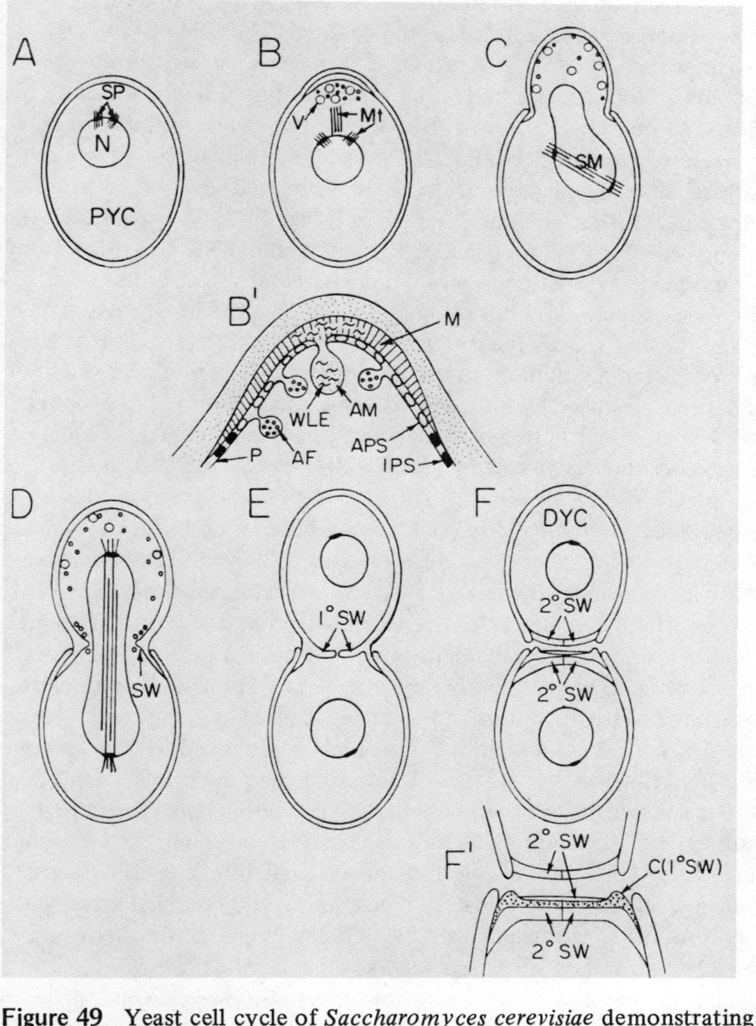

Figure 49 Yeast cell cycle of *Saccharomyces cerevisiae* demonstrating bud emergence (A, B,B′), bud growth (C-E), and cell separation (F,F′). AF, activating factor; AM, amorphous material; APS, activated polysaccharide synthetase; C, chitin; DYC, daughter yeast cell; IPS, inactive polysaccharide synthetase (zymogen); M, microfibrils; Mt, microtubules; N, nucleus; P, plasmalemma; PYC, parental yeast cell; SM, spindle microtubules; SP, spindle plaques; SW, septal wall; V, vesicles; WLE, wall lytic enzymes. (From Ref. 86.)

(SP) on the nuclear envelope is a necessary precursor of bud initiation. This process may be linked to events that trigger bud emergence at the fertile locus of the parental yeast while preventing additional budding from the same cell until completion of cytokinesis [155-157]. As the parental cell wall blows out during early development of the daughter yeast (Fig. 49B), microtubules extend between the nucleus and burgeoning initial [53] and supposedly participate later in migration of a lobe of the dividing nucleus into the newly formed cell (Fig. 49C). Cytoplasmic vesicles are also visible in this region and are suggested to take part in lysis of the preexisting parental cell wall as well as synthesis of the new wall encompassing the yeast initial (Figs. 49B and C). The structural and functional interrelationships of the plasmalemma and fusion vesicles are illustrated in Fig. 49B′. This diagrammatic interpretation is based on the work of Cabib and co-workers discussed earlier in the chapter (Sec. II.A.1). It is proposed that there is a basic similarity between cytological events associated with bud initiation, hyphal tip growth, and early blastic conidium formation (see Figs. 8, 9, and 49B′). As the yeast bud continues to grow, the nucleus begins to migrate into the initial (Fig. 49C), and the spindle fibers subsequently become orientated parallel to the long axis of the proliferating cell (Fig. 49D) [52,235, 265]. The nuclear envelope remains intact until telophase, when it finally separates at the midregion of the nucleus [263,265]. During nuclear migration and division, formation of the primary septum is initiated at the isthmus between the parental and daughter yeast cells (Fig. 49D and E). In *S. cerevisiae,* chitin is found only in the primary septum, which persists after secession as the bud scar on the surface of the parental cell (Fig. 49F and F′) [56,57,60,324]. Such restricted distribution of chitin to the primary septal wall does not seem to be the case for yeast produced by most pleomorphic conidial fungi [79,86,261]. For example, Kanetsuna et al. [191] determined that the N-acetylglucosamine content of the yeast cell wall of *Paracoccidioides brasiliensis* and *Blastomyces dermatitidis* is four times greater than that of the mycelial wall. High chitin content was also reported for the yeast wall of *Histoplasma capsulatum* [104,105,215]. On the other hand, *Candida albicans* has little chitin in its yeast wall and considerably less in the yeast than the mycelial wall. [70]. Yeast morphogenesis in the related fungi, *S. cerevisae* and *C. albicans* [26], shows distinct similarities. Upon completion of the primary septum, the secondary septal walls are synthesized using the former as a template (Fig. 49F and F′) [286]. Composition of the latter in *S. cerevisiae* is unlike the primary septum but comparable to the rest of the yeast wall, consisting mainly of glucans and mannans [55]. Thus, the central region of the bud scar in Fig. 49F′ consists of nonchitinous wall material derived from the secondary septum, while the raised rim is composed of chitin derived from the primary septum [301,302,324]. Although *Saccharomyces* does not undergo yeast-hypha conversion, it demonstrates morphogenetic features during its asexual cell cycle which appear to be common to other yeast-producing fungi. Details of development may differ between *S. cerevisiae* and dimorphic forms, but the morphological, biochemical, and genetical investigations that have been performed using this fungus provide the basis for the design of other experimental systems involving pathogenic yeasts such as those cited below.

The asporogenous or anamorphic yeasts consist of about 300 species which belong to 22 genera [24]. For certain forms (e.g., *Cryptococcus*) the telemorph has been identified, but for the majority of imperfect yeasts there is no known sexual state. Lodder [222] distinguished genera of this group primarily on the basis of morphological and physiological characters. In recent years, yeast systematists have placed increasing emphasis on the taxonomic importance of biochemical characters, such as DNA base sequence relatedness [254], composition of coenzyme Q systems [335], fatty acid content of cells [145], and chemistry of the cell walls [25,330]. Developmental differences between yeasts, on the other hand, have not been extensively explored but may be of systematic value for distinguishing genera as well as major groups of yeasts (i.e., ascomycetous and basidiomycetous yeasts). Von Arx [23,24] has referred to propagules of anamorphic yeasts as conidia and has classified their modes of reproduction as blastic and thallic conidiogenesis. As previously mentioned, cytological data indicate fundamental similarities between processes of conidium formation from fertile hyphae and bud development from parental yeast cells. A single conidiogenous locus of a parental yeast may give rise to a solitary blastic conidium or a plurality of daughter cells. In either case, the outer wall of the parental cell ruptures during conidium initiation and only the newly formed inner wall encompasses the daughter yeast cell (enteroblastic; Fig. 49C). Von Arx and Weijman [25] have reported that formation of enteroblastic conidia in basipetal succession (i.e., phialidic-annellidic development) have been observed only among the heterobasidiomycete-like yeasts. Anamorphic species of the Saccharomycetales demonstrate both enteroblastic (solitary, acropetal) and holoarthric conidium ontogeny. A diagrammatic interpretation of the major kinds of conidiogenesis in yeasts is presented in Fig. 50.

1. *Candida albicans*; Enteroblastic (Multilateral) Conidia

A ubiquitous yeast in nature, *C. albicans* has been isolated from dairy products, sand and soil, air, terrestrial plants, and aquatic flora from inshore freshwater and marine environments [17,18,244,284,290]. It has also been frequently isolated from human feces and is recognized as a common component of the gastrointestinal (GI) microflora [142,244]. In healthy individuals, the presence of *C. albicans* in the GI tract usually poses no immediate problem. An exception is the drunken syndrome, or "meiteisho," which is apparently due to active fermentation by this and other yeast in the GI tract of individuals on carbohydrate-rich diets [187]. Although *C. albicans* is a potentially pathogenic microorganism, the human body has several defense mechanisms at its disposal for keeping the yeast in check. However, every congenital or acquired weakness in the lines of this chain of defense means an increased chance of infection. A patient with one or more disturbances in this defense system is described as "compromised" [240,257]. Generalized systemic infections produced by *C. albicans* have increased in importance in recent years as causes of morbidity and mortality in patients undergoing chemotherapy for cancer [121, 258], immunosuppressive therapy [101], or prolonged antibiotic treatment [280]. It is surprising that less than 15 years ago epidemiology of *Candida* infections, as a concept, did not exist, whereas "today, it is well known in hospitals that *Candida* is an important source of infection to which patients with impaired immunity mechanisms are particularly susceptible" [333]. Passage through the gastrointestinal mucosa of individuals is believed to be a likely portal of entry into the body since the causative yeast has been shown capable of crossing the human bowel walls [211]. Studies of this type of behavior by *C. albicans* in experimental animals have been hampered by a lack of a suitable animal model. Intragastric challenge of infant mice [252] has provided a model to study the events leading to systemic candidosis by dissemination from localized sites of colonization and infections. Figure 51 shows a large number of yeast cells associated with mucin (arrows) and the surface of villi (V) in the ileum of an infant mouse 6 h after inoculation. Using this same model, it has been possible to maintain a gastrointestinal infection for periods up to 6 weeks [123] and such animals may be employed to study the effects of physical and chemical treatments that are known to compromise humans to candidosis.

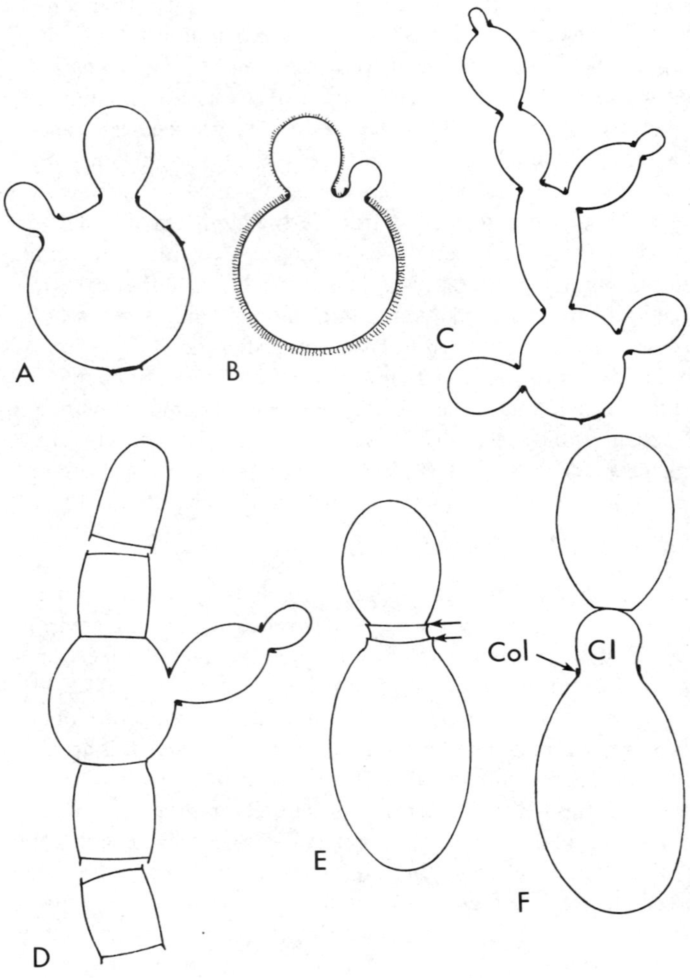

Figure 50 Diagrammatic interpretation of conidiogenesis of anamorphic yeast. Multilateral budding (A), sympodial budding (B), acropetal and multilateral budding (C), holoarthric development and acropetal budding (D), annellidic (E) and phialidic (F) development. Arrows in (E) indicate annellations. CI, conidium initial; Col. collarette.

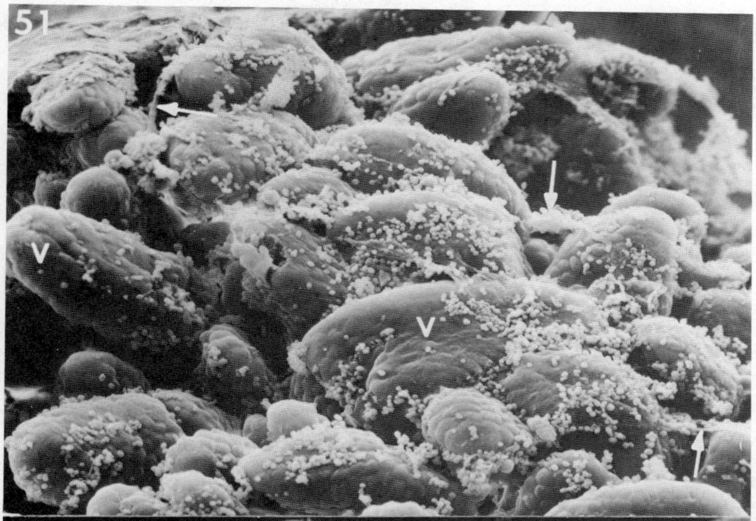

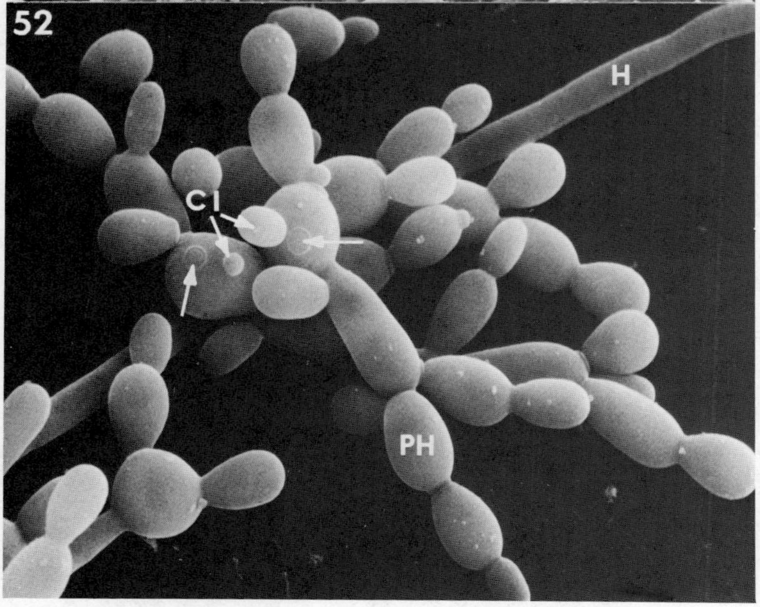

Figure 51 *Candida albicans.* Yeast present on surface of villi (V) and associated with mucin (arrows) in small intestine of an infant mouse 6 h post-inoculation with 5×10^8 colony-forming units in 0.05 ml of saline. 2100X. (From Ref. 252.)

Figure 52 *Candida albicans.* Yeast, pseudohyphae (PH), and hyphae (H) grown on a glass coverslip coated with Sabouraud's dextrose agar. 3000X.

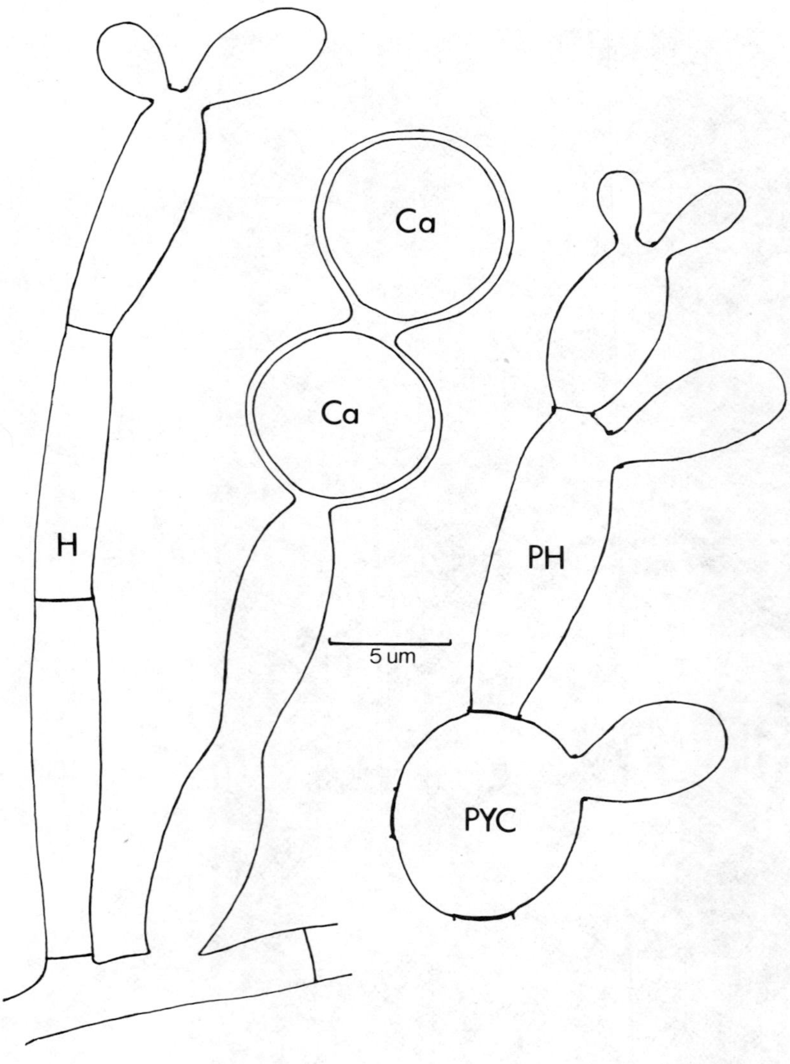

Figure 53 *Candida albicans.* Hypha (H), thick-walled chlamydospores (Ca), pseudohyphae (PH), and parental yeast cells (PYC) which show scars of multilateral budding.

Bud scars (arrows) and conidium initials (CI) are visible on the surface of parental yeast cells in Fig. 52. The pleomorphic nature of *C. albicans* is indicated by this micrograph. Conidia may form by multilateral budding of a single parental cell or may be produced in acropetal chains. Several chains may arise from elongated, ellipsoidal buds which are referred to as pseudohyphae (PH). Conidia may also germinate to form cylindrical germ tubes which continue to elongate at their tips (i.e., polarized growth), become septate, and differentiate into typical hyphae (H). These filaments may later give rise to acropetal chains of conidia [86]. Swollen, thick-walled, blastic cells referred to as "chlamydospores" are also produced by *C. albicans* (Fig. 53). These structures are suggested to be a dormant phase of the fungus [30], and their formation is influenced by various growth factors, such as glucose and nitrogen concentrations in the medium [106,244]. Only one case is known of chlamydospores occurring in infected tissue [165]. Details of wall differentiation during conidium, chlamydospore, and germ tube formation have been examined [69,234,275]. Thin sections have revealed that the outer parental cell wall ruptures during the initiation of conidium ontogeny, as demonstrated by *Saccharomyces cerevisiae* in Fig. 49C (enteroblastic development). Production of germ tubes and chlamydospores by *C. albicans* is frequently used as a criterion for identifying this fungus when cultured from clinical specimens [49,220,310]. However, the reliability of these developmental characters has been questioned in view of the isolation of germ tube positive strains of *C. tropicalis* from patients with pulmonary and oral diseases [228,312] as well as production of chlamydospores in vitro by both *C. stellatoidea* [323] and *C. tropicalis* [159].

2. *Torulopsis glabrata*; Enteroblastic (Multilateral) Conidia

Although this yeast apparently has a low degree of virulence, like *C. albicans* it frequently occurs in association with steroid, antibiotic, and immunosuppressive therapy [117,172, 250]. It has been isolated from soil and water [91] as well as the intestinal and urogenital tracts of humans and other animals [204,299]. However, torulopsosis is generally less understood than candidosis and may more often be overlooked by pathologists in routine analyses [272]

Although pseudohyphae have been observed in histological sections of infected tissue, *T. glabrata* does not produce germ tubes or true hyphae. Yeast budding is multilateral [26] and several bud scars are visible on the surface of the parental cell in Fig. 54.

3. *Cryptococcus neoformans*; Enteroblastic (Sympodial) Conidia

Cryptococcosis has been referred to as a "sleeping giant" [4] because it is thought to be much more prevalent than indicated by numbers of case reports. Kaufman and Blumer [192] pointed out that only 300 cases of cryptococcosis were reported between 1900 and 1955, compared to 1264 proven cases during the 12-year period between 1965 and 1977. Even more dramatic are the following percentage increases of cryptococcosis and other mycoses recorded between 1970 and 1976; cryptococcosis (78%), aspergillosis (158%), and coccidioidomycosis (74%) [127]. Such alarming escalation of these mycotic diseases is due primarily to widespread use of immunological tests for detection of fungal infections, combined with improvements in statistical records of case histories. However, it is reasonable to assume that more extensive use of immunosuppressive drugs in the therapy of various human disorders have presented a larger number of susceptible hosts to these opportunistic pathogens. Migration of susceptible persons into hyperendemic areas and aging of the population have probably also contributed to the increases [127].

The presence of *C. neoformans* in pigeon and canary droppings and nesting areas of these birds is well documented [117,308]. The fungus has been isolated rarely from nature except in areas frequented by pigeons. However, it is not known how the birds are originally contaminated with the yeast. Experimental inoculation of pigeons with virulent strains of *C. neoformans* have demonstrated that the yeast can survive for several months in the crop with no apparent ill effects to the bird. Conidia in the dried feces may be air-dispersed, and thus human contact with this "potential" pathogen commonly involves the respiratory system.

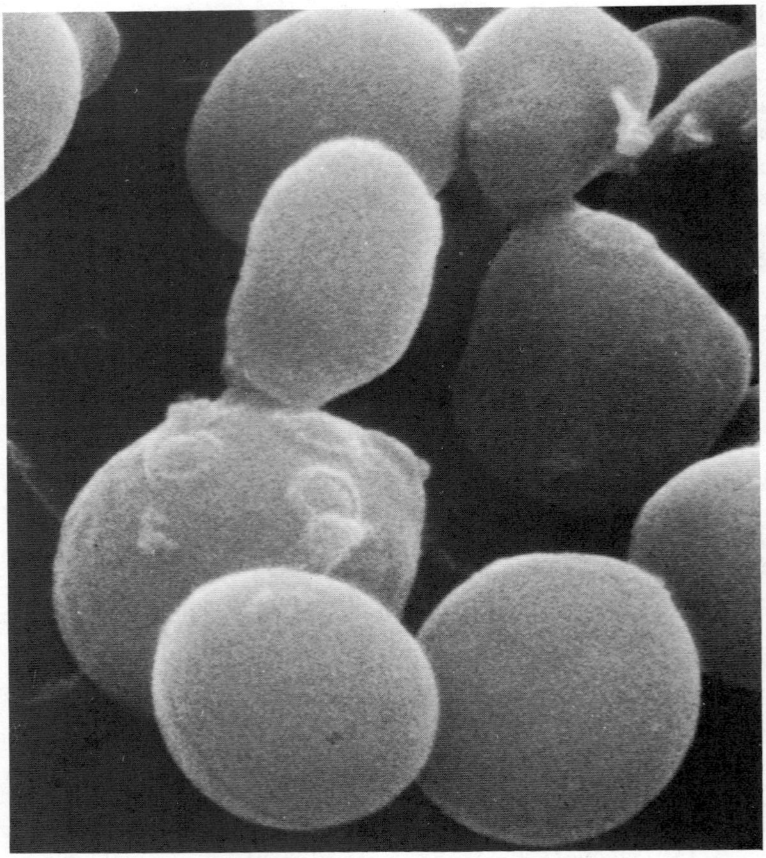

Figure 54 *Torulopsis glabrata.* Parental yeast cell with multiple bud scars. 11,000X.

Conidia are produced by sympodial budding (Figs. 50B and 55) [23]. Mature cells are surrounded by a mucoid polysaccharide capsule which is revealed by the scanning electron microscope as a distinct fibrous layer (Fig. 55). It has been suggested that the capsule is associated with inhibition of phagocytosis by macrophages, which often characterizes infections by virulent strains of *C. neoformans* ;208,209]. The outer polysaccharide layer apparently masks the presence of proteins bound to the cell wall and may present "a new surface to the macrophage for which the macrophage has no receptors. The physicochemical characteristics of the capsule surface that are responsible for this recognition failure are not known" [210].

Kwon-Chung [214] has confirmed earlier observations of hyphal clamp connections produced by *C. neoformans* [285] by demonstrating that this fungus is the anamorph of a basidiomycete which the author accommodated in the new genus *Filobasidiella.* Conjugation involves yeast cells which give rise to typical basidiomycetous hyphae with clamp connections and dolipore septa. Aseptate, slender, flask-shaped basidia (B) arise from the hyphae and produce chains of basidiospores, usually from four separate loci at their apex (Fig. 56) [117,215].

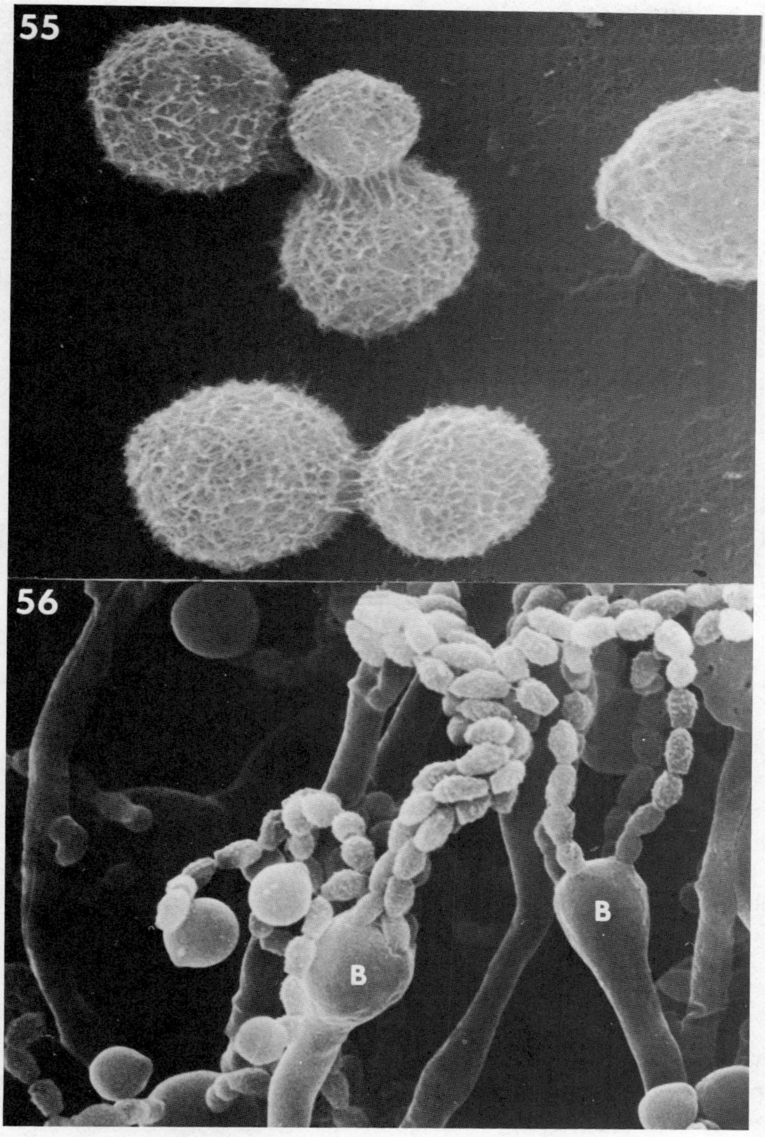

Figure 55 *Cryptococcus neoformans.* Budding cells surrounded by a distinct fibrous outer wall layer which is the mucoid polysaccharide capsule. 7400X.

Figure 56 *Filobasidiella neoformans.* Basidia (B) and chains of basidiospores which arise from four loci at the apex of each fertile cell. 2900X.

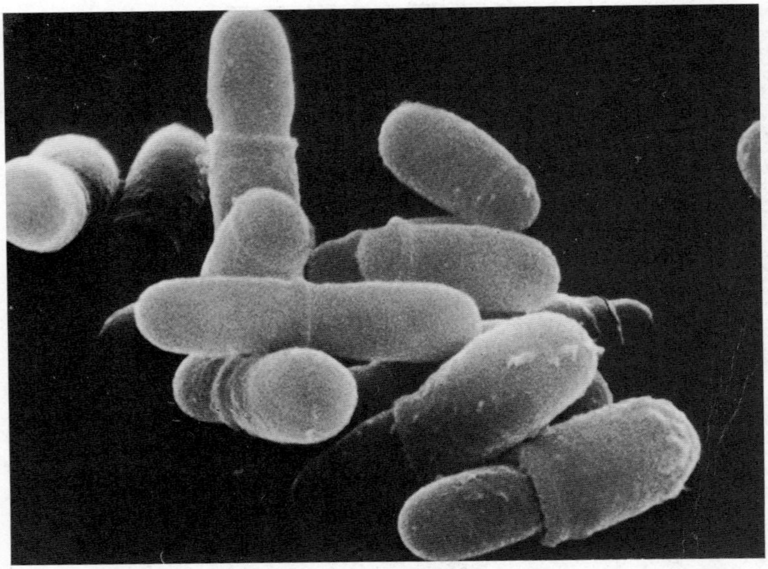

Figure 57 *Pityrosporum malessezii.* Annellate yeast. 9400X.

4. *Pityrosporum malessezii*; Enteroblastic (Annellidic) Conidia

The taxonomy of this yeast has been recently reconsidered. The synonymy of *Pityrosporum* with *Malassezia* is suggested, while *P. ovale* is considered to be a synonym of *P. malassezii* [26]. The fungus is a causative agent of pityriasis versicolor, a skin infection characterized by white, brown, or fawn-colored superficial lesions [117]. The yeast demonstrates unipolar budding and gives rise to a plurality of enteroblastic conidia. Annellations remain at the fertile locus of the parental yeast cell, which is diagnostic of annellidic development (Fig. 57) [236a].

III. SPORANGIOSPOROGENESIS

Benjamin [40] identified seven orders of Zygomycetes, which include Mucorales, Entomophthorales, Zoopagales, Endogonales, Dimargaritales, Kickxellales, and Harpellales. The majority of zygomycetous fungi which are recognized animal pathogens belong to the Mucorales and Entomophthorales. Only these two orders will be considered here. Emmons et al. [117] distinguished the respective diseases produced by certain members of these two orders as mucormycoses and entomophthoramycoses. However, to avoid further proliferation of terminology as additional orders of zygomycetous fungi are revealed to be potential etiologic agents, Ajello et al. [10] have recommended the collective term "zygomycosis" for all infections caused by zygomycetous fungi.

Sporangiospores are simply asexual aplanospores (nonmotile) formed in a sporangium [1], and the mechanism of sporangiospore formation in Zygomycetes is basically the same in all the orders, irrespective of whether the sporangium produces many spores or only one [40,125]. Sporangiosporogenesis is considered to be distinct from conidiogenesis [88,197]. Benjamin [40] has defined sporangiospore formation as the cleavage of cytoplasmic contents of the sporangium . . . without the wall of the sporangium being directly involved in the formation of the spore wall."

A. Mucorales

1. Multispored Sporangia

A classic study of sporangiospore development is Bracker's [45] ultrastructural investigation of the cytological events associated with cleavage in *Gilbertella persicaria* (Fig. 58). In this

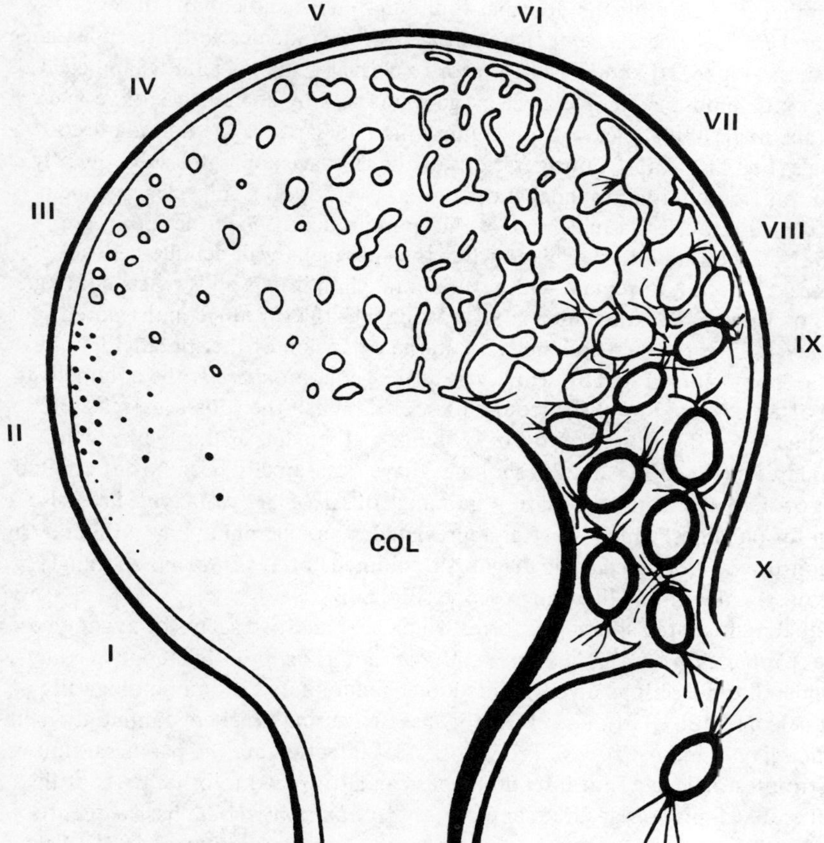

Figure 58 Diagram summarizing sporangiospore formation in *Gilbertella persicaria.* Only cleavage apparatus and walls are indicated. Spatial distribution of stages is for convenience in illustration only, and does not represent spatial progression within a single sporangium. Early precleavage (I,II), mid-precleavage (III), late precleavage (IV), early cleavage (V), mid-cleavage (VI), late cleavage (VII), early postcleavage (VIII), mid-postcleavage (IX), late post-cleavage or spore maturity (X). The discontinuity of the sporangial wall near X represents rupturing of the sporangium as it splits into halves at maturity. COL, columella. (From Ref. 45.)

case, cleavage of the sporangial cytoplasm results in formation of many spores, which characterizes a primitive mechanism of asexual reproduction in the Mucorales [39]. No analogous process of conidiogenesis occurs among the Fungi Imperfecti, with the possible exception of *Coccidioides immitis*. However, wall invagination ("cleavage") associated with endospore formation in the latter (Figs. 35 and 42) is unlike the mechanism of cleavage in mucoraceous fungi described below. Sporangiospore ontogeny involves the differentiation of a cytomembrane system which produces a ramifying, tubular, cleavage apparatus and the eventual separation of spore initials. In the precleavage sporangium (Fig. 58I and II), small vesicles are formed from blebbing cisternal rings [150] which are morphologically similar to the simple dictyosomes found in hyphae and conidia of septate fungi (Figs. 6 and 7). Disappearance of these initial vesicles coincides with the appearance of cleavage vesicles (Fig. 58III and IV). Just prior to cleavage, the nuclei, ER, and cleavage vesicles form a cytomembrane complex. Cleavage is initiated by coalescence of vesicles to form the branching, tubular cleavage apparatus which demarcates the boundaries of the spore initials (Fig. 58V-VII). Configurations of the cleavage apparatus subsequently converge to cut out spore initials as independent cells (Fig. 58VIII). The cleavage membrane is transformed into the plasmalemma of the spore initials. When the spores are mature (Fig. 58IX), they are either released when the sporangial wall deliquesces (e.g., species of *Absidia*; Fig. 59), or remain within the sporangium if its wall is persistent (e.g., certain species of *Mucor*). Most members of the Mucorales that produce multispored sporangia possess a well-differentiated columella, or sterile extension of the sporangiophore apex within the sporangium (Fig. 58). However, in the genus *Mortierella* the columella is indistinct and often obscured by the surrounding spores. When the spores are released from the sporangium of this fungus, a frill of wall material remains at the apex which resembles a collarette (Fig. 60A) and the sporangia have occasionally been misinterpreted as phialides. For similar reasons, the mature sporangia of *Saksenaea vasiformis* have also been mistaken for phialides, which is better appreciated after examining Figs. 60B and 61. However, in contrast to *Mortierella*, the diagnostic columella of *S. vasiformis* sporangia can be clearly observed with the light microscope (Fig. 60B).

The principal multispored sporangial forms which are causative agents of zygomycoses [117] include *Absidia, Mortierella, Saksenaea, Mucor,* and *Rhizopus.* Systemic zygomycoses are considered to be serious mycotic infections in debilitated or immunologically compromised persons [188]. In these cases, the disease causing fungi are commonly members of *Absidia, Mucor,* and *Rhizopus.* Primary sites of infection are the paranasal sinuses, lungs, and gastrointestinal tract, but infectious cells can also spread to other parts of the body by vascular dissemination or direct extension. *Absidia corymbifera* has frequently been identified as one of the causative agents of bovine mycotic abortion [2,183]. *Saksenaea vasiformis,* first isolated and described from forest soils collected in India [271], was identified by Ajello et al. [10] as the disease agent of a patient suffering from a massive cranial infection. A common problem in treating zygomycoses is failure to diagnose the disease in time; identification of the fungal infection is often made only at autopsy [188,218]. Direct examination of tissue will usually reveal broad, aseptate hyphae which characterize zygomycetous fungi. Isolation of the infectious organism in pure culture provides positive identification. However, one should use caution in the isolation procedure to avoid aerial contamination since these etiological fungi commonly occur in the laboratory.

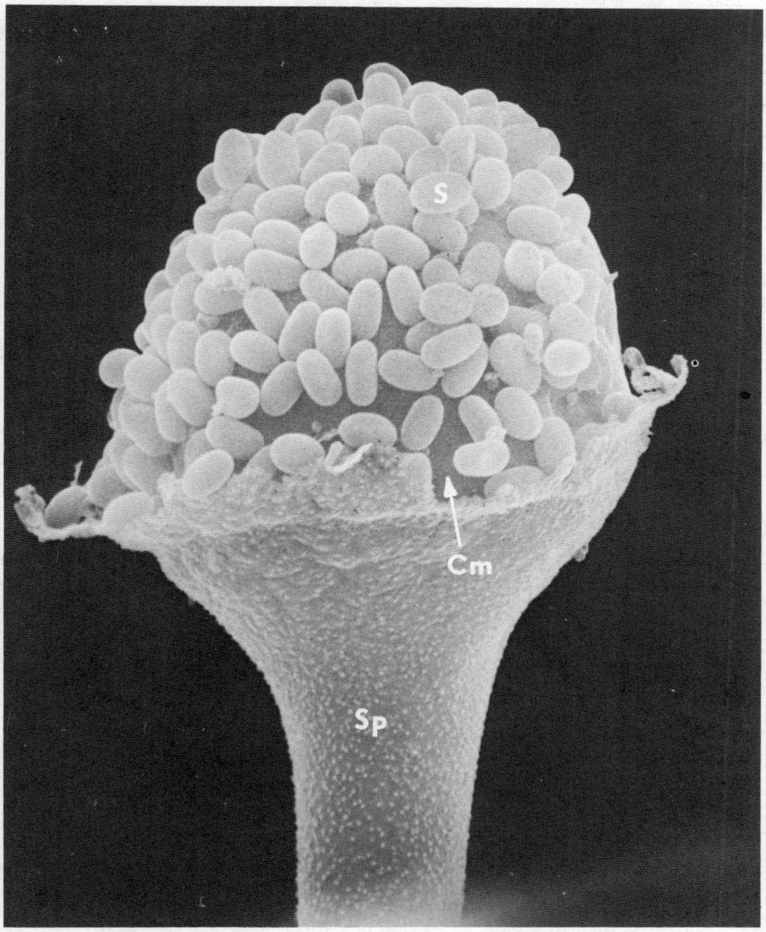

Figure 59 *Absidia corymbifera.* Multispored sporangium (Sp) whose wall has ruptured, revealing the sporangiospores (S) and columella (Cm). 2000X.

2. Monosporous Sporangiola

Certain members of the Mucorales produce reduced sporangia, or sporangiola, which contain a small number of spores. In *Cunninghamella,* a fungus occasionally associated with pulmonary infections in humans [186,198,216], sporangiola are formed by the blowing out of the wall of the sporangiophore vesicle (holoblastic development), and each sporangiolum contains a single spore (Figs. 62-64). The synchronous formation of monosporous sporangiola from fertile vesicles of *C. echinulata* (Fig. 63) is reminiscent of synchronous, botryose conidium formation in the Fungi Imperfecti [88]. In fact, most workers refer to the propagules of *Cunninghamella* as conidia [62,90,117,168,198]. Conidia and monosporous sporangiola, however, are ontogenetically distinct and should be identified as separate kinds of asexual propagules [40]. The basis for this distinction

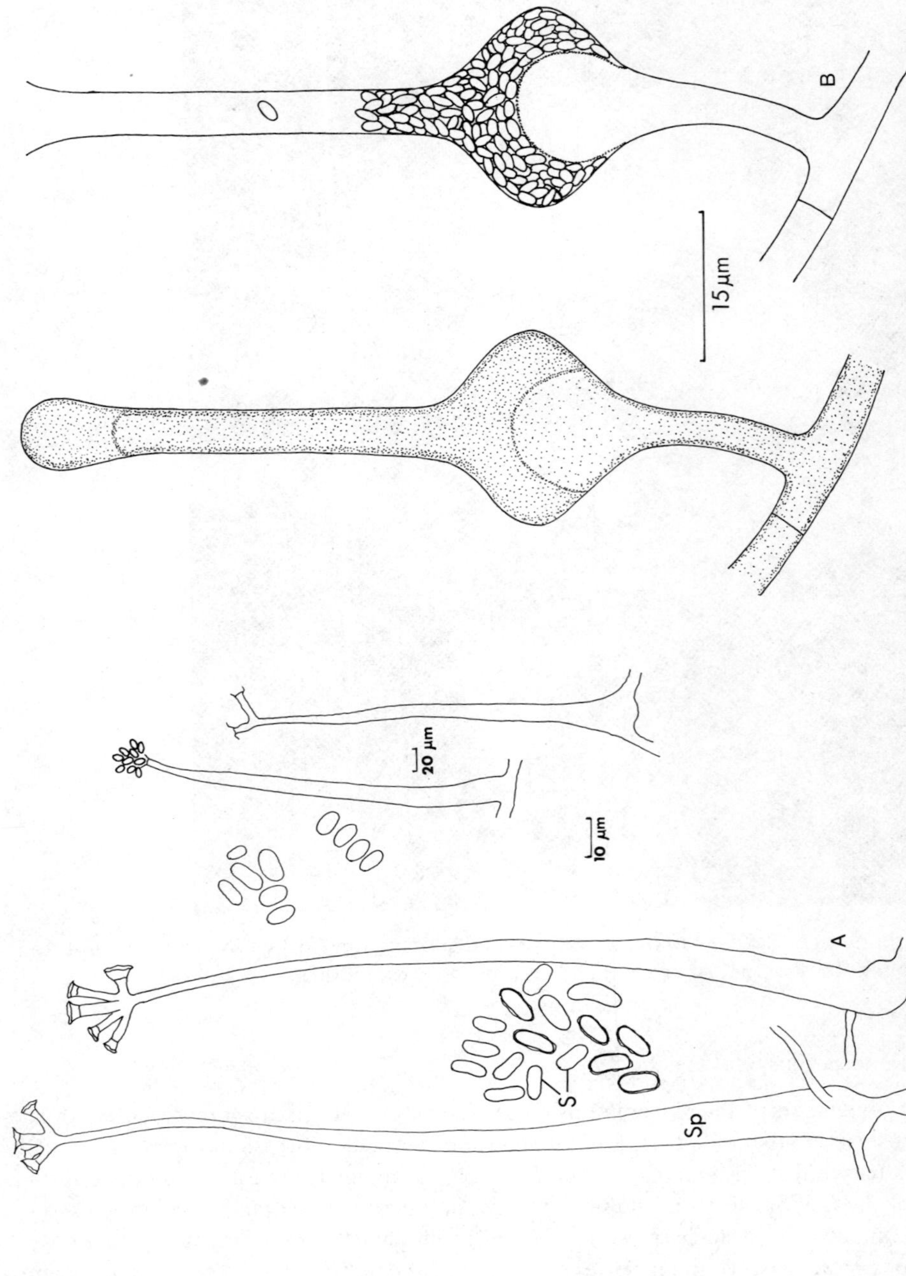

Figure 60 (A) *Mortierella wolfii*; (B) *Saksenaea vasiformis*. S, sporangiospores; Sp. sporangium.

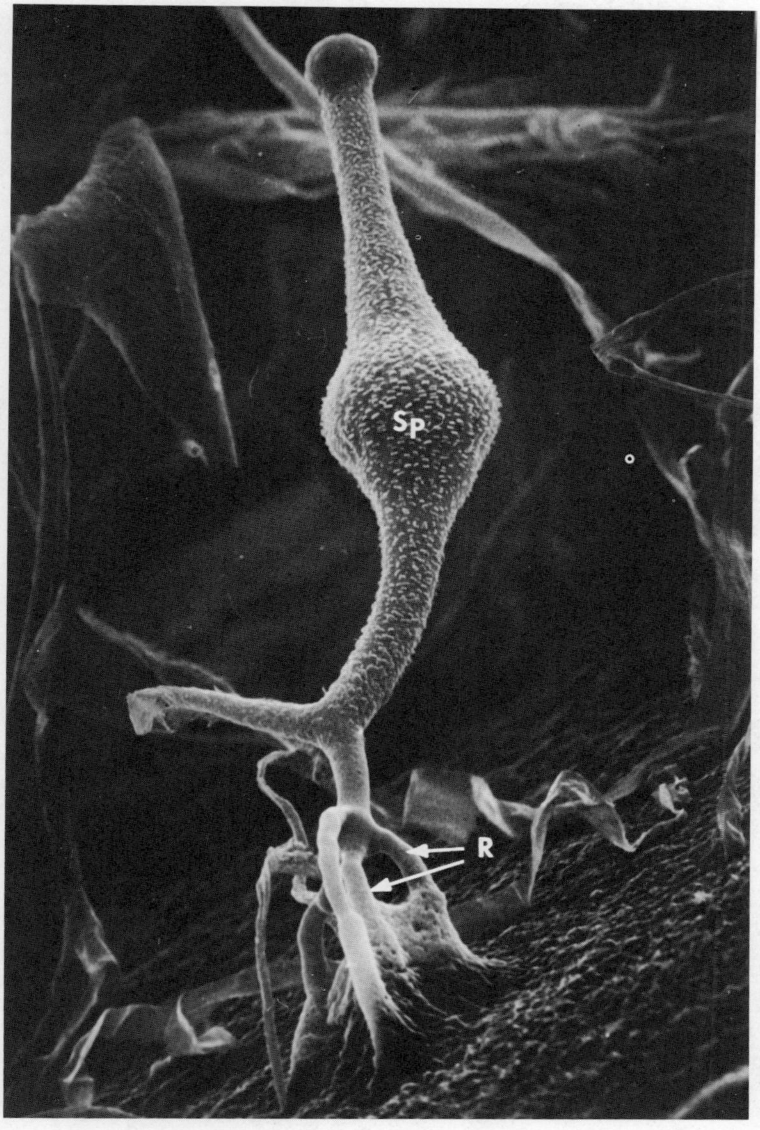

Figure 61 *Saksenaea vasiformis.* Phialide-shaped sporangium (Sp) and basal rhizoids (R). 9500X.

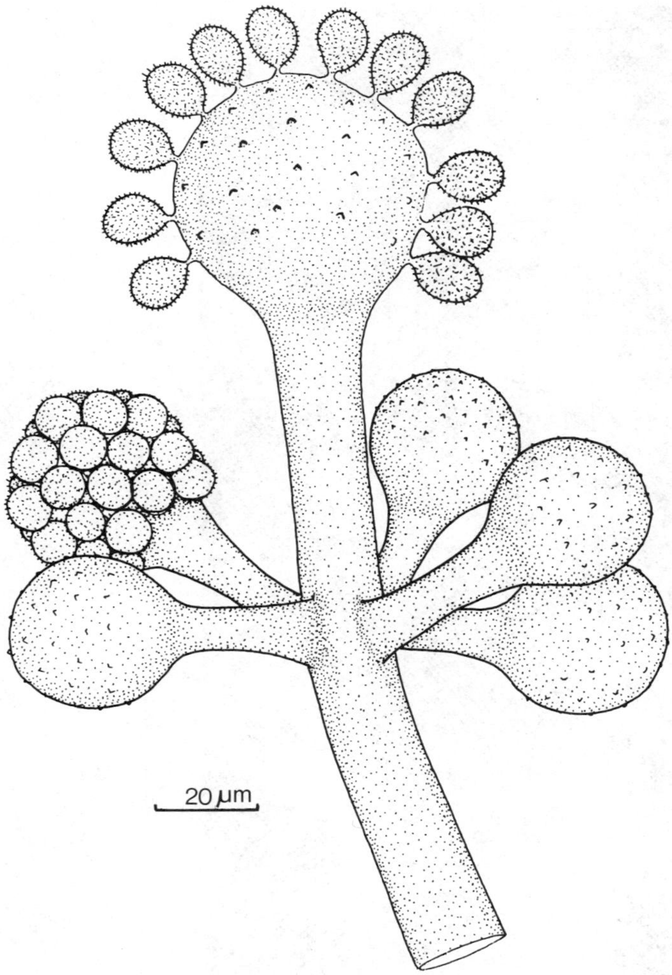

Figure 62 *Cunninghamella elegans.*

is derived primarily from thin-section studies and analyses of wall differentiation during conidiogenesis and sporangiosporogenesis (Fig. 65). Like multisporous sporangia, the single spore is formed by cleavage of the protoplasm contained within the sporangiolum (Fig. 65A-C). Deposition of wall material in the pedicel and occlusion of the cytoplasmic canal occur prior to secession of the sporangiospore (Fig. 65C) [88,196,197]. Surface ornamentation of the outer sporangiolar wall is initiated after cleavage (Figs. 63 and 64) [289]. The inner wall of the holoblastic conidium, on the other hand, is continuous with the basal septum, which develops centripetally (Fig. 65D and E). The latter is usually two-layered with a medial, electron-transparent zone along which schizolysis occurs (Fig. 65F). While the holoblastic conidium is a truly exogenous outgrowth of the conidiogenous cell, the sporangiospore differentiates endogenously within the evaginated sporangiolar wall. The close structural association between the sporangiolar and spore walls, as demon-

Figures 63 and 64 *Cunninghamella echinulata.* Stages of monosporous sporangiolar development. The sporangioles are supported by short pedicels. 4100X and 1600X, respectively. (From Ref. 88.)

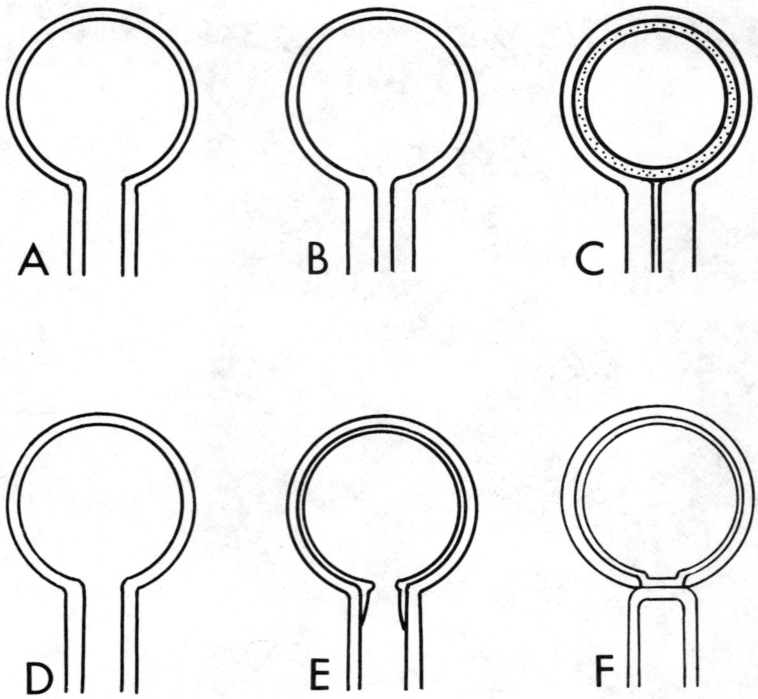

Figure 65 Diagrammatic interpretation of wall differentiation during cleavage and endogenous formation of sporangiospores from monosporous sporangiola (A–C), and exogenous formation of holoblastic conidia (D–F).

strated by *C. echinulata,* presumably evolved concomitantly with reduction from multispored sporangia to monosporous sporangiola.

3. Few-Spored Merosporangia

Several families of mucoraceous fungi produce uniseriately arranged sporangiospores within cylindrical outgrowths of the fertile apices of simple or branched sporangiophores (Fig. 66). Such outgrowths are called merosporangia. In *Syncephalastrum racemosum,* merosporangia develop synchronously from the sporangiophore vesicle, elongating to form tubular structures (Fig. 66A). Cleavage occurs within each merosporangium, resulting in differentiation of 5-10 endogenous sporangiospores [124]. A similar developmental process has been illustrated by thin-section electron microscopy in *Syncephalis sphaerica* [28, 29]. When mature, the spores are free within the tubular structures and may be released by rupture of its outer wall in the region between adjacent spores (Fig. 66B and C). *Syncephalastrum* has been reported as an opportunistic fungal pathogen causing a cutaneous infection in a compormised patient [190]. In this case, the fungus did not cause a systemic mycosis.

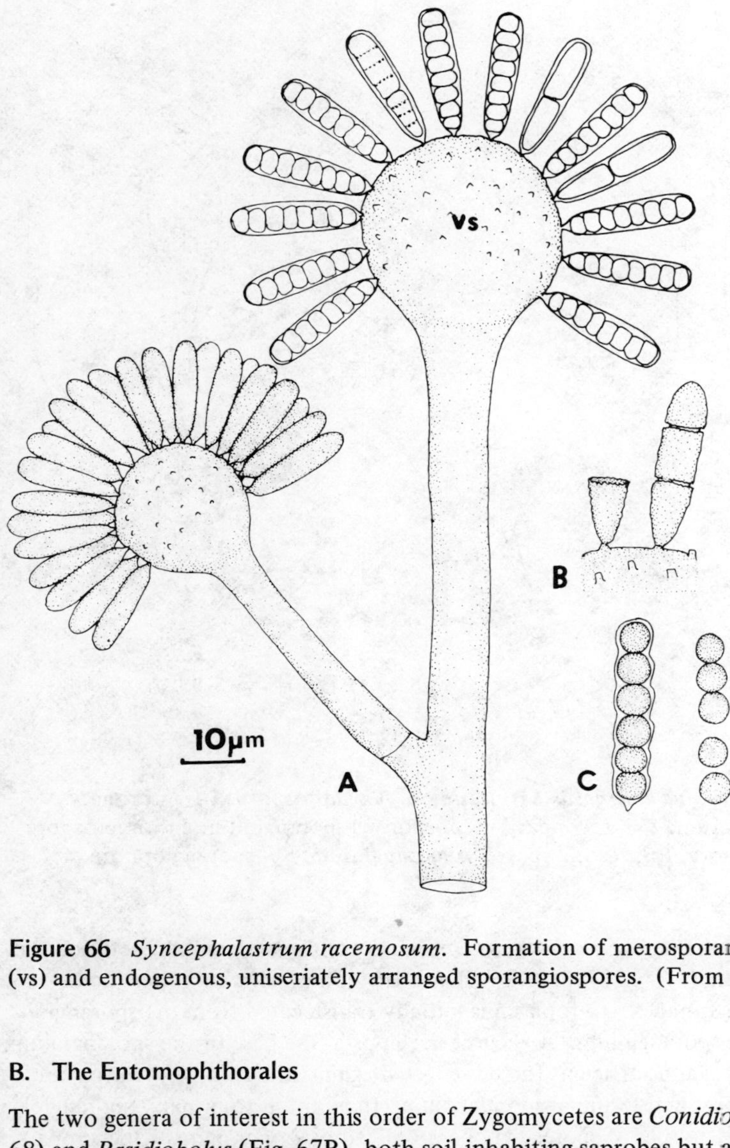

Figure 66 *Syncephalastrum racemosum*. Formation of merosporangia from fertile vesicle (vs) and endogenous, uniseriately arranged sporangiospores. (From Ref. 88.)

B. The Entomophthorales

The two genera of interest in this order of Zygomycetes are *Conidiobolus* (Figs. 67A and 68) and *Basidiobolus* (Fig. 67B), both soil-inhabiting saprobes but also capable of causing severe and massive subcutaneous infections of the face of humans. The principal mechanism of asexual reproduction in these fungi is by formation of single, terminal, uni- or multinucleate, monosporous sporangiola which are forcibly ejected at maturity [40]. In *Basidiobolus haptosporus,* the sporangiolum (S) is subtended by the swollen apex of the sporangiophore (Sp, Fig. 67B). Benjamin [40] has described the mechanism of sporangiolar ejection as follows: the "subsporangial swelling develops a subbasal, circumscissile line of weakness which ruptures and, jet-like, the upper portion of the swelling flies off carrying the sporangiole with it. The detached part of the sporophore, which projects into the base of the sporangiole as a minute columella, usually separates from the sporangiole dur-

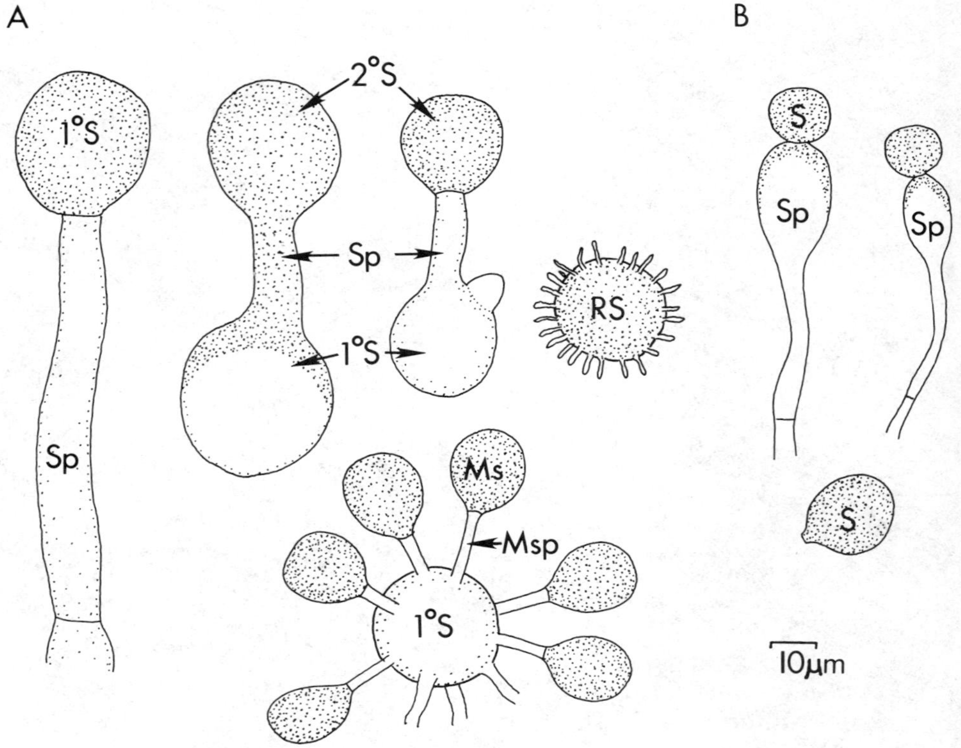

Figure 67 (A) *Conidiobolus coronatus*; (B) *Basidiobolus haptosporus*. 1°S, primary monosporous sporangiolum; 2°S, secondary monosporous sporangiolum; Ms, microspore; Msp, microsporangiophore; RS, resting spore; S, sporangiolum; Sp, sporangiophore.

ing flight." Forcible discharge in *Conidiobolus coronatus* involves interaction between the base of the sporangiolum (1°S or 2°S, Fig. 67A) and the conical-shaped columella at the sporangiophore (Sp) apex. Cytoplasm is initially translocated from the sporangiophore into the growing sporangiolum. A vacuole develops at the base of the sporangiolum, turgor pressure increases, and sufficient force is exerted against the columella to rupture the wall and forcibly eject and terminal sporangiolum. In addition to primary sporangiola, secondary or "accessory" types of spores may be produced by *C. coronatus* which apparently function for short-term survival. A primary sporangiolum (1°S) may germinate to give rise to a secondary sporangiolum (2°S) on a short sporangiophore (Figs. 67A and 68), or it may produce numerous microspores (Ms) at the tips of microsporangiophores (Msp, Fig. 67A). Resting spores (RS, Fig. 67A) are also produced by this fungus, which Benjamin [40] has suggested are necessary for survival of such parasitic forms as *C. coronatus* when a suitable host is not immediately available.

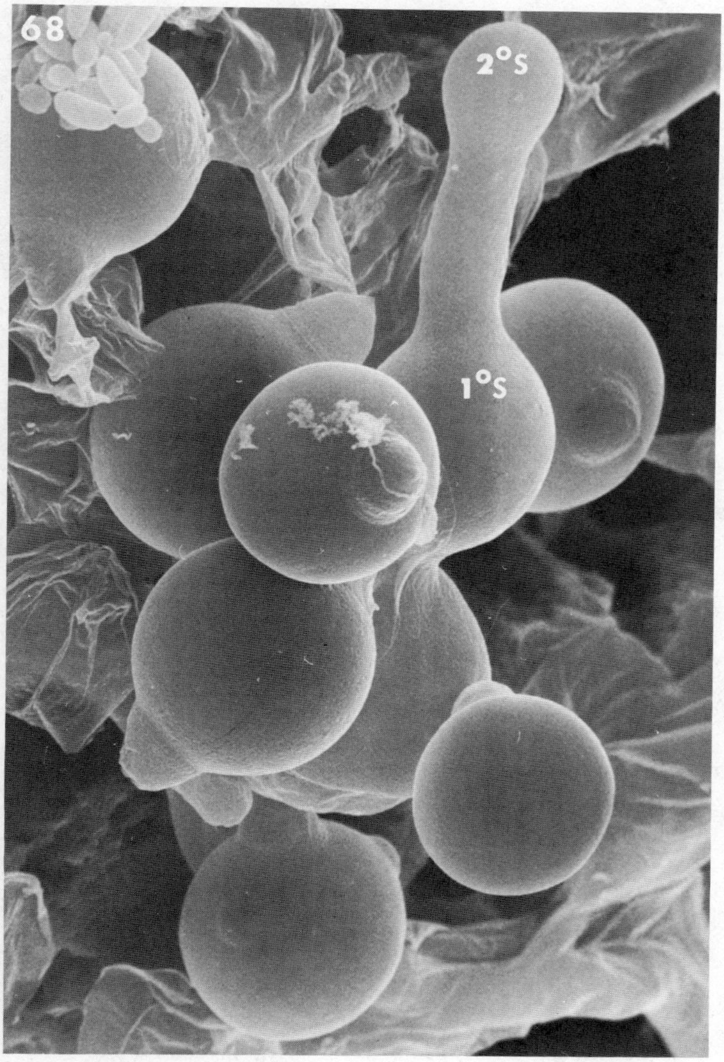

Figure 68 *Conidiobolus coronatus.* Primary (1°S) and secondary sporangiola (2°S). 1800X.

IV. SUMMARY

A summation of the various modes of conidiogenesis, yeast development, and sporangiosporogenesis in fungi cited in this text and their corresponding mycoses is presented in Tables 3-5. Examples have been chosen which illustrate most variations of these develop-

Table 4 Modes of Conidiogenesis in Pathogenic Yeasts and Corresponding Mycoses

Mode of conidiogenesis	Pathogen	Mycosis
Blastic		
Enteroblastic		
Solitary and acropetal	*Candida albicans*	Candidosis
	Candida tropicalis	Candidosis
	Paracoccidioides brasiliensis	Paracoccidioidomycosis
	Blastomyces dermatitidis	Blastomycosis
	Histoplasma capsulatum	Histoplasmosis
	Torulopsis glabrata	Torulopsosis
	Cryptococcus neoformans	Cryptococcosis
	Trichosporon beigelii	White piedra
Annellidic	*Pityrosporum malassezii*	Pityriasis versicolor

Table 5 Modes of Sporangiosporogenesis in Pathogenic Zygomycetes and Corresponding Mycoses

Mode of sporangiosporogenesis	Pathogen	Mycosis
Multispored sporangia	*Absidia corymbifera*	Zygomycosis (systemic) (bovine mycotic abortion)
	Mucor spp.	Zygomycosis (systemic)
	Rhizopus	Zygomycosis (systemic)
	Saksenaea vasiformis	Zygomycosis
	Mortierella wolfii	Zygomycosis
Monosporous sporangiola (Mucorales)	*Cunninghamella echinulata*	Zygomycosis
Few-spored merosporangia	*Syncephalastrum racemosum*	Zygomycosis
Monosporous sporangiola, forcibly discharged (Entomophthorales)	*Conidiobolus coronatus*	Zygomycosis (entomophthoramycosis conidiobolae)
	Basidiobolus haptosporus	Zygomycosis (entomophthoramycosis basidiobolae)

mental processes in pathogenic microfungi, but by no means do these exhaust all the possible methods of conidium or sporangiospore ontogeny [88]. The degree of variability is considerable if one contemplates that any microfungus may be an "opportunistic pathogen" under suitable growth conditions in a compromised animal host [27,117,318]. Ajello [6] has presented a list of the most common pathogenic fungi, which includes 140 species accommodated in the Subdivisions Zygomycotina, Ascomycotina, Basidiomycotina, and Fungi Imperfecti. The different kinds of sporangiospore and anamorphic yeast development represented by members of the author's list have been illustrated in this chapter. However, our discussion of modes of conidium formation by filamentous fungi was restricted to examples from the form-class Hyphomycetes, and did not take into account pathogenic members of the form-class Coelomycetes. The latter are anamorphic microfungi which reproduce by formation of conidia in, or on fructifications called pycnidia, acervuli, and stromata [305]. Most are saprophytes or parasites of vascular plants, but a few pycnidial forms (form-order: Sphaeropsidales) have been reported as human pathogens. Punithalingam [255] has listed 12 coelomycetous fungi belonging to seven genera which have been isolated from humans. Most are associated with cutaneous and subcutaneous infections or mycetomas. However, the author has stated that only a few have been conclusively implicated with mycotic diseases, whereas the others are probably secondary invaders. Isolation of *Pyrenochaeta* spp. and *Phoma* spp. appear to be most common. Few developmental studies of coelomycetous fungi have been conducted, primarily because most of these fungi require their natural substrate for typical growth and differentiation [305]. Conidiogenesis usually occurs within a closed fruiting body, making it difficult to define these developmental processes unless careful sectioning procedures are employed. Such ontogenetic studies of Coelomycetes are in progress [78,100,238].

ACKNOWLEDGMENTS

The authors thank Dr. G. A. de Vries and Ms. L. M. Pope for their suggestions during the preparation of this manuscript. The senior author acknowledges research support from the National Institutes of Health (AI-15583) and the National Science Foundation (DEB78-04263). The authors are grateful to Drs. H. C. Hoch, A. J. Howard, M. Huppert, and S. Sun for submission of previously published micrographs for reproduction in this chapter.

REFERENCES

1. Ainsworth, G. C. 1971. *Ainsworth and Bisby's Dictionary of the Fungi,* 6th ed., Commonwealth Mycological Institute, Kew, Surrey, England.
2. Ainsworth, G. C., and P. K. C. Austwick. 1958. *Fungal Diseases of Animals,* Commonwealth Agricultural Bureaux, Farnham Royal, Slough, England.
3. Ainsworth, G. C., and P. K. C. Austwick. 1973. *Fungal Diseases of Animals,* Commonwealth Agricultural Bureaux, Farnham Royal, Slough, England.
4. Ajello, L. 1970. The medical mycological iceberg, In *Proceedings of the First International Symposium on the Mycoses,* Pan Am. Health Organ. Sci. Publ. No. 205, Washington, D.C., pp. 3-21.

5. Ajello, L. 1975. Phaeohyphomycosis: definition and etiology, In *Mycoses,* Proceedings of the Third International Conference on the Mycoses, Pan Am. Health Organ. Sci. Publ. No. 304, Washington, D.C., pp. 126-130.
6. Ajello, L. 1977. Systemic mycoses in modern medicine. Contrib. Microbiol. Immunol. *3*: 2-6.
7. Ajello, L. 1978. The black yeasts as disease agents: historical perspective, In *The Black and White Yeasts,* Proceedings of the Fourth International Conference on the Mycoses, Pan Am. Health Organ. Sci. Publ. No. 356, Washington, D.C., pp. 9-16.
8. Ajello, L. 1980. Natural habitats of fungi that cause pulmonary mycoses. Zentralbl. Bakteriol. Parasitenkd. Infektionskr. Hyg., Abt. 1, Suppl. *8*: 31-42.
9. Ajello, L. 1980. Chromoblastomycosis, chromomycosis and phaeohyphomycosis—a confusion of terms. In E. S. Kuttin and G. L. Baum (Eds.), *Proceedings of the VII Congress of the International Society for Human and Animal Mycology,* Elsevier/North-Holland, Amsterdam, pp. 187-189.
10. Ajello, L., D. F. Dean, and R. S. Irwin. 1976. The zygomycete *Saksenaea vasiformis* as a pathogen of humans with a critical review of the etiology of zygomycosis. Mycologia *68*: 52-62.
11. Ajello, L., L. K. Georg, R. T. Steigbigel, and C. J. K. Wang. 1974. A case of phaeohyphomycosis caused by a new species of *Phialophora.* Mycologia *66*: 490-498.
12. Ajello, L., M. R. McGinnis, and J. Camper. 1977. An outbreak of phaeohyphomycosis in rainbow trout caused by *Scolecobasidium humicola.* Mycopathologia *62*: 15-22.
13. Allen, E. D., R. Aiuto, and A. S. Sussman. 1980. Effects of cytochalasins on *Neurospora crassa.* I. Growth and ultrastructure. Protoplasma *102*: 63-75.
14. Allermann, K., F. Floto, J. Olsen, O. Sortkjaer, and I. Kier. 1978. Sporulation and the macromolecular composition of the mycelium and arthrospores of *Geotrichum candidum.* Physiol. Plant. *42*: 355-358.
15. Anderson, J. G., and J. E. Smith. 1971. The production of conidiophores and conidia by newly germinated conidia of *Aspergillus niger* (microcycle conidiation). J. Gen. Microbiol. *69*: 185-197.
16. Anderson, J. G., V. Aryee, and J. E. Smith. 1978. Microcycle conidiation in *Paecilomyces varioti.* FEMS Microbiol. Lett. *3*: 57-60.
17. Anderson, J. H. 1979. *In vitro* survival of human pathogenic fungi in seawater. Sabouraudia *17*: 1-12.
18. Anderson, J. H. 1979. *In vitro* survival of human pathogenic fungi in Hawaiian beach sand. Sabouraudia *17*: 13-22.
19. Applegarth, D. A., and G. Bozoian. 1969. The cell wall of *Helminthosporium sativum.* Arch. Biochem. Biophys. *134*: 285-289.
20. Arita, I. 1979. Cytological studies on *Pholiota.* Rep. Tottori Mycol. Inst. *17*: 1-118.
21. Aronson, J. M. 1981. Cell wall chemistry, ultrastructure and metabolism. In G. T. Cole and B. Kendrick (Eds.), *Biology of Conidial Fungi,* Vol. 2, Academic, New York, pp. 459-507.
22. Arx, J. A. von. 1970. *The Genera of Fungi Sporulating in Pure Culture,* J. Cramer, Lehre.
23. Arx, J. A. von. 1979. Ascomycetes as Fungi Imperfecti. In W. B. Kendrick (Ed.), *The Whole Fungus,* Proceedings of the Second International Mycological Conference, Vol. 1, National Museum of Natural Sciences and National Museums of Canada, Ottawa, pp. 201-213.
24. Arx, J. A. von. 1981. Systematics of conidial yeasts. In G. T. Cole and B. Kendrick (Eds.), *Biology of Conidial Fungi,* Vol. 1, Academic, New York, pp. 85-96.

25. Arx, J. A. von, and A. C. M. Weijman. 1979. Conidiation and carbohydrate composition in some *Candida* and *Torulopsis* species. Antonie Leeuwenhoek J. Microbiol. Serol. *45*: 547-555.
26. Arx, J. A. von, L. Rodrigues de Miranda, M. T. Smith, and D. Yarrow. 1977. The genera of yeasts and the yeast-like fungi. *Studies in Mycology,* Vol. 14, Centraalbureau voor Schimmelcultures, Baarn, pp. 1-42.
27. Austwick, P. K. C., and J. W. Copland. 1974. Swamp cancer. Nature (Lond.) *250*: 84.
28. Baker, K. L., and G. R. Hooper. 1979. The relationship between structural development and function in the formation of spore droplets in *Syncephalis sphaerica.* Micron *10*: 159-192.
29. Baker, K. L., G. R. Hooper, and E. S. Beneke. 1977. Ultrastructural development of merosporangia in the mycoparasite *Syncephalis sphaerica* (Mucorales). Can. J. Bot. *55*: 2207-2215.
30. Bakerspigel, A., and S. Burke. 1974. A possible function of the chlamydospores of *Candida albicans.* Mycopathol. Mycol. Appl. *54*: 147-152.
31. Barran, L. R., E. F. Schneider, P. J. Wood, C. Madhosingh, and R. W. Miller. 1975. Cell walls of *Fusarium sulphureum.* I. Chemical composition of the hyphal wall. Biochim. Biophys. Acta *392*: 148-158.
32. Barron, G. L. 1968. *The Genera of Hyphomycetes from Soil,* Williams & Wilkins, Baltimore.
33. Bartnicki-Garcia, S. 1963. Mold-yeast dimorphism of *Mucor.* III. Symposium on biochemical basis of morphogenesis in fungi. Bacteriol. Rev. *27*: 293-304.
34. Bartnicki-Garcia, S. 1973. Fundamental aspects of hyphal morphogenesis. In J. M. Ashworth and J. E. Smith (Eds.), *Microbial Differentiation,* 23rd Symposium of the Society for General Microbiology, 1973, Cambridge University Press, Cambridge, pp. 245-267.
35. Bartnicki-Garcia, S. 1981. Biochemical events and controls: discussant's introduction. In G. Turian and H. R. Hohl (Eds.), *The Fungal Spore: Morphogenetic Controls,* Academic, New York, pp. 457-461.
36. Bartnicki-Garcia, S., J. Ruiz-Herrera, and C. E. Bracker. 1979. Chitosomes and chitin synthesis. In J. H. Burnett and A. P. J. Trinci (Eds.), *Fungal Walls and Hyphal Growth,* Cambridge University Press, Cambridge, pp. 149-168.
37. Beever, R. E., and G. P. Dempsey. 1978. Function of rodlets on the surface of fungal spores. Nature (Lond.) *272*: 608-610.
38. Beever, R. E., R. J. Redgwell, and G. P. Dempsey. 1979. Purification and chemical characterization of the rodlet layer of *Neurospora crassa* conidia. J. Bacteriol. *140*: 1063-1070.
39. Benjamin, R. K. 1959. The merosporangiferous Mucorales. Aliso *4*: 321-433.
40. Benjamin, R. K. 1979. Zygomycetes and their allies. In B. Kendrick (Ed.), *The Whole Fungus,* Proceedings of the Second International Mycological Conference, Vol. 2, National Museum of Natural Sciences and National Museums of Canada, Ottawa, pp. 573-621.
41. Bibel, D. J., D. A. Crumrine, K. Yee, and R. D. King. 1977. Development of arthrospores of *Trichophyton mentagrophytes.* Infect. Immun. *15*: 958-971.
42. Boerema, G. H., and G. J. Bollen. 1975. Conidiogenesis and conidial septation as differentiating criteria between *Phoma* and *Ascochyta.* Persoonia *8*: 111-144.
43. Bowers, B., G. Levin, and E. Cabib. 1974. Effect of polyoxin D on chitin synthesis and septum formation. J. Bacteriol. *119*: 564-575.
44. Bracker, C. E. 1967. Ultrastructure of fungi. Annu. Rev. Phytopathol. *5*: 343-374.

45. Bracker, C. E. 1968. The ultrastructure and development of sporangia in *Gilbertella persicaria.* Mycologia *60*: 1016-1067.
46. Bracker, C. E., J. Ruiz-Herrera, and S. Bartnicki-Garcia. 1976. Structure and transformation of chitin synthetase particles (chitosomes) during microfibril synthesis *in vitro.* Proc. Natl. Acad. Sci. U.S.A. *73*: 4570–4574.
47. Brillinger, G. U. 1979. Metabolic products of microorganisms 181. Chitin synthetase from fungi, a test model for substances with insecticidal properties. Arch. Microbiol. *121*: 71-74.
48. Brunswick, H. 1924. Untersuchungen über Geschlechts und Kernverhaltnisse bei der Hymanomyzetengattung *Coprinus.* In K. Goebel (Ed.), *Botanische Abhandlunge,* Vol. 5, G. Fischer, Jena.
49. Budtz-Jorgensen, E., A. Stenderup, and M. Grabowski. 1975. An epidemiologic study of yeasts in elderly denture wearers. Community Dent. Oral Epidemiol. *3*: 115-119.
50. Bull, A. T. 1970. Chemical composition of wild-type and mutant *Aspergillus nidulans* cell walls. The nature of polysaccharide and melanin constituents. J. Gen. Microbiol. *63*: 75-94.
51. Burnett, J. H. 1979. Aspects of the structure and growth of hyphal walls. In J. H. Burnett and A. P. J. Trinci (Eds.), *Fungal Walls and Hyphal Growth,* Cambridge University Press, Cambridge, pp. 1-25.
52. Byers, B., and L. Goetsch. 1974. Duplication of spindle plaques and integration of the yeast cell cycle. Cold Spring Harbor Symp. Quant. Biol. *38*: 123-131.
53. Byers, B., and L. Goetsch. 1975. Behavior of spindle plaques in the cell cycle and conjugation of *Saccharomyces cerevisiae.* J. Bacteriol. *124*: 511-523.
54. Byers, B., and L. Goetsch. 1976. A highly ordered ring of membrane-associated filaments in budding yeasts. J. Cell Biol. *69*: 717-721.
55. Cabib, E. 1975. Molecular aspects of yeast morphogenesis. Annu. Rev. Microbiol. *29*: 191-214.
56. Cabib, E. 1976. The yeast primary septum: a journey into three-dimensional biochemistry. Trends Biochem. Sci. *1*: 275-277.
57. Cabib, E., and Farkǎs, V. 1971. The control of morphogenesis: an enzymatic mechanism for the initiation of septum formation in yeast. Proc. Natl. Acad. Sci. U.S.A. *68*: 2052-2056.
58. Cabib, E., and F. A. Keller. 1971. Chitin and yeast budding. Allosteric inhibition of chitin synthetase by a heat-stable protein from yeast. J. Biol. Chem. *246*: 167-173.
59. Cabib, E., A. Duran, and B. Bowers. 1979. Localized activation of chitin synthetase in the initiation of yeast septum formation. In J. H. Burnett and A. P. J. Trinci (Eds.), *Fungal Walls and Hyphal Growth,* Cambridge University Press, Cambridge, pp. 189-201.
60. Cabib, E., V. Farkǎs, R. E. Ulane, and B. Bowers. 1973. Yeast septum formation as a model system for morphogenesis. In J. R. Villaneuva, I. Garcia-Acha, S. Gascón, and F. Uruburu (Eds.), *Yeast, Mould and Plant Protoplasts,* Academic, New York, pp. 105-116.
61. Calley, W. A. 1976. Morphological investigations of several human pathogenic fungi under variable conditions of temperature and light in a defined liquid medium. M.A. thesis, University of Texas, Austin.
62. Caretta, G., and E. Piontelli. 1977. *Microsporum magellanicum* and *Cunninghamella antarctica,* new species isolated from australic and antarctic soil of Chile. Sabouraudia *15*: 1-10.
63. Carmichael, J. W. 1957. *Geotrichum candidum.* Mycologia *49*: 820-830.

64. Carmichael, J. W. 1962. *Chrysosporium* and other aleuriosporic Hyphomycetes. Can. J. Bot. *40*: 1137-1173.
65. Carmichael, J. W. 1981. Pleomorphism. In G. T. Cole and B. Kendrick (Eds.), *Biology of Conidial Fungi,* Vol. 1, Academic, New York, pp. 135-143.
66. Carmichael, J. W., A. S. Sekhon, and L. Sigler. 1973. Classification of some dermatophytes by pyrolysis-gas-liquid chromatography. Can. J. Microbiol. *19*: 403-407.
67. Carroll, F. E., and G. C. Carroll. 1974. The fine structure of conidium initiation in *Ulocladium atrum.* Can. J. Bot. *52*: 443-446.
68. Carroll, G. C. 1966. A study of ascosporogenesis in *Saccobolus kerverni* and *Ascodesmis sphaerospora.* Ph.D. dissertation, University of Texas, Austin.
69. Cassone, A., N. Simonetti, and V. Strippoli. 1973. Ultrastructural changes in the wall during germ-tube formation from blastospores of *Candida albicans.* J. Gen. Microbiol. *77*: 417-426.
70. Chattaway, F. W., M. R. Holmes, and A. J. E. Barlow. 1968. Cell wall composition of the mycelial and blastospore forms of *Candida albicans.* J. Gen. Microbiol. *51*: 367-376.
71. Cole, G. T. 1971. The sympodula and the sympodioconidium. In B. Kendrick (Ed.), *Taxonomy of Fungi Imperfecti,* University of Toronto Press, Toronto, pp. 141-159.
72. Cole, G. T. 1973. Ultrastructure of conidiogenesis in *Drechslera sorokiniana.* Can. J. Bot. *51*: 629-638.
73. Cole, G. T. 1973. A correlation between rodlet orientation and conidiogenesis in Hyphomycetes. Can. J. Bot. *51*: 2413-2422.
74. Cole, G. T. 1975. The thallic mode of conidiogenesis in the Fungi Imperfecti. Can. J. Bot. *53*: 2983-3001.
75. Cole, G. T. 1976. Conidiogenesis in pathogenic Hyphomycetes. I. *Sporothrix, Exophiala, Geotrichum* and *Microsporum.* Sabouraudia *14*: 81-98.
76. Cole, G. T. 1977. Conidial development in the Fungi Imperfecti. III. Thallic conidia. 16-mm time-lapse film. Institut für den Wissenschaftlichen Film, Göttingen.
77. Cole, G. T. 1978. Conidiogenesis in the black yeasts. In *The Black and White Yeasts,* Proceedings of the Fourth International Conference on the Mycoses, Pan Am. Health Organ. Sci. Publ. No. 356, Washington, D.C., pp. 66-78.
78. Cole, G. T. 1981. Conidiogenesis and conidiomatal ontogeny. In G. T. Cole and B. Kendrick (Eds.), *Biology of Conidial Fungi,* Vol. 2, Academic, New York, pp. 271-327.
79. Cole, G. T. 1981. Architecture and chemistry of the cell walls of higher fungi. In D. Schlessinger (Ed.), *Microbiology–1981,* American Society for Microbiology, Washington, D.C., pp. 227-231.
80. Cole, G. T., and H. C. Aldrich. 1971. Ultrastructure of conidiogenesis in *Scopulariopsis brevicaulis.* Can. J. Bot. *49*: 745-755.
81. Cole, G. T., and W. B. Kendrick. 1968. A thin culture chamber for time-lapse photomicrography of fungi at high magnifications. Mycologia *60*: 340-344.
82. Cole, G. T., and W. B. Kendrick. 1969. Conidium ontogeny in Hyphomycetes. The phialides of *Phialophora, Penicillium,* and *Ceratocystis.* Can. J. Bot. *47*: 779-789.
83. Cole, G. T., and W. B. Kendrick. 1969. Conidium ontogeny in Hyphomycetes. The annellophores of *Scopulariopsis brevicaulis.* Can. J. Bot. *47*: 925-929.
84. Cole, G. T., and W. B. Kendrick. 1969. Conidium ontogeny in Hyphomycetes. The arthrospores of *Oidiodendron* and *Geotrichum* and the endoarthrospores of *Sporendonema.* Can. J. Bot. *47*: 1773-1780.
85. Cole, G. T., and W. B. Kendrick (Eds.), 1981. *Biology of Conidial Fungi,* Vols. 1 and 2, Academic, New York.

86. Cole, G. T., and Y. Nozawa. 1981. Dimorphism. In G. T. Cole and B. Kendrick (Eds.), *Biology of Conidial Fungi,* Vol. 1, Academic, New York, pp. 97-133.
87. Cole, G. T., and L. M. Pope. 1981. Surface wall components of *Aspergillus niger* conidia. In G. Turian and H. R. Hohl (Eds.), *The Fungal Spore: Morphogenetic Controls,* Academic, New York, pp. 195-215.
88. Cole, G. T., and R. A. Samson. 1979. *Patterns of Development in Conidial Fungi,* Pitman, London.
89. Cole, G. T., T. Sekiya, R. Kasai, T. Yokoyama, and Y. Nozawa. 1979. Surface ultrastructure and composition of the cell walls of conidial fungi. Exp. Mycol. *3*: 132-156.
90. Cook, B. E. 1977. The development of *Oedocephalum*-type conidia in the fungi *Cunninghamella echinulata* and *Fomes annosus.* New Phytol. *73*: 115-130.
91. Cooke, W. B. 1961. Some effects of spray disposal of spent sulfite liquor on soil mould populations. Proc. Indiana Waste Conf. (Purdue Univ.) *45*: 35-48.
92. Correns, C. 1899. *Untersuchungen über die Vermehrung der Laubmoose durch Brutorgane und Stecklinge,* G. Fischer, Jena.
93. Dähn, U., H. Hagenmaier, H. Höhne, W. A. König, G. Wolf, and H. Zähner. 1976. Stoffwechselprodukte von Mikroorganismen. 154. Mitt. Nikkomycin, un neuer Hemmstoff der Chitinsyntheses bei Pilzen. Arch. Microbiol. *107*: 143-160.
94. Da Silva, P. P., and B. Kachar. 1980. Mini review: quick freezing vs. chemical fixation of membrane fusion intermediates. Cell Biol. Int. Rep. *4*: 625.
95. Dean, A. G., J. H. Bates, C. Sorrels, T. Sorrels, W. Germany, L. Ajello, L. Kaufman, C. McGrew, and A. Fitts. 1978. An outbreak of histoplasmosis at an Arkansas courthouse with five cases of probable reinfection. Am. J. Epidemiol. *108*: 36-46.
96. Dean, S. G., and J. E. Smith. 1979. Influence of metabolic inhibitors on microcycle conidiation of *Aspergillus niger.* Trans. Br. Mycol. Soc. *72*: 201-206.
97. Deans, S. G., K. Gull, and J. E. Smith. 1980. Ultrastructural changes during microcycle conidiation of *Aspergillus niger.* Trans. Br. Mycol. Soc. *74*: 493-499.
98. de Hoog, G. S. 1974. The genera *Blastobotrys, Sporothrix, Calcarisporium* and *Calcarisporiella* gen. nov. *Studies in Mycology,* Vol. 7, Centraalbureau voor Schimmelcultures, Baarn, pp. 1-84.
99. de Hoog, G. S., and E. J. Hermanides-Nijhof. 1977. The black yeasts and allied hyphomycetes. *Studies in Mycology,* Vol. 15, Centraalbureau voor Schimmelcultures, Baarn, pp. 1-222.
100. DiCosmo, F., and G. T. Cole. 1980. Morphogenesis of conidiomata in *Chaetomella acutiseta* (Coelomycetes). Can. J. Bot. *58*: 1129-1137.
101. Dillon, M. L. 1975. Clinical aspects of *Candida* infections. In Y. Al-Doory (Ed.), *The Epidemiology of Human Mycotic Diseases,* Thomas, Springfield, Ill., pp. 158-177.
102. Dixon, D. M., H. J. Shadomy, and S. Shadomy. 1980. Dematiaceous fungal pathogens isolated from nature. Mycopathologia *70*: 153-161.
103. Dodge, C. W. 1935. *Medical Mycology,* Mosby, St. Louis.
104. Domer, J. E. 1971. Monosaccharide and chitin content of cell walls of *Histoplasma capsulatum* and *Blastomyces dermatitidis.* J. Bacteriol. *107*: 870-877.
105. Domer, J. E., J. G. Hamilton, and J. C. Harkin. 1967. Comparative study of the cell walls of the yeast-like and mycelial phases of *Histoplasma capsulatum.* J. Bacteriol. *94*: 466-474.
106. Dujardin, L., S. Walbaum, and J. Biguet. 1980. Influence de la concentration du glucose et de l'azote sur la morphologie de *Candida albicans* et la formation de ses chlamydospores dans un milieu de culture synthétique. Mycopathologia *71*: 113-118.

107. Duncan, D. B., J. E. Smith, and D. R. Berry, 1978. DNA, RNA and protein biosynthesis during microcycle conidiation in *Aspergillus niger*. Trans. Br. Mycol. Soc. *71*: 457-464.

108. Duran, A., B. Bowers, and E. Cabib. 1975. Chitin synthetase zymogen is attached to the yeast plasma membrane. Proc. Natl. Acad. Sci. U.S.A. *72*: 3952-3955.

109. Duran, A., F. Uruburu, and J. R. Villanueva. 1973. Morphogenetic and nutritional studies of *Geotrichum lactis* cells. Arch. Mikrobiol. *88*: 245-256.

110. Ebina, K., S. Takashita, A. Kamaguchi, K. Yokota, and O. Sakaguchi. 1978. Comparison of cell body and cell wall compositions between the yeast-like and mycelial phases of *Geotrichum candidum*. Jpn. J. Bacteriol. *33*: 527-538.

111. Ellis, D. H., and D. A. Griffiths. 1977. The fine structure of conidiogenesis in *Alysidium resinae* (= *Torula ramosa*). Can. J. Bot. *55*: 676-684.

112. Ellis, M. B. 1971. Porospores. In B. Kendrick (Ed.), *Taxonomy of Fungi Imperfecti*, University of Toronto Press, Toronto, pp. 71-74.

113. Ellis, M. B. 1971. *Dematiaceous Hyphomycetes*, Commonwealth Mycological Institute, Kew, Surrey, England.

114. Ellis, M. B. 1976. *More Dematiaceous Hyphomycetes*, Commonwealth Mycological Institute, Kew, Surrey, England.

115. Emmons, C. W. 1945. *Phialophora jeanselmei* comb. n. from mycetoma of the hand. Arch. Pathol. *39*: 364-368

116. Emmons, C. W., and W. L. Jellison. 1960. *Emmonsia crescens* sp. n. and adiaspiromycosis (haplomycosis) in mammals. Ann. N.Y. Acad. Sci. *89*: 91-101.

117. Emmons, C. W., C. H. Binford, J. P. Utz, and K. J. Kwon-Chung. 1977. *Medical Mycology*, 3rd ed., Lea & Febiger, Philadelphia.

118. Emmons, C. W., I. G. Murray, H. I. Lurie, M. H. King, J. A. Tulloch, and D. H. Connor. 1964. North American blastomycosis: two autochthonous cases from Africa. Sabouraudia *3*: 306-311.

119. Emyanitoff, R. G., and T. Hashimoto. 1979. The effects of temperature, incubation atmosphere, and medium composition on arthrospore formation in the fungus *Trichophyton mentagrophytes*. Can. J. Microbiol. *25*: 362-366.

120. Endo, A., K. Kakiki, and T. Misato. 1970. Mechanisms of action of the antifungal agent polyoxin D. J. Bacteriol. *104*: 189-196.

121. Eras, P., M. J. Goldstein, and P. Sherlock. 1972. *Candida* infection of the gastrointestinal tract. Medicine *51*: 367-379.

122. Farkăs, V. 1979. Biosynthesis of cell walls of fungi. Microbiol. Rev. *43*: 117-144.

123. Field, L. H., L. M. Pope, G. T. Cole, M. N. Guentzel, and L. J. Berry. 1981. Persistence and spread of *Candida albicans* following intragastric challenge in infant mice. Infect. Immun. *31*: 783-791.

124. Fletcher, J. 1972. Fine structure of developing merosporangia and sporangiospores of *Syncephalastrum racemosum*. Arch. Mikrobiol. *87*: 269-284.

125. Fletcher, J. 1973. Ultrastructural changes associated with spore formation in sporangia and sporangiola of *Thamnidium elegans* Link. Ann. Bot. (Lond.) *37*: 963-971.

125a. Fragner, P., and I. Belsan. 1974. *Scopulariopsis* Bainier as causative agent of onychomycoses. Mycological and clinical study. Part II. Clinical study. Acta Univ. Carol. Med. *20*: 333-353.

126. Franks, F. 1977. Biological freezing and cryofixation. J. Microsc. (Paris) *111*: 3-16.

127. Fraser, D. W., J. I. Ward, L. Ajello, and B. D. Plikaytis. 1979. Aspergillosis and other systemic mycoses. JAMA *242*: 1631-1635.

128. Fuste, F. J., L. Ajello, R. Threlkel, and J. E. Henry. 1973. *Drechslera hawaiiensis*: a causative agent of a fatal fungal menigo-encephalitis. Sabouraudia *11*: 5963.
129. Galgóczy, J. 1975. Dermatophytes: conidium ontogeny and classification. Acta Microbiol. Acad. Sci. Hung. *22*: 105-136.
130. Galgóczy, J. 1978. Conidium ontogeny of dermatophytes. Acta Microbiol. Acad. Sci. Hung. *25*: 55-60.
131. Garrison, R. G., and K. S. Boyd. 1978. Role of the conidium in dimorphism of *Blastomyces dermatitidis*. Mycopathologia *64*: 29-33.
132. Garrison, R. G., and K. S. Boyd. 1978. Electron microscopy of yeast-like cell development from the microconidium of *Histoplasma capsulatum*. J. Bacteriol. *133*: 345-358.
133. Garrison, R. G., J. W. Lard, and M. F. Field. 1970. Ultrastructural changes during the yeastlike to mycelial phase conversion of *Blastomyces dermatitidis* and *Histoplasma capsulatum*. J. Bacteriol. *101*: 628-635.
134. Gaur, P. K., and R. W. Lichtwardt. 1980. Comparative study of a new *Chrysosporium* species with *Histoplasma capsulatum*. Sabouraudia *18*: 105-114.
135. Georg, L. K., B. W. Bierer, and W. B. Cooke. 1964. Encephalitis in turkey poults due to a new fungus species. Sabouraudia *3*: 239-244.
136. Girbardt, M. 1955. Lebendbeobachdungen an *Polystictus versicolor* (L). Flora *142*: 540-563.
137. Girbardt, M. 1957. Der Spitzenkörper von *Polystictus versicolor* (L). Planta (Berl.) *50*: 47-59.
138. Girbardt, M. 1969. Die Ultrastruktur der Apikalregion von Pilzhyphen. Protoplasma *67*: 413-441.
139. Girbardt, M. 1979. A microfilamentous septal belt (FSB) during induction of cytokinesis in *Trametes versicolor* (L. ex Fr.). Exp. Mycol. *3*: 215-228.
140. Gooday, G. W. 1977. Biosynthesis of the fungal wall—mechanisms and implications. J. Gen. Microbiol. *99*: 1-11.
141. Gooday, G. W., and A. P. J. Trinci. 1980. Wall structure and biosynthesis in fungi. In G. W. Gooday, D. Lloyd, and A. P. J. Trinci (Eds.), *The Eukaryotic Microbial Cell*, Cambridge University Press, Cambridge, pp. 207-251.
142. Goodman, N. L., R. B. Otero, and J. C. Parker. 1978. Candidiasis of the brain in the compromised host. In *The Black and White Yeasts*, Proceedings of the Fourth International Conference on the Mycoses, Pan Am. Health Organ. Sci. Publ. No. 356, Washington, D.C., pp. 309-315.
143. Grove, S. N. 1978. The cytology of hyphal tip growth. In J. E. Smith and D. R. Berry (Eds.), *The Filamentous Fungi*, Wiley, New York, pp. 28-50.
144. Grove, S. N., and C. E. Bracker. 1970. Protoplasmic organization of hyphal tips among fungi: vesicles and Spitzenkörper. J. Bacteriol. *104*: 989-1009.
145. Gunasekaran, M., and W. T. Hughes. 1980. Gas-liquid chromatography: a rapid method for identification of different species of *Candida*. Mycologia *72*: 505-511.
146. Gustafson, R. A., R. V. Hardcastle, and P. J. Szaniszlo. 1975. Budding in the dimorphic fungus *Cladosporium werneckii*. Mycologia *67*: 942-951.
147. Hammill, T. M. 1972. Electron microscopy of phialoconidiogenesis in *Metarrhizium anisopliae*. Am. J. Bot. *59*: 317-326.
148. Hammill, T. M. 1972. Fine structure of annellophores. II. *Doratomyces nanus*. Trans. Br. Mycol. Soc. *59*: 249-253.
149. Hammill, T. M. 1974. Electron microscopy of phialides and conidiogenesis in *Trichoderma saturnisporum*. Am. J. Bot. *61*: 15-24.
149a. Hammill, T. M. 1974. Fine structure of annellophores. I. *Scopulariopsis brevicaulis* and *S. koningii*. Am. J. Bot. *58*: 88-97.

150. Hammill, T. M. 1976. Light and electron microscopy of sporangial development in *Mucor* sp. Session 10. Contributed papers. Fungal ultrastructure. Am. Inst. Biol. Sci. Publ., New Orleans.
151. Hammill, T. M. 1977. Transmission electron microscopy of annellides and conidiogenesis in the synnematal hyphomycete *Trichurus spiralis*. Can. J. Bot. *55*: 233-244.
152. Hanlin, R. T. 1976. Phialide and conidium development in *Aspergillus clavatus*. Am. J. Bot. *63*: 144-155.
153. Happ, G. M., C. M. Happ, and S. J. Barras. 1976. Bark beetle–fungal symbiosis. II. Fine structure of a basidiomycetous ectosymbiont of the southern pine beetle. Can. J. Bot. *54*: 1049-1062.
154. Harris, J. L., and W. A. Taber. 1973. Compositional studies of the cell walls of the synnema and vegetative hyphae of *Ceratocystis ulmi*. Can. J. Bot. *51*: 1147-1153.
155. Hartwell, L. H. 1971. Genetic control of the cell division cycle in yeast. II. Genes controlling DNA replication and its initiation. J. Mol. Biol. *59*: 183-194.
156. Hartwell, L. H. 1971. Genetic control of the cell division cycle in yeast. IV. Genes controlling bud emergence and cytokinesis. Exp. Cell Res. *69*: 265-276.
157. Hartwell, L. H. 1974. *Saccharomyces cerevisiae* cell cycle. Bacteriol. Rev. *38*: 164-198.
158. Harvey, I. C. 1974. Light and electron microscope observations of conidiogenesis in *Pleiochaeta setosa* (Kirchn.) Hughes. Protoplasma *82*: 203-221.
159. Hasenclever, H. F. 1971. The consistent formation of chlamydospores by *Candida tropicalis*. Sabouraudia *9*: 164-166.
160. Hashimoto, T., and H. J. Blumenthal. 1978. Survival and resistance of *Trichophyton mentagrophytes* arthrospores. Appl. Environ. Microbiol. *35*: 274-277.
161. Hashimoto, T., J. H. Pollack, and H. J. Blumenthal. 1979. Ultrastructure and chemical composition of *Trichophyton mentagrophytes* arthrospore walls and septa. Abstr. Annu. Meet. Am. Soc. Microbiol., 1979, J 23, p. 92.
162. Hashimoto, T., C. D. Wu-Yuan, and H. J. Blumenthal. 1976. Isolation and characterization of the rodlet layer of *Trichophyton mentagrophytes* microconidial wall. J. Bacteriol. *127*: 1543-1549.
163. Hashmi, M. H., and G. Morgan-Jones. 1973. Conidium ontogeny in Hyphomycetes. The meristem arthrospores of *Wallemia sebi*. Can. J. Bot. *51*: 1669-1671.
164. Heath, I. B., and A. D. Greenwood. 1970. The structure and formation of lomasomes. J. Gen. Microbiol. *62*: 129-137.
165. Heinemann, H. S., E. J. Yunis, J. Siemienski, and A. I. Braude. 1961. Chlamydospores and dimorphism in *Candida albicans* endocarditis. Arch. Intern. Med. *108*: 570-577.
166. Hejtmánek, M. 1976. Scanning electron-microscopy of experimental adiaspiromycosis. Mycopathologia *58*: 91-95.
167. Hess, W. M., M. M. A. Sassen, and C. C. Remsen. 1968. Surface characteristics of *Penicillium* conidia. Mycologia *60*: 290-303.
168. Hesseltine, C. W., and J. J. Ellis. 1973. Mucorales. In G. C. Ainsworth, F. K. Sparrow, and A. S. Sussman (Eds.), *The Fungi*, Vol. IVB, Academic, New York, pp. 187-217.
169. Heuser, J. E., T. S. Reese, and A. M. A. Landis. 1976. Preservation of synaptic structure by rapid freezing. Cold Spring Harbor Symp. Quant. Biol. *40*: 17-24.
170. Hironaga, M., and S. Watanabe. 1980. Cerebral phaeohyphomycosis caused by *Cladosporium bantianum*: a case in a female who had cutaneous alternariosis in her childhood. Sabouraudia *18*: 229-235.

171. Hoch, H. C., and R. J. Howard. 1980. Ultrastructure of freeze-substituted hyphae of the basidiomycete *Laetisaria arvalis*. Protoplasma *103*: 281-297.
172. Holiday, H. D., V. Keipper, and A. B. Kaiser. 1980. *Torulopsis glabrata* endocarditis. JAMA *244*: 2088-2089.
173. Hori, M., J. Eguchi, K. Kakiki, and T. Misato. 1974. Studies on the mode of action of polyoxins. VI. Effect of polyoxin B on chitin synthesis in polyoxin-sensitive and resistant strains of *Alternaria kikuchiana*. J. Antibiot. *27*: 260-266.
174. Horikoshi, K., and S. Iida. 1964. Studies of the spore coats of fungi. I. Isolation and composition of the spore coats of *Aspergillus oryzae*. Biochim. Biophys. Acta *83*: 197-203.
175. Horta, P. 1921. Sombre um caso de tinha preta e um nova cogumelo *(Cladosporium wernecki)*. Rev. Med. Cir. Braz. *29*: 269-274.
176. Howard, R. J. 1980. An ultrastructural analysis of hyphal tip growth. Ph.D. thesis, Cornell University, Ithaca, N.Y.
177. Howard, R. J., and J. R. Aist. 1979. Hyphal tip cell ultrastructure of the fungus *Fusarium*: improved preservation by freeze-substitution. J. Ultrastruct. Res. *66*: 224-234.
177a. Howard, R. J., and J. R. Aist. 1980. Cytoplasmic microtubules and fungal morphogenesis: ultrastructural effects of methylbenzimidazole-2-ylcarbamate determined by freeze-substitution of hyphal tip cells. J. Cell Biol. *87*: 55-64.
178. Hughes, S. J. 1951. Studies on microfungi. XI. Some hyphomycetes which produce phialides. Commonw. Mycol. Inst. Mycol. Pap. *45*: 1-36.
179. Hughes, S. J. 1953. Conidiophores, conidia and classification. Can. J. Bot. *31*: 577-659.
180. Hughes, S. J. 1971. Percurrent proliferation in fungi, algae, and mosses. Can. J. Bot. *49*: 215-231.
181. Hughes, S. J. 1971. On conidia of fungi and gemmae of algae, bryophytes, and pteridophytes. Can. J. Bot. *49*: 1319-1339.
182. Hughes, S. J., and W. B. Kendrick. 1968. New Zealand fungi. 12. *Menispora, Codinaea, Menisporopsis.* N.Z. J. Bot. *6*: 323-375.
183. Hugh-Jones, M. E., and P. K. C. Austwick. 1967. Epidemiological studies in bovine mycotic abortion. 1. Effect of climate on incidence. Vet. Rec. *81*: 273.
184. Hunsley, D., and G. W. Gooday. 1974. The structure and development of septa in *Neurospora crassa*. Protoplasma *82*: 125-146.
185. Hunsley, D., and D. Kay. 1976. Wall structure of the *Neurospora* hyphal apex: immunofluorescent localization of wall surface antigens. J. Gen. Microbiol. *95*: 233-248.
186. Hutter, R. V. P. 1959. Phycomycetous infection (mucormycosis) in cancer patients: a complication of therapy. Cancer *12*: 330-350.
187. Iwata, K. 1972. A review of the literature on drunken symptoms due to yeast in the gastrointestinal tract. In K. Iwata (Ed.), *Yeast and Yeast-Like Microorganisms in Medical Science,* Proceedings of the Second International Specialized Symposium on Yeasts, University of Tokyo Press, Tokyo, pp. 260-268.
188. Jones, K. W., and L. Kaufman. 1978. Development and evaluation of an immunodiffusion test for diagnosis of systemic zygomycosis (mucormycosis): preliminary report. J. Clin. Microbiol. *7*: 97-103.
189. Kamalam, A., and A. S. Thambiah. 1979. Adiaspiromycosis of human skin caused by *Emmonsia crescens.* Sabouraudia *17*: 377-381.
190. Kamalam, A., and A. S. Thambiah. 1980. Cutaneous infection by *Syncephalastrum.* Sabouraudia *18*: 19-20.
191. Kanetsuna, F., L. M. Carbonell, R. E. Moreno, and J. Rodriguez. 1969. Cell wall

composition of the yeast and mycelial forms of *Paracoccidioides brasiliensis*. J. Bacteriol. *97*: 1036-1041.
192. Kaufman, L., and S. Blumer. 1978. *Cryptococcosis*: the awakening giant. In *The Black and White Yeasts,* Proceedings of the Fourth International Conference on the Mycoses, Pan Am. Health Organ. Sci. Publ. No. 356, Washington, D. C., pp. 176-182.
193. Kendrick, W. B. (Ed.) 1971. *Taxonomy of Fungi Imperfecti,* University of Toronto Press, Toronto.
194. Kendrick, W. B., and M. G. Chang. 1971. Karyology of conidiogenesis in some Hyphomycetes. In W. B. Kendrick (Ed.), *Taxonomy of Fungi Imperfecti,* University of Toronto Press, Toronto, pp. 279-291.
195. Kendrick, W. B., and G. T. Cole. 1968. Conidium ontogeny in Hyphomyocetes. The sympodulae of *Beauveria* and *Curvularia.* Can. J. Bot. *46*: 1297-1301.
196. Khan, S. R. 1975. Wall structure and germination of spores in *Cunninghamella echinulata.* J. Gen. Microbiol. *90*: 115-124.
197. Khan, S. R., and P. H. B. Talbot. 1975. Monosporous sporangiola in *Mycotypha* and *Cunninghamella.* Trans. Br. Mycol. Soc. *65*: 29-39.
198. Kiehn, T. E., F. Edwards, D. Armstrong, P. P. Rosen, and I. Weitzman. 1979. Pneumonia caused by *Cunninghamella bertholletiae* complicating chronic lymphatic leukemia. J. Clin. Microbiol. *10*: 374-379.
199. Kier, I., K. Allermann, F. Floto, and O. Sortkjaer. 1976. Changes of exponential growth rates in relation to differentiation of *Geotrichum candidum* in submerged culture. Physiol. Plant. *38*: 6-12.
200. Kier, I., F. Floto, J. Olsen, and K. Allermann. 1980. Macromolecular composition of the wall and protoplasm of mycelia and arthrospores of *Geotrichum candidum.* Trans. Br. Mycol. Soc. *75*: 355-361.
201. King, D. S., and S. C. Jong. 1976. Induction of arthroconidia in *Trichosporon.* Mycopathologia *59*: 61-63.
202. Kirmani, N., C. U. Tuazon, and G. W. Geelhoed. 1980. Disseminated *Trichosporon* infection. Occurrence in an immunosuppressed patient with chronic active hepatitis. Arch. Intern. Med. *140*: 277-278.
203. Kishimoto, R. A., and G. E. Baker. 1969. Pathogenic and potentially pathogenic fungi isolated from beach sands and selected soils of Oahu, Hawaii. Mycologia *61*: 537-548.
204. Knudtson, W. U., G. R. Ruth, C. A. Kirkbride, and M. Tinant. 1976. Pneumonia associated with *Torulopsis glabrata* in an aborted bovine fetus. Sabouraudia *14*: 43-45.
205. Kobayashi, G. S., and P. L. Guiliacci. 1967. Cell wall studies of *Histoplasma capsulatum.* Sabouraudia *5*: 180-188.
206. Kodousek, R., and M. Hejtmánek. 1972. Some morphological observations concerning problems of endosporulation in *Emmonsia crescens*: an etiological agent in adiaspiromycosis. Mycopathol. Mycol. Appl. *47*: 343-348.
207. Kodousek, R., M. Hejtmánek, M. Havelkóv, and Z. Strachov. 1972. The ultrastructure of spherules of the fungus *Emmonsia crescens*—a causative agent of adiaspiromycosis. Beitr. Pathol. *145*: 83-88.
208. Kozel, T. R., and J. Cazin. 1971. Nonencapsulated variant of *Cryptococcus neoformans.* I. Virulence studies and characterization of soluble polysaccharide. Infect. Immun. *3*: 287-294.
209. Kozel, T. R., and R. P. Mastroianni. 1976. Inhibition of phagocytosis by cryptococcal polysaccharide: dissociation of the attachment and ingestion phases of phagocytosis. Infect. Immun. *14*: 62-67.

210. Kozel, T. R., E. Reiss, and R. Cherniak. 1980. Concomitant but not causal association between surface charge and inhibition of phagocytosis by cryptococcal polysaccharide. Infect. Immun. *29*: 295-300.
211. Krause, W., M. Matheis, and K. Wolf. 1969. Fungemia and fungaria after oral administration of *Candida albicans*. Lancet *1*: 598-599.
212. Kuttin, E. S., A. M. Beemer, J. Levij, L. Ajello, and W. Kaplan. 1978. Occurrence of *Blastomyces dermatitidis* in Israel: first autochthonous middle eastern case. Am. J. Trop. Med. Hyg. *27*: 1203-1205.
213. Kwon-Chung, K. J. 1969. *Coccidioides immitis*: cytological study on the formation of the arthrospores. Can. J. Genet. Cytol. *11*: 43-53.
214. Kwon-Chung, K. J. 1975. Heterothallism vs. self-fertile isolates of *Filobasidiella neoformans (Cryptococcus neoformans)*. In *The Black and White Yeasts*, Proceedings of the Fourth International Conference on the Mycoses, Pan Am. Health Organ. Sci. Publ. No. 356, Washington, D.C., pp. 204-213.
215. Kwon-Chung, K. J. 1976. Morphogenesis of *Filobasidiella neoformans*, the sexual state of *Cryptococcus neoformans*. Mycologia *68*: 821-833.
216. Kwon-Chung, K. J., R. C. Young, and M. Orlando. 1975. Pulmonary mucormycosis caused by *Cunninghamella elegans* in a patient with chronic myelogenous leukemia. Am. J. Clin. Pathol. *64*: 544-548.
217. Lacaz, C., P. Minami, and A. Purchio. 1970. *O grande mundo dos fungos*, Editôra da Universidade de São Paulo and Editôra Poligono, São Paulo.
218. Landau, J. W., and V. D. Newcomer. 1962. Acute cerebral phycomycosis (mucormycosis). J. Pediatr. *61*: 363-385.
219. Lane, J. W., and R. G. Garrison. 1970. Electron microscopy of the yeast to mycelial phase conversion of *Sporotrichum schenckii*. Can. J. Microbiol. *16*: 747-749.
220. Law, E. J., O. J. Kim, D. D. Stieritz, and B. G. McMillan. 1972. Experience with systemic candidiasis in the burned patient. J. Trauma *12*: 543-552.
221. Lin, L. P., Y. Y. Lee, J. C. Fong, and S. I. Hsieh. 1975. Ultrastructure of membrane-complex systems in *Agaricus bisporus*. In H. Takeo (Ed.), *Fine Structure of Fungi*, Vol. 2, Proceedings of the First Intersectional Congress of the International Association for Microbiology Society, Tokyo, Science Council of Japan, pp. 136-140.
222. Lodder, J. (Ed.). 1970. *The Yeasts*: *A Taxonomic Study*. North-Holland, Amsterdam.
223. Lottenberg, R., R. H. Waldman, L. Ajello, G. L. Hoff, W. Bigler, and S. R. Zellner. 1979. Pulmonary histoplasmosis associated with exploration of a bat cave. Am. J. Epidemiol. *110*: 156-161.
224. Madelin, M. F., and S. Dorabjee. 1974. Conidium ontogeny in *Wallemia sebi*. Trans. Br. Mycol. Soc. *63*: 121-130.
225. Marchant, R., and R. T. Moore. 1973. Lomasomes and plasmalemmasomes in fungi. Protoplasma *76*: 235-247.
226. Marchant, R., and A. W. Robards. 1968. Membrane systems associated with the plasmalemma of plant cells. Ann. Bot., New Series *32*: 457-471.
227. Marchant, R., A. Peat, and G. H. Banbury. 1967. The ultrastructural basis of hyphal growth. New Phytol. *66*: 623-629.
228. Martin, M. V. 1979. Germ-tube formation by oral strains of *Candida tropicalis*. J. Med. Microbiol. *12*: 187-193.
229. McGinnis, M. R. 1977. *Wangiella*, a new genus to accommodate *Hormiscium dermatitidis*. Mycotaxon *5*: 353-363.
230. McGinnis, M. R. 1978. Human pathogenic species of *Exophiala*, *Phialophora*, and *Wangiella*. In *The Black and White Yeasts*, Proceedings of the Fourth International

Conference on the Mycoses, Pan Am. Health Organ. Sci. Publ. No. 356, Washington, D.C., pp. 37-59.

231. McGinnis, M. R. 1979. Taxonomy of *Exophialia werneckii* and its relationship to *Microsporum mansonii.* Sabouraudia *17*: 145-154.

232. McGinnis, M. R. 1980. Taxonomy of *Exophiala jeanselmei* (Langeron) McGinnis and Padhye. Mycopathologia *65*: 79-87.

233. McGinnis, M. R. 1980. *Laboratory Handbook of Medical Mycology,* Academic, New York.

234. Miller, S. E., B. O. Spurlock, and G. E. Michaels. 1974. Electron microscopy of young *Candida albicans* chlamydospores. J. Bacteriol. *119*: 992-999.

235. Moens, P. B., and E. Rapport. 1971. Spindles, spindle plaques and meiosis in the yeast *Saccharomyces cerevisiae* (Hansen). J. Cell Biol. *50*: 344-361.

236. Moore, P. I., H. L. Bank, N. T. Brissie, and S. S. Spicer. 1976. Association of microfilament bundles with lysosomes in polymorphonuclear leucocytes. J. Cell Biol. *71*: 659-666.

236a. Moore, R. T. 1980. Taxonomic proposals of the classification of marine yeasts and other yeast-like fungi including the smuts. Bot. Mar. *23*: 361-373.

237. Muller, G. H., W. Kaplan, L. Ajello, and A. A. Padhye. 1975. Phaeohyphomycosis in a cat caused by *Drechslera spicifera.* J. Am. Vet. Med. Assoc. *166*: 150-154.

238. Nag Raj, T. R. 1981. Coelomycete systematics. In G. T. Cole and B. Kendrick (Eds.), *Biology of Conidial Fungi,* Vol. 1, Academic, New York, pp. 43-84.

239. Najim, L., and G. Turian, G. 1979. Conidiogenous loss of structure-functional polarity in the hyphal tips of *Sclerotinia fructigena.* Eur. J. Cell Biol. *20*: 24-27.

240. Nauta, E. H. 1979. Infection in the compromised host. In G. Dick (Ed.), *Immunological Aspects of Infectious Diseases,* University Park Press, Baltimore, pp. 343-387.

241. Noguchi, T., Y. Kitajima, Y. Nozawa, and Y. Ito. 1971. Isolation, composition, and structure of cell walls of *Trichophyton mentagrophytes.* Arch. Biochem. Biophys. *146*: 506-512.

242. Nozawa, Y., Y. Kitajima, and Y. Ito. 1973. Chemical and ultrastructural studies of isolated cell walls of *Epidermophyton floccosum.* Biochim. Biophys. Acta *307*: 92-103.

243. Nsanzumuhire, H., D. Vollum, and A. A. Poltera. 1974. Chromoblastomycosis due to *Cladosporium trichoides* treated with 5-fluorocytosine. A case report. Am. J. Clin. Pathol. *61*: 257-263.

244. Odds, F. C. 1979. *Candida and candidosis,* University Park Press, Baltimore.

245. Oláh, G. M. 1972. L'ontogénie sporale et l'ultrastructure de la paroi chez *Paecilomyces berlinensis.* Can. J. Microbiol. *18*: 1471-1475.

245a. Onsberg, P. 1980. *Scopulariopsis brevicaulis* in nails. Dermatologica *161*: 259-264.

246. Oujezdsky, K. B., and P. J. Szaniszlo. 1974. Conidial ontogeny in *Phialophora dermatitidis.* Mycologia *66*: 537-542.

247. Oujezdsky, K. B., S. N. Grove, and P. J. Szaniszlo. 1973. Morphological and structural changes during the yeast-to-mold conversion of *Phialophora dermatitidis.* J. Bacteriol. *113*: 468-477.

248. Palevitz, B. A., and P. K. Hepler. 1975. Identification of actin *in situ* at the ectoplasm interface of *Nitella.* Microfilament chloroplast association. J. Cell Biol. *65*: 29-38.

249. Park, D., and P. M. Robinson. 1969. Sporulation in *Geotrichum candidum.* Trans. Br. Mycol. Soc. *52*: 213-333.

250. Parker, J. C., and G. K. Klintworth. 1971. Miscellaneous uncommon diseases

attributed to fungi and actinomycetes. Torulopsis. In R. D. Baker (Ed.), *Human Infection with Fungi, Actinomycetes and Algae,* Special ed., Handbuch der speziellen pathologischen Anatomie und Histologie, Vol. III, Part 5, Springer-Verlag, New York, pp. 1001-1003.
251. Pfaller, W., and E. Rovan. 1978. Preparation of resin embedded unicellular organisms without the use of fixatives and dehydration media. J. Microsc. *114*: 339-351.
252. Pope, L. M., G. T. Cole, M. N. Guentzel, and L. J. Berry. 1979. Systemic and gastrointestinal candidiasis of infant mice after intragastric challenge. Infect. Immun. *25*: 702-707.
253. Previato, J. O., P. A. J. Gorin, and L. R. Travassos. 1979. Cell wall composition in different cell types of the dimorphic species *Sporothrix schenckii.* Exp. Mycol. *3*: 83-91.
254. Price, C. W., G. B. Fuson, and H. J. Phaff. 1978. Genome comparison in yeast systematics: delimitation of species within the genera *Schwanniomyces, Saccharomyces,* and *Pichia.* Microbiol. Rev. *42*: 161-193.
255. Punithalingam, E. 1979. Sphaeropsidales in cultures from humans. Nova Hedwigia *31*: 119-157.
256. Quinones, F., J. P. Koplan, L. Pike, F. Staine, and L. Ajello. 1978. Histoplasmosis in Belize, Central America. Am. J. Trop. Med. Hyg. *27*: 558-561.
257. Ray, T. L. 1980. Fungal infections in the immunocompromised host. Med. Clin. N. Am. *64*: 955-967.
257a. Reiss, H.-D., and W. Herth. 1979. Calcium gradients in tip growing plant cells visualized by chlorotetracycline fluorescence. Planta *146*: 615-621.
257b. Reiss, H.-D., and W. Herth. 1980. Effects of the broad-range ionophore X-537A on pollen tubes of *Lilium longiflorum.* Planta *147*: 295-301.
258. Remington, J. S., and S. E. Anderson. 1976. Pneumocystis and fungal infection in patients with malignancies. Int. J. Radiat. Oncol. Biol. *1*: 313-315.
259. Riley, O., and S. H. Mann. 1960. Brain abscess caused by *Cladosporium trichoides.* Am. J. Clin. Pathol. *33*: 525-532.
260. Rippon, J. W. 1974. *Medical Mycology; the Pathogenic Fungi and the Pathogenic Actinomycetes,* Saunders, Philadelphia.
261. Rippon, J. W. 1980. Dimorphism in pathogenic fungi. Crit. Rev. Microbiol. *8*: 49-97.
262. Rippon, J. W. 1981. Clinical aspects of medically important conidial fungi. In G. T. Cole and W. B. Kendrick (Eds.), *Biology of Conidial Fungi,* Vol. 2, Academic, New York, pp. 3-35.
263. Robinow, C. G., and A. Bakerspigel. 1965. Somatic nuclei and forms of mitosis in fungi. In G. C. Ainsworth and A. S. Sussman (Eds.), *The Fungi,* Vol. 1, Academic, New York, pp. 119-142.
264. Robinow, C. F., and C. E. Caten. 1969. Mitosis in *Aspergillus nidulans.* J. Cell Sci. *5*: 403-431.
265. Robinow, C. F., and J. Marak. 1966. A fiber apparatus in the nucleus of the yeast cell. J. Cell Biol. *29*: 129-151.
266. Robinson, P. M., and J. M. Smith. 1979. Development of cells and hyphae of *Geotrichum candidum* in chemostat and batch culture. Trans. Br. Mycol. Soc. *72*: 39-47.
267. Romano, A. H. 1966. Dimorphism. In G. C. Ainsworth and A. S. Sussman (Eds.), *The Fungi,* Vol. 2, Academic, New York, pp. 181-209.
268. Ruiz, J. M., E. Arteaga, J. Martinez, E. M. Rubio, and J. M. Torres. 1980. Cutaneous and renal geotrichosis in a giant tortoise *(Geochelone elephantopus).* Sabouraudia *18*: 51-59.

269. Saëz, H. 1969. Formation d'endospores chez *Geotrichum candidum.* Ann. Parasitol. Hum. Comp. *44*: 197-204.
270. Saëz, H. 1970. *Geotrichum loubieri* Morenz 1964, un champignon arthrospore, format également des endospores. Bull. Mens. Soc. Linn. Lyon *9*: 283-288.
271. Saksena, S. B. 1953. A new genus of the mucorales. Mycologia *45*: 426-536.
272. Salfelder, K. 1978. Pathology of cryptococcosis, candidiasis and torulopsosis. In *The Black and White Yeasts,* Proceedings of the Fourth International Conference on the Mycoses, Pan Am. Health Organ. Sci. Publ. No. 356, Washington, D.C., pp. 214-217.
273. Samson, R. A. 1974. *Paecilomyces* and some allied Hyphomycetes. *Studies in Mycology,* Vol. 6, Centraalbureau voor Schimmelcultures, Baarn, pp. 1-119.
274. Samson, R. A., A. C. Stolk, and R. Hadlock. 1976. Revision of the subsection Fasciculata of *Penicillium* and some allied species. *Studies in Mycology,* Vol. 17, Centraalbureau voor Schimmelcultures, Baarn, pp. 1-47.
275. Scherwitz, C., R. Martin, and H. Ueberberg. 1978. Ultrastructural investigations of formation of *Candida albicans* germ tubes and septa. Sabouraudia *16*: 115-124.
276. Schneider, E. F., L. R. Barran, P. J. Wood, and I. R. Siddiqui. 1977. Cell wall of *Fusarium sulphureum.* II. Chemical composition of the conidial and chlamydospore walls. Can. J. Microbiol. *23*: 763-769.
277. Schol-Schwarz, M. B. 1970. Revision of the genus *Phialophora* (Moniliales). Persoonia *6*: 59-94.
278. Schwartz, I. S., and C. W. Emmons. 1968. Subcutaneous cystic granuloma caused by a fungus of wood pulp *(Phialophora richardsiae).* Am. J. Clin. Pathol. *49*: 500-505.
279. Schwartz, J., and K. Salfelder. 1977. Blastomycosis: a review of 152 cases. In E. Grundmann and W. H. Kirsten (Eds.), *Current Topics in Pathology,* Vol. 65, Springer-Verlag, Berlin, pp. 166-200.
280. Seeliger, H. P. R. 1975. Increasing spectrum of opportunistic fungal infections. In E. W. Chick, A. Balows, and M. L. Furcolow (Eds.), *Opportunistic Fungal Infections,* Thomas, Springfield, Ill., pp. 5-21.
281. Sekiguchi, J., G. M. Gaucher, and J. W. Costerton. 1975. Microcycle conidiation in *Penicillium urticae*: an ultrastructural investigation of spherical spore growth. Can. J. Microbiol. *21*: 2048-2058.
282. Sekiguchi, J., G. M. Gaucher, and J. W. Costerton. 1975. Microcycle conidiation in *Penicillium urticae*: an ultrastructural investigation of conidial germination and outgrowth. Can. J. Microbiol. *21*: 2059-2068.
283. Sekiguchi, J., G. M. Gaucher, and J. W. Costerton. 1975. Microcycle conidiation in *Penicillium urticae*: an ultrastructural investigation of conidiogenesis. Can. J. Microbiol. *21*: 2069-2083.
284. Seshadri, R., and J. M. Sieburth. 1975. Seaweeds as a reservoir of *Candida* yeasts in inshore waters. Mar. Biol. *30*: 105-117.
285. Shadomy, H. J. 1970. Clamp-connections in two strains of *Cryptococcus neoformans.* Spectrum Monograph Series, Arts and Science, Georgia State University, Atlanta *1*: 67-72.
286. Shannon, J. L., and A. H. Rothman. 1971. Transverse septum formation in budding cells of the yeastlike fungus *Candida albicans.* J. Bacteriol. *106*: 1026-1028.
287. Shearer, C. A. 1973. Fungi of the Chesapeake Bay and its tributaries. II. The genus *Conioscypha.* Mycologia *65*: 128-136.
288. Shearer, C. A., and J. J. Motta. 1973. Ultrastructure and conidiogenesis in *Conioscypha* (Hyphomycetes). Can. J. Bot. *51*: 1747-1751.
289. Shepton, W. A., and J. A. Lunn. 1980. Sporangiole morphology and species separation in *Cunninghamella.* Trans. Br. Mycol. Soc. *74*: 483-491.

290. Sherry, J. P., J. Kuchma, J. Zarzour, and B. J. Dutka. 1979. Occurrence and significance of *Candida albicans* in Lake Ontario bathing beaches. Scientific Series No. 98, Inland Waters Directorate, National Water Research Institute, Canada Centre for Inland Waters, Burlington, Ontario.
291. Sigler, L., P. K. Gaur, R. W. Lichtwardt, and J. W. Carmichael. 1979. *Renispora flavissima,* a new gymnoascaceous fungus with tuberculate *Chyrsosporium* conidia. Mycotaxon *10*: 133-141.
292. Silva-Hutner, M., and A. L. Carrion. 1975. Differential characteristics of the fungal agents of chromoblastomycosis. In *Mycoses,* Proceedings of the Third International Conference on the Mycoses, Pan Am. Health Organ. Sci. Publ. No. 304, Washington, D.C., pp. 118-124.
293. Sinclair, J. B., and Z. M. El-Tobshy. 1969. Pathogenicity of plant and animal isolates of *Geotrichum candidum* in the turtle. Mycologia *61*: 473-480.
294. Smith, J. E., D. R. Berry, and B. Kristiansen (Eds.), 1980. *Fungal Biotechnology,* Academic, London.
295. Smith, J. E., J. G. Anderson, S. G. Deans, and D. R. Berry. 1981. Biochemistry of microcycle conidiation. In G. T. Cole and B. Kendrick (Eds.), *Biology of Conidial Fungi,* Vol. 2, Academic, New York, pp. 329-356.
296. Smith, J. E., S. G. Deans, J. G. Anderson, and B. Davis. 1977. The nature of fungal sporulation. In J. Meyrath and J. D. Bu'Lock (Eds.), *Biotechnology and Fungal Differentiation,* Academic, New York, pp. 17-41.
297. Smith, J. M., and P. M. Robinson. 1980. Development of somatic hyphae of *Geotrichum candidum.* Trans. Br. Mycol. Soc. *74*: 159-165.
298. Spanoghe, L., A. Devos, and N. Viaene. 1976. Cutaneous geotrichosis in the red flamingo *(Phoenicopterus ruber).* Sabouraudia *14*: 37-42.
299. Stenderup, A., and G. T. Pedersen. 1962. Yeasts of human origin. Acta Pathol. Microbiol. Scand. *54*: 462-472.
300. Stewart, P. R., and P. J. Rogers. 1978. Fungal dimorphism: a particular expression of cell wall morphogenesis. In J. E. Smith and D. R. Berry (Eds.), *The Filamentous Fungi,* Vol. 3, Wiley, New York, pp. 164-196.
301. Streiblová, E. 1971. Cell division in yeasts. In P. Pérez-Miravette and D. Peláez (Eds.), *Recent Advances in Microbiology,* Proceedings of the 10th International Congress for Microbiology, pp. 131-140.
302. Streiblová. E., and K. Beran. 1963. Demonstration of yeast scars by fluorescence microscopy. Exp. Cell Res. *30*: 603-605.
303. Sun, S. H., and M. Huppert. 1976. A cytological study of morphogenesis in *Coccidioides immitis.* Sabouraudia *14*: 185-198.
304. Sun, S. H., S. S. Sekhon, and M. Huppert. 1979. Electron microscopic studies of saprobic and parasitic forms of *Coccidioides immitis.* Sabouraudia *17*: 265-273.
305. Sutton, B. C. 1980. *The Coelomycetes. Fungi Imperfecti with Pycnidia, Acervuli and Stromata,* Commonwealth Mycological Institute, Kew, Surrey, England.
306. Sutton, B. C., and D. K. Sandhu. 1969. Electron microscopy of conidium development and secession in *Cryptosporiopsis* sp., *Phoma fumosa, Melanconium bicolor,* and *M. apiocarpum.* Can. J. Bot. *47*: 745-749.
307. Swenson, F. J., and J. A. Ulrich. 1980. Fatty acids of dermatophytes. Sabouraudia *18*: 1-9.
308. Swinne-Desgain, D. 1975. Strains of *Cryptococcus neoformans* of saprophytic origin. Sabouraudia *13*: 303-308.
309. Szaniszlo, P. J., P. H. Hsieh and J. D. Marlowe. 1976. Induction and ultrastructure of the multicellular (sclerotic) morphology in *Phialophora dermatitidis.* Mycologia *68*: 117-130.

310. Taschdjian, C. L., F. Reiss, and P. J. Kozinn. 1960. Experimental vaginal candidiasis in mice; its implications for superficial candidiasis in humans. J. Invest. Dermatol. *34*: 89-94.
311. Thomas, D. des S. 1978. Cytochalasin effects in plants and eukaryotic microbial systems. In S. W. Tanenbaum (Ed.), *Cytochalasins–Biomedical and Cell Biological Aspects,* Elsevier/North-Holland, Amsterdam, pp. 257-275.
312. Tierno, P. M., and M. Milstoc. 1977. Germ-tube positive *Candida tropicalis.* Am. J. Clin. Pathol. *68*: 294.
313. Ton Than, T. C., and G. Turian. 1978. Ultrastructural study of microcyclic macroconidiation in *Neurospora crassa.* Arch. Microbiol. *116*: 279-288.
314. Trinci, A. P. J. 1978. Wall and hyphal growth. Sci. Prog. (Oxford) *65*: 75-99.
315. Trinci, A. P. J., and A. J. Collinge. 1974. Spore formation in nitrogen and carbon starved cultures of *Geotrichum candidum* and *Mucor racemosus.* Trans. Br. Mycol. Soc. *62*: 351-358.
316. Turian, G., N. Oulevey, and M. Cortat. 1973. Recherches sur la différenciation conidienne de *Neurospora crassa.* V. Ultrastructure de la séquence macroconidiogène. Ann. Microbiol. (Inst. Pasteur) *124A*: 443-458.
317. Ulane, R. E., and E. Cabib. 1974. The activating system of chitin synthetase from *Saccharomyces cerevisiae.* Purification and properties of an inhibitor of the activating factor. J. Biol. Chem. *249*: 3418-3422.
318. Vanbreuseghem, R., and C. DeVroey. 1979. Systemic opportunistic fungal infections. Postgrad. Med. J. *55*: 593-594.
319. Vanbreuseghem, R., J. C. Gentles, D. W. R. Mackenzie, O. Male, and A. Mantoviani. 1980. Nomenclature of mycoses. Sabouraudia *18*: 78-84.
320. Van Emden, J. H., and J. W. Veenbaas-Rijks. 1973. *Cylicognone,* a new genus of Hyphomycetes with unusual conidium ontogeny. Acta Bot. Neerl. *22*: 637-640.
321. Vannini, G. L. 1979. Effects of different temperatures on cell wall formation and polysaccharide arrangement in *Trichophyton mentagrophytes*: considerations from a cytochemical study. J. Ultrastruct. Res. *69*: 157-163.
322. Van Oorschot, C. A. N. 1980. A revision of *Chrysosporium* and allied genera. *Studies in Mycology,* Vol. 20, Centraalbureau voor Schimmelcultures, Baarn, pp. 1-89.
323. Van Uden, N., and H. Buckley. 1970. *Candida* Berkhout. In J. Lodder (Ed.), *The Yeasts. A Taxonomic Study,* North-Holland, Amsterdam, pp. 893-1084.
324. Vršanska, M., Z. Krántky, P. Biely, and S. Machala. 1979. Chitin structures of the cell walls of synchronously grown virgin cells of *Saccharomyces cerevisiae.* Z. Allg. Mikrobiol. *19*: 357-362.
325. Waldrip, D. W. 1974. Isolation of *Dactylaria gallopava* from broiler house litter. Avian Dis. *18*: 445-451.
326. Wang, C. J. K. 1979. Pleomorphic Fungi Imperfecti. In B. Kendrick (Ed.), *The Whole Fungus,* Proceedings of the Second International Mycological Conference, Vol. 1, National Museum of Natural Sciences and National Museums of Canada, Ottawa, pp. 81-91.
327. Ward, J. I., M. Weeks, D. Allen, R. Hutcheson, R. Anderson, D. Frazer, L. Kaufman, L. Ajello, and A. Spickard. 1979. Acute histoplasmosis: clinical epidemiologic and serologic findings of an outbreak associated with exposure to a fallen tree. Am. J. Med. *66*: 587-595.
328. Watts, J. C., C. S. Callaway, F. W. Chandler, and W. Kaplan. 1975. Human pulmonary adiospiromycosis. Arch. Pathol. *99*: 11-15.
329. Weedon, D., and G. Ritchie. 1979. Cystic chromomycosis of the skin. Pathology *11*: 389-392.

330. Weijman, A. C. M. 1979. Carbohydrate composition and taxonomy of *Geotrichum, Trichosporon* and allied genera. Antonie Leeuwenhoek J. Microbiol. Serol. *45*: 119-127.
331. Weisburg, S. H., and G. Turian. 1971. Ultrastructure of *Aspergillus nidulans* conidia and conidial lomasomes. Protoplasma *72*: 55-67.
332. Wilsenach, R., and M. Kessel. 1965. The role of lomasomes in wall formation in *Penicillium vermiculatum.* J. Gen. Microbiol. *40*: 401-404.
333. Winner, H. I. 1977. Recent advances in systemic candidosis. In A. M. Beemer, A. Ben-David, M. A. Klingberg, and E. S. Kuttin (Eds.), *Contributions to Microbiology and Immunology,* Vol. 4: *Host-Parasite Relationships in Systemic Mycoses,* Part II, S. Karger, Basel, pp. 64-76.
334. Wu-Yuan, C. D., and T. Hashimoto. 1977. Architecture and chemistry of microconidial walls of *Trichophyton mentagrophytes.* J. Bacteriol. *129*: 1584-1592.
335. Yamada, Y., and K. Kondo. 1972. Taxonomic significance of coenzyme Q system in yeasts and yeast-like fungi. In K. Iwata (Ed.), *Yeast and Yeast-Like Microorganisms in Medical Science,* Proceedings of the Second International Specialized Symposium on Yeasts, University of Tokyo Press, Tokyo, pp. 63-69.

12

Blastoconidium Germination

Nina Dabrowa and Dexter H. Howard / UCLA School of Medicine, Los Angeles, California

I. INTRODUCTION

The germination of fungal spores has been the subject of a large number of morphological and physiological observations, and the results from such studies have been reviewed extensively on a number of occasions [14,37,82,91,106]. Sometimes, spores of zoopathogenic fungi have been the subjects of investigations on spore germination (e.g., Refs. 5, 6, 28, 39, 106) and, of course, there is nothing unique about the event in such pathogens. It seemed unnecessary, therefore, to consider the germination of the conidiospores, sporangiospores, pycnidiospores, and so on, again in this volume. Other types of spores found among zoopathogens whose germination has been studied to a limited extent include the endospores and arthroconidia of *Coccidioides immitis* [85], and the adiaspores of *Emmonsia parva* and *E. crescens* [85].

One particular type of spore, the blastoconidium (i.e., yeast cell), has not been included in the aforementioned reviews. Yeast cell germination figures prominently in the biology of one very important pathogen, *Candida albicans* [78,92], and is a part of the life cycle of the dimorphic pathogens that have a yeast cell phase of growth, e.g., *Blastomyces dermatitidis, Histoplasma capsulatum, Paracoccidioides brasiliensis,* and *Sporothrix schenckii* [41].

In this chapter we review the literature on blastoconidium germination, especially that of *C. albicans,* which has been studied extensively, and mention briefly the literature on the germination of the yeast cell phase of growth of certain dimorphic pathogens. Since the germination event is an integral part of the behavior of these dimrophic fungi, the subject has also been considered comprehensively in Chap. 10.

II. BLASTOCONIDIUM GERMINATION

A. Dimorphic Pathogens

1. *Histoplasma capsulatum*

The zoopathogenic fungus *Histoplasma capsulatum* has two morphological forms: a yeast form composed of blastoconidia, and a mycelial form consisting of hyphae upon which are borne both micro- and macroconidia [24]. This dimorphism is associated with the parasitic and saprophytic phases of growth of the fungus, but some isolates will express the two distinct morphological aspects of the asexual life cycle in artificial culture media [2, 41,42]. Thus, on common laboratory media incubated at 23-25°C, *H. capsulatum* grows as a mycelium, whereas on suitable media at 35-37°C some isolates of the fungus convert to yeast cells [41,42,113].

The mycelial form of growth of *H. capsulatum* is initiated by yeast cell germination [113]. The morphologic aspects of blastoconidium germination of *H. capsulatum* have been observed by transmission electron microscopy [35] and by scanning electron microscopy [33]. Some of the environmental factors involved in blastoconidium germination have been recorded [113]. Biochemical studies of the event have been limited but are certainly more extensive than are those of other dimorphic pathogens with a yeast cell phase of growth.

It has long been known that maintenance of the yeast form of growth requires sulfur-containing, organic compounds and that cysteine is especially useful in meeting this requirement [12,13,41,98]. Re-examining this fact, Boguslawski and his colleagues confirmed that yeast cells, but not mycelial cells, of *H. capsulatum* (Down's strain) have an absolute requirement for cysteine [13] and that with some but not all strains of the fungus, serine can substitute for cysteine [12,98]. Evidence for a prototrophic yeast phase of *H. capsulatum* has been presented [44] and a need has been expressed for a thorough biochemical and genetical characterization of the many strains of *H. capsulatum* used in various laboratories [12,98]. The nutritional requirement of *H. capsulatum* blastoconidia for cysteine is reportedly based on the fact that they lack the enzyme sulfite reductase so that they cannot reduce inorganic sulfite to sulfide, which is used in the synthesis of the amino acid [12,13]. In contrast, mycelial cells with an active sulfite reductase grow in a minimal medium without a cysteine supplement. The enzyme sulfite reductase is repressed in yeast cells by cysteine and may be expressed under certain conditions, e.g., reduction of the temperature of incubation from 37°C to 23°C [12,42]. Sulfite reductase appears in blastoconidia of *H. capsulatum* approximately 12 h after cultures are moved from 37°C to 23°C, at a time several hours before germ tube emergence [42]. The isolation and preliminary characterization of a mutant of *H. capsulatum* defective in the ability of its blastoconidia to germinate have been reported [42]. Work with this mutant has revealed that the expression of sulfite reductase is temperature dependent, i.e., repressed by cysteine at 37°C [12] but constitutive in mycelial cells and germination-defective yeast cells at 23°C [42]. It should be noted that the yeast cells of some strains are derepressed in the absence of cysteine at 37°C [12,42] and yet cannot grow in minimal media. Thus, additional metabolic defects must exist in the yeast cell phase of growth of *H. capsulatum* [98].

Evidence obtained from previous investigations in our laboratory have revealed that the mycelial-phase cells of *H. capsulatum* accumulate leucine and glutamic acid more rapidly at 23°C than do yeast-phase cells at 37°C [38,99]. Bowdon and Garrison (see Ref. 42) confirmed this observation and reported the same disparity in transport rates for cystine in *Blastomyces dermatitidis*. However, Maresca et al. [66] and Gilbert and Howard (see Ref. 42) reported that the initial rate of cystine transport in blastoconidia of *H. capsulatum* was temperature dependent, being about half as rapid at 25°C as at 37°C. Maresca et al. additionally recorded that mycelial cells and yeast cells at 37°C transport cystine at the same rate and that the rate was halved by incubation of each growth form at 25°C.

When blastoconidia of the fungus grown at 37°C are placed at 23°C, the rate of transport of leucine and glutamic acid is markedly depressed and then gradually recovers as incubation continues at 23°C [99]. Moreover, after recovery the rate of uptake of these same amino acids is more rapid in mycelial cells at 23°C than it is in yeast cells at 37°C [38,49,99]. Under the same condition of temperature shift (37°C to 23°C) yeast cells continue to incorporate guanine and adenine at a constant but lower rate than they do at 37°C [49]. In one study, however, it was reported that guanine uptake was more rapid by mycelial cells at 23°C than it was by yeast cells at 37°C [99]. It is obvious that physiologic and enzymatic differences exist among the several strains of *H. capsulatum* used in various laboratories [12].

It is interesting that the three RNA polymerases of the yeast and mycelial phases are distinctive [50]. The corresponding enzymes from the two phases differ in sensitivity to α-amanitin, ion and salt requirements, temperature sensitivity, and subunit structure.

2. *Paracoccidioides brasiliensis*

The dimorphic fungus *Paracoccidioides brasiliensis* exists in humans and experimentally infected animals as a yeast-like organism. The yeast form can be propagated on synthetic media incubated at 37°C. In fact, temperature seems to be the only factor controlling the dimorphic behavior of this fungus [48]. A mycelial form of growth is initiated by germination of the yeast cells when the temperature of incubation is reduced from 37°C to room temperature (22-25°C). Sequential observations by light microscopy on the germination of blastoconidia of *P. brasiliensis* have been published recently [87].

It is reported that RNA synthesis is inhibited during the early stages of germ tube formation and resumes as true hyphae are formed [87]. On the basis of this observation it has been proposed that RNA synthesis is not required for the early stages of blastoconidium germination but is necessary for production of true hyphae and subsequent mycelial growth [87]. The chemical composition of the cell walls of the two morphologic forms of *P. brasiliensis* are different: an α-1,3-glucan is present in yeast cells, whereas a β-1,3-glucan is the polymer in mycelial cells [87-89]. The synthesis of α-glucan stops and that of β-glucan commences when yeast cells are shifted from 37°C to 22°C. How the mechanism is switched at a molecular level is not understood, but the problem is being investigated [87,88].

3. *Blastomyces dermatitidis*

The dimorphic behavior of *Blastomyces dermatitidis* is similar to that described for *H. capsulatum* and *P. brasiliensis.* Once again the mycelial form of growth is initiated by blastoconidium germination when the temperature of incubation is reduced from 37°C to room temperature [41,47]. As with *P. brasiliensis,* the transition seems to be controlled predominantly by temperature [56].

Differences in the wall composition of hyphal cells and yeast cells have been recorded [47]. Thus, as yeast cells germinate at room temperature, there must be a remarkable shift in cell wall synthesis. However, this aspect of the problem has not been investigated thoroughly (see Chap. 10 for a review of the literature). Some preliminary observations have been made on RNA and protein synthesis in germinating yeast cells of *B. dermatitidis* [102], but much more needs to be done in order to describe changes in these sorts of macromolecules during germination. Changes in the spectrum of isozymes of malate dehydrogenase [86] clearly display the need of active protein synthesis during germination, but study on the timing of the events has yet to be made.

4. *Sporothrix schenckii*

Sporothrix schenckii exhibits a yeast-hyphal dimorphism which is primarily controlled by temperature and CO_2 concentration [41]. As with the other systems so far discussed, the mycelial phase is initiated by blastoconidium germination [41]. The morphological details of phase transition have been extensively studied [34], but biochemical events have received little attention (see Chap. 10).

B. *Candida albicans*

1. General Description

Candida albicans (Robin) Berkhout, 1923, is a commensal of the gastrointestinal tract of humans and other warm-blooded animals [2,36]. The fungus may produce disease under certain conditions [2] and the lesions that it provokes frequently involve the skin and mucous membranes. Occasionally, more serious disease occurs in which any visceral organ may be affected [2]. It is generally agreed that the pathogenicity of *C. albicans* is favored by predisposing factors, and the fungus is clearly among the most prominent of the opportunistic pathogens [23]. The frequency of isolation of different *Candida* species and their relative distribution according to the body sites are summarized in a recent book by Odds [78], and *Candida albicans* was the most commonly isolated of all the species from human sources.

On most standard laboratory media *C. albicans* reproduces by budding, and this form of growth is independent of the temperature of incubation [58,94]. Under certain culture conditions, generally those in which the level of utilizable carbohydrate is very low, a well-differentiated pseudomycelium is formed. This pseudomycelium is comprised of pseudohyphae which are elongated, unseparated yeast cells. Characteristically, the pseudomycelium bears ball-like clusters of blastoconidia and thick-walled chlamydospores [58,78,84, 94,95,107]. The pseudomycelium is generally fostered by certain media, such as cornmeal agar or cream of rice agar, often supplemented with Tween 80 and routinely incubated at 20-25°C [78,95].

Candida albicans can also form a true mycelium [58] which arises from blastoconidium germination [78]. The germ tubes grow by apical extension and septa are produced as the hyphae elongate. The true mycelium may also, like the psyedomycelium, bear clusters of blastoconidia [78]. A bud and a germ tube are readily distinguishable in the early stages of their development. The germ tubes are longer and narrower than the bud initials and there is no constriction at the point of emergence [78]. But, once formed, the pseudo- and true mycelium are not easily distinguished, and investigators have not always been careful to discriminate between the two types of hyphae when describing the filamentation of *C. albicans* (see Ref. 78 for an excellent review of this topic). Moreover, it is somewhat misleading, or at least potentially confusing, to use terms such as "pseudo-germ tube" for the structures that arise from the yeast cells of *C. albicans* by germination [60].

2. Environmental Factors That Induce Germination

a. Introduction: Interest in the origin of the true mycelium was sharply focused when Hu et al. [43] and Johnson et al. [46] reported germination of blastoconidia in serum incubated at 37°C. Reynolds and Braude [83] described the process in greater detail, and various investigators subsequently studied germination in serum and developed from those studies a useful diagnostic test for distinguishing *C. albicans* from other species of *Candida* in the clinical laboratory [9,17,29,31,32,78,101]. Many environmental factors have been identified which foster germination. A comprehensive survey of the literature on the subject is presented by Odds in his monograph (see Tables 7 and 8 of Ref. 78). We present here only a few key references on various features of blastoconidium germination of *C. albicans*.

b. Physical Factors: It has been repeatedly observed that germination is a temperature-dependent event occurring in a variety of substrates at 37°C, but not at 20-25°C [7, 30,30a,51,59,95]. This temperature dependence, together with the fact that other conditions which sponsor germination are like those which obtain in vivo, suggest that a significant portion of filamentation observed in the lesions of disease arises by blastoconidium germination.

Germ tube formation occurs over a fairly wide pH range of 6.0-8.0. Below pH 6.0 blastoconidium growth is usually favored [78]. For example, in one study pH was used as a morphology-controlling factor: stationary phase cultures of yeast cells released into fresh medium at pH 6.5 and incubated at a temperature of 37°C germinated, while cells incubated in the same medium and temperature at pH 4.5 budded [8,70].

Cell density is an important factor and germination is greatly inhibited at concentrations above 5×10^6-1×10^7 cells per milliliter [25,53,60]. This concentration effect has been related to the presence of autoinhibitors [40,57].

c. Nutritional Factors (1) Serum: The initial observations of blastoconidium germination by *C. albicans* were made by investigators using plasma or serum [7,43,45,53,60,71,101]. Therefore, subsequent work focused on manipulation of this substance in the customary ways in order to discover key attributes which sponsored germination. The "transforming factor" (a word used to characterize the "factor" in serum which sponsored germination) was found to be unrelated to *Candida* agglutinins and precipitins in normal human or animal sera. The factor was stable at 56°C, nondialyzable, and its effects could be reproduced by egg albumin or oleic acid albumin (Difco) [17,83]. Landau et al. [53] confirmed these observations and extended them by showing that the filament-fostering factor in serum functioned over the pH range 6.0-8.0 and was absorbed by viable and nonviable yeast cells of *Candida*. Germination was depressed by dilution of serum with saline and by addition of cysteine but was unaffected by glucose [53].

(2) Amino Acids: Taschdjian and Kozinn [100] noted that during blastoconidium germination of *C. albicans* the serum proteins were not broken down, and they postulated that free amino acids provided the nitrogen source for germination. Landau et al. [53] demonstrated that a completely synthetic medium, tissue culture medium 199, would foster germination nearly as well as did serum. Barlow et al. [7], while studying the effects of various serum proteins, noted that a synthetic amino acid mixture consisting of aspartic acid, lysine, histidine, threonine, proline, and B-alanine supported good germination. Others have also devised chemically defined media of amino acid mixtures in which germination readily takes place [51,52,55,65,77].

There is, of course, no assurance that the ingredients of serum that foster germination are the same as those within the several synthetic media recommended by these studies. In fact, some of the work on serum factors, e.g., absorption by whole yeast cells and nondialyzability, argue against a common explanation for germination of *Candida* in serum and in the various amino acid mixtures. But continuation of work with serum was complicated by the diversity of constituents. Moreover, there is a possibility that an important effect of serum is the inhibition of blastoconidium growth [53,54] and that inhibited blastoconidia then respond to other stimuli (e.g., free amino acids) by germinating. Some data to support these conjectures are available [26,53,54].

In any event, the advantages of working with a defined medium were obvious, and it was subsequently reported that amino acids which enter metabolism by conversion to α-ketoglutarate (Table 1) are among the most effective inducers of germ tubes and that proline is one of those which sponsors good filamentation in a large number of strains [26, 51]. The percentage germination of several clinical isolates in various amino acids is recorded in Table 2. It is clear that proline most consistently fosters germination, although several other amino acids also promote it. Considerable differences in response to stimulation by different amino acids by clinical isolates have been observed. Therefore, strain variability may account for some contradictory results from different laboratories. Analogs of proline or the D isomer markedly reduced germination in a proline biotin buffer. These same analogs competitively inhibit uptake of L-proline (unpublished observations from our laboratory). Thus, *C. albicans* seems to transport proline via an imino acid permease, and the inhibition of proline-sponsored germination by analogs may be due to inhibition of uptake of the proline or its subsequent intracellular utilization.

Sulfur-containing amino acids, particularly cysteine [25,75], methionine [3,63-65], and S-adenosylmethionine [3], have been investigated as germination promoters. When added to the medium, methionine induced both germination and a higher level of the enzyme homocysteine methyltransferase [3]. Addition of glycine also promoted germination with increased levels of homocysteine methyltransferase and S-adenosylmethionine [3]. Increased activity of S-adenosylmethionine-splitting enzymes during the germination process may indicate that other methylating enzymes also play a role in filamentation [3]. Post-translational methylation of proteins has been noted during morphogenesis of *Mucor racemosus* [32a]. Also, greater uptake of S-adenosylmethionine in the vacuoles and cell walls of hyphae of *C. albicans* has been demonstrated [4], but the significance of this observation remains unclear.

L-α-Amino-n-butyric acid or glycine are capable of supporting germination in a glucose salts biotin broth when the inoculum was grown in a rich medium (i.e., Sabouraud's agar) but not when it was grown in minimal medium [65]. This result suggests that conclusions about substances which sponsor germination can be influenced by nutritional factors stored up by the cells before harvest and use in experiments.

(3) Carbohydrates: Glucose is generally the carbohydrate selected for media to support germination of *C. albicans* [30,30a]. High levels of glucose have been shown by some to foster yeast cell replication and to suppress germ tube formation [25,30,30a, 103]. Other workers, however, have observed germination in media with a high glucose concentration (e.g., Ref. 51). The discrepancy noted is probably related to variation in the phosphate level in the media used in the different experiments (see Ref. 26 for a discussion of this point).

Chitin is a prominent component of the cell walls of *C. albicans* and occurs more abundantly in mycelial phase cells [15,18,20,90]. It is perhaps not surprising, then, that N-acetyl-D-glucosamine is an inducer of germination [11,18,93]. However, the functioning of this substance is strongly influenced by the phosphate concentration of the medium and the strain studied [104].

(4) Inorganic Nitrogen Source: Ammonium sulfate and ammonium chloride will support a certain amount of germination in a glucose biotin buffer medium [51]. Generally, however, amino acids sponsor more abundant filamentation than do inorganic nitrogen sources, and the interaction of various other factors, e.g., glucose and phosphate levels, are very important in the induction of filamentation by ammonium ions [51,52,108].

(5) Other Factors: Some investigators have induced germination by causing a zinc deficiency [111] that is accompanied by a decrease in RNA levels and a doubling of the inorganic phosphates. Since other germination promoters such as low yeast extract concentration [108] and low biotin levels [109] also reduce the RNA content of the cells, it is suggested that net RNA synthesis is not necessary for the initial stages of the transition from yeast to mycelium [111]. Along these lines Ogletree et al. [79] observed that yeast cells treated with rifampin, which binds mitochondrial RNA polymerase, still germinated [78]. This result suggests that mitochondrial RNA polymerase function is more important to budding and germ tube elongation than to germination.

In another study on the effect of zinc deficiency in *Candida* germination, an interesting dual effect was observed [8]. In these studies a mutant that formed hyphae at 25°C was used. Zinc completely suppressed hyphal formation at 25°C but did not suppress germination in released stationary phase cultures at 37°C. The results suggest alternative pathways for hyphae formation in *Candida*: one sensitive and the other insensitive to zinc [8].

3. Cell Wall Changes in Germinating Blastoconidia

a. Alteration of Ultrastructure: The original studies of Nickerson on the presence of disulfide bonds in and the glucan-mannan composition of the cell wall of *C. albicans* stressed the importance of that structure in dimorphism. The ultrastructure of the two phases of growth have been studied [18]. Five distinct layers can be observed in the yeast cells of *C. albicans* (Fig. 1). The layers undergo modifications in the amount, morphology, and orientations of their components during germination [90,109]. The cell wall of the germ tube is thinner than that of the yeast cells. The outermost layer of the yeast cell wall diminishes during germination [90] and in a biotin-deficient medium may totally disappear [109]. The second and fourth layers of the yeast wall continue into the hyphae, and the fifth layer becomes the innermost layer of the emergent germ tube. The fate of the third layer of the yeast cell wall remains unclear: it either disappears or merges with the fourth layer in the germ tube. The ultrastructural features of the cytoplasm of the yeast cells of *C. albicans* were essentially like those described by other investigators (see Chap. 9 and Ref. 112). Hyphal cells differed most markedly in that they had more frequent and deeper invaginations of the cell membrane, with extensive development of intracytoplasmic membrane system appearing as whorled structures [112]. The endoplasmic reticulum was also better developed in the hyphal cells [112].

b. Alteration in Cell Wall Composition: The cell wall of the germ tube and that of the blastoconidia of *C. albicans* both contain glucans, mannans, and chitin, but the proportion of these polysaccharides differs in the two forms. The cell walls of hyphal cells grown in a biotin-deficient medium had greater quantities of glucans and smaller quantities of

Table 1 Growth Characteristics of *C. albicans* Incubated with Various Amino Acids for 6 h in a Glucose Biotin Phosphate Buffer (pH 7.2)

Nitrogen source	Route of entry into metabolism[a]	Percent filamentation[b]	Increase in growth[c]	pH of culture medium at 6 h
Proline	Glutamate α-ketoglutarate	93 ± 5	6.3 ± 0.06	7.0
Arginine	Glutamate α-ketoglutarate	90 ± 4	4.6 ± 0.05	8.0
Alanine	Glutamate α-ketoglutarate	89 ± 6	4.1 ± 0.065	8.0
Histidine	Glutamate α-ketoglutarate	68 ± 3.6	4.5 ± 0.07	7.4
Thioproline	Glutamate α-ketoglutarate	51 ± 4	6.3 ± 0.04	7.0
Glutamate	Glutamate α-ketoglutarate	38 ± 0.33	6.0 ± 0.051	7.2
Ornithine	Glutamate α-ketoglutarate	14 ± 0.67	6.1 ± 0.08	6.8
Asparagine	Oxaloacetate	44 ± 2.1	6.1 ± 0.056	7.6
Aspartate	Oxaloacetate	40 ± 1	5.5 ± 0.06	7.6
Isoleucine	Succinyl-CoA	70 ± 1	4.3 ± 0.075	8.0
Methionine	Succinyl-CoA	16 ± 0.33	5.2 ± 0.06	8.0
Valine	Succinyl-CoA	20 ± 0	5.5 ± 0.05	8.0
Lysine	Acetoacetyl-CoA	70 ± 3	4.0 ± 0.03	8.0
Tryptophan	Acetoacetyl-CoA	42 ± 2	5.9 ± 0.045	7.6

Phenylalanine	Acetoacetyl-CoA	40 ± 1	6.2 ± 0.066	7.8
Leucine	Acetoacetyl-CoA	36 ± 0.33	4.6 ± 0.08	8.0
Threonine	Acetoacetyl-CoA	31 ± 0.33	5.6 ± 0.02	7.8
Tyrosine	Acetoacetyl-CoA	0	6.1 ± 0.04	8.0
Serine	Pyruvate-acetate	20 ± 0	5.5 ± 0.05	8.0
Glycine	Pyruvate-acetate	8 ± 1	5.7 ± 0.03	8.0
Cysteine	Pyruvate-acetate	3 ± 1	5.5 ± 0.07	7.6
Cystine	Pyruvate-acetate	0	5.0 ± 0.04	7.8
p-NO_2-phenyl proline	Transamination	2 ± 0.67	4.2 ± 0.06	7.2
Benzyl proline	Transamination	0	4.9 ± 0.08	7.2
N-Acetyl proline	Transamination	0	5.1 ± 0.056	7.2
L-Azetidine-2 carboxylic acid	Transamination	0	4.0 ± 0.03	7.0
$(NH_4)_2SO_4$	Transamination	28 ± 0.67	6.7 ± 0.02	6.05
NH_4Cl	Transamination	0	6.8 ± 0.04	4.95

[a]CoA, coenzyme A.
[b]Standard error of the mean of three experiments.
[c]Ratio of dry weight after 6 h of incubation to dry weight of inoculum ± standard error of the mean of three experiments.
Source: Ref. 51.

Table 2 Germination of Clinical Isolates of *C. albicans* in Amino Acids[a]

	Clinical isolate														
Amino acid (0.02 M)	2282	2286	1956	1725A	2087	2065A	1968	2100	9292	2086	734	2059	1792	1966	1580C
L-Lysine HCl	0	2	2	0	0	2	0	0	4	0	1	2	3	0	1
L-Valine	0	0	8	0	0	26	0	0	4	0	1	4	3	0	2
DL-Serine	2	4	10	0	0	12	0	0	4	0	6	5	2	0	0
DL-Methionine	0	0	1	0	0	17	0	0	4	0	2	0	4	0	0
L-Isoleucine	0	2	1	0	0	8	0	0	4	0	3	3	8	0	2
Glycine	0	4	2	0	0	11	0	0	8	0	8	10	2	0	1
L-Cysteine HCl	0	0	6	0	0	0	0	0	3	0	12	15	4	0	2
L-Arginine	2	8	30	0	0	62	24	0	8	4	18	15	31	22	8
L-Phenylalanine	0	2	3	0	0	19	0	0	3	0	8	8	2	5	2
L-Histidine	0	2	3	0	0	2	0	0	2	0	0	1	0	0	0
DL-Threonine	0	1	2	0	0	43	0	0	5	0	9	6	4	1	3
L-Alanine	6	27	56	2	4	29	10	6	32	16	31	15	11	8	17
L-Glutamine	1	2	22	0	0	60	8	1	14	3	28	15	48	5	30
L-Leucine	0	0	3	0	0	24	0	0	0	0	2	6	2	0	2
L-Tryptophan	0	1	2	0	0	32	0	0	0	0	3	5	1	0	0
L-Glutamic acid	2	0	4	0	0	70	10	0	22	3	21	26	18	4	39
L-Aspartic acid	0	0	0	0	0	34	6	0	7	5	2	6	2	0	16
L-Cystine	0	0	2	0	0	34	0	0	4	2	8	23	1	2	0
L-Tyrosine	0	0	0	0	0	15	0	0	4	0	3	6	2	1	2
L-Ornithine HCl	0	0	2	0	0	50	2	0	4	8	14	16	8	0	7
L-Proline	70	49	74	24	46	78	80	16	56	86	53	38	80	60	78
OH-Proline	0	0	14	0	0	64	1	0	16	6	32	26	8	1	10
PB	0	0	0	0	0	4	0	0	1	0	4	2	4	0	0
TC 199	83	68	96	99	81	98	99	92	82	98	70	62	93	99	97

[a]Results shown as percentage of germination under conditions of the test.
Source: Ref. 26.

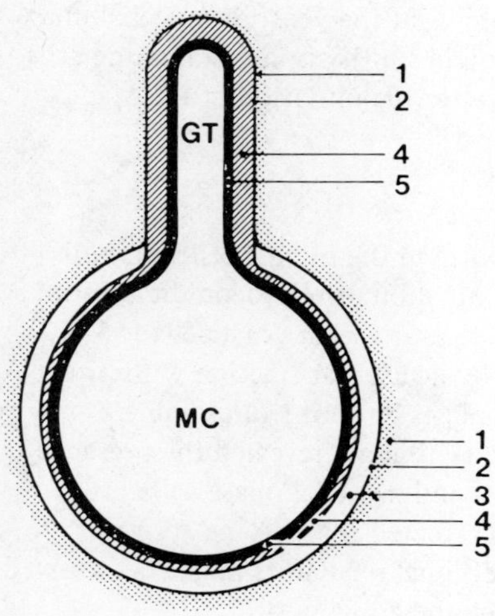

Figure 1 Schematic representation of germ tube formation in *C. albicans.* 1, outer granular layer, strongly diminished in the area of the germ tube (GT); 2, electron-dense amorphous layer, changing smoothly from the mother cell (MC) to the germ tube; 3, main layer in the area of the mother cell, remaining at the starting point of the germ tube; 4, electron-transparent layer forming the major component of the germ tube wall, often not clearly discernible from the third layer in the area of the mother cell; 5, innermost electron-dense layer, immediately adjacent to the plasmalemma. (From Ref. 90.)

mannans than did yeast cell walls [110]. The chitin content of hyphas cells was 3-10 times higher than it was in yeast cells [15,20]. However, the contents of hexoses, hexosamines, and amino acids were similar in the two growth forms [20]. The protein content of hyphal walls was three times less than it was in yeast walls [20]. Enzymatic treatment of cells revealed differences in bonding and spatial distribution of the cell wall components of the two forms [21,22]. There was twice as much carbohydrate in the membranes of the hyphal form as there was in the yeast cells, but the amounts of several enzymes and of lipid contents were about the same in membranes of the two forms [67,68].

4. Biochemical Events Attending Blastoconidium Germination

It is clear that there are a number of factors which may trigger germination of blastoconidia of *C. albicans* [1,78]. A few recent efforts have been made to relate these triggering factors to underlying biochemical events [26,51,52]. Observations of macromolecular synthesis in *C. albicans* have been limited. Yamaguchi [108] described nucleic acid synthesis in yeast and mycelial cells but did not extend his observation to synthetic activities during spore germination. Bhattacharya and Datta [10] have examined the effects of cyclic AMP on RNA and protein synthesis in blastoconidia of *C. albicans.* Dabrowa et al. [27] were perhaps the first to look at macromolecular synthesis during germination of yeast cells. Simple quantitative increases in proteins and nucleic acids were reported. In addition, these workers described qualitative differences in proteins and showed at least one protein unique to germinating cells. In a more recent and more comprehensive study [61,62] no differences were detected in the major cytoplasmic proteins of the yeast or mycelial phase extracts. Wain et al. [105] studied nucleic acid and protein synthesis in germinating cells. From their studies it is clear that DNA synthesis began after the initiation of RNA and protein synthesis [105].

5. Regulation of Germination

a. Chitin Synthetase: The increased amount of chitin in the hyphal cell wall indicates that synthesis of this compound accompanies germination of blastoconidia. Chitin synthetase activity was much higher in the hyphal cells than it was in yeast cells [15]. This enzyme activity was greatly stimulated in hyphae by a vacuolar fraction with protease activity (activator) obtained from the yeast protoplasts but not from hyphal protoplasts [15]. Further studies on the regulation of chitin synthetase revealed the presence of a chitin inhibitor found in supernatants of both yeast and mycelial phase cells [16]. The inhibitor was heat-stable and trypsin-sensitive, and affected the enzyme activator rather than the enzyme itself. A coupler system of additional regulators might be involved in chitin synthesis, at a specific site and time, e.g., below the growing tip [16].

b. Cyclic Nucleotides: Nimi et al. [76] reported that changes in the levels of cyclic AMP but not of cyclic GMP accompanied germination. Moreover cysteine, which is an inhibitor of germination in the system of these investigators, reversed the observed increase in cyclic AMP [76].

c. Autoinhibitors: The role of autoinhibitors in morphogenesis of *C. albicans* has been the subject of a few preliminary observations. Early studies showed that phenethyl alcohol and tryptophol [57] were released by the organism into the media and inhibited filamentation. Recently, a substance unrelated to either phenethyl alcohol or tryptophol which also inhibits germination has been detected in cultures of *C. albicans* [57]. The substance has been called morphogenic autoregulatory substance (MARS) and its release into the culture medium is believed to be the basis of the density inhibition phenomenon, i.e., the fact that the percentage of germination is inversely proportional to the cell density [40].

d. Metabolic and Cell Cycle Controls of Morphogenesis: A theory of morphogenesis in *Candida* was first devised by Nickerson and his colleagues [72-75]. The work upon which the theory rested was a pioneering effort to give a molecular basis to morphogenesis in *C. albicans.* It was postulated that filamentous growth occurs only when cellular division resulting in bud formation is impaired, either permanently by mutation, or temporarily by inhibitors or lack of certain essential metabolites [75]. A special disulfide reductase enzyme which reduces disulfide bonds in the cell wall was necessary for budding and was located in the mitochondria of actively budding cells [74]. Any impairment of the activity of the reductase enzyme resulted in filamentation. A mycelial mutant isolated by Nickerson and Chung [73] apparently was deficient in this division enzyme. Addition of the reducing compounds, such as cysteine or glucose, inhibited germination of blastoconidia [75]. This early work by Nickerson and co-workers provided a basis for subsequent efforts toward a biochemical explanation of morphogenesis in *C. albicans.* Nevertheless, Nickerson's investigations can account for only part of the morphogenetic process. The filamentous mutant used in the studies produced pseudohyphae not true hyphae [78]. Moreover, under condition of proper phosphate concentration, glucose can be incorporated into media that support germination [26,51], and cysteine does not antagonize hyphal growth [78]. Clearly, then, there is little experimental confirmation of some of Nickerson's major predictions regarding control of morphogenesis in *C. albicans.*

Recently, Land et al. advanced the hypothesis that morphogenesis of the mycelial phase of growth of *C. albicans* is correlated with repression of mitochondrial activity [51, 52]. The effect of various electron donors, acceptors, and transport uncouplers on the morphology and oxygen consumption of *C. albicans* is shown in Table 3, taken from the work of these investigators [52]. Competition for electrons, phosphate, and phosphorylated energy or glycolytic intermediates repressed oxidative phosphorylation in the mitochondria and stimulated germination. This generation of internal reducing power or change in the energy flow corresponds to some degree with the Nickerson theory of morphogenesis in *C. albicans.* The increased production of ethanol and CO_2, together with the reduced consumption of oxygen by germinating cells, suggests a sudden change from aerobic to anaerobic metabolism of glucose. The comprehensive study of Land and his co-workers provides a basis for correlating a great many different, and at times seemingly contradictory observations in factors that sponsor germination [51,52].

Chaffin and Sogin [19] presented evidence, derived from a study of yeast cell size, which indicated that germination of blastoconidia could be induced only in stationary phase cells incubated at 37°C in a defined medium consisting of amino acids. Budding cells in a logarithmic phase of growth did not germinate when incubated at 37°C in the same medium. Soll and Bedell [96,97] expanded the observations of Chaffin and Sogin and have established that a "commitment point" is passed by cells growing at 25°C, after which they cannot be induced to form germ tubes by incubation at 37°C until they again enter stationary phase. However, Mattia and Cassone [69] have shown that actively budding cells will germinate when incubated with certain inducers such as serum. Results from our laboratory (unpublished) confirm the observation of both groups of investigators, but it should be noted that a complex medium such as serum may have effects other than induction of germination. For example, serum may inhibit bud formation and thus evoke stationary-phase conditions in the yeast cells, which then respond to the effects of amino acids or other inducers in the medium [25]. The factors involved in germination are diverse and much needs yet to be done to sort out seemingly conflicting observations.

Recent findings [80,81] on the ploidy of *C. albicans* suggest that further work on elucidation of germination regulatory mechanism by production of appropriate mutants will be difficult. Contrary to the general supposition that most laboratory strains and clinical isolates of *C. albicans* are haploid and that appropriate manipulation would provide useful mutations, it has been suggested that most strains are in fact permanent diploids and that therefore the hope for mutants will be difficult to realize [80,81]. Three putative haploid variants studied in our laboratory did not germinate in serum incubated at 37°C (unpublished observation). Thus, the opportunity of exploring the molecular basis of control of dimorphism in *Candida* is, for the time being, frustrated by the uncertain genomic complement of the strains widely available for study.

III. CONCLUSIONS

In conclusion, much has been learned about the biochemistry, physiology, and morphology of the germination of blastoconidia of *C. albicans.* Several hypothesis have been proposed to describe the molecular basis of the event. But a complete description of the control of yeast cell germination remains to be formulated.

Table 3 Effect of Various Electron Donors, Acceptors, or Electron Transport Uncouplers on Morphology and Oxygen Consumption of *Candida albicans*

		NH_4Cl medium		Proline medium	
Category of compound	Metabolite or inhibitor[a]	Percentage filamentation[b]	μl of O_2 consumed $(t_0:t_{120})^c$	Percentage filamentation[b]	μl of O_2 consumed $(t_0:t_{120})^c$
Tricarboxylic acid intermediate or glycolytic products[d]	None	0	0.5:0.5	92	1.3:0
	Ethanol	1	0.5:0.4	6	0.8:0.4
	Acetate	2	0.8:0.4	91	0.8:0
	Citrate	42	0.8:0	87	0.2:0
	α-Ketoglutarate	61	0.2:0	41	0.7:0.1
	Ascorbate	29	0.7:0.1	45	0.9:0.1
	Succinate	56	1.2:0.1	49	0.2:0.1
Glycolysis inhibitors[e]	Iodoacetate	0	0.5:0.2	3	1.2:0.4
	2-DG	1	0.9:0.4	6	0.5:0.2
Electron donors or acceptors[e]	CoASH	82	1.8:0	93	1.8:0
	NAD	62	0.5:0	89	0.5:0
	$NADH_2$	62	0.9:0	92	1.2:0
	NADP	78	1.8:0	98	1.6:0
	FAD	56	1.1:0	97	1.5:0

	Glutathione-SH	44	0.9:0	94	2.0:0
	Methylene blue	0	0.2:0.2	9	0.1:0.1
Nucleotides[e]	AMP	55	0.5:0	99	0.6:0
	ADP	66	0.6:0	100	0.9:0
	ATP	7	0.5:0.3	42	0.9:0.5
	Cyclic AMP	61	0:0	100	0.4:0
Metabolic inhibitors[e]	KCN	35	0:0	62	0:0
	Antimycin A	0	0.5:0.4	92	1.2:0
	Amytal	44	0.5:0	100	1.7:0
	Quinacrine	0	0:0	0	0:0
	Rotenone	52	0.7:0	100	1.3:0
	Dicumarol	89	0.9:0	100	0.9:0
	HOQNO	34	0.6:0	98	1.2:0
	DNP	46	0:0	97	0:0

[a]CoASH, coenzyme A sulfhydryl; NAD, nicotinamide adenine dinucleotide; $NADH_2$, nicotinamide adenine dinucleotide, reduced form; NADP, nicotinamide adenine dinucleotide phosphate; FAD, flavine adenine dinucleotide; AMP, adenosine 5′-monophosphate; ADP, adenosine 5′-diphosphate; ATP, adenosine 5′-triphosphate; DNP, 2,4-dinitrophenol; 2-DG, 2-deoxyglucose; KCN, potassium cyanide.

[b]Determined after 6 h of growth.

[c]Ratio of the number of microliters of O_2 consumed at 2.5 min in culture compared to that consumed at 120 min.

[d]Glucose replaced in the medium with intermediates at 10^{-2} M concentrations.

[e]Concentration (100 μg/ml).

Source: Ref. 52.

REFERENCES

1. Auger, P., and J. Joly. 1977. Factors influencing germ tube production in *Candida albicans.* Mycopathologia *61*: 183-186.
2. Baker, R. D. (Ed.). 1971. *Human Infections with Fungi Actinomycetes and Algae,* Springer-Verlag, New York.
3. Balish, E. 1973. Methionine biosynthesis and S-adenosylmethionine degradation during an induced morphogenesis of *Candida albicans.* J. Gen. Microbiol. *19*: 847-853.
4. Balish, E., and G. Swihla. 1966. Ultraviolet microscopy of *Candida albicans.* J. Bacteriol. *92*: 1812-1820.
5. Barash, I. 1968. Liberation of polygalacturonase during spore germination of *Geotrichum candidum.* Phytopathology *58*: 1364-1371.
6. Barash, I., M. L. Conway, and D. H. Howard. 1967. Carbon catabolism and synthesis of macromolecules during spore germination in *Microsporum gypseum.* J. Bacteriol. *93*: 656-662.
7. Barlow, A. J. E., T. Aldersley, and F. W. Chattaway. 1974. Factors present in serum and seminal plasma which promote germ tube formation and mycelial growth of *Candida albicans.* J. Gen. Microbiol. *82*: 261-272.
8. Bedell, G. W., and D. R. Soll. 1979. Effects of low concentrations of zinc on the growth and dimorphism of *Candida albicans;* evidence for zinc resistant and sensitive pathways for mycelium formation. Infect. Immun. *26*: 348-354.
9. Beheshti, F., A. G. Smith, and G. W. Krause. 1975. Germ tube and chlamydospore formation by *Candida albicans* on a new medium. J. Clin. Microbiol. *2*: 345-348.
10. Bhattacharya, A., and A. Datta. 1977. Effect of cyclic AMP on RNA and protein synthesis in *Candida albicans.* Biochem. Biophys. Res. Commun. *77*: 1438-1444.
11. Bhattacharya, A., M. Puri, and A. Datta. 1974. Induction of N-acetylglucosamine kinase in yeast. Biochem. J. *141*: 593-595.
12. Boguslawski, G., and D. A. Stettler. 1979. Aspects of physiology of *Histoplasma capsulatum* (review). Mycopathologia *67*: 17-24.
13. Boguslawski, G., J. M. Akagi, and L. G. Ward. 1976. Possible role for cysteine biosynthesis in conversion from mycelial to yeast form of *Histoplasma capsulatum.* Nature (Lond.) *261*: 336-338.
14. Brambl, R., L. D. Dunkle, and J. L. van Etten. 1978. Nucleic acid and protein synthesis during fungal spore germination. In J. E. Smith and D. R. Berry (Eds.), *The Filamentous Fungi,* Vol. III: *Developmental Mycology,* Wiley, New York, pp. 94-118.
15. Braun, P. C., and R. A. Calderone. 1978. Chitin synthesis in *Candida albicans:* comparison of yeast and hyphal forms. J. Bacteriol. *133*: 1472-1477.
16. Braun, P. C., and R. A. Calderone. 1979. Regulation and solubilisation of *Candida albicans* chitin synthetase. J. Bacteriol. *140*: 666-670.
17. Buckley, H. R., and N. van Uden. 1963. The identification of *Candida albicans* within two hours by the use of an egg white slide preparation. Sabouraudia *2*: 205-208.
18. Cassone, A., N. Simonetti, and V. Strippoli. 1973. Ultrastructural changes in the wall during germ tube formation from blastospores of *Candida albicans.* J. Gen. Microbiol. *77*: 417-426.
19. Chaffin, W. L., and S. J. Sogin. 1976. Germ tube formation from zonal rotor fraction of *Candida albicans.* J. Bacteriol. *126*: 771-776.
20. Chattaway, F. W., M. R. Holmes, and A. J. E. Barlow. 1968. Cell wall composition of the mycelial and blastospore forms of *Candida albicans.* J. Gen. Microbiol. *51*: 367-376.

21. Chattaway, F. W., S. Shenolikar, and A. J. E. Barlow. 1974. Release of acid phosphatase and polysaccharide and protein containing components of the mycelial and blastospore forms of *Candida albicans* by treatment with dithiothreitol. J. Gen. Microbiol. *83*: 423-426.
22. Chattaway, F. W., S. Shenolikar, J. O'Rilley, and A. J. E. Barlow. 1976. Changes in the cell surface of the dimorphic forms of *Candida albicans* by treatment with hydrolytic enzymes. J. Gen. Microbiol. *95*: 335-347.
23. Chick, E. W., A. Balows, and M. L. Furcolow (Eds.). 1975. *Opportunistic Fungal Infections*, Thomas, Springfield, Ill.
24. Conant, N. F. 1941. A cultural study of the life cycle of *Histoplasma capsulatum* Darling 1906. J. Bacteriol. *101*: 563-579.
25. Dabrowa, N. 1968. Dimorphism in *Candida albicans.* Ph.D. dissertation, University of California, Los Angeles.
26. Dabrowa, N., S. S. Taxer, and D. H. Howard. 1976. Germination of *Candida albicans* induced by proline. Infect. Immun. *13*: 830-835.
27. Dabrowa, N., D. H. Howard, J. W. Landau, and Y. Shechter. 1970. Synthesis of nucleic acids and proteins in the dimorphic forms of *Candida albicans.* Sabouraudia *8*: 163-169.
28. Dill, B. C., T. J. Leighton, and J. J. Stock. 1972. Physiological and biochemical changes associated with macroconidial germination in *Microsporum gypseum.* Appl. Microbiol. *24*: 975-977.
29. Dolan, C. T., and D. M. Ihrke. 1971. Further studies on the germ tube test for *Candida albicans* identification. Am. J. Clin. Pathol. *55*: 733-734.
30. Evans, E. G. V., F. C. Odds, and K. T. Holland. 1975. Resistance of the *Candida albicans* filamentous cycle to environmental change. Sabouraudia *13*: 231-238.
30a. Evans, E. G. V., F. C. Odds, M. D. Richardson, and K. T. Holland. 1974. The effect of growth medium on filament production in *Candida albicans.* Sabouraudia *12*: 112-119.
31. Evans, E. G. V., F. C. Odds, M. D. Richardson, and K. T. Holland, 1975. Optimum conditions for initiation of filamentation in *Candida albicans.* Can. J. Microbiol. *21*: 338-342.
32. Fahlberg, W. J., C. D. Dukes, and R. K. Guthrie. 1957. Rapid classification of *Candida (Monilia) albicans.* J. Invest. Dermatol. *29*: 111-119.
32a. Garcia, R. J., W. H. Hiatt, J. Peters, and P. S. Sypherd. 1980. S-adenosylmethionine levels and protein methylation during morphogenesis of *Mucor racemosus.* J. Bacteriol. *142*: 196-201.
33. Garrison, R. G., and J. W. Lane. 1971. Yeast-like to mycelial phase transformation of *Histoplasma capsulatum* as observed by scanning electron microscopy. Mycopathol. Mycol. Appl. *43*: 185-193.
34. Garrison, R. G., K. S. Boyd, and F. Mariat. 1975. Ultrastructural studies of the mycelial to yeast transformation of *Sporothrix schenckii.* J. Bacteriol. *124*: 959-968.
35. Garrison, R. G., J. W. Lane, and M. F. Field. 1970. Ultrastructural changes during yeast-like to mycelial phase conversion of *Blastomyces dermatitidis* and *Histoplasma capsulatum.* J. Bacteriol. *101*: 628-635.
36. Gentles, J. C., and C. J. La Touche. 1969. Yeasts as human and animal pathogens. In A. H. Rose and J. C. Harrison (Eds.), *The Yeasts,* Vol. 1, Academic, New York, pp. 108-182.
37. Gottlieb, D. 1978. *The Germination of Fungus Spore,* Meadowfield, Durham, England.
38. Gupta, R. K., and D. H. Howard. 1971. Comparative physiological studies of the

yeast and mycelial forms of *Histoplasma capsulatum:* uptake and incorporation of L-leucine. J. Bacteriol. *105*: 690-700.
39. Hashimoto, T., C. D. R. Wu, and H. J. Blumenthal. 1972. Characterization of L-leucine induced germination of *Trichophyton mentagrophytes* microconidia. J. Bacteriol. *112*: 967-976.
40. Hazen, K. C., and J. E. Cutler. 1979. Autoregulation of germ tube formation by *Candida albicans.* Infect. Immun. *24*: 661-666.
41. Howard, D. H. 1962. The morphogenesis of the parasitic forms of dimorphic fungi. Mycopathol. Mycol. Appl. *18*: 127-139.
42. Howard, D. H., N. Dabrowa, V. Otto, and J. Rhodes. 1980. Cysteine transport and sulphite reductase activity in germination defective mutant of *Histoplasma capsulatum.* J. Bacteriol. *141*: 417-421.
43. Hu, F., C. S. Livingood, P. Johnson, and C. M. Pomerat. 1954. Tissue culture studies on human skin. Arch. Dermatol. *17*: 1-15.
44. Jacobson, E., and A. C. Harrell. 1980. Adaptation of yeast phase of *Histoplasma capsulatum* to a very simple medium. Abstr. Annu. Meet. Am. Soc. Microbiol., 1980, F81, p. 333.
45. Jayakumars, S., M. Singh, and R. Prasad. 1978. Characteristics of proline transport system in normal and starved cells of *Candida albicans.* Biochim. Biophys. Acta. *514*: 345-355.
46. Johnson, S. A. M., M. G. Guzmon, C. T. Agurlera. 1954. *Candida (Monilia) albicans.* Effect of amino acids, glucose, pH, chlortetracycline (aureomycin), dibasic sodium and calcium phosphates, and anaerobic and aerobic conditions on its growth. Arch. Dermatol. *70*: 49-60.
47. Kanetsuna, F., and L. M. Carbonell. 1971. Cell wall composition of the yeast-like and mycelial forms of *Blastomyces dermatitidis.* J. Bacteriol. *106*: 946-948.
48. Kanetsuna, F., L. M. Carbonell, I. Azuma, and Y. Yamamura. 1972. Biochemical studies on the thermal dimorphism of *Paracoccidioides brasiliensis.* J. Bacteriol. *110*: 208-218.
49. Knight, R. H., B. A. Body, G. S. Kobayashi, and G. Medoff. 1980. Balanced growth and morphogenesis of *Histoplasma capsulatum* in a defined synthetic medium. Sabouraudia *18*: 39-50.
50. Kumar, B. V., R. A. McMillan, G. Medoff, M. Gutwein, and G. Kobayashi. 1980. Comparison of the ribonucleic acid polymerase from both phases of *Histoplasma capsulatum.* Biochemistry *19*: 1080-1087.
51. Land, G. A., W. C. McDonald, R. L. Stjernholm, and L. Friedman. 1975. Factors affecting filamentation in *Candida albicans:* relationship of the uptake and distribution of proline to morphogenesis. Infect. Immun. *11*: 1014-1023.
52. Land, G. A., W. C. McDonald, R. L. Stjernholm, and L. Friedman. 1975. Factors affecting filamentation in *Candida albicans*: changes in respiratory activity of *Candida albicans* during filamentation. Infect. Immun. *12*: 119-127.
53. Landau. J. W., N. Dabrowa, and V. D. Newcomer. 1965. The rapid formation in serum of filaments by *Candida albicans.* J. Invest. Dermatol. *44*: 174-179.
54. Landau, J. W., N. Dabrowa, V. D. Newcomer, and J. R. Rowe. 1964. The relationship of serum transferrin and iron to the rapid formation of germ tubes by *Candida albicans.* J. Invest. Dermatol. *43*: 473-482.
55. Lee, K. L., H. R. Buckley, and C. C. Campbell. 1975. An amino acid liquid synthetic medium for the development of mycelial and yeast forms of *Candida albicans.* Sabouraudia *13*: 148-153.
56. Levine, S., and Z. J. Ordal. 1946. Factors influencing the morphology of *Blastomyces dermatitidis.* J. Bacteriol. *52*: 687-694.

57. Lingappa, B. T., M. Prasad, Y. Lingappa, D. F. Hunt, and K. Biermann. 1969. Phenethyl alcohol and tryptophol: autoinhibitors produced by the fungus *Candida albicans.* Science *163*: 192-194.
58. Lodder, J. (Ed.). 1970). *The Yeasts. A Taxonomic Study,* North-Holland, Amsterdam.
59. McClary, D. O. 1952. Factors affecting the morphology of *Candida albicans.* Ann. Mo. Bot. Gard. *39*: 137-164.
60. Mackenzie, D. W. R. 1962. Serum tube identification of *Candida albicans.* J. Clin. Pathol. *15*: 563-565.
61. Manning, M., and T. G. Mitchell. 1980. Analysis of cytoplasmic antigens of the yeast and mycelial phases of *Candida albicans* by two dimensional electrophoresis. Infect. Immun. *30*: 484-495.
62. Manning, M., and T. M. Mitchell. 1980. Morphogenesis of *Candida albicans* and cytoplasmic proteins associated with differences in morphology strain or temperature. J. Bacteriol. *144*: 258-273.
63. Mardon, D. N. 1973. In vivo synthesis of sulfur containing amino acids in *Candida albicans.* Can. J. Microbiol. *19*: 155-161.
64. Mardon, D., A. W. Balish, and A. W. Phillips. 1969. Control of dimorphism in a biochemical variant of *Candida albicans.* J. Bacteriol. *100*: 701-707.
65. Mardon, D. N., S. K. Hurst, and E. Balish. 1971. Germ tube production by *Candida albicans* in minimal liquid culture media. Can. J. Microbiol. *17*: 851-856.
66. Maresca, B., B. V. Kumar, J. Medoff, G. Medoff, and G. S. Kobayashi. 1980. Studies on the dimorphism in *Histoplasma capsulatum:* biochemical changes during a differentiation process. Zentralbl. Bakteriol. Parasitenkd., Infektionskr. Hyg., Abt. 1, Suppl. *8*: 17-22.
67. Marriot, M. S. 1975. Isolation and chemical characterization of plasma membranes from the yeast and mycelial forms of *Candida albicans.* J. Gen. Microbiol. *86*: 115-132.
68. Marriot, M. S. 1975. Enzymatic activity of purified plasma membranes from the yeast and mycelial forms of *Candida albicans.* J. Gen. Microbiol. *89*: 345-352.
69. Mattia, E., and A. Cassone. 1979. Inducibility of germ tube formation in *Candida albicans* at different phases of yeast growth. J. Gen. Microbiol. *113*: 439-442.
70. Mitchell, L. H., and D. R. Soll. 1979. Commitment to germ tube formation during release from stationary phase in *Candida albicans.* Exp. Cell Res. *120*: 167-179.
71. Newton, A., E. D. Shaw, C. S. Goode, and F. O. Rigall. 1960. Identification of *Candida albicans.* J. Invest. Dermatol. *35*: 123.
72. Nickerson, W. J. 1948. Enzymatic control of cell division in microorganisms. Nature (Lond.) *162*: 241-245.
73. Nickerson, W. J., and C. W. Chung. 1954. Genetic block in the cellular division mechanism of a morphological mutant of a yeast. Am. J. Bot. *41*: 114-120.
74. Nickerson, W. J., and G. Falcone. 1956. Identification of protein disulfide reductase as a cellular division enzyme in yeasts. Science *124*: 722-723.
75. Nickerson, W. J., and G. Falcone. 1959. Function of protein disulfide reductase in cellular division of yeasts. In R. Benesch (Ed.), *Sulfur in Proteins,* Academic, New York, pp. 409-424.
76. Nimi, M., K. Nimi, J. Tokunaga, and H. Nakayama. 1980. Changes in cyclic nucleotide levels and dimorphic transition in *Candida albicans.* J. Bacteriol. *142*: 1010-1014.
77. Nishioka, Y., and M. Silva-Hutner. 1974. Dimorphism sensitivity to nystatin and acriflavin uptake in a strain of *Candida albicans* grown with glutamate as sole nitrogen and carbon source. Sabouraudia *12*: 295-301.

78. Odds, F. C. 1979. *Candida and Candidosis,* University Park Press, Baltimore.
79. Ogletree, F. F., A. T. Abdelal, and D. G. Ahearn. 1978. Germ tube formation by atypical strains of *Candida albicans.* Antonie Leeuwenhoek J. Microbiol. Serol. *44*: 15-24.
80. Olaiya, A. F., and S. J. Sogin. 1979. Ploidy determination of *Candida albicans.* J. Bacteriol. *140*: 1043-1049.
81. Olaiya, A. F., J. R. Steed, and S. J. Sogin. 1980. Deoxyribonucleic acid deficient strains of *Candida albicans.* J. Bacteriol. *141*: 1284-1290.
82. Orlowski, M. 1980. Cyclic adenosine 3′,5′-monophosphate and germination of sporangiospores from the fungus *Mucor.* Arch. Microbiol. *126*: 133-140.
83. Reynolds, R., and A. Braude. 1956. The filament inducing property of blood for *Candida albicans:* its nature and significance. Clin. Res. Proc. *4*: 40.
84. Ridley, M. F. 1960. A comparison of methods for identification of *Candida albicans.* Aust. J. Dermatol. *5*: 209-213.
85. Rippon, J. W. 1980. Dimorphism in pathogenic fungi. Crit. Rev. Microbiol. *8*: 49-97.
86. Roy, I., and J. W. Landau. 1971. Polymorphism of malate dehydrogenase in cytoplasmic and mitochondrial fractions of *Blastomyces dermatitidis.* Sabouraudia *10*: 107-110.
87. San-Blas, F., G. San-Blas, and D. Inlow. 1980. Dimorphism in *Paracoccidioides brasiliensis.* Zentralbl. Bakteriol. Parasitenkd. Infektionskr. Hyg., Abt. 1, Suppl. *8*: 28-78.
88. San-Blas, G. 1979. Biosynthesis of glucans by subcellular fractions of *Paracoccidioides brasiliensis.* Exp. Mycol. *3*: 249-258.
89. San-Blas, G., and F. San-Blas. 1977. *Paracoccidioides brasiliensis:* cell wall, structure and virulence. Mycopathologia *62*: 77-86.
90. Scherwitz, C., R. Martin, and H. Ueberberg. 1978. Ultrastructural investigations of the formation of *Candida albicans* germ tubes and septa. Sabouraudia *16*: 115-124.
91. Schmit, J. C., and S. Brody. 1976. Biochemical genetics of *Neurospora crassa* conidial germination. Bacteriol. Rev. *40*: 1-41.
92. Sherr, G. H., and P. H. Weaver. 1953. The dimorphism phenomenon in yeasts. Bacteriol. Rev. *17*: 51-92.
93. Simonetti, N., V. Strippoli, and A. Cassone. 1974. Yeast-mycelial conversion induced by N-acetyl glucosamine in *Candida albicans.* Nature (Lond.) *250*: 344-346.
94. Skinner, C. E. 1947. The yeast like fungi *Candida* and *Brettanomyces.* Bacteriol. Rev. *11*: 227-274.
95. Skinner, C. E., and D. W. Fletcher. 1960. A review of the genus *Candida.* Bacteriol. Rev. *24*: 397-416.
96. Soll, D. R., and G. W. Bedell. 1978. Bud formation and inducibility of pseudomycelium outgrowth during release from stationary phase in *Candida albicans.* J. Gen. Microbiol. *108*: 173-180.
97. Soll, D. R., M. Stassi, and G. W. Bedell. 1978. The regulation of nuclear migration and division during pseudomycelium outgrowth in the dimorphic yeast *Candida albicans.* Exp. Cell Res. *116*: 207-215.
98. Stettler, D. A., and G. Boguslawski. 1979. Cysteine biosynthesis in a fungus, *Histoplasma capsulatum.* Sabouraudia *17*: 23-34.
99. Tang, S. L., D. H. Howard, and R. K. Gupta. 1978. Uptake of metabolites by germinating blastospores of *Histoplasma capsulatum.* Sabouraudia *16*: 271-278.
100. Taschdjian, C. L., and P. J. Kozinn. 1961. Metabolic studies of the tissue phase of *Candida albicans* involved in vitro. Sabouraudia. *1*: 73-82.
101. Taschdjian, C. L., J. J. Burchale, and P. J. Kozinn. 1960. Rapid identification of

Candida albicans by filamentation on serum and serum substitutes. Am. J. Dis. Child. *99*: 212-215.

102. Taylor, J. J. 1961. Nucleic acid and dimorphism in *Blastomyces*. Exp. Cell. Res. *24*: 155-158.

103. Ton, S. H., and M. S. Kaminairatnam. 1976. Activities of some carbohydrate metabolising enzymes in both the yeast like and mycelial like forms of *Candida albicans*. Southeast Asian J. Trop. Med. Public Health *7*: 72-76.

104. Wain, W. H., A. R. Brayton, and R. A. Cawson. 1974. Variation in the response to N-acetyl glucosamine by isolates of *Candida albicans*. Mycopathologia *58*: 27-29.

105. Wain, W. H., M. F. Price, A. R. Brayton, and R. A. Cawson. 1976. Macromolecular syntheses during the cell cycles of yeast and hyphal phases of *Candida albicans*. J. Gen. Microbiol. *97*: 211-217.

106. Weber, D. J., and W. M. Hess (Eds.). 1976. *The Fungal Spore*, Wiley, New York.

107. Wickerham, L. J., and E. F. Rettinger. 1939. A taxonomic study of *Monilia albicans* with special emphasis on morphology and morphological variation. J. Trop. Med. Hyg. *42*: 174-216.

108. Yamaguchi, H. 1974. Dimorphism in *Candida albicans*. I. Morphology dependent changes in cellular contents of macromolecules and respiratory activity. J. Gen. Appl. Microbiol. *20*: 87-99.

109. Yamaguchi, H. 1974. Effect of biotin insufficiency on composition and structure of cell wall of *Candida albicans* in relation to mycelial morphogenesis. J. Gen. Appl. Microbiol. *20*: 217-228.

110. Yamaguchi, H. 1974. Mycelial development and chemical alteration of *Candida albicans* from biotin insufficiency. Sabouraudia *12*: 320-328.

111. Yamaguchi, H. 1979. Control of dimorphism in *Candida albicans* by zinc: effect on cell morphology and composition. J. Gen. Microbiol. *85*: 370-372.

112. Yamaguchi, H., Y. Kanda, and M. Osumi. 1974. Dimorphism in *Candida albicans*. II. Comparison of fine structure of yeast like and filamentous phase of growth. J. Gen. Appl. Microbiol. *20*: 101-110.

113. Yen, C. M., and D. H. Howard. 1970. Germination of blastospores of *Histoplasma capsulatum*. Sabouraudia *8*: 242-252.

13
Metabolite Transport

Rishab K. Gupta / UCLA School of Medicine, Los Angeles, California

Dexter H. Howard / UCLA School of Medicine, Los Angeles, California

I. INTRODUCTION

Transport of metabolites by fungi has been studied for several decades, but relatively little has been reported about the process in pathogenic fungi. Most of the work on transport has been done with *Saccharomyces cerevisiae* [35,46,65,106,113,191,208], and *Neurospora crassa* [51,121,142,162,201,202,206,207,251], or other filamentous fungi [3,4, 19,33,43,77,93,94,101,105,151,181,209,214,227,234,248,262]. The only pathogenic fungi that have been studied with any intensity are *Histoplasma capsulatum* [63,76,90, 236], *Candida albicans* [224,244,261], *Blastomyces dermatitidis* [12] and some of the agents of aspergillosis [139,167,168,184,188,189]. This dearth of observations is certainly not because of a lack of enthusiasm on the part of medical mycologists, but rather is due to the slow development of methods whereby pathogenic fungi can be handled safely in the experiments required to perform the transport studies.

II. METHODS OF STUDY

Transport of metabolites into cells of pathogenic fungi has been studied in several ways. Of particular importance are flux measurements, kinetic and competitive inhibition studies, and use of mutants.

A. Flux Measurement

For the measurement of fluxes, the cells, harvested at appropriate stages of the growth cycle, are washed and resuspended in appropriate medium containing the solute. Samples of the suspension are withdrawn at predetermined times and filtered through a membrane filter (0.45 μm). The filtrate and/or the cells retained on the membrane are analyzed for the amount of solute. A change in the amount of the solute in the specimens compared to a zero-time sample is a measure of the influx of the solute. Obviously, the rate of efflux of solutes (movement from cells to the environment) may also be measured in this manner.

The amount of solute can be determined (1) by analytical methods, (2) by use of radioisotopes, or (3) by electrical measurements. The transport rates or flux measurements of metabolites are generally denoted in terms of moles per unit dry weight and time rather than per unit area. This is because of difficulties involved in assessing the cellular surface area.

1. Analytical Methods

These methods are generally used for flux measurements of ions and include: (1) flame photometry (e.g., sodium and potassium); (2) atomic absorption spectrophotometry (e.g., calcium, rubidium); (3) chemical reduction (e.g., nitrate); and (4) colorimetry (e.g., nitrite, phosphate, ammonia, etc.). Methods such as these are often cumbersome and time consuming.

2. Radioisotopic Methods

The use of isotopically labeled substrates has revolutionized the field of biochemistry and physiology, and has fostered studies on the transport of metabolites. This is because of the simplicity of detection and quantitation of radioisotopes. Appropriately prepared cells are suspended in radiolabeled substrate and the amount of label removed from a filtrate or the amount accumulated by the cells is a measure of uptake.

For efflux measurements the procedure outlined above is reversed. The cells are pretreated with radioactive solute and resuspended in a medium supplemented with nonradioactive solute. Analysis of supernatant or membrane filtrate at sequential time intervals for radioactivity provides the information about efflux rate of the solute which occurs by exchange with external nonradioactive solute. If the cells are resuspended in a solute-free medium, any release of radioactivity would be due to leakage of the solute from the cells.

3. Electrical Measurements

In a few experiments, electrodes have been inserted into hyphal cells for electrical measurements. Electrogenic pumps operative in the transport of ions have been demonstrated in a few saprophytic fungi by means of this novel technique of electrode insertion [43-45, 215-218].

B. Kinetics and Competitive Inhibition of Carrier-Mediated Transport

Studies such as these are performed to elucidate the specificity of transport systems. The uptake of a metabolite is studied in the presence of other substances. If these other substances are able to decrease the initial rate of uptake of the metabolite under study, they are considered to be transported by the same system. However, it may be that the competitive substances are actually transported by their own specific systems, but under experimental conditions they may also enter the cell by the transport system designed for the metabolite under study. The competitive inhibition studies are based on the assumption that the transport or carrier systems behave like enzymes and express specificity with respect to the metabolite. This assumption is based on the fact that carrier-mediated transport of metabolites is pH and temperature dependent.

Changes in pH can modify the ionizable groups of a metabolite. Alternatively, pH may affect the site or mechanism of metabolite transport rather than the metabolite itself. Moreover, these two possibilities are not mutually exclusive; i.e., pH can simultaneously influence ionization of metabolite and the cellular site concerned in the metabolite transport. Similarly, estimation of energy of activation of various transport systems made either graphically (Arrhenius plot) or by mathematical computation reveals them to be of the same magnitude as those of catalytic or enzyme-like systems. The energy of activation is computed with the following equation:

$$E_a = 2.303R\left(\frac{T_1 T_2}{T_2 - T_1}\right)\log\frac{V_2}{V_1}$$

where R is the gas constant expressed in cal deg^{-1} mol^{-1} (1.987); T_1 and T_2 are temperatures in degrees absolute; V_1 and V_2 are rates of uptake of the metabolite at T_1 and T_2, respectively; and E_a is the energy of activation.

The concept of an enzyme-like transport system includes a carrier system that binds the metabolite at the outer surface and releases it at the inner surface of the cell. Thus, the transport process can be described by Michaelis-Menten kinetics; i.e., a Michaelis-Menten constant (K_m) for the metabolite transport can be obtained by determining the influence of the external concentration of the metabolite on the rate of its uptake. If the transport process is carrier mediated, a plot of metabolite uptake rate as a function of external concentration results in a typical saturation curve as depicted in Fig. 1. Such a curve indicates that the metabolite uptake is limited to a finite number of sites (carrier sites) at the cell surface and the carrier system is of limited capacity. The linearity of a double reciprocal plot of the data shown in Fig. 1 by the method of Lineweaver-Burk [125] indicates the formation and dissociation of an intermediate complex (Fig. 2). If the curve bends toward the origin (Fig. 2), involvement of more than one transport or carrier systems is indicated at high external substrate concentration. In general, the higher the affinity of a metabolite for the carrier site, the smaller is the K_m value.

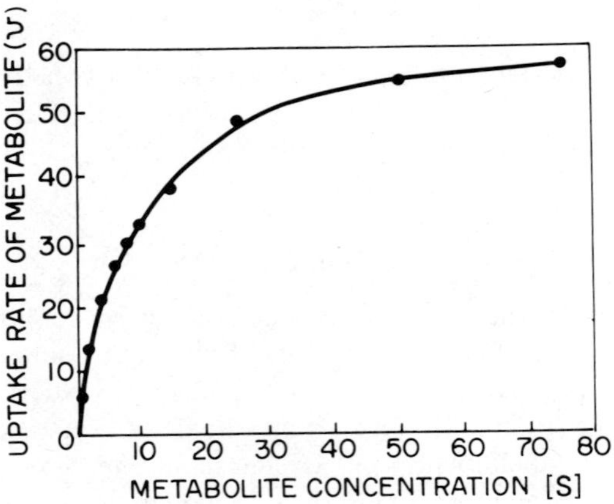

Figure 1 Representative plot of metabolite uptake rate as a function of external concentration. This plot indicates that the metabolite uptake is limited to a finite number of carrier sites and at high external metabolite concentration the carrier system becomes saturated. As a result, there is no net increase in uptake rate with increasing external metabolite concentration after reaching a saturation point. (From Ref. 77.)

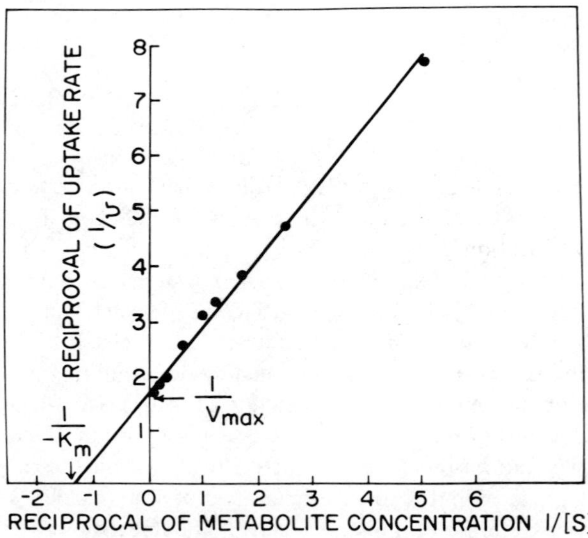

Figure 2 Representative double reciprocal plot of uptake rate and external metabolite concentration. (Data plotting method from Ref. 125; data from Ref. 77.)

The following mathematical relationship between the rate of reaction of an enzyme and the number and uniformity of binding sites that function at a locus for the reaction has been established:

$$\log \frac{v}{V_{max} - v} = n \log [S] - \log K$$

where v is the rate of reaction at substrate concentration [S], V_{max} is maximum rate of reaction reached when all binding sites are saturated, K is the product of the dissociation constant of binding sites with different affinities for the substrate, and n is the number of such sites. A plot of log $[v/(V_{max} - v)]$ as a function of log [S] would result in a straight line with a slope equal to n. This plot is known as a Hill plot, the slope of which is a function of the number of binding sites (Fig. 3). When a Hill plot does not result in a straight line, it indicates involvement of additional but different binding sites which become functional as the external concentration of the substrate increases. When all the binding sites are similar in their affinity for the substrate, the slope usually approximates unity. Since carrier-mediated metabolite transport follows enzyme-like kinetics, its characteristics by these mathematical formulations can be evaluated.

Differences in transport sites exist for the uptake of different metabolites, and these sites vary with different species. Some are highly specific, whereas others do not differentiate between large groups of metabolites. Kinetic analyses provide a rapid means of determining if the affinity of a binding site for the two isomers of a metabolite is equal or preferential.

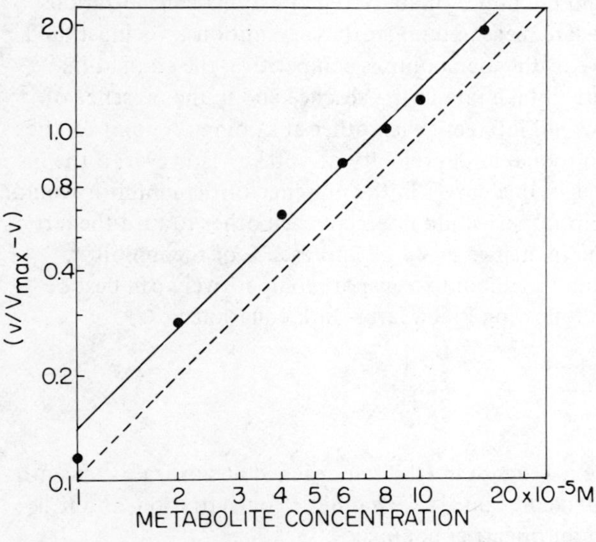

Figure 3 Representative Hill plot to determine the uniformity of carrier sites for a metabolite. Dashed line, theoretical; solid line experimental. v, uptake rate at a given metabolite concentration; V_{max}, maximum uptake rate attainable at high external metabolite concentration.

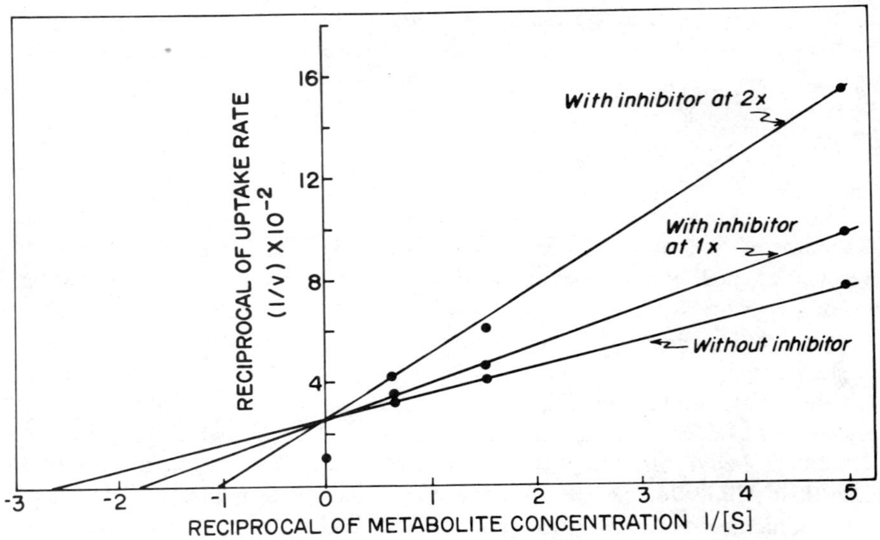

Figure 4 Double reciprocal plot of uptake rate as a function of external metabolite concentration in the absence and in the presence of various amounts of the inhibitor. (From Ref. 77.)

When a structural analog or other isomer of a metabolite is used as a competitive inhibitor, both the test metabolite and the inhibitor are tested at varying concentrations and uptake rates of the metabolites are measured under these conditions. As illustrated in Fig. 4, if the inhibition of uptake of the metabolite is competitive, the straight-line curves, each representing metabolite uptake rate in the absence and in the presence of various amounts of the inhibitor, would intercept each other at a common point on the vertical axis that would correspond to maximum velocity of uptake. However, if the inhibition is noncompetitive, the straight-line curves in the presence of the inhibitor would not intercept on the vertical axis but rather would intercept each other toward the left of the vertical line. This would be due to change in V_{max} in presence of the inhibitor.

The ability of substance to inhibit metabolite transport competitively can best be described by the value of K_i in the following Lineweaver-Burk equation:

$$\frac{1}{v} = \frac{K_m}{V_{max}\,[S]}\left(1 + \frac{[I]}{K_i}\right) + \frac{1}{V_{max}}$$

where v is uptake rate, V_{max} is the maximum uptake rate reached at substrate concentration [S], K_m is the Michaelis-Menten constant, [I] is inhibitor concentration, and K_i is the dissociation constant of the inhibitor-carrier complex.

As an alternative, when the V_{max} of metabolite transport is well established, one can determine K_i by observing the uptake rate of the metabolite at one single concentration in the presence of various amounts of the inhibitor. In such instances, the inverse of uptake rate of the metabolite is plotted as a function of the inhibitor concentration (Fig. 5). The

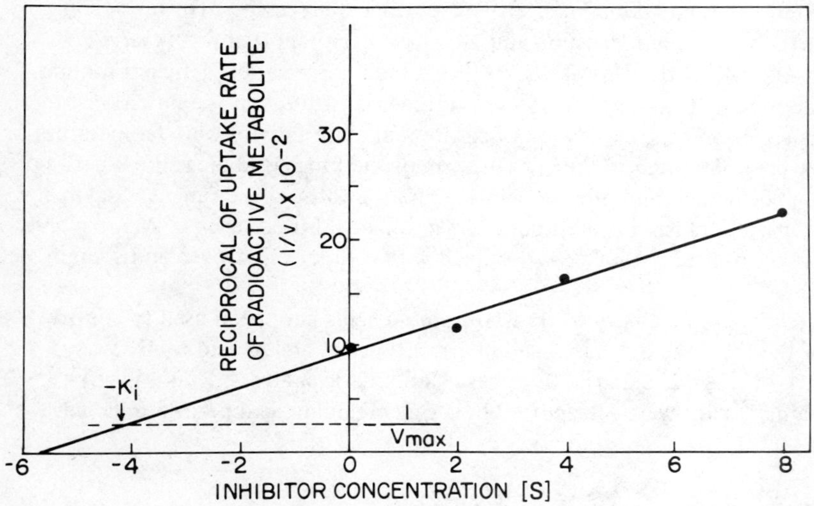

Figure 5 Representative Dixon plot to determine inhibition constant (K_i) of a competitor for transport system of the metabolite. (From Ref. 77.)

point of intersection of the straight line with one horizontally drawn at the height of $1/V_{max}$ is equal to $-K_i$.

As has been pointed out by Jennings [100], the only limitation, if any, of competition studies is that they focus attention on the one metabolite whose uptake is measured radioisotopically and only the effects of structurally related compounds on the uptake of the labeled metabolite are studied.

C. Use of Mutants in Carrier-Mediated Transport Studies

Carrier-mediated transport of metabolites is under genetic control, and the properties of enzyme-like carriers are determined by certain genetic loci [117,136,150,165,171,180]. A number of possible functions can be outlined for these genetic loci: (1) to control the synthesis of a specific protein (permease) for the transport of an individual metabolite, (2) to regulate the synthesis of a permease that is specific for one or more groups of metabolites, (3) to regulate the enzymes that are involved in the metabolism of the transported metabolite, and (4) to regulate enzymes that affect energy systems coupled with the metabolite transport. These genetic loci may contain regulatory genes that could control the level of one or more transport systems [240]. Thus, mutants unable to grow in the presence of a certain metabolite, but that contain all the enzymes necessary to metabolize the substance internally, may be considered as deficient in the transport system for that metabolite. The presence of a functional endogenous pathway of nicotinamide adenine dinucleotide (NAD) biosynthesis from tryptophan in a dimorphic fungus *Candida albicans* has been demonstrated by Chaffin et al. [36]. However, the organism is unable to utilize nicotinate, one of the biosynthetic precursors. The failure to utilize nicotinate is due to a defect in transport. Nonmetabolizable substrates can also be used to detect mutants defective in transport systems. Pateman and Kinghorn [166] have postulated that most mutants affecting transport in fungi are recessive. However, some dominant or partially dominant transport mutants have been reported for *Aspergillus nidulans* [110,213]. Recently, Howard et al. [90] used a germination-deficient mutant of the dimorphic fungus *Histoplasma capsulatum* to demonstrate that the nutritional requirement for cysteine and the expression of sulfite reductase activity were temperature dependent and were not related to the growth form of the fungus, whereas the kinetics of cysteine transport depended on both temperature and the form of the fungus. Slayman et al. [219,223] used a transport mutant and the normal wild-type strain of *Neurospora crassa* to show that potassium uptake was coupled with efflux of hydrogen and sodium ions.

Mutants defective in the ability to transport amino acids have been used to confirm the existence of the three (acidic, basic, and neutral) types of amino acid transport systems in fungi [97,110,187,238,254]. By genetic and kinetic analyses, Rytka [198] showed that D-amino acids were transported by a general amino acid permease in yeast.

III. HOW METABOLITES ENTER CELLS

Movement of a solute across the cell membrane will take place only under the influence of a driving force, and permeation can result from any of a variety of mechanisms. These mechanisms include (1) free diffusion, (2) diffusion through pores, and (3) carrier-mediated transport.

A. Free Diffusion

Because of biochemical composition of the membrane, metabolites of a lipid nature are partitioned from an aqueous environment and enter the cell under the influence of a concentration gradient. If no other forces intervene, the movement will cease when the concentration on both sides of the membrane is equal, i.e., the distribution of the substance has reached equilibrium. Thus, it appears that for this particular type of transport the cell membrane acts like a dialysis bag. As a result, the rate of diffusion obeys Fick's first law, i.e., the steady-state rate of diffusion is proportional to the concentration difference. Structural analogs do not compete with the metabolite, and the diffusion rate of such analogs is independent of one another or that of the metabolite. No stereospecificity is observed, i.e., the L isomer of the metabolite penetrates the membrane at the same rate as the D isomer. At equilibrium, the external concentration of the metabolite is equal to that of internal concentration.

B. Diffusion through Pores

Water and water-soluble materials move into cells through hypothetical channels or pores in the membrane, and are driven by a concentration gradient or an electrical potential or carried by the flow of water (solvent drag). The radius of these pores has been estimated to vary from 4 to 8 Å [164]. The limiting sizes of molecules that can penetrate the intact cell and protoplast of *Saccharomyces cerevisiae* have been determined by Scherrer et al. [203]. In the yeast protoplasts the diffusion-exclusion threshold is about 110 daltons and the Einstein-Stokes hydrodynamic radius is 0.42 nm. The rate of diffusion in this situation may be dependent on pH. For example, organic acids at a lower pH will enter the cell at a higher rate than they do at a higher pH.

C. Carrier-Mediated Transport

Water-soluble substances that do not enter cells through pores are assisted in their passage by carrier substances. The carrier itself does not provide the energy required for translocation. Energy is derived from any of several possible sources, and depending on the nature of the driving force, one distinguishes among (1) facilitated diffusion, (2) exchange diffusion, and (3) active transport.

1. Facilitated Diffusion

In this type of transport process, the movement of a metabolite is controlled by a concentration gradient, and the transportability of a metabolite is determined by the pore size of the cell membrane. The net movement of a metabolite ceases after reaching equilibrium. However, unlike free diffusion, facilitated diffusion exhibits stereospecificity and it is several times faster than free diffusion. The rate of diffusion may not be strictly proportional to the concentration difference, i.e., at exceedingly high external metabolite concentrations, the rate of transport may reach a limit. Thus, the process of facilitated diffusion is saturable. However, metabolic energy is not required to drive the process.

2. Exchange Diffusion

This form of transport is a specialized case in which metabolites are eventually translocated without the need of energy expenditure. Here, the influx or inward movement of a metabolite is coupled with the outward movement of another metabolite.

3. Active Transport

This is a carrier-mediated process which often proceeds against a chemical or electrical gradient. Active transport requires an expenditure of energy on the part of the cell. Such metabolic energy must be coupled to the transport process. The criteria for active transport are (1) substrate movement against a concentration gradient, (2) inhibition of substrate accumulation by inhibition of energy-yielding reactions, (3) the occurrence of saturation, and (4) competitive inhibition of transport by compounds that are related structurally.

As outlined in Fig. 6, during the process of active transport a metabolite, A, combines with a carrier, X, to form a complex AX, at the surface of the cell. The complex moves from the exterior to the interior surface of the cell membrane due to confirmational or other changes, and then dissociates, generating an inactive carrier, X′, and releasing the metabolite, A, to a metabolic pool. The overall process and the reactivation of the carrier (transformation of X′ to X) require energy. In effect, the conversion of adenosine triphosphate (ATP) to adenosine diphosphate (ADP) and inorganic phosphate, P_i, provides the force necessary to transport and accumulate the metabolite against a concentration gradient. Since inactive carrier, X′, cannot combine with the metabolite, and since transformation of X′ to X occurs on the outer surface of membrane, there is no loss of the pooled metabolite and the internal concentration may be 2.5×10^4 times greater than the external concentration [104].

D. Demonstration of Carrier or Transport Proteins

Genetic and biochemical approaches to a study of metabolite uptake in fungi have provided evidence that in the majority of cases, transport occurs by means of carriers which exhibit enzyme-like properties. These carriers (permeases) are constitutive in most cases;

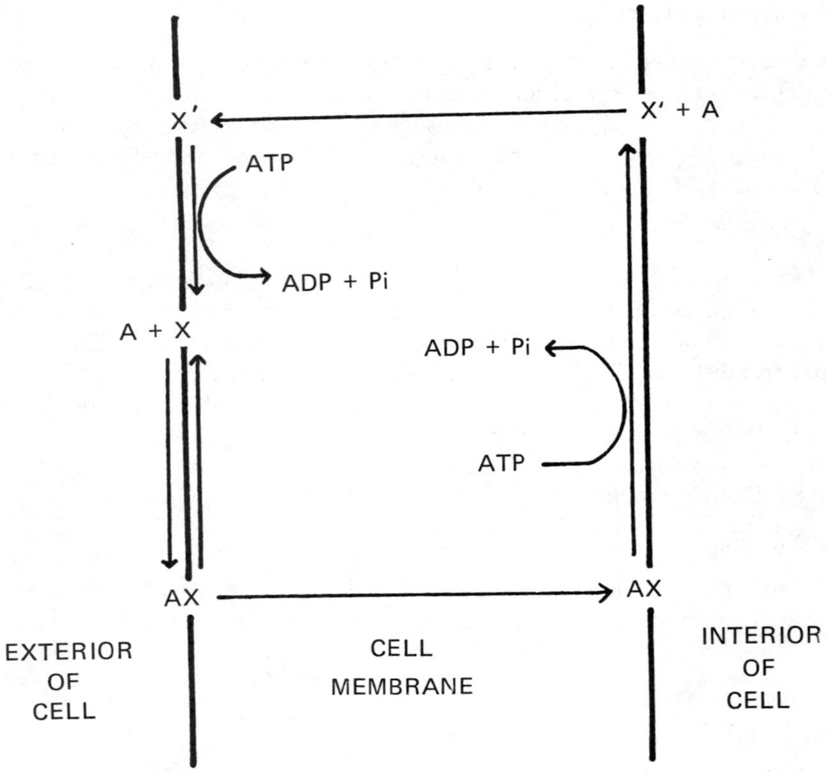

Figure 6 Hypothetical model for active transport of metabolite across fungal cell membrane. X, carrier; X′, inactive carrier; A, amino acid, AX, amino acid-carrier complex. (From Ref. 77.)

however, some are inducible. The binding proteins isolated from bacterial cells are considered to be components of the carrier-mediated transport systems for a variety of amino acids, sugars, and ions. So far the evidence that binding proteins are components of transport systems is indirect. Oxender, in his review [157], has summarized these indirect lines of evidence as follows:

1. The transport activity is associated with the cell membranes.
2. Treatment of cells with hypotonic solutions results in loss of transport activity and at the same time binding activity can be recovered from the hypotonic solution in which the cells were treated.
3. There is a parallelism between transport activity and the ability to synthesize the binding protein.
4. Binding protein and cellular transport systems exhibit similar kinetic constants.
5. Reversion of transport-negative and binding protein-deficient mutants to transport-positive mutants consistently results in appearance of binding protein activity.

Stimulation of amino acid transport in *Saccharomyces cerevisiae* by metabolic inhibitors has been proposed to be caused by retarded degradation of the transport proteins [88].

Wiley [250] has isolated and purified a tryptophan-binding protein from germinated conidia of *Neurospora crassa* by cold osmotic shock treatment. The cells lost 90% of their ability to transport tryptophan without losing viability. The dissociation constant of binding protein recovered from the shock fluid for tryptophan was quite comparable to the K_m value of tryptophan transport by the intact whole cells (8×10^{-5} M versus 5×10^{-5} M, respectively). The cells repressed for the tryptophan transport system yielded less transport protein into the shock fluid. The binding activity of the protein for various amino acids decreased in the following order: phenylalanine > tryptophan > leucine. This order was found to be consistent with the competitive order of tryptophan transport by the whole cells. Thus, it was concluded that the binding protein represented a component of the neutral amino acid transport system of *Neurospora crassa.*

Two glycoproteins with a relatively high affinity for L-arginine have been isolated from *Neurospora crassa* by extraction with potassium chloride [229]. This amino acid does not readily enter an artificial film of *Neurospora crassa* conidial lipids. However, addition of glycoprotein extracts increases the entry of arginine 1000-fold [230,232]. Stuart and DeBusk [231] have been able to purify one of the two glycoproteins. This purified glycoprotein contained eight different types of sugar residues and 113 amino acids per monomer.

Thiamine-binding proteins have been solubilized from the membrane of *Saccharomyces cerevisiae* by Iwashima et al. [96]. A methionine-binding protein has been partially purified from *Aspergillus nidulans* by affinity chromatography [227]. Bayer et al. [13] attempted to localize and quantitate the biotin transport components in yeast by tracing the fate of a radioactive affinity label, [^{14}C] biotinyl-p-nitrophenyl ester, through various stages of subcellular fractionation, and by use of a biotin-derivatized, impermeable, electron-dense, affinity-cytochemical label (ferritin-biotin conjugates) for subsequent visualization by electron microscopy. Values of approximately 8000 and 4000 transport sites per cell, respectively, were observed by the two methods.

IV. ION TRANSPORT IN FUNGI

Pathogenic fungi have not yet been examined with respect to ion transport. Most of our knowledge in this area comes from studies on filamentous and unicellular saprophytic fungi [29,103,140,141,146,210,222].

Fungi, like other microorganisms, discriminate between sodium and potassium. The net uptake of potassium is, in general, balanced by the outward flow of sodium and hydrogen ions. The process of potassium uptake is energy dependent and obeys saturation kinetics. Rubidium can partially substitute for potassium [1]. In yeast, the affinity series for various monovalent cations has been outlined as follows: $H^+ > K^+ > Rb^+ > Cs^+ > Na^+ > Li^+$ [196].

The transport systems for bivalent cations have been little studied. Fuhrmann and Rothstein [59] have advanced the following affinity series for bivalent cation transport system: $Mg^{2+}, Co^{2+}, Zn^{2+} > Mn^{2+}, Ni^{2+} > Ca^{2+} > Sr^{2+}$. It has been stated that each cation displayed a saturation phenomenon, and in pairs they competed with each other for the transport system. The K_m for Mg^{2+}, Co^{2+}, and Zn^{2+} was on the order of 10^{-5} M. For the uptake of each bivalent cation, two K^+ or two Na^+ ions were extruded. Thus, uptake of bivalent cations required electrical balance. Once taken up, the bivalent cations were nonexchangeable. Although monovalent cations such as K^+ did not compete with bivalent cations, Mg^{2+} could enter the cell via the K^+ transport system. However, the affinity of Mg^{2+} for the K^+ transport system was extremely low [41].

Rodriguez-Navarro and Sancho [186] reported that environmental Mg^{2+} influenced the K^+/Na^+ exchange rate in metabolizing yeast. A calcium transport system in *Achlya* has been shown to be composed of two components: a low molecular weight Ca^{2+}-binding glycopeptide present within the cell wall matrix and a Ca^{2+} uptake component associated with the cell membrane [122]. In *Saccharomyces cerevisiae,* uptake of Ca^{2+} and Sr^{2+} is energy dependent and shows a deviation from simple Michaelis-Menten kinetics [193]. The uptake rate is influenced by pH of the cell and of the medium. The inhibition of uptake of Ca^{2+} and Sr^{2+} by monovalent cations is independent of the bivalent cation concentration. The inhibition shows saturation kinetics. This inhibition by monovalent cations is, perhaps, related to depolarization of the cell membrane [193]. Stroobart and Scarborough [228] reported that the active transport of Ca^{2+} across the plasma membrane of *Neurospora* occurs via a Ca^{2+}/H^+ "antiporter," and this antiporter functions to pump Ca^{2+} out of the intact cell.

The uptake of Mn^{2+} by *Saccharomyces cerevisiae* is an active process which is dependent on the stage of culture. Okorokov et al. [156] reported that the uptake rate of Mn^{2+} is maximum during the midexponential growth phase of the yeast. Oligomycin at a concentration of 25 μg/ml inhibits the uptake rate by 50-85%.

Because of unavailability of iron for transport in an ionic form, microorganisms, including fungi, have evolved special mechanisms to transport this ion from the environment. Siderochromes, special molecules that chelate iron, are produced and released by the cells [34,55]. These ferric chelators are then transported into the cell by a special high-affinity system. In fungi, hydroxamates of the type of ferrichrome, ferrioxamine, and rhodotorulate have been shown to be produced [9,55,122,249]. A pronounced specificity of chelate-iron uptake has been observed among various species of *Aspergillus* [249].

Sulfate transport has been studied in *Neurospora, Aspergillus, Penicillium,* and yeast [26,129,143,148,241,262]. In many cases, the sulfate transport was energy dependent and showed saturation kinetics. However, in *Penicillium* and in a transport-negative mutant of *Aspergillus nidulans,* the uptake of sulfate was nonsaturable, energy independent, temperature independent, and unaffected by pH [148,262]. The sulfate uptake system has been shown to transport chromate, thiosulfate, selenate, and molybdate [241]. Uptake and efflux of sulfate is inhibited by p-chloromercuribenzoate. The efflux of sulfate occurs by exchange reaction, i.e., exit of sulfate is coupled with the entry of either sulfate or related ion [143]. In *Saccharomyces cerevisiae* uptake of sulfate is accompanied by a net proton influx of $3H^+$ and an efflux of K^+ for each sulfate ion taken up. Bivalent cations stimulate sulfate uptake at low concentrations of sulfate [192].

The phosphate transport system in yeast and other fungi is saturable, energy dependent, inhibited by metabolic poisons, and competitively inhibited by arsenate [23,66,197]. *Neurospora crassa* is able to maintain the same rate of phosphate uptake in media containing widely different concentrations of phosphate [18]. This is because of the existence of two uptake systems, one of high affinity and the other of low affinity. The activities of these two systems vary with growth conditions and these variations are enough to maintain a constant uptake rate of phosphate [30,31]. Phosphate uptake in various mutants of *N. crassa* has been studied by Lehman [121]. Roomans and Borst-Pauwels [191] reported that phosphate in *Saccharomyces cerevisiae* is taken up by two transport systems, one Na^+ dependent and the other Na^+ independent. The Na^+-dependent transport system at pH 7.2 is stimulated by bivalent cations, whereas the Na^+-independent transport system is inhibited. On the contrary, phosphate uptake via the Na^+-independent system is stimulated by bivalent cations at pH 4.5. The effect of bivalent cations on phosphate uptake has been postulated due to decrease in surface potential.

A nitrate transport system in *N. crassa* has been described by Schloemer and Garrett [204]. They observed that this system is induced by either nitrate or nitrite. It is absent in cells grown in ammonia or casamino acid-supplemented media. The induction of the nitrate transport system can be prevented by inhibitors of protein synthesis such as cyclohexamide, puromycin, or 6-methylpurine. Nitrate uptake is inhibited by 2,4-dinitrophenol, cyanide, and antimycin A, and noncompetitively by ammonia and nitrite. With the use of nitrate reductase deficient and nitrate nonutilizing mutants (*nit-1, nit-2,* and *nit-3*) it has been documented that nitrate transport is not dependent on nitrate reduction [204]. Although, like the nitrate transport system, the nitrite uptake system is inducible in *N. crassa,* it is not inhibited by ammonia, nitrite, or casamino acids [205].

V. TRANSPORT OF ORGANIC METABOLITES

There have been numerous studies on organic metabolite transport in microorganisms and animal cells, and some excellent reviews are available in the literature [100,114,157,158, 179,195,242,247,253]. Most studies on organic metabolite transport in fungi have been concerned with sugar and amino acid transport.

A. Sugars

With the exception of glucose, sugar transport systems in microorganisms are inducible [139, 149]. Phosphate esters of sugars are not transported by microbial cells [157]. In *Neurospora crassa,* glucose appears to be transported by two permeases [154,201,202,206,207]. Both uptake systems are temperature dependent and inhibited by metabolic poisons such as sodium azide and 2,4-dinitrophenol. However, transport system I has a high K_m whereas system II has a low K_m. The low-K_m system is repressed by glucose. The high-K_m system is constitutive. The low-K_m system is insensitive to fructose, galactose, and lactose. But glucose is able to inhibit noncompetitively the uptake of fructose and galactose. The efflux of the nonmetabolizable substrate, 3-O-methyl-D-glucose, occurs only in the presence of metabolic inhibitors [206]. The high- and low-affinity transport systems are under different enzymatic controls and they can be genetically dissected [79]. The existence of at least two transport systems for the uptake of trehalose by *Saccharomyces cerevisiae* has been reported by Kotyk and Michaljanicova [116]. A transport system in *Candida parapsilosis* used in uptake of D-glucose, D-mannose, D-galactose, and maltose but not of fructose, sucrose, or pentose, which is most active at pH 4.5-8.5, has also been described by these investigators [115].

Hofer [85,86] studied the transport of monosaccharides in *Rhodotorula gracilis* under conditions of blocked energy supply. He observed that the plasma membrane of the yeast was impermeable to monosaccharides when the energy-requiring transport system was rendered inoperative by metabolic poisons. Later, Hofer and Misra [87] provided evidence that the active transport of monosaccharides in *Rhodotorula gracilis* is accompanied by proton uptake. The system ("H^+-symport") is energized by the electrochemical gradient of H^+ across the plasma membrane of the yeast. The maximum velocity and apparent affinity of the membrane-bound monosaccharide transport system is temperature dependent [84]. Kinetic studies of D-xylose transport in this fungus by Alcorn and Griffin [2] revealed the presence of at least two carriers: one with low K_m and the other with a high K_m. The low-K_m carrier was repressed in rapidly growing cells and derepressed by starvation. 2-Deoxy-D-glucose competitively inhibited both of the D-xylose carriers. On the contrary, D-galactose was able to inhibit the low-K_m carrier only.

Romano and co-workers [28,139,188,189] have provided evidence that phosphorylation, which is a necessary event during sugar transport by bacteria, is not required for such transport by the filamentous fungus *Aspergillus nidulans.* A specific fructose uptake system has been described by Rand and Tatum [176] in *N. crassa.* The uptake is carrier-mediated and is not inhibited by glucose, mannose, or 2-deoxy-D-glucose. It is competitively inhibited only by L-sorbose (K_i = 9 mM compared to K_m = 0.4 mM for glucose). In *A. nidulans,* the fructose transport system is highly specific [139].

B. Nucleic Acid Bases

Despite the potential importance of analogs of nucleic acid bases as mutagens, studies on the uptake of the bases themselves are few. Tang et al. [236] studied uptake and incorporation rates of purine bases in *H. capsulatum.* Little or no uptake of radiolabeled adenine, thymidine, or uracil by yeast or mycelial cells of *H. capsulatum* was observed. The uptake of guanine by mycelial cells was faster than that by yeast cells. Other investigators [37,111] have also observed limited uptake of nucleic acid precursors by this fungus.

In *N. crassa* adenine, guanine, and hypoxanthine are transported by one transport system, and this system is insensitive to adenosine, cytosine, and purine base analogs [132, 172]. There is another adenine transport system, separate from the hypoxanthine uptake system, which is not inhibited by guanine or hypoxanthine [131].

In *S. cerevisiae* a number of transport systems for nucleic acid bases have been reported. One system transports adenine, hypoxanthine, and cytosine [174,178], while another transports guanine and hypoxanthine [174]. Yet another can transport adenine, guanine, hypoxanthine, and cytosine after preincubation of cells in glucose [178]. The mechanism of substrate binding to this purine-transport system has been postulated by Foret et al. [57]. The carboxyl group (pK_a = 5.1) of the transport system, a proton and a substrate molecule, arrange to form an uncharged ternary complex. Losson et al. [128] reported that the uptake of cytosine, uracil, and uridine is mediated by three distinct permeases. Uptake of adenine in yeast is optimum at pH 5.6-6.0 [147]. A common transport system for cytosine, adenine, and hypoxanthine in *Candida albicans* has been described by Polak and Grenson [175].

In mutants of *Aspergillus nidulans* insensitivity to purine analogs has been postulated to be due to defective hypoxanthine-adenine-guanine and xanthine-uric acid transport systems [50]. Fink and Hygaard [56] reported than exogenous purine nucleosides were readily taken up and metabolized in *Physarum polycephalum,* whereas free purine bases were hardly taken up at all.

Tang and Howard [235] studied the uptake and incorporation of radioactive adenine in *Cryptococcus albidus.* These investigators documented that uptake and incorporation of adenine were not substantially different at 37°C and 23°C. The uptake of the metabolite at both temperatures was a linear function of time. By 2.5 h, 82-90% of the total metabolite in the medium was accumulated by the cells. The majority of the accumulated metabolite was incorporated into RNA at both temperatures. Its incorporation into DNA at 37°C was only 40% of that at 23°C. It was postulated that the inability of *C. albidus* to incorporate adenine into DNA might have occurred through damage to the proteins involved in DNA synthesis or to the DNA itself at 37°C.

Dalke and Magill [49] have reported that pyrimidine (uracil) uptake in *Neurospora* occurred via a specific system which recognized certain molecular features of uracil. Inhibition studies in which various analogs were used revealed that:

1. The tautomeric form of uracil analogs was a predominant determinant recognized by the uptake system.
2. Substitution at the five positions of the pyrimidine ring imposed steric hindrance on binding.
3. The presence of negative charge resulted in the loss of recognition.
4. The double bond between the carbon atoms 5 and 6 appeared to be important for recognition.
5. Purine bases did not inhibit uracil uptake.

Studies on nucleoside transport in *N. crassa* by Dunaway-Mariano and Magill [52] strongly suggested the presence of a general transport system which accepts both purine and pyrimidine nucleosides. The general structural features of the nucleosides which are involved in binding to the transport system are as follows:

1. Presence of a purine of pyrimidine as the heterocyclic ring
2. Presence of an unfunctionalized ribose or 2′-deoxyribose as the sugar unit
3. Presence of a β-configuration about the anomeric carbon
4. Absence or substituents at C_8 in the purine series and at C_5 and C_6 in the pyrimidine series
5. Presence of a double bond between carbon atoms 5 and 6 in the pyrimidine series
6. The absence of a charge on the heterocyclic ring

C. Urea

In *Aspergillus nidulans* urea and thiourea are transported by a specific permease [53]. Cooper and Sumrada [42] have reported that uptake of urea in *S. cerevisiae* occurs in two ways: (1) by an active transport system which is dependent on metabolic energy and sensitive to nitrogen repression, and (2) by passive or facilitated diffusion.

D. Amino Acids

1. Types of Amino Acid Transport Systems

Amino acids probably constitute the largest groups of metabolites which are commonly taken up by active transport system in living cells. The other means by which amino acids can enter cells include passive and facilitated diffusion. In fungi, a number of active transport systems for amino acids have been observed. Such systems can be grouped roughly into four categories: (1) those for neutral amino acids, (2) those for basic amino acids, (3) those for acidic amino acids, and (4) those which admit a large number of different sorts of amino acids (e.g., general transport systems). Systems specific for a given amino acid, such as methionine, lysine, arginine, and so on, have also been observed. In addition, a given amino acid may be transported into fungal cells by more than one system. Thus, systems of both high and low specificity have become recognized. Analyses of the kinetics of transport and genetic studies of transport-defective mutants have contributed a great deal to the understanding of the specificities of amino acid permeases in fungi.

a. Acidic Amino Acid Transport Systems: Uptake of the acidic amino acids, aspartate, glutamate, and α-aminoadipate, was observed to occur via a single transport system in *S. cerevisiae* by Joiris and Grenson [102]. These investigators obtained a mutant that was unable to utilize external α-aminoadipic acid. This mutant had a decreased rate of uptake of the acidic amino acids. Hunter and Segel [91] reported a specific transport system for L-glutamic acid and L-aspartic acid in *Penicillium chrysogenum.* Uptake pf acidic amino acids in *A. nidulans* occurs by a common transport system that is specific for those amino acids [95,108,110,167,183,184]. A mutant of *A. nidulans* defective in L-glutamate and L-aspartate uptake has been described by Kinghorn and Pateman [108,110]. Pall [161] reported an uptake system in *N. crassa* which transported both D- and L-acidic amino acids. Wolfinbarger and Kay [257] reported that acidic amino acids were also transported by neutral or general amino acid transport systems. In addition, these investigators also reported a dicarboxylic acid transport system in *N. crassa* that could transport L-aspartic acid [256].

b. Basic Amino Acid Transport Systems: A number of investigators [11,160,187] have identified a basic L-amino acid transport system in *N. crassa.* A number of basic amino acids, such as L-arginine, L-ornithine, L-lysine, and canavanine, are transported by this system (termed system III) into mycelial cells and germinated conidia [11,160]. A number of mutants of *N. crassa* defective in basic amino acid transport systems have been identified. These mutants have been labeled *bat* [160,238], *hlp-1* [187], *bm-1* [38], *bas-9* [200], *CR-10* [134], and *PmB* [187,254]. Normally, basic amino acids are transported by a general system (designated system II) and by the basic amino acid system (system III), but in the *bat* mutant any amino acid can completely inhibit arginine and lysine uptake.

Evidence for the presence of a constitutive basic amino acid transport system in *P. chrysogenum* has been provided by Hunter and Segel [91]. This system has a low affinity for 2,4-diaminobutyric acid, L-ornithine, and L-histidine; however, it has high affinity for L-lysine and L-arginine ($K_m = 6 \times 10^{-6}$ M). A low-affinity system for L-histidine has been reported in *N. crassa* [160]. A system for L-arginine and L-lysine in *A. nidulans* has been observed by Cybis and Weglenski [47]. This transport system is present in young mycelial cells and transports L-arginine preferentially over L-lysine.

Two active lysine transport systems in *S. lipolytica* have been described by Beckerich and Heslot [16]. The kinetic properties of one of these systems revealed it to be a high-affinity lysine permease which was competitively inhibited by arginine and its analogs, L-canavanine and L-ornithine, and by analogs of lysine, L-5-aminoethylcysteine and L-4,5-transdehydrolysine. The inhibition by other amino acids was noncompetitive. Thus, this lysine permease appeared to be similar to the basic amino transport system described in other filamentous fungi. A similar permease common for basic amino acids, L-lysine, L-arginine, and L-ornithine, has been reported in *S. cerevisiae* [69].

c. Neutral Amino Acid Transport Systems: *Neurospora crassa* has been studied extensively with respect to neutral amino acid transport systems. Both aromatic and aliphatic neutral amino acids are transported by this system [27,51,123,159,225,251]. It has also been observed that certain acidic amino acids (aspartic and glutamic acids) at a pH where they are neutrally charged can be transported into growing mycelial cells [255-257]. A number of mutants defective in the ability to transport neutral (aliphatic and aromatic) amino acids have been obtained in several laboratories [123,134,225,254,259]. The neutral amino acid transport system has been designated as system I by Pall [159].

No distinct neutral amino acid transport system has been recognized in *P. chrysogenum* and *A. nidulans.* However, in a dimorphic, pathogenic fungus, *H. capsulatum,* L-leucine at low substrate concentration ($< 5 \times 10^{-5}$ M) was transported into both yeast and mycelial cells by a neutral amino acid permease system. Since the uptake of L-leucine was inhibited by neutral aliphatic-side-chain amino acids and not by neutral aromatic-side-chain amino acids or the D isomer of leucine, it was concluded that the transport system was stereospecific with high specificity for neutral, aliphatic-side-chain amino acids [76]. The exact mechanism by which L-leucine uptake was inhibited by neutral aliphatic-side-chain amino acids and not by aromatic neutral amino acids is not known. However, it can be speculated that the electronegative nature of the aromatic ring of the aromatic neutral amino acids enabled them to compete for L-leucine. This speculation assumes that L-leucine interacts with the permease by three-point attachment. This probably is the reason for the inability of basic and acidic amino acids to compete with L-leucine for the transport site. This charge (negative on acidic amino acid side chains and positive on basic amino acid side chains) interferes with binding of the metabolite to the permease. The fact that glycine (a neutral amino acid without any side chains) could not inhibit L-leucine uptake suggests that the aliphatic side chain is one of the attachment points. L-Proline (an imino acid) competitively inhibited the L-leucine uptake. This could be because of structural confirmation of proline in solution, where its buckled ring provides all the points necessary for interaction with the permease [76].

d. General Amino Acid Transport Systems: General amino acid transport systems with similar specificities have been documented in various filamentous and unicellular fungi [67,72,77,92,233,235,251]. The general permease system can transport a wide range of amino acids, which include acidic, basic, and neutral. Proline and hydroxyproline are generally not transported [77]. This system is usually found in older cells which have been starved of carbon and/or nitrogen, and Pall [159] has suggested that such a system may act as a scavenger which takes up any remaining exogenous amino acids. Tang and Howard [235] reported that glutamic acid in *Cryptococcus albidus* is transported by a general system similar to that of other fungi. Table 1 lists the organisms in which a general amino acid transport system has been studied comprehensively.

The general amino acid transport system is constitutive and its activity is inhibited by the presence of ammonium ions [51,71]. The existence of general amino acid transport systems has been confirmed by studies on mutants of *N. crassa* [133,177,226] and *S. cerevisiae* [198]. The *Su-mtr* mutant of *N. crassa* [226] has increased activity of the transport system II of Pall [159]. The *PmG* mutant is defective in general amino acid transport system [177]. The double mutant *his-3,neu*^{a}*,bas*^{a}, obtained by a cross between

Table 1 Fungi Studied with Respect to General Amino Acid Transport System

Transport system for:	Organism	Competitive inhibition caused by	Reference
Phenylalanine	*Neurospora crassa*	Methionine Leucine Tryptophan Tyrosine Cysteine Alanine Serine Asparagine Threonine Valine Isoleucine Histidine Glycine Ornithine Arginine	51
Arginine Lysine Citrulline Tryptophan	*Saccharomyces cerevisiae*	Glycine Alanine Valine Leucine Tryptophan Glutamate Citrulline	72
Phenylalanine	*Geotrichum candidum*	Alanine Arginine Aspartate Cysteine Glutamate Lysine Histidine Leucine Methionine Tryptophan Tyrosine Valine	151
Valine	*Arthrobotrys conoides*	Leucine Isoleucine Alanine Phenylalanine Methionine Serine Glycine Cysteine β-Alanine	77

(continued)

Table 1 (continued)

Transport system for:	Organism	Competitive inhibition caused by	Reference
		α-Aminobutyrate	
		Aspartate	
		Glutamate	
		Asparagine	
		Glutamine	
		Histidine	
		Lysine	
		Arginine	
Glutamic acid	*Cryptococcus albidus*	Glutamine	235
		Aspartate	
		Arginine	
		Lysine	
		Histidine	
		Tryptophan	
		Phenylalanine	
		Leucine	
		Valine	
		Isoleucine	

his-3,bas[a] and *his-3,neu*[a], lacks the basic and neutral amino acid transport systems. The substrate specificity of the general transport system present in the double mutant of *N. crassa* has been fully characterized. Although this system has a broad specificity for neutral, hydrophobic, or hydrophilic amino acids, and for those with positively charged side chains, the acidic amino acids aspartate and glutamate are only moderate inhibitors [133]. This may be due to the fact that the side chains of acidic amino acids (glutamate and aspartate) are folded so that the carboxyl groups form hydrogen bonds with the α-amino group, thereby greatly reducing the positive charge on the α-amino group. This may result in a decreased affinity of acidic amino acids for the general transport system. It is well documented that a positive charge at the α position is essential for binding of amino acids.

By kinetic and genetic analyses, Rytka [198] showed that D-amino acids enter the cells of *S. cerevisiae* by a general transport system that accommodates both D and L stereoisomers. An amino acid analog, N-δ-chloroacetyl-L-ornithine (NCAO), has been used by Larimore and Roon [120] as a site-specific reagent for the general amino acid transport system in this yeast.

e. Specific Amino Acid Transport Systems: Fungi, like other microorganisms, contain several distinct amino acid transport systems, each of which serves for a limited number of structurally related amino acids. For example, proline transport in many fungi and other microorganisms is highly specific, and this permease system can transport only proline and hydroxyproline. Although *S. cerevisiae* possesses a general amino acid transport system, it also displays specific permeases for arginine [69], lysine [68], and methionine [65]. Table 2 represents a partial list of specific amino acid transport systems that have been reported in fungi.

Table 2 Partial List of Specific Amino Acid Transport Systems That Have Been Recognized in Filamentous Fungi and Yeast

Transport system specific for:	Organism	References
Cystine	*Histoplasma capsulatum*	63
	Penicillium chrysogenum	91, 214
Cysteine	*Penicillium chrysogenum*	91, 214
Methionine	*Saccharomyces cerevisiae*	65
	Penicillium chrysogenum	20
	Neurospora crassa	162
Arginine	*Penicillium chrysogenum*	91
	Saccharomyces cerevisiae	22, 47, 69
Lysine	*Saccharomyces lipolytica*	16
	Saccharomyces cerevisiae	68, 72
	Penicillium chrysogenum	91
Phenylalanine	*Penicillium chrysogenum*	20
Leucine	*Penicillium chrysogenum*	20
Histidine	*Saccharomyces cerevisiae*	46
Proline	*Candida albicans*	98, 99
	Penicillium chrysogenum	91
	Aspergillus nidulans	8
	Saccharomyces cerevisiae	24, 25
	Saccharomyces chevalieri	130
	Histoplasma capsulatum	48

When the specificity of the arginine transport system was measured, Grenson [69] found that many basic amino acids (L-lysine, L-ornithine, L-histidine, and D-arginine) were inhibitory. All these amino acids caused competitive inhibition, with the exception of histidine, which showed a mixed type of inhibition. Mutants resistant to L-canavanine were isolated. In these mutants, uptake of L-arginine was specifically impaired. Genetic analysis showed that the mutation was recessive. It presented a gene-dose effect and segregated normally as a single gene. The conclusion that the system is specific for arginine even though lysine is a competitive inhibitor was based on the fact that the K_i of lysine inhibition of arginine uptake is much higher than the K_m for lysine uptake, suggesting that arginine uptake system does not normally serve for lysine transport in yeast. Lysine at high external concentrations, which generally enters the cell by its specific permease, may also, in part, be transported by the arginine system when the lysine system is overloaded. On the other hand, the highly specific lysine-permease system of yeast cannot transport arginine and loss of its activity by mutation affected lysine uptake exclusively [68].

The arginine transport system in isolated vacuoles of *S. cerevisiae* has been characterized by Boller et al. [22]. The advantages of using isolated vacuoles for transport studies are obvious in that they represent a much simpler system than intact cells. The vacuolar isolates consist of simple membrane bags which are not ruptured in the course of isolation, so that the membranes are oriented as they are in vivo. When isotopically labeled arginine is added to the medium in which vacuoles are suspended, the amino acid is transported into the vacuoles by exchange specifically with the arginine already present inside. Such an exchange does not take place in intact spheroplasts. The exchange transport in vacuoles occurs simply because they lack a principal energy-producing system. The amino acid transport shows saturation kinetics with an apparent K_m 20 times higher than that in the spheroplasts. Although the arginine transport in vacuoles by the highly specific transport system is inhibited by histidine and canavanine, their K_i values are substantially higher (13- and 20-fold, respectively) than the K_m for arginine uptake. On the contrary, Nakamura and Schlenk [153] and Christensen and Cirillo [40] were unable to detect uptake of any of several labeled metabolites by isolated vacuoles of *Candida utilis* and *S. cerevisiae.* This difference from the studies of Boller et al. [22] and of Merkel et al. [145] could be due either to the different species used, to the different procedure (osmotic shock) used for isolating vacuoles, or to the different conditions used in transport experiments, e.g., low pH and addition of ATP, glucose, and lysate of spheroplasts. Transport of sugars (glucose, L-sorbose, and D-galactose) and purine compounds into vacuoles (vesicles) of *S. cerevisiae* has been observed by Christensen and Cirillo [40], Nagy [152], and Wehrli et al. [246].

Methionine enters cells by at least two distinct permeases in *S. cerevisiae.* Mutants that have lost the methionine permease with a high affinity for methionine and retained the methionine permease with a low affinity have been isolated by Gits and Grenson [65]. These mutants are resistant to low concentrations of L-methionine. Competitive inhibition studies revealed that the permease with high affinity for methionine is highly specific.

Gilbert and Howard [63] described a specific transport system for cystine in the yeast phase of *H. capsulatum.* The kinetics of uptake of the amino acid showed a saturation at 70 μM and an average K_m value of 3×10^{-5} M. The optimal pH and temperature for cystine uptake were 6.5 and 37°C, respectively. The calculated energy of activation was 14.1 kcal/mol with a temperature coefficient of 2.1. The uptake of cystine was not inhibited by any of the 14 other amino acids tested. This could be because of the dimer nature of the amino acid, which contains a disulfide bond. The disulfide bond might provide a specific point of attachment to the carrier site. Alternatively, the disulfide form of the amino acid might be associated with a reductase, which would allow the sulfhydryl form to cross the membrane. If this is the case, one would expect inhibition of cystine uptake by cysteine; however, the effect of this amino acid on cystine uptake was not reported. A 32% inhibition of cystine uptake in presence of 0.3% (w/v) yeast extract might be attributable to cysteine content of the yeast extract. A cysteine and cystine-specific transport system in *P. chrysogenum* has been described by Skye and Segel [214].

A distinct proline transport system has been detected in *P. chrysogenum* [91], *S. cerevisiae* and other yeasts [24,25,130], *A. nidulans* [8], and *C. albicans* [98,99]. The imino acid transport system of *S. chevalieri* has the highest affinity for L-proline and sarcosine (K_m = 25 and 31 μM, respectively); however, a distinct inhibition of proline uptake by this system was observed when alanine and glycine were used as inhibitors. It was postulated that this might have been due to the small molecular nature of these two amino acids, which may "mimic" the imino acid structure [130]. Brandriss and Magasanik [24] reported that a proline transport system is fully expressed by *S. cerevisiae* even when grown on a poor nitrogen source (urea, proline) but that the system does not require proline as an inducer. This transport system is repressed by ammonia. These observations were confirmed by the use of proline nonutilizing mutants [25]. Conversely, in addition to the constitutive proline transport system [98,211,212], an inducible system in *C. albicans* and other species has been reported by Jayakumar et al. [98,99]. When *C. albicans* cells were grown in the presence of proline as the sole nitrogen source or preincubated with proline, a more efficient proline transport system was induced [99]. Although the K_m values of both the constitutive and the inducible transport systems were similar, the uptake rate of proline by these systems was different, i.e., the cells exposed to proline, before the ^{14}C-labeled proline uptake rate was measured, had a 6-12 times higher uptake rate than those cells that were not preincubated with proline. This transport system was induced by proline only and not by other amino acids. The induction of the transport system could be inhibited by inhibitors of protein synthesis. Furthermore, ammonium ions did not repress the inducible proline transport system of *C. albicans.*

In *A. nidulans,* one of the genes located in linkage group VII has been shown by Arst and McDonald [7] to specify a proline-inducible proline transport system. This permease is responsible for most of the proline uptake under induced conditions. Another genetically undefined permease accounts for a small but significant proportion of proline uptake. This minor proline transport system does not respond to proline induction but is subject to repression and/or inhibition by ammonium ions [7]. The repression by ammonium ions occurs at 37°C but not at 25°C [8].

f. High-Affinity Versus Low-Affinity Transport Systems: Like phosphate [18] and sugar [79,154,202,206,207] transport systems, a number of amino acid transport systems have been observed to be of two affinities: high and low. When a double reciprocal plot of the initial rate of uptake versus amino acid concentration is made, a sharp break in the curve at a given amino acid concentration may suggest that the amino acid is transported by two different carrier sites, one with higher affinity (lower K_m value) and the other with lower affinity (higher K_m value). Tang and Howard [235] encountered an example of this phenomenon in their studies on the uptake of L-glutamic acid by *Cryptococcus albidus.* Observations on the initial rate of uptake at different external concentrations of the amino acid failed to reveal a complete saturation phenomenon. The rate of uptake of the amino acid increased with increasing external substrate concentration up to 5×10^{-5} M, but thereafter, any increase in external concentration resulted in a slow but linear increase in the initial rate (Fig. 7). The biphasic nature of the saturation kinetics suggested that the amino acid was transported by at least two distinct systems. The presence of two types of transport is further indicated by the discontinuity in the Lineweaver-Burk plot shown as an insert to Fig. 7. By extrapolation of the two linear portions of the double reciprocal plot, one can obtain two K_m values. Such a simple extrapolation cannot distinguish between saturable and simple diffusion unless the straight line observed after the break in the double reciprocal plot does not pass through the origin. If it does pass through the origin, entry of the substrate into cells by simple diffusion is indicated. Furthermore, complete inhibition of uptake even at high external substrate concentration by appropriate concentrations of the energy-yielding metabolic inhibitors would signify the uptake of the substrate by two active transport systems: one of high affinity (operative at low substrate concentration) and the other of low affinity (operative at high substrate concentration). As illustrated in Fig. 7, the uptake of L-glutamate occurred by two saturable transport systems. The high affinity system had a K_m of 1.15×10^{-5} M with maximum uptake rate (V_{max}) of 0.049 μmol per milligram of cells (dry weight) per hour, and the low-affinity system had a K_m of 2.5×10^{-3} M with V_{max} of 3.61 μmol per milligram of cells (dry weight) per hour.

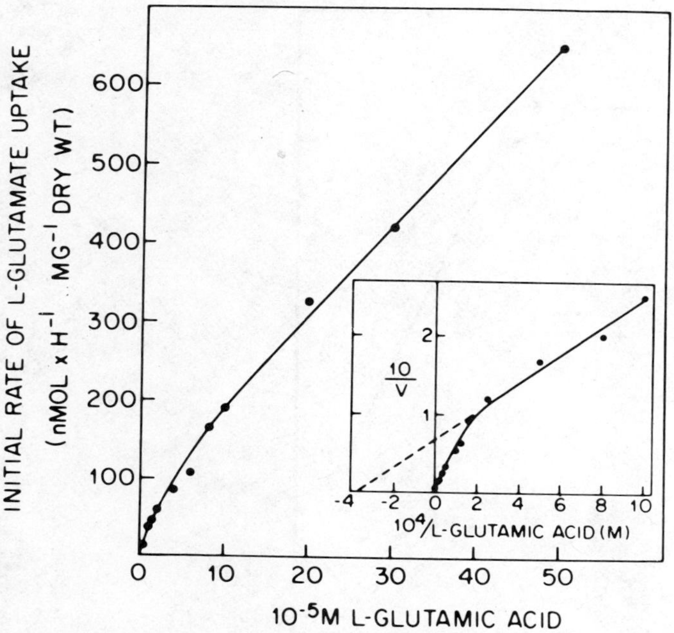

Figure 7 Graphical analyses (saturation and double reciprocal plot) of transport data to determine the involvement of more than one transport system for an amino acid. (From Ref. 235.)

The low-affinity transport system may represent the uptake of the substrate by other transport system(s) not meant for the substrate. However, because of structural similarity, this substrate at high external concentration may gain excess to the other transport system(s). This is a situation analogous to lysine transport in *S. cerevisiae* [68]. Thus, theoretically it is possible to get an array of K_m values with increasing substrate concentrations, each representing involvement of additional transport system. Such nonlinear, double reciprocal plots are characteristics of many amino acid transport systems in bacteria and fungi [5,58,65,68,73,74,76,77,80,107,235,243,260].

In yeast, uptake of lysine [68], methionine [65], and histidine [46] has been found to have two distinct transport systems. In each case, at low external concentration, the high-affinity transport system was very specific; but at high external concentration, the low-affinity transport system was relatively nonspecific. In the uptake of histidine, however, both high- and low-affinity systems appeared to be specific. A number of mutants of fungi have been obtained that are deficient in high-affinity transport systems [65].

2. Variations During Morphogenic Events

Certain pathogenic fungi exhibit two morphological forms: a parasitic form which exists in the tissue of a host and a saprophytic form which occurs in nature. Infection of a host by one of these fungal forms is accompanied by conversion of the organism from its saprophytic to its parasitic form of growth. Pathogens that exhibit this type of dimorphism include *Histoplasma capsulatum, Blastomyces dermatitidis, Coccidioides immitis, Sporo-*

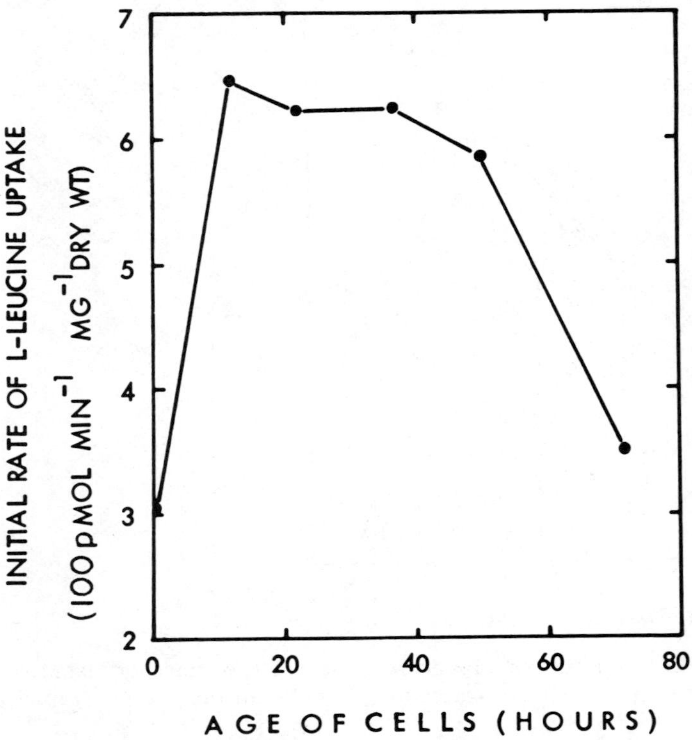

Figure 8 Initial rate of L-leucine uptake by yeast cells of *Histoplasma capsulatum* of various ages during growth. Initial rate of L-leucine uptake per minute per milligram dry weight of cells was computed from observations made at 5-, 10-, and 15-min intervals. The reaction mixture contained 1.0 mg of yeast cells (dry weight) per milliliter suspended in 0.02 M phosphate buffer (pH 7) and 10^{-5} M [^{14}C] L-leucine (approximately 10,000 counts per minute per milliliter). The system was incubated at 37°C. Uptake was measured by filtering the sample through a membrane (0.45-μm pore size) filter and analyzing the 0.5-ml aliquot of the filtrate for ^{14}C activity. The decrease in radioactivity at various time intervals compared to the whole suspension represented the uptake of L-leucine.

thrix schenckii, Emmonsia parva, E. crescens, and the various dematiaceous, imperfect fungi which cause chromomycosis [89].

It is obvious that a change in morphology must be preceded by changes in biochemical and physiological processes. Thus, there are numerous comparative studies on the morphology and biochemistry of dimorphism in the literature. Such studies are summarized and discussed elsewhere in this volume (Chap. 10). However, there are only a few studies of the changes in uptake rates of various metabolites that occur during morphogenic events among dimorphic, pathogenic fungi.

Studies performed in our laboratory on the kinetics of accumulation and incorporation of radioactive L-leucine by the yeast and mycelial forms of *H. capsulatum* have provided some insight into the physiological changes that occur during transition between the two morphological forms. The age of yeast phase cells is critical in achieving maximum uptake rate of L-leucine. The data depicted in Fig. 8 show that the uptake rate of L-leucine by the yeast cells was lowest during the stationary phase of growth (denoted by zero hour of age of the cells). The cells were cultured in Salvin's medium [199] and as they entered the logarithmic phase of growth, the uptake rate increased until a maximum was reached after 10 h at 37°C. The high uptake rate remained more or less constant until the cells were 56 hours old (representing the phase of active growth). Further aging of yeast cells resulted in a rapid decrease in the initial rate of the amino acid uptake (initiation of the stationary phase of growth). Similar studies were performed with the mycelial form of *H. capsulatum.* These cells were obtained by shifting cultures of yeast cells from 37°C to 23°C. The time when the culture was shifted to the lower temperature was considered as zero time. The maximum uptake rate of L-leucine was observed when the cells were 60-72 h old. During the period of 60-72 h of incubation at 23°C about 85% of the yeast cells germinated, thereby representing the mycelial form of growth.

The time course of L-leucine uptake by the yeast and mycelial form in the presence and absence of glucose has been published [76]. A summary of the data is shown in Fig. 9. The uptake of the amino acid was initiated immediately by both morphological forms of *H. capsulatum* after adding the radiolabeled L-leucine to the system. Most of the external amino acid (75-87%) was accumulated at a linear rate within 7-10 min by the mycelial form. The accumulated radioactivity appeared to leak out from these cells during the next 10-30 min and then it was taken up again within the experimental time of 80 min. Addition of 10^{-3} M glucose enhanced the rate of uptake, and the maximum accumulation of L-leucine was reached about 2 min earlier with glucose than without glucose. Furthermore, glucose suppressed the outflow of ^{14}C activity accumulated by the cells. On the contrary, the uptake of L-leucine by the yeast form of *H. capsulatum* was linear with time for 20-30 min, then decreased as the external supply of the amino acid diminished, reaching a plateau after about 50 min. The efflux of accumulated radioactivity was insignificant. Addition of 10^{-3} M glucose resulted in an increased rate of L-leucine uptake. In contrast to the situation with mycelial phase cells, the total accumulation of the amino acid by yeast cells was slightly higher in the presence of glucose than it was in the absence of glucose.

Figure 9 clearly indicates that the initial rate of L-leucine uptake by mycelial cells was approximately 5-10 times faster than it was by yeast cells under similar conditions. Moreover, the maximum accumulation of the amino acid was reached within 7-10 min in mycelial cells, whereas about 50 min was required by yeast cells to achieve maximum accumulation. These results suggested that the amino acid uptake system of mycelial cells was more efficient than was that of yeast cells.

Neither of the two morphological forms of *H. capsulatum* showed a complete saturation phenomenon [76]. At low external concentrations of L-leucine the rate of uptake of the amino acid by yeast phase cells increased rapidly with increasing external concentration. This rapid increase started to diminish when the external concentration of L-leucine was approximately 5×10^{-5} M, and beyond this point any increase in the external concentration resulted in a slow but linear increase in the initial rate of L-leucine uptake. A similar trend of the saturation curve was observed with mycelial cells except that the initial rate of L-leucine uptake by mycelial cells was always higher at any point than it was by yeast cells. It was concluded that at high external concentrations of leucine ($> 5 \times 10^{-5}$ M) the amino acid entered the cells through simple diffusion, in addition to the saturable system observed at low concentrations of substrate ($< 5 \times 10^{-5}$ M).

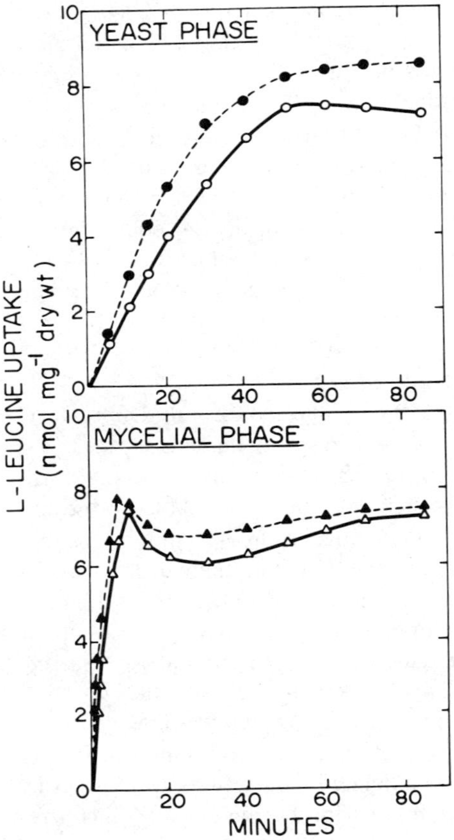

Figure 9 Time course of L-leucine uptake by mycelial and yeast phases of *Histoplasma capsulatum* in the presence and absence of glucose. Triangles, mycelial phase; circles, yeast phase. Cells were incubated in the presence of 10^{-3} M glucose at 37°C for 15 min before L-leucine was added to the system (represented by dashed line). Cells used to study the uptake of L-leucine in the absence of glucose were incubated similarly before the amino acid was added (represented by solid line). After the addition of L-leucine, standard procedures were used to measure the amino acid uptake as described in the legend of Fig. 8 at the determined intervals.

The effect of some well-known metabolic inhibitors on L-leucine uptake by yeast and mycelial phases has been studied [76]. Poor inhibition of L-leucine uptake by these metabolic inhibitors was observed and the authors suggested that this was due to the fact that when permease-like transport system was rendered partly inactive, the amino acid entered the cells via a diffusion-like process.

The effect of other amino acids on the initial rate of L-leucine uptake by yeast and mycelial forms of *H. capsulatum* established that the permease-like system specifically bound only aliphatic-side-chain amino acids, and that removal of negative charge from side chains of anionic amino acids did not enable the amino acids to inhibit L-leucine uptake, especially by yeast phase [76]. The puzzling fact that proline inhibited L-leucine

Table 3 Apparent K_i Values for Amino Acids That Competitively Inhibited L-Leucine Uptake by Yeast and Mycelial Phases of *Histoplasma capsulatum*

	K_i^a ($\times 10^{-5}$ M)	
Competitive inhibitor	Yeast phase	Mycelial phase
L-Alanine	5.4	3.5
L-Serine	5.3	3.5
L-Methionine	6.0	3.8
L-Valine	4.2	2.5
L-Proline	6.2	5.0
L-Isoleucine	3.4	1.5
N-Monomethyl-L-leucine	6.5	5.0

[a]The K_i values were determined as described in the text. The range of K_m values for L-leucine in yeast and mycelial phases were 2×10^{-5} to 4.2×10^{-5} and 2.4×10^{-5} to 4×10^{-5} M, respectively.

uptake was resolved, after a series of observations, with the suggestion that the buckled ring of proline acted as an aliphatic side chain [76]. A summary of the inhibition data from the work of Gupta and Howard is shown in Table 3. It was of interest to note that all the compounds with aliphatic side chains were competitive inhibitors for L-leucine uptake. Moreover, the K_i values for these amino acids were within the range of the K_m for L-leucine or were fairly close to that range. Clearly, the L-leucine transport system of yeast and mycelial forms of *H. capsulatum* operative at low substrate concentrations were identical. The permease-like L-leucine transport system has a limited specificity with respect to side chains in that only aliphatic neutral amino acids can be transported by this system with an equal affinity.

As an extension to the study of amino acid transport in the dimorphic fungus, *H. capsulatum,* Dabrowa and Howard [48] reported on the kinetics of L-proline uptake in the yeast phase of this organism. These investigators observed that the uptake of L-proline occurred by at least two types of transport: one by an active transport system with a K_m of 1.7×10^{-5} M and the other by simple diffusion. Certain analogs inhibited L-proline uptake and prevented incorporation of the amino acid into cellular constituents. Furthermore, the active transport system was stereospecific and was restricted to neutral aliphatic-side-chain amino acids. The inhibition of L-proline uptake by L-leucine was competitive, an observation that confirms an earlier conclusion [76] that the buckled ring of L-proline, in solution, acts as an aliphatic side chain and that this cyclic amino acid is transported by a system more or less specific for amino acids with neutral aliphatic side chains.

Starting from the observations of Garrison et al. [62], Maresca et al. [138] assumed that respiration was a good measure of the active uptake of cysteine and cystine in *H. capsulatum.* Maresca and his colleagues reported that the addition of cysteine or cystine to a medium stimulated the respiration of yeast but not that of the mycelia. Furthermore, when yeast cells were shifted down from 37°C to 25°C, the respiration rate decreased immediately and returned to baseline within 8 h. Shifting up to a temperature of 37°C caused an immediate increase in the respiration rate. Addition of cysteine accelerated the stimulus to respiration which a temperature shift up caused. When the temperature was shifted up to 37°C in the presence of cyclohexamide or chloramphenicol, the cysteine-induced stimulation in respiration still occurred. Mycelia incubated at 25°C did not respond to cysteine or cystine, but when the cells were switched to 37°C, they became responsive to these amino acids within a few minutes. Thus, it was concluded that the change in uptake of cysteine and cystine did not require new protein synthesis and that the transport protein(s) may have been present on cell membranes of both phases of *H. capsulatum,* but active only at 37°C.

Tang et al. [236] studied the uptake of metabolites by germinating yeast cells of *H. capsulatum.* Metabolite (L-leucine, L-glutamic acid, and guanine) uptake was faster in mycelial cells at 23°C than it was in yeast cells at 37°C. During the yeast-to-mold transition, the initial rate of uptake was first markedly decreased and then increased and achieved a level characteristic of the mycelial phase of growth.

In an attempt to characterize the role of cystine and cysteine in the phase transition, Maresca et al. [137] analyzed the uptake and reduction of cystine in relation to the morphological transition. They observed that the two morphological forms of *H. capsulatum* concentrated cystine at comparable rates and that mutants defective in the uptake of the amino acid were still able to undergo the transition normally. Therefore, the uptake of cystine was not a requirement for transition to or maintenance of the yeast phase. The uptake rate at 23°C by each phase was about half of that at 37°C. A similar reduction in uptake rate of cystine by yeast cells of *H. capsulatum* incubated at 23°C was previously reported by Gilbert and Howard [63].

In a recent study, Howard et al. [90] used a mutant of *H. capsulatum* that was defective in the ability of its blastoconidia to germinate. These investigators documented the fact that the nutritional requirement for cysteine and the expression of sulfite reductase activity were temperature dependent and not related to the growth form of the fungus. However, the kinetics of cysteine transport were dependent on both temperature and the form of the fungus. The initial rates of uptake of cysteine by the mutant (NG-2) and the parental wild type were the same at 37°C. However, when cells were tested at 23°C, there was a depression in the initial rate of uptake. Mycelial cells were unable to transport cysteine as efficiently as yeast cells, regardless of the temperature of incubation [90].

A number of low molecular weight organic sulfur-containing compounds were tested for their effect on the respiratory activity of yeast-like and mycelial forms of *H. capsulatum* by Garrison et al. [62]. They found that L-cysteine gave the maximum stimulatory effect on yeast-like phase respiration. The D and meso isomers of cysteine as well as substituted derivatives were much less effective. Respiration of homologous mycelial cells was depressed in the presence of L-cystine. Studies on transport of ^{35}S-labeled L-cystine and methionine by yeast cells of *H. capsulatum* revealed that these compounds but not [^{35}S]β-mercaptoacetate were actively transported. The uptake of ^{35}S-labeled L-cystine by yeast and mycelial phases of *Blastomyces dermatitidis* and *Sporothrix schenckii* was much lower than that by yeast cells of *H. capsulatum.* Garrison et al. [62] concluded that significant differences existed between yeast and mycelial forms of *H. capsulatum* in the uptake and subsequent intracellular metabolism of low molecular weight sulfur-containing α-amino acids and related compounds.

In a subsequent study, Bawdon and Garrison [12] observed that L-leucine and L-cystine were transported by yeast and mycelial forms of *B. dermatitidis* by a permease-like system at low external concentrations. However, at high concentrations, the amino acids entered the cells by simple diffusion as well. These observations confirmed earlier reports by Gupta and Howard [76] on transport of L-leucine into yeast and mycelial cells of *H. capsulatum.*

A specific inhibition of amino acid transport in *H. capsulatum* by a lysosomal extract from rabbit alveolar macrophages has been reported by Calderone and Peterson [32]. The degree of inhibition depended on the amino acid used. Uptake of sugars, nucleotides, or rubidium was not affected by the extract. The amino acids tested were leucine, proline, and glutamic acid. The ability of lysosomal extracts of alveolar macrophages to inhibit selectively amino acid transport in *Candida albicans* has been implicated as a mechanism by which growth and protein synthesis of the yeast is inhibited following phagocytosis by macrophages [173].

E. Peptide Transport

The transport of peptides in bacteria and animal cells has been studied by various investigators [10,39,54,144,169,170], but such studies are rare in fungi. Transport of methionine-containing peptides in *S. cerevisiae* was first reported by Becker et al. [15]. Recently, Marder et al. [135] confirmed that the Z1-2D strain of *S. cerevisiae* contains a functional peptide transport system. However, several structural requirements were observed. Di- and tripeptides of similar composition but different sequence varied in their affinity for the peptide transport system. A free carboxyl terminal was not necessary for the uptake of a leu-leu peptide. Peptides composed of proline (imino acids) were transported. A protonatable amino terminus played a role in peptide transport. The amino acid side chains of a peptide determined its effectiveness as an inhibitor. The peptide transport system of the strain Z1-2D had a higher affinity for hydrophobic peptides; a similar observation has been made in other strains of *S. cerevisiae* [14,15] and of *C. albicans* [124].

Size limits to peptide uptake could result from either a sieving action of the cell envelope or from the failure of the peptide transport system to bind peptides above a certain size [10,258]. *Neurospora crassa* mutants containing porous cell walls were used to show that the peptide transport system was incapable of handling peptides of more than five residues [258]. A mutant, Z1-2D Eta^r, of *S. cerevisiae* has been isolated which lacks the cellular mechanism to transport peptides to the site of peptidase activity. With the use of this mutant Marder et al. [136] confirmed that di- and tripeptides share a common mode of entry into yeast.

A detailed study of peptide transport in a pathogenic yeast, *C. albicans,* has been conducted by Logan et al. [127]. The peptide transport system is saturable and is temperature and pH dependent. It is affected by inhibitors of metabolic energy production. Whether or not uptake of peptides occurred against a concentration gradient could not be determined because of degradation of accumulated substrate (trimethionine) by intracellular peptidase. The K_m (3.3×10^{-5} M) of trimethionine uptake by *C. albicans* was comparable to those reported in *S. cerevisiae* [14] and in *N. crassa* [258] for gly-leu-tyr uptake. Free methionine did not affect the trimethionine uptake to a significant extent. Dimethionine competed only slightly and this slight competition was considered insignificant. Tripeptides of following composition: leu-leu-leu, gly-met-gly, acetyl-$(met)_3$, D-met-L-met-L-met, and met-met-ile effectively competed with trimethionine (met-met-met) uptake; but lys-lys-lys, L-met-L-met-D-met, D-met-D-met-D-met, $(met)_3$ methyl ester, and ala-ala-ala did not compete for trimethionine. In view of a number of limitations Logan et al. [127] have examined a number of features of the tripeptide transport system. These include:

1. The carboxyl terminal is not required as an attachment point but must be present as free carboxyl group.
2. Hydrophobic residues are desirable in the side chain.
3. D isomers may be included only at certain positions; however, the carboxyl terminal residue must be an L isomer.

VI. EFFECT OF POLYENES AND n-ALKANES ON MEMBRANE TRANSPORT

It is well documented that polyenes interact with sterols present in the membranes of microorganisms [81,118] and thereby bring about changes in permeability which lead to leakage of the cellular constituents and ultimately to death of the cells [60,61,82,126]. Destruction of selective permeability by polyenes allows cations to leak into the environment and permits the entry of protons (hydrogen ions) with acidification of the internal environment [83]. However, it is conceivable that polymers also bring about alterations in membrane functions that are responsible for the transport of certain essential metabolites from the environment. In this regard, certain imidazole, antimycotic compounds, e.g., clotrimazole and miconazole, have been shown to decrease the rate of uptake of leucine and lysine by *C. albicans* [261]. Bedicevsky and Grossowicz [17] reported that the polyene antibiotic amphotericin B inhibited phosphate uptake in *C. albicans* more strongly than it inhibited growth. Recently, Obukhovskaya et al. [155] observed that inhibition of the transport of neutral amino acids with hydrophobic chains (leucine, phenylalanine) by the polyene antibiotic levorin in *C. albicans* was due mainly to redistribution of monovalent cations caused by the antibiotic. In an acidic medium, levorin also caused a decrease of proton release from the cells which was accompanied by inhibition of leucine transport. Uptake of neutral amino acids with short side chains, such as alanine, including glycine and proline, was inhibited by the polyene antibiotic regardless of the ionic composition or pH of the medium.

The exposure of log-phase yeast cells of *H. capsulatum* and *B. dermatitidis* to amphotericin B and hamycin results in detectable alterations of the plasma membrane. These alterations are compatible with the concept of binding of the polyene with membrane sterols and subsequent damage due to changes in permeability [119].

Timberlake and Griffin [239] reported that cyclohexamide directly inhibited the uptake of proline by *Achlya*. This drug affected the transport of metabolites in other fungi either by inhibiting the protein synthesis and therefore inhibiting the permease synthesis, or by causing efflux of calcium ions from the cell membrane [33,70,94,252].

Factors other than sterols have recently been implicated in rendering the fungal cell sensitive to polyene antibiotics [112,212]. The magnitude of the nystatin effect in terms of proline uptake and potassium release by *C. albicans* was more pronounced in cells having a high ergosterol level than in those with a low level; thus, a correlation between nystatin sensitivity and ergosterol content of *C. albicans* was suggested [212]. However, von Hedenstrom and Hofer [245] reported that the effect of nystatin on active transport in *Rhodotorula glutinis* is restricted to its interaction with the plasma membrane, whereas the energy metabolism in mitochondria remains unaffected by the antibiotic.

The unique property of *C. albicans* of growing on n-alkanes of different chain lengths has been utilized by Singh et al. [211] to alter the lipid composition of plasma membrane in a predictable manner. In alkane-grown cells, the transport of lysine, proline, serine, and methionine was reduced, but there was no effect on the transport of glycine, glutamate, phenylalanine, and leucine. As a result, it was concluded that different permeases responsible for amino acid transport respond differently to the change of lipid environment [211]. The effect on glucose uptake of growing cells of *Candida* sp. on alkanes was investigated by Gill and Ratledge [64]. The functional differences between glucose- and alkane-grown cells with respect to the effect of n-decane or n-undecane on glucose uptake was related to a structural difference between the two types of cells. Using *Cladosporium resinae,* Teh and Lee [237] concluded that the rapid loss of selective permeability of cells grown in n-alkanes was due to a disorganization of the cell membrane rather than removal of membrane lipids such as sterols and phospholipids by the n-alkanes.

VII. REGULATION OF TRANSPORT SYSTEMS

Although most transport systems are constitutive, a number of regulatory mechanisms may control their operation. Transport of a number of metabolites in fungi is known to be inhibited by ammonia [6,20,21,71,78,167,168,184,185,190,200]. Roon et al. [194] postulated that the inhibition of amino acid transport by ammonium ion in *S. cerevisiae* results from the uptake of this metabolite into the cell and its subsequent incorporation into the α-amino group of glutamate and other amino acids. No significant decrease in amino acid transport in a mutant strain of yeast, which lacked the nicotinamide adenine dinucleotide phosphate (NADP)-dependent (anabolic) glutamate dehydrodenase, was observed when ammonium sulfate was added to the nitrogen-starved cells. However, another double mutant, which lacked the NADP-dependent enzyme and had a depressed level of NAD-dependent (catabolic) glutamate dehydrogenase, was equally as sensitive to ammonium ion as was the wild-type strain. An association of glutathione synthetase deficiency and diminished amino acid transport in yeast has been reported by Mooz [150].

Grenson et al. [72] have shown that the constitutive general amino acid transport system of *S. cerevisiae* is inhibited by ammonium ions. The wild-type strain (1278b) transported amino acids 5 (methionine) to 460 (citrulline) times faster when grown in proline minimal medium than it did when grown in ammonium minimal medium. The transport rate of these and other amino acids was the same in a mutant *(gap)* whether it was grown in ammonium medium or in proline medium, suggesting that only the general amino acid permease system was inhibited by ammonium ions. Furthermore, ammonium ions do not inhibit amino acid uptake by inhibiting the synthesis of the permease or by competing for the recognition site of the transport system [72]. On the contrary, Beckerich and Heslot [16] reported that ammonium ions inhibit the uptake of lysine in *S. lipolytica* by at least two mechanisms: (1) external ammonium acting as a competitive inhibitor, and (2) internal ammonium affecting the transport by inhibition of the carrier or by repression of its synthesis. The inhibition or repression of the carrier was suggested to occur indirectly through NADP-glutamate dehydrogenase. This mechanism involving NADPH-specific glutamate dehydrogenase has also been described by Grenson and Hou [71] in *S. cerevisiae.*

Even though the amino acid permease is not degraded after synthesis in *A. nidulans,* ammonium inhibits its activity and represses further synthesis. It is the intracellular ammonium and not glutamate which is important in regulation of the acidic amino acid permease [182]. Hynes [95] provided evidence that glutamate uptake in *A. nidulans* is controlled by ammonium ions and glucose. The *amdT* gene is involved in glucose repression of glutamate uptake. A lesion *(amdT102)* leads to increased sensitivity to glucose repression of glutamate uptake, but this lesion does not alter the response to ammonium ion repression [95]. The ammonium regulation mechanisms are not operative where the substrate of the ammonium repressible system is used as the carbon and nitrogen source [109].

In *Penicillium chrysogenum,* the general amino acid transport is also inhibited by ammonium ions. In this case, intracellular ammonium ions act as feedback inhibitors of the transport system [21]. According to Rytka [198], ammonium ions in combination with the gene product (*amc* product in case of *S. cerevisiae*) are responsible for production of some ammonium metabolite, which affects the general amino acid transport system.

The inhibition of amino acid transport by intracellular amino acid pools has been demonstrated for several transport systems [46,92,162,163]. This type of inhibition is known as transinhibition. The uptake of L-histidine is regulated by L-histidine accumulated within cells [45]. Rytka [198] suggested that in yeast the uptake of other amino acids, not only of L-histidine, is regulated by transinhibition. Inability of tryptamine-grown cells of *A. fumigatus* to transport tryptophan is due to severe feedback inhibition of the tryptophan transport system, and not by repression of the formation of a transport component [75].

For transport systems that are shared by more than one metabolite, any of the accumulated metabolites can cause transinhibition of other metabolites. For instance, L-α-aminoadipate is a better transinhibitor of the acidic amino acid transport system than is the metabolizable substrate, e.g., glutamate or aspartate [91]. Furthermore, the transinhibition is most pronounced under conditions in which the metabolite is transported much faster than it is metabolized. Thus, regulation of transport would be subjected to metabolite concentration in the free pool [67]. Therefore, care should be exercised when protein synthesis inhibitors are used in the investigation of amino acid transport because the inhibition of protein synthesis can cause a buildup of free amino acid pools which may result in transinhibition [100].

VIII. CONCLUSIONS

Even though studies on metabolite transport in pathogenic fungi are few in number, the fungi, both filamentous molds and unicellular yeasts, have been studied to an appreciable extent with respect to both inorganic and organic solute uptake. From kinetic and genetic analyses, it is obvious that diverse and complex metabolite transport systems exist in fungi. A limited number of studies have provided a description of some of the membrane components involved in transport processes. Most organic metabolites, i.e., amino acids and sugars, are taken up by an active process that requires expenditure of energy by the cells. Among various metabolites, the transport of amino acids by fungi has been studied most extensively. From these studies it is now evident that specificities of the amino acid transport systems may vary from fungus to fungus. A given amino acid within a single species of fungus may be transported by more than one system. The types of transport systems that may operate in a fungus may change depending on the age of the culture and growth conditions. Although the majority of amino acid transport systems are constitutive, some are inducible. Fungi may possess both a constitutive and an inducible transport system for the same amino acid.

The regulation of amino acid transport is quite complex and more than one factor appears to play a role in the control mechanisms. Ammonium ions appear to regulate amino acid transport by several mechanisms, e.g., by regulation of metabolic enzyme, by repression of transport systems, by acting in combination with gene products, and by binding with the transport system. Transinhibition caused by free amino acid pools, apparently, is another regulatory mechanism of amino acid transport.

Although the knowledge gained from the studies on transport mechanisms among saprophytic fungi may hold true for pathogens, observations on these pathogens are warranted, nevertheless. Since the pathogens in the host are surrounded by a unique environment and since they are under constant pressure from the host's defense mechanisms, study of metabolite transport is likely to unveil unknown features of transport mechanisms. For example, changes that occur during transition of one phase to another in dimorphic pathogenic fungi may be initiated at the level of the membrane that directly controls traffic of metabolites from external environment to internal environment, and vice versa.

REFERENCES

1. Aiking, H., and D. W. Tempest. 1977. Rubidium as probe for function and transport of potassium in the yeast *Candida utilis* NCYC 321, grown in chemostat culture. Arch. Microbiol. *115*: 215-221.
2. Alcorn, M. E., and C. C. Griffin. 1978. A kinetic analysis of D-xylose transport in *Rhodotorula glutinis.* Biochim. Biophys. Acta *510*: 361-371.
3. Allaway, A. E., and D. H. Jennings. 1971. The effect of cations on glucose utilization by, and on the growth of, the fungus *Dendryphiella salina.* New Phytol. *70*: 511-518.
4. Allen, R. D. 1974. Some new insights concerning cytoplasmic transport. Symp. Soc. Exp. Biol. *28*: 15-16.
5. Ames, G. F. 1964. Uptake of amino acids by *Salmonella typhimurium.* Arch. Biochem. Biophys. *104*: 1-18.
6. Arst, H. N., and D. J. Cove. 1969. Methylammonium resistance in *Aspergillus nidulans.* J. Bacteriol. *98*: 1284-1293.
7. Arst, H., and D. W. McDonald. 1975. A gene cluster in *Aspergillus nidulans* with an internally located *cis*-acting regulatory region. Nature (Lond.) *254*: 26-31.
8. Arst, H. N., D. W. McDonald, and S. A. Jones. 1980. Regulation of proline transport in *Aspergillus nidulans.* J. Gen. Microbiol. *116*: 285-294.
9. Atkin, C. L., J. B. Neilands, and H. J. Phaff. 1970. Rhodotorulic acid from species of *Leucosporidium, Rhodosporidium, Rhodotorula, Sporidiobolus,* and *Sporobolomyces,* and a new alanine-containing ferrichrome from *Cryptococcus melibiosum.* J. Bacteriol. *103*: 722-733.
10. Barak, Z., and C. Gilvarg. 1975. Peptide transport. In H. Eisenburg, E. Katchalski-Katzir, and L. A. Manson (eds.), *Biomembranes,* Vol. 7, Plenum, New York, pp. 167-218.
11. Bauerle, R. H., and H. R. Garner. 1964. The assimilation of arginine and lysine in canavanine resistant and sensitive strains of *Neurospora crassa.* Biochim. Biophys. Acta *93*: 316-322.
12. Bawdon, R. E., and R. G. Garrison. 1974. The uptake and incorporation of leucine and cystine by the mycelial and yeast like phases of *Blastomyces dermatitidis.* Mycopathol. Mycol. Appl. *54*: 421-433.

13. Bayer, E. A., E. Skutelsky, T. Viswanatha, and M. Wilchek. 1978. Specific localization and quantitation of biotin transport components in yeast by use of a biotin-conjugated, impermeant, electron dense label. Mol. Cell. Biochem. *19*: 23-29.

14. Becker, J. M., and F. Naider. 1977. Peptide transport in yeast: uptake of radio active trimethionine in *Saccharomyces cerevisiae.* Arch. Biochem. Biophys. *178*: 245-255.

15. Becker, J. M., F. Naider, and E. Katchalski. 1973. Peptide utilization in yeast: studies on methionine and lysine auxotrophs of *Saccharomyces cerevisiae.* Biochim. Biophys. Acta *291*: 388-397.

16. Beckerich, J. M., and H. Heslot. 1978. Physiology of lysine permease in *Saccharomyces cerevisiae.* J. Bacteriol. *133*: 492-498.

17. Bedicevsky, I., and N. Grossowicz. 1977. Effect of a polyene antibiotic on growth and phosphate uptake by *Candida albicans.* J. Gen. Microbiol. *102*: 229-304.

18. Beever, R. E., and D. J. W. Burns. 1977. Adaptive changes in phosphate uptake by the fungus *Neurospora crassa* in response to phosphate supply. J. Bacteriol. *132*: 520-525.

19. Bellinger, N., P. W. Nissen, T. C. Wood, and I. H. Segel. 1968. Specificity and control of choline-o-sulphate transport in the filamentous fungi. J. Bacteriol. *96*: 1574-1584.

20. Benko, P. V., T. C. Wood, and I. H. Segel. 1967. Specificity and regulation of methionine transport in filamentous fungi. Arch. Biochem. Biophys. *122*: 783-804.

21. Benko, P. V., T. C. Wood, and I. H. Segel. 1969. Multiplicity and regulation of amino acid transport in *Penicillium chrysogenum.* Arch. Biochem. Biophys. *129*: 498-508.

22. Boller, T., M. Durr, and A. Wiemken. 1975. Characterization of a specific transport system for arginine in isolated yeast vacuoles. Eur. J. Biochem. *54*: 81-91.

23. Borst-Pauwels, G. W. F. H., J. K. Peter, S. Jager, and C. C. B. M. Wijffels. 1965. A study of the arsenate uptake in yeast cells compared with phosphate uptake. Biochim. Biophys. Acta *94*: 312-314.

24. Brandriss, M. C., and B. Magasanik. 1979. Genetics and physiology of proline utilization in *Saccharomyces cerevisiae*: enzyme induction by proline. J. Bacteriol. *140*: 498-503.

25. Brandriss, M. C., and B. Magasanik. 1979. Genetics and physiology of proline utilization in *Saccharomyces cerevisiae*: mutation causing constitutive enzyme expression. J. Bacteriol. *140*: 504-507.

26. Breton, A., and Y. Surdin-Kerjan. 1977. Sulphate uptake in *Saccharomyces cerevisiae*: biochemical and genetic study. J. Bacteriol. *132*: 224-232.

27. Brink, N. G. 1972. Tryptophan transport in *Neurospora crassa* by various tyeps of *mtr* revertants. J. Gen. Microbiol. *73*: 153-160.

28. Brown, C. E., and A. H. Romano. 1969. Evidence against necessary phosphorylation during hexose transport in *Aspergillus nidulans.* J. Bacteriol. *100*: 1198-1203.

29. Budd, K. 1969. Potassium transport in non-growing mycelium of *Neocosmospora vasinfecta.* J. Gen. Microbiol. *59*: 229-238.

30. Burns, D. J. W., and R. E. Beever. 1977. Kinetic characterization of the two phosphate uptake systems in the fungus *Neurospora crassa.* J. Bacteriol. *132*: 511-519.

31. Burns, D. J. W., and R. E. Beever. 1979. Mechanism controlling the two phosphate uptake systems in *Neurospora crassa.* J. Bacteriol. *139*: 195-204.

32. Calderone, R. A., and E. Peterson. 1979. Inhibition of amino acid uptake and incorporation in *Histoplasma capsulatum* by a lysosomal extract from rabbit alveolar macrophages. J. Reticuloendothel. Soc. *26*: 11-19.

33. Cameron, L. E., and H. B. LeJohn. 1972. On the involvement of calcium in amino acid transport and growth of fungus *Achyla.* J. Biol. Chem. *247*: 4729-4739.

34. Carrano, C. J., and K. N. Raymond. 1978. Coordination chemistry of microbial iron transport compounds: rhodotorulic acid and iron uptake in *Rhodotorula pilimanae.* J. Bacteriol. *136*: 69-74.
35. Carter, B. L. A., and H. O. Halverson. 1973. Periodic changes in rate of amino acid uptake during the yeast cell cycle. J. Cell Biol. *58*: 401-409.
36. Chaffin, W. L., R. A. Barton, E. L. Jacobson, and M. K. Jacobson. 1979. Nicotinamide adenine dinucleotide metabolism in *Candida albicans.* J. Bacteriol. *139*: 883-888.
37. Cheung, S. C., G. S. Kobayashi, D. Schlessinger, and G. Medoff. 1974. RNA metabolism during morphogenesis in *Histoplasma capsulatum.* J. Gen. Microbiol. *82*: 301-307.
38. Choke, H. C. 1969. Mutants of *Neurospora crassa* permeable to histidinol. Genetics *62*: 725-733.
39. Christensen, H. N., and M. Liang. 1966. Transport of diamino acids in the Ehrlich cell. J. Biol. Chem. *241*: 5542-5549.
40. Christensen, M. S., and V. P. Cirillo. 1972. Yeast membrane vesicles: isolation and general characteristics. J. Bacteriol. *110*: 1190-1205.
41. Conway, E. J., and H. M. Gaffney. 1966. The further preparation of inorganic cationic yeasts and some of their chief properties. Biochem. J. *101*: 385-391.
42. Cooper, T. G., and R. Sumrada. 1975. Urea transport in *Saccharomyces cerevisiae.* J. Bacteriol. *121*: 571-576.
43. Cowan, M. C., B. G. Lewis, and J. F. Thain. 1972. Uptake of potassium by the developing sporangiophore of *Phycomyces blakesleeanus.* Trans. Br. Mycol. Soc. *58*: 113-126.
44. Cowan, M. C., J. F. Thain, and B. G. Lewis. 1972. Mechanism of translocation of potassium in sporangiophores of *Phycomyces blakesleeanus* in an aqueous environment. Trans. Br. Mycol. Soc. *58*: 91-102.
45. Cowan, M. C., J. F. Thain, and B. G. Lewis. 1972. Electrical mobility of potassium in the sporangiophores of *Phycomyces blakesleeanus.* Trans. Br. Mycol. Soc. *58*: 161-164.
46. Crabeel, M., and M. Grenson. 1970. Regulation of histidine uptake by specific feedback inhibition of two histidine permeases in *Saccharomyces cerevisiae.* Eur. J. Biochem. *14*: 197-204.
47. Cybis, J., and P. Weglenski. 1969. Effects of lysine on arginine uptake and metabolism in *Aspergillus nidulans.* Mol. Gen. Genet. *104*: 282-287.
48. Dabrowa, N., and D. H. Howard. 1976. Uptake of L-proline by *Histoplasma capsulatum.* Can. J. Microbiol. *22*: 1188-1190.
49. Dalke, P., and J. M. Magill. 1979. Specificity of uracil uptake in *Neurospora crassa.* J. Bacteriol. *139*: 212-219.
50. Darlington, A. J., and C. Scazzacchio. 1967. Use of analogues and the substrate-sensitivity of mutants in analysis of purine uptake and breakdown in *Aspergillus nidulans.* J. Bacteriol. *93*: 937-940.
51. DeBusk, B. G., and A. G. DeBusk. 1965. Molecular transport in *Neurospora crassa.* I. Biochemical properties of a phenylalanine permease. Biochim. Biophys. Acta *104*: 139-150.
52. Dunaway-Mariano, D., and J. M. Magill. 1978. Specificity of nucleotide transport in *Neurospora crassa.* J. Bacteriol. *136*: 924-928.
53. Dunn, E., and J. A. Pateman. 1972. Urea and thiourea uptake in *Aspergillus nidulans.* Heredity *29*: 129.
54. Elliot, K., and M. O'Connor. 1972. *Ciba Foundation Symposium on Peptide Transport in Bacteria and Mammalian Gut,* Elsevier, Amsterdam.

55. Emery, T. 1974. A model for carrier-mediated iron transport. Biochim. Biophys. Acta *363*: 219-225.
56. Fink, K., and P. Nygaard. 1979. Purine metabolism in microplasmedia of *Physarum polycephalum*. Biochim. Biophys. Acta *563*: 269-277.
57. Foret, M., R. Schmidt, and U. Reichert. 1978. On the mechanism of substrate binding to the purine-transport system of *Saccharomyces cerevisiae*. Eur. J. Biochem. *82*: 33-43.
58. Friede, J. D., D. P. Gilboe, K. C. Triebwasser, and L. M. Henderson. 1972. Transport of lysine and hydroxylysine in *Streptococcus faecalis*. J. Bacteriol. *109*: 179-185.
59. Fuhrmann, G. F., and A. Rothstein. 1968. The transport of Zn^{2+}, Co^{2+}, and Ni^{2+} into yeast cells. Biochim. Biophys. Acta *163*: 325-330.
60. Gale, E. F. 1974. The release of potassium ion from *Candida albicans* in the presence of polyene antibiotics. J. Gen. Microbiol. *80*: 451-465.
61. Gale, E. F., and J. P. Folkes. 1967. The effect of lipids on the accumulation of certain amino acids by *Staphylococcus aureus*. Biochim. Biophys. Acta *144*: 461-466.
62. Garrison, R. G., H. T. Todd, and J. W. Hamilton. 1970. The uptake of low molecular weight sulfur containing compounds by *Histoplasma capsulatum* and related dimorphic fungi. Mycopathol. Mycol. Appl. *40*: 171-180.
63. Gilbert, B. E., and D. H. Howard. 1970. Uptake of cystine by the yeast phase of *Histoplasma capsulatum*. Infect. Immun. *2*: 139-144.
64. Gill, C. O., and C. Ratledge. 1972. Effect of n-alkanes on the transport of glucose in *Candida* sp. strain 107. Biochem. J. *127*: 59P-60P.
65. Gits, J. J., and M. Grenson. 1967. Multiplicity of amino acid permeases in *Saccharomyces cerevisiae*. III. Evidence for a specific methionine-transporting system. Biochim. Biophys. Acta *135*: 507-516.
66. Goodman, J., and A. Rothstein. 1957. The active transport of phosphate into the yeast cell. J. Gen. Physiol. *40*: 915-925.
67. Greasham, R. L., and A. G. Moat. 1973. Amino acid transport in a polyaromatic amino acid auxotroph of *Saccharomyces cerevisiae*. J. Bacteriol. *115*: 975-981.
68. Grenson, M. 1966. Multiplicity of amino acid permeases in *Saccharomyces cerevisiae*. II. Evidence for a specific lysine-transporting system. Biochim. Biophys. Acta *127*: 339-346.
69. Grenson, M. 1967. Multiplicity of amino acid permeases in *Saccharomyces cerevisiae*. I. Evidence for a specific arginine transport system. Biochim. Biophys. Acta *127*: 325-338.
70. Grenson, M., M. Crabeel, J. M. Wiame, and J. Fechet. 1968. Inhibition of protein synthesis and stimulation of permease turnover in yeast. Biochem. Biophys. Res. Commun. *30*: 414-419.
71. Grenson, M., and C. Hou. 1972. Ammonia inhibition of the general amino acid permease and its suppression in NADPH specific glutamate dehydrogenase-less mutants of *Saccharomyces cerevisiae*. Biochem. Biophys. Res. Commun. *48*: 749-756.
72. Grenson, M., C. Hou, and M. Crabeel. 1979. Multiplicity of the amino acid permeases in *Saccharomyces cerevisiae*. IV. Evidence for a general amino acid permease. J. Bacteriol. *103*: 770-777.
73. Gross, W., and K. Ring. 1971. Active transport of glutamate in *Streptomyces hydrogenans*. I. Studies on uptake and pool size, and their interrelationship. Biochim. Biophys. Acta *233*: 652-665.
74. Guardiola, J., M. De Felice, T. Klopotowski, and M. Iaccarino. 1974. Multiplicity of isoleucine, leucine, and valine transport systems in *Escherichia coli* K-12. J. Bacteriol. *117*: 382-392.

75. Gupta, A. R., and K. K. Rao. 1977. Inhibition of tryptophan uptake in *Aspergillus fumigatus* by tryptamine. Experentia *33*: 853-854.
76. Gupta, R. K., and D. H. Howard. 1971. Comparative physiological studies of the yeast and mycelial forms of *Histoplasma capsulatum*. J. Bacteriol. *105*: 690-700.
77. Gupta, R. K., and D. Pramer. 1970. Amino acid transport by the filamentous fungus *Arthrobotrys conoides*. J. Bacteriol. *103*: 120-130.
78. Hackette, S. L., G. E. Skye, C. E. Burton, and I. H. Segel. 1970. Characterization of an ammonium transport system in filamentous fungi with methylammonium ^{14}C as the substrate. J. Biol. Chem. *245*: 4241-4250.
79. Halaban, R. 1975. Glucose transport deficient mutant of *Neurospora crassa* with an unusual rhythmic growth pattern. J. Bacteriol. *121*: 1056-1063.
80. Halpern, Y. S., and A. Evans-Shoshan. 1967. Properties of glutamate transport system in *Escherichia coli*. J. Bacteriol. *93*: 1009-1016.
81. Hamilton-Miller, M. T. 1974. Fungal sterols and mode of action of polyene antibiotics. Adv. Appl. Microbiol. *17*: 109-134.
82. Hammond, S. M., P. A. Lambert, and B. N. Kliger. 1974. The mode of action of polyene antibiotics: induced potassium leakage in *Candida albicans*. J. Gen. Microbiol. *81*: 325-330.
83. Hammond, S. M., P. A. Lambert, and B. N. Kliger. 1974. The mode of action of polyene antibiotics: induced entry of hydrogen ions as a consequence of polyene action of the cell membrane of *Candida albicans*. J. Gen. Microbiol. *81*: 331-336.
84. Heller, K. B., and M. Hofer. 1978. Temperature dependence of the apparent affinity and the maximum velocity of the membrane-bound monosaccharide transport system in the yeast *Rhodotorula gracilis*. Biochim. Biophys. Acta *514*: 172-177.
85. Hofer, M. 1971. Transport of monosaccharides in *Rhodotorula gracilis* in the absence of metabolic energy. Arch. Mikrobiol. *80*: 50-61.
86. Hofer, M. 1971. A model of the monosaccharide uphill transporting cell membrane system in yeast. J. Theor. Biol. *33*: 599-603.
87. Hofer, M., and P. C. Misra. 1978. Evidence for a proton/sugar symport in the yeast *Rhodotorula gracilis (glutinis)*. Biochem. J. *172*: 15-22.
88. Horak, J., A. Kotyk, and L. Rihova. 1978. Stimulation of amino acid transport in *Saccharomyces cerevisiae* by metabolic inhibitors. Folia Microbiol. (Prague) *23*: 286-291.
89. Howard, D. H. 1962. The morphogenesis of the parasitic forms of dimorphic fungi: a review. Mycopathol. Mycol. Appl. *18*: 127-139.
90. Howard, D. H., N. Dabrowa, V. Otto, and J. Rhodes. 1980. Cysteine transport and sulfite reductase activity in a germination-defective mutant of *Histoplasma capsulatum*. J. Bacteriol. *141*: 417-421.
91. Hunter, D. R., and I. H. Segel. 1971. Acidic and basic amino acid transport systems of *Penicillium chrysogenum*. Arch. Biochem. Biophys. *144*: 168-183.
92. Hunter, D. R., and I. H. Segel. 1973. Control of the general amino acid permease of *Penicillium chrysogenum* by transinhibition and turnover. Arch. Biochem. Biophys. *154*: 387-399.
93. Hunter, D. R., and I. H. Segel. 1973. Effect of weak acids on amino acid transport by *Penicillium chrysogenum*: evidence for a proton or charge gradient as the driving force. J. Bacteriol. *113*: 1184-1192.
94. Hunter, D. R., C. L. Norberg, and I. H. Segel. 1973. Effect of cyclohexamide on L-leucine transport by *Penicillium chrysogenum*: involvement of calcium. J. Bacteriol. *114*: 956-960.
95. Hynes, M. H. 1973. Alteration in the control of glutamate uptake in mutants of *Aspergillus nidulans*. Biochem. Biophys. Res. Commun. *54*: 685-689.
96. Iwashima, A., H. Nishimura, and Y. Nose. 1979. Soluble and membrane bound

thiamine-binding proteins from *Saccharomyces cerevisiae.* Biochim. Biophys. Acta *557*: 460–468.

97. Jacobson, S. E., and R. L. Metzenberg. 1968. A new gene which affects uptake of neutral and acidic amino acids. in *Neurospora crassa.* Biochim. Biophys. Acta *156*: 140-147.

98. Jayakumar, A., M. Singh, and R. Prasad. 1978. Characteristics of proline transport in normal and starved cells of *Candida albicans.* Biochim. Biophys. Acta *514*: 348-355.

99. Jayakumar, A., M. Singh, and R. Prasad. 1979. An inducible proline transport system in *Candida albicans.* Biochim. Biophys. Acta *556*: 144-150.

100. Jennings, D. H. 1976. Transport and translocation in filamentous fungi. In J. E. Smith and D. R. Berry (Eds.), *The Filamentous Fungi,* Vol. 3, Wiley, New York, p. 35.

101. Jennings, D. H., J. D. Thornton, M. F. J. Galpin, and C. R. Goggings. 1974. Translocation in fungi. Symp. Soc. Exp. Biol. *28*: 137-154.

102. Joiris, C. R., and M. Grenson. 1969. Spécificité et régulation d'une perméase des acides amines dicarboxyliques chez *Saccharomyces cerevisiae.* Arch. Int. Physiol. Biochim. *79*: 154-156.

103. Jung, C., and A. Rothstein. 1965. Arsenate uptake and release in relation to the inhibition of transport and glycolysis in yeast. Biochem. Pharmacol. *14*: 1093–1112.

104. Kaback, H. R., and E. R. Stadtdman. 1960. Proline uptake by an isolated cytoplasmic preparation of *Escherichia coli.* Proc. Natl. Acad. Sci. U.S.A. *55*: 920-927.

105. Kackett, S. L., G. E. Skye, C. Burton, and I. H. Segel. 1970. Characterization of an ammonium transport system in filamentous fungi with methyl-ammonium ^{14}C as the substrate. J. Biol. Chem. *245*: 4241-4250.

106. Kalsow, C. M., and R. J. Doyle. 1972. Facilitated diffusion of monosaccharides in *Saccharomyces cerevisiae.* I. Theoretical considerations of asymmetry. J. Theor. Biol. *34*: 125-133.

107. Kay, W. W., and A. F. Gronlund. 1969. Proline transport by *Pseudomonas aeruginosa.* Biochim. Biophys. Acta *193*: 444-455.

108. Kinghorn, J. R., and J. A. Pateman. 1972. L-glutamate and L-aspartate transport in *Aspergillus nidulans.* Heredity *29*: 128.

109. Kinghorn, J. R., and J. A. Pateman. 1974. The effect of the carbon source on ammonium regulation in *Aspergillus nidulans.* Mol. Gen. Genet. *128*: 95-98.

110. Kinghorn, J. R., and J. A. Pateman. 1975. Mutations which affect amino acid transport in *Aspergillus nidulans.* J. Gen. Microbiol. *86*: 174-184.

111. Knight, R. H., R. A. Brady, G. S. Kobayashi, and G. Medoff. 1980. Balanced growth and morphogenesis of *Histoplasma capsulatum* in a defined synthetic medium. Sabouraudia *18*: 39-50.

112. Kobayashi, G. S., and G. Medoff. 1977. Antifungal agents: recent developments. Annu. Rev. Microbiol. *31*: 291-308.

113. Kotyk, A. 1972. Transport of non-electrolytes in yeast. Biochem. J. *127*: 51P-52P.

114. Kotyk, A., and K. Janacek. 1970. *Cell Membrane Transport,* Plenum, New York, p. 233.

115. Kotyk, A., and D. Michaljanicova. 1978. Transport kinetics of 6-deoxy-D-glucose in *Candida parapsilosis.* Folia Microbiol. (Prague) *23*: 18-26.

116. Kotyk, A., and D. Michaljanicova. 1979. Uptake of trehalose by *Saccharomyces cerevisiae.* Proc. Natl. Acad. Sci. U.S.A. *76*: 1814-1818.

117. Kurtz, M. B., and S. P. Champe. 1979. Genetic control of transport loss during development of *Aspergillus nidulans.* Dev. Biol. *70*: 82-88.
118. Lampen, J. O. 1966. Polyene antibiotics and membrane function. Symp. Soc. Gen. Microbiol. *16*: 111-130.
119. Lane, J. W., R. G. Garrison, and D. R. Johnson. 1972. Drug-induced alterations in the ultrastructural organization of *Histoplasma capsulatum* and *Blastomyces dermatitidis.* Mycopathol. Mycol. Appl. *48*: 289-296.
120. Larimore, F. S., and R. J. Roon. 1978. Possible site-specific reagent for the general transport system of *Saccharomyces cerevisiae.* Biochemistry *17*: 431-436.
121. Lehman, J. F. 1973. Regulation of phosphate metabolism in *Neurospora crassa*: characterization of regulatory mutants. Genetics *75*: 61-73.
122. LeJohn, H. B., L. E. Cameron, R. M. Stevenson, and R. U. Mauser. 1974. Influence of cytokinins and sulfhydryl group-reacting agents on calcium transport in fungi. J. Biol. Chem. *249*: 4016-4020.
123. Lester, G. 1966. Genetic control of amino acid permeability in *Neurospora crassa.* J. Bacteriol. *91*: 677-684.
124. Lichliter, W. D., F. Naider, and J. M. Becker. 1976. Basis for the design of anticandidial agents from studies of peptide utilization in *Candida albicans.* Antimicrob. Agents Chemother. *10*: 483-490.
125. Lineweaver, H., and D. Burk. 1934. The determination of enzyme dissociation constants. J. Am. Chem. Soc. *56*: 658-666.
126. Liras, P., and J. O. Lampen. 1974. Sequence of candicidin action on yeast cells. Biochim. Biophys. Acta *372*: 141-153.
127. Logan, D. A., J. M. Becker, and R. Naider. 1979. Peptide transport in *Candida albicans.* J. Gen. Microbiol. *114*: 179-186.
128. Losson, R., R. Jund, and M. R. Chevallier. 1978. Properties of three distinct pyrimidine transport systems in yeast. Biochim. Biophys. Acta *513*: 296-300.
129. Lukaszkiewicz, Z., and A. Pazewski. 1976. Hyper-repressible operator type mutant in sulphate permease gene of *Aspergillus nidulans.* Nature (Lond.) *259*: 337-338.
130. Magana-Swheneke, N., J. Kuznar, and J. Schwencke. 1973. Imino acid transport in yeast: the uptake of sarcosine. Biochim. Biophys. Acta *318*: 281-288.
131. Magill, J. M., and C. W. Magill. 1973. Correlation of growth inhibition patterns to nucleoside transport models in *Neurospora crassa.* J. Bacteriol. *116*: 1071-1074.
132. Magill, J. M., and C. W. Magill. 1975. Purine base transport in *Neurospora crassa.* J. Bacteriol. *124*: 149-154.
133. Magill, C. W., S. O. Nelson, S. M. D'Anbrosio, and G. I. Glover. 1973. Histidine uptake in mutant strains of *Neurospora crassa* via the general transport system for amino acids. J. Bacteriol. *113*: 1320-1325.
134. Magill, C. W., H. Sweeney, and V. W. Woodward. 1972. Histidine uptake in strains of *Neurospora crassa* with normal and mutant transport systems. J. Bacteriol. *110*: 313-320.
135. Marder, R., J. M. Becker, and F. Naider. 1977. Peptide transport in yeast: utilization of leucine- and lysine-containing peptides by *Saccharomyces cerevisiae.* J. Bacteriol. *131*: 906-916.
136. Marder, R., B. Ross, J. M. Becker, and F. Naider. 1978. Isolation of a peptide transport-deficient mutant of yeast. J. Bacteriol. *136*: 1174-1177.
137. Maresca, B., E. Jacobson, G. Medoff, and G. S. Kobayashi. 1978. Cystine reductase in the dimorphic fungus *Histoplasma capsulatum.* J. Bacteriol. *135*: 987-992.
138. Maresca, B., G. Medoff, D. Schlessinger, and G. S. Kobayshi. 1977. Regulation

of dimorphism in the pathogenic fungus *Histoplasma capsulatum*. Nature (Lond.) *266*: 447-448.

139. Mark, C. G., and A. H. Romano. 1971. Properties of the hexose transport systems of *Aspergillus nidulans*. Biochim. Biophys. Acta *249*: 216-226.
140. Marzluf, G. A. 1970. Genetic and metabolic controls for sulfate metabolism in *Neurospora crassa*: isolation and study of chromate-resistant and sulfate transport-negative mutants. J. Bacteriol. *102*: 716-721.
141. Marzluf, G. A. 1970. Genetic and biochemical studies of distinct sulfate permease species in different developmental stages of *Neurospora crassa*. Arch. Biochem. Biophys. *138*: 254-263.
142. Marzluf, G. A. 1973. Regulation of sulfate transport in *Neurospora* by transinhibition and by inositol depletion. Arch. Biochem. Biophys. *156*: 244-254.
143. Marzluf, G. A. 1974. Uptake and efflux of sulfate in *Neurospora crassa*. Biochim. Biophys. Acta *339*: 374-381.
144. Matthew, D. M., and J. W. Payne (Eds.). 1975. *Peptide Transport in Protein Nutrition*, American Elsevier, New York.
145. Merkel, G. J., F. Naider, and J. M. Becker. 1980. Amino acid uptake by *Saccharomyces cerevisiae* plasma membrane vesicles. Biochim. Biophys. Acta *595*: 109-120.
146. Miller, A. G., and K. Budd. 1971. Chloride uptake by mycelium of *Neocosmospora vasinfecta* and its inhibition by glucose. J. Gen. Microbiol. *66*: 243-245.
147. Mills, D. 1972. Effect of pH on adenine and amino acid uptake during sporulation in *Saccharomyces cerevisiae*. J. Bacteriol. *112*: 519-526.
148. Moore, B., and B. Spencer. 1972. Non-specific uptake and specific binding of sulfate by *Aspergillus nidulans*. Biochem. J. *127*: 72P.
149. Moore, D., and M. S. Devadatham. 1979. Sugar transport in *Coprinus cinereus*. Biochim. Biophys. Acta *550*: 515-526.
150. Mooz, E. D. 1979. Association of glutathione synthetase deficiency and diminished amino acid transport in yeast. Biochem. Biophys. Res. Commun. *90*: 1221-1228.
151. McEvoy, J. J., and J. R. Murray. 1972. Amino acid transport in germinated arthrospores of *Geotrichum candidum*. Arch. Mikrobiol. *86*: 101-110.
152. Nagy, M. 1979. Studies on purine transport and on purine content in vacuoles isolated from *Saccharomyces cerevisiae*. Biochim. Biophys. Acta *558*: 221-232.
153. Nakamura, K. D., and F. Schlenk. 1974. Examination of isolated yeast cells' vacuoles for active transport. J. Bacteriol. *118*: 314-316.
154. Neville, M. M., S. R. Suskind, and S. Roseman. 1971. A derepressible active transport system for glucose in *Neurospora crassa*. J. Biol. Chem. *246*: 1294-1301.
155. Obukhovskaya, A. S., E. B. Lishnevskaya, and I. M. Tereshin. 1979. Mechanism of the effect of levorin on transport of amino acids in *Candida albicans*. Curr. Microbiol. *2*: 219-222.
156. Okorokov, L. A., V. M. Kadomtseva, and B. I. Titovskii. 1979. Transport of manganese into *Saccharomyces cerevisiae*. Folia Microbiol. (Prague) *24*: 240-246.
157. Oxender, D. L. 1972. Membrane transport. Annu. Rev. Biochem. *41*: 777-814.
158. Oxender, D. L., and S. C. Quay. 1976. Regulation of leucine transport and binding proteins in *Escherichia coli*. J. Cell. Physiol. *89*: 517-521.
159. Pall, M. L. 1969. Amino acid transport in *Neurospora crassa*. I. Properties of two amino acid transport systems. Biochim. Biophys. Acta *173*: 113-127.
160. Pall, M. L. 1970. Amino acid transport in *Neurospora crassa*. II. Properties of a basic amino acid transport system. Biochim. Biophys. Acta *203*: 139-149.
161. Pall, M. L. 1970. Amino acid transport in *Neurospora crassa*. III. Acidic amino acid transport. Biochim. Biophys. Acta *211*: 513-520.

162. Pall, M. L. 1970. Amino acid transport in *Neurospora crassa.* IV. Properties and regulation of a methionine transport system. Biochim. Biophys. Acta *233*: 201-214.
163. Pall, M. L., and K. A. Kelly. 1971. Specificity of transinhibition of amino acid transport in *Neurospora.* Biochem. Biophys. Res. Commun. *42*: 940-947.
164. Pappenheimer, J. R., S. R. Heisey, E. F. Jordan, and A. K. Solomon. 1963. Perfusion of the cerebral ventricular system in unanesthetized goats. Am. J. Physiol. *203*: 763-774.
165. Pateman, J. A., E. Dunn, J. R. Kinghorn, and E. C. Forbes. 1974. The transport of ammonium and methylammonium in wild type and mutant cells of *Aspergillus nidulans.* Mol. Gen. Genet. *133*: 225-236.
166. Pateman, J. A., and J. R. Kinghorn. 1976. Nitrogen metabolism. In J. E. Smith and D. R. Berry (Eds.), *The Filamentous Fungi,* Vol. 2, Wiley, New York, p. 187.
167. Pateman, J. A., J. R. Kinghorn, and E. Dunn. 1974. Regulatory aspects of L-glutamate transport in *Aspergillus nidulans.* J. Bacteriol. *119*: 534-542.
168. Pateman, J. A., J. R. Kinghorn, E. Dunn, and E. Forbes. 1973. Ammonium regulation in *Aspergillus nidulans.* J. Bacteriol. *114*: 943-950.
169. Payne, J. W. 1976. Peptides and microorganisms. Adv. Microb. Physiol. *13*: 55-113.
170. Payne, J. W., and C. Gilvarg. 1971. Peptide transport. Adv. Enzymol. Relat. Areas Mol. Biol. *35*: 187-244.
171. Pendyala, L., J. Smyth, and A. M. Wellman. 1979. Nature of 6-methylpurine inhibition and characterization of two 6-methylpurine-resistant mutants of *Neurospora crassa.* J. Bacteriol. *137*: 248-255.
172. Pendyala, L., and A. M. Wellman. 1978. Purine base transport in *nit-2* mutants of *Neurospora crassa.* J. Bacteriol. *133*: 401-402.
173. Peterson, E. M., and R. A. Calderone. 1978. Inhibition of specific amino acid uptake in *Candida albicans* by lysosomal extract from alveolar macrophages. Infect. Immun. *21*: 506-513.
174. Pickering, W. R., and R. W. Woods. 1972. The uptake and incorporation of purines by wild type *Saccharomyces cerevisiae* and a mutant resistant to 4-amino-pyrazolo (3,4-d)pyrimidine. Biochim. Biophys. Acta *264*: 45-58.
175. Polak, A., and M. Grenson. 1973. Evidence for cytosine, adenine and hypoxanthine in *Saccharomyces cerevisiae* and *Candida albicans.* Eur. J. Biochem. *32*: 276-282.
176. Rand, J. B., and E. L. Tatum. 1980. Fructose transport in *Neurospora crassa.* J. Bacteriol. *142*: 763-767.
177. Rao, E., and A. G. DeBusk. 1975. A mutant of *Neurospora* deficient in the general (PmG) amino acid transport system. Neurospora Newsl. *22*: 12-13.
178. Reichert, U., and M. Winter. 1974. Uptake and accumulation of purine bases by stationary yeast cells pretreated with glucose. Biochim. Biophys. Acta *356*: 108-116.
179. Rickenberg, H. V., G. N. Cohen, G. Buttin, and J. Monod. 1956. La galactoside-perméase d'*Escherichia coli.* Ann. Inst. Pasteur *91*: 829-857.
180. Roa, T. K., and A. G. DeBusk. 1973. Characteristics of a transport deficient mutant *(nap)* of *Neurospora crassa.* Biochim. Biophys. Acta *326*: 619-626.
181. Robertson, L. W., J. E. Robbers, and H. G. Floss. 1973. Some characteristics of tryptophan uptake in *Claviceps* species. J. Bacteriol. *114*: 208-219.
182. Robinson, J. H., C. Anthony, and W. T. Drabble. 1972. Regulation of acidic amino acid permease of *Aspergillus nidulans.* Proc. Biochem. Soc. *127*: 71P.

183. Robinson, J. H., C. Anthony, and W. T. Drabble. 1973. The acidic amino acid permease of *Aspergillus nidulans.* J. Gen. Microbiol. *79*: 53-63.
184. Robinson, J. H., C. Anthony, and W. T. Drabble. 1973. Regulation of the acidic amino acid permease of *Aspergillus nidulans.* J. Gen. Microbiol. *79*: 65-80.
185. Robinson, J. H., C. Anthony, and W. T. Drabble. 1974. The utilization of nitrogen sources by *Aspergillus clavatus.* J. Gen. Microbiol. *85*: 23-28.
186. Rodriguez-Navarro, A., and E. D. Sancho. 1979. Cation exchange of yeast in the absence of magnesium. Biochim. Biophys. Acta *552*: 322-330.
187. Roess, W. B., and A. G. DeBusk. 1968. Properties of a basic amino acid permease in *Neurospora crassa.* J. Gen. Microbiol. *52*: 421-432.
188. Romano, A. H., and H. L. Kornberg. 1968. Regulation of sugar utilization by *Aspergillus nidulans.* Biochim. Biophys. Acta *158*: 491-493.
189. Romano, A. H., and H. L. Kornberg. 1969. Regulation of sugar uptake by *Aspergillus nidulans.* Proc. R. Soc. Lond., Ser. B *173*: 475-490.
190. Romkes, S. C. E., and M. J. Lewis. 1971. Some factors which affect amino acid uptake by *Saccharomyces carlsbergensis.* Appl. Microbiol. *21*: 799-805.
191. Roomans, G. M., and G. W. Borst-Pauwels. 1979. Interaction of cations with phosphate uptake by *Saccharomyces cerevisiae.* Effect of surface potential. Biochem. J. *178*: 521-527.
192. Roomans, G. M., G. A. Kuypers, A. P. Theuvenet, and G. W. Borst-Pauwels. 1979. Kinetics of sulfate uptake by yeast. Biochim. Biophys. Acta *551*: 197-206.
193. Roomans, G. M., A. P. Theuvenet, T. P. van den Berg, and G. W. Borst-Pauwels. 1979. Kinetics of Ca^{++} and Sr^{++} uptake by yeast. Effects of pH, cations and phosphate. Biochim. Biophys. Acta *551*: 187-196.
194. Roon, R. J., F. Larimore, and J. S. Levy. 1975. Inhibition of amino acid transport by ammonium ion in *Saccharomyces cerevisiae.* J. Bacteriol. *124*: 325-331.
195. Roseman, S. 1972. Carbohydrate transport in bacterial cells. In L. E. Howkins (Ed.), *Metabolic Pathways,* Vol. VI, Academic, New York, pp. 41-89.
196. Rothstein, A. 1972. Ion transport in microorganisms. In L. E. Howkins (Ed.), *Metabolic Pathways,* Vol. IV, Academic, New York, p. 21.
197. Rothstein, A., and K. Donovan. 1963. Interaction of arsenate with the phosphate-transporting system of yeast. J. Gen. Physiol. *46*: 1075-1085.
198. Rytka, J. 1975. Positive selection of general amino acid permease mutants in *Saccharomyces cerevisiae.* J. Bacteriol. *121*: 562-570.
199. Salvin, S. B. 1950. Growth of yeast-like phases of *Histoplasma capsulatum* in a fluid medium. J. Bacteriol. *59*: 312-313.
200. Sanchez, S., L. Martinez, and J. Mora. 1972. Interactions between amino acid transport systems in *Neurospora crassa.* J. Bacteriol. *112*: 276-284.
201. Scarborough, G. A. 1970. Sugar transport in *Neurospora crassa.* J. Biol. Chem. *245*: 1694-1698.
202. Scarborough, G. A. 1970. Sugar transport in *Neurospora crassa.* II. A second glucose transport system. J. Biol. Chem. *245*: 3985-3987.
203. Scherrer, R., L. Louden, and P. Gerhardt. 1974. Porosity of the yeast cell wall and membrane. J. Bacteriol. *118*: 534-540.
204. Schloemer, R. H., and R. H. Garrett. 1974. Nitrate transport system in *Neurospora crassa.* J. Bacteriol. *118*: 259-269.
205. Schloemer, R. H., and R. H. Garrett. 1974. Uptake of nitrite by *Neurospora crassa.* J. Bacteriol. *118*: 270-274.
206. Schneider, R. P., and W. R. Wiley. 1971. Kinetic characteristics of the two glucose transport systems in *Neurospora crassa.* J. Bacteriol. *106*: 479-486.

207. Schneider, R. P., and W. R. Wiley. 1971. Regulation of sugar transport in *Neurospora crassa.* J. Bacteriol. *106*: 487-492.

208. Seaston, A., C. Inkson, and A. A. Eddy. 1973. The absorption of protons with specific amino acids and carbohydrates by yeast. Biochem. J. *134*: 1031-1034.

209. Shere, S. M., and L. Jacobson. 1970. Mineral uptake in *Fusarium oxysporium* f.sp. *vasinfectum.* Physiol. Plant. *23*: 51-62.

210. Shere, S. M., and L. Jacobson. 1970. The influence of phosphate uptake on cation uptake in *Fusarium oxysporium* f.sp. *vasinfectum.* Physiol. Plant. *23*: 294-303.

211. Singh, M., A. Jayakumar, and R. Prasad. 1978. The effect of altered lipid composition on the transport of various amino acids in *Candida albicans.* Arch. Biochem. Biophys. *191*: 680-686.

212. Singh, M., A. Jayakumar, and R. Prasad. 1979. Lipid composition and polyene antibiotic sensitivity in isolates of *Candida albicans.* Microbios *24*: 7-17.

213. Sinha, U. 1969. Genetic control of the uptake of amino acids in *Aspergillus nidulans.* Genetics *62*: 495-505.

214. Skye, G. E., and I. H. Segel. 1970. Independent regulation of cysteine and cystine in *Penicillium chrysogenum.* Arch. Biochem. Biophys. *138*: 306-318.

215. Slayman, C. L. 1965. Electrical properties of *Neurospora crassa.* Effect of external cations on the intracellular potential. J. Gen. Physiol. *49*: 69-92.

216. Slayman, C. L. 1965. Electrical properties of *Neurospora crassa.* Respiration and intracellular potential. J. Gen. Physiol. *49*: 93-116.

217. Slayman, C. L., C. Y.-H. Lu, and L. Shane. 1970. Correlated changes in membrane potential and ATP concentrations in *Neurospora.* Nature (Lond.) *226*: 274-276.

218. Slayman, C. L., and C. W. Slayman. 1962. Measurement of membrane potentials in *Neurospora.* Science *136*: 876-877.

219. Slayman, C. L., and C. W. Slayman. 1965. Membrane studies on mutants of *Neurospora crassa.* Abstr. 23rd. Congr. Physiol. Sci., Tokyo, item 129.

220. Slayman, C. L., and C. W. Slayman. 1968. Net uptake of potassium in *Neurospora.* Exchange for sodium and hydrogen ions. J. Gen. Physiol. *52*: 424-443.

221. Slayman, C. W. 1970. Net potassium transport in *Neurospora.* properties of a transport mutant. Biochim. Biophys. Acta *211*: 502-512.

222. Slayman, C. W., and C. L. Slayman. 1970. Potassium transport in *Neurospora.* Evidence for a multisite carrier at high pH. J. Gen. Physiol. *55*: 758-786.

223. Slayman, C. W., and E. L. Tatum. 1965. Potassium transport in *Neurospora.* III. Isolation of a transport mutant. Biochim. Biophys. Acta *109*: 184-193.

224. Sreedhara-Swamy, K. H., M. Sirsi, and G. Ramananda-Rao. 1974. Studies on the mechanism of action of miconazole: effect of miconazole on respiration and cell permeability of *Candida albicans.* Antimicrob. Agents Chemother. *5*: 420-425.

225. Stadler, D. R. 1966. Genetic control of the uptake of amino acids in *Neurospora.* Genetics *54*: 677-685.

226. Stadler, D. R. 1967. Suppressors of amino acid uptake of *Neurospora.* Genetics *57*: 935-942.

227. Stephien, P. P. 1976. Partial purification of methionine binding protein from *Aspergillus nidulans* by affinity chromatography. Biochim. Biophys. Acta *439*: 154-159.

228. Stroobart, P., and G. A. Scarborough. 1979. Active transport of calcium in *Neurospora* plasma membrane vesicles. Proc. Natl. Acad. Sci. U.S.A. *76*: 3102-3106.

229. Stuart, W. D., and A. G. DeBusk. 1971. In vitro studies of isolated glycoprotein subunits of the amino acid transport system of *Neurospora crassa.* Arch. Biochem. Biophys. *144*: 512-518.

230. Stuart, W. D., and A. G. DeBusk. 1973. Transport of arginine by an in vitro system. Biochem. Biophys. Res. Commun. *52*: 1046-1050.
231. Stuart, W. D., and A. G. DeBusk. 1975. Genetic alternatives of ribonuclease-sensitive glycoprotein subunits of amino acid transport system in *Neurospora crassa* conidia. Arch. Biochem. Biophys. *166*: 213-222.
232. Stuart, W. D., and D. O. Woodward. 1975. Effects of ribonuclease A on amino acid transport in *Neurospora crassa*. J. Bacteriol. *122*: 41-46.
233. Surdin, Y., W. Sly, J. Sine, A. M. Bondes, and H. deRobichon-Szulmajster. 1965. Propriétés et controle génétique de système d'accumulation des acides amines chez *Saccharomyces*. Biochim. Biophys. Acta *107*: 546-566.
234. Taber, W. A. 1971. Uptake of ^{14}C-labelled succinate, L(+)-dihydroxy-succinate, L-monohydroxy-succinate, citrate, alpha-ketoglutamate and D-glucose by washed mycelium of *Calviceps purpurea*. Mycologia *63*: 290-307.
235. Tang, S. L., and D. H. Howard. 1973. Metabolism, macromolecular synthesis, and nuclear behavior of *Cyrptococcus albidus* at 37°C. J. Bacteriol. *115*: 574-581.
236. Tang, S. L., D. H. Howard, and R. K. Gupta. 1978. Uptake of metabolites by germinating blastospores of *Histoplasma capculatum*. Sabouraudia. *16*: 271-278.
237. Teh, J. S., and K. H. Lee. 1974. Effects of n-alkanes on *Cladosporium resinae*. Can. J. Microbiol. *20*: 971-976.
238. Thwaites, W. M., and L. Pendyala, 1969. Regulation of amino acid transport in a strain of *Neurospora crassa* lacking basic amino acid transport activity. Biochim. Biophys. Acta *192*: 455-461.
239. Timberlake, W. E., and D. H. Griffin. 1973. Direct inhibition of the uptake of proline by cyclohexamide. Biochem. Biophys. Res. Commun. *54*: 216-221.
240. Tsao, T. F., and G. A. Marzluf. 1976. Genetic and metabolic regulation of purine base transport in *Neurospora crassa*. Mol. Gen. Genet. *149*: 347-355.
241. Tweedie, J. W., and I. H. Segel. 1970. Specificity of transport processes for sulfur, selenium, and molybdenum anions by filamentous fungi. Biochim. Biophys. Acta *196*: 95-106.
242. Ullrich, K. H. 1979. Sugar, amino acid, and Na^+ cotransport in the proximal tubule. Annu. Rev. Physiol. *41*: 181-195.
243. Utech, N. M., K. G. Reid, and J. T. Holden. 1970. Properties of a dicarboxylic amino acid transport-deficient mutant of *Streptococcus faecalis*. J. Biol. Chem. *245*: 5273-5280.
244. Van Den Bossche, H. 1974. Biochemical effects of miconazole on fungi. 1. Effects on uptake and/or utilization of purines, pyrimidines, nucleosides, amino acids and glucose by *Candida albicans*. Biochem. Pharmacol. *23*: 887-889.
245. von Hedenstrom, M., and M. Hofer. 1979. The effect of nystatin on active transport in *Rhodotorula glutinis (gracilis)* is restricted to the plasma membrane. Biochim. Biophys. Acta *555*: 169-174.
246. Wehrli, E., C. Roehm, and G. F. Fuhrmann. 1975. Yeast plasma membrane vesicles suitable for transport studies. J. Bacteriol. *124*: 1594-1597.
247. West, I. C. 1980. Energy coupling in secondary active transport. Biochim. Biophys. Acta *604*: 91-126.
248. Whitaker, A., and A. G. Morton. 1971. Amino acid transport in *Penicillium griseofulvum*. Trans. Br. Mycol. Soc. *56*: 353-369.
249. Wiebe, C., and G. Winkelmann. 1975. Kinetic studies on the specificity of chelate-iron uptake in *Aspergillus*. J. Bacteriol. *123*: 837-842.
250. Wiley, W. R. 1970. Tryptophan transport in *Neurospora crassa*: a tryptophan binding protein released by cold osmitic shock. J. Bacteriol. *103*: 656-662.

251. Wiley, W. R., and W. H. Matchett. 1966. Tryptophan transport in *Neurospora crassa.* I. Specificity and kinetics. J. Bacteriol. *92*: 1698-1705.
252. Wiley, W. R., and W. H. Matchett. 1968. Tryptophan transport in *Neurospora crassa.* II. Metabolic control. J. Bacteriol. *95*: 959-966.
253. Wilson, D. B. 1978. Cellular transport mechanisms. Annu. Rev. Biochem. *47*: 933-965.
254. Wolfinbarger, L., and A. G. DeBusk. 1971. Molecular transport. II. In vivo studies of transport mutants of *Neurospora crassa* with altered amino acid composition patterns. Arch. Biochem. Biophys. *144*: 503-511.
255. Wolfinbarger, L., and A. G. DeBusk. 1972. The kinetics of L-aspartate transport in *Neurospora crassa* conidia. Biochim. Biophys. Acta *290*: 355-367.
256. Wolfinbarger, L., H. H. Jervis, and A. G. DeBusk. 1971. Active transport of L-aspartic acid in *Neurospora crassa.* Biochim. Biophys. Acta *249*: 63-68.
257. Wolfinbarger, L., and W. W. Kay. 1973. Acidic amino acid transport in *Neurospora crassa.* Biochim. Biophys. Acta *330*: 335-343.
258. Wolfinbarger, L., and G. A. Marzluf. 1975. Size restrictions on utilization of peptides by amino acid auxotrophs of *Neurospora crassa.* J. Bacteriol. *122*: 949-956.
259. Woodward, C. K., C. P. Read, and V. W. Woodward. 1967. *Neurospora crassa* mutants defective in the transport of amino acids. Genetics *56*: 598-610.
260. Yabu, K. 1970. Amino acid transport in *Mycobacterium smegmatis.* J. Bacteriol. *102*: 6-13.
261. Yamaguchi, M., and K. Iwata. 1979. Effect of two imidazole antimycotics: clotrimazole and miconazole on amino acid transport in *Candida albicans.* Sabouraudia. *17*: 311-322.
262. Yamamoto, L. A., and I. H. Segel. 1966. The inorganic sulfate transport system of *Penicillium chrysogenum.* Arch. Biochem. Biophys. *114*: 523-538.

Author Index

Numbers in brackets are reference numbers and indicate that an author's work is referred to although his name may not be cited in the text. Italicized numbers give the page on which the complete reference is listed.

B

C

D

E

H

K

L

M

O

Q

R

T

Y

Z

Subject Index

Illustrations are indicated by italicized page numbers.

B

E

G

H

R

S

U

V

W

Y

Z